Getting shot hurts. Still my fear was growing because no matter how hard I tried to breathe it seemed I was getting less & less air. I focused on that tiled ceiling and prayed. But I realized I couldn't ask for Gods help while at the same time I felt hatred for the mixed up young man who had shot me. Isn't that the meaning of the lost sheep? We are all Gods children & therefore equally beloved by him. I began to pray for his soul and that he would find his way back to the fold.

I opened my eyes once to find Nancy there. I pray I'll never face a day when she isn't there. Of all the ways God has blessed me giving her to me is the greatest and beyond anything I can ever hope to deserve.

All the kids arrived and the hours ran together in a blur during which I was operated on. I know it's going to be a long recovery but there has been such an outpouring of love from all over.

The days of therapy, transfusion, intravenous etc. have gone by — now it is Sat, April 11 and this morning I left the hospital and am here at the W.H. with Nancy & Patti. The treatment, the warmth, the skill of those at G.W. has been magnificent but it's great to be here at home.

Whatever happens now I owe my life to God and will try to serve him in every way I can.

THE REAGAN DIARIES

Contents

TO NANCY REAGAN

HarperCollins books may be purchased for educational, business, or sales promotional use. For information, please write: Special Markets Department, HarperCollins Publishers, 10 East 53rd Street, New York, NY 10022.

FIRST EDITION

Designed by William Ruoto

Library of Congress Cataloging-in-Publication Data is available upon request.

ISBN: 978-0-06-0876005
ISBN-10: 0-06-087600-X

07   08   09   10   11   NMSG/RRD   10   9   8   7   6   5   4   3

THE
REAGAN
DIARIES

RONALD REAGAN

EDITED BY DOUGLAS BRINKLEY

HarperCollins*Publishers*

Introduction

When I was first asked to edit President Ronald Reagan's diaries, I was, of course, flattered and excited, not to mention terribly curious. What would I find? Would the "real" Reagan finally emerge from the man's personal journals? How would the diaries alter perceptions of his White House tenure?

To my knowledge only four presidents other than Ronald Reagan maintained written diaries on a consistent basis: George Washington, John Quincy Adams, James K. Polk, and Rutherford B. Hayes. (Others, like Harry Truman and Dwight Eisenhower, kept them sporadically.) Reagan had never kept a diary before entering the White House, which was something he and Nancy later regretted. "The Sacramento years flew by so quickly; we both wish we had kept diaries," she would tell friends. "But the kids were younger then, and we just did not seem to have time." Reagan made a serious commitment to maintaining one as president, starting with his inauguration on January 20, 1981. Unlike so many new diarists who trail off after the first few weeks, he took his task seriously, and in eight years he never neglected a daily entry, except when he was in the hospital.

When I first saw the White House diaries at the Ronald Reagan Presidential Library in Simi Valley, California, I was astounded. Lined up on a research table, all five volumes resemble a handsome half-set of an encyclopedia. For a moment, I just looked at them. In physical appearance, the diaries were hardcover books, 8½ by 11, bound in maroon and brown leather with the presidential seal embossed on the center of the front and the name RONALD WILSON REAGAN in gold lettering at the bottom right. No words appeared on the spines. The inside cover boards all had elegant designs, with intertwined brown autumnal leaves or swirling red-blue-beige paisley or some other exquisite pattern.

Within the pages themselves, you see immediately that Reagan had neat, rounded handwriting, done in ink that is variously blue or black. It is a welcoming script, easy to read. Cross-outs are rare. Economical to the core, Reagan filled every page to the very bottom. Occasionally, he inserted auxiliary material into the diary pages: comic photographs, a picture of a young child wearing a REAGAN straw hat, or newspaper clippings pertaining to U.S. soldiers killed abroad.

Reagan was a master of the art of summary. As an orator, he was known to keep notes in shorthand on cards that he kept in his breast pocket. He wrote in the diary in a similar fashion. The very act of composition helped him organize his thoughts, as it had since his boyhood. Over the decades, he wrote his own speeches, radio broadcasts, and newspaper columns. He once claimed that the creative act gave him "great clarity." As president, he made the time to write (or revise) many of his own speeches. He answered his own mail. "You specified that you wanted to hear from me personally," Reagan wrote a citizen from the White House in 1981, "so here I am." He enjoyed reading books of all sorts—if the writing was inspiring. To judge by the liveliness of many diary entries, from the first to the very last, keeping a daily log was, for Reagan, anything but a chore.

Nancy Reagan explains that her husband kept the diaries in his second-floor White House study next to their bedroom. When traveling, he'd bring a volume with him, often writing while on Air Force One. "We just wanted a way to capture the moment and our feelings before we were whisked on to the next day," Nancy explained, "so we could savor it a little more." For several years after their return to California, the Reagans would often sit together in their den after dinner, reading aloud from their diaries and reminiscing about their White House years.

As soon as I started reading the daily entries the president had written, I could almost hear his voice. It was Reagan, circa the 1980s. The familiar, plain-spoken, direct tones were back. It was as if he were talking just to me. And I found that I was fascinated by what he had to say.

In these writings Ronald Reagan's true nature is revealed. His uncomplicated and humble notations are on display in these pages: genuine, thoughtful, and caring. They are an extension of an honest man who loved freedom but hated communism, inflation, and … taxes. The Reagan appeal, evident in each volume, helps to explain why he never outlasted his welcome with the American people. They, and millions of others around the world, regarded him as a powerful leader, and, at the same time, a trusted neighbor. Those who made Reagan a hero in his own lifetime even saw something of themselves reflected in him—a modern American unashamed of the nation's majesty and his own pride in time-honored traditions. It's all here in his own hand, in his own words: from the everyday musings of ordinary thoughts to the detailing of diplomatic summits, Reagan's entries reveal the hidden rhythm of life as a president, a husband, a father, and a friend.

Many have referred to Ronald Reagan as "larger than life." The national outpouring of grief when he died in 2004 was proof positive that he was no ordinary politician. But now, with the publication of *The Reagan Diaries*, there is even more: an immense presidential legacy is extended.

Ronald Reagan wrote about seemingly everything and everyone in the diaries—from his first inauguration in 1981 to his very last day in the White

House in 1989. Arms reduction, Reaganomics, the military, Lebanon, Iran-Contra, Camp David, Rancho del Cielo, and much more are recorded in considerable detail. Queen Elizabeth II, Mohammar al-Qaddafi, Pope John Paul II, Mikhail Gorbachev, Margaret Thatcher, Tip O'Neill, George H. W. Bush, Fidel Castro—they are all here, too.

And rarely can one turn a page without realizing that Ronald Reagan's marriage to Nancy was the cornerstone of his life. Nothing was more important to Ronald Reagan than his marriage. His deep and enduring love for her was a foundation for him. In the George Washington University Hospital after being shot, he wrote, "I opened my eyes once to find Nancy there. I pray I'll never face a day when she isn't there. Of all the ways God has blessed me giving her to me is the greatest and beyond anything I can ever hope to deserve." He felt a real sense of loss whenever his wife was away. On one occasion, when she left for a ship christening, he asked himself, "Why am I so scared when she leaves like that?" Later, when she went on a trip to New York, he contracted one of his typical bouts of White House loneliness. "I don't like it here by myself," he wrote.

Like his marriage to Nancy, his strong relationship with God was of paramount importance in Reagan's life. Consistently, he thanks God for allowing him to be physically fit and for sparing his life from Hinckley's bullet spray, after which he recalls lying on a bed in the emergency room: "I focused on that tiled ceiling and prayed. But I realized I couldn't ask for God's help while at the same time I felt hatred for the mixed-up young man who had shot me. Isn't that the meaning of the last sheep? We are all God's children & therefore equally beloved by him. I began to pray for his soul and that he would find his way back to the fold."

Another personality trait that frequently surfaces in passages is Reagan's enormous empathy for citizens with physical disabilities. At one juncture or another, he comments on blindness, epilepsy, sickle cell anemia, deafness—the list goes on and on. He learned as much as possible about those afflicted with conditions such as muscular dystrophy or cancer, and reported their stories to his diary. On many occasions, as he freely admitted, he "puddled up" in the face of human tragedy and the courage that often accompanied it. "Francis Albert (Sinatra) came by with the Multiple Sclerosis Mother and Father of the year," he wrote. "He's heading up their fundraising drive—'FS for MS.' What a cruel disease." Observations such as this one abound.

Looking through the pages of Reagan's White House diaries, we see daily events and historical moments as he saw them. We see the nation, too, through his eyes. With the immediacy of the moment, these diaries present history as it unfolded.

Interest in forming personal bonds was especially distinct in his meetings with world leaders. He chafed at the layers of communication that he felt complicated international relations. In the diary, we see him searching for a way to improve his contact with the successive general secretaries of the Soviet Union: Brezhnev, Andropov, Chernenko, and Gorbachev. While Reagan was always guarded in his attitude toward the Soviets, he believed that progress would be made if he could communicate directly with them—by letter, telephone, or in person. And he was right.

The president was, by nature, anything but a Washington insider. Yet he had a remarkably sure understanding of his relationship with Congress. It was the rare week that went by without his personal negotiations over legislation or an outright campaign in anticipation of a coming vote. Some of the struggles could be rough, as any president can attest. Reagan, however, had the admirable quality of being able to rise above policy disagreements and set aside the day's battles for friendly socializing. Some of his best friends, according to the diary, were those Democrats with whom he had had the most bitter differences politically. He was especially fond of Speaker of the House Thomas "Tip" O'Neill (D-MA), with whom he could match Irish stories by the hour. "Tip is truly a New Deal Democrat," he wrote while in the middle of a budget crisis feud. "He honestly believes that we're promoting welfare for the rich."

History regards Reagan as an important president with lasting influence. How he accomplished as much as he did can be better understood with the publication of *The Reagan Diaries*. Prime Minister Margaret Thatcher of Great Britain once said that Reagan "won the Cold War without firing a shot." Reagan's own entries through the years explain how he fought his most critical adversary, the USSR. It may surprise even Cold War scholars. As he wrote of April 6, 1983: "Some of the N.S.C. staff are too hard line & don't think any approach should be made to the Soviets. I think I'm hard line & will never appease but I do want to try & let them see there is a better world if they'll show <u>by deed</u> they want to get along with the free world." In foreign relations, Reagan knew innately when to show power and when to be subtle. The personal relationship he developed with Gorbachev, in fact, as recorded in these diaries, was of a deeply historic nature. "There is no question in my mind," he wrote at one juncture in 1988, "but that a certain chemistry does exist between us."

As with the man himself, there is an appealing earnestness to the diaries, an unvarnished accounting of his days in office. The entries don't dazzle in a self-congratulatory fashion. Nor do they consciously attempt to spin history in his favor. They are prosaic, not grandiose. The power of the diaries is in their cumulative effect. His attitude seemed to be that it would be left to the echo of history to decide whether he was right or wrong. Nowhere in the entries did the president bask in glory, savor the misfortune of adversaries, or wallow in his own defeats. More often than not, he is self-deprecating. The furthest that Reagan reached into hyperbole was along the lines of "all in all, a good day" or "things might have been better."

Because the complete Reagan diaries would fill two or three fat volumes, I had to be selective in deciding what to choose to include in this abridged version. Heavy cuts were made. My objective was to combine both the most intriguing historical material and a healthy sampling of the more mundane, day-to-day realities of his center-stage life in the 1980s, (or the Reagan Era, as it is frequently called). Great effort was made not to lose the rhythm of his economical prose. Unlike the wording often found in his speeches, here he assiduously avoids flowery, adjective-driven observations.

All the annotations in brackets throughout the text are mine. They are an editorial attempt to earmark what topical events or material was excised. In the book's back matter I have included a glossary, where readers will find capsule biographies for people Reagan mentions in the diary. Occasionally in the text I've placed asterisks to indicate a footnote on the page's bottom. The National Security Council read all five diary volumes and redacted only about six pages of material for national security reasons. In addition, Nancy Reagan requested that a few entries be edited out for personal reasons. I respected her wish. The locations of all redactions within the text are indicated by ellipses within brackets: [. . .]

The most profound achievements of the Reagan administration changed the world. When Reagan took office, the Soviet Union expected war imminently, according to a report the president received. By the time he left the White House, the threat of nuclear war with the USSR was reduced to a memory, as the Cold War was fast coming to a close. Domestically, Reagan left as strong a legacy. His personality, no less than his policies, led to a resurgent belief in the power of the individual, competition in business, and private sector initiative to solve domestic problems. Such successes were no accident; Reagan worked hard, coming fully prepared to each difficult decision. His executive ability, and the work habits that went along with it, are evident throughout the diary.

It was quickly apparent that my initial curiosity about reading these diaries was well founded. Far from being disappointed or feeling intrusive, I was intrigued by the unprecedented look behind the scenes of one of the most significant presidencies of modern times. Even though the historical events described are very familiar, I could not read the diaries fast enough. In many ways, though, this is a whole new story, because it is told from the unique perspective of the man at its center.

The Ronald Reagan you'll encounter in these pages is principled, confident, happy, free of ego, and devoted to his wife. And, most of all, a man who understood instinctively that he did not "become" president, but was given "temporary custody" of an office that ultimately belongs to the people. By sharing these stream-of-consciousness diaries with us, the Reagans have given a gift to history that will endure forever.

THE REAGAN DIARIES

CHAPTER I

—

1981

1981

The Inaugural (Jan. 20) was an emotional experience but then the very next day it was "down to work." The first few days were long and hard—daily Cabinet meetings interspersed with sessions with Congressional leaders regarding our ec. plan.

Monday, January 26

A meeting on terrorism with heads of F.B.I.—S.S.—C.I.A. Sec's of St., Defense & others. Have ordered they be given back their ability to function. Next a Cabinet meeting on the deal with Iran. We just may not implement some of the Carter executive orders on grounds they violate our own laws. Hostages will arrive in country tomorrow. It seems some of them had some tough questions for Carter in Germany as to why they were there so long and why they were there to begin with. Rest of day meeting committee chairmen & Sens. on raising the debt ceiling.

Tuesday, January 27

Ceremony on S. Lawn to welcome hostages home. Thousands of people in attendance. Met the familys earlier. Now we had in addition the familys of the 8 men who lost their lives in the rescue attempt. One couple lost their only son. His widow was also here. I've had a lump in my throat all day.—Evening 1st white tie reception for the diplomatic corps.

Wednesday, January 28

Visit by P.M. Seaga of Jamaica, his wife & members of his admin. Our 1st state luncheon. He won a terrific election victory over a Cuban backed pro-communist.

I think we can help him & gradually take back the Caribbean which was becoming a "Red" lake.

Thursday, January 29

Nancy had a great triumph with committee which rides herd on White

House (to preserve its history). They were enthusiastic about what she has already done to upgrade the 2nd & 3rd floors.

[Received cable from Mike Mansfield, U.S. ambassador to Japan.]

Friday, January 30

More meeting with Cong. leaders on trying to get debt ceiling lifted. If don't we'll be out of money by Feb. 18. Cong. recessing from 5th to 12th. Must get passage of bill by Fri. the 5th.

Short day in office—left for 1st weekend in Camp David. It was great to be in a house with the knowledge you could just open a door and take a walk outdoors if you wanted.

Saturday, January 31

Had a before lunch walk (it was cold). Spent afternoon in front of fire reading intelligence reports & Briefing papers for visit by Pres. Chun (Korea). We have definite evidence Nicaragua transferring hundreds of tons of arms from Cuba to El Salvador. P.M. ran a movie—"Tribute"—Jack Lemmon. He is truly a great performer.

[Sunday, February 1: took walk; returned to W.H.]

Monday, February 2

What's getting to be routine—full day in Oval office.

Tuesday, February 3

The arrival of Pres. Chun, his wife & staff. These meetings through an interpreter which can become a strain. Good meetings though—assured him we would not withdraw our troops from Korea

Wednesday, February 4

Cabinet discussion of grain embargo. I've always felt it hurt our farmers worse than it hurt Soviets. Many of our allies?? filled the gap & supplied Soviet. But now—how do we lift it without sending wrong message to Soviets? We need to take a new look at whole matter of strategy. Trade was supposed to make Soviets moderate, instead it has allowed them to build armaments instead of consumer products. Their socialism is an ec. failure. Wouldn't we be doing more for their people if we let their system fail instead of constantly bailing it out?

[Compliment from Weinberger on cabinet meeting.]

Thursday, February 5

[President's prayer breakfast; meetings with Boy Scouts and high school students.]

Lots of phone calls—Sen. Robt. Byrd (D) is playing games with bill to raise debt ceiling. Has held vote over til tomorrow.

Friday, February 6

My birthday. Nancy, Tip O'Neil, Paul Laxalt, Tom Evans & Cong. Wright from Texas surprised me (all duly recorded by Cap. Press Corps) with a beautiful cake. Tip gave me a tie & the flag which flew over the Cap. on Jan. 20. We wheeled the cake into another room where it was cut up by about 200 of our staff.

That afternoon received a great present—our own Sens. who had held out on debt. ceiling turned around and we carried the day.

[Surprise birthday party with California friends.]

P.S. During day discovered my Ambas. appts. were processed by State Dept. They take forever. I want Bill Wilson cleared by them before 26th so he can meet Pope (he's to be Ambas. to Vatican) in Alaska. Told Penn James to tell the guy at State that was advising him to get off his A-- & do it.

[Saturday, February 7: photo sessions and dinner party.]

Sunday, February 8

Thank you letters for gifts we found on 2nd floor Fri. night. It took entire Sat. morning to open. Just had a call from Al Haig. I had asked that we quietly have Swiss [. . .] tell Iranians if they did not free Mrs. Dryer (Am. woman they had charged with being a spy & imprisoned) we might find it difficult to implement the terms of the Carter hostage agreement. Mrs. Dryer is coming home. She was turned over to the Swiss. Word [. . .] is that last 2 weeks of hostage negotiations were completely dominated by Iranian fear they'd have to negotiate with our admin. I couldn't be happier.

Monday, February 9

Started the day learning Mrs. Dryer did not leave Iran—some snafu with paper work. Hopefully tomorrow.

[Meetings on timing of tax cuts, and with groups of state legislators; signed citation for Vietnam veterans.]

Tuesday, February 10

This was a day. I was wired for sound. David Brinkley is doing "a day in my life" for TV showing Fri. His cameras catch me in every meeting etc. and I turn on the sound for those things suitable & turn it off for balance of meetings. Began with Brkfst. with labor leaders who supported me in campaign; Teamsters, Merchant Marine, Marine engineers & Air Controllers.

[Meetings on economy with governors and with labor leaders, including Lane Kirkland of the AFL-CIO and Douglas Fraser of the UAW.]

An early hurried dinner then to Kennedy Center for Harlem Ballet. This group represents effort of one man (Mitchell) to take kids off the streets of Harlem & make something of them and he has. They are really good and it was an enjoyable evening. Cong. Tom Evans & his wife, Cong. Rangel & wife whose district includes Harlem, the Bushes etc. We discovered some of the niceties that go with this job. At intermission in the room behind the Presidential box we found a W.H. usher on hand serving snacks & drinks. Went back stage & met the dancers. Found out no other Pres. had even gone to one of their performances.

Came home to the news that Mike was being charged with fraud in connection with his "gas a hol" business. I suspect this is plain pol. and aimed at me.

Wednesday, February 11

High spot a Nat. Security Council meeting. We have absolute proof of Soviet & Cuban activity in delivering arms to rebels in El Salvador—Also their worldwide propaganda campaign which has succeeded in raising riots & demonstrations in Europe & the U.S.

Intelligence reports say Castro is very worried about me. I'm very worried that we can't come up with something to justify his worrying.

[Thursday, February 12: *meeting with cabinet regarding economic program; dinner with the Bushes.* Friday, February 13: *left for Camp David.]*

Saturday, February 14

Slept in—took a morning walk then spent the afternoon with desk work. We had both sneaked out & bought valentines for each other & believe it or not we did surprise each other. Ran a movie & had Dan Ruge & the others who have to go with us over for it. It was a comedy (Jane Fonda, Dolly Parton & Lilli Tomlin) "Nine To Five." Funny—but one scene made me mad. A truly funny scene if the 3 gals had played getting drunk but no they had to get stoned on pot. It was an endorsement of Pot smoking for any young person who sees the picture.

Sunday, February 15

I was at Camp David but I was indoors all day writing my speech for Wed. nite in Cong. Nancy spent the day interior decorating—very well I might add in the other cabins "Dogwood"—"Birch" etc. Finished speech by bedtime.

Monday, February 16

Walked to cabins to help Nancy with more picture hanging—an early lunch & back to Wash. A meeting regarding the speech—a few minor changes. Dinner here with Tip & Mrs. O'Neill—Ed, Jim B. & Susan & the Friedersdorf's.

[Several of the guests arrived late.]

It was a nice evening but maybe Tip & I told too many Irish stories.

Tuesday, February 17

Met with Congressmen & Sens. on grain embargo. Those from farm states want it lifted. I explained we'd made no decision but while I was against the embargo we had to worry about making a concession to the Soviets without some Quid Pro Quo. It might send a wrong message.

Tip had last word & it was a good one. He told me I was Pres. and had to think of all the states. The gist was—was lifting the embargo good for the U.S. and our security vis-a vis the Soviets.

Wednesday, February 18

This was the big night—the speech to Cong. on our ec. plan. I've seen Presidents over the years enter the House chamber without ever thinking I would one day be doing it. The reception was more than I'd anticipated—most of it of course from one side of the aisle. Still it was a thrill and something I'll long remember.

[Initial telegrams favored the speech. **Thursday, February 19:** *met with editors from all over the U.S.; left for California.* **Friday, February 20–Sunday, February 22:** *vacation at ranch; returned to Washington.* **Monday, February 23:** *speech at National Governors Conference.* **Tuesday, February 24:** *ceremony at Pentagon; state dinner for governors.]*

Wednesday, February 25

A cocktail reception for the 14 Cong'men. & wives who 1st came out for me and campaigned enthusiastically. Then an evening of trying to catch up with paper work. Maybe our drive to reduce govt. paper work should begin with us.

Thursday, February 26

P.M. Margaret Thatcher arrived. A most impressive ceremony on So. Lawn—review of troops etc. We had a private meeting in Oval office. she is as firm as ever re the Soviets and for reduction of govt. Expressed regret that she tried to reduce govt. spending a step at a time & was defeated in each attempt. Said she should have done it our way—an entire package—all or nothing.

[State dinner.]

Friday, February 27

P.M. getting great press. Went up to the hill and was literally an advocate for our ec. program. Some of the Sen's. tried to give her a bad time. She put them down firmly & with typical British courtesy.

Dinner at British Embassy—truly a warm & beautiful occasion. I believe a real friendship exists between the P.M. her family & us—certainly we feel that way & I'm sure they do.

Saturday, February 28

Saw the Thatchers off at 11 AM after coffee together. Mexican Ambassador brought message from Pres. Lopez Portillo to Dick Allen & Jim Baker. The Pres. is willing to go forward (with Venezuela) in trying to negotiate an end to El Salvador problem. I'm all for it. A call in evening reported a boat load of Haitians approaching our shores. I'm all for opening the door to refugees from totalitarianism but this is more complicated. These are just people who believe they can have a better life here. They are in fact illegal aliens. We'll have to deport them but it's a long & complicated business due to our own laws.

Had the Wilsons & Smiths to dinner & showed a movie in the "family theatre"—"The Black Stallion."

Sunday, March 1

[Awakened by false fire alarm.]

Church at 11—Nat. Presbyterian—very nice.

Took our 1st walk on W.H. grounds—saw the pool, dog runs & tennis court.

[Monday, March 2: addressed local government officials; new poll indicates support for budget cuts.]

Tuesday, March 3

[Meetings with economist Arthur Burns regarding China, and with President Ford on economy; message from Pope John Paul II expressing general greetings; VFW reception for Senator Laxalt (R-NV); dinner party.]

During day I did a 1 hr. interview with Walter Cronkite—his last for CBS. He spent the 1st 20 min's. on El Salvador. He didn't throw any slow balls but the reaction was favorable. Because of our dinner we couldn't watch the show but I was treated to another W.H. service. They taped the program & played it back to us later in the evening.

Wednesday, March 4

Our wedding anniversary. 29 years of more happiness than any man could rightly deserve. Nancy came down to the office & we met the Easter Seal girl for 1981 Colleen Finn, a charming 6 yr. old in a wheelchair.

[Met with freshmen Republican congressmen.]

A Pakistani plane was hijacked and landed in Kabul. The Russians are holding it & 3 or 6 of the passengers are American. We haven't been able to learn which figure is right but we're going to let the Soviets know we won't put up with their games.

Tonite—a dinner party (old friends from home) at the Jockey Club. I'm really looking forward to it.

Home—a wonderful evening—Alfred & Betts, Smiths, Wicks, Lee A. & Ted G.

[**Thursday, March 5**: *breakfast with conservative Democratic congressmen; photo session with children afflicted with cystic fibrosis; meeting with Louisiana congressmen; lunch with Vice President Bush; cabinet meeting; taped video appearances for various groups.* **Friday, March 6**: *met with governors regarding automobile industry problems; press conference; went to Camp David.* **Saturday, March 7**: *responded to news of the execution of a church worker in Bolivia.*]

Sunday, March 8

[Returned to Washington; dinner party.]

During the day watched 2 panel shows, Issues & Answers & Face the Nation. On one the Soviet Consul proved I was right when I said Communists reserved the right to lie. On the other Walter Heller proved ecs is the dismal science. He was against our ec. program but had nothing to say about 3 decades or more of his ec. principles had led to the disastrous situation we are in.

Monday, March 9

[Interviewed by Frank Reynolds, termed "objective and he's also a Gentleman of great integrity"; met with West German foreign minister Hans-Dietrich Genscher.]

Had meeting with Jewish Repub. leaders. They were concerned about sale of F-15's to Saudi Arabia. I think we made it alright when I told them we had discussed the matter with the Israeli foreign minister and that we were going to increase help to Israel.

Meeting in situation room—approved some covert operations. I believe we are getting back on track with a proper approach to "intelligence" under Bill Casey.

Tuesday, March 10

Off to Canada with Sec's of State, Commerce, Treasury & Special trade rep. Bill Brock.

[Formal arrival ceremony at airport.]

It was a warm welcome with Canadians lining the streets cheering & clapping. Quartered at Rideau house which is truly a magnificent old mansion—except that Nancy & I were in separate rooms—1st time in our marriage.

Went to Parliament hill to meet P.M. Trudeau. Discovered I liked him. Our meetings were very successful. We have some problems to be worked out having to do with fishing, energy & environment but I believe we've convinced them we really want to find answers.

[Gala evening and dinner.]

Wednesday, March 11

Met with Consv. leader Joe Clark—then to Parliament for speech which was very well received. Sandy Vanocur called it best summit he's ever covered.

Stopped by American Embassy to speak to staff there then on to the airport.

In a large hangar faced honor guard while band played the Nat. Anthems of both countries. Somehow the Star Spangled Banner when you hear it in another country brings a tear to the eye. As we turned to leave the band played Auld Lang Syne—then we were really undone.

[*Thursday, March 12: breakfast with Republican congressmen; lunch with Bush; visited Senator Dole in hospital; reception for National Newspaper Association; received set of bronzes from businessman and former ambassador Walter Annenberg.*]

Friday, March 13

Nancy off to N.Y. I'm to join her there. The Rev. Jesse Jackson announces he's staging a march on Wash. Mon. demanding help for Atlanta in this 19 month tragedy of the murders of black children which has so far numbered 20 with 1 missing. Atlanta has a financial problem due to $100,000 a month overtime for police. What Jesse apparently doesn't know is that we've already given Atlanta about $1 mil. and have roughly 40 F.B.I. agents in there on the case. Today I went to the press briefing room and read a statement detailing this & then announced we were giving another $1½ million. & that V.P. G. Bush was going personally to Atlanta.

Finished calling a number of St. Dept. Professionals appointing them to Ambassadorial posts—no Pres. has ever done this personally before. I'm enjoying doing some of the things I'm then told Pres's. haven't done before.

Flew into LaGuardia then helicoptered around the lower tip of Manhattan and landed on a heli-pad on the lower west side. The usual lengthy motorcade with police closing off intersections and taking us traffic free through town. One thing was unusual and very humbling; the streets were lined with people as if for a parade all the way to the Waldorf. They cheered & clapped and I wore my arms out waving back to them.

I keep thinking this can't continue and yet their warmth & affection seems so genuine I get a lump in my throat. I pray constantly that I won't let them down.

Saturday, March 14–Sunday, March 15

Met with the editorial board of the Daily News. It was a good & friendly meeting concerned mainly with our ec. program.

Then went to lunch in Little Italy with Sen. D'Amato & Cong. Molinari & their wives. It was a wonderful experience—again the streets were jammed. Met with Mayor Koch later and ironed out some misunderstandings about our program. That night saw "Sugar Babies." Again a great reception from the people in the theatre & from the cast when we went backstage to see Mickey R. and Ann Miller.

At the final curtain Mickey asked the audience to stay in place until we exited. As we went up the aisle the audience began singing "America the Beautiful." We went to supper with Alfred & Betts, Claudette Colbert, Jerry Z. the Cowles etc. & all in all had a wonderful time.

Sunday—went to the Met for the Joffrey Ballet. Ron was flown in from the road company for the performance. I think I held my breath until he finished but he was good. Flew back to Wash.

Monday, March 16

Jerry Lewis came to the Oval office with the M.D. poster girl—a beautiful child who I learned would probably die before the next year ends. What a terrible disease. Jerry has for 29 years worked to raise more than $400 mil. for research of M.D.

Paul Laxalt came by & played Santa Claus with gifts he'd received for me— including the most beautiful western belt buckle I've ever seen. He had a letter from an Irishman in Nev. who complained because he didn't think I knew the R.W. Service poem "The Shooting of Dan McGrew." We put in a call to Nevada and after I convinced him I really was who I said I was I recited the poem to him. He's a Dem. who I think may now turn Repub.

[Appearance at convention of general contractors. Mrs. Reagan had lunch with Mrs. Sadat; president met with congressmen. **Wednesday, March 18:** *saw teacher who once taught Reagan children; breakfast with freshmen senators; met Jewish woman and the Dutch citizen who saved her from the Nazis.]*

Thursday, March 19

The auto task force met with Cabinet—still some disagreement about any quotas on Japanese imports. Some even with regard to a Japanese voluntary cutback. The V.P. summed it up nicely. He said we're all for free enterprise but would any of us find fault if Japan announced without any request from us that they were going to reduce their export of autos to America?

There was no dissent. I told them I'd heard enough I would make a decision. Privately I told Al Haig to call Amb. Mike Mansfield and have Mike advise "Ito" before his visit that we were threatened by a bill in Cong. to set a quota. An announcement by Japan of a voluntary cutback could head that off. We'll see what happens.

Al then told me he felt he was being undercut by other agencies etc. I worry that he has something of a complex about this. Anyway I've arranged that he & I meet privately 3x a week.

Kennedy Center in the evening for "Little Foxes" starring Liz Taylor. She was darn good—so was the show.

Friday, March 20

[*Meetings with Senator Jacob Javits (R-NY) and with former Japanese prime minister Takeo Fukuda.*]

Forgot to say in yesterday's account that I met with Chai the Amb. P.R. of China. With him a man from Beijing. The Amb. got on the Taiwan subject—promised peaceful resolution of the problem, then made it plain he didn't feel we should sell weapons to Taiwan. I made it just as plain that we had an obligation we would fulfill. No harm done.

[*Met with state legislators; swearing-in ceremony for Lee Annenberg as chief of protocol; spoke at Conservative Dinner.*]

Saturday, March 21

Quiet day at home until 5:10. Then a reception for the entertainers who performed later at Ford Theatre.

Stopped by Georgetown Club a roast being given by press to Jim Brady. Jim is well liked & deservedly so.

On to Ford Theatre for 2 hour benefit for support of the theatre. Our 1st visit—there is a definite feeling when you see the flag draped Presidential Box where Booth shot Lincoln.

[*Sunday, March 22: a quiet day.*]

Monday, March 23

East Room in the morning to meet with Jim Roosevelt & all the March of Dimes poster children from the beginning until the present—many of course now grown up. This years little charmer drops her crutches in a second to be picked up and then she promptly kissed me & told me I was cute.

[*Presented award to veteran civil servant.*]

Ec. Council meeting and meeting with Al Haig. The grain embargo is coming to a head. We shouldn't lift it unless the Russians show some signs of being decent—still it's hurting our farmers I fear worse than it's hurting the Russians.

Tuesday, March 24

[*Breakfast with congressional leaders regarding economic package; met with Japanese foreign minister Masayoshi Ito.*]

Nat. Security Council meeting in Situation Room. We adopted a plan to persuade African States of our desire to help settle the Namibian question—an election after a const. is adopted. At the same time we would urge Angola's govt. to oust the Cubans at the same time we helped Savimbi. Our hope being that with the Cubans out NATO & Savimbi could negotiate a peace.

Later in day a call from Al Haig, all upset about an announcement that George B. is to be chairman of the Crisis Council. Historically the chairman is

Nat. Security Advisor (Dick Allen). Al thinks his turf is being invaded. We chose George because Al is wary of Dick. He talked of resigning. Frankly I think he's seeing things that aren't there. He's Sec. of St. and no one is intruding on his turf—foreign policy is his but he has half the Cabinet teed off.

Wednesday, March 25

Awakened at 7:30 by Mike D.—Al Haig again. He called me I told him to meet me in the Oval O. at 8:45. Mike told me Al was talking resignation etc. He arrived and was very calm—never mentioned resigning but had a statement he wanted me to make. It wasn't bad but it wasn't good. After he left I drafted a statement of my own—short, simple & it did the job.

Met with our volunteer ec. advisors chaired by Geo. Schultz. They are opposed totally to any import limit on Japanese cars.

Went for my 1st horseback ride at Quantico. It felt great. We should do this often.

[*Thursday, March 26: reception for Young Republicans; dinner for broadcast correspondents.*]

Friday, March 27

A Honduran plane hijacked—Americans aboard. It landed in Managua Nicaragua—the usual demand—release of some prisoners in Honduras. Nicaragua's Sandinista govt. has been trying to convince us they'll quit trying to help the Cuban sponsored revolution in El Salvador. Well now's their chance.

[*Hosted luncheon in honor of National Baseball Hall of Fame.*]

Laxalt & Maureen for dinner & a movie—"Tess."

Saturday, March 28

A day at the desk upstairs doing remarks for Sat. night—the "Gridiron" & Mon. afternoon address to the AFL-CIO—Building & Construction Trades. Conf.

The "Gridiron" was very good and a good time was had by all. I was able to announce the Honduran Hijackers were in custody in Panama & all hostages safe.

Sunday, March 29

Church at St. Johns—a beautiful Spring Day and at the service the Navy Choir from Annapolis. They looked & sounded so right you have to feel good about our country.

[*Report on hijacking in Thailand; conferred with decorator regarding Oval Office.*]

Monday, March 30

My day to address the Bldg. & Const. Trades Nat. Conf. A.F.L.-C.I.O. at the Hilton Ballroom—2 P.M. Was all dressed to go & for some reason at the last min. took off my really good wrist watch & wore an older one.

Speech not riotously received—still it was successful.

Left the hotel at the usual side entrance and headed for the car—suddenly there was a burst of gun fire from the left. S.S. Agent pushed me onto the floor of the car & jumped on top. I felt a blow in my upper back that was unbelievably painful. I was sure he'd broken my rib. The car took off. I sat up on the edge of the seat almost paralyzed by pain. Then I began coughing up blood which made both of us think—yes I had a broken rib & it had punctured a lung. He switched orders from W.H. to Geo. Wash. U. Hosp.

By the time we arrived I was having great trouble getting enough air. We did not know that Tim McCarthy (S.S.) had been shot in the chest, Jim Brady in the head & a policeman Tom Delahanty in the neck.

I walked into the emergency room and was hoisted onto a cart where I was stripped of my clothes. It was then we learned I'd been shot & had a bullet in my lung.

Getting shot hurts. Still my fear was growing because no matter how hard I tried to breathe it seemed I was getting less & less air. I focused on that tiled ceiling and prayed. But I realized I couldn't ask for Gods help while at the same time I felt hatred for the mixed up young man who had shot me. Isn't that the meaning of the lost sheep? We are all Gods children & therefore equally beloved by him. I began to pray for his soul and that he would find his way back to the fold.

I opened my eyes once to find Nancy there. I pray I'll never face a day when she isn't there. Of all the ways God has blessed me giving her to me is the greatest and beyond anything I can ever hope to deserve.

All the kids arrived and the hours ran together in a blur during which I was operated on. I know it's going to be a long recovery but there has been such an outpouring of love from all over.

The days of therapy, transfusion, intravenous etc. have gone by—now it is Sat. April 11 and this morning I left the hospital and am here at the W.H. with Nancy & Patti. The treatment, the warmth, the skill of those at G.W. has been magnificent but it's great to be here at home.

Whatever happens now I owe my life to God and will try to serve him in every way I can.

Sunday, April 12

The 1st full day at home. I'm not jumping any fences and the routine is still one of blood tests, X rays—bottles dripping into my arms but I'm home. With the let up on antibiotics I'm beginning to have an appetite & food tastes good for the first time.

Monday, April 13

I'm beginning to have a work schedule. We meet in the mornings in the treaty room. It feels good to be whittling at the problems. Afternoon is still nap time and bed time follows dinner by about ½ an hour.

Tuesday, April 14

Our astronauts landed and what a thrill that was. I'm more & more convinced that Americans are hungering to feel proud & patriotic again.

The circus came to town and paraded in front of the White House & put on a show. They had a big get well banner. I waved from the biggest window I could find & thank Heaven they saw me.

*[***Wednesday, April 15***: met with Haig and Weinberger; Mrs. Reagan has call from Queen of Jordan regarding president's health.* **Thursday, April 16**: *met with secretary general of NATO; press flap about pardon of FBI agents involved in Weathermen surveillance.]*

Friday, April 17

Good Friday and while representatives of the Nat. Co. of Churches paraded (carrying the cross) in front of the W.H. protesting our effort to help El Salvador Cardinal Cooke of N.Y. came to visit us in the W.H. He's a good man and doing magnificent work in ed. in the inner cities.

Sometimes I think (forgive me) the Nat. Co. believes God can be reached through Moscow.

Later in the day talked by phone with Billy Graham. He knows the family of the young man who did the shooting. They are decent, deeply religious people who are completely crushed by the "sickness" of their son.

Saturday, April 18

A nice quiet day—no emergencies, slept in late but still managed an afternoon nap. Wrote a draft of a letter to Brezhnev. Don't know whether I'll send it but enjoyed putting some thoughts down on paper. 9 P.M. and we're off to bed.

Sunday, April 19—Easter

A beautiful Easter morning. In the afternoon Rev. Louis Evans & his wife called and brought us communion. They made it a most meaningful day.

Watched some TV in bed and saw Gloria Steinem take me over the coals for being a bigot and against women. Either she is totally ignorant of my positions which I doubt or she is a deliberate liar.

*[***Monday, April 20***: described Easter Egg Roll, but refrained from attending for security reasons; Ron, Doria, and C.Z. arrive.* **Tuesday, April 21**: *meetings with staff and with*

Republican governors; comment on penal regulations; note on progress of letter to Brezhnev through State Department channels.]

Wednesday, April 22

Won part of the battle with the diplomats. They drafted the letter to Brezhnev along usual lines but included major positions of mine. We sent it back for a re-write including more of mine.

More phone calls to Congressmen. I'm having more luck with Demos. than Repub's. Asked Tip O'Neill if I could address a joint session next week. He agreed.

Did an interview with AP & UPI on the shooting. Just learned they have to re-operate on Jim Brady. No word yet on what this means to the otherwise optimistic prognosis.

Thursday, April 23

Met with Sen. Baker re the AWAC's sale to Saudi Arabia & the fuss being kicked up about it. He made a sound suggestion that we don't present the proposal to Cong. until after next weeks vote on the ec. prog.

I'm disturbed by the reaction & the opposition of so many groups in the Jewish community. First of all it must be plain to them, they've never had a better friend of Israel in the W.H. than they have now.

We are striving to bring stability to the Middle East and reduce the threat of a Soviet move in that direction. The basis for such stability must be peace between Israel & the Arab nations. The Saudi's are a key to this. If they can follow the counsel of Egypt the rest might fall in place. The AWAC's wont be theirs until 1985. In the meantime much can be accomplished toward furthering the Camp David format.

We have assured the Israelis we will do whatever is needed to see that any help to the Arab States does not change the balance of power between them & the Arabs.

Last night watched the NBC special on our first 100 days. If only Bobby Burns had waited for TV he would never have written "If only we had the gift to see ourselves as others see us."

Attached is a script of letter I wrote to Brezhnev by hand.

My Dear Mr. President

I'm sorry to be so long in answering your letter to me and can only offer as an excuse the problems of settling into a routine after my hospitalization. I ask your pardon.

I won't attempt a point by point response to your letter because I agree with your observation that these matters are better discussed in person than in writing. Needless to say we are not in agreement on a number of points raised in both my letter & yours.

There is one matter however which I feel I must bring to your attention. All information having to do with my govt's practices & policies past & present is available to me now that I hold this office. I have thoroughly investigated the matter of the man Scharansky an inmate in one of your prisons. I can assure you he was never involved in any way with any agency of the U.S. govt. I have seen news stories in the Soviet press suggesting that he was engaged in espionage for our country. Let me assure you this is absolutely false.

Recently his wife called upon me. They were married and spent one day together before she emigrated to Israel assuming that he would follow shortly thereafter. I believe true justice would be done if he were released and allowed to join her.

If you could find it in your heart to do this the matter would be strictly between us which is why I'm writing this letter by hand.

While on this subject may I also enter a plea on behalf of the two familys who have been living in most uncomfortable circumstances in our embassy in Moscow for three years. The   family & the   are Pentacostal Christians who feared possible persecution because of their religion. Members of that church in America would, I know, provide for them here if they were allowed to come to the U.S.

Again as in the case of Scharansky this is between the two of us and I will not reveal that I made any such request. I'm sure however you understand that such actions on your part would lessen my problems in future negotiations between our two countries.

Sincerely.

Friday, April 24
1st day in oval office. Staff meeting at 9:30 then met with Al Haig & Jack Block about lifting grain embargo. I'm reluctant about it but think it will reassure our allies that while we're hard nosed about the Russians we aren't refusing to talk. At 11 AM met with Cabinet—1st since shooting. They greeted me with applause and I was very moved. What a team they've turned out to be & how proud I am of them. (I told em so) Big issue for discussion was how to fund a strategic oil reserve. No decision yet.

[Report on poll numbers; visit from father-in-law, Dr. Loyal Davis. **Saturday, April 25–Monday, April 27:** *quiet days at Camp David.]*

Tuesday, April 28

Spent the morning in the office—1st time. Met with Congressional (Repub.) leadership. All of us agreed we shouldn't be over confident.

Rec'vd. a mailgram from the Hinckleys—parents of the boy with the gun.

Evening was the big part of the day. Addressed a joint session of Cong. re the ec. pckg. I walked into an unbelievable ovation that went on for several minutes. Frank Reynolds says the speech was interrupted 14 times (3 of them standing ovations). In the 3rd of those suddenly about 40 Democrats stood and applauded. Maybe we are going to make it. It took a lot of courage for them to do that and it sent a shiver down my spine. Except for that all the applause came from the Repub. side, the Demos. just sit on their hands—except for the greeting.

The response from the public has been overwhelming.

Wednesday, April 29

Again spent morning in the office—returned to residence for lunch & nap. I must be getting better—my naps are getting shorter.

We had a Nat. Security Council briefing (meeting tomorrow). Several situations in the world are worsening—particularly Lebanon.

Donn Moomaw came by & also the Ted Cummings. Recv'd. word that Jules Stein died last night. That is truly the end of an era.

Thursday, April 30

[Met with Republican congressmen; ceremony remembering Holocaust.]

An N.S.C. meeting devoted mainly to charting our course—a double track course of getting NATO agreement on installing Theatre Nuclear weapons and opening arms limitation talks with the Soviets. This latter is most important politically to our allies.

Spent evening at home watching Patti on the Merv Griffin Show.

Friday, May 1

More meetings with Congressmen. Highlight was noon visit by Prince Charles. He's a most likeable person. The ushers brought him tea—horror of horrors they served it our way with a tea bag in the cup. It finally dawned on me that he was just holding the cup & then finally put it down on a table. I didn't know what to do. Mike escorted him back to the W.H. and apologized. The Prince, "I didn't know what to do with it." Betty Wilson & the Prices came by in the afternoon. Why does that just make me more homesick?

Saturday, May 2

This was the big night—our non-state, upstairs dinner for the Prince—32 in all. It was a great success & I feel everyone present had fun. We had 4 tables (of 8) in the family dining room after cocktails in the Oval room. Then back to the Oval room for music by Bobby Short & a special song by Sammy Kahn.

[Praise for Mrs. Reagan's redecoration of the W.H. **Sunday, May 3**: called man who helped subdue John Hinckley during assassination attempt.]

Monday, May 4

[Meeting with Democratic congressmen.]

It was a beautiful spring day. We lunched in the solarium & then I sun bathed for about an hour.

Heard from Haig & Dick Allen—approved closing the Libyan embassy here. One of their officials linked to a terrorist killing in Chi. Our embassy there was burned down & there has never been compensation or rebuilding.

Tuesday, May 5

P.M. Begin sent a letter by way of Israeli Ambass. in answer to my request that Israel hold off on the Syrian Missiles in Lebanon while we tried to bring about a settlement. His letter gave us the go ahead. Then I met with Habib. We asked him to come out of retirement and undertake a special mission to Syria, Lebanon & Israel. He's on his way.

Al Haig back from a most successful NATO meeting in Rome. They have agreed to * accept Theatre Nuclear Weapons on their soil & are convinced we are willing to negotiate with the Russians. Clincher Al said was when he showed them a copy of my handwritten letter to Brezhnev.

Spent rest of my time meeting with Demo. Congressmen.

Wednesday, May 6

More meetings with Cong. These Demos. are with us on the budget and it's interesting to hear some who've been here 10 years or more say it is their 1st time to ever be in the Oval office. We really seem to be putting a coalition together.

[Working lunch to plan stance for meeting with Japanese prime minister Zenko Suzuki.]

Thursday, May 7

P.M. Suzuki & party arrive—full mil. honors on S. Lawn. I'd been told he

---

* At this point in the diary, five lines were crossed out inconsequentionally by President Reagan: it seems that pages were stuck together, and the president started writing on those pages, then realized he was wasting paper, so he crossed out the lines and went back and rewrote them two pages earlier.

was nervous about the visit. He's not the usual political type and is new in the job—that makes 2 of us.

He and I met in the Oval office for more than half an hour. It was a good meeting. I had the interpreter send a copy of my hand written letter to Brezhnev. In the cabinet room later in the larger meeting he told them I had made him feel that he could pick up the phone and call me personally about any problems that might come up. Tonite the State Dinner.

But 1st this was the big day. The Budget bill passed 253 to 176. All the Repubs. stayed together & 63 Demos. voted with us. We never anticipated such a landslide. We felt we were going to win due to the conservative block of Demos but expected R. defectors so that we might win by 1 or 2 votes. It's been a long time since Repubs. have had a victory like this.

[Considers meeting with Suzuki a success.]

Friday, May 8

Spent the morning—or most of it—in a meeting with the P.M. his ministers and my cabinet members. We got down to specifics. He opened the whole subject of defense etc. and then stated our position on the Soviet U. and what we intended to do and that we would consult with Japan as our senior partner. There would be no pressure or arm twisting by us—just frank discussion of how we can help each other.

Saw him off and then took off for Camp David.

*[*Saturday, May 9–Sunday, May 10: *Ron, Doria, Patti, and family and friends celebrate Mother's Day weekend.* Monday, May 11: *meeting with task force on Social Security; reception for congressmen who favored budget cuts.* Tuesday, May 12: *meetings with Senate Republican leaders and with governors regarding block grants.]*

Wednesday, May 13

Word from Habib in Israel holds out a little hope that we can avert a nose to nose between Syria & Israel.

[Met with Democratic senators regarding crime.]

Word brought to us of the shooting of the Pope. Called Cardinal Cooke & Cardinal Crowell—sent message to Vatican & prayed.

Thursday, May 14

Cabinet met 90 min.—working lunch. Subject—the way the Bureaucrats are slipping in cost overruns which are increasing the '81 bud. deficit. There is no question but it is the usual warfare the perm. structure wages to protect its turf. It's a war we must win.

5:45 a reception for the Sens. D. & R. who voted for our budget. Then upstairs to dress for my 1st time to go out on the town. The Tom Jones' had a dinner at Georgetown Club for all the Californians (& others) who were in town for the Wrathers premiere of "The Legend of the Lone Ranger."

It was wonderful to see all of them and get homesick all over again. I was surprised to find my back tingled a bit when I got out of the car & went up the steps with a crowd of people across the street in the dark.

Friday, May 15

Met with Botha—F. Minister S. Africa. Then with the F.M. of Romania. An N.S.C. meeting with Lebanon on the table. Latest message from Habib does not sound good although he said to make no decision for a few days. Begin was more flexible than Assad of Syria (bolstered by Soviets) Habib going to Saudi Arabia to see if they'll lean on Assad. Sometimes I wonder if we are destined to witness Armageddon.

[Mrs. Reagan's plans for U.S. Navy ship christening.]

Saturday, May 16—Armed Forces Day

Nancy up at the crack of dawn to leave for Miss. & the launching.

Why am I so scared always when she leaves like that? I do an awful lot of praying until she returns. She returned and I've said my thank you. While she was gone I got a call that Prince Turki [al-Faisal] of Saudi Arabia was in N.Y. If we could persuade him to come here & meet me (he's a nephew of Crown Prince Sahd) maybe we could persuade him to fly back to Saudi A. and see if they would lean on Syria while we continue to talk to Begin. Meeting a success we're sending him back by mil. aircraft.

Sunday, May 17

Change signals on Prince Turki—he left for Saudi Arabia today by commercial plane.

An easy morning with the Sunday papers—my exercises and then boarded the chopper & Air Force 1 for South Bend Ind. Had lunch on board with Walter & Lee Annenberg. Father Hesburgh met us at the airport and we drove to Notre Dame. It was commencement for 2000 graduates but there must have been 15,000 all told in the auditorium. Pat O'Brien was there also to get an honorary degree. It really was exciting. Every N.D. student sees the Rockne film and so the greeting for Pat & me was overwhelming. Speech went O.K. and I was made an honorary member of the Monogram Club. When I opened my certificate I thought they'd made 2 copies—they hadn't, the 2nd was to "The Gipper." He died before graduation so had never been made a member.

Got back to the plane wringing wet—Cap & Gown plus an "iron" vest makes for heat. Discovered a service I hadn't been aware of—a change of clothing is always carried when I go on a trip. Change in this case meant a welcome dry shirt.

[Watched 60 Minutes; called Senator Dole (R-KS) about tax plan.]

Monday, May 18

We may have a chance in Lebanon of heading off a war. The Saudis sent an emissary to Syria. Our problem now is 2 pol. leaders Assad & Begin & finding a way that saves both their faces & doesn't look like the U.S. managed this.

Met with Volcker of the Fed. Reserve. Int. prime rates went to 20 today. This is "chicken little" stuff in the money mkt. based on pessimism that Cong. won't give us what we're asking so inflation will go up.

Bill Smith came in with a task force report on immigration. Our 1st problem is what to do with 1000's of Cubans—criminals & the insane that Castro loaded on refugee boats & sent here.

Finally an interview with Teddy White. Yes now I'll be in a book called "The Making of the Pres."

Tonite we went to the Lisner Auditorium for the benefit performance by the Joffrey Ballet. Ron is featured in much of it—the old show biz story. One of the principles strained a muscle & Ron replaced him. He was darn good. He has a grace that is remindful of Fred Astaire—a little extra flair that makes it all look easy.

[Tuesday, May 19: luncheon with NASA astronauts. Wednesday, May 20: meetings in anticipation of Ottawa summit and visit from German chancellor Helmut Schmidt; official send-off for group of U.S. ambassadors; reception for National Advertising Council; birthday party for Ron. Thursday, May 21: meeting with Schmidt, regarded as a show of friendship. Friday, May 22: finds agreement with Schmidt on "future course with regard to Russia." Leaves for vacation at ranch in California.]

Saturday, May 23–Tuesday, May 26

The weather was beautiful (most beautiful this morning when we had to leave). We rode in the mornings—Little Man, Fancy & Gwalianko. Nancy of course on "No Strings." Fancy is the new mare Mrs. Weldon gave us. She's a beautiful chestnut—A little uptight at trail riding because she's spent her life in the show ring.

In the afternoons (short ones Nancy saw to that) we hacked away at the snowstorm debris—broken limbs & downed trees. We cut out the wood suitable for fuel and then piled the brush for burning next winter. Dinner by the fireside & early to bed.

Now we're back in Wash. & off to West Point in the morning.

Wednesday, May 27

Up at 7:15 & on Marine 1 at 8:15. Boarded A.F.1 at Andrews & flew up to West Point where we helicoptered & drove to the stadium. Can there be anything more stirring than a West Point graduation? The speech was well received and I shook 900 hands. We then returned to the White House & problems. We've called Habib home for consultations. He felt he'd be sitting in one or the other of the 3 capitals doing nothing which might make our peace efforts look like a failure. This way it gives an appearance of continued action which we hope will keep all sides from taking a precipitate flyer. He's done a great job and kept things that were about to blow, quiet for 3 weeks.

[Received word of a bomber crash on the deck of the USS Nimitz.*]*

Thursday, May 28

Latest on Nimitz disaster 14 dead 44 injured. Had an N.S.C. meeting. Gave the go ahead to a plan for the Caribbean & Central Am. to help the smaller cos. economically, social reforms etc. plus mil. aid to protect against Cuban exported radicals. We'll involve or ask to be involved Canada, Mexico, Venezuela & others. I proposed that the answer to the Haitians continuing to come here as illegal aliens for ec. reasons should be a program to upgrade Haiti's economy so it could provide jobs for its people.

Cabinet meeting. Demo's. finally have come up with a counter proposal to our tax program. They want to include a reduction of the inc. tax rate on unearned income from 70% to the 50% top rate on earned income. We wanted that in the 1st place but were sure they'd attack us as favoring the rich. Several of their other proposals are things we wanted. Decision is, I'll reluctantly give in provided they'll accept the 3 yr. across the board cut which will be 5–10 & 10 instead of the 10–10 & 10 we originally proposed. I'll hail it as a great bipartisian solution. H--l! It's more than I thought we could get. I'm delighted to get the 70 down to 50. All we give up is the 1st year 10% beginning last Jan. to 5% beginning this Oct. Instead of 30% over 3 yrs. (36 mos.) it will be 25% over 27 months.

[Dinner party at the home of Dillon Ripley, wildlife expert and secretary of the Smithsonian Institution.]

I almost forgot to mention the 1st meeting of the day. Mrs. Scharansky & a young Jewish refugee from the Gulag (10 yrs. in prison) came to see me. Mrs. S. married her husband 10 yrs. ago. She had a visa to leave Russia for Israel the day after the wedding. She left—the authorities told her the Groom would be allowed to leave very soon to join her. He is in the Gulag—was never allowed to leave—is said to have been an American spy which he never was.

D--n those inhuman monsters. He is said to be down to 100 lbs. & very ill. I promised I'd do everything I could to obtain his release & I will.

Friday, May 29

[Photo session with diplomatic appointees.]

Then met with Am. Habib who has been performing miracles in holding off a Syrian-Israeli war for 3 weeks. He has the Sandis involved pressuring the Syrians to be reasonable.

He'll go back after the weekend.

Saturday, May 30

How good to sleep in and not get up til the papers have been read.

Al Haig called—the Saudis have persuaded a combined force of the moderate Arab nations to go into Lebanon as a deterrent in place of the Syrians. This could give the Syrians a face saving way of withdrawing and that would mean peace.

Tonite a dinner for 18 here at the W.H. It was very nice—Frank & Barbara, the Michels, Isaac Sterns, Geo. Murphy, the Kissingers etc., Bill Paley, The Wyeths.

[Sunday, May 31: lunch with friends in Virginia.]

Monday, June 1

Quite a day! Met with Sen. Simpson re immigration. He had a great collection of fraudulent S.S. cards, drivers licenses, union cards etc.—Oh also phony food stamps. How can we curb this kind of counterfeiting? Then it was Sen. Ted Kennedy's turn. He's interested in gun control but not as rabid about it as I thought.

[Negotiations with congressional leaders over tax bill.]

Later an N.S.C. meeting. I okayed reversing the Carter policy against selling arms to our friends. Venezuela wants to buy F16 fighter planes. Also said yes to a plan to enlist Canada & Mex. in aiding Central America & the Caribbean nations. Ec. help, social reform & help in resisting revolutions exported from Cuba.

Moving to the middle east. I approved naval maneuvers in Mediterranean waters that Khadafi of Libya has declared are his territorial waters. I'm not being foolhardy but he's a madman. He has been harassing our planes out over international waters & it's time to show the other nations there Egypt, Morocco, et al that there is different management here.* Nancy went to Wolf Trap—Bob Hope performing—a big benefit opening. Security wouldn't OK it for me. She came home with evidently "a bug" that's been going around—a kind of 1 day intestinal disorder. She was up most of the night & thoroughly miserable. That rules her out for visiting Jim Brady tomorrow.

---

* *In 1973, Libyan dictator Mohammar al-Qaddafi declared territorial waters in excess of the internationally recognized twelve-mile limit, such that his claim encompassed the entire Gulf of Sidra in the Mediterranean Sea. He threatened to attack any planes or ships crossing what he called the "Line of Death." Previous U.S. presidents had respected the line.*

Tuesday, June 2

[Meeting with Republican leadership, mainly on tax plan and block grants.]

Cabinet meeting re the selling of 100,000 tons of butter from govt. stockpile. Only real customer would be Russia. The world price is $1 a lb.—Ams. are paying $1. I've proposed finding what we can do toward giving it to hunger areas instead. I haven't given my answer yet but feel it will be against selling even though we'd get a couple of hundred million $.

Met with Jim Brady at the hospital. He looks well and has muscle movement in his left arm & leg. He got emotional a few times but they say that is normal at this stage. The prognosis continues to be that he'll have good recovery.

Nancy went to Wolf Trap opening Mon. Security said no to me going.

[Wednesday, June 3: Reagan feels ill early in the day, rallied later; met with people with multiple sclerosis, escorted by Frank Sinatra, and with the NSC; reception for European civic leaders.]

Thursday, June 4

A very good day. Met with Dem. Conservatives. Believe a majority of the 33 will support our tax program.

[NSC meeting on Cuba and upcoming Haig trip to the People's Republic of China; received ambassadors; met with senators on tax bill; greeted old friends in the family quarters.]

High spot of the day however was lunch. Our guest was Mother Theresa. What an inspiration this quiet little Albanian born Nun is. She radiates joy because God, as she says, has given her the opportunity to serve the lepers, the poverty stricken & the hopeless.

Friday, June 5

A hectic ¾ of a day—one meeting or appointment on top of another with no time in between to shift gears.

Did however give a medal to Ethel Kennedy honoring Robt. K. It was voted by Cong. in 1978 and the former Pres. never presented it.

Off to Camp David.

Saturday, June 6

Had planned to have our 1st horseback ride at Camp David but it rained so I did my homework for the meeting with Pres. Lopez Portillo—Mexico.

Sunday, June 7

Weather broke bright & clear so finally our ride. I wanted to see the trail at least once because the Pres. & I are scheduled to ride on Monday.

Got word of Israeli bombing of Iraq—nuclear reactor. I swear I believe Armageddon is near.

Returned to W.H. at 3 P.M. More word on bombing. P.M. Begin informed us after the fact.

Monday, June 8

Pres. Lopez Portillo and party arrived. We had the Marine band on hand for the welcome. Then off by helicopter for Camp David. Learned he was looking forward to complete informality so told him we'd change when we arrived. He & I met for more than an hour discussing our idea of U.S., Mex. & Canada launching a program to put the Caribbean nations & Central America on a solid Ec. base & correct the social inequities which make them ripe for Cuban inspired revolution: Then a working lunch—a dozen of us out on the terrace.

[Horseback ride with President José López Portillo; barbecue dinner.]

Tuesday, June 9

[Further meetings with President José López Portillo and Mexican officials; state luncheon.]

Ended day with an N.S.C. meeting re the bombing of Iraq. P.M. Begin insists the plant was preparing to produce nuclear weapons for use on Israel. If he waited 'til the French shipment of "hot" uranium arrived he couldn't order the bombing because of the radiation that would be loosed over Baghdad.

I can understand his fear but feel he took wrong option. He should have told us & the French, we could have done something to remove the threat.

However we are not turning on Israel—that would be an invitation for the Arabs to attack. It's time to raise H--l worldwide for a settlement of the "middle-east" problem. What has happened is the result of fear & suspicion on both sides. We need a real push for a solid peace.

Jus. Dart called—a solid offer has been made for our home. We said yes but we'll cry when it goes.

Wednesday, June 10

More meetings about the Israeli bombing. Under the law I have no choice but to ask Cong. to investigate & see if there has been a violation of the law regarding use of Am. produced planes for offensive purposes. Frankly, if Cong. should decide that I'll grant a Presidential waiver. Iraq is technically still at war with Israel & I believe they were preparing to build an atom bomb.

Henry K. came to see me at 5:30 to report on his European & Middle East trip. He has an amazing knowledge of all the "players" and his report was most enlightening & helpful.

Thursday, June 11

I feel like I've spent the day in the U.N. The morning was pretty well taken with a labor leader meeting—those leaders who support our Ec. program. Then it was the Ec. Advisory Council. Art Laffer dropped a grenade on his colleagues when he said we weren't going to solve the fiscal program until we returned to convertibility of money for gold. I would like to have heard the discussion among those economists after I left.

Then after lunch the U.N. phase started; 5 Arab St. Ambassadors, followed by the Israeli Amb. The 5 of course protested we hadn't been hard enough on Israel. The Israeli deplored the harsh action we'd taken. I told all of them our goal was peace in the middle east. The bombing was the result of the suspicions & hostilities that exist there. We are going to keep on trying to bring peace to Lebanon & then between Arabs & Israelis.

The truth is the Arab indignation on behalf of Iraq is a waste. Saddam Hussein is a "no good nut" and I think he was trying to build a nuclear weapon. He has called for the destruction of Israel & he wants to be the leader of the Arab world—that's why he invaded Iran.

[Reception for supporters on economic package; ceremony at Marine barracks. **Friday, June 12:** *received new ambassadors; signed POW/MIA Recognition Day proclamation; lunch with Vice President Bush; reception for Republican National Committee.* **Saturday, June 13:** *dinner party with the Regans and others.* **Sunday, June 14:** *church services; barbecue with* Washington Post *publisher Katherine Graham and others.* **Monday, June 15:** *met with Charlton Heston and members of the Task Force on Arts and Humanities; discussion with congressmen on tax bill.]*

Tuesday, June 16

[Meeting with Republican leadership on Democratic opposition to budget proposal.]
Today was Press Conference. I don't think I was very scintillating.

[Recognized Canadian help in Iranian hostage crisis; reception for eleven senators and their wives.]

It's lonely. Nancy is in N.Y. for the Royal Ballet. I don't like it here by myself.

We have just learned that Israel & the previous Admin. did communicate about Iraq & the nuclear threat & the U.S. agreed it was a threat. There was never a mention of this to us by the outgoing admin. Amb. Lewis cabled word to us after the Israeli attack on Iraq & now we find there was a stack of cables & memos tucked away in St. Dept. files.

*[***Wednesday, June 17:*** *award from the Broadcast Pioneers; send-off for ambassadors; horseback ride.]*

Thursday, June 18

Called Tip O'Neil—there is no question but that games are being played.

The Dem. dominated Committees have put together their plans for implementing the Gramm-Latta bill. Some did alright but a number of them claimed savings by putting in unrealistic figures they knew would not hold up. For example one of them called for eliminating ⅓ of the P.O.'s in the U.S. Now they know we'd have to ask for replacement of those which means their claimed savings doesn't exist. I called attn. to this one in the Tues. Press Conf. & the next day Pickle called a comm. meeting to change it & stated publicly it was because of my press conf.—in other words he confessed they'd been caught.

Tip was bluster on the phone & accused me of not understanding the const.—separation of powers etc. I was asking only that he allow an amendment to be presented on the floor for correction of the phony comm. recommendations. He won't allow that of course. He is blocking our move to consolidate categorical grants into Block grants. Claims Cong. would be abdicating its responsibility. In truth Wash. has no business trying to dictate how States & local govts. will operate these programs. Tip is a solid New Dealer and still believes in reducing the states to admin. districts of the Fed. govt. He's trying to gut our program because he believes in big spending.

[Friday, June 19: *meeting with Prime Minister Lee Kuan Yew of Singapore; left for Camp David.* **Saturday, June 20:** *Ron and Doria arrive with others for Father's Day.*]

Sunday, June 21

[Relaxed with family and friends.]

Negotiations are still going on to try and head off tomorrow's illegal strike by the air controllers. I told Drew L. to tell their Union Chief I was the best friend his people ever had in the W.H. but I would not countenance an illegal strike nor would I permit negotiations while such a strike was in process.

Monday, June 22

Nancy left for Calif. at noon & I'm already lonesome. The morning was mainly briefing on our hassle on the bill. The House leadership is revealing it really doesn't want to cut spending or taxes.

[Meetings with governors regarding block grants and with Congressman Bob Michel (R-IL).]

Bill Smith came in on the Sup. Ct. I believe if the check up goes well we should go with the lady in Ariz. to replace Potter Stewart to the Supreme Ct.

Ended afternoon at reception for champion athletes & familys. It was fun but somehow moving to meet the children of men like Jesse Owens, Joe Louis etc.

Tuesday, June 23

This was a day to end all days. I guess they figure they had to make up for the fact I'll be gone until next Tues. I speak tomorrow in San Antonio—the Jaycee's, Thurs. in L.A.—then the ranch for the weekend & Mon. the N.A.A.C.P. in Denver.

I started with the usual staff briefing at 9 A.M. then breakfast with the Dem. conservatives who helped us with the Gramm-Latta bill. A Nat. Security session, more staff, interview with U.S. News & World Report, Ben Hooks & Margaret Bush Wilson of N.A.A.C.P., saw 3 Ambassadors off. Lunched with Jacque Cousteau who is a true ecologist who like me is disgusted with the eco. freaks. Meeting with Sen. Finance Comm. (Repubs.). They've done a good job & will have a tax bill on the Sen. floor tomorrow. A Cabinet meeting, President's Advisory Comm. on Federalism, session with Pen James, then Bill Casey, addressed the Teenage Repubs. & a reception for the 190 Repub. Reps. from the House. It was non stop all day. Photographed all day by U.S. News.

Tip O'Neil is getting rough. Saw him on TV telling the United Steel workers U. I am going to destroy the nation.

Wednesday, June 24–Monday, June 29

[Addressed National Jaycees Convention in San Antonio; flew to L.A. and Century Plaza Hotel; made calls necessary before the following day's congressional vote on budget.]

There I was in Calif. & never left the hotel room except for a speech next day (noon) in the hotel dining room. Just before going down learned we had won the 1st big vote. It meant some quick changes in the speech. Back to the room for an afternoon on the phone. We won the next two votes & the victory was ours.

Finally Fri. about 11 A.M. got out of the hotel & on to the ranch. The weather was beautiful & so was Rancho del Cielo. Patti arrived Sat. morning. We all rode in the A.M. After lunch Geo. & Barbara Bush came up to brief me on European trip on their way to the Marcos' Inaugural in Manila.

[Relaxed at the ranch. **Monday, June 29:** *flew to Denver to address convention of the NAACP. Returned to Washington.]*

Tuesday, June 30

[Signed order ending quotas on shoe imports; met with Prime Minister Malcolm Fraser of Australia; state dinner.]

This afternoon met with Cap. W. on the B-1. He leans toward going for the "Stealth" which would leave a several year gap with only the aging B-52's. I tend to favor filling the gap with B-1's while we develop the "Stealth." The Israel election returns coming in looks like a dead heat between Begin & Peres which means problems whichever one wins.

Wednesday, July 1

A half day in the office with nothing of great importance but spent the entire afternoon in my study catching up—reading, intelligence reports, the options for a new immigration policy, etc.

The press keeps score on office hours but knows nothing about the never ending desk & paperwork that usually goes on 'til lights out.

Thursday, July 2

Met with the task force David Rockefeller has put together to stimulate investment in Jamaica. P.M. Siaga is trying to recover from the ec. damage done to his country by the Marxist Mandel who preceded him. I'm greatly encouraged.

[Meeting with former Republican state chairmen; wrote letters to three mothers of child rape/murder victims; appeared at dinner to honor economist and author Bill Simon; Mike Abrams, fitness expert, staying at W.H. **Friday, July 3**: *preparations for July 4; gym set up at W.H.* **Saturday, July 4**: *birthday party for Mrs. Reagan.* **Sunday, July 5**: *used gym; took call from King Juan Carlos of Spain requesting visit; desk work.]*

Monday, July 6

Now it's Nancy's birthday. A lengthy N.S.C. meeting re trade with Soviet U. & what to do about the Soviet pipeline to W. Europe. A dozen options ranging from almost total trade with restrictions only on extremely sensitive mil. technology to almost total boycott. I have to choose one of the options. Our allies of course lean toward trade.

Lunch with Phil Habib—he's a real pro & he's kept us from having a war break out between Israel & Syria for a long time now. He's going back & our prayers go with him.

Called Judge O'Connor in Ariz. & told her she was my nominee for Supreme Ct. Already the flack is starting & from my own supporters. Right to Life people say she's pro-abortion. She declares abortion is personally repugnant to her. I think she'll make a good Justice.

Tuesday, July 7

This morning I announced my nominee for Supreme Court, Mrs. O'Connor of Ariz. I made some calls because someone has started a bonfire among the Right to Life people. Apparently it all started with a woman—Dr. Gersten in Phoenix. Her claims don't match the record we have of O'Connor's voting record when she was a state Sen. But she's spread her message far & wide. A full Cabinet meeting on the Ft. Chafee problem (Cubans—criminal & mentally deranged dumped on us by Castro). We believe we can move them to an unused mil. base in Maryland which can be developed also as a holding area for legitimate refugees in the future.

[Fund-raising dinner in Chicago; returned to Washington.]

Wednesday, July 8

Met with the 11 handicapped young people (Blind, some deaf, an Epileptic & one with an artificial leg) who climbed Mt. Rainier July 4. We'd given them a flag to carry up there—they presented it to me. Nat. Geographic presented me with a map case & maps of every section of the world.

Bill Clark & I went riding at Quantico. Back here to a 5:30 concert by the "Most of Mozart" Orch. from Lincoln Center. They played the recently discovered symphony of Mozart written when he was 9 years old.

[Thursday, July 9: received fly rod from Alaskan congressman; met quadriplegic athlete; announced new comptroller general; NSC still undecided on Soviet pipeline; met African American supporters; two W.H. receptions.]

Friday, July 10

P.M. Trudeau arrived on his way back from trip to Europe. We discussed his Ottawa Summit meeting coming up in about 9 days. I believe it will be a general meeting to discuss ec. issues, N.-S. relationships & some East-West trade matters.

At lunch we finally got around to 3 bilateral issues. On their side they want action on acid-rain which they believe results from air pollution on our side of the border. I assured him we are researching that because we have people in Vermont who think they are getting acid-rain from Canada. Second issue is the gas pipe line to be built from Prudhoe Bay down to Calgary to join existing pipe line to U.S.

The whole project was conceived during the energy crunch. Carter pledged we'd go through with it even though the Alaska portion was & is supposed to be built by pvt. enterprise. The problem is that now no one (but Canada) thinks or at least is sure it should be built. There have been new gas discoveries down here and it might be more practical for Alaskan gas to be liquefied & sold to Japan.

Our issue had to do with a discriminatory tax policy on Am. owned business in Canada while we do nothing of that kind to Canadian owned business on our side of the border. He grew very defensive. I think our problem is that he leans toward outright nationalization of industry.

He left & I then met with board of Black State Legislators organization. A good meeting—better than the N.A.A.C.P. in Denver. I think they discovered I might be different than the image they've had.

Final meeting with Repub. Reps & Senators on defense matters. Much of it had to do with whether we should go for the B-1 or wait for the newer "stealth" bomber (which may not fly for 10 yrs. or so if at all). M.X. was also a problem particularly having to do with how it will be deployed. No decision from Defense (or me) yet but I lean against the proposed "race track" system.

Abt. 4:30 we left for Camp David. Almost upon arrival the relaxing began. What a joy that beautiful spot is getting to be.

[Saturday, July 11–Sunday, July 12: swimming and relaxing; returned to Washington. Monday, July 13: reviewed budget projections; met with Polish American and Italian American leaders; considered problem of returning undesirable Cubans; met with Gaston Thorn, president of the Commission of the European Communities; entertained Mrs. Reagan's parents.]

Tuesday, July 14

We are still meeting & stewing about East-West trade and now we must take on the problem of what to do or if to do something to help the Polish people. Their economy is going bust. Here is the 1st major break in the Red dike—Poland's disenchantment with Soviet Communism. Can we afford to let Poland collapse? But in the state of our present economy can we afford to help in a meaningful way? We can't alone but if our NATO allies could really unite we could handle it.

Met with Willie Stargell (Pitts. Pirates—most valuable player '79 World Series). He has created a foundation for research in Sickle Cell Anemia. He's quite a guy.

[Invited to speak at Eureka College on fiftieth anniversary of graduation; attended ceremony at Irish Embassy in honor of father-in-law, Dr. Loyal Davis, a respected neurological surgeon.]

Wednesday, July 15

Jimmy Roosevelt came in to see me about the Centennial Birthday of F.D.R. next year. I agree there should be appropriate ceremonies.

Began the briefing for the Ottawa Summit. A lot to do & I wonder if it's worth it.

Had to get dressed up for a drop by at the reception for top labor leaders.

Thursday, July 16

Most of the day given over to Ottawa & Cabinet meeting on our immigration problems. The Haitians and the criminal Cubans Castro sent us mixed in with the refugees our 2 greatest problems.

Went to a small but pleasant dinner at Ex. Sen. John Sherman Cooper's home. A nice evening—interesting talk with Lady Henderson (Ambas's. wife).

Friday, July 17

Signed bill stretching out compliance time, clean air act for steel industry. This should be a big help.

[Met with congressmen on tax bill and with veterans; prepared for Ottawa summit.]

Rev. & Mrs. Billy Graham came for dinner & overnight. A really fascinating conversation. He's well informed about world situation & knowledgeable about world leaders thru personal acquaintance.

Saturday, July 18

Sens. Baker & Dole, Dave Stockman, Max, Mike D., Ed Meese & Dick Darman came up to the study. It seems Jim Wright is playing games.

On the House floor he introduced a resolution that the sense of the Congress be that we should drop from our budget cuts the elimination of the minimum Soc. Sec. benefit. It is presently in both the House & Senate versions. It is a pol. trick aimed solely at creating a 1982 election year issue for the Dem's.

[Reviewed Social Security benefit program.]

Sen. Baker said I could fight or retreat—he's not sure we can win either on the floor or in the conference committee. If I retreat the gains we've made in lowering inflation will be lost. It will be taken as a test of my determination & looked upon as a sign that I'll back down on the tough decisions. Well I have no intention of retreating; I've sent a letter to our leaders on the hill informing them that early in August I'll go on TV to discuss the Soc. Security problem. I think this will shake Jim Wright more than a little.

Tomorrow off to Ottawa.

Sunday, July 19

Air Force One to Ottawa & then Marine One to Montebello—the largest log cabin in the world. The hotel is a marvelous piece of engineering, totally made up of logs.

Had a one on one with Chancellor Schmidt. He was really down & in a pessimistic mood about the world.

Following—met with Pres. Mitterrand—explained our ec. program & that high interest rates were not of our doing.

Dinner that night was just the 8 of us. The 7 heads of State & the Pres. (Thorn) of the European Community. It became a really free wheeling discussion of ec. issues, trade etc. due largely to a suggestion by P.M. Thatcher.

Monday, July 20–Tuesday, July 21

I had one on ones (breakfast meetings with Margaret Thatcher & P.M. Suzuki. Also meetings with P.M. Sabatini of Italy. I hope he can hang on. I think he might do well with Italy.

The schedule was heavy—plenary meetings morning & afternoon—always preceded by sessions with my own people—then working dinners just with the principals.

[Concluding social events; flew back to Washington.]

It was a successful summit—not divisive although it could have been with regard to our interest rates. I think we held our own very well and opened the door to a meeting on trade with the Soviet U.

Margaret Thatcher is a tower of strength and a solid friend of the U.S.

We held off what could have been a ganging up on Japan.

Wednesday, July 22

A half day with an N.S.C. meeting on our grain sales to the Soviet. I agreed to a 1 yr. extension—our agreement expires Sept. 30. We'll let them wonder for awhile whether there can be a long term deal. Frankly I want some give from them on their obvious expansionism.

[Considered federal assistance for synthetic fuel plants; luncheon for editors; horseback ride with Vice President Bush.]

Six thirty a drop by at the annual "Gym Dinner" of the House—Carter never went.

Thursday, July 23

Saw "Mommie" off for London & the Royal Wedding. I worry when she's out of sight 6 minutes. How am I going to hold out for 6 days. The lights just don't seem as warm & bright without her.

Went to the Oval office for the morning brief but my mind wasn't on it. Met in the State Dining Room with St. Legislators, Mayors & Co. officials from a dozen states re the tax bill.

[Visits by coach from Eureka football days and by former colleague from Sacramento; meetings separately with Republican and Democratic congressmen on tax bill.]

Al Haig called & said we may have a "cease fire" tomorrow between Israel & the P.L.O.

Watched the "Waltons" on T.V. and so to bed.

Friday, July 24

[Meetings with 191 Republican representatives on tax bill and with New Zealand prime minister Robert Muldoon.]

Met with Congressmen who want the "Legal Assistance program" continued. We have proposed reducing it. We don't object indeed we support legal aid for people too poor to hire a lawyer when they need one, but we object to the way it has become a lobby for all sorts of causes—class action suits, etc. which are the lawyers' idea, not that of a client.

Went to Smithsonian to see the "Catlin" exhibition. He was an early painter of the American Indian.

Al Haig has brought me a case of our Ambassador who is undercutting him (Al) both with the govt. of the country in which he serves and indeed our foreign policy itself. In this case I believe Al is right. I'm going to have to have a nose to nose with Dick Allen (who picked the guy) and back Al.

[Dinner party and showing of film Victory *(1981).]*

Saturday, July 25

The cease fire in the Middle East seems to be holding although some bands of P.L.O. may not have the word.

Off to Camp David with Mike D. & the Bakers. This time bad weather—drizzly rain & fog. Not bad for me—I had a speech to write. Watched a movie—"Arthur."

Sunday, July 26

Weather still bad so I finished the speech (for TV Mon. night). At noon 15 Dem. Congressmen (1 woman) came for a BarBQ & a pitch by me for our tax program. I think we did some good. Weather cleared some—the others got in some tennis. Left for W.H. about 5. Now here I am and it's d--n lonesome. I did get a glimpse of Nancy on TV at the Polo match.

Monday, July 27

[Meetings all day: corporate lobbyists, trade associations, congressmen regarding tax program, Governor Fob James (D-AL), and Philip Habib, who has helped prolong a cease-fire in the Mideast.]

Now I've just finished my broadcast. The D's are already screaming. Calls in the 1st hour were 629 pro—148 against.

Tuesday, July 28

The morning after—W.H. received more calls than on any other appearance they ran better than 6 to 1 in favor of our tax bill.

The schedule however was unbelievable. Throughout the entire day I was meeting with Cong'men. singly & in groups soliciting support. All of them told the same story—their phones in their districts & in the Cap were ringing off the wall. Tomorrow is the day & it's too close to call but there is no doubt the people are with us.

Met with Prince Bandar (Saudi Arabia). The last time I saw him he was learning to be a jet fighter pilot in Alabama. Met also with Jeanne Kirkpatrick. She says the U.N. is a worse can of worms than even she had anticipated. We've agreed the U.S. has got to get tough and maybe walk out a few times.

[Met with WWII veterans; dinner party hosted by Mr. and Mrs. James Baker.]

Wednesday, July 29

The whole day was given to phone calls to Congressmen except for a half dozen to name Ambassadors. I went from fearing the worst to hope we'd squeak through. As the day went on though somehow there was a feel that something good was happening. Then late afternoon came word the Senate had passed its tax bill (ours) 89 to 11. Then from the house where all the chips were down— we won 238–195. We got 40 Dem. votes. On final passage almost 100 joined the parade making it 330 odd to 109 or thereabouts. This on top of the budget victory is the greatest pol. win in half a century.

Tip O'Neil & his leadership called me and with complete graciousness congratulated us on our win.

Now we must make it work—and we will.

[Thursday, July 30: *spoke to National Association of State Legislatures; Mrs. Reagan is home.*]

Friday, July 31

A lengthy working lunch. N.S.C. briefing on Naval games in the gulf of Sidra. We'll find out how serious Khadaffi is about claiming half the Mediterranean for himself.

Big discussion about proceeding with the tactical 8 inch weapon the "Enhanced radiation warhead." Al worries about Europe's reaction. My decision we're not deploying them, we're stockpiling a defensive weapon. We don't announce it but if word leaks as it will we point out the Carter Admin. had ordered the key part to be made but not assembled into the shell. In other words, we stockpiled 2 parts of the shell separately. We're just putting them together.

Off to Camp David—late as usual.

Saturday, August 1–Sunday, August 2

Two days of loafing by the pool & swimming. Weather wonderful. Rec'd word our home has gone into escrow. Also learned the Air Controllers will probably strike Mon. morning. That's against the law. I'm going to announce that those who strike have lost their jobs & will not be re-hired.

Monday, August 3

The strike was called for 7 A.M. I called the press corps together in the Rose Garden & read a statement I'd written yesterday. I included in it a paragraph from the written oath each employee signs—"that he or she will not strike against the U.S. govt. or any of its agencies." I then announced they would have 48 hrs. in which to return & if they don't they are separated from the service.

By afternoon an estimated 29% of the 17,000 were at work. Have my fingers crossed for tomorrow.

[*Cabinet meeting regarding the budget.*]

An N.S.C. meeting concerned mainly with the defense plans. Our task force has reported to Cap that we need a world wide communications system for reaching our Subs, missile sites & Air Forces.

They recommend going ahead with 100 B.1's while we develop the ATB plane. Go for the new DS missile for our Tridents & go for the MX without the race track but put some in Silos & in airplanes. There is a possible new plane—very slow & easy on fuel which could stay aloft 5 or 6 days. We could have them carrying MX's on a round the clock airborne schedule, thus making them invulnerable to Soviet missile attacks.

[Civil defense program discussed; Senate passed tax bill.]

Tuesday, August 4

Today the House passed the tax bill 285 to 95. It's mine to sign now.

As of mid-day 35% of the Air Controllers were on the job and flying was 65% of normal.

[Received check for funds saved by government agency; lunched at Star *newspaper during its last week; revolutionary situation in Gambia discussed at W.H.]*

Wednesday, August 5

Deadline day for the air controllers strike. About 38% are working—air travel is at 75% of normal. We've learned we've had about 6000 more air controllers than we need.

In Chi. Lane Kirkland (Pres. of AFL-CIO) & the exec. comm. joined the picket line. How do they explain approving of law breaking—to say nothing of violation of an oath taken by each a.c. that he or she would not strike.

Pres. Sadat arrived. We had good meetings. I told him of our naval maneuvers in the Mediterranean on Aug. 18 & that our ships & planes would cross the Kadaffi line into the Gulf of Sidra. He almost shouted—"magnificent!"*

The State Dinner was a success. I'm encouraged that between us maybe we can do something about peace in the middle east.

Thursday, August 6–Thursday, September 3

[Saw Sadat off; meetings on defense and upcoming speech; flew to California for vacation; barbecue for Seabees; press event for bill signing.]

Each day we rode in the A.M. then in the afternoon Barney, Dennis & I with vol. help from Dr. Ruge & our mil. aides cut up downed limbs & built brush piles for rainy season burning. Ten days of this—our longest single stretch at the ranch. The S.S. agents have caps titled U.S.S.S. Mounted Patrol. They have become enthusiastic horsemen.

---

* On August 19, with the U.S. Sixth Fleet on maneuvers in the gulf, two Libyan jet-fighters attacked and were shot down.

Met with Haig the afternoon of the 16th. On the 17th back to L.A. & the Century Plaza. Every night was a dinner out—with old friends. It was really enjoyable. The days were rather busy. Meetings—Cap., N.S.C., Charles Wick. A pol. reception on the 17th. More meetings on following days on Budget, with Sen. Tower & Cong. Dickinson on MX, the A.M.A. representatives, etc.

[Visit to USS Constellation*; fund-raiser in Orange County.]*

August 22nd back to the ranch. We had finally made a dream come true. Bud Bredall made part of the barn into a tack room. We now have a dozen saddles & they were being stacked on top of each other. It turned out great.

We're waiting for word our house is sold so we can put up a guest house & a house to replace Lee's trailer. Our buyer is strange and each day Escrow is to close making the sale final, a new excuse pops up delaying it.

Cap & Ed came up for a meeting. Went to Santa Barbara one evening for a reception we gave the press corps, a Repub. fundraiser and finally a quiet dinner with the Deutches at the Tuttles.

Back to the ranch for the last few days and then to L.A. A few more meetings and then to Chi. on Sept. 2nd. Back into the "iron vest." We have an intelligence report that Quadaffi has told others he intends to have me assassinated. It's a strange feeling to find there is a "contract" out on yourself.

Sept. 3rd addressed about 5,000 Carpenter Union delegates at McCormack Place in Chi. Expected possibly a cool reception but it turned out fine including my remarks about the PATCO strike.

Back in Wash. and we begin again.

Friday, September 4

[Oval Office redecoration complete.]

Not a big day. Met with Al Haig—the world is still exploding. The French Ambass. to Beirut was gunned down by terrorists doing the Syrians' work.

Met with Jim Watt. He's taking a lot of abuse from environmental extremists but he's absolutely right. People are ecology too and they can't forage for food and live in caves.

Saw a film on Begin, a kind of character study. He'll be here Wed.

Had a pleasant evening with a stack of horse and western magazines.

[Saturday, September 5: photo session; desk work; dinner party and play starring Claudette Colbert.]

Sunday, September 6

I know we should have gone to church & yet I'm concerned about the elaborate security measures which have me arriving in a siren screaming motorcade. I don't know what the answer will be but one must be found.

Spent the day at my desk then after dinner we ran a movie—a bloody one called "The Fan."

[Monday, September 7: in New York to initiate Westway highway project; labor reaction to president mixed; back to Washington.]

Tuesday, September 8

Back on schedule—9 A.M. in the office. First of what will be several budget meetings. Hi interest rates are going to force more budget cuts or we won't meet our 82, 83 & 84 targets.

I've given the order—we'll cut & we'll meet the goals we set for ourselves. We have to convince the money mkt. that we mean it & that means some cuts in defense. But we have to do that in such a way that the world sees us as keeping our word to restore our defense strength. It can be done.

[Briefing for meeting with Israeli prime minister Menachem Begin; talk with Republican congressional leadership.]

Bill Smith came over to ask for more money for the Justice Dept. This is a tough one—Carter had cut his budget in real dollars 40% in 4 years. Crime is a number 1 concern.

Wednesday, September 9

The P.M. (Begin) arrived & we had a beautiful day for it—the So. Lawn ceremony. In the Oval office we did some getting acquainted & surprisingly it was very easy.

A.W.A.C.'s came up in the cabinet room session. He of course objects to the sale. I told him how strongly we felt it could help bring the Saudis into the peace making process. I assured him we (Israel & U.S.) were allies. That the partnership benefited us as much as it did Israel & that we would not let a risk to Israel be created.

While he didn't give up his objection he mellowed. By the time the meetings and the State Dinner ended he said this was the warmest reception he'd ever had from a Pres. of the U.S.

I think we're off to a good start in the difficult business of peace in the Middle East. My own feeling is that it should come through bi-lateral agreements just as it did with Egypt. That's why we want to start with Saudi Arabia.

Thursday, September 10

[Future meetings planned and then Begin leaves.]

Dropped in on meeting of Ec. Advisors—A roomful of our country's greatest economists. None of them could explain why interest rates are so high.

[Dedication of corridor at Pentagon in honor of General Douglas MacArthur; per-

formance on South Lawn by Lionel Hampton and Pearl Bailey; phone calls on additional budget cuts.]

Friday, September 11–Sunday, September 13

Cap, Dave & I settled Defense cuts at $13 Bil. over 3 yrs., 2 Bil. of it to come off the '82 spending. These figures are for outlays not obligational authority which would probably be about $21 Bil. Gave bravery & service awards to 9 young Americans. They had performed the deeds in 75 & 78 but J.C. wouldn't give out the medals. The law says only the Pres. can do that. Friday afternoon off to Camp David with the Deavers & the Bakers—children & all. A nice weekend—mainly by the pool with one try at skeet shooting—my 1st attempt. A very humbling experience even if I did get a couple of birds.

Rcv'd a call while there—our consulate in Frankfurt Germany was fire bombed—no casualties.

Monday, September 14

[Ceremonial appearances; meetings with "Boll Weevil" Democratic congressmen and then a bipartisan group of senators.]

First meeting was re the budget. The 2nd had to do with Saudi AWAC's sale. Was annoyed to learn that P.M. Begin had gone to the hill & had lobbied against the sale after leading me to believe he wouldn't.

Bill Clark is going to meet him in N.Y. if possible before he leaves tomorrow to tell him this may affect our relationship unless he undoes what he's done.

Tuesday, September 15

[Meetings regarding budget with Republican leadership in Congress; displeasure with Senator Bob Packwood (R-OR) over AWAC position; ceremonial appearances; lunch with presidents of traditionally black colleges, signed executive order for greater federal involvement in funding; met with Weinberger.]

An N.S.C. meeting on Poland. I cannot see helping the govt. with cash but I'm all for getting surplus food to Polish people.

[Met with governing mayor of West Berlin Dr. Richard von Weizsäcker; reception for supporters.]

We've had disturbing news that P.M. Begin was planning to invade Lebanon. Of course I've been upset by Sen's. tales that he lobbied hard against AWAC's when he'd told me he wouldn't.

Al H. flew up to N.Y and saw him before he left for home. He promised no invasion unless some act forced his hand. Denied he broke his word & Al is convinced he really wants to get along with us. He told Al he'd never felt as close to a U.S. Pres. as he does to me.

Wednesday, September 16

[Ceremonial appearances, including greeting Boys Club "Boy of the Year."]

State Dining Rm. lunch for Nat. Hispanic leaders. Like yesterday's lunch this was most enjoyable plus which they made it plain they support fiscal conservatism such as our program. I believe their long loyalty to the Dem's is ending.

Staff meeting with Stockman on strategy for additional cuts in budget '82 & for '83 & '84. A good plan & I believe we'll make it.

[Reception for Republican donors; met with leader of Maronite Christians in Lebanon.]

Thursday, September 17

[Cabinet meeting on budget cuts; discussion with Secretary of Energy James Edwards and Secretary of Education Terrel Bell on which of their two departments should be dissolved first.]

Leaving at 12:35 for Grand Rapids, Mich.—opening of Ford library & museum.

Betty & Jerry Ford met us at the plane. At the hotel I went into meetings almost immediately with former Pres. of France, then P.M. Trudeau & last Pres. Lopez Portillo. The P.M. & I had a go around about his energy policy which has American investors in Canada quite upset. Then Jose & I had a real set to about El Salvador. He has evidently believed that we were on the verge of sending in the Marines. His whole demeanor changed when I told him we'd never entertained such a thought.

[Gala evening taped for Bob Hope television special. **Friday, September 18:** *tour of the museum and dedication ceremony; flew to Denver for speech before National Federation of Republican Women; returned to Washington.* **Saturday, September 19–Sunday, September 20:** *relaxed at Camp David; returned to W.H.]*

Monday, September 21

I don't like Mondays. This morning I was given the news that our successful coalition in Cong. is breaking up in the face of our need to get additional budget cuts. I think they have '82 election jitters. We have to have those cuts & we have to solve the Soc. Security problem. The D. leadership has refused to tackle the problem in an effort to find a bi-partisan solution. They are staking it out as an '82 campaign issue in which we'll be portrayed as trying to pinch the Sr. Citizens. The fact is the "trust fund" will be out of money by election time.

[Met with business people about privatization of government social initiatives; met with congressmen on issues, including AWACs program and high interest rates; received resolution of support on budget from 152 Republican representatives. **Tuesday, September 22:** *signed petition for reapportionment in California; met with Ralph Abernathy, civil rights leader; worked on upcoming speech on budget; three receptions.]*

Wednesday, September 23

Finally a meeting on the budget that made me feel good. It's going to be a tough fight & we won't get all we ask for but we'll do fairly well I'm sure. I'm withdrawing Soc. Security from consideration & challenging Tip & the Dems. to join in a bipartisan effort to solve the fiscal dilemma of S.S. without all the politics they've been playing. Since I'm inviting him on nat. TV, he might be persuaded to go on at 9 P.M. tomorrow night & I haven't finished writing the speech yet. Saw old friends from Dixon who are involved in fixing up my boyhood home.

[Meeting with Roy Brewer, early ally and labor leader in Hollywood during the 1940s; barbecue for congressmen.]

Thursday, September 24

[Met with governors concerned about budget cuts; send-off for three ambassadors.]

Prince Bandar of Saudi Arabia came in with Sen. Baker. Sen. Glenn is influencing many Sens. against the A.W.A.C.'s deal on proposition that we should have joint crew operation if they are sold. This may be our chance. Prince Bandar believes Crown Prince Fahd may agree to this. We in return would give them the enhanced planes with the top level equipment.

Tonight I did the speech live. Everyone seems to think it went well—I hope so.

*[***Friday, September 25:*** lunch with President Daniel Arap Moi of Kenya; swearing-in of Sandra Day O'Connor; went to Camp David. **Saturday, September 26:** worked on upcoming speech. **Sunday, September 27:** calls to senators regarding extensions of debt ceiling.]*

Monday, September 28

Brkfst. on A.F.1—headed for New Orleans. Had teleprompter problems that almost threw me for about 7 min. in front of several thousand Police Chiefs and Sheriffs (International meeting). Finally got going. Rcv'd a tremendous reception. Later a Repub. fund raiser—noon time $1000 per head. I remember going to La. in '63 to campaign for Charlton Lyons who ran for Gov. only to establish a 2 party system in that state. Well it's 2 party now with a Repub. Gov. & a few Congressmen.

Back to Wash. and what I've been told may be the most momentous decision any Pres. has had to make (that's hard for me to believe as I review our history). It was to OK the strategic missile & bomber buildup for our future defense needs. One hundred MX missiles & 100 B-1 bombers (to start with).

Word from Calif. that we may have sold our house at last.

Tuesday, September 29

Addressed the annual meeting of the World Bank & International Monetary fund. While only 125 delegates from nations all over the world, the actual crowd numbered more than 1000. All had staff, finance ministers, etc. I was amazed to learn our S.S. agents had to disarm a great many until after I spoke. It seems they bring bodyguards because of the fear of kidnapping or assassination.

Day ended happily when at 10 P.M. I learned the Sen. had passed the debt limit extension so we wouldn't go out of business Oct. 5.

Wednesday, September 30
Supposed to be a light day—but the schedule underwent some changes. Lunch at home with Mommie became lunch alone in the office.

Met with Repub. leadership of the House & Senate. This is a good practice & one we'll continue. I mainly listen & that's not bad.

Briefings for tomorrow's press conference—took hours.

Dinner for quite a group & a movie—a beautiful & unusual one. "The French Lt.'s Woman."

[Signed debt-limit extension. **Thursday, October 1:** *briefings for press conference; change in secretarial staff; press conference; meetings with senators on AWACs.]*

Friday, October 2
Met press again in East Room with Cap W. to give decision on strategic arms. We're going with 100 MX missiles—1st 36 in Minute Man & Titan silos—no multiple shelter plan. We'll study in the meantime a basing mode for the rest. Also building 100 B-1 bombers while research continues on the "Stealth."

Then off to Camp David.

Saturday, October 3
Windy & cold but we rode & for 1st time—went out a back gate into the Catoctin Nat. Forest. Followed an old road that led to a stone ruin of what was a summer hotel in 1933 when the lady owner Bessie Darling was murdered by her Dr. boyfriend. During WWII it was used for briefing foreign intelligence officers. It's a tumbled ruin now.

Sunday, October 4
[Rode horseback; enlisted Bill Verity Jr., chairman of the U.S. Chamber of Commerce, to head task force on Private Sector Initiatives.]

It bothers me not to be in church on Sunday but don't see how I can with the security problem. I'm a hazard to others. I hope God realizes how much I feel that I am in a temple when I'm out in his beautiful forest & countryside as we were this morning. Back to the W.H.

Monday, October 5

I hate Mondays. This was one that didn't stop. First off—over to the Exec. Off. Bldg. to do a Q&A with 70 out of town—in fact from all over the country—Editors & Radio news editors. Then to the Sheraton to address the Nat. Assn. of Bus. on Volunteerism. They were enthusiastic. I think we'll really take the lead in this.

Lunch was a 1st for the W.H. There were 28 former Nat. Security Officials from Admins. all the way back to Truman all came to join in support of A.W.A.C.'s. Henry Kissinger & former Sec. of Defense Harold Brown (under Carter) both spoke to the press endorsing the sale.

A moving ceremony in the First Lady's Garden. I signed the bill making Raoul Wallenberg the 2nd honorary citizen in our history. He is the Swede who saved 100,000 Hungarian Jews from the Nazis. Then with the war over the Soviets kidnapped him & evidence is he's still alive in the Gulag. Now that he's a citizen of the U.S. we intend to ask for his return. His brother & sister were present. Cong. Tom Lantos was there—he wrote the bill. Tom was one of those saved by Wallenberg—he was 16 at the time.

Finally a meeting on Cancun and how to keep from being the villain there. We do more for the under developed nations than anyone in the world but they act as if we're out to destroy them and they never say boo to the Soviets.

Tuesday, October 6

Most of the days event cancelled. Awakened at 7:25 by a phone call from Al Haig with news that Anwar Sadat had been shot along with many others. All morning we waited for news—receiving word only that he was not fatally wounded & was in surgery. Then came the word he was dead.

It's hard to describe the shock & the sorrow we both felt. Even though their visit was short we discovered we had a deep feeling of friendship for them. Maybe it has to do with a state visit. You start out with knowledge of each other & immediately get into the problems you mutually want to solve.

Anyway, we both feel a great sense of loss. He was truly a great man, a kind man with warmth and humor.

I'm trying not to feel hatred for those who did this foul deed but I can't make it. Qadhafi gloating on TV, his people jubilantly celebrating in the streets. He is beneath contempt.

He goes on radio (clandestine) and began broadcasting propaganda, calling for a holy war, etc. before Sadat's death was confirmed. This material had to have been already prepared. In other words, he knew it was going to happen.

[**Wednesday, October 7:** *met with senators regarding AWACs plan; discussion of job placement with two governors; memorial ceremony for Sadat.*]

Thursday, October 8

A busy day—signed pledge for Fed. govt. charity fund with Mac Baldridge & others present. Cabinet Council on energy meeting—went forward with new nuclear policy which should get 33 Nuclear Power plants licensed. They've been sitting idle while we kept on importing more oil for power.

[Discussion of wealth sharing in anticipation of Cancún summit; farewell for eleven ambassadors.]

Pres.'s Nixon, Ford & Carter arrived at the W.H. for a brief send off by helicopter from S. Lawn. They are representing us at Sadat's funeral. All wanted to go—all knew him. It's the 1st time 4 Presidents have ever been in the W.H. together.

*[***Friday, October 9:*** *Senator Thad Cochran (R-MS) pledged support for AWACs plan; meeting on education; greeted National Epilepsy Foundation poster child; lunch for Medal of Freedom awards; Columbus Day proclamation; went to Camp David.* **Saturday, October 10–Monday, October 12:** *rode horseback; called Margaret Thatcher regarding Mideast; enjoyed comedy films* Paternity *(1981) and one he called "The Great Divide," though he probably meant* Continental Divide *(1981); back to W.H.]*

Tuesday, October 13

King Juan Carlos & Queen Sophia arrived today. The usual ceremonies. They flatteringly but nicely remembered our having tea with them in Madrid in 1972.

[Good rapport with King.]

The State dinner was most entertaining. Ella Fitzgerald sang—great as always. Then we danced—my 1st dance ever with a Queen.

This afternoon met with J.C. [Jimmy Carter]. I expected the worst but he was cordial, friendly and just exchanged views on the Middle East, etc.

Wednesday, October 14

Another busy day. I've discovered I hate those days where I have one meeting right after the other with no time to collect my thoughts between meetings & I'm supposed to make remarks or a short speech at every one.

[Press announcement of Congressman Eugene Atkinson (D-PA) switching parties; lunch with Arts and Humanities task force; meetings with senators regarding AWACs; noted Senator Harry Byrd (I-VA) is considering joining GOP; called friend whose wife was diagnosed with terminal cancer.]

Thursday, October 15

Short morning in the office—greeted newly appointed members of the Export board. Then off to Phil. & N.J.—airports too small to handle A.F.1 so took a small job—a Gulf Stream I think. It doesn't take many trips in A.F.1 to spoil you.

Addressed Foreign Relations Council in Phil. The speech was really meant for the nations going to Cancun to plant the idea we weren't going to buy their idea of a new international bureaucracy empowered to share the wealth.

[Speech well received; met with leaders of volunteers groups; fund-raisers in New Jersey.]

We lost the vote on AWAC's in Foreign Relations Comm. 9 to 8—all D's but one—Rudy Boschwitz.

Back on the plane & a box lunch. Ky. fried chicken and home 10 minutes ahead of schedule.

Friday, October 16

This was a day that didn't stop—1 on 1 with 8 different Sens. on AWAC's. A meeting with the P.M. of Mauritius. Then one with Dr. Kohl leader of the opposition party to Helmut Schmidt. He said the 250,000 demonstrators in Bonn against the U.S. came from all over Europe and it was an affair orchestrated by the Soviet U. He made a good point—that propaganda is painting us as a militaristic people when the truth is we are the most moral & generous people on earth. We should be appealing to the world on the basis of morality.

Lunched with a group of editorial page editors from all over the U.S. Did a Q&A.

An NSC meeting that has left me with the most profound decision I've ever had to make. Central America is really the world's next hotspot. Nicaragua is an armed camp supplied by Cuba and threatening a communist takeover of all of Central America. More meetings & finally home with an arm full of homework & my problem.

Saturday, October 17

Over to the Cabinet room for a 10 AM meeting. Another bomb—the latest figures on deficit projections—bad. It seems our success is actually hurting us. Inflation is a tax. We have brought inflation down so much faster than we anticipated that tax revenues will be lower than we figured. We force the prospect of low inflation & lower interest rates—all of which is good—but gigantic deficits & that's bad.

[Senator Jennings Randolph (D-WV) pledged support for AWACs plan.]

Ten minute interview with Trudy Feldman. Back to the house & our guest arrived—Bob Hope. Tonite the big U.S.O. dinner honoring Bob. I made him an Ambassador (WorldWide) of Good Will. It was a glittering affair.

[Sunday, October 18: meetings with French president François Mitterrand in Virginia; state dinner. Monday, October 19: commemoration of Cornwallis's surrender at Yorktown; return to Washington. Tuesday, October 20: campaigned for AWACs plan with senators; meeting with American Business Council; Mrs. Reagan left for New York City; briefings for Cancún summit.]

Wednesday, October 21

Departed for Cancun—breakfast on A.F.1—3 hour flight. Arrived to a full military reception—greeted by Pres. Lopez Portillo. 21 gun salute—reviewed troops, etc. Helicoptered across Yucatan Peninsula to Cancun—a beautiful white beach & the bluest, clearest water I've ever seen. The Sheraton is an attractive hotel.

Held meetings (1 on 1) with Indira Ghandi who was not what I expected. She is tiny and seemed very reasonable and moderate. Met with Pres. of Nigeria, P.M. People's Republic of China & Pres. Herrera of Venezuela. Then had my 1st swim in the Caribbean. Water warm but rough with a strong undertow—too rough to catch a wave for body surfing.

[Dinner; report of unrest in Nicaragua.]

Thursday, October 22

Met with Pres.'s (1 on 1) of Austria & Yerge Sloveci [Kraigher]. Then 1st session devoted to speeches by each of 22 delegates. I know everyone was waiting for mine—possibly with chip on shoulder. We fooled them—it was well received.

Didn't get lunch til about 3 P.M. & back into session at 5 until 7:30.

This session was a free wheeling discussion of food & agriculture. I suggested task forces of agri. experts to visit the hunger countries & recommend new procedures, etc. I volunteered to provide such volunteers. Sec. Gen. of U.N. expressed great appreciation.

Pleasant dinner at 9 P.M.

Friday, October 23

Plain by now things going very well. Started day with 1 on 1—Pres. Bendjedid of Algeria & Pres. Nyere of Tanzania.

These meetings dealt with energy, trade & financial help—paused at noon for group photo.

Dinner on beach—fireworks and a wrap up of a successful meeting. I believe any hostility toward U.S. is gone.

[Farewell ceremonies; good feeling on homebound plane, except for concern over AWACs; Mrs. Reagan and Patti were waiting on balcony at W.H.]

Sunday, October 25

A quiet day at home catching up on homework—principally the extent to which the Soviets are ahead of us on mil. capability in every line.

Had a talk with Patti—told her what a bum rap her mother was taking from a few bitchy columnists.

Al Haig called—there has been some shooting in the streets of Cairo. He says not serious or widespread.

Dinner party & ran "Ragtime." Both Jimmy Cagney & Pat O'Brien were here. Jimmy now uses a wheel chair since his stroke & has trouble hearing. He was great in the movie.

[Monday, October 26: meetings with senators; working lunch with cabinet; NSC discussion of plan for Cuba and Central America; received ambassadors. **Tuesday,** *Octo-ber 27: meetings with senators over AWACs vote, president "cautiously optimistic"; cam-paign rally in Virginia.]*

Wednesday, October 28

This is get ready for physical exam day. It began with a laxative—all meals will be liquid, etc. All in all—yes I'd rather be in Phil.

Another day at home (thank goodness) meeting with Sens. The debate is on & at 5 P.M. they vote on AWAC's.

Met with Zorinsky—didn't get a commitment but believe he's leaning.

It is 5:15 and we won—52–48. Zorinsky was with us so was Bill Cohen of Maine. What a victory this is—and what it means world wide. Our Sen. leaders came over from the hill happy & proud.

Thursday, October 29

Met with Huang Hua. There is a real push going on. China is virtually delivering an ultimatum re arms to Taiwan. I don't like ultimatums. We have a moral obligation & until a peaceful settlement is reached between the mainland & Taiwan we're going to meet that obligation.

Now to Bethesda for our check up & tomorrow to Camp David.

I won a purple heart. A Fleets E. before boarding Marine 1 and upon arrival at the Naval Hospt. right in for the Barium E. That is an experience I've never had before—Oh, I've had B.E.'s but they have a new procedure that's worse than March 30th.

One of the young ladies (Navy) made me a paper purple heart & I wore it.

Finally our 1st meal in 2 days.

[Friday, October 30: routine medical tests; Mrs. Reagan had a cold; left for Camp David; saw the film Rich and Famous *(1981) and considered it pornographic.* **Saturday, October 31:** *desk work; saw* Heartland *(1979) and thought it strange.]*

Sunday, November 1

More desk work—Nancy not much better & now back at the White House.

Al Haig called. Jack Anderson had a column set to go that I was about to fire Haig—quoting an unnamed W.H. source. I called Jack told him it was false—he graciously agreed to pull the column & quote me instead—that Al was the best Sec. of State we've had in a long time. Of course he wouldn't reveal the W.H. source.

Monday, November 2

Day really started with the arrival ceremony for King Hussein & Queen Noor. I'd received a letter from Margaret Thatcher telling me the King was very nervous about the meeting.

In the Oval office after the ceremony I told him where we stood and the climate got better. I think my remarks at the ceremony declaring our friendship with Jordan were well received by him. He ad-libbed a finish to his own remarks that was very sincere and expressed appreciation for what I'd said.

It's going to be touchy with Israel but we're going to provide defensive weapons for Jordan. If we don't they'll turn to the Soviets for them because of their fear of Syria.

Later in the day we had a legislative planning session in which a very dark picture economically was painted for 83 & 84. I will not give in and raise taxes.

The State Dinner was a smash from 1st cocktail to the last dance. Benny Goodman entertained with 4 of the top men in the field including Buddy Rich on drums. Everyones foot was tapping.

I believe we've really established a bond of friendship. His Majesty's response to the toast was obviously sincere, warm & moving.

*[**Tuesday, November 3**: meetings with former senator Edward Brooks (D-MA) on voting rights bill; farewell to King Hussein; working lunch on budget; met with Primate of Greek Orthodox Church in U.S.; meetings on the state of the Merchant Marine, and on decontrol of natural gas; spoke to Southern senators about voting rights bill. **Wednesday, November 4**: meetings with legislators; with cabinet over voting rights bill, which president saw as flawed but considered some action necessary; noted election results in Virginia and New Jersey; went horseback riding with Senator Paul Laxalt (R-NV).]*

Thursday, November 5

Not too busy a day but a serious one. A very secret paper somehow made its way to the N.Y. Times. We have to find a way to end this endless & dangerous leaking.

Called in Dick Allen & Al Haig and ordered a halt to the sniping—wherever it's coming from so we can stop this press obsession that we are having chaos & feuding in the admin.

[Met with professional money managers regarding economic program.]

A long afternoon N.S.C. meeting. We plan to talk to the world via TV just prior to Breshnev's visit to Germany to announce that in the INF arms limit talks we'll ask for total elimination of Theatre Nuclear weapons in Europe.

The Central American situation is going sour. Nicaragua is becoming an armed camp with the most modern weapons & an army disproportionate to its size or its needs.

Friday, November 6

We're going to have to build in more open time around meetings with Congressional leaders. Met 1st with Sen. leaders and meeting went 45 min. over time. Then the House leadership which of course started late but ended later.

The meetings were about the economy. With our plan barely started unforeseen things such as the high interest rates, etc. have increased the estimated deficit & make a balanced budget by '84 look unlikely. On the hill they automatically start thinking of tax increases. We differ & I think with good reason. I believe we reduced the differences between us but the press is going wild with its usual irresponsibility.

[Fund-raiser in New York City. **Saturday, November 7:** *desk work all day;* La Bohème *at Kennedy Center in company with Beverly Sills.* **Sunday, November 8:** *desk work all day; watched* 60 Minutes; *studied briefing book for Tuesday press conference.]*

Monday, November 9

[Ceremonial appearances; prepared for press conference; met child star Ricky Schroder.]

Jim Brady appeared with us for the opening of the rebuilt press room. It was a most heart warming event. All the interviews seemed to go well.

George & Barbara Bush came over for dinner. It was an early but pleasant evening.

*[*Tuesday, November 10: *met freshman New York congressman Peter A. Peyser about use of Fort Drum to house Haitian refugees; reception for Vietnam veterans; press conference "went well"; NSC meeting regarding Central America and Cuba; presented with gifts from constituents; reception for donors.]*

Wednesday, November 11

[Veterans Day: instructed not to go to Arlington National Cemetery for security reasons.]

Spent the day at my desk upstairs and managed to get a little drawer cleaning done in addition to homework.

Now I'm reading an article about Dave Stockman supposedly telling all to a reporter in the Atlantic Monthly. If true, Dave is a turncoat—but in reality he was victimized by what he'd always thought was a good friend.

Thursday, November 12

It looked like everything was going wrong today. The "Big 3" were waiting with a "what to do about Stockman" question. Before we could get into that—had a meeting with leadership. Repub. of the House & Sen. on the budget—Stockman present. He asked for the floor—got up & told them he'd made a stupid mistake, etc. & they applauded.

Back in the Oval Off. met with staff, George B. & Don Regan. I didn't go along with one or two who wanted to fire Dave—nor did Don R. or George B.

Dave came over he and I had lunch. I had lunch—he couldn't eat. He stood up to it and then tendered his resignation. I got him to tell the whole thing about his supposed friend who betrayed him, then refused to accept his resignation. Told him he should do a "mea culpa" before the press and clear the misconception that had been created by the story. He was all set to go & did—taking their questions head on.

[Used influence to help family stymied by federal hospital regulations; received Senator Ted Kennedy (D-MA) with his mother, Rose; met with army rodeo group and with NSC regarding arms reduction talks with Soviets.]

Friday, November 13–Sunday, November 15

Nancy left for Calif. to get "Scopus Award." I was in meetings until early afternoon & then off to Houston Texas for Repub. fund raiser. It was honoring Jim Baker and all the Texans now in our admin.—(quite a few of them). A big affair—Gov. Clemens, John Connally, several Congressmen (wives for all just named), Eddie & Fran Chiles—he was dinner chairman. Billy Graham dropped in to do a benediction. It was a warm and friendly affair.

The next morning off to San Antonio, then by chopper to ranch for our 2 days of Turkey hunting. The ranch house is small—really evidently just a cabin for this sort of thing.

Hunting Turkeys provides some good nap time. You sit in a blind—in our case 3 hrs. each day. The 1st day Turkeys finally came in but they were all hens so no shooting. The second day—3 hours and then had to get ready for home but saw no Turkeys at all.

Came home in the 747 Doomsday plane. Was briefed on its capabilities. It's like being in a submarine—no windows. [. . .]

Monday, November 16

Staff meeting re the Allen $1000 episode. I don't think it amounts to much. It's hard to believe a fellow didn't forget the money since he left it in the safe of the office he was moving out of.

It goes back to a Japanese mag. interview with Nancy. It seems they give cash gifts to people they interview. This took place clear back at Inaugural time. Dick intercepted the envelope to avoid embarrassing Nancy or the Japanese since she obviously could not accept it. It was then given to his Sec. to put in the safe. Both forgot & it was found after he'd moved out.

We had an F.B.I. investigation—everybody was all clear but we didn't know that under the 1978 "Special Prosecutor Act"—several Justice Depts. must successfully review the findings. Someone in Justice after all these months leaked it & I suspect bureaucratic sabotage. In other words our admin. not Dick Allen is the target.

An N.S.C. meeting following an interview with D. Lambro. This has to do with Central Am. We have decided on a plan of covert actions, etc. to block the Cuban aid to Nicaragua & El Salvador. There is no question but that all of Central Am. is targeted for a Communist takeover.

[Tuesday, November 17: official visit by President Luis Herrera Campins of Venezuela.]

Wednesday, November 18

Today was the big day—the speech to the world at the Nat. press club. It really was to the world. I'm told it was the largest network ever put together—all of Europe, China & I don't know how many other places. It has been wonderfully received worldwide except for Russia—Tass is screaming bloody murder. I asked Russia to join us in total elimination of all medium range nuclear weapons in Europe. Funny—I was talking peace but wearing a bullet proof vest. It seems Kadaffi put a contract on me & some person named Jack was going to try for me at the speech. Security was very tight.

[Farewell to President Luis Herrera Campins; received traditional gift of Thanksgiving turkey; two receptions.]

Thursday, November 19

An NSC meeting on what we can do to further help El Salvador by way of the C.I.A.

Reports still coming in on the Speech. It was telecast in 50 countries with an estimated audience of 150 to 200 mil.

[Met with Jewish groups to discuss sale of AWACs to Saudis; saw film From Mao to Mozart: Isaac Stern in China *(1981) and thought it "wonderful."* **Friday, November 20:** *appearance at Library of Congress in honor of James Madison; met with Haig to discuss Israeli defense minister Ariel Sharon; saw President Gaafar Nimeiri of Sudan; Midwestern governors regarding unemployment and Idaho congressmen about silver sales by Treasury; interviewed by People magazine; filmed campaign spot for General Paul Calvo of Guam.]*

Saturday, November 21

Home at my desk most of the day cleaning up odds & ends. The House, Sen. Conf. committee is slugging it out on the "Continuing Resolution" which was to have started at midnight last night. Still no budget of course. Neither house gave us the $8 Bil in cuts we want but the Sen. bill was closest & would have given us about half that. The Dem's are fighting for less. It seems at 5 PM they were about ½ bil $ apart on the Dem. side. That isn't enough to justify a veto since I indicated I'd settle for the Sen. version. I'm still waiting to learn what happens.

Don Regan home from China—says in 4 days of meetings with all the top people, Taiwan wasn't mentioned once.

Helmut Schmidt called to see if I had any ideas about what to say when he meets Brezhnev tomorrow night. I said to tell him I really mean it about the SS20's.

A rub down & home for the evening.

Sunday, November 22

[Photo session.]

The Cong. is still meeting. Finally adopted a conference report I said I'd veto. I've suggested they pass a 15–20 day extension of the present "continuing res." go home for Turkey day & then come back & take up where they left off. If they don't & I have to veto—we won't board A.F.1 for the ranch and the battle will go on.

[Unofficial visit by King and Queen of Sweden; reception in honor of PBS young artists series.]

Then the long evening wait for the vote in Cong. The house went against us by only 11 votes. I said "veto."

Monday, November 23

At 8 A.M. I read the veto message to the press. The House was convening at 10, the Sen. at 12. Finally the House voted and we won with every Repub. & 43 Dems. The afternoon went on and finally came the Senate vote 88 to 1 in our favor. The 1 negative was Alan Cranston. Now I'm waiting to sign the extension and then it's A.F.1 & Calif.

Arrived in dense fog—had to motorcade to ranch. Barney & Anne on hand.

Tuesday, November 24–Monday, November 30

[Horseback riding; interview with Barbara Walters.]

Thanksgiving turned out rainy but inside dinner went well with Bess & Moon, Maureen & Dennis, Patti & Peter Straus about whom she is very serious.

[Riding and chores; phone calls to key advisors.]

Security very tight even at the ranch. A "hit band" is supposed to have crossed into the U.S. by way of Canada with me, George B., Al Haig & Cap W. (any or all) as targets. (Our friend in Libya).

[Flight home with fund-raising stop in Cincinnati.]

Tuesday, December 1

[Met with Teamsters national board, received their support.]

Pres. Mobutu of Zaire came in. We're trying to help him but his economy is in bad shape—partly due to corruption which he's promised to clear up.

[Planned fund-raiser for 1982 campaigns.]

At noon addressed the Conf. on Aging. To my surprise they were most friendly giving me ovations about 6 times during the speech & at the finish. An N.S.C. meeting re Central Am. We're proceeding with covert activity in Nicaragua to shut off supplies to the Guerillas in El Salvador.

[SCT; dinner at home.]

Wednesday, December 2

Finally the meeting with the whole AFL-CIO Board. Lane Kirkland led off with why they don't like our ec. program. To hear him it is responsible for the recession. Never let it be said that 21½% interest rates last year, double digit inflation & factories closing down played any part in our troubles. It's easier to blame a program which has barely started & already interest rates are down to 15¾—inflation almost cut in half & while there is great unemployment there are also 266,000 more people working than there were last Jan. 20th.

He wants us to take back the Patco strikers & embark on a plan of government spending that sounds like the "new deal" all over again—and it didn't work. He chided me for not communicating with labor. I politely reminded him of the approaches we'd made—all of which he turned down and I said, "Lane—you slammed the door, not me."

[Meetings with farm leaders and task force on private initiative.]

Nancy went to Arlington to Tex Thornton's funeral. Security said no to me. It seems Kadaffi has enlarged his hit list to include even Sen. Baker—but I'm still target No. 1.

Thursday, December 3

A lengthy interview with several editors of "Reader's Digest." I think it went well—most of the Q's were on foreign policy. A long day filled with meetings & appointments. A rumor started somewhere that I had a heart attack. It was just like the phony rumor during the campaign. We traced that one to a brokers office on Wall Street.

N.S.C. meeting—I approved starting a Civil Defense buildup. Right now in a nuclear war we'd lose 150 mil. people. The Soviets could hold their loss down to less than were killed in W.W. II.

Friday, December 4

An easy day. Meeting with leaders of Steel industry. Persuaded them to let us deal with Europe's dumping of 3½ mil. tons of subsidized steel on our mkt. They wanted to go at them with law suits.

[Meeting on economic program and maintaining the country's faith in it.]

A most enjoyable dinner (about 12) mainly old Hollywood friends on hand for the annual awards Sun. at Kennedy Center—the Stewarts, Grants, Fairbanks, etc. After dinner gave them a W.H. tour. They were enthused about all that Nancy & Ted have accomplished.

[Saturday, December 5: met with Bill Casey on Allen case; ran Reds *(1981) for guests—which included the film's director, Warren Beatty—and thought it "a most imaginative job."]*

Sunday, December 6

Sun. but hardly a day of rest. A pic. at 11 AM for Time magazine. 2 P.M. meeting with Al Haig. Then changed clothes for another pic. This one downstairs with the Christmas tree. Nancy & Ted have done a beautiful job of decorating the entire downstairs. Then change into Black Tie for reception for the annual winners of honors as entertainers; Helen Hayes, Cary Grant, Count Bassie, Jerome Robbins & Rudolph Sarkin (pianist). We shook 450 hands in a receiving line. But it was a beautiful & warm affair amid all the Xmas lights & decorations.

Then upstairs to snatch a quick dinner and off to Kennedy Center for the show (taped to go on CBS) honoring the 5. This too was a wonderful event.

Monday, December 7

[Meeting on government efficiency; received visits from astronauts, grassroots Republican workers, and celebrities Douglas Fairbanks Jr. and Tommy Lasorda.]

An NSC meeting re Quadaffi. No conclusions reached except that we can't do anything until we find an answer to the 1700 Americans still working in Libya.

[Inducted as an honorary colonel in the Army of Texas.]

Had a long talk with Bill Clark about some in house problems that must be solved.

Tuesday, December 8

A full day. First a meeting to hear the 1st 1983 budget review. We who were going to balance the budget face the biggest budget deficits ever. And yet percentage wise they'll be smaller in relation to G.N.P. We have reduced Carter's 17% spending increase to 9%. The recession has added to costs & reduced revenues however so even with that reduction in govts. size we face a large deficit.

Then an N.S.C. meeting—2 of them—to discuss Quadaffi and Global negotiations on helping the 3rd World. We are sending a secret or private warning from me to him that harm to any of our people by his terrorist goons will be considered an act of war. We're looking at other actions that must however come only after we've tried to get our people out of Libya. We also debated the O.M.B. difference in budget for foreign assistance and State's request $1.1 Bil. bigger. I think this time I came down on States side.

The press is beginning to charge that we are making up the Quadaffi threat because we won't tell them the sources of our information. I've come to the conclusion they are totally irresponsible and won't be satisfied (if then) until someone is gunned down by the "hit men."

I've told our people not to tell the press we're going to Camp David. As it is we're changing our route for security reasons.

[Lee Annenberg resigned for personal reasons.]

Wednesday, December 9

Made decision to ask for waiver allowing striking air controllers to apply for govt. jobs (not including air control) without waiting 3 yrs. Also decided to OK sale of Caterpillar Pipe Layers to Soviet U. Told Cong. Repub. leaders of that decision this morning. Saw some Ambassadors off & met with March of Dimes Poster Child for those born with defects. Richard Wagner was a joy. He uses leg braces & crutches but plays basketball, soccer & swims—5½ yrs. old. Bright as a whip & happy.

Rcvd. Blackthorn walking sticks as gifts for Nancy & me along with Tip O'Neil. Lunch with him, Mike, Max, Ken. It's Tips birthday & we had a good time telling stories—Irish stories.

[Desk work; heard Navy choir sing carols; dinner with Annenbergs.]

Thursday, December 10

Met with Council of Ec. advisors. While one or two spoke of possible tax increases after 1982 the others (majority) said no. Tax increases don't eliminate deficits they increase govt. spending. The general consensus was that our plan is the proper medicine for the recession and we should stick to it. That's what I intended to do all the time. In N.S.C. decided to go forward with a $100 mil. sale of corn to Poland (on credit).

[Met with David Rockefeller and committee to assist Jamaica; cabinet meeting on difficulty of further budget cutting.]

Today the House signed a continuing resolution I can sign.

Friday, December 11

Off to Camp David by a quite different Helicopter route. Our Libyan friends are thought to possess heat seeking missiles that can be hand launched.

[Considered various budget cuts; met with President Ford and discussed Qaddafi case.]

Saturday, December 12—Camp David

A long walk in the morning then an afternoon at the desk. Somehow it seems easier there. Word received that Poland has moved on Solidarity. Leaders have been arrested, union meetings & publications banned, martial law declared. Our intelligence is that it was engineered & ordered by the Soviet. If so, and I believe it is, the situation is really grave. One thing certain—they won't get that $100 mil. worth of corn.

Sunday, December 13

Many calls all re Poland. Returned to W.H. at 1:30—a brief security council meeting on the latest info. A quiet evening with "60 Minutes."

Monday, December 14

[Spoke to group of conservative state legislators.]

Met with Paul Volcker. I'm not sure he sees the need to let the increase in money supply go forward in the upper range of their moderate schedule. This recession is because they slammed the door in April & kept it closed until Sept.—almost Oct.

Our plan will get the Ec. moving again only if the Fed. allows—not an upward surge—but a moderate growth geared to Ec. growth.

Called the Pope re Poland. Expressed our concern & intention to keep close tabs on what is going on.

This is take my picture day—a half dozen including a portrait by Arnold Newman.

Then a meeting with Papal Sec. of State Cardinal—whoops! Wait a minute—that's tomorrow. Todays meeting with 4 professors—the Pope's Vatican study team on Nuclear War. Their findings—we must not have one.

Al Haig got back, met with him in the Study. He says Europe's leaders are on Cloud 9 over my speech about zero missiles in Europe.

Cap came by—may have an answer re the Allen situation. Even if cleared, I don't believe he can resume as W.H. advisor. Cap has 2 proposals—either one of which would be good for him if he'd take it.

The W.H. Christmas party is now over—that is the 1st shift. Tomorrow night we'll have one for the other half. Everyone seemed very pleased with it.

Tuesday, December 15

Just up from the 2nd Christmas party. It was a success & every one seemed to have a warm, wonderful time. Nancy has her wish—it is snowing.

Today was a tough one—2 major decisions facing me and I don't have the answers yet. One is a debate between OMB and EPA (Ann Gorsuch) on the '83 figure for EPA. Even though I want to save money I lean toward Ann. She's a good manager & has done a great job of cutting back.

Second problem worse. I promised to do away with the Energy Dept. Jim Edwards (Sec.) has carried this out. The problem is do the functions which must continue go to Interior or Commerce. Both want it & I think they should have it but only 1 can get it.

Lunched with Cardinal Casaroli—Sec. of St. to the Vatican. Most of the talk was on Poland. This thing going on there was no sudden reaction as the Communist Govt. would have us believe. The operation is so smooth it must have taken weeks for planning. Solidarity was going to demand a vote by the people as to whether they wanted to continue under Communism. That the Commies can never permit.

Wednesday, December 16

[Visit from Charles Price, ambassador to Belgium.]

Made decisions on Energy–to Commerce. And came down in favor of Ann Gorsuch of the E.P.A. budget with a slight nod to O.M.B. She's a good manager & if money can be saved, she'll do it.

[Interview with PBS; receptions for Gannett employees, and for congressmen after vote on farm bill.]

Thursday, December 17

[Conflict between OMB and HUD over budget figures; press conference.]

Learned a Major Gen. of ours—Dozier by name, was kidnapped in Italy—probably by the Red Brigade. I fear for his life.

[Interview with Tom Brokaw of NBC; reception for Christmas tree lighting; dinner with Nofzigers and Laxalts.]

Friday, December 18

Had to be the heaviest schedule & most frustrating day yet. One long series of meetings on bud. on Nat. Security and half a dozen other things. Broke the routine in late afternoon to go over to the W.H. for W.H. staff party. It was nice. All these good people who take care of us & tend the place with their families.

Most important meeting with heads of Auto Co's. That industry is facing total disaster. I'm recommending that we tell Japan we are going to impose same restrictions on their exports as they do on import of Am. cars in Japan. Auto makers need passage of Clean Air Act changes which can reduce price of Am. cars.

Met with Sen. leaders who are beginning to panic on taxes. They want us to raise or impose new ones. I'm resisting. D--n it our program will work & it's based on reduced taxes.

[Saturday, December 19: lunch with Senator Paul Laxalt (R-NV); visit with Henry Kissinger; report that Polish ambassador to U.S. defected.]

Sunday, December 20

First thing this A.M. a message from Israel—an angry denunciation of us with charges of betrayal & anti-semitism. Obviously it is intended for the public to arouse our own Jewish community against us. We'll meet tomorrow on this but I'm trying to put together a possible, personal letter to Begin.

[Reception with entertainment by young opera singers, taped for PBS.]

Dinner then back down for the Senior Staff party. It was a fun time. Mike D. teamed up with the piano player and we had a 2 piano concert & sang Christmas Carols.

Monday, December 21

Another non-stop day with virtually no time between meetings. Most important was NSC meeting re Poland. I took a stand that this may be the last chance in our lifetime to see a change in the Soviet Empire's colonial policy re Eastern Europe. We should take a stand & tell them unless & until martial law is lifted in Poland, the prisoners released and negotiations resumed between Walesa (Solidarity) & the Polish govt. We would quarantine the Soviets & Poland with no trade, or communications across their borders. Also tell our NATO allies & others to join us in such sanctions or risk an estrangement from us. A TV speech is in the works.

We also touched on Israel. There are signs Begin went too far both here & at home. His country is divided (even the Cabinet) on his berating of us. And here the Jewish community is resentful of what he's done.

There were budget meetings. We're coming down to the wire although there is wide disagreement on what kind of deficits we forecast for 5 yrs. ahead.

[Farewell party for Max Friedersdorf, counsel to Bahamas.]

Tuesday, December 22

I think we're packing everything we've missed into these last several days of the year. By the end of the 1st meeting I was half an hour behind & it stayed that way all day.

A budget meeting. We've finally come together on the cuts—probably won't get all we ask for from Congress. They're so used to spending (for votes) they're getting edgy with '82 being an election year.

The recession has worsened, throwing our earlier figures off. Now my team is pushing for a tax increase to help hold down the deficits. I'm being stubborn.

I think our tax cuts will produce more revenue by stimulating the economy. I intend to wait and see some results.

Met with the Polish Ambassador & his wife. It was an emotional meeting. They have asked for asylum here. He is defecting because of what the Polish govt. (ordered by the Soviets) are doing to the Polish people.

An NSC meeting on what we are going to do about the situation. I go on TV tomorrow nite 3 networks. It's supposed to be a Christmas message but I intend to deliver a message to the Soviets & the Pols. We can't let this revolution against Communism fail without our offering a hand. We may never have an opportunity like this one in our lifetime.

We have just learned we have to make public every gift we've received—even from friends we've exchanged gifts with for years—the price & who gave them. A little inheritance from J.C. and a d--n gross invasion of privacy.

Wednesday, December 23

Meetings, pictures & year end interviews—one with Roger Mudd who gave a biased account of my last press conference. I was a good boy though & didn't ding him.

In NSC worked out final touches for speech tonite on all networks. OK'd letters to Brezhnev & Jaruzelski. Said more in the letters than I will in the speech.

Speech went OK—about 14 minutes. First ½ hour calls at W.H. running about 6 to 1 in favor.

Thursday, December 24

Christmas Eve.—Ron & Doria arrived in time for lunch—the others later in the day. To the Wicks for dinner at Watergate & a typical Wicks Xmas Eve. Lots of singing—carols, etc.

Friday, December 25

[Opened gifts in the morning.]

Tonite the Wicks & Ruges came here. It was a wonderful Christmas dinner. Earlier a letter arrived from Brezhnev. It seems we're intervening in Poland and he's upset about it. I've suggested that in our reply we remind him that we are only suggesting the Polish people be allowed to have a voice in the kind of govt. they want. Under the Yalta Pact the Soviets agreed they & others would be allowed to do this. The Soviets have never honored that promise. We should also agree—we won't intervene if they won't.

Saturday, December 26

Dick & his family left this morning. Doug & C.Z. came over for lunch.

Patti & Peter leave tonite. I believe, Ron & Doria in the morning. We leave for Calif. at 11 AM. We were all in the gym exercising this A.M.

Sunday, December 27–January 3, 1982

I won't do a day by day but it was a wonderful trip. Our first few days at the Century Plaza. On Mon. Barney & I went up to the ranch. It was a beautiful day. Most of our time was spent cutting up fallen limbs for firewood & brush to burn. That night at the hotel—dinner with Patti, Maureen & Dennis.

There was some work of course—bill signings, intelligence reports etc. To Palm Springs on the 30th. Went by Marine 1. The whole holiday season there was great—all our friends for a wonderful New Year's Eve—then the same group for 2 additional parties. Golf in the mornings & on New Year's Day football on TV all day. Met with Al Haig, Bill Clark, Mike Deaver. I think Bill will come over as National Security Chief. The problem is how to treat the Allen problem so as not to hurt him. He has good offers in the private sector but I don't want to leave him with any onus. The press has really been a lynch mob & I don't think they'll stop which is why he can't be back in N.S.C.

Left Sunnylands on Jan. 3rd—took A.F.1 out of Palm Springs. Learned before leaving that "Little Man" may have a fractured skull. Vet has him for a few days of observation. Evidently he's in bad shape. Is it the end of an era? His mother became mine in '46—then I rode his half sister Nancy D & now him—last of the line.

We're back in the W.H. and the whole trip seems like a beautiful dream from which we've just awakened.

Chapter 2

—

1982

Monday, January 4

An easy day except for the Allen case. Dick has been cleared. He came in & of course said he'd like to continue. Ed has already told him as I had to that we've learned the press is not going to let up if he's in that job and some Dem. Congressional committee chairmen are ready to start investigations—make that lynchings. He agreed to resign—we at same time will make him a consultant to P.F.I.A.B. and look for other assignments.*

Tuesday, January 5

Helmut Schmidt arrived. We did a last minute switch in what to say based on his remarks to Al & Don R. last nite on arrival. Our press had distorted Germany's & for that matter NATO's position with regard to our sanctions against Soviets & Poland. While they might not be able to join us in canceling their own trade they openly support us & will not undercut our sanctions. It was a good day. Helmut wants me to meet Brezhnev. He says he is very curious about me & doesn't know what to expect. Also says B. truly fears war. Maybe our disarmament talks might work after all.

An N.S.C. meeting to confirm our actions & some photos by Bachrach & Karsh. Why do I hate photo sessions—I always did even when I was in movies.

Wednesday, January 6

An easy day—gave Al H. his way on Stoesell [Stoessel] as replacement for Bill Clark. Some staff objections on grounds it should be a pol. appt instead of career. I've given my word to Al.

[Meeting with productivity task force; photo session.]

Rumor that Gen. Dozier has been killed—no confirmation.

---

*Allen was appointed National Security Advisor at the beginning of the Reagan administration. In November 1981, he was accused of taking a $1,000 bribe from a Japanese magazine for arranging an interview with Mrs. Reagan. The charges were never proven, nor did they go away. His resignation was accepted January 4, 1982.

Thursday, January 7

Dispatched (top secret) a specialist mil. team at Italian's request. Has to do with a possible informer regarding whereabouts of Gen. Dozier.

Laid down the law to Cabinet on leaks. Had Ed M. tell press my decision to extend draft registration—that leaked before he got to the press at 4:30 P.M. to tell them.

We have a drastic plan to phase a transfer over years of most of our grant programs to the states & local govts. along with a transfer of several excise taxes to fund them. It will be the biggest turn around in govt. since Wash. started usurping local & state rights.

[Visited with L.A. county supervisor, a supporter.]

Met with Paul Nitze & Eugene Rostow who are here for a few days before going back to the arms reduction talks in Geneva. I told them even 1 nuclear missile in Europe was too many and that if anyone walked away from the table it would have to be the Soviets.

Friday, January 8–Sunday, January 10

[Complains of leak to press about secret meeting on Federalism.]

A team is off to Peking to tell them at 1st hand we're going to sell F5E's & some F104's to Taiwan. We're softening things a little. Taiwan really wants the more advanced F5G's—They'll have to come later.

Off to Camp David where it was only 8 degrees. We ran an old movie, "One Way Passage."—Kay Francis, Wm. Powell. It was wonderful. Sat. took a long walk in a snow storm then spent the afternoon getting a lot of mail done & tidying up a speech.

I've learned there is a China Lobby and it has its moles in the State Dept. The Post had a story on why we should cling to the P.R.C. & never mind Taiwan. A cartoon carried out the same theme. The timing is amazing because no word has been spoken about plans & I've told no one what my decision will be.

Sun. a short walk—with wind chill it was 52 degrees below zero.

Monday, January 11

Repub. House leaders came down to the W.H.—Except for Jack Kemp they are h--l bent on new taxes & cutting the defense budget. Looks like a heavy year ahead.

Al Haig seems to have won a moral victory in Brussels on NATO vs. Poland & the Soviets.

Press running wild with talk that I reversed myself on Taiwan because we're only selling them F5E's & F104's. I think the China Lobby in State Dept. is selling this line to appease the P.R.C. which doesn't want us to sell them anything.

The planes we're offering are better than anything the P.R.C. has. Later on if more sophistication is needed we'll upgrade & sell them F5G's.

[Reception for the Hoover Institute board of governors.]

Tuesday, January 12

A busy day mainly of meetings including a working lunch & the cabinet. In the afternoon had some desk time. Met with Senators about legislation re tax exemption for schools practicing segregation. I've had complaints because the I.R.S. passed its own civil rights regulations & began harassing schools & colleges that didn't segregate but simply on the grounds of the I.R.S. wasn't satisfied with the number of minority students they had. In my view it was a classic case of bureaucracy pre-empting the function of Congress. So we shut I.R.S. off with the intention that Cong. would amend the tax laws. The press has carried on that I was reversing the govts. stand on civil rights. When we announced we wanted Cong. to handle it the press announced that as a retreat under fire. I'm burned up.

Wednesday, January 13

[Troubled by leak regarding possible shipment of jet fighters into Cuba by USSR; telephoned Senator Barry Goldwater (R-AZ) regarding Taiwan; received gift of drawings by editorial cartoonists.]

At 3:45 pm a 737 with about 70 people aboard crashed into the 14th Street bridge taking off from Nat. airport in a blinding snow storm. I've been watching the rescue efforts on TV. The plane crushed several cars then sank leaving some passengers in the ice filled Potomac. What tragedy.

I've called Pres. Cheek of Howard U. & Ralph Abernathy about our Civil Rights situation. They were very understanding.

Thursday, January 14

Flew over the scene of the crash (Marine 1). Cranes & barges on hand—no evidence they've raised the fuselage which is in 25 ft. of water.

[Trip to New York City in support of civic-business partnership program.]

Met new Sec. Gen. of U.N. from Peru and then home—landed at Andrews in a blinding snow storm. This morning before leaving had an N.S.C. meeting on planes in Cuba—no decision. I lean toward a giant propaganda campaign aimed at the Cuban people urging them to disavow Soviet U. and become once again a part of the Latin Am. community.

Jeanne [Jeane] K. tells me a Soviet ship designed to neutralize efforts to detect presence of Soviet Nuclear Subs has reached Cuba. First time such a ship has ever been in Am. waters. This definitely is an offensive weapon—thus a violation of the '62 agreement.

Friday, January 15

An N.S.C. meeting re Cuba. My own thought is that we should create a plan to urge Cuba & yes Castro to come back into the orbit of the Western Hemisphere. Castro is in trouble—his popularity is fading, the ec. is sinking and Soviets are in no position to help. We could start a campaign to persuade him & the disenchanted Cubans to send the Russians home & once again become a member of the Latin Am. community.

Met with Repub. Sen. leadership. They are uptight about budget, taxes, etc.

Wm. Raspberry & J. J. Kilpatrick both took me on in their columns re the I.R.S. tax exemption decision. But they came at me from opposite sides. I called each of them & think I defanged them. The most famous & best loved entertainer in the Black community—name of Brown came to see me. To my surprise he is a devoted follower & pledges his all out support.

Wash. is beautiful today—everything covered with several inches of fresh snow & a bright sun & blue sky.

Saturday, January 16

Al Haig home with bad news about the middle east. It's possible Mubarak will abandon the Camp David Accords & settle down with his Arab brothers once he gets the Sinai back. At the same time Begin may renege on the Sinai although he swears he won't. Sharon is the bad guy who seemingly looks forward to a war. Al will be going back.

On Cuba we're sending Gen. Walters in to open talks with Castro. It's just possible we could talk him into moving back in to this hemisphere. His ec. is flat and he's losing popularity.

[Attended banquet. **Sunday, January 17:** *slept late, quiet day.]*

Monday, January 18

[Met with representatives of Fellowship of Christian Athletes; conference with U.S. Ambassador to Japan Michael Mansfield and Japanese trade minister.]

Then the day turned bad. Lt. Col. Ray, our attaché in Paris was gunned down & killed as he left for work. I called Mrs. Ray—but what can one say? The other attaches & their wives were with her. I told her we were meeting with the French govt. to see if we couldn't provide some kind of security & we're reviewing our own efforts.

A Tex Saunders called—said he'd been a grip on some of our "Death Valley Days" shows. He's in town for his son's funeral. His son is an A.F. Lt. who was knocked off the bridge and killed by the 737 that crashed in the Potomac. I talked to his widow.

[Report of Air Force. Thunderbirds training accident; mix-up in date for dinner with Frank Reynolds and his wife.]

Tuesday, January 19

Most of the day was spent in preparing for the 2 PM press conference. It's always an experience to walk into the East room jammed to the walls with the press corps. The questions covered a wide range. I'd been told they were surly and wanted blood. I think it was my day though.

[Taped appearances for various groups; phone call from son Mike, whose mother-in-law died.]

Wednesday, January 20

[First anniversary of inaugural; received mementoes from friends.]

The day however was a tough one. A budget meeting and pressure from everyone to give in to increases in excise taxes tied to Federalism program. I finally gave in but my heart wasn't in it.

[Banquet to celebrate anniversary.]

Thursday, January 21

Supposed to helicopter to Baltimore for some appearances & lunch re the enterprise zone program. Baltimore on its own had done a remarkable job combining city, state & private effort. Trip cancelled—no fly weather. The Mayor and his group drove here for lunch.

Early in day met with U.S. Chamber group. They made an impassioned plea that I not raise any taxes. They were touching a nerve when they said I would look as if I were retreating from my own program. That's exactly how I feel. After meeting told Ed, Jim & Mike we had to go back to the drawing board. I just can't hold still for the tax increases. We'll go at it again in the morning.

At N.S.C. meeting we decided to go back to the law of the sea negotiations but make it plain we could not accept their proposals on sea bottom mining, etc.

Friday, January 22

Lyn N. & his staff came over for a farewell picture. I told him not to go too far away.

I told our guys I couldn't go for tax increases. If I have to be criticized I'd rather be criticized for a deficit than for backing away from our Ec. program.

Left for Camp David to work on the St. of the U. speech. Saw an old movie "Adam's Rib."—Hepburn & Tracy.

[Saturday, January 23: worked on speech, watched 1932 movie Grand Hotel; Sunday, January 24: watched Walter Wriston, a member of the president's Economic Policy Advisory Board, and Senator Howard Baker (R-TN) on television shows, and the Super Bowl. Monday, January 25: met with small-business representatives; looked over com-

ments on speech by cabinet members; dinner with congressional leaders, with entertainment by Vince Dowling.]

Tuesday, January 26

Have been getting calls from Al Haig in Geneva. Gromyko declares he won't talk about Poland. Al says that's what he's going to talk about. Last call came after 7½ hour meeting. Al says Gromyko is more nervous than he's ever seen him. He talked arms limitation. Al talked Poland & Cuba. Gromyko tried to play down the big arms buildup they are doing in Cuba.

[Met with Republican congressional leaders on returning programs to state and local governments.]

At noon a working lunch with the Cabinet. They now know what's in the St. of the U. address (I'm writing this before leaving for the Capitol). I wonder if I'll ever get used to addressing the joint sessions of Cong.? I've made a mil. speeches in every kind of place to every kind of audience. Somehow there's a thing about entering that chamber—goose bumps & a quiver. But it turned out fine—I was well received & I think the speech was a 4 base hit. We'll know more tomorrow.

Wednesday, January 27

A lot of briefing for a 2 hr. special with Dan Rather. Saw the show & wonder why we bothered.

[Visit from Senator Strom Thurmond (R-SC) and Clemson University champion football players.]

Thursday, January 28

Awakened at 6:45 A.M. by Bill Clark telling me Gen. Dozier has been rescued in Padera Italy. Before morning was over I'd called Pres. Pertini to thank him for all his country had done. Their elite anti-terrorist force did a brilliant piece of work. Then I called the General. A lot of prayers were answered this day.

Nancy & I visited the F.D.R. exhibit at the Smithsonian. Then back to the W.H. (East room) for a luncheon which brought together all the Roosevelt family plus many who had served in his admin. & current politicos of both parties. The press is dying to paint me as now trying to undo the New Deal. I remind them I voted for F.D.R. 4 times. I'm trying to undo the "Great Society." It was L.B.J.'s war on poverty that led to our present mess.

[Met with African American Republican leaders, Republican mayors, medical advisory group, and the cabinet regarding unemployment; dinner at the home of George Will.]

Friday, January 29

A lengthy cable from P.M. Thatcher re Poland. She is working hard to get allies to be more forceful in their actions.

Met with Sens. Hatch & Thurmond on Voting Rts. Bill and "Tax Exemption Legis." It was mainly lawyer talk so I moved them on to settle the legal points.

John Jacobs, Pres, of Urban League came in. I believe we can continue communicating & establish a good relationship.

Sen. Hayakawa came in to tell me he will not run for re-election. I think that's good.

We're going to present a program to Congress calling for reforms—not cuts in tax regulations to tighten loopholes particularly for business. They'll gain about $12 Bil. in revenue but they are truly reforms & all in all well justified.

Saturday, January 30

Solidarity Day. Charles Wick has really created a great international telecast studded with celebrities & heads of state proclaiming solidarity with the Polish people.

Al Haig & Bill Clark came in for report on Al's trip. He believes he did some good with Israel although he fears they may—on slightest provocation—war on Lebanon. [. . .] Our choice—to go it alone with harsher steps against Poland and risk split in the alliance or meet with E. alliance on things we can do together. The latter is my choice. The plain truth is we can't—alone—hurt the Soviets that much. The Soviets will however be disturbed at evidence their attempts to split us off from allies have failed.

[Concern for Pentecostal Christian woman on hunger strike at U.S. Embassy in Moscow. **Sunday, January 31:** *noted dinner the night before with Polish ambassador and others, watched 1981 Polish film* Man of Iron *and said, "It was most moving & made all of us more determined than ever to help these people."* **Monday, February 1:** *met with Elizabeth Dole, assistant to the President for Public Liaison, and then "hard nosed conservative friends" concerned about direction of administration; dinner with Mr. and Mrs. Frank Reynolds.]*

Tuesday, February 2

An Ec. meeting to hear about the ec. The banks have upped the interest rates. Don R. says this is typical fluctuation when you are working your way out of a recession. He thinks it will be temporary. He also believes however, we've got to have more cooperation from the Fed. Reserve Bd.

Addressed about 100 young people—participants in the Youth Senate program sponsored by the Hearst foundation. I told the B-17 story and again choked up so I could hardly finish.

[Reception for Council on Physical Fitness; prepared for visit by President Hosni Mubarak of Egypt.]

Wednesday, February 3

A rainy day so Pres. Mubarak's greeting was in the East room. Our 1st meeting following the ceremonial was just the 2 of us in the Oval office. I had a whole program of subjects I intended to bring up. He began explaining his situation with the Soviets and I listened. But I guess the script will be taken up by others because he has meetings scheduled with Al, Cap, Don R., Bill Casey & others. Tonite the State Dinner and an Israeli violinist for entertainment.

[Answered mail; state dinner.]

Thursday, February 4

[Attended Annual Prayer Breakfast.]

Back to the W.H. for a last meeting with Pres. Mubarak. We decided to make $400 mil. of the $1.3 bil. mil. weapons for Egypt an outright grant. We'd been prepared to grant $200 mil. but politically he needed to take something back symbolizing a triumph resulting from his visit. I think he's going to do darn well. We're pushing for Suez passage for our nuclear vessels. He's reluctant because the Soviets will want the same thing.

Day was filled with meetings, N.S.C. etc. We have problems with El Salvador—the rebels seem to be winning. Guatemala could go any day & of course Nicaragua is another Cuba. Lots of options but no decisions. Mid afternoon I could hardly keep my eyes open—in fact I didn't.—Tonite early to bed.

Friday, February 5

[Meeting on Federalism with congressmen; ambassadorial formalities.]

The Sr. staff plus Stu Spencer, Dick Wirthlin, et al, are up at Camp David strategizing on the political situation. I'm not sure I'm happy about that. I don't want them to be eternal optimists but sometimes they decide to fix things that aren't broke.

["Cleaned up" speeches for delivery the following week.]

Saturday, February 6

Birthday phone calls—delivery of gifts, etc. Worked on speeches for Iowa, Ill. & Ind. the 8th & 9th & did thank you notes. Bill Clark called—trouble brewing in the Middle East. Israel on verge of major invasion of So. Lebanon because of P.L.O. mil. build up. We are trying to persuade them they must not move unless there is a provocation of such a nature—the world will recognize Israel's right to retaliate. Right now Israel has lost a lot of world sympathy.

A very nice dinner with Geo. Murphy & friend, the Mike Cowles, Annenbergs & Wirthlins. Finished up with a 1911 Port.

[Sunday, February 7: a quiet day and some more desk work.]

Monday, February 8

Met with bi-partisan leadership on budget then with Repubs. alone. Then signed the bud. and sent it to the hill. Tip O'Neil still thinks I'm depriving the needy. Told the press I associate with the country club crowd. He plays golf & I don't.

[Lunch with star hockey players; flew to Minnesota for political rally; flew to Des Moines. **Tuesday, February 9:** *spoke at fund-raiser and before state legislature on Federalism; flew to Indianapolis for similar appearances; returned to Washington and spoke to religious broadcasters convention.]*

Wednesday, February 10

Short day but long NSC meeting. Subject should I or shouldn't I make a Nat. speech re our Central American problems. Worry of the "no's" is that I'll sound like we're going to war over El Salvador or Cuba. I finally said I thought we should do a speech about need to bring the Americas together in a solid alliance & I'd like to see a suggested draft on such a speech.

At noon had about 300 of the gal's who've been appointed in govt. at the W.H. to a stand up lunch.

Last night ABC News had a story of an L.A. family about to be evicted. They bought a home in 1954 for $10,000 paid off the mortgage in 1965. In '71 new street lamps were installed. Their assessment was $51. They never received a notice. Their home was sold for $170, then resold twice—again without any notice to them. All this time they've paid the taxes. Now a man appears with a court order claiming ownership & $6,100 rent due for the years they've lived there. I called Frank Reynolds & got all the dope—turned it over to Bill Clark. There must be something we can do about this.

[National Press Club dinner.]

Thursday, February 11

Sens. Baker, Laxalt & Cong. Bob Michel came over this morning. There is unrest among the troops over the budget problem. The mystery of Sen. Hollings proposal to freeze spending, taxes, etc. at this year's level for '83 was explained. Sen. Baker persuaded him to do that to drop a bomb on his own party. They'd love to go for the tax & the spending list—for Defense only—but they have to be against him for suggesting lower social reform spending than we did.

N.S.C. briefing. Castro is very nervous. We'll try to keep him that way.

[Ray Donovan presented proposal for job training program to replace Comprehensive Employment and Training Act (CETA).]

I've agreed to a committee to review Fed. govt.'s property holdings (⅓ of the land in the U.S.) and come up with a plan to sell unneeded property & use the money to pay off the debt.

*[Accepted resignation of Joe Canzeri, assistant to the President, over apparent finan-
cial impropriety; state dinner for foreign diplomats.]*

Friday, February 12

*[Signed proclamation for National Day of Prayer; spoke to appointees about cam-
paigning; dentist visit; left for Camp David.]*

Still snow up here & more promised tonite. Ran a movie called "Missing."
It is a pretty biased slam at Chile and our own government.

Saturday, February 13

A nice long walk in the fresh snow. They've put a couple of bird feeders (at
Nancy's request) outside the windows. It's a contest between the squirrels & the
birds but I guess they all do fine.

Ran "Inchon"—it is a brutal but gripping picture about the Korean War
and for once we're the good guys and the Communists are the villains. The producer was Japanese or Korean which probably explains the preceding sentence.

Sunday, February 14

We're back in the W.H. now. Nancy is off to Fla. tomorrow. I don't want
to think about that.

Monday, February 15—Washington's Birthday

Stayed home except for session in Situation Room. Briefing by C.I.A. on
El Salvador. The Guerillas really have a sophisticated set up. I'd never suspected
their organization & communications. They have divided the country into sec-
tions with a separate command group for each section and a network of permanent camps, well fortified. [. . .] Now we must find a way to counter it. In
the meantime, Ed Asner & some performers show up in Wash. with $25,000
they've raised for the Guerillas.

Met with Paul Volcker. I think we've broken ground for a new and better
relationship aimed at getting interest rates down. He thinks we can get short
term rates down 3 or 4 points by June. Long term will take longer.

I miss Mommie. She called from Fla.—will be home tomorrow nite.

*[**Tuesday, February 16:** received new ambassadors, including one from Israel upset
over supposed remark by Caspar Weinberger; Mrs. Reagan returned; eye exam showed need
for new contact lenses; noted George Bush appearance at meeting of antagonistic AFL-CIO
leaders, saying, "Funny—I'm the 1st Pres. never invited to the AFL-CIO conventions & I'm
the only Pres. who is a lifetime member of an AFL-CIO union." **Wednesday, February
17:** agreeable meeting with Wilfried Martens, prime minister of Belgium and chairman
of the European Council; prepared for press conference; cocktails with friends and Prime
Minister Martens.]*

Thursday, February 18

Press conference day. I think a good one but the "pack" was blood thirsty. Had a briefing on Soviet Arms. It was a sobering experience. There can be no argument against our re-arming when one sees the production complex they have established for the mfg. of every kind of weapon and war machine. Their sophistication is frightening.

Friday, February 19

Should have added yesterday our dinner for the Diplomatic Corps. This time Dobrynan was here as Dean of the Corps. Everything we've heard is true—they are a most likeable couple. In fact so much so you wonder how they can stick with the phoniness of the Soviet system.*

Now—Friday—a quiet day. Came home early. Mr. and Mrs. Jack Lemmon and a friend came by for cocktails and it was fun.

[Saturday, February 20: *briefing by Mike Deaver on upcoming European trip; screened film by Charlie Wick on Poland.*]

Sunday, February 21

To "Christs Church" in Alexandria—sat in Geo. Washington's pew. It was good to be back in a church. We haven't gone since the Libya threat. I didn't even mind the bullet proof vest.

Rest of the day and evening at home. Geo. Murphy came by. He's worried about the hard core conservatives rebelling. The "Conservative Digest" never has me captured by the moderates.

Monday, February 22

Helicoptered out to Mt. Vernon—250th B.D. of George Washington. A wreath on the tomb & then a brief speech which was live radio & carried in pub. schools across the country.

Lunch on issues. I'm convinced of the need to address the people on our budget & the economy. The press has done a job on us and the polls show its effect. The people are confused about ec. program. They've been told it has failed and it's just started.

[Spoke at two conventions of government officials regarding Federalism; met with Phil Habib.]

---

* Reagan reiterated his impression of the meeting in an insert for the diary written four days later: "Truth is he and his wife are most likeable and very much in love with each other after 40 years of marriage."

Tuesday, February 23

A short night. We were sound asleep last nite when it sounded like the War of 1812. It seems some group had scheduled a Wash. B.D. fireworks display down on the mall at midnight.

Met this A.M. with our Cong. leaders. They are really antsy about the deficit and seem determined that we must retreat on our program—taxes and defense spending. Yet they seem reluctant to go for the budget cutting we've asked for.

Capital Historical Society presented me with a silver medal—struck for Geo. W's B.D.

Amvets came by with Gold helmet award I was to have received last April before the interruption.

Lunched with George Will—hope I reassured him that I'm not a false front.

Met with special emissary of P.M. Suzuki—told him Japan would have to do better regarding trade restrictions.

Cab. Council on Food & Agri. Jack Block wants a tough anti-grain embargo statement to reassure our farmers. Can't do completely what he wants because it is a threat now putting pressure on Soviets.

[Met with Governor Bill Clements (R-TX) regarding the dissolution of the Energy Department; phoned presidents José López Portillo (Mexico) and Luis Herrera Campins (Venezuela) and Prime Minister Pierre Trudeau to brief them on the Caribbean plan to be introduced in speech before the Organization of American States (OAS); state dinner for governors.]

Wednesday, February 24

Day started early—briefing Cong. delegation—Dem. & Repub. on the Caribbean Initiative. I think we may have support on this—Jim Wright seemed darn right enthused.

Then the address to the O.A.S. It was extremely well received and remarks from Ambassador relayed to me afterward were to the effect it was most impressive presentation ever made to the O.A.S.

[Addressed employees of the Voice of America on its fortieth anniversary.]

Thursday, February 25

A jammed up day with meetings running one on top of the other. Everything from briefing Repub. Leadership on the O.A.S. speech at the same time I was trying to finish writing the speech which I deliver tomorrow noon; to meeting with representatives of various groups and causes.

Friday, February 26

Series of N.S.C. meetings re sanctions on Soviets, what to do with 3000 jailed Cubans. Castro infiltrated with the Mariel refugees. These have criminal records and history of mental problems. They are truly violent and were evidently released from prison and hospitals in Cuba just to be dumped on us.

A judge threatens to release them from our jails and turn them loose on society. The problem—as yet unsolved is how to return them.

Richard Viguery held press conference along with John Lofton and blasted me as not a true conservative—made me wonder what my reception would be at the Conservative Dinner. I needn't have worried—it was a love fest. Evidently R.V. & J.L. don't speak for the rank and file conservatives. Speech was well received.

[Saturday, February 27: worked at home, mainly on foreign relations. Sunday, February 28: noted morning snow storm. Monday, March 1: worked on speeches for upcoming trip; Mrs. Reagan left for L.A. Tuesday, March 2: morning in the office then flew west for rallies in Cheyenne and Albuquerque; day ended at Century Plaza Hotel in L.A. with Mrs. Reagan and Patti.]

Wednesday, March 3–Sunday, March 7

[Met with businessman Peter Grace on recent trip to Japan; spoke at rally as guest of the L.A. County Board of Supervisors.]

Back to hotel—Bill Wilson came by with an account of a secret meeting in Rome where Italian businessmen (2) brought a Libya official to see him who claims Quadaffi wants to restore good relations with us. Turned info over to Ed Hickey.

That nite an anniversary (our 30th) dinner at the Wilsons. All our old friends—a wonderful evening. Thurs. 4th thru to Monday March 8 at the ranch. Wonderful as always. Construction started on Lee's house. "Little Man" is out to pasture—end of an era. There was his mother 1st—"Tar Baby" then his sister, "Nancy D." Now end of line. Sunday a fantastic affair with Merle Haggard and his band at a neighbor's ranch—BBQ and all as part of series on presenting young future stars on P.B.S. Several hundred present who have helped us over the years.

Monday, March 8

As always—the most beautiful day of the trip. So we got on A.F.1 and returned to Wash. Doria and Ron are staying at the W.H. with us.

Tuesday, March 9

[Met with Republican senators to reestablish support for proposed budget, and later with Foreign Minister Hans-Dietrich Genscher of West Germany.]

Then heard Gen. Walters story of his meeting with Castro. Walters does one h--l of a job. He's going back again and maybe we'll be sending Castro back his jailbirds and maniacs he hid in the Marial boat lift. He says Castro really sounds like he'd like to make up. Walters let him know we have a price.

[Wednesday, March 10: answered mail; attended Joffrey Ballet.]

Thursday, March 11

[Meetings with NSC; the National Association of Black County Officials; local officials regarding Federalism plan.]

Situation room for a hush hush on Cuba. They are uptight thinking we may be planning an invasion. We aren't but we'll let them sweat. We're concerned that some place in the N.S.C.—State or C.I.A. we have a mole—there is too much information leaking.

Friday, March 12–Sunday, March 14

[Met with leaders of Business Round Table; generally supportive, but concerned about worsening inflation.]

Pres. Mitterrand of France flew in just for the day—on the Concorde. We had good discussion and I convinced him to suspend shipment of weapons to Nicaragua until his and our Ambassadors there can get together. Ours will brief him on our evidence of Soviet and Cuban involvement there.

We got along fine but the Press tried to make it sound as if we are at odds.

[Went to Camp David; finished speeches for upcoming trip; felt that it "was nice to sleep in and keep no schedule"; saw films Shoot the Moon *(1982) and* Evil Under the Sun *(1982); helicopter was diverted on the way back to Washington because of security concerns.* **Monday, March 15:** *flew to Montgomery, Alabama, to address legislature and Nashville, Tennessee, for fund-raisers.* **Tuesday, March 16:** *in Oklahoma City for speech before legislature; made last-minute detour to Fort Wayne, Indiana, site of flooding.]*

Wednesday, March 17

P.M. Haughey of Ireland and his wife arrived. We had a big luncheon in the East room. It was a good get-together. Frank Patterson, the great Irish Tenor and his harpist wife Eily entertained. He explained what "Danny Boy" was before he sang it and there wasn't a dry eye in the house. I never knew it was a mother saying goodbye to her son who was going off to war.

Thursday, March 18

Herr Straud, leader of the W. German Conservative Party came in. We had met with him in Munich in 1978. He's a solid citizen who believes Schmidt's party is falling apart—a leftist wing getting too strong.

[Spoke to National Association of Manufacturers meeting and then publishers of black periodicals; reception for supporters of a congressional committee.]

Friday, March 19

[Met with builders groups looking for government subsidies for home mortgages.]

Staff meeting on strategy to break log jam on budget with Dems. Part hinges on Cap W. finding some defense savings that he can say will not delay our defense rearming program. Also finding some unnamed sources that will not curtail our tax incentive plan. For all this we must get Dem. willingness to cut entitlements on a bi-partisan basis. Off to Camp David—rainy—saw "Das Boot" a German movie of a WWII submarine crew. Very good but strange to find yourself rooting for the enemy.

[Saturday, March 20: work on speeches and mail; watched Deathtrap (1982); Sunday, March 21: returned to W.H.; vetoed bill to grant president allocation of fuel controls and signed a disaster order for Fort Wayne, Indiana, hit by floods; dined with Burt Reynolds, noted, "Had not known him well in Hollywood so surprised at his seriousness and his sincere crusade spirit against drugs, etc."]

Monday, March 22

A busy Mon.—one of those pile-up days but finally cleared the debris away by late afternoon. Addressed farm paper editors. Met with Dem. & Repub. chairmen of committees on farm problems. I share their concern about the farmer's plight but was surprised at how many of them had no concept of where grain sales could affect the world situation with the Soviets. I suggested that yes we could negotiate a long term grain contract with the Soviets but shouldn't we get some concessions re Latin Am., Afghanistan, etc. The Soviets are, after all, in deep ec. trouble & need our help desperately.

Met with Paul Laxalt's Comm. on Federalism. Some want perfection before we start.

Norman Podhoretz sent me his book on "Why We Were In Vietnam" personally inscribed: He wrote "To Pres. RR—Who always knew and still knows why we were in Vietnam and why it was indeed 'a noble cause.'"

Tuesday, March 23

[Desk work in the morning, then flew to New York City for the National Conference of Christians and Jews dinner, received an award presented by Henry Kissinger; large demonstrations outside.]

Wednesday, March 24

Met with business leaders on private initiative program; cabinet council meeting on matters including drug problem.

Prosecutors of Hinckley called on me. They want me to testify in the trial. No President has ever done that before. Staff lean very much against—I'm not as sure as I was that it would be wrong.

[Noted arrival of Betsy Bloomingdale as a W.H. guest.]

Thursday, March 25

Pres. Pertini of Italy arrived—1st visit to U.S. He's 84 and a terrific old gentleman. We had the full ceremony under a warm sun. Good meeting. He loves America—very touching moment on the way into the W.H. He paused by the Marine holding our flag and kissed it.

N.S.C. meeting to hear Sen. Buckley's report on Europe trip trying to get them to join us in shutting off credit to Soviets. I proposed a plan for consideration—that we explore if time hadn't come to confront the Russians and tell them all the things we could do for them if they'd quit their bad acting and decide to join the civilized world.

State dinner a huge success—a really fun time and great entertainment by Perry Como and Frank Sinatra.

Friday, March 26

Briefing on Soviet Ec. They are in very bad shape and if we can cut off their credit they'll have to yell "Uncle" or starve. We had cabinet meetings on our own economy. It is imperative that we get further budget cuts. So far the Dems. aren't budging. Our strategy is to move them toward bipartisan agreement on the entitlements—the only real savings must be found there.

[Telephoned Ted Cummings, U.S. ambassador to Austria, regarding his health.]

Called Geo. Schultz in S.A. and asked him to do a mission for us in Europe and Japan. What a nice man—busy as he is he agreed.

Jim Brady came in—we had a small wine taste of some 19th century wines brought by a friend of his—a dealer.

Al and I are on opposite sides I'm afraid about China. He wants to make concessions which in my view betray our pledge to Taiwan. We'll have to work it out.

Had lunch with about 80 or 100 Black Clergy. I think they understand us better now.

Saturday, March 27

Sat. off—except that Bill Clark and others didn't take the day off. A steady stream of manila envelopes kept coming to my desk. Spent the day reading.

Come evening, White Tie and Tails—The Gridiron dinner. I think it was the best yet. The smash of the evening was Nancy. This had to be one of the best kept secrets. The Repub. skit was a choral rib of Nancy about clothes. She had left the head table supposedly for the powder room. Suddenly there she was on stage—even the cast wasn't in on the secret. She was dressed in a hokey costume and sang a parody of "Second Hand Rose"—Second hand clothes. She was carrying a plate which she smashed as a finale. It brought the house down. She got 2 standing ovations. Maybe this will end the sniping.

Sunday, March 28

Lunch at the Bushes with Billy Graham, the Sulsbergers, Thompsons (L.A. Times) and others. Spent most of my time with the Sulsberger children talking movies. Didn't get in on George and Nancy's tête-à-tête with Billy but they are going to fill me in.

Back to the W.H. and the last in the series of Young American Artists for P.B.S. Gene Kelly brought some wonderful young dancers from Martha Graham, and the S.F. Ballet.

Monday, March 29

[Addressed National Association of Realtors.]

Meetings with Al Haig about China and Taiwan. State wanted to send a paper to the P.R.C. and letters from me because we are about to send mil. equipment and spare parts to Taiwan. I objected to some of the terms they wanted in these papers—the note of almost apology to the P.R.C. I'm convinced the Chinese will respect us more if we politely tell them we have an obligation to the people and Taiwan and no one is going to keep us from meeting it. We didn't send the papers.

[Tuesday, March 30: noted the "anniversary of another March 30th"; Cabinet Council meeting on mandatory retirement age, coming out against it; visit from Chief Rabbi of Israel, who was against the establishment of a Palestinian state; photo session; meeting with Senators Bob Dole (R-KS) and Russell B. Long (D-LA) on deficit and taxes.]

Wednesday, March 31

[Visited by National Arthritis Foundation poster girls; spoke to winners of high school democracy contest.]

Nancy Kassebaum and the Cong. group who went to El Salvador to observe the election reported in. They told the most inspiring stories about the people standing in line 10 to 12 hours in order to vote. I liked it best when they said the people chanted at the press (our press) "Tell the Truth."

Spent 2 hours in a Q&A with my gang preparing for the press conf. It's over now—held it in East room, 8 P.M. live. I think I won. Dr. Burton Smith in town—dropped by with Dan R. He's checking my plumbing tomorrow.

Thursday, April 1

Met with Gordon Luce & reps. of "Savings & Loan" industry. They are really in trouble and again because of the high interest rates. We are trying to help in several ways.

[Medical tests; results normal; Mrs. Reagan in Atlanta.]

Friday, April 2

A couple of Rose Garden signing ceremonies on the 2nd nice day in a row. Met with U.S. Canada Border Commission. Had Cabinet Council meeting on Prayer in Schools and Busing. We'll try for a Const. Amendment on prayer.

Paul Laxalt and I lunched on the patio. A good talk on our budget problems. He's on the task force (bi-partisan) that's trying to work out a compromise. He seems pretty optimistic. Dick Wirthlin came in—more bad polling news. I'm slipping badly in the polls. I think it reflects the constant media drumbeat of biased reporting against what we're trying to do. After the press conference the calls were heavy in my favor and a goodly number jumped all over the press.

Worked on 1st 5 min. radio spot—do tomorrow. Nancy came home—now the place seems better. Last night I called the Pres. of Argentina—talked for 40 min. trying to persuade him not to invade the Falkland Islands (property of U.K. since 1540 or so). Argentina has been trying to claim them for 149 years. I got nowhere. This morning they landed some 1000 or so men. Population of islands is only 2000 almost all English. Now we learn there is a possibility of oil there.

Saturday, April 3

Oval office and did 1st of weekly 5 min. radio broadcast live—we're going to do. Talked about recovery from recession and why our ec. program is best answer.

The argument about a nuclear freeze is heating up. Big issue is freeze now with Russians out ahead, or negotiate a substantial verifiable reduction—then freeze. I support the latter. Some critics are upset because in the recent press conf. I flatly stated Russians are ahead of us.

[Went to Camp David despite poor weather; enjoyed the movie The Awful Truth *(1937).* **Sunday, April 4:** *alternating sun and snow; returned to W.H.]*

Monday, April 5

[Breakfast for Republican freshmen congressmen.]

Met with Nat. Security Planning group re a former Nicaraguan rebel leader who has left the govt. there and wants to head up a counter revolution.

Tried something new with press. Had them come into the Oval Office and gather round the desk. It worked darn well. I took advantage of the occasion to announce I'd be speaking to U.N. Arms Limitation Conf. in June after return from Europe. Also suggested Brezhnev should do the same. That was the evening news.

Went to Hilton Hotel—to speak to same group I addressed a yr. ago on

March 30. They didn't exactly fall all over me. Their business—bldg. trade is hurting and they'd really like some govt. handouts. Cap. W. home from Asia—now we'll try to get the MX back in the budget.

Met with Lyn N.—he drops in every once in awhile.

Tuesday, April 6

Fairly quiet day except in the So. Atlantic. The Royal Navy is sailing toward the Falkland Islands to oust Argentina. Both sides want our help. I'm leaving Al Haig home from Barbados to cope with it. We have to find some way to get them to back off.

Wednesday, April 7

8:10 A.M. meeting on the So. Atlantic problem then off to Barbados—well Jamaica 1st at 10 AM.

Sunday, April 11—Easter

6 P.M. arrived back at the W.H. Hard to sum up the days since Wed. without writing a book. Wed. we arrived at Jamaica and went right into meetings with P.M. Seaga. Amazed at the warmth of the people and their sincere affection for Americans. Met Gov. Gen. and his lady—a state dinner at his residence. Next morning on to Barbados—again a series of meetings this time with P.M.'s of 6 Island Nations. An evening reception and then to Claudette's (Claudette Colbert) for dinner. The Buckleys have joined us.

Friday our one really clear day off was great. We went down to Claudette's—all 4 of us—swam in the ocean, lunched, swam some more and then had dinner. She is a wonderfully warm and charming person. Sat. spent morning on homework. Talked to Al Haig in Buenos Aires on the Falkland Islands problem—did my radio broadcast and then to Claudette's. The tide was high and the ocean rough with waves—surf breaking right on the beach. I swam anyway. Coming in was a chore against the undertow.

Sunday—church at the oldest church in the Caribbean. Lunch again at Claudette's then back to Wash.

Monday, April 12th

Hard to get back in the swing after being a beach bum for a day and a half.

Al Haig is not too optimistic—called from London. Since he left Argentina things have deteriorated there. The Soviets have a number of ships in the vicinity of the U.K. fleet. [. . .]

[Met with Jewish leaders regarding the situation in the Middle East, stressed support of Israel; attended Easter Egg Roll with Mrs. Reagan; named Lionel Hampton Ambassador of Music for upcoming overseas tour.]

Tuesday, April 13th

Al Haig is on his way home. Chances for a peaceful settlement of the Falklands dispute seems very dim.

Cabinet Council meeting on dairy products. Because of the ridiculous subsidy of dairy products the govt. has enough butter, cheese and dried milk to feed every human in the U.S. for a year.

Lunch with religious leaders brought in for Bill Veritys Pvt. Initiative program.

Signed a bill to authorize charitable gifts be allowed to help with Jim Brady's medical bills. Heretofore, this was illegal for govt. employees.

[Visit by Henry Kissinger on utilizing businessmen to improve East-West relations; met with fund-raisers for the Reagan Scholarship at Eureka College.]

Wednesday, April 14

Al Haig is back. We really have a tough problem and it hasn't been made any easier by the press. In what I think is a most irresponsible act—engineered by Bernstein of the Post, they have charged that we are lending aid to Britain's Navy in the Falkland dispute. This of course has set the Argentinians on fire. The charge is false. We are providing Eng. with a communications channel via satellite but that is part of a regular routine that existed before the dispute. To have cancelled it would have been taken as supporting the Argentine.

We're still in the game as to trying for a peaceful settlement & Al will be going to Argentina.

[Reaches compromise position on military budget.]

Thursday, April 15

Income tax day. People actually demonstrating against paying their tax and Dem. Congressmen are demanding that we rescind our tax cuts or at least part of them.

[Visited Chicago and announced plan for tuition tax credit at convention of Catholic educators.]

Flew back to Wash. & spent half an hour on the phone with the Pres. Galtiere of Argentina. He sounded a little panicky and repeated several times they want a peaceful settlement of the Falklands problem.

Friday, April 16

A day with a schedule so full I couldn't read the memos on my desk. Spoke to and lunched with broadcasters from all over the country. Met with 6 Islamic country Ambassadors. They are all fired up about the tragedy in Jerusalem. A demented young man opened up on the Mosque full of worshippers killing a number of them. This started 3 days of bloody riots. I plead with them to recognize this as the act of a deranged individual not the act of a govt. I think I made some headway.

Meetings of every kind went on all day. Finally 2 back to back receptions for "Eagles." Sen. Packwood made a speech in Oregon picking on me as the enemy. And he's the chairman of our Sen. campaign committee.

Saturday, April 17

Departed for Camp David—Ron & Doria were here but he's playing in Maryland. Broadcast my radio spot from C.D. Afternoon interrupted by phone calls—Bill Clark—re the Falkland situation. Al Haig is there and as of noon things looked hopeless. I called Margaret Thatcher to tell her I'd cabled him to return home if there was no break in the Argentine position. He used that message however and there must be some movement—he's still there.

[Sunday, April 18: received message of support from Pope John Paul II regarding peace efforts in Falklands dispute; Haig feeling optimistic about negotiations; rode horseback in the afternoon; returned to W.H.]

Monday, April 19

Al H. is on his way home. Argentina made some concessions but not nearly enough. Apparently Pres. Galtiere is overruled by the junta. He would try to work things out with Al, then he'd meet with the junta and everything would be back to ground zero.

I don't think Margaret Thatcher should be asked to concede anymore.

Queen Beatrix of Holland and Prince Claus arrived on this beautiful spring day. She is a warm, charming lady. This time, no meetings to haggle over issues—theirs is a monarchy in which she is similar to the Queen of Eng. or the King of Spain.

The state dinner was very nice and entertainment was by George Shearing and Brian Torf.

Tuesday, April 20

Met with Repub. Cong. Leadership. The budget was the subject. I think they were relieved to learn that I'm willing to compromise some in return for a bi-partisan program. I called Tip O'Neill—I'm not sure he's ready to give. Tip is truly a New Deal liberal. He honestly believes that we're promoting welfare for the rich.

[Noted Haig's latest, pessimistic report on Falklands; was moved by observation of Holocaust Day; press availability in Rose Garden.]

Our new Repub. N.J. caretaker Sen. Nicholas Brady was sworn in today. He replaces Williams (Abscam scandal).

Wednesday, April 21

An N.S.C. meeting on "START" our idea for Strategic Arms Reduction Talks with the Soviets. We've had a team working on this. Some of the journalists who write so easily as to why we don't sit down and start talking with the Soviets should know just how complicated it is. Our team is doing a good job. Israel bombed a P.L.O. base in Lebanon. There have been some provocations and an unfounded report that an Israeli plane was shot down.

Took the afternoon off on a beautiful spring day and went down to Quantico for a horseback ride.

[Overnight guests, Marion and Earle Jorgenson.]

Thursday, April 22

[Report on Falklands outlook, diplomatic and military; violence in Israel and Lebanon, but truce is maintained.]

Last night C.B.S. did a special 1 hour documentary (Bill Moyers) on 4 cases of poverty and illness they laid to our ec. program. It was a thoroughly dishonest demagogic, cheap shot. They may have gone too far—some columns already have taken them on. We're going to reply—forcefully.

A reception for the Folger Library. I'm now an honorary member. This was its 50th anniversary.

Friday, April 23

Both the Falklands and the Budget battle seem to be coming to a moment of truth. Defense Minister Pym (U.K.) is on his way back to London with what could be the last proposal we can cook up. The shooting could start—it would be a war mainly because an Argentine General, President (result of a coup) needed to lift his sagging pol. fortunes.

The group debating the budget seems unable to arrive at any kind of consensus. If we can't get a Bi-partisan agreement to act together in the face of the projected deficit—then I take to the air (TV) and there will be blood on the floor.

[Visited mail office volunteers; signed executive order addressing problems of crime victims. **Saturday, April 24:** *radio broadcast; W.H. correspondents dinner.]*

Sunday, April 25

First news—British Helicopters damaged an Argentine Sub of New Georgia. Next—the Sinai has been returned to Egypt. Mubarak & Begin together said "Peace Forever." I called Mubarak and Begin. The call to Menachem was most important because this has been a traumatic experience. Many of his people resisted leaving the Sinai and his army had to physically eject them. The army went in unarmed and did a magnificent job. I pledged to both that we will continue to help in any way we can to further the peace process. Rest of the day tried to catch up with some of the upstairs desk work.

Monday, April 26

It was raining—Mommie left for Calif. (for 5 days) and it was Mon.

Then at 10:15 I went to Const. Hall to address 2000 delegates of the U.S.C. of C. convention. What a shot in the arm. They interrupted me a dozen times or more. I was talking about the budget and our tax program.

[British send fleet to Falklands; hoping Haig can return to Argentina for further negotiations.]

A Budget meeting—tomorrow the gang goes back at it with the Demos. tomorrow. I felt they were tiring a bit and the Demos are holding out for more concessions. The D's are playing games—they want me to rescind the 3rd yr. of the tax cut—Not in a million years!

[Tuesday, April 27: budget talks stalled, meeting scheduled with Tip O'Neill; no word from Argentina on renewing diplomatic efforts; NSC meeting on "a master strategic plan" submitted by Tom Reed (no description of it given); meeting with leaders of building trade.]

Wednesday, April 28

The big thing today was a meeting with Tip, Howard Bohling, Jim Wright, Jim B., Ed M. Don R. and Dave S. the result of my call to Tip. The "gang of 17" had come fairly close together on the budget and revenue package.

Their combination of cuts and revenue increases came to $60 bil. on the R. side to $35 bil. on the D. side. It should have been straight haggling especially since our original budget package called for $10 bil. I didn't try to start bargaining from that figure but started at the $60 bil. Three hours later we'd gotten nowhere. Finally I said I'll split the difference with you and they refused that. Meeting over.

[Thursday, April 29: lunch with Prime Minister Muhammad Mzali of Tunisia, who thought Yasser Arafat was ready to recognize Israel's right to exist; television address on budget.]

Friday, April 30

Al H. has announced the turn down of our plan by Argentina and that we now must come down on the side of Britain.

[Messages to W.H. showed support for speech; met with congressmen negotiating budget.]

Saturday, May 1

Off to Knoxville and the opening of the World Fair. Also to meet Nancy who has been gone since Mon. I've been lonesome.

[Spoke at opening ceremonies and at U.S. Pavilion; appeared at political fund-raiser.]

Helicoptered up into the foothills to Sen. Baker's place. Had a nice afternoon and very informal BBQ dinner with them and their friends and neighbors and his daughter Cissy who is running for Congress. I think she'll make it—I hope so. Joy and Howard are gracious and comfortable hosts. We overnighted there.

[Sunday, May 2: attended Presbyterian church; back to W.H.]

Monday, May 3

Met with Repub. Leadership. Also, Rep members of Sen. Budget Committee. While we have some disagreements as to the makeup of the budget I think we'll be together. There is a feeling of optimism and an esprit de corps.

Read this morning of a black family—husband and wife both work in govt. printing office. They live in a nice house near U. of Maryland. They have been harassed and even had a cross burned on their lawn. It was all on the front page of the "Post." I told Mike and Jim I'd like to call on them. We cleared the last part of the afternoon schedule and Nancy & I went calling. They were a very nice couple with a 4 year old daughter—grandma (a most gracious lady) lived with them. Their home was very nice and tastefully furnished. They were very nice about our coming and expressed their thanks. The whole neighborhood was lining the street—most of them cheering and applauding us. I hope we did some good. There is no place in this land for the hate-mongers and bigots.

[Bloodshed in Falklands War with loss of an Argentine cruiser; no progress in efforts to stop it.]

Tuesday, May 4

[British take losses in war.]

Turned State dept. down on a message they wanted me to send our Ambassador in China urging him to have informal talks with Chinese preceding George Bush's arrival. The talks were to soften the Taiwan issue some more. We can't do that—the Taiwanese have proven their friendship.

[Cabinet meeting regarding sugar imports; met with Republican members of the House Budget Committee.]

Dick Wirthlin reported in. My speech was to a smaller TV audience than usual. Probably due to its last min. nature and no promotion. But people are confused on the whole budget issue and my ratings are way down.

[Dinner for the Congressional Campaign Committee.]

Wednesday, May 5–Friday, May 7

Never got to this journal until today (Fri). Wed. things turned up. We got together with Sen. Pete Dominici's Budget committee & worked out a compromise budget pckg. It passed out of committee by a straight 11 to 9 party line vote. It's a good budget and will trim the projected deficits down to $106 bil. next year, $69 bil. in '84 and $39 bil. in '85. The Demos. are screaming and lying like bandits charging us with cutting Soc. Security—we aren't touching Soc. Security.

[Tentative cease-fire in the Falklands; meeting with congressmen on defense budget; went riding with Bill Clark and Paul Laxalt. **Saturday, May 8:** *deskwork; new diplomatic initiative in Argentina.* **Sunday, May 9:** *visited Eureka College for fiftieth reunion; in commencement address introduced START program for nuclear weapons reduction; left for Chicago.* **Monday, May 10:** *fund-raising and civic appearances in Chicago; returned to W.H.* **Tuesday, May 11:** *considered Democratic budget proposal; sought support on budget in meetings with business groups; noted that "Dem. Phil Gramm is coming up (as he did last year) with his own plan which may even be better than ours."]*

Wednesday, May 12

[Visit from President João Baptista Figueiredo of Brazil.]

Al H.- George Bush and I lunched on the patio and heard about George's trip to China. The P.R.C. is really obsessed with our continuing to be friendly to Taiwan. Well there isn't going to be any change on our part.

Francis Albert (Sinatra) came by with the Multiple Sclerosis Mother and Father of the year. He's heading up their fundraising drive—"F.S. for M.S." What a cruel disease—a young mother getting around in a walker—has a handsome little boy. Louis Unser—brother of the Indianapolis 500 champ—(Louis was a race driver too), he's in a wheelchair. He has a lovely little girl.

[State dinner; President João Baptista Figueiredo said that he'd heard British intended to attack Argentine mainland bases.]

Thursday, May 13

I called P.M Thatcher [. . .]. I talked to Margaret but don't think I persuaded her against further action.

[Press conference. **Friday, May 14:** *political fund-raisers in Pennsylvania; returned to W.H.; Mrs. Reagan in Chicago to accept honor for her father at Northwestern University.]*

Saturday, May 15

Held morning meetings with George Schultz and Phil Habib. George met with heads of state I'll be meeting with in Versailles. It's a little un-nerving to find they expect me to be the leader in solving our world wide ec. problems.

Phil thinks our Lebanon cease fire is on thinner ice than it has been in 9 months. A radical wing of the P.L.O. wanting to take over from Arafat is out to provoke Israel into action. This will rid them of Arafat who is more moderate & will rally Arab & Soviet support to the radicals. We're trying to make Israel see this.

Did my radio script on the 33rd observance of Armed Forces Day. We've really made the volunteer military work. Enlistments are up, so are re-enlistments but most important the intelligence and ed. level of the enlistees is higher than it's ever been.

Long call to Ron. He wants to Sign off Secret Svc. for a month. S.S. knows he's a real target—lives in a N.Y.C. area where the Puerto Rican terrorist group is active. In fact he's on a hit list. He thinks we're interfering with his privacy. I can't make him see that I can't be put in a position of one day facing a ransom demand. I'd have to refuse for reasons for the Nation's welfare.

[Sunday, May 16: *went swimming; false report that British would attack Argentina overnight.* Monday, May 17: *visit from Prime Minister Malcolm Fraser of Australia, both agreed Australia should be included in European summit at Versailles; calls to congressmen on budget; Nancy left for New York City for the opening of Broadway musical* Annie, *watched* Marco Polo *on television.*]

Tuesday, May 18

Lunched with Joint Chfs. of Staff and Cap W. It was a good meeting with a sound discussion of strategic problems—for example—the importance of the sea above Norway and what Iceland means in the Navy strategy should these be conflict on the NATO front.

[*Greeted Baseball Hall of Famers Walter Alston and Al Lopez, German publisher Axel Springer; cabinet meeting on proposal for a gas tax and highway-building program; noted progress on budget talks; possibly imminent bombing raid on Port Stanley by British bombers; Prime Minister Menachem Begin's discomfort with the PLO in southern Lebanon.*]

Wednesday, May 19

[*Official visit from King Hassan of Morocco; talks and then riding at the Bill Marriott estate in Shenandoah Valley.*]

Nancy is very depressed about her father's health and understandably so. I want so much to speak to him about faith. He's always been an Agnostic—now I think he knows fear for probably the 1st time in his life. I believe this is a moment when he should turn to God and I want so much to help him do that.

Thursday, May 20

Ron's birthday. But here a frustrating day, much of it spent on the phone with members of Cong. on the budget. Most are going to be with us but some I talked to are dedicated (they say) to what we're trying to accomplish but they won't vote with us because the budget compromise isn't exactly to their liking.

A compromise is never to anyone's liking—it's just the best you can get and contains enough of what you want to justify what you give up.

[Offer of help from the King of Spain on Falklands diplomacy; meeting with economic advisors—"as usual they are not in total agreement—economists never are"; attended fundraiser for Howard University; met with John Davis Lodge, interested in ambassadorial posting.]

Friday, May 21

The British have landed in the Falklands & the fighting has started.

My fight has started but it isn't bloody—just messy. I'm on the phone all day calling Congressmen. The Senate is supposed to pass a budget—not too different from what I wanted—tonight. The House has 5 budgets before it—one of which resembled the Sen. Bill. It's called the "Bipartisan Recovery Budget." They'll be voted on next Mon.

[Briefings on upcoming European trip; Don Regan called with inflation numbers.]

Saturday, May 22

Last nite the Senate Budget bill passed 49 to 43. Now it's up to the House where there are 7 Budgets only one of which is any good—our "Bipartisan Budget Proposal." They'll begin voting on all of them Mon. Ours first, but then according to the rules, any other proposal that passes after ours has been voted on even—if it passes—will be replaced by the last one to pass.

[Desk work.]

Sunday, May 23

At home all day. Ron came down from N.Y. He's a little rebellious and wanted us to sign off S.S. protection for a month. He's the only one of the kids who is on the hit list of groups like FALN etc. Ed Hickey came over and it's all straightened out.

I've been working on the speech for the British Parliament. Tomorrow Nancy leaves for Phoenix to see her father who apparently is desperately ill.

Monday, May 24

[Busy with meetings, briefings, and interviews.]

Nancy's on her way to Phoenix & L.A.

We had a session on Sanctions, limiting Soviet Credit & the Versailles meeting. There was a lot of talk about not having a set to with our allies. I finally said to h--l with it. It's time we tell them this is our chance to bring the Soviets into the real world and for them to take a stand with us—shut off credit, etc.

Day finally ended and tomorrow it's Calif. for me too.

[**Sunday, May 30:** *reported on week's activities, starting with Tuesday speech at California aerospace plant; Wednesday to Saturday at the ranch; budget stalled in Congress; returned to W.H.*]

Monday, May 31

Easy day—to Arlington for Decoration Day Ceremony. It was very moving and inspiring.

Back to office to meet with Jeanne [Jeane] Kirkpatrick. She and Al H. have been at each others throats. Later in day met with Al. Bill Clark at both meetings—thank Heaven. As it turns out there is some right and wrong on both sides. I think we can get a lid on it with no further damage.

Talked by phone to P.M. Thatcher. We had a new proposal to get or try to get a settlement without the U.K. scoring a total victory which could topple the govt. of Argentina. The P.M. is adamant (so far). She feels the loss of life so far can only be justified if they win. We'll see—she may be right.

[**Tuesday, June 1:** *briefings for summit; told Republican congressional leaders, "I d--n well want a budget."*]

Wednesday, June 2–Friday, June 11

We flew out Wed. morning—I was loaded with briefing books until my head was stuffed with more than I needed to know. Flying into night we arrived in Paris at near midnight their time but only 6 P.M. ours. Met by Van & Bootsie Galbraith our Ambassador plus a welcoming French group. Put up at the Ambassadors residence which is a magnificent palace built in 1840. Suddenly the W.H. looked like an ordinary residence. I had meetings the next day with P.M. Susuzki & Thatcher—plus briefings and that night—Thurs a black tie dinner for Pres. Mitterrand. Fri. visited Mayor Chirac. Every place we drove the streets were lined with friendly crowds.

Left Nancy in Paris and flew to Versailles for 1st dinner with the Summit group.

Sat. the plenary sessions began. There were differences over East-West trade, Global negotiations and credits to Soviet U. plus how to stabilize currency exchange rates. We more than held our own and I believe came out on top.

Sundays meetings were topped with a dinner in the Hall of Mirrors—the table had to be at least 100 feet long—an unbelievably beautiful sight. Nancy had joined me. Then we were entertained by the Paris Opera and later a recital in the Palace Chapel.

Up early Mon. morning and off to Rome for meeting with Pope. A group of young American priests broke into song—"America the Beautiful." Both of us were crying. Lunched with Pres. Pertini—then called on P.M. Spadolini. Met the young men (police) who had saved Gen. Dozier. Again, the streets were lined with cheering people. We were told that had never been done for any other head of state.

Flew out for London & helicoptered to Windsor Castle. This was a fairy tale experience. Black tie dinner with the Queen and Prince Phillip plus family—the Queen Mother et al. The next day a ride with the Queen. Nancy and the Prince did a carriage of 4 (horses.) We then left for London—I addressed Joint session of Parliament. First Pres. ever to do so. Lunch with P.M. Thatcher then back to Windsor for a White tie dinner—almost as spectacular as Versailles—1 long table for about 150 guests.

Bonn and Berlin were again a series of palaces. The NATO meeting was a success in every way. We were housed in an old castle complete with moat. Thurs. night a dinner hosted by Pres. and Mrs. Carstens of Germany—they are very nice. Fri. to Berlin—Checkpoint Charley and a heartwarming welcome by hundreds of our mil. personnel and their familys. Final event a speech in Charlottenburg Castle gardens to 25,000 Germans waving American flags.

Almost forgot—1st day in Bonn addressed the Bundestag (Parliament). It was hailed as a great success.

While in Bonn learned the House had passed a budget—we're on our way. Also learned though that Israel had invaded Lebanon. I'm afraid we are faced with a real crisis.

Saturday, June 12

Spent the entire day at Camp David working on a speech for Houston Tuesday. It was foggy and rainy—saw the world's worst movie called "Health."

[Sunday, June 13: desk work.]

Monday, June 14

Finally the sun and a horse back ride. Bill C. and Mike D. came up to Camp David. The Al H. situation is coming to a head. I have to put an end to the turf battles we're having and his almost paranoid attitude.

George B. came by at 5 P.M.—same subject. Then at 6 P.M. we had an N.S.C. meeting on the Lebanon situation. There is a possibility the separate Lebanese factions can unite—get Syrians and Israelis out of their country and disarm the P.L.O. Al H. made great good sense on this entire matter. It's amazing how sound he can be on complex international matters but how utterly paranoid with regard to the people he must work with.

[Tuesday, June 15: met with Capitol pages; cabinet meeting on summit; flew to Houston for major fund-raiser; wrote that "Going to Texas can be good for the morale"; report that task force on fraud saved $5.8 billion in six months.]

Wednesday, June 16

Back to Wash. and a hectic afternoon. Met with Deputy P.M. Ali of Egypt on the Lebanon mess. Spoke to Nat. Assn. of broadcasters. Met with Al H. and Bill Clark on Lebanon. We're walking on a tightrope. Some 6000 armed P.L.O. are holed up in Beirut. Pres. Sarkis of Lebanon can't say openly but he apparently wants Israel to stay near until the P.L.O. can be dis-armed. Then he wants to restore the Central govt. of Lebanon—allow Palestinians to become citizens and get all foreign forces to withdraw from Lebanon. The world is waiting for us to use our muscle and order Israel out—we can't do this if we want to help Sarkis but we can't explain the situation either. Some days are worse than others.

Thursday, June 17

This was a day in N.Y. This morning I addressed the U.N. General Assembly. Ambassador Gromyko did not applaud. I said some blunt things about the Soviet U. that needed saying. They were not well received by the large segment which usually votes against the U.S. and with the U.S.S.R. On the other hand, I think my talk added to the results of the trip to Europe & was a plus.

[Report on Argentine army, possible improvement in Lebanon.]

Friday, June 18

[Reason for optimism regarding Lebanon; report that Vatican working to reestablish relations with Libya.]

Met to finally decide whether to lift sanctions on pipe line material to Soviets. Cabinet very divided. I ruled we would not remove sanctions. There hadn't been the slightest move on the Soviets part to change their evil ways.

Barry Goldwater came to see me—he's upset by rumors that I'm going to dump Taiwan. I convinced him there is no way I'll ever do that.

[NRC reception in the East Room; left for Camp David; impressed by Clint Eastwood's work on film Firefox *(1982).* **Saturday, June 19–Sunday, June 20:** *riding and swimming; prepared for upcoming meeting with Prime Minister Menachem Begin; returned to W.H.; dinner with Maureen and Dennis.]*

Monday, June 21

Whatever else happened this was P.M. Begin day. He and I had almost an hour 1 on 1 with first our 2 Ambassadors present. I was pretty blunt regarding whether even a savage assassination attempt which will probably turn out to be successful (if not the Ambas. will be paralyzed) warranted the retaliation which has taken so many lives in Lebanon.

He came back with a defense based on the shelling of Israeli villages by P.L.O. elements in Lebanon.

It's a complex problem. While we think his action was overkill it still may turn out to be the best opportunity we've had to reconcile the warring factors in Lebanon and bring about peace after 7 years.

In the larger meetings with his people and ours we went at it again. He's adamant against our proposal to sell arms to Jordan. My argument is we're trying to create more "Egypts" who'll make peace with Israel. He refuses to believe another Arab state will do what Egypt did. Frustrating.

Tuesday, June 22

[House passed the conference budget bill; meeting with President Luis Alberto Monge of Costa Rica, agreed to cooperate on Caribbean Basin Initiative (CBI), a program of economic assistance to Caribbean and Latin American nations in an effort to draw the region together under free-market, democratic values; then met educators regarding tuition tax bill; Senators Richard Lugar (R-IN) and Jake Garn (R-UT) unsuccessfully sought support for bill to subsidize home mortgages.]

Met 125 candidates for Cong. and had photo with each. I hope they make it. I wonder what it would be like to have a Repub. House & Senate?

Yesterday Hinckley was found innocent by reason of insanity. Quite an uproar has been created.

Wednesday, June 23

Went over to C.I.A. and spoke to all the employees. It was wonderful to see their high morale. I signed the bill which now prevents people like Agee and others from exposing and identifying our agents. Some have been killed because of this.

Lunch with V.P. Bush. Then 2 receptions for the Ground Floor Committee with an hour in between in which I met with P.M. Thatcher.

Thursday, June 24

Met with Al H. this A.M. He and Bill C. had met for 2 hours last night. I was prepared for Al to resign. He was ready but then said it was over policy. Bill had the impression he really wanted to leave but he launched into an attack on staff at W.H. some Cabinet and said he couldn't function. He had a bill of particulars—so I have it to read and told him I'd talk to him later after I had.

Met with task force on drugs. I think we can get a handle on this problem with the group now working on it.

This evening the 4th reception in 3 days. I'm ready for Camp David.

Friday, June 25

Today was the day—I told Al H. I had decided to accept his resignation. He didn't seem surprised but he said his differences were on policy and then said we didn't agree on China or Russia, etc. I made a simple announcement to the press and said I was nominating George Shultz for the job. I'd called him and like the patriot he is he said "yes."

This has been a heavy load. Up to Camp David where we were in time to see Al read his letter of resignation on TV. I'm told it was his 4th re-write. Apparently his 1st was pretty strong—then he thought better of it. I must say it was O.K. He gave only one reason and did say there was a disagreement on foreign policy. Actually the only disagreement was over whether I made policy or the Sec. of State did.

Saturday, June 26–Sunday, June 27

Good weather—2 morning rides. Ed M., Jim B., & Bill C. helicoptered up with Mr. and Mrs. George Shultz who had just arrived by Concorde from London. We had a good working lunch. I think things are going to work out fine. Al will stay on for the transition.

Sun. before leaving C. D. Bill & Bud McFarland called. The Israeli cabinet has submitted a new idea. Calls for a Lebanese army to disarm the P.L.O. in Beirut. All of the armed P.L.O. will then be sent out of Lebanon to Syria. The odd note was the Israeli call then for the U.S. to join in the negotiations with the P.L.O. Heretofore, we have refused to do any talking to the P.L.O. until and unless they agreed to recognize Israel's right to exit, which they've never done. We're having about 30 people in for dinner at the W.H. and then running Spielberg's movie, "E.T."

Monday, June 28

N.S.C. briefing on space. No question that Soviets have moved to a military priority in space. We must not be left behind.

The Haig-Shultz situation is still a top press item. The time table for Al to leave is moved up—we'll have an interim with Stoussel [Stoessel] as Acting Sec. George is handling himself with great class and dignity.

We have a leak and someone in the W.H. is stirring the pot with an animus toward Al that has me d--n mad.

[Reception for donors to Reagan Scholarship Fund for Eureka College.]

Tuesday, June 29

Decided in NSC meeting—will not sign "Law of the Sea" treaty even without seabed mining provisions.

Possible agreement in Lebanon regarding removal of P.L.O. Even in Israel there is opposition to Sharon's attacks on Beirut. So far the ceasefire is holding.

[Signed extension of the Voting Rights Act; economic indicators showing gradual improvement. **Wednesday, June 30:** *meeting with President Sékou Touré of Guinea, commented, "He's come all the way from Karl Marx to free enterprise"; press conference.*

Thursday, July 1: *left for California; situation in Lebanon dimming.* **Thursday, July 1–Sunday, July 11:** *vacation interrupted only by two-day trip to L.A. for meeting with L.A.* Times *editors and appearances with civic groups; birthday party for Mrs. Reagan; trip to Edwards Air Force Base to watch landing of the shuttle; returned to W.H.]*

Monday, July 12

A typical Mon.—nothing but problems and the kind you can't solve without a lot more study. Farmers are in trouble. By Sept. we have to decide whether to negotiate a long term sales agreement with the Soviets. If we don't we bankrupt a lot of American farmers—if we do we look like we're caving in on our sanctions.

We announced our go ahead on getting an Amendment to the Const. requiring a balanced budget.

[Pessimistic report on Lebanon; lifted sanctions on Argentina; videotaped messages for conventions. **Tuesday, July 13:** *trip to Baltimore; addressed the National Association of Counties on the Federalism program; returned to W.H.; briefed congressional leadership on international issues.]*

Wednesday, July 14

Met with Repub. leaders on Lebanon, Iran-Iraq and C.B.I. I believe we have their support on C.B.I.

Pres. Suaza of Honduras, a Dr. and quietly religious came in for an hour's meeting. He is first democratically elected Pres. in 10 years. He needs help, his economy is in a shambles from years of mis-management and he's being harassed by Nicaragua. We got along fine.

Met with House Foreign Affairs Committee on C.B.I.—again it would seem we have bi-partisan support.

Upstairs for a desk full of mail.

Thursday, July 15

[Day crowded with meetings: task force on cost control, Senator Larry Pressler (R-SD) on farm problems, Amway business leaders, President-Elect Jorge Blanco of Dominican Republic, Vice President Bush, Republican congressmen on Soviet grain sales, cabinet, photo session.]

It made for a busy day—the push for extension of grain sales to Soviet U. is heavy. My own belief is we should do a 1 yr. extension with an increase in the minimum they have to buy and then hold out chance of a long term deal if they'll do something like relax emigration freeze on Jews and Christians who want out of Russia. Our farmers are hurting too much to order an embargo.

The Pres. Elect Blanco sounds alright—but he needs ec. help. Cabinet meeting also dealt with grain and the problem of Fed. product liability standard to replace 50 dif. standards in the states. We'll have a piece of legislation drafted and discuss that one again. Even though I'm for Federalism—I believe this is a call for a nat. regulation.

[Staff resignations for personal reasons; George Shultz unanimously confirmed as Secretary of State.]

Friday, July 16

[George Shultz was sworn in; NSC meeting.]

I must make a decision on grain sales to the Soviets. Our agreement expires in Sept. My inclination (very strong) is to extend for 1 year—raise the minimum they must purchase and get the word to them that we could talk a long term agreement if they'd perform a few good deeds like allowing emigration of people now being held because of their religion, etc.

*[Left for Camp David. **Saturday, July 17–Sunday, July 18:** swam and rode; returned to W.H.; dinner party.]*

Monday, July 19

[Announced "Captive Nations Week" in Rose Garden.]

N.S.C. meeting re nuclear testing—ruled we'd keep on doing what we were doing but no point in announcing that in the face of all the anti-nukes. We're doing what we've been doing since 1975.

Up to Capital Hill & addressed several thousand enthusiastic supporters of Const. Amendment requiring balanced budget. Spoke from W. steps of Capitol. Even with my iron T-shirt on—not as hot as in the Rose Garden—the sun went under a cloud.

[Meetings with congressmen and senators. Called Anne Higgins, director of the White House Office of Correspondence, in hospital with lung cancer; reception for champion skiers and tennis players.]

Tuesday, July 20

Today we met with Prince Saud (Saudi Arabia) and the Foreign Minister of Syria. They were an odd combination because the Saudis and Syrians are not exactly friends. The F.M. is abrasive and obviously a hater of Israel. Still I think we made some progress—we'll know in a few days. We submitted a proposition that Syria take the top rank 1000 or so of the P.L.O. in Beirut. The rest 5 or 6000 to be moved to N. Lebanon until they can be distributed to Arab States. Then at Lebanon's request the Syrians and Israelis go home. They are taking this back to the Arab League.

[Cabinet meeting on progress in drug problem.]

A reception this afternoon to launch the new James S. Brady Presidential Foundation. It is to raise funds for Jim & others who suffer as he has as a result of assassination attempts on government officials.

Wednesday, July 21

[Lunch with Prime Minister Lee Kuan Yew of Singapore; meeting with Caribbean leaders.]

A really tough problem not yet resolved has to do with defense budget and the projection Dave Stockman must give to Congress re deficits for next 5 yrs. Cutting defense sends a message I don't like to allies & enemies alike. But Dave's report, if deficits are too high, sends a shock wave to the world just when we seem to be gaining ground.

*[Reception at Organization of American States. **Thursday, July 22:** meeting with Weinberger on budget; flew to St. Louis; attended opening of the Mathews-Dickey Boys Club; fund-raiser for Eureka College; Olympic Committee dinner honoring August Busch III; Mike and Colleen present. **Friday, July 23:** returned to W.H.; heard of the death of Ed Meese's son in an auto accident; left for Camp David. **Saturday, July 24–Sunday, July 25:** rode horses; talked with journalist Larry Barrett of Time; returned to W.H.; dinner party.]*

Monday, July 26

A meeting to resolve the budget update due before Cong. O.M.B. et al were trying to get an additional cut in defense spending because they said new deficit figures for 5 yrs. would show an increase instead of the decline we had earlier projected as a result of our budget figures. Cap & Geo. Shultz & I must say we were opposed. The new budget forecasts would, it's true send a msg. to the money markets that would be devastating. On the other hand defense cuts would send an equally disastrous message to our allies & the Soviets. I said there had to be a 3rd choice that will foreclose sending either message—and we worked one out. This problem has been haunting me for a week.

[Met with minister of defense of Indonesia, then met with president of Cameroon; attended Meese boy's funeral; met with Private Sector Initiative task force.]

Tuesday, July 27

N.S.C. to O.K. some covert operations [. . .].

[Met Future Farmers of America.]

We're having to gig the St. Dept. to let us announce the extension of grain sales to Soviets in time for me to announce it at press conference tomorrow night.

[Prepared for press conference.]

Wednesday, July 28

Press conference day & Indira Gandhi tomorrow. The "Conservative Digest" came out—an entire issue devoted to cutting me up down and crosswise. John Lofton and his compatriots seem to be determined to paint me as a turncoat conservative. The tone is one of devoted but now disillusioned followers.

H--l, in 1980 they held a secret meeting trying to persuade Al Haig to run against me.

[Press conference seemed successful.]

Thursday, July 29

Indira Gandhi day. The weather turned out beautiful. Our meetings were most successful. I think we've established a rapport. She's said to be aloof to the point of arrogance. I think she's shy—once past that she's warm, generous and has a sense of humor.

The state dinner tonite was a real success with music outdoors by the N.Y. Philharmonic directed by Zuban Mehta.

[Ambassadorial formalities.]

The Arab League has told the P.L.O. to leave Beirut—this could be a break-through.

Friday, July 30

Met with group of Sen's. who wanted to be re-assured that I was not abandoning Taiwan. Kasten was the main spokesman. I think they are now satisfied.

Did a video tape for King Hussein—celebration of his 30th year as King.

Most important of all—a meeting with deputy F.M. of Egypt Ali. He and his associates wanted a commitment from us to recognize and negotiate with the P.L.O. We pointed out that while they asked that of us—so far Egypt has refused to take any of the P.L.O. fighters in Beirut. It is Egypt as well as some other Arab states who are destroying the P.L.O. by refusing to take them in even temporarily while we settle the Lebanon problem.

Rain & thunder but we made it to Camp David where it was also foggy and raining. Saw "The World According to Garp."—good show.

Saturday, July 31–Sunday, August 1

Cancelled riding but both days turned out fine. We were able to lunch outside and sat by the pool. Calls and cables back and forth with Lebanon. U.N. with us supporting voted 15 to 0 for a ceasefire and U.N. observers on the scene. Israel will scream about the latter but so be it. The slaughter must stop.

Back to Wash.

Monday, August 2

Met with F. Minister of Israel—Shamir. I was rather severe regarding Israels continued shelling and bombing of W. Beirut and in effect delivered an ulti-matum. Another cease fire is in effect and things look better. Then off to Des

Moines where it was 95 degrees. Addressed 12,000 of the Nat. Corn growers Assn. Very well received. Helicoptered out to a Hog farm—the Dee family. It was a great afternoon—homemade ice cream and lemonade. Saw their operation and then did Q&A with about 35 of their friends. Back to Des M. for a meeting with execs. of the Farm Bureau—friendly but are going to oppose the tax bill. Then met with Agri. Council of State Repub. party. All in all a good day.

[Tuesday, August 3: *flew to Hartford to address Knights of Columbus on their one hundredth anniversary; returned to W.H.*]

Wednesday, August 4

Awakened about 6:30 AM. by Bill Clark. The Israelis had moved to new positions within W. Beirut and were shelling the city. NSC meeting at 9:30—we sent reply to King Fahd re getting the P.L.O. to quit stalling & get out and a msg. to Begin from me that a ceasefire was necessary and continued offense could bring a drastic change in our future relationship.

Then I moved to a full day of meetings—1 Cabinet but the rest with various groups of Reps. and Sens. re the tax bill and 1 on the resolution regarding a nuclear freeze.

We won on the Const. amendment to balance the budget—69 to 31. Russell Long gave us the deciding vote.

Met with Jack Kemp (alone) & then in leadership meeting. He is adamant that we are wrong on the tax increase. He is in fact unreasonable. The tax increase is the price we have to pay to get the budget cuts.

Thursday, August 5

Nancy left for Des Moines Iowa and Dallas Texas—back tomorrow night. But this place seems lonely as h--l when I know she isn't here.

Long Cabinet meeting on policy planning. We really found out why we came here. We saw and heard the impossible management structure of government. It is by any standard a cumbersome, costly incompetent monster.

Rest of day met with various congressional groups on the tax bill and the nuclear freeze battle. Meantime the Sen. passed $12 bil. in spending cuts.

I had to be a little stern in some of the meetings but it paid off. So did the phone calls on the nuclear freeze. We've just won it 204 to 202.

[Friday, August 6: *vacation plans changed due to upcoming votes in Congress; meetings with congressmen on taxes and with cabinet, discussed steel import issue; dentist visit; intention to veto spending bill; Mrs. Reagan returned; telephoned local officials eliciting support for Federalism plan.* Saturday, August 7: *worked on speech explaining support of tax increase bill; dinner party and movie,* In Name Only *(1939).*]

Sunday, August 8

Again at the W.H. More of Saturdays work plus a long letter I feel I have to write to Loyal. I'm afraid for him. His health is failing badly.

Tonite the Laxalts and we're seeing "Stagecoach."

Monday, August 9

[Phoned movie producer Doug Morrow, facing heart surgery at the Cleveland Clinic.]

A lengthy cable from Habib outlining a complete plan for withdrawal of the P.L.O. from Beirut. I have my fingers crossed. Sharon cabled wants to come here Thurs. to see me. George S. cabled back that I would not undercut Habib by doing that & that he should talk to Habib, not me.

[Met with congressmen regarding the tax bill; with others on tuition tax credit; Senator John Warner (R-VA) on MX missile program; farewell party for staffer; NSC meeting on START negotiating strategy; report of terrorist attack against Jews in Paris.]

Nancy in tears when I came home—her father is back in the hospital. She's going out there Wed. I wish I could bear her pain myself.

Tuesday, August 10

Things continue to look better in the Middle East.

Met with Israeli opposition leader Shimon Peres of the Labor party. He's quite a contrast to Begin and believes once P.L.O. leaves Beirut Israel should leave Lebanon. Believes also we must resolve the Palestinian problem. Surprisingly he wants us to continue befriending the Arabs and wants Jordan brought into the peace process.

Most of day spent in meetings with Congressmen (& ladies) re the tax bill. We're up against a strange mix—it isn't going to be easy.

Sen. Bill Roth came in—unlike Jack K. he's all for us and wants to help.

Met with Chryslers Lee Iacocca. He's all for our program and wants to help. He has reduced Chryslers cost of production 50%. No wonder they show a profit.

Rcv'd. letter from Richard Viguerie with copy of Conservative Digest. He tried to write in sorrow, not anger about my betrayal of the conservative cause. He used crocodile tears for ink.

[Wednesday, August 11: flew to Billings, Montana, for civic celebration and fundraiser; returned to Washington for rally; Mrs. Reagan traveling to Phoenix to see her ailing father.]

Thursday, August 12

Met with the news the Israelis delivered the most devastating bomb & artillery attack on W. Beirut lasting 14 hours. Habib cabled—desperate—has basic agreement from all parties but can't arrange details of P.L.O. withdrawal because of the barrage.

King Fahd called begging me to do something. I told him I was calling P.M. Begin immediately. And I did—I was angry—I told him it had to stop or our entire future relationship was endangered. I used the word holocaust deliberately & said the symbol of his war was becoming a picture of a 7 month old baby with its arms blown off. He told me he had ordered the bombing stopped—I asked about the artillery fire. He claimed the P.L.O. had started that & Israeli forces had taken casualties. End of call. Twenty mins. later he called to tell me he'd ordered an end to the barrage and pled for our continued friendship.

Spent rest of day meeting with Congressmen on Tax bill.

Friday, August 13

George Shultz came for a 1 hr. meeting on entire Middle East situation. Then Congressional meetings and a press availability in press briefing room. Finally off to Camp David. Mike, Lyn & Bonnie came along. We had a movie, "Officer and a Gentleman." Good story spoiled by nudity, language & sex.

[Saturday, August 14: horseback riding; meeting on Middle East with members of the NSC. Sunday, August 15: worked on speech; lunch with group of congressmen on tax bill; spoke with Mrs. Reagan, upset over the failing condition of her father; returned to W.H. Monday, August 16: met with Shultz regarding Taiwan; telephoned congressmen on tax bill; nationally televised speech on tax bill.]

Tuesday, August 17

P.L.O. out of Beirut looks much closer. Press and TV with a leak from State Dept. has gone crazy declaring our joint communiqué with P.R.C. of China is a betrayal of Taiwan. Truth is we are standing with Taiwan and the P.R.C. made all the concessions.

Met and lunched with Chairman Moe [Samuel K. Doe]—head of Liberia. Spent rest of day calling & meeting with people (Congress) on tax package.

[Stock market responded positively to speech; meeting in order to calm concerns of congressmen and senators anxious over relations with Taiwan.]

Wednesday, August 18

Tomorrow is D-Day in the House. Most of the day with Congressmen—in groups & singly. Met with the hard nose conservatives in the State Dining Room. Met with about 20 "Boll Weevils." They are pretty much with us. Then a series of one on one's. Interesting photo opportunity in Rose Garden with Tip O'Neil, Bolling and our leaders as a bipartisan for the tax bill.

Tonite a dinner in the W.H. for a group of undecideds.

Last night or the night before Nancy says Loyal asked for the chaplain at the hospital in the middle of the night.

Thursday, August 19

Dr. Loyal died this morning. Nancy wasn't alone—thank Heaven Ron & Doria were there. But it seemed awful to be here and not be with her. All day I sat at my desk phoning Congressmen on the tax bill and tonite it passed with 103 Repub. votes and more than half the Dems. 226 to 207. Tip O'Neill made a speech to Repubs. telling them why they should support me. It seemed strange—both of us on the same side. The Sen. took it up tonite and it won 53 to 47. Again some of our ultra pure conservatives deserted.

Now I'm packing to leave for Phoenix and my Sweetheart.

Friday, August 20–Tuesday, September 7

Phoenix for the sad occasion—a short but moving service then back to Deedee at the house. Many friends were there. I overnighted and then left for Calif. & straight to the ranch. Nancy followed & arrived about 6 P.M.

[Fund-raisers in L.A.; televised speech on the Middle East; two weeks of riding and ranch work; awarded Medal of Freedom to Phil Habib.]

Wednesday, September 8

What a 1st day back! Correspondence, bills, memos, reports, background-ers—and all of it stayed on my desk all day because meetings were scheduled all day. One meeting and a state luncheon was with a group of charming people—Princes and Princesses from 3 Royal Scandanavian families and the Pres. (woman) of Iceland. All were here for a cultural tour called "Scandinavia Today."

Had ec. meetings re the bank situation and possible disaster if a half dozen or more countries including Mexico should default on the bils. of $ they've borrowed from Am. banks.

[Met with congressmen regarding the balanced budget amendment, and with busi-nessmen financing the campaign to pass it.]

Thursday, September 9–Friday, September 10

Off to Kansas and Utah. Spoke at Alf Landon lecture series K.S.U. About 16,000 present—students and towns people. Alf is 95—We all sang "Happy Birthday." Sens. Nancy Kassebaum & Bob Dole were there. Speech well received. Then went to a fund raiser for Repubs. I think maybe I rambled on too long. Boarded A.F.1 for Utah—Orrin Hatch aboard. Had a round table meeting with western St. Repub. Chmn.

Spent the evening on the phone to Senators re the override of my veto of an appropriations bill. Then until midnight did some cards for speech on Fri.

to Repub. fundraiser. We put Orrin on a plane back to Wash. that evening so he could vote.

[Visited Mormon cannery; spoke at rally; returned to Camp David. **Saturday, September 11–Sunday, September 12:** *relaxed by pool at Camp David; returned to W.H.]*

Monday, September 13

Another typical Mon. Only 1 hr. in the whole day scheduled for time to read, write, etc. & that got eliminated by phone calls, a stack of things to sign and an extra meeting.

There was a meeting with Joint Chiefs of Staff. It was inspiring. We've really turned the mil. around morale wise & every other way. Use of Pot is down from 50% to 16%.

[Cabinet meeting on gloomy situation in international finance; National Security planning group meeting on INF; met Prince Bandar of Saudi Arabia; appeared on show hosted by Jack Anderson; taped greetings for conventions; worked out in gym.]

Tuesday, September 14

Met briefly with a group of anti-abortion leaders. They recognize the effort I've made to end the Weicher-Packwood filibuster of the Helms amendment.

Word came that a bldg. had been bombed in Beirut where Pres. elect Bashir Gemayel was speaking. Later word came he was alright. Much later word came he'd been killed. We're not sure at this point what we should believe, but fear the worst.

[Met with senators on tuition tax credit bill.]

Dr. Teller came in. He's pushing an exciting idea that nuclear weapons can be used in connection with Lasers to be non destructive except as used to intercept and destroy enemy missiles far above the earth.

[Met with businessmen to encourage interest in the private initiative program; launched the Combined Federal Campaign charity; met with legislators and businessmen on textile imports from Chinese; charity dinner with Pat Boone, Efrem Zimbalist Jr., "and several other old friends"; report of the death of Princess Grace of Monaco.]

Wednesday, September 15

[Met with Republican congressional leadership, then with seven mayors.]

Bill Verity gave a sum up of achievements & future plans for the Pvt. Initiatives program. He's performed miracles. Lunch with George S. and Bill C. Think we clarified things for George on what his turf was and what I wanted.

[Reception for Hispanic Heritage Week and one for launch of USA Today; *dinner for a conference of Black Republicans; anticipated visit by Ferdinand and Imelda Marcos of the Philippines; trip by Mrs. Reagan to funeral for Princess Grace.]*

Thursday, September 16

The Israelis moved into W. Beirut following assassination of Bashir and fight between leftist Muslims & Lebanese Arm.

Pres. Marcos & Imelda arrived—the usual mil. ceremony which he said was more impressive than anything he'd seen here before. We had a good meeting & then a beautiful state dinner in the Rose Garden & entertainment "the 5th Dinner on the So. Lawn." I think they will go home reassured and confident of our friendship.

[Friday, September 17: Mrs. Reagan headed for funeral in Monaco; flew to New Jersey for two fund-raisers for Millicent Fenwick; met with Eastern Republican State Chairman; returned to W.H.]

Saturday, September 18

With Nancy gone the only event of the day should have been my noon radio broadcast. Unfortunately, things changed. In Beirut, Haddads Christian Phalangist Militia entered a Palestine refugee camp and massacred men, women and children. The Israelis did nothing to prevent or halt it. George S. and I met and agreed upon a blunt statement which he delivered to the Israeli Ambassador. It is a sad day and one which may very well set our peace efforts back.

Nancy came home about 6 P.M.—thank Heaven. We watched tapes of Princess Grace's funeral.

Sunday, September 19

A busy day for N.S.C.—State and Defense & Staff. I attended meeting in the morning re the Beirut massacre. The Israelis did finally attempt to oust the killers. They have proclaimed their outrage. I finally told our group we should go for broke. Let's tell the people we are in at the request of the Lebanese—sending the multi-national force back in. Italy has agreed and we believe the French will, too. We are asking the Israelis to leave Beirut. We are asking Arabs to intervene and persuade Syrians to leave Lebanon at which time we'll ask Israelis to do likewise. In the meantime, Lebanon will establish a govt. & the capability of defending itself. No more half way gestures, clear the whole situation while the M.N.F. is on hand to assure order.

George S. and Jeanne K. were enthusiastic about the idea and apparently there was no disagreement. The wheels are now in motion.

[Monday, September 20: met with NSC planning group to frame speech on Lebanon; cabinet meeting on Native American issues, railroad strike, and need for emergency powers to force workers to return; met Sickle Cell Anemia poster girl. Tuesday, September 21: meetings with beauty pageant winners, aerospace business leaders, Jaycees leaders, Senator Barry Goldwater (R-AZ) regarding pro-China influence of State Department; sent disarmament negotiators on trip to Geneva; Elizabeth Dole reported on her role in the

administration fighting discriminatory laws and regulations against women; fund-raiser for Senate candidate from Maine. **Wednesday, September 22:** *addressed black college presidents in East Room and pledged support; NSC meeting on continuing sanctions on Russian gas pipeline; visit from Boys Club members; tapings.]*

Thursday, September 23

Another crowded day. Met with Sec. Shultz re his upcoming meeting with Gromyko. Decided he should low key and with regard to a summit agree in principle but say we'd have to see some action 1st—permission for Jews to emigrate, let the Pentecostals out of our embassy in Moscow. Seven of them have been trapped there for 4 years. Then there is always Afghanistan & Poland.

Had an ec. rundown in Cabinet—not bad, not good. One thing sure the recession has bottomed. Price index for Aug. $\frac{3}{10}$ of 1% that would be an annual rate of less than 3.3.

Lunch for Jim Brady Foundation. A success—people from all over the country—including Lew Wasserman.

[Met with business leaders on summer jobs for teenagers, and with an African American minister on proposal to grant tax incentives to businesses for hiring disadvantaged youths; interview with Trudy Feldman; reception for Hispanic young people; attended farewell for Walter Stoessel Jr., Undersecretary of State for Political Affairs; report of torpedo incident involving Soviet and U.S. subs.]

Friday, September 24

Finally an easy day. An N.S.C. Planning Group meeting. Subject—Latin America. It's possible the Soviets will ship "Mig" fighter planes into Nicaragua. George S. Monday will let Gromyko know we'll take very seriously any overt moves by them or their stooge Cuba toward any part of Latin Am. We have contingency plans leading all the way up to troop involvement if Cuba should send troops to stir the pot in Central Am.

[Lunch with trade magazine journalists; ambassadorial formalities.]

Then photos with some of our Sen. candidates & John Palmer and his wife, Nancy (she's with child). He's leaving the W.H. to anchor the Today Show. Chris Wallace comes here.

*[***Saturday, September 25:*** *radio broadcast; attended rally for prayer in school; reception for Ford Theatre Foundation, with performances by Liza Minnelli, Ben Vereen, and others.]*

Sunday, September 26

A day at home. Camp David is better on a weekend. Our Marines were to have landed in Lebanon but now it will be Tues. It is Yom Kippur so the Israelis won't be out til Tues. but they have agreed to withdraw.

Yesterday we lost 2 officers in Lebanon. They were assigned to the U.N. observers. They and an Irish and Finnish officer were in a vehicle on the road to Damascus—hit a land mine—all dead.

Monday, September 27

Nancy up early and off to Miss. Ala. and Ark.—part of drug program.

Most of my day spent on homework for press conference tomorrow night. It should be free wheeling what with Lebanon and all. Yom Kippur will be over tomorrow. Our troops will probably go ashore Wed.

It's lonesome here.

Tuesday, September 28

Cap signed order for Marines to go ashore Wed. A.M. Israelis will withdraw to So. of the airport—Marines will be stationed at the airport.

[Spoke to business leaders.]

Spent afternoon getting ready for press conf. Have just had it and made it through O.K. Everyone says "best yet." My favorite ans. was to Sam D. who asked if I didn't think I had something to do with our ec. problems which I'd been laying on the Dems. over the past years. I said: "Oh my yes. I share the responsibility—I was a Dem. for years."

Wednesday, September 29

Left 11:40 A.M. for Richmond Va. with Cong. Paul Trible, cand. for US. Sen. Before leaving met Miss America, a Calif. girl by way of "Cut and Shoot" Texas. She flattered me by presenting a card from "Man Watchers, Inc.," making me an awardee of "Well Worth Watching."

Rally at Richmond was heartwarming—over 5000 many of them young people—(not the most traditional thing at Repub. Rallys).

[Fund-raiser and grant presentation to Hampton Institution; returned to W.H.; photo session; dinner with friends.]

Thursday, September 30

Cabinet council meeting on crime. We're really launching an attack on organized crime and the drug epidemic. It's astounding to learn that 243 drug addicts over an 11 yr. period committed 500,000 crimes.

[Congress brought the balanced budget to the floor; spoke to members; barbecue dinner at W.H. with entertainment by Roger Williams, Tammy Wynette, and the Army chorus; report of four Marines killed in Beirut.]

Friday, October 1

Lunch in the St. Dining room for the Supreme Ct.—very enjoyable. They were gratified. Tradition has it that they report their return to Wash. to the Pres. as court convenes for the Oct. session. I guess few, if any Presidents in recent years have formalized it with a luncheon.

Supposed to leave early for Camp David but House was playing games with balanced bud. amendment. First they introduced a straight statute which was overwhelmingly defeated as it should have been. Then came the Const. amendment which required a ⅔ vote. It got a sizeable majority but failed to make the ⅔. I appeared in the press room and laid it on the House Dem. leadership where the blame properly lies. Prior to that met with Amb. (to Russia) Hartman. It was a fruitful meeting. He confirmed much of what I feel about Soviets—their economy is staggering, there is corruption & rampant cynicism re communism & the govt. Finally at 5:30 off to Camp David.

[Saturday, October 2–Sunday, October 3: *rode horseback and relaxed.* Monday, October 4: *flew to Ohio for appearances; accepted resignation of director of Veterans Administration; returned to W.H.; Mrs. Reagan on trip to Phoenix.*]

Tuesday, October 5

[*Listed day's meetings with Al Spiegel, Bill Clements, and Patriarch of Greek Orthodox Church in Jerusalem; taped message for Marines' anniversary.*]

Al is concerned about a growing anti-semitism and I concede anger at Begin and Sharon has made that possible. We'll do all we can to stop it. The patriarch awarded me the "great cross" as the foremost peace maker in the world. It contains a sliver from the cross upon which our Lord was crucified.

The news—C.B.S. & Wash. Post described my trip to Ohio as a chilly reception. That is blatant falsehood. I've never been greeted with greater warmth and enthusiasm. I think their bias toward the Dems. is showing.

Wednesday, October 6

I've had a top secret briefing on our ability at covert operations abroad. They made it plain we had lost this capacity under the previous admin. If our people only knew the heroism of unsung Americans risking their life every min. of every hour around the clock, they'd be as proud as I am.

Had an unusual experience in the East Room. I was speaking to our Congressional candidates plus quite a group of Congressmen and women. Suddenly a young bearded man from Calif. stood up and started a diatribe against me. It was as if he were reciting that terrible edition of the "Conservative Digest." I got a little angry a la N.H. and the debate. The whole press corps was covering the event. I teed off and answered his charges with facts and told him to shut up. I got the most heartwarming ovation from all present. Everyone seems to think I won a great victory.

[*Cabinet meeting on anti-drug campaign.*]

Thursday, October 7–Monday, October 11

[Flew west for appearances in Nevada, California, Mexico, and Texas.]

Nevada was great—the big rally was at the U. of Nev. Reno. I'm not used (after the 60's) to having college students whooping it up for me. Thurs. nite the Joffrey Ballet performed in open air at night—the Murdock's home in Bel Air. It was beautiful and they raised $1,700,000.

Fri. to Long Beach for a public signing of the "Export" bill and a speech to a large crowd on the pier—many of them unemployed dock workers. Jesse Calhoun—Pres. of the Union is truly on our side.

Then we flew to TiJuana—met with Pres. elect de la Madrid at Del Coronado. We got along fine. I think he'll be good for Mexico.

[Arrived at the ranch Friday night; saw Mrs. Reagan; she left Sunday for Washington; "It's the 1st time I've ever been at the ranch without her. I don't like it"; noted fires in L.A. County; left Monday for fund-raiser in Dallas; returned to W.H.]

Tuesday, October 12

1st day back—seemed like a Monday. Pres. & Mrs. Soharto of Indonesia arrived—usual ceremony—good meetings and magnificent State dinner. Entertainment a very charming, opera singer who captivated everyone—Fredericka Von Stabe.

[Lunch with Vice President Bush; signed the Omnibus Crime Bill and Missing Children's Act; met with cabinet on agricultural problems; Secretary of the Interior James Watt present for signing of repeal of 1902 water bill.]

Wednesday, October 13

Met in W.H. Treaty room with Alan Greenspan & several Cabinet members and staff re the Soc. Security Commission. We have put the figures together and Soc. Security is a disaster beyond our worst imaginings. We meet here in W.H. to avoid press ever knowing we are meeting.

Met with Don Rumsfeld who is going abroad to meet with our allies and try to persuade them to join us in rejecting the deep sea mining provisions of the Law of the Sea treaty.

An N.S.C. meeting—then preparation for broadcast speech on ec. Demos furious—ABC refuses to carry it. NBC & CBS are at 7:30. Back home heard the speech went well. Now to wait for the returns. Market up again today—Dow Jones above 1000. Prime rate at 12.

[Thursday, October 14: addressed group at the Justice Department on anti-drug program; viewed CIA movie on Soviet space program, regarded as "much further ahead than most people realize and their main effort has been military"; visits from soccer stars Pelé and Steve Moyers, from California Republican politician Elsa Sandstrom, and then Justin Dart (disabilities superactivist and presidential appointee to the Board of Directors

of the Communications Satellite Corporation); made appearance at conference of business leaders; addressed Republican rallies nationwide via closed-circuit television.]

Friday, October 15

Did a radio address to farm network telling them on grain sales we guaranteed sanctity of contract—meaning once they made the deal to sell we would not put an embargo on.

Signed bill in Rose Garden—Sen. Jake Garn & several Congressmen and Cabinet members present as well as bank & savings & loan leaders from all over the country. It had to do with untangling and removing some of the restrictive regulations on such institutions.

An N.S.C. meeting. George S. has made some progress on negotiations with our European cousins. If we can get enough agreement on credit restrictions, etc. to the Soviet U. we could lift the sanctions on the pipeline.

[Visit from Florida Republican fund-raisers; went to Camp David.]

Saturday, October 16–Sunday, October 17

All set for a ride on Sat. then it started to rain—even a snow flurry so we sent the horses home. Then the sun came out and Sunday was also beautiful although crisp and cool.

Saturday's radio broadcast was on the economy. I cited F.D.R. and said our greatest problem now was fear.

*[**Monday, October 18**: met with Senator Nicholas Brady (R-NJ), who predicted prime interest rate would drop by year's end; received citation from National Venture Capitalists Association; met with commission on military recruiting; tea with King of Norway; teleconference with Republicans in fifteen cities.]*

Tuesday, October 19

Brkfst. with Pres. of Lebanon—Amin Gemayel & then a meeting with him and his ministers and our NSC and state people as well as some Cabinet members in the Cabinet Room. I'm impressed, he is modest and apparently totally dedicated to doing his job for the people of Lebanon. He wants us to help him set up his admin.

Lunch with George Bush—he's really been out on the trail campaigning for our cand's. (about 5 days a week).

Jimmy Lyons and Ted Stivers came in with an ambitious plan for holding our team together. They've talked to Bill Wilson about taking it on & he's willing apparently to give up the Vatican role to do it. I'll have to talk to him. First, he's very good at that—2nd, they would have him a deputy to me complete with staff, at Cabinet level etc. I think it would cause chaos in a well organized set up.

Taped commercials for cand's.—then a reception in tent on S. Lawn for N.R.C. staff. A pleasant affair. They've already got bumper stickers reading "shut-up."

[**Wednesday, October 20:** *flew to Illinois for campaign rally on farm; flew to Peoria for rally with Charlton Heston and Pat Boone; met with grateful Lebanese Americans; flew to Omaha.* **Thursday, October 21:** *signed grain alcohol bill; receptions for local candidates; returned to Washington; met Mrs. Reagan, who had been at funeral for Bess Truman.*]

Friday, October 22

The big day! King Hassan (Morocco) and representatives of the Arab League including one from Syria & one representing the Palestinians. This was no negotiating session. They were here to exchange views with the express idea there would be nothing calling for a yes or no answer. We met in the Cabinet room. It was apparent the King in his presentation as spokesman had to be careful to represent a consensus view. Still it gave room for optimism. There seems to be a general feeling that King Hussein (Jordan) should be the negotiator with Israel.

Then the King & I went in for a one on one in the oval office. There we got down to the real meat—the P.L.O. He had met with Arafat. He accepted my view that Arafat would have to recognize Israel's right to exist as a nation before we could have any contact with him. He offered a sample of what Arafat should say and I agreed it was good enough. Then he indicated he thought he could deliver that in 3 weeks or a month.

We broke for lunch—a very pleasant one in the State Dining room. The King introduced us to tea with Am. mint. It was very tasty.

A few office chores then off to Camp David.

Saturday, October 23

Cold & bright—took a short horseback ride. Late afternoon Doria & Ron arrived for a family pow-wow. He'd been rude to Nancy on a phone call and when I phoned him about it he said he thought we needed to clear the air.

It wasn't the greatest meeting but still I think it opened the door to a closer relationship. He seemed to be carrying water for Patti who has a kind of yo-yo family relationship. She's either warmly attentive or very distant & Nancy seems to bear the brunt of it.

Sunday, October 24

Saw them off and back to the W.H. Having a dinner with sports flavor. Father Hesburgh of N.D. is presenting us with a print of "Knute Rockne, All American." I've never owned one. So dinner with a lot of sports figures & a running of the film.

It was quite an experience to see it again after all these years and to see it in such company—Sonny Jorgenson, Gale Sayers, Merlin Olson, Joe Theisman—the Notre Dame Coach and many, many more. For me it was a truly nostalgic evening.

Monday, October 25

Brkfst. with Jerry Ford. He's been campaigning all over the country. He attended our N.S.C. briefing, also.

Dick Wirthlin came in with poll results—some good, some not so good. He gives us 8 chances out of 10 of holding the Sen. It will be disastrous if we don't.

Had a Cabinet meeting about high technology & Japans move to grab that off as they did the auto market. I've OK'd a study as to how we resist.

Some bill signings, tapings and a meeting with our commission on women and how we can really help achieve equality.

[Tuesday, October 26: flew to North Carolina for campaign appearances; presented singer Kate Smith with Medal of Freedom; fund-raiser; returned to W.H.; Mrs. Reagan in New York in connection with her book; report on economic indicators.]

Wednesday, October 27

Winners of N.Y. Marathon male and female came by with lovely young lady who finished the 26 miles in 11 hours. She has Cerebral Palsy and did the race on crutches. ABC covered her entire 11 hours. I could hardly keep from crying at the bravery of this girl. She has a twin sister also suffering from C.P. also on crutches.

Then lunch with all the relatives alive of T. Roosevelt. It's his 125th birthday. He won 1st Nobel Peace Prize to an Am. They've presented it to the W.H.—1st time it's ever been on public display. It was a wonderfully warm and enjoyable occasion.

[Ambassadorial formalities; opened savings bond drive.]

Rcv'd. a letter from Mrs. Sharon Gardner out in Hammond, Ind. She had written before. She is 42, divorced, a victim of Spinal Bifida—no legs left—in a wheel chair. No complaint—she's on Soc. Security disability. Her 17 yr. old son, Kevin sounds like a great young man. He's working to help out and getting an education at the same time. When I answered in July I enclosed a $100 check made out to him. This latest letter was one of warmest thanks and all, but some banker had told him I probably didn't intend him to cash the check, so he still has it. I phoned Mrs. Gardner and told her to cash it and I'd send the cancelled check back as a souvenir. It was a lump in throat call for me. She sounds like the nicest kind of person. It's been a nice day.

[Thursday, October 28–Friday, October 29: flew to western states for campaign

appearances; returned East to Camp David, noting "Nancy was watching 'Meet Me in St. Louis.' It was good to be there holding hands with her again."]

Saturday, October 30–Sunday, October 31

Did radio show I'd written on the plane—an open letter replying to Don Riegel's ridiculous message to me. Ride in the afternoon. I rode a Stud named Bull. Jackpot died a few days ago. Ran a mystery movie—not a naughty word in it. Today, Sunday watched TV panel shows all talking about the election. Took a walk after lunch—a beautiful Indian Summer weekend. Now back at W.H.

Monday, November 1

An easy day in that there were no meetings other than the issues briefing lunch and the usual staff meetings, NSC etc. The press is really revealing its bias on this election eve. After lunch went to G.W. Hospital to see Ed Rollins in for a clot in his carotid artery. He looks fine. His parents were there—a very nice couple. It did seem a little strange walking in there. Incidentally Hinckley was caught in an escape attempt today.

Back at the W.H. Came to the study and worked all afternoon on letters Anne Higgins thought I'd like to see. She was right. They are so upbeat, from all kinds of people—many with problems but they write to say they'll handle them without govt. help. Tonite over to Mike Deaver's for a farewell part for Morgan Mason who's going to work for Rogers & Cowan in L.A. King Vidor died to-day—97. Nancy got word her mother has gall stones. This has been a terrible year and a half for Nancy, beginning with my mishap March 30, 1981.

Tuesday, November 2

Election day! Harry Byrd yesterday virtually endorsed our candidate as his successor. I'd called him to ask if he would. He explained his son was working our man Tribble as were all of his associates and his papers had endorsed him. But he didn't think it was right for him to—in a sense name his successor. Well he must have changed his mind.

I didn't have time today to think about the election. An NSC briefing by Tom Reed on the MX. Then an NSC meeting on the new credit lines etc. we're trying to get our allies to go along. If they do we can lift much of the pipeline sanction as unnecessary in light of there new restrictions. We're close but France is holding out on one paragraph.

Briefing on our upcoming budget situation. We really are in trouble. Our one time projections—pre recession are out the window & we look at $200 bil. deficits if we can't pull some miracles. Speaking of miracles & such the mkt. went up more than 20 pts.

[Briefing with Cabinet Council on positive economic news; lunch with Senator Paul Laxalt (R-NV).]

Tonite a gang coming in for buffet & to watch the election returns. It was a nice evening & turned out alright—some disappointments & some bell ringers. Lost 25 in house—had to expect that & it could have been worse. Held the Senate 54–46. Millicent Fenwick lost—I'm sorry. But high spot—we won Gov. & Senator in Calif. Bye Bye Brown.

Wednesday, November 3

Did a press appearance in Rose Garden re the election returns. I'm sure they were sorry I was so happy.

[Visit from Prime Minister Giovanni Spadolini of Italy, termed Italy under him "a dependable ally"; budget meeting; met Don Hodel, new secretary of Energy; called candidates.]

Thursday, November 4

Continued making calls to winners & losers. No call to Jim Thompson Gov. Illinois. His election is tied up with Dem. controlled Chi. finding empty ballot boxes, ballots in car trunks and you name it. We may not know for a year who won. Had a Nat. Security Planning Group meeting re activities in Nicaragua & Poland. It was reassuring to find out how effective we can be.

Pres. Monge of Costa Rica came in. He's a good man & has been doing his best to pull his country out of an economic crisis. We then lunched in the East room with delegates from all over the free world here for the Conference on Elections. This grew out of my speech to Parliament last summer in London—start of a program to sell democracy to the world.

[Paul Laxalt accepted chairmanship of the Republican National Committee (RNC); met with own regional political directors; farewell party for staffer.]

ABC news had story about a woman with a country restaurant in Texas along a hi-way in a Nat. Park. The Park service wants to bulldoze her café to improve the view. I called Jim Watt (in California). He's going to look into it.

Friday, November 5

An easy day for some reason—except for the 1¼ hours I spent in the dentist chair having my choppers cleaned.

Before that an N.S.C. meeting then a meeting with Jeanne [Jeane] Kirkpatrick. Jeanne says the U.N. is a miserable place. I thought she was about to resign—but, no, she'll stay with it. Expressed a thought though that deserves real consideration. We're the only major country that dispenses aid on the basis of need instead of using it to reward friends. We, therefore have no voting block in the U.N. beholden to us and loyal.

[Visit from a family that adopted fourteen handicapped children; lunch with Nancy and Bill Buckley; deskwork all afternoon. **Saturday, November 6:** *radio broadcast; lunch with regional party Republican chairmen to announce that NRC should run reelection campaign.* **Sunday, November 7:** *coffee with Prince Philip and UK ambassador and wife; PBS "Young American Talent" program, with Itzhak Perlman hosting.* **Monday, November 8:** *meeting with Cabinet Council on economics regarding high unemployment and massive deficit; met with Supreme Allied Commander of NATO; conference on falling exports.]*

Tuesday, November 9

Most of the day spent in preparing for budget discussions. More than 3 hours. A long NSC meeting also. Geo. Shultz has worked out the agreement with our allies re the Soviets—trade, credits etc. This is more effective than our pipe line sanctions so I'm going to lift those sanctions. The agreement is what we tried for before the sanctions—the sanctions have done their job.

Lunched with V.P.—he and Barbara are off for Africa to visit a few states we need to handhold a bit.

Wednesday, November 10

Bud. Meeting—1st Drew Lewis on subject of 5 cent gas tax pledged to repair of hi ways and bridges. Then Dick Schweiker on medicaid and medicare. Hospital care is still on an inflation rate of 15%. These programs must be reformed or in a few years they will dominate the budget. In the afternoon 2 hours of preparing for press conference. Then at 5 P.M. Nancy and I went to the Nat. Cathedral where all day the reading of the names of those who died in Vietnam was going on. It was a moving experience—the familys, parents, widows and orphans go at the time they know the name of their loved one will be read.

[Ceremony honoring Marine Corps anniversary. **Thursday, November 11, Veterans Day:** *presented medal to holiday activist; budget session, more interest in new highway program; Phil Habib to return to Lebanon; report that Brezhnev died; Regan and Shultz presented plan for budget; press conference.]*

Friday, November 12

A busy day. Met with an assorted group of families—wonderful people in the state dining room to sign the proclamation of family week—Nov. 21–28.

Meeting in the Oval office with P.M. Ali of Egypt. He brought a good letter from Pres. Mubarak. The letter makes it plain that relations between Egypt & Israel are badly strained. My meeting with Begin had better result in his agreement to stop being so intractable.

Lunch with Jack Kilpatrick. He tried to tell me of his visit to the new Vietnam Vets. memorial and started to cry. Had a meeting in the East room with a group of youth volunteers from all over the country. They were wonderful kids & lifted my spirits.

Spent most of afternoon in another budget meeting. It's going to be a battle but we must make deeper budget cuts.

Rcvd. a warm message from Margaret Thatcher about the agreement we think we've reached with our allies re East-West trade. If nothing happens between now and tomorrow noon, I will announce the agreement—also the lifting of the pipe line sanctions.

The Soviets have told us our funeral delegation should only be 3 people. So it will be the V.P., Secretary of State and our Ambassador. Incidentally, our allies have followed my lead—no heads of state will attend.

Saturday, November 13

To the Soviet Embassy to sign the condolence book for Pres. Brezhnev. There's a strange feeling in that place—no one smiles. Well that is except Amb. Dobrynin. Back to the Oval office to do the Sat. broadcast. Then an emergency. With all 7 nations agreed on a uniform policy on E-W trade, something we've been after for a year and a half—we get word that Mitterrand has some objection. My script was written as an announcement of our agreement and that as a result I was lifting the pipeline sanction. The State dept. chickened and wanted me to go with a backup script on crime. I put in a call to Mitterrand—he was unavailable. I had in my hand Chancellor Kohl's and Margaret Thatcher's messages of joy over the agreement. I said to h--l with changing and did the announcement. Maybe Francois Mitterrand will get the message—& maybe the striped pants types at State will, too.

Now we are off to Chi. for the memorial service to Loyal by the Am. College of Surgeons.

It was a nice ceremony with several former associates, students etc. of Loyal's speaking of him. It wasn't easy for Nancy. [. . .] Got word that George Shultz had swallowed something that stuck in his throat. He was hospitalized but word is he is alright. Also received word—Reuters is running a story that France has repudiated the E.W. trade agreement we've all (including France) just agreed to.

[Sunday, November 14: watched morning panel shows; read Lou Cannon's book, caught "a pot full of inaccuracies. All in all though his editorializing was friendly"; anticipated visit from German chancellor Helmut Kohl.]

Monday, November 15

More flack from Paris but we're not answering. We've told them if they are reneging for any reason about the E.W. Trade agreement take it up with all of us—not just the U.S.

Briefing for the Kohl visit. This will be my 5th meeting with him but now he is Chancellor of the Fed. Repub. of Germany.

We had the full ceremonial on a raw, windy day. Our meeting was good. He is entirely different than his predecessor—very warm and outgoing. Mrs. Kohl is the same and very charming.

We did hit it off and I believe we'll have a fine relationship. No state dinner but a dinner for about 40 upstairs in our dining room. They felt very good about that and accepted it as something special.

During the day I sandwiched in meetings with Lyn N. and another with John Tower re the MX. No doubt we're going to have trouble—the Dems. will try to cancel out the whole system. It will take a full court press to get it. If we don't I shudder to think what it will do to our arms reduction negotiations in Geneva.

Tuesday, November 16

A busy morning and then off to New Orleans to address the Nat. Savings and Loan convention. But 1st a briefing by Geo. Shultz on his meeting with Andropov at the Brezhnev funeral. He agrees now I was right not to go.

[Flew to Miami. **Wednesday, November 17**: *meeting with Florida supporters; cited Coast Guard and civic anti-drug efforts; returned to W.H.]*

Thursday, November 18

Met with Senator Baker and Rep. Bob Michel. They weren't as upset as I was prepared to see them re the ec. situation and the election. Still they predicted tough going for much of what I feel we have to do in view of the continuing unemployment & the projected budget deficits.

Then an N.S.C. meeting about the MX and its basing mode. I know what my decision is but if I reveal it there will be a leak (d--n it) so I've taken it under study.

Had my yellow fever shot for the S.A. trip after Thanksgiving.

Friday, November 19

Another budget review then met with members of the "General Advisory Comm. of the U.S. Arms Control and Disarmament Agency." It was more ceremonial than anything—Judge Clark swore them in. I then told them we would not settle for an arms treaty just to have a treaty. If it wasn't a good one, I'd told our team to walk away & leave it. Went over to the Chamber of Commerce studio & did one of their hook-ups (satellite TV) to 37 meetings across the country of builders & industrial leaders. Spoke to them and then took questions from several of the groups—all had to do with unemployment.

Back to a Cabinet meeting on the budget. Our deficits are structural as well as recession caused. We have a built in increase in the budget which is automatic—we must deal with it. Finished up talking about trade and the coming G.A.T.T. meeting in Europe. There is a threat of protectionism widespread here & abroad. We must oppose it & strive for free trade.

Ended the afternoon (early) on the S. Lawn with an exhibition by riders from the Spanish Riding School in Austria—with those beautiful and especially trained Lippizaner horses.

One a stallion was presented to me—I accepted on behalf of the Am. People. The Austrian Ambas. & his wife & the Pres. of the Austrian C. of C were on hand. The Pres. of C. of C. Mr. Salinger is the actual donor of the horse. The whole thing was televised live to Austria.

I felt sorry for the trooper whose horse I was accepting. The rider & horse are permanently assigned to each other so this has been his horse for almost 8 yrs.

We then left for Camp David—had to finish the trip by car because of fog. Saw Clint Eastwood's picture, "Honky tonk man." He dies of T.B. in the end. Sat. night saw a picture in which the little girl dies at the end of Leukemia. It's still too early for Nancy to be seeing things like that—the wound hasn't healed. On my Saturday broadcast or before it I should say I did a voice check & remarked the economy was in a h--l of a mess. The mike was opened to the press room—I'd been told it wasn't. That night on the news—there I was being quoted on NBC, even played a sound tape. It was a cheap shot on their part.

[Sunday, November 21: *read most of the day; returned to W.H.*]

Monday, November 22

A hectic day probably because I'm leaving tomorrow for Thanksgiving at the ranch. Nancy left today to spend a couple of days in Phoenix. I'm lonely.

Started the day with last details on MX statement. Met with Cong. bipartisan leadership on MX. Only outspoken against was Rep. Addabbo. He'll cause trouble if he can. Tip was non-committal but Dem. Sen's. like Scoop Jackson were supportive. We're going to base the missile in Wyoming in what's called Dense Pack.

[*Meeting regarding Reagan scholarships at Eureka College; televised speech on peace and arms reduction.*]

Tuesday, Nov. 23 is getaway day for ranch but before going I have asked our people to look into the 10% inc. tax cut slated for July 1st and see if I could order (without Cong.) cutting withholding by 5% on January 1st as an economic stimulant. It wouldn't change the tax cut but would get some of it in the people's hands early. Chase Manhattan lowered prime rate to 11½.

Tuesday, November 23–Monday, November 29

Off to the ranch. Nancy is in Phoenix—will join me Wednesday. Got to the ranch—it had rained but seemed to be clearing. Up Wednesday for a ride—had to stay on the roads—ground soft from rain. Nancy late getting to ranch—about 4 P.M. The day had been beautiful. Rode Thurs.—stayed on the roads again but the day was beautiful. Moon & Bess, Maureen & Dennis came for Thanksgiving dinner—no fireworks. Rode Fri. & Sat. mornings then Dennis LeBlanc & Barney and I worked in the afternoons cutting up a downed tree & splitting wood—lots of it. Sunday turned up foggy and wet—no ride—more wood cutting. Mon. sun came out but we were on our way to Wash. after I spoke to Nat. League of Cities in L.A. Sun. night Clint Eastwood came to ranch with a man named Gordon Wilson who is part of a small group trying to get P.O.W.'s out of Laos. Clint has contributed to this. I'm checking the group out with Bill Clark—right now it looks a little questionable.

Tuesday, November 30

[Flew to Brasilia, Brazil.]

Pres. Figuereda was there to meet us with full mil. honors—review the troops etc. The motorcade to the Palacio de Alvarado was interesting. We drove on a deserted boulevard past miles of landscaped open space seeing city lights all around in the distance but no feeling of being in a city of 1½ million people.

The Palacio was meant to be the President's residence but he lives at his ranch. I don't blame him. The Palacio is like a public building, not a home. You could believe it was an insurance co. H.Q.—extremely modern.

Wednesday, December 1

To the Palacio de Planalto for the meeting with the Pres. We hit it off very well as I knew we would from our meeting in Washington. Ambassador Motley interpreted. I told the Pres. I was there to seek his advice as to how we could realize my dream of an accord between the nations of S.A. & N.A. He responded instantly. [. . .]

We moved into a meeting with all his ministers & our Cabinet members. Arranged to have ministerial level meetings re all our problems on trade, nuclear, military, etc.

Changed into riding clothes & went to his ranch. His home there is modest but his stables and horses are fantastic. He is a collector & showed great ingenuity in displaying sculpture, antique spurs, swords, etc.

I rode a cross between a T.B. and a German Hanover—18 hands. A great horse—twice Nat. Champion in Brazil. He insists on giving it to me & I'm not about to refuse. Lunch was a B.B.Q. All in all, a great day.

That night a reception for members of Cong. & then dinner.

[Thursday, December 2: ceremonies and final meeting with president; flew to São Paulo; addressed business leaders; learned that MX bill passed in congressional committee.]

Friday, December 3

Off to Bogota Colombia. Had an advance copy of toast he would deliver at lunch. It was a little harsh with a tone of resentment toward the "Colossus of the North." He has talked of recognizing Cuba & not being a satellite to U.S. etc.

[Full honors in reception at airport.]

At the palace he & I (plus interpreters) met alone. He spoke of previous visits by Am. Presidents (J.F.K. last) and how nothing came of it. This gave me a chance to say I came to ask questions—what were our problems and how could we solve them. Told him of my dream for the western hemisphere: That we had a common heritage of coming to this land from all over the world and that from pole to pole we worshipped the same God. He told me of his poor beginnings and I told him of mine. Then I pointed out—we were now the Presidents of our countries; that I wanted this kind of opportunity for everyone in all the Americas. By the time we did the farewell ceremonies at the airport, I believe we were real friends.

We flew on to San Jose, Costa Rica. Same formal ceremonies. This was my 3rd meeting with Pres. Monge. I went on to Hotel Cariari where I had a meeting with Pres. Magana of El Salvadore. We spoke of need to emphasize their pursuit of human rights so we could continue to help.

Saturday, December 4

Up at 7 A.M. Helicoptered to Casa Presidencial for a very constructive meeting with Pres. Monge. Costa Rica is the oldest real democracy in S.A. They have no armed forces.

On to Nacional Theatre where we each addressed an audience of civic and governmental leaders. When I started to speak a man in one of the boxes stood up and began to read a speech. The crowd called to him to stop but he kept on. The Pres. told me he was a communist deputy in their Parliament.

Since I had a mike and he didn't, I spoke. I told them I understood he was expressing the communist view point and this proved Costa Rica's devotion to democracy that he could do this but, we, in turn would not be allowed to do the same in any country with a communist government. The crowd cheered and clapped and he sat down.

The Pres. & I signed a new extradition treaty on stage. Then we went to a very nice lunch. Flew out that P.M. to Honduras where we were met by Pres. Suazo (had met him in Wash.) Had a meeting with him in the Base commanders office (We had landed at an A.F. base). Then he & I read statements to the press.

Back to office & met with President Rios Montt of Guatemala. They are getting a bum press rap as a mil. govt. which makes it hard for us to help even though they are under constant terrorist attacks.

I was greatly impressed by his presentation. We are going to give it real study. Then—home and mother (3 hrs. & 45 min. later).

[Sunday, December 5: quiet day; reception for Kennedy Center Awards honoring George Abbott, Gene Kelly, Lillian Gish, Eugene Ormandy, and Benny Goodman.]

Monday, December 6

Spent almost entire day in meetings with Congressmen on MX—tomorrow is the day. Patti is here and I promised I'd meet Dr. Caldicott—the lady is a round the clock anti-nuke lecturer & writer. She seems like a nice, caring person but is all steamed up and knows an awful lot of things that aren't true. I tried but couldn't get through her fixation. For that matter, I couldn't get through to Patti. I'm afraid our daughter has been taken over by that whole d--n gang.

Tuesday, December 7

The weather turned out fine for the official greeting ceremony for Pres. Zia of Pakistan. We got along fine. He's a good man (cavalry). Gave me his word they were not building an atomic or nuclear bomb. He's dedicated to helping the Afghans & stopping the Soviets. After our meetings—spent the day on the phone re the MX. No use—we took a beating 245 to 176—50 Republicans defecting. We had rabbits when we needed tigers. We still have a chance in the Sen. & then a conference committee. I really resent Chmn. Adabbo of N.Y. who honchoed killing the MX. No matter what he says he's just plain against re-arming the U.S.—unless the weapons are built in New York.

Then the State dinner—a fine toast by Pres. Zia. All in all a very pleasant evening.

Wednesday, December 8

Had to move out of the Oval office—a van full of dynamite pulled up at the Wash. Monument. A character against nuclear power threatened to blow it up. Security figured the windows here might go so everyone worked in the other side of the building. Held lunch for Bill Veritys task force in the entry lobby of the W.H.

[Meetings with representatives of the Council on Humanities and later with House Ways and Means Committee regarding the Caribbean Basin Initiative.]

Watched Cong. on TV—Addabbo couldn't kill the B-1 but he tried. I've never heard such pious hypocrisy from him & his storm troopers—especially Jim Wright. Late word—they killed the man in the van & disarmed the charges. We'll sleep in our own bed.

Thursday, December 9

[Meeting with Dick Wirthlin on public perception of administration; Cabinet Council meeting on a plan to aid farmers using surplus grain.]

Lunch with Habib—I told him he'd have to let Begin & Sharon know that their intransigence could cost them our long time relationship and support. Lebanon has asked them to leave so there is in effect an occupation army against the wishes of Lebanon.

[Gave HUD awards to local officials for using partnerships to spawn programs; interviewed by two reporters from People, *"Thought they'd be birds of prey—they seemed very nice and it was a pleasant time"; met March of Dimes poster boy; Senate dinner at the Library of Congress.]*

Friday, December 10

A busy day—signed two human rights proclamations before a group in East Room—rather emotional—one re Poland.

[Taped a Christmas story before a group of children; met with General Edward Rowny, chief negotiator and head of the U.S. Delegation for Arms Control Negotiations, with Senators Henry Jackson (D-WA) and John Stennis (D-MS) on the MX; group from University of Alabama establishing Reagan chair in communications; press availability; conference with three Republican governors on budget; went to Camp David. **Saturday, December 11:** *spent most of the day in front of the fire; radio broadcast; photo session; watched* Sophie's Choice *(1982).* **Sunday, December 12:** *phoned speech to Bridgeport, Connecticut; returned to W.H.; attended Christmas show starring Dinah Shore.]*

Monday, December 13

[NSC briefing; visit from Prime Minister Poul Schlüter of Denmark, "a good man and more akin to us in thinking than his predecessor"; issues briefing lunch; met with Drunk and Drugged Driving task force.]

Called Geo. Shultz in Rome to wish him a happy Birthday. He says Pres. Pertini is still talking about Nancy.

Had a hush hush meeting with NSC to complete report on covert activities by C.I.A. These had almost ceased under the previous admin. I was greatly reassured to learn what we are doing now—it's quite extensive and effective.

Met with about 30 Hearst editors in a Q&A session—told them I'd answer all the Q's we didn't get to by mail.

Don Rumsfeld came to report on his mission to steer our allies off the Law of the Sea treaty. He did a good job, Japan, U.K., F.R.G., Italy etc. all joined us in not signing. The treaty would turn the entire oceans of the world over to an international body with supreme power even to tax.

Saw the Mexican Ambassador off to Mexico—he is now the Foreign Minister replacing Castaneda whom I never trusted. His son-in-law is a member of the Cuban Polit. Bureau.

Early dinner & downstairs to the Cong. Christmas reception where we shook hands in an hour long reception line.

Tuesday, December 14

Held an MX meeting with Dem. & R. Sens. Consensus was we should go for a compromise they've worked out to appropriate money but fence it in until both houses can agree on a basing mode. Not good but best we can do.

[Meeting with Republican leadership on MX, gas tax; then business executives in National Security Telecommunications Advisory Committee.]

Spent most of afternoon in a dismal ec. briefing about the deficits and the little chance we have of getting further budget cuts. Then on to Soc. Security. I'm afraid our bi-partisan commission has failed us.

[Photo session. **Wednesday, December 15:** *visit from Prime Minister Francisco Pinto Balsemão of Portugal, who expressed doubt that Angola and Mozambique would continue to maintain relationships with the USSR; budget review; calls seeking support for MX bill; dinner and overnight with the Reverend and Mrs. Billy Graham.* **Thursday, December 16:** *staff meeting, NSC briefing; cabinet meeting regarding changes in education grants; meeting with telecommunications executives; lunch with Vice President Bush; met with Eagles, Republican loyalists; NSC regarding overall policy toward Soviet Union; interview with* Washington Post; *photo sessions with sports stars in anti-drug message, and with Future Farmers of America, Mormon Youth Choir; ambassadorial formalities; Christmas tree lighting; holiday reception for W.H. press corps; "Hanging over the whole afternoon & evening—A B52 crashed on take off in Sacramento—loss of crew"; another plane crash in Cincinnati, which killed four FBI agents.]*

Friday, December 17

Cabinet council meetings—re trade matters and moving "off bud." loan programs on top the bud. They are part of this whole Mickey Mouse budgeting system the Congress insists on preserving. They are responsible for 10% of the Trillion $ debt. Did some taping and met with reps. of the Minority business Assn. Took an hour out and went riding at the Park Police Indoor riding hall. Then in late afternoon dropped in on Bill Clark's NSC party at Blair house.

I also today phoned the widow & children of the 4 F.B.I agents who died in yesterday's plane crash. They tore my heart out.

Saturday, December 18

Radio broadcast & a round table live interview by representatives of 8 independent radio networks. It went well and I had a chance to get in some licks about the Dems. so-called "jobs" programs—also re the charge that we discriminate against minorities.

3:30 the military visit to the White House. It was frustrating—the crowd was so thick in every room and the hall that we could only stand on the stairway and yell Merry Xmas.

6:09 P.M.—the Secret Service reception—took pictures with all of them & their wives or girlfriends.

Ron and Doria arrived—will stay through Sunday. Doria is doing a review of some performance at Kennedy Center.

Howard Baker with Bob Dole & Paul L. called (phone) to tell me they would keep the Sen. in session all night & take up the "gas tax" bill in effort to break the filibuster by Helms, Humphrey & Nichols if I said O.K.—I said it.

Sunday, December 19

Snowing but I'm afraid it's too warm—37 degrees for it to last. Nancy is yearning for a White Xmas. Quiet day—the 4 of us lunched in the solarium. At 3 P.M. we have to go downstairs for the W.H. and Exec. O. Bldg. reception. We went and every room was jammed. We shook as many hands as we could through the State Dining Room, the Red, Blue, Green & East Rooms & back to the staircase.

Word from the hill indicates we may get the gas tax and the Continuing Res. in a form I can sign.

Tonite Ron told us he wants to quit Ballet & go to writing. I can't say I'm sorry although he worked hard & was getting along well—but there isn't much of a future and it is a short career.

Monday, December 20

Only one thing on my mind today. Nancy had a small growth removed from her upper lip. They took a biopsy but we won't have word for 24 hours. I know she's worried and so am I.

Part of the day was spent by me phoning the "Hill" on all the unfinished business going on in this lame duck session. Even the members of both Houses are going on the air declaring that the game playing & politics is a disgrace.

Tuesday, December 21

Word came that the growth was a form of skin cancer caused by the sun. But we are both re-assured. It is an easily cured form & in her case it was cleanly excised—no further treatment needed. Dan Ruge says he had one removed 10 years ago—no problem.

King Hussein of Jordan arrived. I really like him. He is our hope to lead the Arab side and the P.L.O. in negotiating with the Israelis. He has some problems in order to keep the trust of other Arab states & right now Israel is proving difficult. I told the King this was a top priority of mine & we'd go all the way to bring peace to the Middle East & we'd stand by Jordan. They'll be here until Thurs. so his team & ours will meet tomorrow & he & I will have breakfast Thurs. A.M.

Up on the Hill it looks like we won't get the C.B.I. or the gas tax but we have the continuing resolution so the govt. won't be shut down.

[Call to woman with rare disease.]

This mornings paper, N.Y. Times had a story of a 29 yr. old man (out of work for 12 months), wife & 8 kids, had applied for a job in N.Y. On the way home a 75 yr. old blind man broke his cane & fell between the cars of the subway. Young Mr. Andrews without hesitation leaped off the platform between the cars & rescued him. I called him—from his voice I knew he was black. I asked if he'd gotten the job. He said they'd called him to come for an interview & he was on his way there. I called the company—the operator said the manager & his mother were both on the phone. I asked if they'd read the story—they had. Andrews has a job.

Wednesday, December 22

[Noted that business executive also helped the New York hero Reginald Andrews.]

Gave out medals today to teenagers for service and for heroism in 1980 & one for 1981. Choked up on one—an 18 year old went down a pipe into a tank to try & save 2 paramedics who had passed out from noxious fumes. I had to give the medal to his parents—he died in the attempt.

[Reiterated support for Caribbean Basin Initiative before regional ambassadors.]

Another Bud. meeting—there'll be blood on the floor with this one—we're going to try and reform the entitlement programs.

Thursday, December 23

A busy hectic day—but the Senate voted cloture & then passed the gas tax highway bill. Thank Heaven the Congress has now gone home.

Final meeting with King Hussein. I believe we've made great progress and, unless the Israelis throw sand in the gears, he should be back here in a few weeks to announce he'll negotiate with them.

A day full of meetings on Federalism, Women's issues, the Private Initiative program, etc.

Friday, December 24

Nancy didn't get her "White Christmas Eve." In fact it is almost a shirt sleeve day. We've spent it quietly here in the White House. Patti arrived at 6 P.M. and at 7 we were on our way to the Wicks for the Christmas eve dinner that has become tradition with our families. It was as usual—fun evening— Cindy played Santa this time and we all sang Xmas Carols with Charles at the piano.

Saturday, December 25

We opened presents and at noon I called Mr. Ottman, V.P. of Sheraton Corp. He opened the ballroom at the Sheraton Central Hotel in N.Y. to provide

Christmas dinner for 500 homeless people. His employees gave up their Xmas to serve & work.

The Wicks, Deavers & Ruges came for a 5 o'clock dinner—a most enjoyable time. Lucy Fisher and Doug Wick, houseguests overnight.

Sunday, December 26

Quiet day—guests are gone. Patti to the movies. The brutal overthrow in Surinam by a small officer clique in the army. Wants slaughter of about 30 civic leaders and total takeover of all media. We believe Dutch who were contributing $100 mil. a year have shut the spigot. I advocate approaching the Dutch about sending in their 4 or 500 marines who are in the Caribbean. We could help legislatively and block out Cuba intervention.

Monday, December 27–Sunday, January 2, 1983

[Left for western trip; visited Mrs. Reagan's mother in Phoenix, her health improved "due at least in part to the change in doctors"; attended recommissioning ceremony for battleship New Jersey in Long Beach, California; traveled Wednesday to Palm Springs for golf and socializing.]

Fri. the weather turned beautiful—played 11 holes & surprised myself with a few pars. New Years Eve at Sunnylands & as always a wonderful evening with old friends. Sat. New Years day—beautiful but spent half the day in meetings & the rest watching the bowl games on TV. Dinner party at a new club—given by the Wrathers & Darts. Mike D. proposed & I agreed to go by the La. floods on way home.

And Sun. on the way back (Nancy staying in Calif. for a few days doing some charity events) set down at Monroe, La. It really is devastated by the floods but the people were out in droves and their spirits are high. I shoveled a few sand bags for the cameramen—toured the area with Governor Dave Treen & a good boll-weevil Congressman. Then went on radio and called on others to help the flood victims. Promised I'd approve Governor's request for disaster aid.

Now it's bedtime and tomorrow is Mon.

CHAPTER 3

—

1983

Monday, January 3

[Staff and NSC meetings; calls from congressional leadership.]

A tough budget meeting & how to announce the deficits we'll have—they are horrendous & yet the Dems. in Cong. are saying there is no room for budget cuts. Met with a group of young Repub. Congressmen. Newt Gingrich has a proposal for freezing the budget at the 1983 level. It's a tempting idea except that it would cripple our defense program. And if we make an exception on that every special interest group will be asking for the same.

[Signed bill subsidizing phone aids for the hard of hearing; dinner with congressional leadership.]

Oops—almost forgot—signed order launching our commission to study the MX landing system. Former General Scowcroft is Chmn.

Dinner very pleasant—Eddie Albert entertained—songs & poetry. He was great.

Tuesday, January 4

Brkfst. with G.O.P. leaders (Sen.). Gave them bad news about deficits. They agree the law that says we must project 5 yrs. ahead is crazy but we still have to do it. No economist can predict more than 1 yr. ahead (if that) with any degree of accuracy. Nat. Security Planning Group—met re Surinam. Press is neglecting this place possibly because the blood thirsty dictator is on the left. He had about 30 labor leaders, academics & civic leaders executed and is seeking alliance with Cuba. This must not be allowed. We have to find a way to stop him. The Marines could do it but we'd lose all we've gained with the other Latin Am. countries.—Working lunch with cabinet on medicare and medicaid. I'll have a decision to make there. Started briefing for tomorrow night's press conf.

[Visit from President and Mrs. Yitzkah Navon of Israel; rehearsal for press conference; introduced Elizabeth Dole as secretary of Transportation; difficult questioning at press conference.]

Thursday, January 6

Signed the Gas Tax bill, had lunch with the V.P., met with House Repub.

leaders. Talked budget with them. Didn't get any hint of panic & believe we can come up with some saleable ideas. My own unspoken idea is that we should give 2 sets of projections—one if recovery is low say 3% & one if it is at 6% as many economists predict.

[Cabinet Council meeting, discussed using unemployment insurance in relation to job training; photo session; gym; dinner for new members of Congress.]

Friday, January 7

Met with Sens. re the crime bill. It doesn't have all of the things we wanted & does have one thing we don't want—appt. of a drug Czar. I'll decide Mon. whether I can sign it Monday. If I sign we can make the Att. Gen. the Czar and maybe down the line we can legislate some changes.

Lunch with Bill C. and Henry K. He's very much in support of what we are doing in the Middle East. In fact, he was most complimentary.

[Signed bill on nuclear waste disposal; meeting with groups sympathetic to Lebanon, Danny Thomas present.]

Bill Clark talked to me about leaving—problems with his ranch & his family. I didn't talk or try to talk him out of it because of his problems. Then I talked to Helene who knows he is torn and really would prefer to stay. Lord knows I want him to stay. So I told him to take a month's leave—go to the ranch & return. That's what he'll do.

Off to Camp David.

[Saturday, January 8–Sunday, January 9: radio broadcast Saturday; worked in front of the fire; returned to W.H.]

Monday, January 10

Wet weather & a typical Mon. in every other way—a jam packed schedule with no time to catch up on memo etc. Main part of day spent on budget. We're close to agreement on a lean budget including $11.3 bil. in defense cuts. Cap isn't happy but says he can do it without wiping out new weapons systems. Part of it will be—no pay increase for 1984. But then we're asking for that across the board for everyone.

Met with Phil Habib—he's on his way back to Lebanon. The msg. for P.M. Begin is that I want action. Another meeting was with NSC planning group re our strategy with the Soviet Union. Geo. S. thinks our re-direction since we've been here is a success.

[Met with Governor Thomas Kean (R-NJ) with plan for revolving loan fund, authorized by Congress. Tuesday, January 11: flew to Dallas to address National Farm Bureau; met with supporters from 1980 campaign; visited food bank; noted prime interest rate was lowered; Weinberger agreed to budget cuts; returned to W.H.; received gift of electric train set from hobbyists.]

Wednesday, January 12

Supposed to be a half day but it turned out to be a full one. Met with Sen. R. Byrd to do a little stroking. A long Cab. meeting on options for treating with the unemployed. Bill Casey came by—feels strongly we should have a p.r. office to get our story to the country. We certainly can't depend on the news media. I agree with him.

A lot of N.S.C. business—Eugene Rostow resigned as head of "arms control." We've had trouble there. We're appointing a new man also to head up our conventional arms reduction negotiators in Geneva. I think some of our friction in the top 3 staff is cleared up. Announced Margaret Heckler as Secretary of HHS.

Thursday, January 13

A drop by—Christian Dem. party chairman of Italy, Mr. Mata. Then a visit by Mayor Jacque Chirac of Paris. He's a good, sound & charming man with solid ideas. An N.S.C. meeting re our arms negotiations—we'll stick with our zero option plan. Found I was wishing I could do the negotiating with the Soviets—They can't be any tougher than Y. Frank Freeman & Harry Cohen. A phone call from Malcolm Fraser of Australia—P.M. I think he'll visit us in the Spring. He's a good man. I called 3 N.Y. police officers wounded by terrorist bombing in N.Y. One lost a leg, another an eye and one has serious wounds.

Friday, January 14

A long NSC briefing re the visit of P.M. Nakasone of Japan. Amb. Mike Mansfield was on hand. Following this, a meeting with U.N. Sec. Gen. Perez Cuellar. We'd decided I should have a press availability today to straighten out the girls and boys of the press who are all preaching that our "arms policy" is in disarray over the resignation (firing) of Eugene Rostow. I opened by using all their words about them—that they were in Disarray, chaotic about to unravel into complete shambles. They got the msg. & seemed a little self-conscious. It was a good session. I enjoyed myself.

[Photo session; phoned televised job fairs around the country. **Saturday, January 15:** *met with staff regarding Social Security Commission; radio talk; concert by Harlem Boys Choir and reception to honor memory of Martin Luther King Jr.; Social Security proposal accepted by Democratic congressional leadership.]*

Sunday, January 16

Spent most of day at desk, then at 6:30 guests arrived for a stag-bull session dinner. Geo. Shultz, Geo. Bush, Bryce Harlow, Lee Iacocca, Geo. Will and Irving Kristol. Most of the conversation centered around the auto industry. Lee discussed the Japanese problem & their restrictive trade practices, including under valuing the Yen. It was most informative & useful since we'll be dealing with P.M. Nakasone Tues.

[**Monday, January 17**: *called Tip O'Neill and other congressional leaders to discuss accord on Social Security; meeting with CEOs of industrial companies and trade groups for "a solid discussion of Japan's gimmicks to pretend free trade but practice protection-ism."*]

Tuesday, January 18

P.M. Nakasone (Japan) arrived. We met privately. I acknowledged what he had done already about reducing tariffs & upping the mil. bud. but told him more was needed. I recognized his pol. problems but told him we had some too—a drive in Cong. to turn to protectionism, which was growing in strength. I believe he will take action. We moved to a larger meeting in the Cabinet room & continued discussion there & lunch in the W.H. I invited him & his wife & daughter to brkfst. tomorrow. Cabinet met & heard last word on budget. In constant $ it will be equal to '83 bud. Allowing for inflation it will [be] 4½% bigger the exact amt. of present inflation rate.

Charley Wick brought in Rupert Murdock for a meeting. He is supportive of me which means some of the press is with us—his press.

[**Wednesday, January 19**: *farewell breakfast with Nakasones; desk work in morning; flew to Chicago; visited high school; political fund-raiser; visit with radio personality Paul Harvey; Republican dinner, noted demonstration outside.* **Thursday, January 20**: *noted second anniversary of inauguration; Mrs. Reagan in Arizona; NSC briefing; press conference; lunch with Vice President Bush; honored youth exchange program; annual Executive Forum; photo session.*]

Friday, January 21

Sen. Bennett Johnson brought the Mardi Gras King and Queen in for pictures—the Queen is his 17 yr. old daughter Sally.

Met with Nitze and Rowny who are headed Back to Geneva for the arms negotiations. The Soviets are pulling out all the propaganda stops trying to turn off the allies. I think we can top them.

Went over to exec. office bldg. and did a TV show in which I faced a group of high school kids from all over the nation & fielded their questions. I really enjoyed it. Then we took a half hour or so and dropped in on some offices I've never visited before. That was fun too. Final meeting was with pro-life reps. for planning how we are going to get rid of this abortion monster.

Waiting for Nancy to come home so we can go to Camp David. She's here and we're on our way.

[**Saturday, January 22**: *radio broadcast; photo session; worked on State of the Union Address; called Super Bowl football teams.*]

Sunday, January 23

More speech work. Had to drive down because of fog & rain. "60 Minutes" surprised me by taking on the Nat. Council of Churches & the World Council for their quiet spending in support of leftist forces and governments in the world.

Monday, January 24

Nancy is off to N.Y. Comes back tomorrow with Ron & Doria. N.S.C. briefing—not good news from Lebanon. The Israelis are becoming more intransigent. Habib is on his way back here. Met with Ec. Policy Advisory Board—as usual there were differing ideas about our budget & tax policies but altogether were more supportive of what we are doing than an outside group—Pete Peterson, John Connally, McNamara and Bill Simon who were all over the airwaves with proposals which are impossible to realize. Mr. Simon did not attend the meeting—he's a member—he's also a loose cannon.

[Visit from Muhammad Ali, "gave me an autographed Muslim Prayer Book"; new budget figures show sharp increase in projected deficits, blamed recession for declining revenues.]

Tuesday, January 25

Met with House & Sen. Repub. leadership re tonite's St. of the U. and next Monday's budget message. It was the best meeting we've ever had and all pledged loyalty. Then a meeting preliminary to the V.P.'s visit to Europe and Geo. S.'s visit to Asia. George is going to read a msg. from me publicly in Berlin re the I.N.F. talks in Geneva. We intend to steal back the high ground the Russians are trying to occupy. In China Sec. Shultz will resist any effort by P.R.C. to oust Taiwan from the Asian Bank. A lengthy meeting of NSC re the "Start" talks which are kind of stalled over the Phase I Phase II framework—Russians want to make them all Phase I. I think we've found a framework.

[Lunch with television anchormen; Mrs. Reagan home with Ron and Doria; State of the Union Address; ABC poll showed 15 percent rise in approval afterward. **Wednesday, January 26:** *met with West German foreign minister Hans-Dietrich Genscher, who pledged support for administration strategy in INF talks; flew to Boston to visit schools and factories and an Irish pub; met with Republican leaders; returned to W.H.]*

Thursday, January 27

President Mubarak of Egypt arrived. We had good meetings and reaffirmed our solid relationship. He's very concerned about Israels stubbornness about getting out of Lebanon & fears our peace initiative may simply die on the vine because of them. I share his concern. He believes they & the Syrians may be playing a game with an eye to both, even though they are hostile toward each other, cutting up Lebanon between them.

[Visit from Christian Democratic leader of Austria; Republican party reception; dinner in honor of National Gallery of Art benefactor Paul Mellon.]

Friday, January 28

Spoke to 700 people—the familys of those who are still listed as Missing in Action in Vietnam. An emotional experience. We are going all out to follow up persistent leads that some Americans are still P.O.W.'s in that d--n Communist sink hole. Later spoke to the Vietnam Vets who have formed a task force to help establish those vets who have been unable to get back on their feet. These fellows are all very successful & now want to help the others.

[Farewell reception for Secretary of Transportation Drew Lewis. Ron and Doria arrived for weekend. **Saturday, January 29:** *radio broadcast; taped segment for NBC program featuring questions from children; spoke at Alfalfa Club; Don Regan reported hearing from business leaders that the recovery had begun.* **Sunday, January 30:** *watched* Gandhi *(1982), called it "a truly great movie"; watched Super Bowl.]*

Monday, January 31

Bi-partisan Cong. leaders came to oval office re the budget—which was just being sent to Cong. today. Tip and I get into a donnybrook—I really had my dander up. The worst of it was Tip didn't have the facts of what is in the budget—besides he doesn't listen. Then we moved to the St. Dining room where Committee Chmn. were assembled. I made opening remarks & turned meeting over to Don Regan, Dave Stockman & Marty F. Early afternoon picked up Nancy & we went to the Shoreham Hotel for convention of religious broadcasters. Nancy received an award for her humanitarian causes. I addressed the group (4000). I've never had such a reception. Billy Graham called this evening and said they are still talking about it—called it the greatest declaration for the Lord any Pres. has ever made. I feel very humble.

At Six thirty took Marine 1 to Dulles & met homecoming Redskins—Nat. Champs.

*[*Tuesday, February 1:** *flew to St. Louis to visit resurgent Chrysler plant with Lee Iacocca and governor; speech to four thousand small-business owners; returned to W.H.; met with Republican campaign leaders.]*

Wednesday, February 2

[Met with Jewish leaders briefly; cabinet meeting on Federal Charity Campaign; report from the Department of Agriculture refuting charges of deterioration of nutrition standards.]

Finally had a visit with 6 Afghan Freedom Fighters here in this country to tell of inhumanity of the Soviets. One's wife had been executed while her 2 children watched. One was a young, pretty and tiny lady—held in custody for

4 months—tortured daily. She had been a medical student. Her crime? Caught looking at a "freedom" leaflet. The others were from a village where the Russians had burned 105 people alive with gasoline & chemicals in an irrigation tunnel. Jack Hume came by with a plan for getting our story to the people. Looks like a good idea.

Tonite 6:30 P.M., another Stag dinner, Laxalt, Rostenkowski, Cheney, R. Long, Scoop Jackson & Pickle. Good time had by all & some brownie points chalked up.

Thursday, February 3

Left W.H. at 7:25 A.M. for Nat. Prayer Brkfst. As always, it was a warm, heartening experience. Campbell College N.C. choir sang. Gen. Vessey gave a firm talk. Back to oval office for an N.S.C. briefing on C.I.A. covert activities. We've made great progress in re-establishing the C.I.A.'s effectiveness under Bill Casey.

[Chancellor Bruno Kreisky of Austria arrived, observed of him, "he is not a well man"; farewell party for Health and Human Services Secretary Richard Schweiker.]

Friday, February 4

Met with High School Honor students from all over the country who are winners in a competition sponsored by Hearst people. Did Q&A. I hated to leave. They had chosen unanimously 97 of them to have a girl present me with a group photo autographed by all of them. The girl was on crutches—has only one leg. Found out later it was a recent operation—bone cancer—she is probably terminal. She insisted on making the trip. Final 2 hour briefing—Phil Habib is on his way back to Israel. I asked him to let me know if a phone plea to Begin would help. If not we'll just have to separate ourselves from Israel.—Off to Camp David.

[Saturday, February 5: desk work; radio broadcast; noted birthday; returned to W.H.; reception with birthday cake for Young College Republicans. **Monday, February 7**: *breakfast with Republican freshmen congressmen; received delegation from Louisiana expressing gratitude for assistance after flooding the previous December; met with local TV anchors, and with Air and Space employees; swearing-in ceremony for Elizabeth Dole; ceremonial visits; Mrs. Reagan away in New York.]*

Tuesday, February 8

An NSC meeting about Surinam. [. . .]

Dropped in on Edit. writers lunch & took some questions.

Cabinet meeting on deregulating gas. Hodel is going to work with our leaders in Congress to come up with a package to deregulate & at same time protect consumers. Under the present (Dem.) control plan there are 28 dif. classifications of gas price & it has still gone up about 50%.

Staff and D. Stockman came in about accelerating pub. works already in budget to head off a Dem. job's program.

Bill Clark is off to Geneva to huddle with our negotiators.

[Wednesday, February 9: met with Commission on Strategic Forces, related directly to deployment of the MX missile, a group commonly known as the Scowcroft Commission after its chair, Lieutenant General Brent Scowcroft; also met with journalists ("Did Q&A without leaving any time bombs around"); answered mail.]

Thursday, February 10

Should have saved the mail til today—had a lot of open space in the schedule. The Israeli Cabinet meeting again—don't know what they'll decide about Sharon & the investigation report calling for his ouster from govt. Cabinet meeting on Davis Bacon Act.* A court ruled we can't change the regulations—do we try for legislation or appeal? I decided appeal—we couldn't get a bill against the labor lobby through this Congress.

[Met with CEOs regarding budget; photo sessions; met with Builders Association.]

Friday, February 11

Winter is back—virtually a blizzard. Camp David is out—we're in for the weekend. Started snowing last night. By noon we'd sent almost everyone home before they got snowed in.

[NSC planning group meeting, report from Shultz and Bush on their respective trips abroad.]

An almost 2 hr. lunch with Joint Chiefs of staff. Most of time spent on MX & the commission, etc. Out of it came a super idea. So far the only policy worldwide on nuclear weapons is to have a deterrent. What if we tell the world we want to protect our people, not avenge them; that we're going to embark on a program of research to come up with a defensive weapon that could make nuclear weapons obsolete? I would call upon the scientific community to volunteer in bringing such a thing about.

[Saturday, February 12: some staffers snowed-in at the W.H.; Vice President Bush and Deputy National Security Advisor Bud McFarlane brought report that Qaddafi was allegedly planning attack against Sudan; dinner with Secretary and Mrs. Shultz. **Sunday, February 13:** *caught up on reading.]*

Monday, February 14

Valentine's Day. Had a brief on the West Bank. There can be no question but that Israel has a well thought out plan to take over the W.B.

---

* *The Davis-Bacon Act, signed into law in 1931, ensured the payment of locally prevailing (typically union) wage scales on federal projects.*

[Met congressmen Jack Kemp (R-NY) and Jerry Lewis (R-CA) regarding their concern over multinational bank funds, defended their usefulness; teleconference with Chamber of Commerce event. Did a closed-circuit telecast to a meeting of the Young Presidents Organization in California; "On way back stopped and bought some valentines"; interview with journalist Allan Ryskind of Human Events, "maybe this will help them know I'm not betraying the Conservative Cause."]

N.S.C. meeting about Khadafy and a report that he's going to send bombers to help a coup in Sudan. Some question as to whether the report is valid. We are providing AWAC's to aid Egypt fighters (planes) who will intercept Lybian planes.

[Dinner in honor of five presidential historians.]

Tuesday, February 15

[Meeting with Republican congressional leadership, reports of Shultz and Bush trips, discussion of economy; intelligence briefing on Palestinian situation, heard of Palestinians scattered all over the Middle East, regarding Israel and the West Bank as their homeland.]

Did a Q&A in the family theatre preparing for the press conf. tomorrow night. Home to wood shed for that exam.

Almost forgot—Geo. Shultz sneaked Ambassador Dobrynin (Soviet) into the W.H. We talked for 2 hours. Sometimes we got pretty nose to nose. I told him I wanted George to be a channel for direct contact with Andropov—no bureaucracy involved. George tells me that after they left the Ambassador said "this could be an historic moment."

Wednesday, February 16

Most important event—the Press conf. By the time I got to it I was mad as h--l. I'd watched the news and seen the witch hunt that is on for Ann Gorsuch at E.P.A. The media is a lynch mob that thinks it smells blood. Then I saw 2 Repub. Sens. Mathias & Pressler join the Dems. on the Foreign relations committee to oppose Adelman (a d--n good man) for appointment to the Arms Control Commission. Maybe being mad helped—it was a good conference & some said the best.

They were snooping over another leak—this time the leaker could be risking lives. Khadafy, we've been tipped, is planning an air attack on the Sudan as part of a coup to overthrow Pres. Neimieri. We've flown AWAC's into Egypt ostensibly to engage in a training mission. They are there to help direct Egyptian fighter planes if he tries it. The date is supposed to be the 18th.

Thursday, February 17

The leaks re Lybia are all over the place. We're stonewalling.

Met with Ann G. She's really uptight. I'm reassured there has been no hanky panky because she wants us to give up on exec. privilege & turn the documents in question over to Congress. We are trying to work out a deal whereby members of the Committee can see them but we preserve the principle of E.P. so we don't set up a pattern where they'll think all they have to do is raise h--l & presidents will cave in.

Jeanne Kirkpatrick reported on her trip to Central America. A grim story. Our Ambas. Hinton under the direction of the same kind of St. Dept. bureaucrats who made Castro possible are screwing up the situation in El Salvador. I'm now really mad. Bill C. is bringing George S. up to date & then I'm determined heads will roll, beginning with Ambas. Hinton.

Friday, February 18

We haven't settled with Cong. yet on E.P.A. Poor Anne G. is really strung out. Now some zany on the hill is saying $50 mil. in the superfund is "unaccounted for."

P.M. Willoch of Norway arrived. He's a fine man and a good friend of the U.S.A. We had a good meeting on a number of subjects. He's worried that we should move to a lesser position in the I.N.F. talks before we begin to look inflexible. He has a "peace" mob in his country too. All in all though we were together on most things.

The N.S.P.G. meeting was something. It was on El Salvador & Central Am. We have been so restricted by Cong. in our help to the El S. govt. that they are losing to the Rebels. I finally told the Council to come up with how we could be of more help, not whether we should. We (I learned) are tracking by radar the planes flying ammo & guns to the guerrillas but we're forbidden to tell our friends so they can intercept them. That has to stop.

[Mrs. Reagan left for Phoenix for a week's visit with her mother; attended American Conservative Union dinner, commented that "evidently the Right Wing Rebels have had little effect. I was warmly received." **Saturday, February 19:** *felt lonely without Mrs. Reagan; made radiocast; edited speech; "stag dinner" with senators Jim McClure (R-ID), Al Simpson (R-WY), and Howell Heflin (D-AL) and representatives Henry Hyde (R-IL), John Breaux (D-LA), and Jim Martin (R-NC).* **Sunday, February 20:** *cleaned desk, read; noted lack of call from Mrs. Reagan, who was helping her mother move; wondered if phones were working.]*

Monday, February 21

Holiday. Maureen & Dennis arrived early morning—took the red eye plane. They are now sacked out. I'm getting a real hang up on the press. Front page of Wash. Times (which is becoming as R. Wing as the Post is L. Wing) had a story that the Conservative Union was cool to me Fri. night. I was interrupted a dozen times with applause and got a lengthy standing ovation.

[Reception sponsored by the National Review; *appearance at Society of Cincinnati*

gathering. **Tuesday, February 22:** *met with Prime Minister Edward Seaga of Jamaica, who was working to unravel "left wing pro Cuban policies"; addressed American Legion convention; cabinet meeting on division of responsibilities on trade; dinner with members and contributors of conservative groups; dinner with Maureen and Dennis.]*

Wednesday, February 23

[Breakfast with press.]

Met with Ambassador Ahrens—he's leaving to become Israel's minister of defense. I hope I made him aware of how serious we are about Israel getting out of Lebanon.

Soares of Portugal (probably be P.M. after election) came by. He's a socialist but entirely anti-Communist & pro-America.

Lunch for a dozen recipients receiving Pres. Medal of Freedom. About 350 people in all. After lunch, Dan R. caught me for an electrocardiogram & tests before and after the treadmill. Home for the afternoon. Called a Mrs. Nelson, Canton Ga.—her son, a Green Beret trainee on a training mission in Panama was killed. He was 25 and told her he was in the service because he wanted to serve me. I had a newspaper clipping with his picture before me. He was strikingly handsome. How I wish there were words that could bring comfort to his mother. When she told me what he had said about me I felt very humble. And oh how I pray I can even be a little deserving of that.

[Thursday, February 24: closed-circuit telecast for international business leaders; met with local officials to explain block grants; Cabinet Council on deregulation of natural gas, "I've decided to go for it"; National Security Planning Group (NSPG) meeting on El Salvador, decided to increase aid; spoke to American Legion Auxiliary about parental notification of birth-control requests by teenage girls; Mrs. Reagan in L.A.]

Friday, February 25

An interesting day. Anne (Gorsuch) Burford came in. She's taken a beating due to a cluster of Demagogs—Dem. Congressmen trying to scare up a "sewergate" re the E.P.A. She wants to turn over all papers to the committees on the hill with no regard to what it does to "Executive Privilege." She may have a point. Aided by the press they've created a false image making it look as if there is a scandal cover-up in E.P.A. I've set our people & I called Bill Smith to get into this and see if we shouldn't do as she asks but at the same time, make it publicly clear we are doing it only to head off a lynch mob.

[Closed-circuit appearance with high school students.]

Saturday, February 26

[Radiocast announced deregulation of natural gas, generating controversy over potential for price fluctuation.]

Since compromise to make documents available to Cong. on E.P.A., the supposed scandal has moved to page 11. Congressional ardor has cooled now that they face about 800,000 documents which turn out to be bland, run of the mill reports with no smoking guns.

Tomorrow we start trying to convince Congress we must have more money for El Salvador. We have an entire plan for bolstering the govt. forces. This is one we must win.

Took time off to run slides & sound tape on "Shroud of Turin." I'm convinced it is the burial cloth of Jesus and it certainly gives credence to the bodily ascension.

Sunday, February 27

St. Dinner tonite for Nat. Gov's. Conf. Watched them this afternoon on TV. A half dozen Dem gov's. kept sounding off on how our programs were unfair and favored the rich—paralyzed the poor. Their ans? cut defense spending. Our Bud. for H.H.S. is the biggest & highest % of Bud. in history. It is 36.7% of the bud. Defense is only 26.7%. Food stamps are budgeted at $4 bil. more than in 1980 and 3 mil. more people are getting them. As for taxes favoring the rich—they are 25% across the board and indexing the tax brackets does nothing for the rich—they are already in the top tax bracket. I'm host tonight—I'll try not to get in an argument. Pearl Bailey entertains.

[*Monday, February 28: flew to California, met by Mrs. Reagan.* **Tuesday, March 1:** *Queen Elizabeth and Prince Philip visited, Mrs. Reagan left with them on the royal yacht, headed for San Francisco.* **Wednesday, March 2:** *ranch work.* **Thursday, March 3:** *in L.A. for address to Olympic fund-raisers; flew to San Francisco.* **Friday, March 4:** *meeting with Shultz, Clark, and Weinberger regarding El Salvador and Middle East, commented, "We're stepping up the action"; enjoyable formal dinner on royal yacht.* **Saturday, March 5:** *flew to Oregon for speech to lumber-business group; met with Republican leaders; returned to W.H.* **Saturday, March 6:** *noted speech to San Francisco Commonwealth Club on previous Friday, with many old colleagues present; report that three Secret Service agents were killed in car accident en route to protecting royal entourage; phoned widows.*]

Monday, March 7

Wash. Post carried leaks on meetings we held in S.F. last Friday. Met with delegation of Mayors—didn't complain too much but of course want more money. Young high school science winners came in—amazing but majority were Japanese Am. Another meeting was with leaders of groups & org's. joined in a cause called "Peace Through Strength"—they are (I'm happy to say) a counter lobby to the Nuclear Freeze people.

Late afternoon met with Repub. Sens. & Reps.—only a few of leadership, & mainly armed services committees. The closest we've been to a real split in the ranks. The issue—the defense budget. They want still more cuts. We've done our own cutting and are to rock bottom. They say our Repub. Sens. won't support present budget. I'm going to take our case to the people only this time we are declassifying some of our reports on the Soviets and can tell the people a few frightening facts. The d--n media has propagandized our people against our defense plans more than the Russians have. We are still dangerously behind the Soviets & getting farther behind.

Tuesday, March 8

In office early 8:30 for staff & N.S.C. daily meetings then a session with Bi-partisan Congressional leaders before leaving for speeches in Orlando Fla. Naturally, there is division on subject of aid to El Salvador—still I think we gained some points and cleared the air that we aren't just seeking mil. aid.

[Flew to Orlando, Florida, for meetings with high school students at Disney World and with Evangelical clergy; political fund-raiser.]

Back to Wash. where it was raining cats and dogs & lots of lightning which doesn't reassure you on the helicopter ride to the W.H.

Lyn N. has evidenced a desire to come back to us. He wants to report directly to me—this, of course, is upsetting to Jim B. I'd like Lyn back but it's a touchy thing to work out.

Wednesday, March 9

Meeting with Repub. Cong. leadership. Discussed strategy for all the major items on the legislative plate—El Salvador, Soc. security, withholding of inc. tax on Int. & dividends, our job bill. They seemed a little more upbeat than they've been. An N.S.C. meeting on El Salvador. It looks like we'll have to redistribute funds from other countries to increase El S's grant. Other actions such as providing AWAC's info to them on ammo supply flights from Nicaragua to the rebels are proceeding as scheduled.

Met with exec. comm. of Nat. Repub. Federated Women. A good but brief time. These ladies are so supportive.

Ended afternoon by calling Mr. and Mrs. Carl Rossow in Conn. This couple had adopted 12 children—all as deformed or handicapped babies—the kind who usually are victims of infanticide. I talked to them & to all the children. You could feel the love & joy that is in that household. If ever one needed proof that God has a purpose for each one of us, let them meet the Rossows.

Thursday, March 10

A really hectic day with a lot of courtesy and ceremonial but I suppose necessary chores. High spot was a speech to the N.A.M. I suppose I fooled them somewhat by talking on El Salvador & why it was necessary that we help.

The Soc. Security bill passed both houses with an amendment by Jake Pickle to increase age to 67 in year 2000. Today was the aftermath of an E.P.A. 3rd act curtain last evening. Anne Burford resigned on her own so the agency could get back to work. This whole business has been a lynching by headline hunting Congressmen. I can't wait to get a question from the Press so I can say that.

[Met a kidney transplant recipient and his father (the donor); met with publishers of small newspapers.]

Friday, March 11

Brkfst. with our Soph. class (my class) of Congressmen. They are a fine young group who've been very supportive. I needed them this morning after reading Lou Cannon's story in the Washington Post. It was a vicious series of falsehoods and I was mad as h--l. An N.S.C. meeting—Phil Habib is back. F.M. Shamir (Israel) is coming as is the F.M. of Lebanon. We have hopes that maybe Israel is softening on the withdrawal—we'll know by Monday.

[Reception for National State Legislators conference; press availability.]

I'm waiting for Nancy—coming in from California and we'll be off to Camp David. And here she is.

*[*Saturday, March 12: *recorded strong wind, answered mail.* Sunday, March 13: *phone calls and mail, returned to W.H.; report from Shultz regarding meetings with two foreign ministers, Israel's Yitzhak Shamir and Lebanon's Elie Salem; watched* 60 *Minutes.]*

Monday, March 14

[Memorial service for federal law-enforcement agents killed in the line of duty.]

Meeting with Shamir (F.M. of Israel) and his party. I put in my pitch as best I could. George S. passed me a note telling me I'd been terrific.

Finished afternoon meeting 150 local and state government people in the East Room. Did Q&A. They seemed very friendly. There were other meetings which kept me busy. Did some phone calling to Senators on the jobs bill. Senator Kasten has hung an amendment on having to do with repealing withholding tax on interest & dividends. I'm trying to get it off—let it stand by itself—so we can get on with the jobs bill.

Tuesday, March 15

Met with Pete Domenici & Howard Baker. Howard said he'd failed to get Pete to delay budget markup in committee until after Easter. We want to delay to see if we can't find a legitimate defense savings in MX plan when commission reports. Otherwise they'll go ahead now & slash the budget in a harmful way. I made my pitch that had a possibility of an accepted cut if he'd give us time. He then asked if he could bring the ranking Dem. Senator on the committee in to see me. Howard left. Our Dem. arrived. I made my pitch. Before day was out we'd won—they've postponed.

[Visit from Prime Minister Rudolphus Lubbers of the Netherlands regarding Interme-diate-Range Nuclear Forces Treaty (INF); met with state attorneys general on administra-tion crime package; with congressmen on Zabloski nuclear freeze bill, promoted substitute bill; saw Jerry Falwell—"he is a good friend & highly supportive"; reception for Country Music Association; noted perceived negative reporting on CBS, saying "CBS evening news an almost total attack on our admin. They are beginning to look like a deliberate cam-paign."]

Wednesday, March 16

Meeting with our team, Phil Habib et al, and F.M. Salem & former P.M. and now elder statesman Salem of Lebanon. I assured them we'd stay in there helping to get foreign forces out of Lebanon. The old gentleman told me I was much loved in Lebanon. He said previous experience with Am. Presidents was that they advanced so far and then retreated. I told him I didn't have any reverse gear.

Met with Repub. Congressmen regarding Sen. Roth's proposal for a new Cabinet dept. of Trade. This is a sticky one. Dropped in on a meeting with sev-eral dept. heads from E.P.A. We're trying to boost their morale.

Took the afternoon off & rode in Rock Creek Park. First time on horseback in a long time—not since last Fall.

Tonite Constitution Hall for the Country Music Assn. anniversary. A great show & a wonderful warm evening. Minnie Pearl. Roy Acuff, Tennessee Ernie Ford, Gene Autry—well on and on. High spot—a piano & signing duet—Ray Charles and a young (also blind) artist who said Ray had been his inspiration. They drew a standing ovation that came from the heart.

Thursday, March 17

St. Patrick's Day. A shamrock tie from Margaret Heckler & one from Tip O'Neil. Lunch on Capitol Hill as Tips guest. About 30 people including F.M. Barry and Ambass. O'Sullivan of Ireland. Tip is a true pol. He can really like you personally & be a friend while politically trying to beat your head in. After lunch—to the Irish Embassy for a reception. I'm talking with a brogue already. From all that to a meeting with Al Seigle and Max Fisher on Israel problems and our own Jewish community. Max is a regular visitor to the Middle East. I've ar-ranged for him to report to me on return from those trips.

[Ambassadorial formalities and photo sessions.]

Friday, March 18

Interview with Henry Brandon—his swan song with the London Times.

Met with House Repubs. re the Dem. budget that came out of comm. on a party line vote. Our guys were all gung ho & ready to do battle. The Dems. must know they can't get their budget but Tip said it was a moral statement of Dem. beliefs. I think they've handed us an issue. They would reduce or repeal everything we have done. They'd cut defense more than $100 bil. over 5 yrs. but they'd increase social spending $181 bil. and increase taxes $315 bil. On Mon. Commerce will announce G.N.P. growth for 1st Quarter '83 as 4%—highest since 1st Q. of 81. Our program is working even better than we expected 2 months ago.

An N.S.P.G. meeting on I.N.F. and Lebanon. I'm of the mind we should tell Nitze to offer an interim missile reduction plan to the Soviets while still claiming zero is our ultimate goal. On Lebanon it's still Israel dragging their feet.

[Met with small-businessmen's group; signed annual report; made plan to recruit William Ruckelshaus, senior vice president of Weyerhaeuser, as head of EPA.]

Saturday, March 19–Sunday, March 20

Camp David—rain until noon today (Sunday) then it became beautiful just as we had to leave. Ron came down for a talk at the W.H. Same old problem—his itchiness at having S.S. protection. Wants to sign off for a week trip to L.A. I wish he'd be more thoughtful of what it means to me to have him do that.

Monday, March 21

[Noted constantly changing weather; staff and NSC meetings, recorded opinion that Moshe Ahrens as Defense Minister of Israel was an improvement over Ariel Sharon; radio taping for Afghanistan Day; speeches at Department of Agriculture.]

They had a Hereford calf in a small pen in the room by the dais. It was the 4H. project of 2 small boys. About ⅔ of the way through my talk the 2 boys entered the pen. Suddenly I was talking to an audience of profiles. Everyone was watching the boys & their calf.

I couldn't resist—I said—"You know I once learned you should never do a scene with kids or an animal." It got a laugh—and the audience back. Back to the W.H.

The usual issues lunch—then a meeting with some top business people hosted by Charles Wick. Propose to enlist the pvt. sector in our campaign to sell democracy worldwide. Later Charles told me he had contributions of $3 mil. to help in the effort.

Phyllis Schafley came by. She's darned effective. Her plan to counter the new E.R.A. campaign in Cong. is brilliant. The women's groups backing E.R.A. are really only trying again as a fundraising Dem. party gimmick. They will also use it to embarrass Repub. candidates in '84. Her idea is to say that if Cong. is going to put that Const. Amendment on the floor, we want other amendments to also be given a chance—abortion, bussing, prayer in schools etc. I think it would give Tip et al fits.

[Shultz concerned about insert (unidentified in diary) planned for upcoming speech on defense; decided to rewrite section.]

Then 2 meetings with different Repub. Congress. groups. Same subject for both—the ridiculous Dem. bud. passed out of committee. It cancels most of the savings we've made in the last 2 yrs.—raising taxes, cuts defense spending, increases domestic spending by bils. of dollars and ends up adding $8 bil. to the deficit. They've called it a return to Dem. tradition. And it is—all the things that messed up our economy to begin with. And today the Dept. of Commerce released the fig. on growth of G.N.P. for the 1st Q. of '83—up 4%. We ourselves had only dared predict 1½ %.

[Reception for the National Republican Congressional Leadership Council, contributors group.]

Called Chuck Percy about holding off a vote tomorrow in his committee re the El Salvador funding we want. He's going to stall because the appropriations comm. is voting on it and we think we'll get it. His Foreign relations comm. could louse that up if the voted no first.

Tuesday, March 22

Another day that shouldn't happen. On my desk was a draft of the speech on defense to be delivered tomorrow night on TV. This was one hassled over by NSC, State & Defense. Finally I have a crack at it. I did a lot of re-writing. Much of it was to change bureaucratic into people talk. But all day there were meetings—with Congress, with our volunteer leaders from the business world, unscheduled meetings having to do with problems & finally a trip to the Capital Club to address a few hundred Repub. donors, Congressmen etc. Finally back to the W.H. to catch up with all the memos & reports that had piled up on my desk since early morning.

During the day speaking to our Cong. Repub. leadership & I blasted the Dem. budget with the press in attendance. It was a good pitch exposing the ridiculous irresponsibility of the phony budget—but on the evening news they showed a quick 20 seconds in which I wasn't saying anything about the tax increase, the increase in spending etc.

Wednesday, March 23

The big thing today was the 8 P.M. TV speech on all networks about the Nat. Security. We've been working on the speech for about 72 hours & right down to deadline. We had a group in for dinner at the W.H. I didn't join them except before dinner a few words of welcome. Nancy & I then dined early upstairs. The group included several former Secs. of State, Nat. Security Advisors, distinguished Nuclear scientists, the Chiefs of Staff etc. I did the speech from the Oval office at 8 & then joined the party for coffee. I guess it was O.K.—they all praised it to the sky & seemed to think it would be a source of debate for some time to come. I did the bulk of the speech on why our arms build up was necessary & then finished with a call to the Science community to join me in research starting now to develop a defensive weapon that would render nuclear missiles obsolete. I made no optimistic forecasts—said it might take 20 yrs. or more but we had to do it. I felt good.

Thursday, March 24

[Called widow of Barney Clark, recipient of first artificial heart.]

Our women representatives (Repub.) in the House came over for a meeting on womens legislation. We (our admin.) have already done more to correct inequities than any other admin. before us. A couple of the gals are pretty aggressive sounding.

[Messages to W.H. indicate positive response to speech; heads of Hispanic organization pledge support for '84 campaign; met young boy representing Easter Seals.]

This evening we went out to the Library of Cong., for the Cowboy Exhibit. It was interesting & we enjoyed it. Then we went to the Hilton not to stay for the dinner of the TV & Radio Correspondents, but to make remarks—supposed to be funny. Finally home.

[Friday, March 25: NSC briefing; met with speechwriters regarding Saturday radiocast; new poll numbers indicate improvement in job approval rating on economy, but reduction in regard to foreign policy, blamed dip on "the drumbeat of anti-defense propaganda"; discussed upcoming Williamsburg summit; lunch with Vice President Bush, Shultz, and New York Times publisher Arthur Ochs "Punch" Sulzberger; met with high school students; visit from Governor Lamar Alexander (R-TN) pledging support in '84; met with Young Republicans; hour-long meeting with Shultz—"just the 2 of us to talk about our quiet diplomacy efforts with Dobrynin"; expressed hope of freeing Pentecostals from U.S. Embassy in Moscow.]

Saturday, March 26

Radio broadcast re the Dem. phony budget. Early afternoon went over to the Hilton to rehearse for the Gridiron banquet. They had an idea that after Nancy's big surprise smash last year I should be a surprise guest this year. No Pres. has ever done this before. And for me it's another first. They have me singing a parody as Nancy did. It's to the tune of "Mañana." My 1st time ever singing a song on a stage. I was nervous. But it went off O.K.—White tie & tales under a serape & a Mexican sombrero. Of course I then had to go back to the head table & wind up with some funnies in the grid iron tradition. Much of the day was spent in culling oneliners.

[Sunday, March 27: slept late; watched the panel shows; edited upcoming speech; answered mail.]

Monday, March 28

Usual briefings—only specific of concern was a setback in Lebanon. Israel is demanding Haddad be commander in So. Lebanon—Lebanon says no—Lebanon is right.

The Murdocks came by—our former Calif. finance chairman, his wife Gabrielle & son Justin.

In the afternoon, met with Nat. coalition trying to do something about pornography. I'm with them.

[Mrs. Reagan narrates a philharmonic performance. **Tuesday, March 29:** *interview with panel of journalists; Mrs. Reagan in Phoenix.]*

Wednesday, March 30

Pres. Kaunda of Zambia arrived. A good meeting & lunch. I think he feels good about the trip. We made clear we detest Apartheid but believe we can do better with S. Africa by persuasion.—Left for L.A. Met Nancy at Century Plaza late afternoon. Maureen came by re Mike. She's really being a trooper & solid sister about the problem.

Thursday, March 31

Met with Doug Morrow re some ideas he has about taxes. Then on to the World Affairs Council for a speech & Q&A. Speech went well—it was on entire matter of disarmament. In Q&A, one Q. led to my response we could not under the law deliver 75 F16s to Israel so long as they remained in Lebanon. This became the feature story. Maybe it was a necessary signal to Israel. Afternoon went to Ranch. It's beautiful—windy but sunny & warm.

[Friday, April 1: went riding; Ron, Doria, and Patti arrived. **Saturday, April 2:** *ranch work.* **Sunday, April 3:** *church services for Easter; returned to W.H.]*

Monday, April 4

Big thing of this day was the traditional Easter Egg roll. Nancy & I went out on the lawn at 11 A.M.—there were more than 30,000 people—mainly children. Then I had to go back in the Cabinet room for a meeting with the Joint Chiefs of Staff. They are all on board for the Commission report on M-X. There were meetings all day on the upcoming Summit in Williamsburg, defense budget etc. A very top meeting in the situation room had to do with Surinam and possible plans to oust the dictator who seized power & murdered about 15 top civic leaders. Some of the plans I had to veto. We're going to see if Venezuela & Brazil might help.

[Ambassadorial formalities.]

Tuesday, April 5

[Visit from Muscular Dystrophy poster boy and comedian/fund-raiser Jerry Lewis; economic briefing, troubled by the difficulty of gauging money supply due to bank deregulation; visits from Eisenhower Fellows; then the Rossows, adoptive parents of twelve handicapped children.]

I had an unsatisfactory meeting with Repub. Sens. on the budget committee. They are determined we must cut defense spending & increase domestic in order to get a budget passed. I'm opposed. We have rabbits where we need Tigers.

Joe Coors came by for a drink & some talk. He's a fine man.

Wednesday, April 6

Learned in office George S. is upset—thinks NSC is undercutting him on plans he & I discussed for "quiet diplomacy" approach to the Soviets. They have let Lydia—the young hunger striker member of the family that's been living in the embassy basement in Moscow for 4 yrs. go. She is in Vienna as of today. We had a meeting later in the day with George & cleared things up I think. Some of the N.S.C. staff are too hard line & don't think any approach should be made to the Soviets. I think I'm hard-line & will never appease but I do want to try & let them see there is a better world if they'll show by deed they want to get along with the free world.

[Flew to Pittsburgh, visited job-training school; spoke to conference of business leaders and local officials from all over the country; returned to Washington for entry in PBS's Young Artists program.]

Thursday, April 7

A day that never should have happened. It was a displaced Monday. A 9:30 breakfast with the 16 Sens. who came into office when I did. A pleasant enough affair. Then an NSC briefing re Surinam. A solid plan is offered if Venezuela & Brazil will provide combat forces—only a few hundred needed. We provide Naval & Air support. Bill C. is off tomorrow for S.A. to personally present plan to both countries.

Met briefly with St. & Local officials, did a Q&A on Federalism plan. Lunch with V.P. A strain has developed between George S. & Bill C. I've invited the Shultz's to Camp D. for the weekend. I don't think the problem is serious.

[Brief appearance at National Catholic Education Conference, reiterated support for tuition tax credits; Cabinet Council meeting on coal slurry pipelines and clean-water poli-

cies; visits from leader of the Antiochian Orthodox Church, Senator Sam Nunn (D-GA) with honored disabled veteran; Olympic representatives; appearance at National Eye Care Project meeting; ambassadorial formalities.]

Interrupted to make a call to Sen. Pete Domenici. Budget committee is debating defense budget. I asked them to hold off voting til Monday so we could discuss a reasonable proposal. He refused & I got mad. These supposed to be Republicans went ahead and cut in half the increase we'd asked for. The Russians must be very happy tonite.

This evening Ron called all exercised because S.S. agents had gone into their apartment while they were in Calif. to fix an alarm on one of the windows. I tried to reason with him that this was a perfectly O.K. thing for them to do. [. . .] I told him quite firmly not to talk to me that way & he hung up on me. End of a not perfect day.

Friday, April 8–Sunday, April 10

[President Osvaldo Hurtado Larrea of Ecuador visited; reception for Eagles Republican loyalists; reception for thirty-one female POWs from WWII.]

Then off to Camp David. We had the Shultz's with us as guests & had to drive because of the fog & rain. It stayed foggy & rainy until Sun. Still we had an enjoyable time. They are nice to be with. George & I had hours of discussion of all our international problems. On Sunday we got word that the P.L.O. had rejected Arafat's agreement with King Hussein & offered a counter proposal which must have been written in Moscow. Hussein has refused it. I phoned him. Word had it that King Fahd (Saudi Arabia) & Hassan of Morocco favored the new plan of the radicals. I called Fahd—he's on our side now that he's seen the new draft. I'm waiting now to talk to Hassan. We're back at the W.H. & the weather is now sunny & bright.

Monday, April 11

N.S.C. meeting—Bill C. back from quick secret trip to Venezuela & Brazil. We had a plan which required their cooperation to take Surinam back into the family of Am. States before it becomes a Cuban patsy. Venezuela couldn't go along. Pres. of Brazil had an idea somewhat different than ours but I believe different. So operation "Guiminish" is born. We'll know before the month is out whether it has succeeded.

[Received MX plan from Commission on Strategic Powers, deemed it "a good plan"; meeting with National Federation of Independent Unions, received support; report from secretary of Agriculture on trip to North Africa and Middle East; photo sessions; Republican "Inner Circle" fund-raiser.]

Then on to the Holocaust meeting—some 16,000 people—they had to turn 10,000 more away—no room.

Practically all these people—all but their children, are survivors of the Nazi death camps. It was an emotional experience for them & for us. I know I choked up a couple of times while I was addressing them. Back at the W.H. & have a call in to Mubarak who is in Pakistan—it's about the Arafat-Hussein affair.

Tuesday, April 12

His Majesty King Qabbos [Qaboos bin Said, Sultan of Oman] arrived— mil. reception on the S. Lawn. State Dinner tonite. He gave some good advice about the P.L.O. statements. He says let them sweat for awhile—we've made them feel more important than they are. So we'll stop chasing them.

[Met with Democratic congressmen regarding the nuclear freeze resolution; visit from California investment banker Albert E. Schwabacher.]

A Cabinet briefing about the move to lift a rule & let networks buy & own syndication rights to shows. I'm afraid it will be a monopoly to the detriment of the independent producers.

Met with about 20 bankers who are against our withholding tax on interest. Don Regan carried most of the argument on our side. No agreement reached—it's a testy issue.

Today the Pentecostals left the Am. Embassy basement in Moscow where they've lived in the basement for 4 yrs. They left at our request. We think—well more than that we're sure we have a deal that they will be allowed to emigrate.

[State dinner successful; operatic performances by Robert Merrill and Anna Moffo.]

Wednesday, April 13

Another group of bankers re withholding. These were on our side. As one would expect—their statements to the press did not appear on the TV evening news.

Called Colleen who has given us another grandchild—Ashley Marie R. Lunch in the East room to present awards to 20 winners of private initiative volunteer efforts across the country.

Out to Rock Creek park to ride Giminish for ½ hour in the indoor riding hall. Back at the W.H. Ed Meese came by to give me an update on our W.H. problem of leaks & in house rivalry in the staff. I'm afraid it has reached a point where the axe must fall.

Nancy is in Pittsburgh filming another program re drugs.

Thursday, April 14

Signed our income tax forms. The Pres. should do something about taxes. The Defense Minister of S. Korea (Yoon) came by—brief meeting & picture. NSC meeting on the MX commission report. I'll announce my approval Tue. A work session on preparing for the Summit—subject trade. Then another group in on the Int. Withholding tax. About 50 top business & finance people—almost all very supportive of our position. Somehow they never seem to appear on the TV news. The bankers who were opposed did.

[Cabinet Council meeting on economic affairs, decision in favor of extension of revenue sharing for local governments, discussion of health care for the unemployed; meeting with OAS ambassadors in recognition of Simon Bolivar's birthday; announced a scholarship program for Caribbean youth; press availability.]

Meeting with George S, Bill Clark et al re Israel. I think the time has come to approach P.M. Begin with a palm leaf. We're going to offer some helpful measure through Defense Minister Ahrens.

Friday, April 15

Tax day & our returns are on all the front pages with emphasis on our supposed wealth—as usual the stories are slanted.

Chancellor Helmut Kohl—W. Germany arrived. Good meetings. Ended with lunch in the St. Dining room. We have established a fine relationship. I asked him to establish an embassy in El Salvador. They have one in Nicaragua. He's waiting for Duarte to visit Germany. Duarte will be a cand. in coming election. Kohl favors him & will give him the embassy to take home to El S.

[Met with muscle-builders; left for Camp David in the rain. **Saturday, April 16– Sunday, April 17:** *deskwork.]*

Monday, April 18

Awakened with word a car bomb did great damage to our embassy in Beirut—killed scores of people including 5 of our Marine guard detail. First word is that Iranian Shiites did it—d--n them.

A Budget briefing by Dave S. If the Dems. have their way the recovery will be over before it starts. They must give us the spending cuts we want or we face a trillion dollar deficit over the next 5 yrs. In the Rose garden I presented peace corps awards to 6 outstanding volunteers. One a nun from Ghana whispered to me she needed flour for her country. Before day ended, Dick Meyer of Seabord Corp. agreed to send 3000 lbs. of flour to her.

Cabinet meeting was on budget—it's a grim picture.

[Observance of National Crime Victim Week; live telecast to North Carolina state champion basketball team; went to Kennedy Center for opening night of Frank Sinatra show.]

Just before show started—got a call from Jim Baker. He & Ed were with Paul Laxalt & Bob Dole. We don't have the Sen. votes to keep them from rolling over us on the withholding of tax on Int. & dividends. In fact we can't keep them from over riding my veto. They've come up with a substitute as a possible compromise & needed my OK to try to put it together. Keep withholding of dividends & postpone Int. withholding on Int. for 2 yrs. It will get us about 80% of what we want in revenue. I bit hard & said O.K.

Barry G., Howard Baker & John Tower were our guests in the box so at intermission I cornered them & told them—they said it was the thing to do—so I'll sleep better tonight.

Tuesday, April 19

Cong. leadership re the Commission Strategic Force report. One present— Sen. Alan Cranston is a big bleeding heart for nuclear freeze. He said nothing to us in the meeting but tore us apart outside on the lawn to the press. He only favors Soviet arms, not U.S.

Then went over to the East Room to formally announce my approval of the commission report.

An unpleasant briefing—N.S.C. on the Beirut bombing. Lord forgive me for the hatred I feel for the humans who can do such a cruel but cowardly deed.

[Report from American Medical Association on efforts to provide medical care for the uninsured and unemployed; reception for Building Trade Unions et al.; private dinner for HRH Alexandra of Britain and her husband, Angus Ogilvy, entertainment by songwriter Cy Coleman.]

Wednesday, April 20

A cold grey day. NSC briefing on Beirut. We lost [. . .] our top research man on Middle East. The bodies will return Sat. I called our Ambassador—he's a fine man.

Outdoors on S. Lawn, a big ceremonial signing of the Soc. Security bill. Cong. committees, Cabinet, Task force & about 1000 people—all freezing to death. Then in to meet with the Ec. Advisory Board. All seemed certain recovery is happening but all agreed if we don't stay on course it will be aborted.

[Ceremony honoring volunteerism. Lunch with regional campaign leaders; president's former radio producer, Harry O'Connor, visited, awaiting confirmation for Public Broadcasting board; disaster proclamation for Louisiana flood damage; visited volunteers in W.H. mail room.]

Thursday, April 21

N.S.C. Briefing—Brazil is holding Lybian transport planes that were headed for Nicaragua. Their cargo—medical supplies turned out to be weapons.

[Visit from President Gaston Thorn of European Commission; cabinet meeting, approved a phase-out of farm-marketing orders, discussed health insurance for unemployed, "Interesting thing is the way the medical profession has already moved to provide free care for these people"; gathering for local television stations that staged job-a-thons; video tapings for various conventions.]

Friday, April 22

N.S.C.—mainly on Beirut. Our honored dead will be arriving at Andrews A.F. Base tomorrow evening.

Met with speechwriters on T.V. speech re Nicaragua & El Salvador. An N.S.P.G. meeting on Middle East & George Shultz's trip. I approved it. It's time we tried to get our relationship back on track. George feels the same. Phil Habib has made great progress regarding Israeli's withdrawal but there are still some sticking points.

[Press availability; swearing-in ceremony for Ken Adelman as director of the U.S. Arms Control and Disarmament Agency; Cabinet Council meeting on trade problems with Japan, intention to negotiate agreement on machine tools; reception for Small Business Advisory Council.]

Saturday, April 23

Merv Griffin came by for lunch. We left at 6:15 for Andrews A.F. Base—the ceremony for the 16 who died in Beirut. It was a moving experience. Nancy & I met individually with the families of the deceased. We were both in tears—I know all I could do was grip their hands—I was too choked up to speak. Then home to change clothes & off to the W.H. Correspondents dinner. I was supposed to do a routine of jokes etc. a la the Gridiron. I couldn't change gears that swiftly. So as not to put a damper on the evening & comedian Mark Russell, I waited til the last & then asked their pardon for not "singing for my supper" because of our sad journey to Andrews A.F. Base. I asked for a rain check & said I'd save my remarks for then.

[Sunday, April 24: worked on speech planned for Wednesday; visit from Secretary and Mrs. Shultz, on their way to Egypt.]

Monday, April 25

An easy day except for getting up—the 1st day of daylight savings. I've told Ed & Mike they & Jim B. & Bill C. should get together & lay it out on the table. A real split has developed & the press is making much of it due to leaks probably by subordinates in behalf of their bosses.

[Department of Energy ceremony; met with Jim Coyne, special assistant and director of the Office of Private Sector Initiatives, on publicizing private programs that are "doing things Congress thinks only govt. can do"; interview with German journalist; visit from actress and old Hollywood friend Joy Hodges-Scheiss with her husband.]

Tuesday, April 26

Sec. Gen. Luns of NATO came by. He believes strongly in our disarmament proposals & in the need for us to deploy intermediate missiles in Europe.

I was scrambling all day to get the speech on Central Am. in shape for Weds. nite. State Dept., N.S.C., Defense & C.I.A. all are putting an oar in. I had 3 scripts on my desk at one time. I'd already turned in my own version.

Met with selected members of the House Intelligence Committee. Cong. has eroded away much of the Const. authority of the Presidency in foreign affairs matters. They can't & don't have the information the Pres. has & they are really lousing things up.

[Met with Governor John Sununu (R-NH) regarding his suspicion of Comprehensive Employment and Training Act (CETA) bureaucrats; report from National Commission on Excellence in Education contending that federal interference has undermined educational system; interview with USA Today.*]*

Early dinner for about 40 people honoring Lawrence Olivier—then we saw his picture, "King Lear." It was a tour de force for him.

Bill & Pat Buckley stayed the night. It was good to see them.

Wednesday, April 27

[Flew to New York with the Buckleys; worked on the plane; visited police precinct to meet private citizens who have stopped crime; addressed newspaper publishers association; returned to W.H.]

8 P.M.—addressed Joint Session of Cong. & gave speech we've all been working on. Got 3 standing ovations with some Demos. on 2, & all of them on the 3rd. That was on the line that we had no intention of sending troops to Central America. I think we scored well with the TV audience.

Thursday, April 28

Nancy left for Phoenix—we'll be apart until Monday. I don't like this but I know it can't be helped. The wires, phone calls & mailgrams are running about 3 to 1 in favor on the speech.

Today named Dick Stone, former Dem. Senator as personal envoy to Central America. Sen. Dodd & other far out liberals & left wingers are all over the tube screaming foul. Dodd calls me ignorant. His claim to expertise on Central Am. is 2 yrs. as a peace corps vol. many yrs. ago in Dominica.

P.M. Trudeau of Canada here for meetings & lunch. It turned out alright & he's pleased with our plans for the Summit.

[Phoned twelve-year-old former Cambodian refugee on winning a spelling bee; two medal ceremonies.]

Patti called—she's been looking for work quite awhile—needed to borrow some money.

*[*Friday, April 29: *flew to Houston; visited Cenikor Foundation, rehabilitation center; two party fund-raisers; minor motorcade accident.* Saturday, April 30: *met University of Houston basketball team; returned to W.H.]*

Sunday, May 1

Went to church. We kept it a secret until the very last minute. It felt good. A little girl gave me a small safety pin on which small beads had been threaded. I learned it's all the thing with kids today—a friendship pin. On way home stopped & put flowers at the Vietnam memorial. It's quite a place—a very impressive & moving experience.

Nancy phoned—very upset. Ron casually told the S.S. he was going to Paris in a few days. I don't know what it is with him. He refuses to cooperate with them. [. . .] I'm not talking to him until he apologizes for hanging up on me.

*[***Monday, May 2:*** *presented with new Bible edition containing his frontispiece declaring 1983 the Year of the Bible; addressed various local leaders from Mississippi and Ohio; preparation meeting for upcoming summit; met with International National Private Enterprise Task Force; interviewed by* Family Weekly; *received letters of support from congressmen regarding tax cuts; taped appearances.]*

Tuesday, May 3

Opened day with Repub. Cong. leaders—subjects the budget & the MX vote plus the interference with Presidential authority in international affairs. Two House committees are putting together measures to limit our help to El Salvador & the freedom fighters in Nicaragua. My case on the budget was that the House Dem. budget was so far out of line, there was no way I could accept any split or compromise with the Sen. version. Howard Baker came up with the idea of passing the Sen. version—letting the Dems. in the House pass theirs & since the budget resolution is a non binding thing not a bill we'd have 2 and I would only sign appropriation bills conforming with the Sen. resolution. It would be quite a campaign plus—the Dem. & Repub. versions of govt. spending & the contrast.

Cable traffic from Beirut was ring of optimism. George S. apparently believes they are close to an agreement on pull out. Foreign Minister Lee Bum Suk—S. Korea came by for a short visit. He's a personable & very intelligent man & completely sincere about S. Korea friendship with us.

Had a good Cab. Council meeting on Drugs & our help to our historic Black Colleges & U's. On drugs we really have a great program going to steer kids away from them. There is real progress & Nancy is a big part of that.

Presented the Am. Eagle award to Marvin Stone—editor of U.S. News & World Report—their 50th anniversary.

Tonite the annual reception for the Diplomatic Reception.

Wednesday, May 4

Paul Laxalt came by to talk a little politics. He's being pushed by friends who want to start fundraising etc. for '84. We agreed it was not a good idea.

NSC briefing not as optimistic as the last few days re the middle east. Syria is poisoning the well & the possibility of an Israeli-Syrian (plus Soviet) confrontation cannot be ruled out. Armageddon in the prophecies begins with the gates of Damascus being assailed.

[Press ceremony with U.S. World Cup soccer team.]

Congressman Carol Campbell brought in a young lady who is being honored as the handicapped person of the year. Mrs. Patti Just Long. Polio at age 4 & now in a wheel chair, she is one of the prettiest, most vivacious likeable persons I've ever met. Harold Russell was with her—he's the soldier who lost both arms in W.W.II & won an Oscar in "Best Yrs. of our Lives."

Was interviewed by 6 press members on behalf of entire press corps. Think I got through it unbloodied—I'll know when the evening news comes on.

Dick Wirthlin with poll taken after April 27 speech. It did swing some people our way on Central America.—but it was astonishing how few people even know where El Salvador & Nicaragua are.

Spent balance of afternoon calling Sens. on Intelligence Committee re El Salvador & our aid to the freedom fighters in Nicaragua.

[Met with early campaign supporters in Congress.]

Trip—Thursday, May 5 through Monday, May 9

A meeting with members of Cong. re the MX—this one was preaching to the choir. Then time for a photo with Mrs. Clark, the widow of the Dr. who died after months of living with an artificial heart. She's a remarkable & likeable lady.

Off to San Antonio for the Cinco de Mayo celebration. A big outdoor event, a largely Hispanic audience & I believe a fruitful day. Then back on A.F.1 & on to Phoenix. Checked in at Ariz. Biltmore & over to Nancy's mothers apt. for dinner. She was very much with it & we had a good visit. The next day I addressed the Sun City Posse & was made an honorary member. This is a group of vol. senior citizens who patrol that Sr. Citizens community & have markedly reduced crime there. A bite of lunch at the hotel & off to address the nat. convention of the Nat. Rifle Assn. A crowd of thousands—a 20 min. speech ran 35 mins. because of applause including a standing ovation. Back to Deedee's—Nancy had arrived & we left for Pt. Mugu & Marine 1 & on to the ranch. Weather beautiful. Sat. like old times except that we rode after lunch because of the timing on some phone calls I had to make in the A.M.

[Ranch work. **Sunday, May 8:** *ranch work; lunch with Mike Deaver.* **Monday, May 9:** *flew to Ashland, Ohio, for speech, commenting, "It's wonderful to get out in the country & meet the people of this land"; returned to W.H.]*

Tuesday, May 10

Back to the routine. NSC meeting to settle some matters about the START talks. We're going to emphasize "warheads" as a negotiation base not launchers or missiles. Over to C. of C. for a closed circuit T.V. program including Mac Baldridge in Paris & meetings of small businessmen + women in 42 locations. Everything went well.

The problem of the Big 4—the press stories that a feud exists between them (partly true) is intolerable. Tomorrow I'm telling them they must get together, lay everything out on the table & heal their wounds. They have to restore trust in each other. The big split is between Jim B. on one side & Ed M. & B.C. on the other. Lately the leaks have been aimed at Bill & I think they come from the 2nd echelon in Jim's department.

Later in day I called the young Sec. of Cong. Don Young, Fla. He had told me of her dying mother—to whom I'd written. She died 4 weeks later. Beverly (his Sec.) wrote me the nicest letter. I felt I had to call her—glad I did.

Tonite dinner W.H. with the 19 R. Sens. who are up for election in '84. It was a pleasant evening even if a few of them have strayed from the party path.

Wednesday, May 11

[Busy schedule; gave small-business award to owners of Mexican food business.]

Cardinal Krol came in. He will be with the Pope in Poland. We have made surplus food available at concession prices to the Catholic relief for the people of Poland. They have shipped $87 mil. worth this year. The Cardinal will now be on our Pvt. Initiative Council.

[Met Ray Cave, new managing editor of Time *magazine; lunch with John Naisbitt, author of* Megatrends; *meetings with senators on MX bill; signed proclamation for amateur baseball month, watched game on South Lawn, "I also signed a jillion baseballs."]*

Late afternoon George Shultz came in—just back from the Middle East. He did a fantastic job & is not too discouraged about Syria's Soviet backed troublemaking.

At last meeting with several D. & R. Congressmen on MX & at last Home & Mother.

Thursday, May 12

Peter Grace, Chmn. of our private survey team (1500 Vol. Business men & w.) came in with a 1st report. I wish he'd been delivering it to those big spenders on the hill. It shows that if we don't continue reducing spending—the deficits will be $500 bil. a year before 1990.

Then met with our citizens commission P.F.I.A.B. which evaluates intelligence. We haven't one Am. employee in the Soviet Embassy in Wash. They have 209 in our embassy in Wash [Moscow].

[Met Prime Minister George Price of Belize.]

Met with Dr. Nitze who leaves Saturday for Geneva to resume the Nuc.

Reduction talks we call I.N.F. Like me he believes the Soviets won't move until & unless we display our intermediate missiles in Europe. A Cabinet meeting to settle a few matters—one, whether to rush to complete filling our salt dome storage areas with crude oil or wait for possible lower prices. We'll talk some more about that. The same with some changing Food stamp regulations. The cheating in that program is a scandal. Finally agreed to urge State Unemp. Ins. boards to permit unemp. funds to be used for training in colleges for the unemployed.

[Reunion of White House Fellows; reception for party supporters.]

Friday, May 13
N.S.C.—discussed Soviet Summit & how to handle if they broached it. There is possibility Andropov might come to the U.N. If so we should invite him to Wash. & will.

Met with speechwriters—told them shorter sentences & single syllable words wherever they can be used.

Intelligence oversight board reported in (every ½ year). They told me they can not find that I'm breaking any laws regarding intelligence operations or covert in Central Am.

Met with some bipartisan leaders of the House re MX. I think we did some good.

[Lunch with Jim Baker and Senator Paul Laxalt (R-NV) on budget bill; met with representative of Catholic League for Human Civil Rights; with Lane Kirkland and foreign labor leaders, "Somehow they think we can solve the world's ec. problems by increasing pay to labor"; visit by Mr. and Mrs. Charlton Heston regarding television syndication rights.]

Off to Camp David. Weather beautiful. Cocktails outdoors. I have a load of homework with me.

Saturday, May 14–Sunday, May 15
Began to cloud up. We scheduled a ride for 2 P.M. It was my 1st ride at Camp David on Giminish. We had to cut it short. He was either spooked by the new territory or he's got some problems we don't know about. In the arena (Rock Creek park) when I rode him he was perfect. He began acting up on the trail & became darn near impossible. I felt a little loose in the saddle a few times.

Went back to the house & got a lot of work done.

[Senator Paul Laxalt (R-NV) called to withdraw his support of budget compromise, agrees with president to "stay the course"; returned to W.H.]

Monday, May 16
His R.H. Prince Hassan of Jordan arrived. He believes the Arab states

will pressure Syria into withdrawing from Lebanon after some face saving exchange.

[Addressed home-builders convention, criticized Congress on budget progress; economic briefing regarding foreign debt; rehearsal for press conference; speech to business leaders on MX; swearing-in ceremony for new ambassador to Austria.]

Tuesday, May 17

Started day with Cong. Repub. leadership meeting. Main item was budget. I told them I wanted to stick with our budget. They want to try for a budget resolution which means a compromise with the Dems. I think it would be a disaster. We'll probably wind up with a year of me vetoing appropriations.

We had an N.S.C. briefing then right into another meeting with a group of Dem. & Repub. House members on MX. I think we did some good. Later in the day the House Appropriations Committee approved the MX 30 to 26 (11 Dems. with us). It will be a close battle on the floor.

[Luncheon for the Council on the Arts and Humanities; rehearsal for press conference and then the conference; noted that Israel and Lebanon signed agreement.]

Wednesday, May 18

Signed a joint resolution before guests in the Rose Garden. The resolution declared May 21st Andrei Sakharov day. Among the guests was his daughter & her husband. I'm kind of sorry about the whole thing. We've been working behind the scenes to get him released—quiet diplomacy. This kind of public demand puts the Soviet politics in a corner where they lose face if they give in.

Cabinet meeting—Dave Stockman reported on budget deficits. He paints a dismal picture which I can't help but suspect was designed to convince me we'd have to have tax increases. Frankly, I didn't understand his figures & since George Shultz didn't either, I'm still a hold out on taxes.

[Lunch with leaders from 1980 campaign.]

N.S.C. meeting on how to push the Soviets on the conventional arms reductions talks. We're going to see if they will talk verification—on right inspection etc. If they won't, there's not much point in continuing the negotiations.

Swearing in of Don Rumsfeld as head of E.P.A. He graciously stated that when E.P.A. started 13 yrs. ago with him as head then We in Calif. were out in front of the nation on environmental protection.

Taped an interview with Jess Marlowe on El Salvador. Tonight 9 P.M. dropped by the W.H. photographers dinner. A few minutes of quips & home.

Thursday, May 19

Met with Cong. leadership. Tip a no show. Maybe because on the Today Show he kicked my brains out. I'm a liar, cruel etc., etc.

Jim Wright in the meeting criticizes me for being partisan. I told him I'd seen (TV) his speech on the Floor about me & I thought I had a couple of licks coming.

Lew Lehrman & Jack Hume came by. They have a great plan for getting our supporters organized at the Cong. District level. Lunched with George B. Did an interview with Helen Thomas. A Cabinet Council meeting. Don Regan joined me in the office about Ron & his paranoia about S.S. protection. I think he's being ridiculous & d--n unfair to the guys who are trying to protect his hide. That is settled—we let him sign off permanently—no protection.

A wonderful family (Filipino) came to the office. Their teenage son has won the nat. science award. He has invented a foil for airplane wings that can result in a 22% savings on fuel.

[**Friday, May 20:** *flew to Miami to address Cuban American group; returned to Washington; USO special in celebration of Bob Hope's eightieth birthday.*]

Saturday, May 21

Off early to Seton Hall U. in N.J. to speak at commencement. We left Wash. in a downpour. Fortunately in N.J. it was just a grey foggy day. The ceremony was out on the Athletic field—1900 graduates & many thousands more in attendance. I was on edge because I didn't think my prepared remarks were very impressive. I've never been able to or wanted to use a graduation ceremony as a forum to get off some speech I wanted to make on an issue important to me. I think the graduates should have a speech directed toward their accomplishment & their day. Well, with the help of some ad-libbing, it turned out fine. Mr. Nardina (A.B.C.), Pearl Bailey & I all received honorary degrees.

Flew back to Andrews A.F. Base—Nancy met me & we took Marine 1 to Camp David. A downpour but it was good to sit in front of a fire & get some homework done.

[**Sunday, May 22:** *returned to W.H.; desk work.* **Monday, May 23:** *met with Shultz regarding relationship with the Soviets; commented, "I thought we've come to a point where we should include Bill Casey & Cap W. in some of our decisions"; dropped in on meeting with local representives; lunch at Vice President Bush's house with top high school students; gave awards to businesses promoting exports; met with eight physically or mentally handicapped Girl Scouts; photo sessions; met with school chess champions and with local officials from North Carolina who built their City Hall with private help, longtime supporter Representative William Lipinski (D-IL); dinner in support of MX for sixty congressmen.*]

Tuesday, May 24

Our newest appointees came to the W.H. I met with them at 9:45 A.M. Met with Sen. Gordon Humphrey re the MX. He's stubborn & I don't think I moved him. He's stuck on the idea it is vulnerable & won't listen to any arguments. Met then with 4 Sens.—Jackson, Tower, Nunn & Warner on the "Confidence Building Measures" with the Soviet U. to lessen chances of accidental war or crisis.

Went to the East Rm. & participated in giving 12 science awards.

Lunched with Henry K. & Bill C. Henry gave me some food for thought re the Soviets. We've been trying to loosen them with some little deeds. We've told them we need deeds not words to see if they really want a good relationship. He thinks we should put down a marker on something big—like blockading Nicaragua.

Had briefing on summit—I'll be in charge & frankly I'm a little edgy. Made some calls on MX—the vote comes this afternoon. Cabinet meeting on women's affairs & things we can do to eliminate or reduce discrimination.

[Met New York Islanders hockey team, then with Japanese philanthropist; noted that MX bill passed in the House.]

Wednesday, May 25

Early meeting with George S. & staff plus N.S.C. Reps. George declared need for a better set up to deal with Central Am. He says there is no set or proper organization to centralize command. He's right. I OK'd a plan. We are bringing Motley—Ambas. to Brazil back to be in charge. We'll offer Tom Enders Amb. to Spain.

Spent virtually entire day in Cabinet meeting doing dry runs on the Summit meetings. Various people played the parts of the other heads of State. The young fellow who played P.M. of England—Margaret Thatcher was a stranger to me. When I called on him I told him his gown was lovely.

Word came in afternoon—the Senate passed the MX 59 to 39. The House later passed the Pershing II missile. Home & Mother.

[Thursday, May 26: NSC meeting regarding Navy commander murdered in El Salvador; television interviews with foreign correspondents; visit from Prime Minister Amintore Fanfani of Italy and from movie executives who renovated White House theater; "another hour briefing on the Summit. Enough already"; had a haircut; taped radiocast; phone calls to various gatherings.]

Friday, May 27–Tuesday, May 31

The summit in Williamsburg on all our minds but 1st on Fri. had a press interview & then met with Prime Minister Nakasone of Japan—1st name Yasu. We met in office & then lunch in the St. dining room. He impresses me more every time I see him. At lunch we surprised him with a birthday cake. I can't believe he's 65. I had him pegged for 45. He's off to Johns Hopkins U. for a speech & then on to the summit.

I finished the day in Williamsburg with a 2 hr. prep meeting for the Summit.

Sat. 28th—A full day what with touring all the facilities arranged for the Summit. Our table—at least 40 feet long was hand built by a craftsman as a gift to the govt. for use at the Summit.

Had bilaterals with Pres. Mitterrand (France) & P.M. Margaret Thatcher, then a dinner meeting with the other 6 heads of state & the Pres. of the European Council, Gaston Thorn. I opened the subject of the I.N.F. deployment. After full discussion it was agreed that we'd have foreign ministers draw up a statement of approval of deployment & of negotiations to reduce & hopefully eliminate all such intermediate range weapons. We met on Sunday morning & out of the blue both Mitterrand & Trudeau said they couldn't support such a statement. The discussion grew very brisk with Margaret, Helmut, Yasu & Amantore (Italy) all having at them. I got angry & did about 20 minutes. We were one hour late for lunch. In the afternoon meeting we started again on a new draft that tried to meet some of their language complaints without weakening the statement.

While Ministers were working on drafts, we took up matter of an ec. statement & believe it or not the same 2 had objections to that. We stood firm—I thought at one point Margaret was going to order Pierre to go stand in a corner. It was hard to remember we had started the day with a prayer service in the tiny church. Maybe that's what did it because we closed the day with both issues resolved, cordially restored & no winners or losers.

Sunday night dinner was very pleasant at the old Royal Gov's. palace. We discussed the middle east but didn't make it an agenda issue.

Monday morning meeting very productive—agreed to do more to cooperate on medical research etc. Rallied around our full ec. statement. At an outdoor lunch we met exchange students—1 from each of our countries. I filled our leaders in on Central Am. Later in day, met with Helmut Kohl—he's solid & with us all the way. Monday night dinner (Nancy had arrived, thank Heaven) was something of a banquet. Press already hailing the Summit as a success. Tues. saw each of the heads of state off. (Mitterrand, Thatcher & Kohl) had left the night before. Thank you's all around & back to the W.H.

Wednesday, June 1

[Met with leaders of the Red Cross, which was suffering financial problems.]

Met with Gen. Rowny—he leaves in a week to resume the "Start" talks in Geneva. I've given him some needed flexibility.

[Lunched with space shuttle crew, including Dr. Sally Ride, called her "a charming, likeable gal"; cabinet meeting regarding equal rights for women, Social Security, disability and hospice care; visit from leaders of Jewish War Veterans; diplomatic formalities; visit from former governor Mel Thompson (R-NH); photo session; videotapings; reception for David Rockefeller and investors in Latin America.]

Thursday, June 2–Sunday, June 5

[Memorial service for W.H. staffer; left for Camp David.]

Thurs. afternoon at Camp D. we rode. After my last experience with Giminish I didn't know what to expect, but God Bless Dennis Ayres (Park Mounties). My horse was entirely different & we had a nice ride. Giminish stayed a little tense but no acrobatics. Fri. morning we rode again & this time he was relaxed & just a darn good ride. Whatever it was I think he's cured. We decided (correctly) not to ride Sat. based on weather reports & sure enough the skies opened Sat. afternoon plus lightning & thunder. We both got a lot of homework done. Sun. of course was magnificent—85 degrees & bright sun. Of course we left right after lunch & here we are at the W.H. Tonite to the Shultz's for a B.B.Q. It was a very pleasant evening & we learned something. A stripper steak was covered with salt & laid in the fireplace on hot coals. Turned over once & it was ready for eating.

Monday, June 6

A busy day. A moving ceremony in the Oval office presenting a gold medal posthumously to a man who has been known for more than a year as "the unknown hero." The Coast Guard has finally established that the man who kept passing the helicopter rescue rope to others in the "Air Florida" crash into the Potomac—Jan. 13 1982 was Arlin Williams Jr. Five people's lives were saved but when the Copter went back for him he was gone. His mother, father, son (18) & a daughter (16) were present for the ceremony.

[Met with finalists in National Spelling Bee; received report on American place in world marketplace; discussion of loan request from Nigeria; PBS's Young Talent program, cohosted by Mary Martin and John Raitt.]

Before that I met with Paul Volcker—? do I reappoint him as Chmn. of the Fed. Aug. 1 or change. The financial mkt. seems set on having him. I don't want to shake their confidence in recovery. Cliff White came over from R.N.C. to talk about our staff battles. It's a tough problem made worse by 2nd echelon spokesmen leaking stories in behalf of their top men.

Tuesday, June 7

N.S.C. meeting on Start negotiations. We've opened them up & given our negotiators more flexibility. We're going to try & reduce warheads & missiles & throw weights.

[Visit from President and Mrs. Houphouët-Boigny of the Ivory Coast; met with congressmen on the START talks; visit from Stu Spencer, campaign strategist, and from Max Fisher, oil magnate and advisor, with Jewish leaders, all offering support.]

Don Regan came in—just between us for now he wants to resign. His family is pressing for it & I know it's concern for his health. He puts in a 60 or 70 hour week. I think we'll re-appoint Paul Volcker for about a year & a half. He doesn't want a full term.

Tonite—State Dinner but before that—a short meeting with Alexandre de Marenches—one time head of French Intelligence. He travels extensively & had depressing news about a number of areas & Soviet subversion in them. He picks Iran as a real danger spot.

[State dinner.]

Wednesday, June 8

Sen. Malcolm Wallop came to the Oval office. He makes a good case that we are actually further ahead on laser technology & could carry out my idea of a defense against nuclear missiles much earlier if there wasn't bureaucratic foot dragging.

Met with bipartisan leadership of both houses to report on Wmsburg & to discuss "start" negotiations. Tip & Bob Byrd conspicuous by their absence. Good meetings—then adjourned to Rose Garden where I made statement on changes in our negotiations in "Start." A lot of Ambassadors, the Scowcroft Commission, Joint Chiefs of Staff etc. present. It was broadcast world wide.

[Received NBA champion Philadelphia 76ers; lunch with educators, supportive of administration findings; cabinet meeting on farm problems, worker safety; new poll shows improvement in job approval and other areas; Maureen staying at W.H.]

Thursday, June 9

Ambas. Hinton just relieved as Ambas. to El Salvador, stopped by. He's a good man & did a fine job under extremely difficult circumstances. I hope he can convince some of our left leaning Congressmen how wrong they are.

Geo. Bush reported on his phone calls to J. Loeb, our Ambas. to Denmark. We are replacing him—the news had leaked to him already—he was climbing the wall. He really had to be replaced. Finally on 3rd call Geo. told him I wanted him to serve as a delegate to the U.N. in N.Y. That calmed the angry waters.

[Introduced new staffer, Lee Verstandig; flew to Minneapolis, briefing en route by head of Commission on Excellence in Education; visited high school; political dinner, noted hostile demonstration outside; chastised mayor of Minneapolis for "hosting a seminar of Soviets & Am's. who see no threat from the left"; returned to the W.H.]

Friday, June 10

A short day. An N.S.C. meeting re security for our nuclear installations. Chmn. Dingell (D. Congman) has been trying for headlines on that subject.

[Lunched with business leaders; left for Camp David; sat by pool.]

Saturday, June 11–Sunday, June 12

Did my radio broadcast—lunch & then a ride. Giminish was a perfect gentleman. Wore my new field boots for the 1st time. We've been working on the ankles with Lexol. They've been darn hard to break in but we're getting there. Spent rest of afternoon pool side. Had some homework with me which took more time than I'd anticipated. Had some calls to make—Anti-Defamation League, George Bush—Birthday & Charlie Price in Brussels who had some good suggestions. Discovered for 1st time the embassy doesn't have a secure phone—he has to go to NATO H.Q.

Early leaving on Sunday because of Special Olympics on W.H. lawn. The Special Olympics are for retarded & otherwise handicapped children. It was a truly grand evening—special events by these children, then a picnic dinner in a huge tent & finally entertainment by the Beach Boys. We had one of the athletes, a young girl (gymnast) at our table. She couldn't keep from hugging Nancy who hugged right back. Eunice Shriver heads up this special activity—she was at our table with her husband & the Giffords. Chris Reeves (Superman) was on hand as honorary chief coach. He's done some acrimonious interviews about me being a cold fish with a heart only for the rich. I'm just optimist enough to think he might have changed his mind. Eunice told someone her brother (J.F.K.) would never give as much time to something like this as we did.

Monday, June 13

Visited by Bob Hawke, new P.M. of Australia. We got along fine & quietly just between the 2 of us he let me know he did not represent the left wing of the labor party.

[Cabinet meeting, approved reform in food stamp program; meeting with congressmen touting various pet bills; signed proclamation for Baltic Freedom Day; dinner party. **Tuesday, June 14:** *received gift of giant flag; flew to Tennessee, visited high school and Meharry Medical School, a traditionally black institution, commented, "A short time ago it looked like it was going belly up—we did some things that cleared it's debt & saved it"; met with party leaders; flew to Albuquerque; reception for party leaders; called to follow up on intervention effort to cut red tape and bring a hospitalized girl home to her parents.]*

Wednesday, June 15

This was the morning I addressed the Nat. Convention of the P.T.A. All the local papers were telling me how unwanted I was. The P.T.A. is opposed to Tuition Tax Credits. The young D. Gov. of N.M. was all over the news blasting me for cuts in ed. funding—which we've never made. They had ordered 2000 green buttons to wear showing they were against the tax credits. Incidentally I had nothing on that subject in my speech.

I was braced for trouble. I was warmly received & was interrupted a number of times by enthusiastic applause & when I got to "prayer in school" got a standing ovation—which the network TV news failed to mention.

Back to the W.H. & a meeting with staff, Cap W. & George S. He's meeting with Dobrynin & Gromyko & wanted to check with us on subject matter & positions. We were all in agreement that we be firm, willing to hold out a hand same time let them know we d--n well want them to stay away from Central America.

Thursday, June 16

Wrong start to a pretty good day. One of our Presidential Scholars—high school seniors chosen for outstanding scholarship was a visitor before the ceremony. A young lady named Ariela Gross came by to give me a petition she had persuaded 14 of the more than 140 to sign advocating a nuclear freeze. I should have asked her questions & then caught her up when she was wrong. I didn't—feeling pressed because I only had a few minutes with her. I tried to tell her why a freeze was ridiculous. She was unreachable & more than a little arrogant. I spoke to ears that refused to hear.

[*NSC meeting on the danger of the rising debt in developing countries; lunch with Vice President Bush; economic briefing on the stock market; Don Regan unusually optimistic; visits by United Negro College Fund supporters, by leader of Armenian Apostolic Church, by committee working on restoration of Statue of Liberty, by cancer researchers who received award from GM; presented with Maine salmon; ambassadorial formalities; dropped by dinner for Jesse Helms.*]

Friday, June 17

President Magana arrived—a good man. When we went out for the photo op in the Rose Garden, Sam Donaldson (who else) started yelling a question at me—each time I said "Photo Opportunity." I'd explained beforehand to the Pres. that we had a rule against Q's. at a photo op. Sam then yelled a Q. at President Magana. The Pres. said—"Photo Opportunity." That finished Sam. We had good meetings. Selwa Roosevelt brought her family in for a photo. Archie gave me a small bronze that once belonged to Teddy R. I went over to the E.O.B. to speak to the board of the Knights of Columbus. We took a round about way to get there—seems one of the sniffer dogs showed some excitement about a parked car along the usual route. Did some tapings & we were off to Camp David—where again we spent a rainy weekend. Oh! Forgot on S. Lawn addressed the Presidential Scholars & families. One—a girl came to the office 1st to present a nuclear freeze petition she'd gotten 14 others to sign. She really had a closed mind.

Saturday, June 18

Played a game on the leakers & the press. About 11 A.M. I phoned Paul Volcker in N.Y. & asked him to accept re-appointment as Chairman of the Fed.

Then I called Don Regan to tell him. Then I did my radio broadcast—I boiled down the prepared script & opened with my now handwritten announcement about Paul's appt.

[Answered mail. **Sunday, June 19:** *desk work; returned to W.H.; dinner with Wicks family.]*

Monday, June 20

Met with Habib & Dick Stone. Habib's was kind of a final sum up on where we are in the Middle East—Dick is going to Central Am. & try to meet with the El Salvador bandits to see if they'll meet with the El Salvador Peace Commission to discuss a pol. settlement. He has to do this to appeal to the hard heads in Congress who are trying to dictate foreign policy.

[Met with private pilots who avoided a crash of their disabled plane; spoke to congressional pages; flew to Jackson, Mississippi, for political fund-raiser, rally; returned to W.H.]

Tuesday, June 21

[Met with congressmen regarding 1983 supplemental budget.]

An N.S.P.G. meeting on how to move Syria out of Lebanon. I've asked for more specifics.

P.M. (Pres.) Gonzalez of Spain arrived. He's sharp, a bright, personable, young moderate & pragmatic socialist. I think we hit if off pretty good which was what he wanted. I did lecture him a little on Central Am.

[Cabinet Council meeting on synthetic fuels and farm subsidies; photo session; visits by winner of wheelchair marathon, by retiring Washington Times *reporter Carlyle Reed.]*

Wednesday, June 22

Finally got everything cleared away as to Dick Stones trip to Central Am. Last night got a call that 3 Am. Press men had been killed on the border of Honduras. By morning we knew it was only 2—3rd man was an Honduran driver. They were killed by a rifle grenade fired across the border by Nicaraguan soldiers. The bodies couldn't be recovered til night fall because the Sandinista forces kept shelling the area. They can't lay it to a mistake—the vehicle was painted white.

[Met with new commission to improve Japan-U.S. relations, chaired by David Packard, cofounder of Hewlett-Packard; spoke at convention of independent business owners; answered mail.]

The Dems. are trying to pick up a scandal saying someone gave us or we stole their scenario for the Jimmy Carter campaign debate which I won. Evidently someone was given something that was merely a compilation of what they thought were achievements. I'd never heard of it until it broke in the press yesterday. It certainly was no strategic plan of any kind & I've never seen it.

Thursday, June 23

Every time I got within half a block of the press corps they were yelling "what about the (& they call it) Stolen Carter document." I haven't replied & won't until next weeks press conference. This was the day for Chi.—arrived at O'Hare still fairly early in the day—helicoptered to the H.Q. Polish American Assn. Spoke to a rather small group of invited leaders—well received. They are most supportive of what we've done for the Polish people.

[Addressed American Medical Association; returned to W.H.; plan to go to Florida for shuttle landing canceled due to delays caused by uncertain weather.]

Friday, June 24

Prince Bandar (Saudi Arabia) came by with message from the King re the Syrian situation. They are most anxious that we do something to rid the Middle East of the Soviet influence. Bandar believes Syria would like to trade the Soviets for U.S. help & influence.

Met with Repub. plus 1 Dem. Bud Roemer re El Salvador. Many of the group had just come back from there. Not all had been on our side before they went—they are now. The stories they told all added up to the fact that El Salvador is only the current battlefield. What is going on is a general revolution aimed at all of Central Am. & yes, Mexico.

[Lunch with 1980 campaign supporters; interviewed by Malcolm Forbes and Forbes reporters, commented, "They too are supportive"; desk work; ran an unnamed movie; phoned astronauts in California, where shuttle ultimately landed.]

Saturday, June 25

No Camp David—but weather wonderful. After my broadcast we went over to pool—had a swim & lunched there with the Deavers, Jim Baker. Very pleasant day. We're being harassed by a charge that 3 yrs. ago we obtained a campaign strategy book from Carter campaign & used it to win debate. No one can remember anything other than a bunch of pages citing Carter admin. accomplishments. I'd never even heard of that until this story broke. It resulted from Larry Barrett's book on me & sounds like a book promotion stunt but a Dem. Congmn.—Chairman of a sub comm. thinks he's struck pay dirt. He's struck dirt alright—campaign mud.

Sunday, June 26

Church in the A.M.—same plan as Easter—didn't put it on the schedule—told press pool at last minute. It felt good to be there—I've missed it.

Afternoon by pool—Bill Clark & Bill Casey came over for discussion of Central Am. We're losing if we don't do something drastic. Those in Cong. who are dribbling out about ¼ of what we ask for & need could be playing politics.

They'd like to give enough money to keep us in the game but El Salvador bleeds to death. Then they call it my plan & it lost Central Am. We have to take this to the people & make them see what's going on. If the Soviets win in Central Am. we lose in Geneva & every place else.

[Swam; Marine ceremony after dinner.]

Monday, June 27
Nancy left this A.M. for Phoenix. I won't see her until Wed. night—or is it Thurs. night—I'll have to check. Anyway I still don't like this place without her.

Met with Cardinal Krol who has returned from being with the Pope in Poland. He believes there may well be some easing. I believe also that Walesa may not be as much of a force as he once was.

[Photo session; issues lunch; videotapings; rehearsal for press conference.]

Tuesday, June 28
N.S.C.—In the Sudan—rebels picked up several missionary types, 2 of whom are Americans—they've suggested ransom. I wish to h--l we had forces in the area who could move in. We don't of course but I'm fed up with this kind of thing.

[Photo sessions with Bay Buchanan and with Wayne Newton; met with former secretaries of the Treasury and current secretary Regan in support of appropriation for International Monetary Fund (IMF); NSC briefing on the Middle East and Central America; lunch with Council on Private Sector Initiatives; last rehearsal for press conference.]

The news conference dealt almost entirely on the "Carter papers." There is no question but that the Wash. press is going to try for another lynching. I think it went pretty well though & I did get in some good licks—particularly on the few Q's that didn't dig at the papers—which incidentally I had never heard about until the press started in last week or something that was supposed to have happened 2½ yrs. or more ago.

[Wednesday, June 29: flew to Louisville, Kentucky; addressed vocational students and teachers; flew to Kansas City; visited meeting of student leaders; flew to California, met Mrs. Reagan at Beverly Wilshire Hotel.]

Thursday, June 30
Helicoptered out to Whittier to a high school we've recognized as outstanding for excellence in Ed. Presented a flag as symbol of that to student body & townspeople. It was another of our regional seminars with the Excellence in Ed. Commission present. It was a great morning—78% of student body are Hispanic. Back to the hotel & in early afternoon met with Ansel Adams & his wife.—He is the great nature photographer. He has expressed hatred for me be-

cause of my supposed stand on the environment. I asked for the meeting. I gave him chapter & verse about where I really stand on the environment & what our record is. All in all the meeting seemed pleasant enough & I thought maybe I'd taken some of the acid out of his ink. Then I read the story of the meeting as he'd given it to the press. I'm afraid I was talking to ears that refused to hear.

[*Addressed party fund-raiser in Long Beach, toured* Spruce Goose *airplane; went to ranch.* **Friday, July 1:** *ranch work.* **Saturday, July 2:** *birthday party for Mrs. Reagan.* **Sunday, July 3:** *ranch work.* **Monday, July 4:** *riding and desk work.*]

Tuesday, July 5

Into L.A. in the A.M., addressed the Am. Fed. of Teachers. About 150 greeted me by walking out. They had "get out of El Salvador" placards. I ignored them & addressed the roughly 2000 delegates re our ed. plan. Was interrupted 22 times by applause.

Here we are back in the W.H. & homesick for the ranch already.

Wednesday, July 6

Nancy's Birthday! Life would be miserable if there wasn't a Nancy's Birthday. What if she'd never been born. I don't want to think about it.

A really open day at the office. Met with new man who will work with Motley on Central Am. Amb. Charley Price came by. Outside of that I had the day to catch up on reading reports etc. Have written an idea for a reply to Andropov letter in which he pleads for an end to war & nuclear weapons.

[**Thursday, July 7:** *latest poll numbers show a drop in perceived trustworthiness, commented, "No question but that reflects the media blitz over the so-called Carter papers—about which I knew absolutely nothing"; interview with* Sports Afield *journalist; photo session; lunch with public relations executive Dick Whalen; cabinet meeting about new legislation regulating banks; visit by daughter of Eureka College alumna; received gift of quill pen set; presented staff award for fighting waste and fraud; visit by Dolores Ballachino, personal correspondent for more than forty years; met with editor of* London Times *and owner Rupert Murdoch.*]

Friday, July 8

—Surprised staff at 8:15 by dropping in & telling them to answer any & all Q's F.B.I. had on Carter papers.

An N.S.C. meeting on Central Am. & whether to abide by Congress's request to formal bipartisan commission to study area between now & end of year. I suggested we go along—provided that Congress provide funds we've requested between now & end of year.

Met with Scowcroft commission again—explained our flexibility on arms reductions.

Geo. Shultz is back (as of 3 A.M.) but he came in with report on trip. A good & fruitful trip except where Syria is concerned. We're going to have to find a way to get them out of Lebanon.

[Photo session with actor James Mason; left for Camp David. **Saturday, July 9–Sunday, July 10:** *rode and swam; returned to W.H.; polls indicate public not concerned with scandal involving Carter papers; noted that before leaving for the weekend, he ordered a complete FBI investigation of Carter papers.]*

Monday, July 11

Ambas. John Gavin came by. He's a darn good Ambassador. Had the usual "issues lunch." Later met with A.W. Clausen now head of the World Bank. He'd like us to increase our contribution to the bank but there's no way we could get an increase through Cong. Foreign Minister Genscher of W. Germany came to report on Chancellor Kohl's Moscow visit. The Chancellor really stood firm on our NATO unity & that we were going to deploy intermediate range missiles in Europe on schedule in Dec. No question but the "Russkys" are upset about this.

Kase Bendtson, Bill Wilson, Jack Hume, Joe Coors & Dr. Edward Teller came by to press me on setting up a "Manhattan" type project to have a crash program on finding a defensive weapon against nuclear missiles. I have to agree with them it's the way to go.

[Marine band concert.]

Tuesday, July 12

Met with G.O.P. leadership. Geo. S. reported on Middle East & Central Am. We lobbied for our defense program as necessary to Arms Reductions. Told them we'd go for their proposal to have a commission set up on Central Am. for six months but only if Congress (who wants the commission) will give us the money for Central Am. that we're asking for so that our allies will still be there 6 months from now.

N.S.C. meeting to O.K. plan for pre-positioning mil. equipment in Israel as well as Arab countries as hedge against Soviet push.

Had lunch with Gen. Sec. of O.A.S. Orfila. He says I'm only one who can bring about an alliance between all the Am. nations. Had a not too pleasant Ec. briefing & then a Cabinet meeting to hear Dave Packard report on our research labs. We have some work to do in that department. Some of them were started for a particular purpose but the purpose no longer exists.

[Visit from Y. K. Pao, Hong Kong businessman; met with clergy on school prayer amendment; photo session; poster boy for asthma fund; ambassadorial formalities.]

A new scandal?? Some cockamania lawyer claims he had a video tape of several "govt. figures & friends of mine in a sex orgy" with the Morgan girl who was

Al Bloomingdale's mistress. She has been murdered—her live-in boyfriend has confessed to the murder. The lawyer claimed he would only give the film to me. Fred Fielding on phone told him to turn it over to the D.A. Then the character said it's been stolen. I don't think there ever was such a film but the press had a field day linking my name to the supposed orgy.

Wednesday, July 13

Met with Repub. women of Congress on legislation we're sponsoring to improve collection of child support from runaway husbands.

[Met with supporting congressmen on probable vetoes they intend to uphold; with Jack Kemp on vagary of using money supply as means of fostering recovery; lunch with Senator Paul Laxalt (R-NV) and James Baker on potential 1984 campaign.]

Geo. Shultz & Max Kampelman met to discuss our position re the Madrid meeting on human rights etc. The Soviets are being devious about their promise to let Scharansky go. We're going to hold them to it.

Thursday, July 14

Gov. Pierre DuPont came by. I'm trying to convince him to run for Bidens Sen. seat. He doesn't want to & I can't blame him. I wouldn't have wanted to be a Senator after my 8 yrs. as Gov. He's been a d--n good Gov. too. Met with group of young Repub. Congressmen to sell them on voting for a $8.5 bil. quota in the Inter Nat. Monetary Fund. I think we sold them. Sen. Denton came by with a plan or idea of getting a combine of Latin Am. countries & ourselves to quarantine Nicaragua & stop flow of arms to govt. there.

[Lunch with Vice President Bush; calls to senators encouraging support of B-1 bomber; cabinet meeting; met with group of congressmen on International Monetary Fund (IMF); with Sir Geoffrey Howe, British foreign minister; photo session; received award for use of radio; greeted new members of Washington press corps; met with ninety-six-year-old African American National Guard officer Colonel West A. Hamilton, honored with rank of brigadier general; met with poster child for disabled organization; videotapings; dinner for Scowcroft Commission; noted House vote in favor of Caribbean Basin Initiative (CBI).]

Friday, July 15

A hectic ½ day plus—Began with House leader Jim Wright. Jim has kicked my head off on domestic issues but been pretty good on foreign policy—worked hard to get C.B.I. passed. Had him in to solicit help on MX. Couldn't get a commitment although he didn't say a flat no.

Sens. Laxalt, Thurmond & Hatch in re the hearings on my appointments to Civil Rts. Commission. The Dems. are being vicious & unprincipled. Orin has been working too hard—he's very emotional & on a thin edge.

N.S.P.G. meeting on the middle east. Didn't come up with an answer to the Syrian situation but found a lot of areas of agreement. We'll be back at that again before Pres. Gemayel of Lebanon & Israeli P.M. Begin arrive.

[Lunch with Business Round Table, asked for advice; poll shows further support regarding scandal surrounding Carter campaign papers; left for Camp David.]

Saturday, July 16–Sunday, July 17

Weather close to 90 degrees both days. Horseback on Sat. plus poolside & swimming. Spent Sunday by the pool—broke in a pair of blue jeans. Swam in them & then let them dry on me—it took hours.

Called Henry Kissinger & asked him to chair the Central Am. task force.

Monday, July 18

Met briefly with Ambas. Stoesel who is back from a quick visit to 9 of our allies in Europe—1 day to each Capitol. Mission was to discuss Soviet violations of Human Rights. He met with solid agreement & a desire to keep in close consultation on the subject.

[Flew to Florida to address International Longshoreman's Association; met with group of Jewish leaders, noted, "I think we're on track even though many of them are Dems"; returned to W.H.; Maureen visiting.]

Started making phone calls re the MX vote. Announced today in Florida that Henry Kissinger will be chairman of the Commission on Central Am. One of press yelled a Q. at me that Sen. Dodd says commission is a ploy to get around Cong. I wonder really which side Dodd is on. He comes down against the U.S. on almost every issue.

Tuesday, July 19

Hot as an oven but cloudy & grey—we worried all morning about rain possibly messing up the formal ceremonies for the Emir of Bahrain. Everything went smoothly. He's a very personable & likeable man & expressed great appreciation for our reception of him.

[Addressed gathering for anniversary of Captive Nations Week; met with education leaders; met Mayor Tommy Thompson of Tampa, switching party affiliation from Democratic to Republican.]

More N.S.C. meeting, a haircut & home to get ready for the St. dinner. It was a warm & pleasant affair. Byron Janis, the pianist entertained—so did his Highness. His remarks at dinner were very humorous. We made a friend.

Wednesday, July 20

A short but busy day that started with the tragic news that our friend Frank Reynolds died at 12:40 A.M. He was a very decent man & as a journalist a man of great fairness & integrity. I signed a waiver permitting his burial in Arlington.

Back to the Exec. bldg. for a short speech & Q&A with the Central American Outreach group. Had a Q. about the Sandinista treatment of the Miskito Indians in Nicaragua. I gave a full account of their torture, confinement, persecution etc. Then someone said there was a Miskito in the audience. He stood up so I asked him if my account was accurate. He replied in Spanish. Two people interpreted. He had said I was completely accurate.

[Met with economic advisors, agreement on contingency tax "to go into effect in '86 if certain conditions prevailed at that time"; presented funds to Mayor Marion Barry for summer youth employment in city of Washington, D.C.]

A Nat. Security Planning Group meeting on the Lebanon situation. It is really a can of worms. How to get Syria out is the problem. Incidentally the other Pentecostal family—15 people has been allowed to leave the Soviet U. Quiet diplomacy is working.

Thursday, July 21

Held a mini press conf. in press room to announce new ec. figures. Growth in G.N.P. in real $ was 8.7% for 2nd Quarter. Not one question about the Carter debate papers. Lunch with V.P. The Multiple Sclerosis foundation brought in the Father of the Year & the Mother of the Year & her little boy. Both were in wheelchairs. The father is totally paralyzed from the neck down. His family was with him. His 3 daughters wrote the nominating letter that won him the title. Cabinet Council meeting took up matter of lumber companies who contracted for Nat. Forest Timber. Their bids were based on expectations of continuing inflation. Lumber prices now make it a loss for them to cut the timber. Some of our team think we should extend their contracts but make them pay a penalty. I disagree—this would just make it even a greater loss or raise the price of lumber which would come out of the pockets of future home builders. I say, extend time, no penalty but make them agree to a timetable for cutting. Finished the day with 1½ hours in the dentist chair—teeth cleaning time.

Friday, July 22

Dr. Dodge (Pres. of Am. U. in Beirut) kidnapped 1 yr. ago is safely back in Conn. The Syrians got him out. He had been kidnapped by the Iranians who smuggled him through Damascus drugged & in a box. He is well but was constantly hassled & threatened with death.

Today was Pres. Gemayel (Lebanon) day. We had a good meeting & lunch. I think he is reassured that we are not going to abandon them. While we were meeting word came that Beirut was under rocket attacks by the Syrians. We are going to send them the latest in Radar Art which can zero in on exactly where the rockets are coming from.

No Camp David until tomorrow—Frank Reynolds' funeral at 11 A.M. tomorrow—we'll attend.

[Saturday, July 23: funeral for Frank Reynolds; went to Camp David.]

Sunday, July 24

Spent day by & in the pool. For awhile could almost forget Bud McFarlane is replacing Phil Habib in the middle east. Phil has really worn himself out & has served above & beyond the call of duty. Could almost forget too the block of Congressmen who are determined to stop our aid to Central Am. & our own defense buildup. It's hard for me to believe this is just partisan politics. Whenever it is us versus the Soviets or Cubans they always come down on the wrong side. Back to the W.H.

Monday, July 25

Met with Henry Kissinger, now chairman of our commission on Central Am. We are agreed their purpose is to come up with a long range plan for Central Am.

[Photo session; lunch with Bob Michel (R-IL, House Minority Leader) and Howard Baker (R-TN, Senate Majority Leader) with W.H. staff to plan legislative agenda; presented National Wildlife Awards.]

A momentous meeting in the Oval Office with George Shultz, Bill Clark, V.P., Ed Meese & Jim Baker. The press has been dishing up stories all describing a situation in which Bill C. & George S. are battling for supremacy. It is totally false but George thought he might be so tarnished that he was a liability to me. I told him he had my confidence & that it would be a disaster for all of us if he left. I think the meeting was a great plus & that we go forward from here, conscious of what press is trying to do & all determined to refute their crusade.

[Rehearsal for press conference.]

Tuesday, July 26

Former President of France, Giscard D'estang came by. We had a good visit. He is solidly behind us on the I.N.F. deployment—was most interested in our economic recovery & thinks it bodes well for Europe.

[Appearance in honor of seventy-fifth anniversary of FBI; gathering of 1980 campaign supporters.]

A legal meeting on matter of loans to Ed Meese & Mike Deaver. It seems their accountant arranged the loans & later on we appointed him to the Postal Board. Once again the "small minds" are tearing at their betters. These 2 came to govt. at great sacrifice to themselves. They are paying 18% interest on these loans & there is no connection between their accountant arranging the loans & then accepting an appt. to govt. service himself.

Spent rest of afternoon woodshedding for the press conf. It came off alright but their questions were such iffy inconsequential things that it was like trying to get your teeth into whipped cream. I think it was a dull half hour.

Wednesday, July 27

Maybe I'm wrong about press conference—all the calls etc. indicate we won—I say that because press conferences anymore are an adversary contest. The press isn't after news—they want to trap you into a goof.

A short day in the office but a long afternoon in my study. Don Regan called re the International Monetary Fund (I.M.F.) Jack Kemp who told me he couldn't vote for it but wouldn't actively oppose is working his head off to torpedo it. Tip O'Neil is licking his chops. He's told Don Regan that if we can't deliver 80 Repubs. he'll take it off the calendar so that Repubs. will take the blame for it's failure. I'm about as mad as I've been in these 2½ yrs. P.M. Thatcher & Chancellor Kohl have both succeeded in getting it through their Parliaments & here I am threatened with defeat by my own party.

Had a meeting with some of our leaders but honestly don't know that we did any good.

Top two men in V.F.W. came by. We've fixed the "snafu" about speaking to their convention—I'll be on hand in New Orleans August 15 at 10 AM. I'm glad that's fixed—they've been very supportive & to be turned down by our schedules was a great mistake. They were just back from El Salvador & Nicaragua & are now totally in agreement with what we're trying to do.

Thursday, July 28

Jack Kemp now knows I'm mad. He's against us on the IMF increase but promised he wouldn't work against us. He's been working his head off.

[Telephoned W.H. electrician whose baby daughter received liver transplant; noted that previous week's radiocast mentioned need for organ donors; established crime commission.]

Met with Minister Shamir & Arens of Israel. I believe they will endorse calling their redeployment the 1st phase of total withdrawal of forces from Lebanon. Lunch with Cabinet Council on the pleas for protectionism in steel & textiles. We just can't give in without setting off a wave of retaliation.

[Received visits from Eureka fraternity brother, from mother who adopted handicapped Vietnamese baby, from Mr. and Mrs. Red Skelton; received $100,000 donation for Statue of Liberty from American Hellenic Educational Progressive Association; visit from cook from Eureka college days; addressed National Council of Negro Women, commented, "I believe I've found friends to counter Ben Hooks"; farewell party for Edwin Harper, assistant to the President for Policy Development.]

Friday, July 29

Brkfst. with McLaughlin TV cast—Pat Buchanan, Bob Novak, Georgie Anne Geyer, Jack Germond & of course John McLaughlin. It was a freewheeling & on the record. Most of it had to do with Central Am. It was interesting to see their show on Sat. & hear them discuss the breakfast. Pat (to my surprise) & Bob did seem to remember things in a different way from how I'd said them. Georgie Anne brought them up several times with on the nose corrections.

N.S.P.G. meeting on possible covert operations [. . .] in Central Am. They'll need further discussion.

[Appeared before Future Farmers of America.]

Dick Wirthlin with some poll results. We're not getting across our Central Am. position. House voted against covert aid but I think Senate will kill their bill—if not, I'll veto.

[Spoke to elementary and secondary school principals; off to Camp David. **Saturday, July 30:** *radiocast; ride and swim.* **Sunday, July 31:** *report from Bill Clark that Libyan forces entered Chad civil war on the side of the anti-government rebels, noted, "We're trying to get France to move. They've let us know Chad is in their sphere of influence." Back to the W.H.]*

Monday, August 1

Off to Atlanta & the Am. Bar Assn. Before I left got word things are looking up in Chad. Just in case, we are conducting air drills from 2 carriers in the Gulf of Sidra off Libya. One of them was supposed to be on its way to the Naval exercises off El Salvador.

[Addressed American Bar Association; met with party leaders in Georgia.]

Back at the W.H. by 2:45 & a meeting with Sam Pierce. Sam once again is demanding an appointment of his choice over someone we've picked & who we feel is better for a top spot at H.U.D. I overruled him.

Five P.M. in the study met with Ambas. John Louis. He's upset because we want to replace him. I explained it's part of a general shuffle & not due to any failure on his part. He'll stay on til Nov. 7. I would have preferred he go earlier but I'd also like to have no wounded feelings.

Tuesday, August 2

Where do these days come from—hardly a minute between gatherings. G.O.P. Cong. leadership to start with—mostly on Central Am. I think we won some points. A little time on I.M.F. & Jack Kemp now knows I'm teed off with him. A briefing by Joint Chiefs of Staff again on Central Am. They had a captured map—the kind that's done in the field for a particular mission. It was the map the Guerillas had for the action that resulted in the execution of 34 El Salvador soldiers/prisoners. Significant however was the fact that the symbols used on the map were those the Soviets use.

[Issues lunch with report of Grace Commission on reforms to save an estimated $300 billion over three years; met with Cabinet Council on crime issues; photo session; met with an epileptic man who walked across the country; with executives of Mars candy, supporters of Olympics; met Miss Teenage America; received gift of gratitude from Cardington, Ohio, which received federal aid after tornado damage; met with Senate Republican Steering Committee.]

The real happening of the day was when I learned that hundreds of members of the International Fed. of Business Women (56 countries) had been scheduled to tour the W.H. & some bureaucratic bungle denied them entrance. I called their Am. Pres. Polly Maddonwald & told her I was going to jump off the roof. I'm going to their convention in the A.M. & apologize.

Wednesday, August 3

Went to the International Fed. of Businesswomen with our apology. They were wonderful. I was warmly received & left to quite an ovation. Then came an aftermath. Back in the office a few hours later I was handed a press release given out by Polly Maddonwald. She opened it by accepting my apology in a ho hum way & then took off on me as being lighthearted about the problems of women, not serious about them & that they should remember in November etc. I guess taking her at face value was a mistake. She really convinced me she was legit.

Met with about 90 men & women & many of them old friends who are here because of Jack Hume's idea for organizing movers & shakers in every Cong. District to carry the word about what we're really doing here—which the Nat. press does not do.

From there to lunch with the top 20 or so people in Aero-Space for a work session on how to get the private sector involved in the space program on a solid commercial basis. The whole concept is very exciting.

Dick Stone came in to report on Central Am. He's very upbeat & says the tide has turned in El Salvador & that means improvement in Nicaragua.

Tonite dinner here at the W.H. with the Shultz's & Regans. The TV evenings news played up the episode of my apology but refused to show the applause I received & played it as a great embarrassment for me.

Thursday, August 4

The morning papers were worse than the TV news. I reached the boiling point. Then as the day wore on I received a wire from the Pres. Maxine Hayes, who had so graciously introduced me. She apologized for her colleague & said the women were honored to have me & appreciated my coming to apologize. About that time word came that a local radio station was polling its listeners on the incident & the great majority came down on my side.

[Briefing on November trip to Asia; met with President Joseph Mobutu of Zaire, noting, "He's very impressive"; lunch with Vice President Bush; met with congressmen from

steel-manufacturing districts; with Lieutenant Governor Bob Cashell (D-NV), who had decided to switch parties; with Senator Orrin Hatch (R-UT) and thirteen-year-old boy from Utah who formed a public library; with student athlete; with representative of U.S. Marshals; with Industrial Competitiveness Commission; reception for recipients of Reagan Scholarship to Eureka College.]

Friday, August 5

N.S.C.—Gen. Rowny came by to report on "Start" talks—not much movement on part of Soviets but believes this is due to their preoccupation with the I.N.F. talks.

[Signing ceremony for National Child Support Enforcement Month; met with Shultz and Weinberger on affairs in East Asia, commented, "We are all agreed that much of the future lies out across the Pacific"; lunch in State Dining Room with Hispanic leaders from all over the country; in-depth briefing on Soviet Union; photo session; met with Congressman Henry Hyde (R-IL), a supporter of administration's Central America policy; left for Camp David.]

Saturday, August 6

A day by the pool but also a call on the secure phone. We are sending Awacs & fighter escorts to Sudan. Awacs will keep tabs on Libyan planes in Chad. French asked for this presumably because they are going to give help to govt. of Chad. They've been pretty fuzzy on this even though they claim Chad as a former colony is in their sphere of interest.

Sunday, August 7

A day in the sun & back to the W.H. Bill C. came over to deliver a reply from Andropov to my handwritten letter. Expressed a desire to continue communicating on private basis & wants to talk about main issues.

Monday, August 8

Congress has gone home—it was a pleasant & unhurried day. NSC to start the day—subject Chad. Our AWAC's & fighter cover are in the Sudan—but no French A.F. in Chad. Same time however the Libyan planes seem to have stopped bombing. It's possible the French quietly made a deal. Of course, the Press now portrays us as having sent combat air forces to intervene in Chad but the French refused to intervene.

[Latest poll numbers reflect no response to the U.S. fleet presence in Nicaraguan waters.]

Had a briefing on the Soviets & Space. There is no question but that they are working (twice as hard as us) to come up with a military superiority in outer space.

Had an interview with a young lady from M Magazine. She seemed very nice—now we'll see if she really was when the article comes out in Sept.

[Plans for ranch improvements; rise in prime interest rate.]

Tuesday, August 9

Had a new briefing NSC on upcoming meeting with Pres. of Mexico. La Paz is about 110 degrees this time of year—happily he & I have come to agreement we'll wear the traditional Mexican shirt & slacks—no coats. We'll have lots to talk about. They are in an economic tailspin—we've been helping & will do more but they must tidy up their own kitchen. The Pres. is trying with some austerity moves that are not being well received. I'll try to strengthen his hand.

An Ec. briefing about our own problems. The banks just raised the "prime" to 11%. The stock mkt. nose dived. The Fed. had let monetary growth get a little out of hand—now they are trying to level off. I've suggested to Don R. that he have a little talk with the chairman.

[Lunch with Hispanic leaders concerned over administration's immigration bill and unemployment along the Texas border; staff meeting ended with decision to send a team to Texas to direct employment programs; met with new members of W.H. press corps and with W.H. Fellows; haircut.]

Wednesday, August 10

I'm alone. Nancy left this morning for Denver, Phoenix & the ranch. I'll catch up with her there on Monday. I go by way of Tampa, El Paso, La Paz Mexico, New Orleans.

Well Faza Largeau in Chad has fallen to the rebels & Quadafi's forces. France who says Chad is her special interest did nothing. A squadron of planes could have made the difference.

[Lunch with President Abdou Diouf of Senegal, who made a positive impression.]

A brief one on one after the Diorf lunch with Geo. Shultz. He's been taking a vicious beating in the press. He & I are going to have regular meetings every week. Part of the press attack is to the effect that he has lost standing with me. The h--l he has.

A cable from Lebanon tells us the Druze are shelling the Beirut Airport where our Marines are based—no casualties. The world must have been simpler in the days of gunboat diplomacy.

I don't like an empty W.H.

Thursday, August 11

The situation in Chad worsens. Faza Largeau has definitely been taken by the rebels & the Libyans. The French have sent 500 paratroopers to the capitol of Chad. Still no French air power leaving the air to the Libyans. They may be talking a "deal" with Quadafi—Libya is a big customer of France.

[Met with U.S. ambassador to Romania over problems with State Department.]

Met with our Hispanic appointees. They are a great gang & it was a good meeting. I think they'll do something to rebut the misinformation being peddled to Hispanics.

First meeting with our Commission on Central Am. I'm very hopeful about this group.

[Signed proclamation for celebration on October 3 of the two-hundredth anniversary of the Signing of the Paris Peace Treaty; photo session; received St. John Neumann Award.]

Some tapings—United Way, Willy Mays Day, Pete Wilson etc. Then questioned by F.B.I. re the Carter papers.

I'm packed & ready to head west tomorrow. First will have a big meeting on our I.N.F. strategy. Some want us to make some new movements just for the P.R. of it. I think to h--l with that—let's settle on the bottom line & fight it out.

Friday, August 12

Off to California by way of Tampa, El Paso, La Paz Mexico, New Orleans. Made it all the way then back to Wash.—3 days early.

Friday, September 2

In Tampa 3 wks. ago the Hispanic Business People were wonderfully warm in their reception. In El Paso the same was true of the Hispanic veterans even though the Gov. of Texas & others tried to steam them up to give me trouble. A good meeting in La Paz with Pres. de La Madrid. I left some powerful briefing material with him & hope he will see the Sandinista regime in Nicaragua more clearly. Then back to New Orleans for a speech to the V.F.W. convention the next day. On to the ranch & joined Barney & Dennis in building our new fence. It turned out great.

[Ranch work and relaxation for six days; to L.A. for five days, including day trip to Seattle to address American Legion convention; party fund-raiser, also speech in San Diego before convention of Republican women; public contretemps regarding Justice Department official who spoke out against administration policy regarding women's rights; returned to ranch; report of two Marines killed in Beirut.]

Then as the week went by the Soviets shot down a Korean Airliner with 269 passengers—53 of them Americans including Congressman Larry McDonald. The traffic in conference calls got heavy. We were due to return to Wash. on Labor Day but realized we couldn't wait so we left on Fri. It was heartbreaking—I had really looked forward to those last 3 days. When we got in Fri. I went directly to an N.S.C. meeting re the Soviet affair. We're going to try & persuade our friends to join us in banning Aeroflot flights & in demanding reparations for the victim's familys.

I've overlooked the trip down to Santa Barbara to Fess Parker's ranch where we entertained the press. I've also failed to mention the death—a real loss—of Scoop Jackson.

Saturday, September 3

An NSC meeting, this time on Beirut. Bud MacFarlane is back. His report held out some hope but there would be more if we could persuade the Israelis to stay several days longer in the "Shuf." Gemayal has finally, officially asked Syria to leave Lebanon. So I called Begin, he switched us to Defense Minister Arens. No soap. It was too late—the Israelis are already on the move.

We've held the carrier Eisenhower off Beirut & brought 2000 Marines from Egypt to standby—off shore as protection for the M.N.F. in Beirut. The Lebanese force trained by us has done a d--n good job in quieting things in Beirut but are still needed there & can't be pushed into the Shuf just yet. The Druse have already massacred about 40 men, women & children & we fear there may be more when the Israelis pull out. Paul Nitze is off tomorrow for Geneva & a resumption of the I.N.F. talks.

Sunday, September 4

To the Oval Office for a meeting with Congressional leadership—Dems. & Repubs. Met with our team at 9:30 A.M.—general meeting at 10 A.M. Meeting was very good—ran til 1 P.M. Dealt 1st & longest with Korean plane. Ran a tape of conversation between 2 Soviet pilots including the one who stated he had locked on his radar guided air to air missiles, launched them & "target destroyed." I'm going on air 8 P.M. tomorrow night to tell the story & announce our plans. Strom Thurmond made a great suggestion. We know the whereabouts of many K.G.B. agents [. . .]. We're looking into the practicality of this. [. . .] That would be shooting our selves in the foot.

Second half of meeting was on Lebanon. I believe we have Dem. support for what we're doing there. We may be on the verge of showing the world a truly united front.

Monday, September 5—Labor Day

Only thing scheduled for the day was lunch at the pool with the Wicks & at 8 P.M. a TV speech on the Korean airliner massacre. Well I put on my trunks but the speech draft arrived at 9:30 A.M.—in fact 2 drafts. I didn't like either one so I spent the day til 5:15 P.M. rewriting. It turned out OK & everyone seems to think it was A. O.K. I spent the day in my trunks sitting on a towel in my study but changed into a blue suit for the speech. It went well & everyone seemed pleased.

Tuesday, September 6

N.S.C. meeting with Geo. S. to discuss his meeting with Gromyko. Some are opposed but I think George is right—he should see Gromyko & eyeball him on the Korean plane shoot down. There were 61 Americans on that plane. This could be the 1st time Grimace has been put on the defensive.

A 2nd & larger NSC meeting discussed whether we should give a little on our position in the MBFR. force negotiations in Vienna. I think there is a better option but now is not the time to switch. Lebanon was next on the agenda. We lost 2 more Marines last night in Beirut. The Civil War is running wild & could result in collapse of the Gemayal govt. & the stuff would hit the fan. I called the parents of the 2 Marines—not easy. One father asked if they were in Lebanon for anything that was worth his son's life.

[Issues lunch; meeting with CEO of B. Dalton booksellers regarding adult literacy; meeting with leaders of United Jewish Appeal; appearance at birthday party for Deputy Counselor to the President Jim Jenkins.]

Wednesday, September 7

N.S.C. briefing: George S. is working on foreign ministers in Madrid—re the KAL shoot down. More on Lebanon—we have to show the flag for those Marines. I can't get the idea out of my head that some F14s off the Eisenhower coming in at about 200 ft. over the Marines & blowing hell out of a couple of artillery emplacements would be a tonic for the Marines & at the same time would deliver a message to those gun happy middle east terrorists.

[Launched an initiative on adult literacy; met Chicago teacher Marva Collins; met former governor Arch Moore (R-WV), encouraged him to run for Senate.]

Thursday, September 8

[Noted that U.S. forces in Beirut returned fire and silenced a Druse artillery post.]

Jean Kirkpatrick came in with a pitch for my addressing the opening session of the U.N. There is reason to believe the Soviets may be planning a surprise—like Andropov showing up. King Hassan of Morocco told her Jumblat (Druse) is not really in charge—his family is in Syria & they (the Syrians) give him orders.

Big Cabinet meeting on our program through Justice dept. to wipe out legal discriminations against women. We've changed 27 laws, have 60 more in process & today approved some more. Routine personnel meeting. Then some visitors—the 2 heroic pilots who safely landed their plane saving the passengers even though they themselves were on fire.

[Visit from Joanna Stratton, author of Pioneer Women.]

Talked to Geo. S. in Madrid—he terminated the meeting with Gromyko who insisted on repeating the Soviet lies about the Korean Plane Massacre. George says our allies may be hanging with us on taking more action against the Soviets. We'll know more tomorrow.

Friday, September 9

An interview for the Mon. edition of "Time." All of it in the Korean plane shoot down. I have no idea how it will turn out—Larry Barrett who did a thoroughly inaccurate book on me in the interview.

[Memorial service for the "KAL massacre."]

The Soviets have stepped up their propaganda drive to point us as the villains, the KAL as a spy plane & themselves as protecting their rights.

[Closed-circuit TV meeting with Republicans in Arizona; phoned Senator Daniel Evans (R-WA).]

Saturday, September 10

Met with George S 10:30 A.M. Nancy had left for Phoenix—back Sun. Nite. George reported in full on meeting with Gromyko. No doubt Gromyko was on the defensive & "discombobulated." I think it was our round. We've learned by continuing to electronically process the tapes to bring out the few unintelligible lines that a Soviet pilot did report firing his canon. We don't know if that was at the KAL or as a signal—"traces." The Japanese tapes of the Korean transmissions give no hint that the pilot was aware of the Soviet planes even being in the air. We made this new information public.

Our main meeting was on Lebanon. The situation is worsening. We may be facing a choice of getting out or enlarging our mission. Chfs. of Staff want to send the New Jersey. I'm concerned as to whether that won't have a bad morale effect on our friends in Central Am. We're going to move her thru the canal & off shore in the Atlantic before seeing whether she should head for Lebanon.

Sunday, September 11

NSC is meeting without me on Lebanon re a new Cable from Bud Mac-Farlane. Troops obviously PLO & Syrian have launched a new attack against the Lebanese army. Our problem is do we expand our mission to aid the army with artillery & air support. This could be seen as putting us in the war. George S., Bill C. & Jim Baker have just left me at 2 P.M. to get more info. on what is happening and where our partners in the MNF stand. Contingent on what they learn I've ordered use of naval gunfire. My reasoning is that this can be explained as protection of our Marines hoping it might signal the Syrians to pull back. I don't think they want a war with us. If it doesn't work then we'll have to decide between pulling out or going to the Congress & making a case for greater involvement. NSC will meet again at 6 P.M. Meeting didn't change anything so I've called for use of navy fire power & air strikes if needed.

[Monday, September 12: NSC meeting on Lebanon, situation there quieter; addressed local politicians group, and then Hispanic Heritage gathering; appointed Katherine Ortega as U.S. Treasurer; met with commission on competitiveness in international trade, Japanese business leaders; cabinet meeting on unitary tax policy on international companies; approved extension of unemployment insurance; reception for Republican contributors from Ohio and Michigan.]

Tuesday, September 13

[Visit from Senator Daniel Evans (R-WA) and family.]

P.M. Mugabe of Zimbabwe arrived. I was worried that he tended to filibuster. He talked a monologue for 20 min's. Got around to our—as he put it—intervening in Angola's affairs because of our effort to get Cubans out of Angola. Then moved to El Salvador & Nicaragua. I caught him taking a breath & interrupted. We can only persuade So. Africa to allow Namibia's independence if we get the Cubans out. Then I gave a history lesson on Central America. He's a very opinionated man & his country has voted against us in the U.N. even more times than the Soviets.

[Cabinet meeting for Shultz report on Korea, Lebanon, and other issues (no details recorded in diary); other reports on drought in the Midwest and farm economy; met with educators; reception for Republican Eagles.]

Wednesday, September 14

N.S.C.—Saudi Arabia may exert more pressure on Syria. Things are a little suspended in Lebanon.

Met with Eliz. Dole & our F.A.A. people who are going to the ICAO meeting in Canada. They intend to rally the rest of the 25 nations there to taking action against the Soviets in civil air affairs.

[Interview with journalists from labor, Hispanic, and religious publications; new poll showed improvement; meeting with Phil Gramm and others on appropriations strategy in Congress.]

A round table on the latest violation of classified info. The week-end meetings on Lebanon strategy appeared in the press a day later hurting our strategy there. I've ordered Justice to investigate. Tonite—speak to Hispanic Repubs. at a dinner.

[Thursday, September 15: meeting for Pres. António dos Santos Eanes of Portugal, who maintained that former Portugese colonies Mozambique and Angola were ready to move away from Soviet influence; economic briefing to discuss differences of opinion over

need to raise taxes; report on acid rain; farewell for speechwriter Aram Bakshian; photo
sessions and visits, including one by Clare Boothe Luce.]

Friday, September 16

[Report of improvement in Central American situation; ceremony honoring contribution of Hispanics to military; met coalition for tuition tax credits; interview with News-week on Lebanon and Marines; met Belgian prince studying at Stanford; lunch with Republican National Committee.]

Met with Geo. Shultz briefly & then proceeded to a Nat. Security Planning Group meeting. We approved some re-writing of the directive to C.I.A. for Covert activities in Nicaragua.

Off to Camp David.

Saturday, September 17–Sunday, September 18

It was good to see the Camp. Weather was good & we rode Sat. We also had a talk about "to run or not to run." I think we've agreed that I'll have a go at it. I think I offer the best chance of winning—but we both agree it won't be easy. Word came that Migs flown by Afghans—at least we're supposed to believe that—bombed a village in Pakistan.

The Sun. panel shows were frustrating to me. Everyone talks about what should or should not be done but they don't know the real facts. Their opinions are based on their theories. I'm really upset with Geo. Will. He has become very bitter & personal in his attacks—mainly because he doesn't think I've done or am doing enough about the Russians & the KAL007 massacre. He also believes I should ask for increased taxes.

[Returned to W.H.; attended PBS's "Young Talent" series, with Leontyne Price and singers from the Metropolitan Opera.]

Monday, September 19

NSC—Our Navy guns turned loose in support of the Lebanese army fighting to hold a position on a hill overlooking our Marines at the Beirut airport. This still comes under the head of defense. To allow those who have been shelling our Marines to take that position would have made the Marine base untenable.

[Telephoned football coach George Allen and spoke at a physical fitness gathering; observation of the tenth anniversary of Executive Women in Government.]

We met later—George S., Cap etc on a compromise we hope Cong. will agree to regarding the War Powers Act & Congressional approval of the Marines being in Lebanon. Sen. Baker thinks he can get it through the Sen. Among other things it would settle their presence in Lebanon for 18 months.

[Met with Senator Alan Simpson (R-WY) and promised support for his immigration *bill; met with two Democratic representatives regarding disarmament; Maureen staying at W.H., discussed gender gap over dinner.]*

Tuesday, September 20

An N.S.C. briefing on Cap's trip to China. I believe relations have definitely improved even though they still feel like they have to make a little noise now & then about Taiwan.

Off to Columbia, S. C. Announced to the press 1st that we had reached an agreement with a bi-partisan group in Cong. on Lebanon & the War Powers Act. If passed we'll all stand together for an 18 month extension of the Marines stay in Lebanon. I will—in signing—voice my reservations about the constitutionality of the War Powers Act.

[Flew to South Carolina, received honorary degree; reception for Senator Strom Thurmond (R-SC); political dinner; report of shelling of U.S. ambassador's residence in Beirut, no casualties; returned to W.H.]

Wednesday, September 21

[NSC briefing, but no new reports on Beirut shelling.]

Worked last night on draft of U.N. speech (next Mon.). Met with speech writers to see if we can't re-focus on the idea that people don't start wars—govt's. do. If govt's. met—all determined there must not be a war everything including arms reduction would fall into place.

[Presented award for Boys Club; lunch with regional broadcasters; private meeting with Shultz, mainly on Lebanon (no other details recorded); met Pres. Roberto Suazo Cordova of Honduras.]

Finished day with some tapings. Upstairs to learn that Patti has been storming at Ed Hickey about S.S. protection again. We've agreed that if she wants out—after she's heard their estimate of risks then let her out. She's nuts but we can't battle this way forever. It isn't fair to the agents.

Thursday, September 22

More on Lebanon in NSC briefing—still no cease fire. French however took shells on their positions & suffered more casualties. They for 1st time scrambled 8 planes from their carrier & bombed PLO batteries in the mountains.

[Addressed W.H. Conference on Productivity; signed proclamation for Business Women's Day; desk work, including U.N. speech; visit from Special Olympics poster child, then from Epilepsy poster child; met boxer Hector Camacho; accepted copy of book Keepers of the Sea *by Fred J. Maroon and Edward Latimer Beach; ambassadorial formalities; met with W.H. Fellows; reception for American Women in Radio and TV; reception for board of Smithsonian; home to Mrs. Reagan and Maureen.]*

Friday, September 23

NSC on Central Am. Things are looking much better in El Salvador which is probably why it's no longer on the TV evening news or the front pages. We've trained more units & their army is in the field not the barracks. Nicaragua is also feeling the squeeze.

[*Teleconference with regional party leaders; lunch with John and Sandra Day O'Connor; private meeting with Shultz on Lebanon, report that President Hafez al-Assad of Syria had two new demands; taped* The Merv Griffin Show *with Mrs. Reagan; swearing-in of appointee; reception for National Aquarium Society, which saved capital's aquarium by making it a private institution; desk work upstairs.* **Saturday, September 24:** *photo session in gym; radiocast; attended rodeo.*]

Sunday, September 25

[*Flew to New York City for United Nations opening; checked in to Waldorf-Astoria; lunch with Secretary General Javier Pérez de Cuéllar, noting, "He's a good man & really trying to make the U.N. effective."*]

After the lunch Nancy & I visited Cardinal Cooke who is dying of Leukemia. A brief prayer ceremony had been scheduled in his private chapel—without him of course. But then we went to his bedroom & concluded the final prayer with him. He is a brave & good man & has been most supportive of what we're doing.

[*Meetings at Waldorf with Prince Sihanouk of Kampuchea (Cambodia) and with President Samuel Doe of Liberia; reception at U.N. hosted by Jeane Kirkpatrick.*]

Monday, September 26

Put on my iron undershirt & off to the U.N. Gen. Assembly. It seems many of the members carry weapons. The speech was very well received. Mrs. de Cuellar said it got the most applause of any U.S. presidential address to the U.N. The theme was arms reduction & peace.

[*Spoke briefly at hotel to meeting of Women's Sports Foundation.*]

Up to the suite & working lunch with King Hassan of Morocco. It was a most informative meeting with the King holding forth on the Middle East. His history of the Palestinians presented a view I'd never thought of. The problem is greater than the refugees in Lebanon & what to do with them. There are Palestinians in every Arab nation, 5 million of them unassimilated & wanting their own homeland. But where is the real estate for such a homeland.

Prime Minister Gandhi of India came for a meeting. Then an interview with the N.Y. Post & back to Wash.

In the East Room a reception for heads of Black colleges & U's. I signed a bill—just passed today that will be of great help to those inst's.

Tuesday, September 27

Turned out to be an easy day. After staff & NSC meeting—the latter had our Ambas. to S. Africa as a visitor. He brought us up to date on S. Africa's effort to changed Const. & give full equality to the "Colored" population. Also on efforts to get Cubans out of Angola.

[Address joint meeting of International Monetary Fund (IMF) and World Bank; lunch with President Mauno Koivisto of Finland; scheduled activity was postponed, so president took afternoon off; Mrs. Reagan suffering from a cold; fund-raising dinner for Senate Campaign Committee, noting, "Henry K. as toastmaster was very funny." **Wednesday, September 28:** *meetings with congressional leadership and with members concerned about vote for further military presence in Lebanon; administration bill on Lebanon passed later in the day; met with educators honored for excellence; report from Cabinet Council on Management and Administration, on schedule to reduce number of federal employees by seventy-five thousand; met with Mark Fowler, chairman of FCC on syndication rights for network; honored America's Cup crews.]*

Thursday, September 29

Prince Bandar reported in with message from King Fahd of Saudi Arabia. He has done a great job in helping bring about the Lebanon cease fire. He is about to be the Saudi Ambassador to the U.S.

P.M. Margaret Thatcher arrived. She & I had an hours talk before lunch mainly about the Soviets & what it would take to get back into some kind of relationship. Then we continued on a number of subjects at lunch. I don't think U.S.-U.K. relations have ever been better. After she departed we had a Nat. Security Planning Group meeting re the "Start" talks. We agreed on some strategy as Gen. Rowny prepares to leave for Geneva.

[Rose Garden ceremony for professional and amateur hockey teams, with square of imitation ice, goal, and goalie, noted, "They asked me to hit a puck & believe it or not I scored a goal"; visited by couple from Dixon, Illinois (boyhood home); photo session; visited by ballet dancers who won competition in Bulgaria; Willa Shalit took mold for face mask; noted Senate passed bill allowing Marines to stay in Beirut for eighteen months.]

Friday, September 30

The usual staff & N.S.C. meetings. On NSC it seems there was a breakdown in communication and an issue came in for decisions by me that had not included the legis. section in its meeting regarding options. Naturally I should have some input as to what the situation might be when it went to the Congress. I don't know if this was accident or a little "turf" battle.

Pres. of ABC TV news brought Peter Jennings in to see me. We had never met.

Major part of day was spent with Pres. Mubarak of Egypt. We covered a lot of territory. He has some real ec. problems. We're trying to help with that as well as the modernization of his military. He's really turned off on the Soviets. I think we have a solid friendship going there.

Later Jerry Carmen brought former N.H. Gov. Mel Thompson in for a brief visit. We were pretty close as Gov's. together. Now he is convinced I've abandoned my conservative principles. We discussed a few points then he left me with a lengthy letter listing my sins. I'll reply of course but not sure I can convince him. I'm afraid he gets some of his ideas from the Howie Phillips crowd.

Dick Wirthlin came by with the tracking surveys. I'm up on job rating, the ec. etc. But on foreign policy—Lebanon I'm way down. The people just don't know why we're there. There is a deeply buried isolationist sentiment in our land.

Decision was reached today to postpone the Philippines, Thailand, Indonesia trip—probably til Spring & just do Japan & Korea in Nov. Congress was supposed to go home in Oct. but now they are staying thru Nov. There's no way I could be gone for 3 weeks now. I've written Marcos explaining & Mike D. is going out there on the weekend to meet them all personally & explain.

Saturday, October 1

Radio broadcast—our job training program officially begins. Golf game laid on at Andrews A.F. Base—Don Regan & Geo. B. Then dinner party for Princess Margaret Rose.

I've never played worse golf. For the 1st time I had actually forgotten the basics which like swimming I'd always assumed would never leave me. I bent my left arm on the back swing, kept my weight on my right leg & had a loose grip on the club. Well at least I realized all that before the game was over.

The dinner party was great—everyone had a great time. Marvin Hamlish entertained. He was brilliant. The Wyatts & Claudette Colbert were our house guests.

[Sunday, October 2: *quiet day.* Monday, October 3: *swearing-in ceremony for Katherine Ortega as U.S. Treasurer; met with President Aristides Pereira of Cape Verde; issues lunch; addressed group for Minority Business Week; met with news directors; launched Combined Federal Charity Campaign in Rose Garden ceremony; met with Scowcroft Commission; with senators and congressmen on START strategy; reception for Supreme Court justices; worked out in gym; spoke at Heritage Foundation dinner.*]

Tuesday, October 4

[*Visit from President Karl Carstens of West Germany.*]

After our meeting the day began to resemble yesterday. I met with the Am. Bus. Conf.

Then saw Ed Rowny off to Geneva in a Rose garden ceremony. Secretary

Hodel came in to present a copy of the annual energy report required by Cong. A Cab. Council meeting on the problem of phone rates now that the F.C.C. has broken up the Bell system. Pretty soon the discussion was based totally on pol. considerations. I finally reminded everyone we came here to do what was right not what was politically expedient. That's the way it's going to be. Then a taping session for 6 dif. affairs. Up to exercise & get into black tie for the Carstens state dinner. I left out a Cong. meeting. A large group of Dem. & Repub. Reps. & Sen's. in the Roosevelt room on the issue of the arms reduction talks. There is still that question of my sincerity about the talks. I guess I got carried away & did a speech on how I really feel. Chalk up another 1st in my experience in Wash. They came to their feet in a standing ovation.

Wednesday, October 5

GOP leadership to start the day. They have a slightly different set of priorities than I do on what they should be trying to pass before recess but we agree on some of the most important items.

[Visit from Oklahoma City patrolman hurt in accident while escorting the president; ceremony for Caribbean Basin Initiative (CBI); interviewed by AP board of directors; spoke to Department of Labor employees regarding Job Partnership Training Act; desk work. Thursday, October 6: presented medals to two heroic young people; interviewed by TV executives; lunch with Vice President Bush; interviewed by Washington Times columnist Donald Lambro; economic briefing; personnel meeting, noted, "Bill Smith has a man he wants desperately & we have another candidate. I think I'll have to defer to Bill"; gave medal to aircraft designer Kelly Johnson; military ceremonies; report from Commission on Small and Minority Business; barbecue for Congress members, entertainment by the Oakridge Boys. Friday, October 7: flew to Louisville, Kentucky; spoke to National Federation of Republican Women; two fund-raisers for congressional candidate and former baseball pitcher Jim Bunning; returned to W.H., met by Mrs. Reagan, Ron, and Doria; went to Camp David.]

Saturday, October 8–Sunday, October 9

Rode both afternoons. The weather was magnificent. The usual movies at night. On Sun. Ron & Doria left by car to fly back to N.Y. We stayed on because Mon. was Columbus Day. Sun. evening Jim Watt called from Calif. & tendered his resignation as Sec. of Interior. I accepted with real regret. He's done a fine job. True he has an unfortunate way of putting his foot in his mouth but he's really the victim of a 2½ year lynching. He knows he no longer can be effective with Congress so he's being a bigger man than his detractors.

Monday, October 10

Columbus Day. In the morning at Camp D. I ran the tape of the movie

ABC is running on the air Nov. 20. It's called "The Day After." It has Lawrence, Kansas wiped out in a nuclear war with Russia. It is powerfully done—all $7 mil. worth. It's very effective & left me greatly depressed. So far they haven't sold any of the 25 spot ads scheduled & I can see why. Whether it will be of help to the "anti nukes" or not, I can't say. My own reaction was one of our having to do all we can to have a deterrent & to see there is never a nuclear war.

Back to W.H.

Tuesday, October 11

Congress in recess for the week—it about seems as if you can feel it in the air.

[Mrs. Reagan in NYC for appearance on Good Morning America; *NSC briefing; met Ambassador to the Vatican Bill Wilson, in U.S. for funeral of Cardinal Cooke.]*

Foreign Minister Wu (P.R.C.) in for a visit. Everyone felt he probably wouldn't raise the Taiwan question. That's the 1st thing he did raise. I repeated our position & we went on from there. All in all a rather pleasant visit—no strain.

After an issues lunch came upstairs to Study & caught up on necessary reading—secret reports on Chinese nuclear capability, defense study on possible cooperation from Israelis if Soviets attacked Middle East. A few other items I could never made much headway with in the Oval office.

[Attended World Series in Baltimore.]

Wednesday, October 12

NSC & more on the Arms negotiations. Rumor has it the Soviets may take a walk in a last effort to sway public opinion in Europe. Some on our side want us to come up with an additional proposal. That is lousy negotiating strategy. It's time for the Soviets to come up with a proposal of their own. We can't keep changing our proposals every time they say "nyet."

[Columbus Day lunch at Italian Embassy.]

Geo. S. & I met in office after lunch. Israel is facing real ec. problems & is talking of switching it's money to U.S. Dollars.

[Taped message for Time *magazine's sixtieth birthday; received autographed baseball bat from Tommy Lasorda.]*

Now I'm up in my study waiting for Nancy to come home from N.Y.

Thursday, October 13

N.S.C.—report on South Korea. Pres. Chun is resisting pressure by his people to take some military action against N.K. in retaliation for the bombing in Burma believed to have been done by N.K. terrorists. It killed 16 people, many of them high S. K. govt. officials & Cabinet members. Pres. Chun is following the right course.

We had our 1st Budget overview—an optimistic sum up of what we've accomplished so far in savings. Figured in real dollars we have reduced domestic spending 10% over the 4 years ending with '84.

Good lunch with George Bush. We talked of some personnel problems.

Sen. Laxalt came by—I agreed to formation of campaign committee which technically makes me a candidate.

[Announced federal partnership with schools, pairing departments and offices with specific schools; ambassadorial formalities; received award from Hugh O'Brien's youth foundation; appeared before meeting of Evangelist Christian Women; made surprise announcement of Bill Clark as new secretary of Interior, noted, "Bill I think just finally got caught up with the 18 hour days and 7 day a week routine at N.S.C."]

Friday, October 14

Met with George S. & Bud MacFarlane re Lebanon. It really is a nest of adders & we have some tough decisions to make.

Jim Baker wants to take the NSC post. I was willing but then found great division & resistance in certain quarters. I finally decided that to ignore this & go ahead anyway would leave me with a permanent problem. Mike D. was going to move up to Jim's job. When I put everything on hold, Jim took it well but Mike was pretty upset. It was an unhappy day all around.

[Signed proclamation for World Food Day; National Security Planning Group (NSPG) meeting on Lebanon—no decisions, just a listing of the problems; interview with religious writer; left for Camp David; commented, "Not a pleasant evening what with all the hassle over the NSC spot."]

Saturday, October 15–Sunday, October 16

Friday evening I called Pat O' Brien in St. Johns hospital in S. M., Calif. He's just had a prostate removal. He sounded fine & told me a joke—Irish of course. On Sat. I called his widow Heloise. He had died suddenly of a massive heart attack.

The weekend weather was magnificent. We rode Saturday afternoon. The phone was busy. I'm being lobbied on behalf of Jeanne Kirkpatrick for the NSC job. I fear there would be bad chemistry with Geo. S. & State. I lean toward Bud MacFarlane. Some lobbying is going on for Brent Scowcroft. I'll have to decide tomorrow. Jeanne wants out of the U.N. & I don't want to lose her to the admin. She may have her heart set on Bills job. I'm going to try & find something else & then ask Jim Baker if he'd like the U.N. job.

I spent some of my time picking up acorns—a big bag of them. I'm going to give them to the squirrels outside the Oval office.

Monday, October 17

N.S.C.—mostly about how to announce appointment of Bud MacFarlane as Nat. Security Advisor.

Sen. Laxalt & the basic campaign committee came in for letter signing which makes me a candidate in the eyes of the Elec. Commission.

Cab. meeting was on legis. having to do with reforms in tax structure mainly on women's issues such as tax credits & deductions for child care centers etc.

Lunch with Geo. Will. I think we are straightened out on some differences about taxes & the Soviets. We parted on very good terms.

Jeanne Kirkpatrick came in. We talked for an hour. She has wanted to come back to Wash. from the U.N. for some time & hoped she would be given Bill's job. I offered her a position as Counselor to the Pres. on International Policy. I couldn't convince her it was a job where she'd have a real voice in determining policy. Finally left it with me not accepting her "no" & her promise to think about it during the 2 months or more she'll still be Ambas. at the U.N.

[Photo session; announced appointment of Bud McFarlane as National Security Advisor; phoned parents of Marine killed in Beirut; videotapings; dinner with Maureen and friend. **Tuesday, October 18:** *morning briefing on negotiations over auto quotas in upcoming Japan trip; National Security Planning Group (NSPG) meeting on Lebanon; lunch with editors of ethnic papers; rehearsal for press conference.]*

Wednesday, October 19

Discussion about a successor to Bud MacFarlane in Beirut. It won't be a problem we have several good choices. Presented Presidential awards to selected Math & Science tchrs. from all over Am.

Jim Watt came by. He's in good spirits. He knew that in carrying out my policies his days would be numbered. Actually he thought he would have had to leave sooner than he did. He gave me a report on his stewardship & it reveals the hypocrisy of the Environmental lynch mob. I don't think the Dept. of Interior or our Nat. Parks & wild lands have ever been in better shape.

[Greeted business leaders volunteering in bond sale campaign.]

After lunch went to the Air & Space Museum to celebrate the "Nasa" 25th birthday. Saw a film on the "Shuttle" in a process called "IMAX." It is the most spectacular thing of it's kind I've ever seen.

[Two-hour briefing for press conference, then staged the conference.]

Thursday, October 20

P.M. Craxi of Italy for a lengthy meeting & lunch. He's a different kind of Italian official. He's socialist but totally anti-communist. He's been very supportive of us & firm on the Pershing II deployment in Europe. A Cabinet Council meeting on 2 subjects—no decision. One was whether to sell Alaska oil to Japan. It would put a hole in our relations with the maritime industry & union. The other issue was the issue raised by the F.C.C. as to whether the TV networks should be allowed to own syndicated rights to programs. I believe it would amount to a monopoly by the networks.

[Presented the National Security Medal to Richard Helms; send-off for Olympic Ski Team.]

Friday, October 21

A Budget overview meeting. We've actually achieved more savings in domestic spending than we're given credit for. We've only been given about 50% of the spending cuts though that we've asked for.

[Met with business leaders regarding natural gas price deregulation; interview with David Hartman for Good Morning America.*]*

Henry Kissinger & the Commission on Central Am. came by. It's amazing how much consensus there is now that they've been there, that what we're doing is right & Nicaragua is the real villain.

I taped my Sat. radio bit & we left for Augusta, Ga. We all stayed in the Eisenhower cottage at Augusta Country Club—home of the Masters golf tourney. There were the Shultzs, the Don Regans & the Nick Bradys. We turned out to be a fun group. Dinner at the Club then home to bed. About 4 A.M. or so I was awakened by Bud McFarlane. I joined him & George S. in the living room. We were on the phone with Wash. about the Grenada situation. I've OK'd an outright invasion in response to a request by 6 other Caribbean nations including Jamaica & Barbados. They will all supply some forces so it will be a multinational invasion. Finally back to bed for a short while & then up for golf. I was better than at Andrews but still not good. I guess you have to play more than 4 times in almost 3 yrs.

We reached the 16th hole & suddenly were stopped. A man with a gun was holding hostages in the golf shop demanding to talk to me. I got on the car phone & tried 5 times to talk to him but he always hung up on me. David F. was a hostage but talked his way out on the grounds that he could get the message to me. Lanny Wiles remained a hostage for almost 2 hours before he made a break & got away. One by one the hostages got away one way or another. The gunman was taken into custody. Meanwhile we had taken a back road & reached Eisenhower cottage.

And so to bed—after a pleasant dinner.

Saturday, October 22–Sunday, October 23

About 2:30 in the morning awakened again: This time with the tragic news that more than 100 Marines in Beirut had been killed by a car bomb driven by a suicide driver who drove the truck right into the H.Q. building & blew up with it. All our plans changed—we arranged to depart the cottage at 6:30 A.M. & go back to Wash. Of course by this time it was Sun.

Oct. 23. I've spent the day in meetings on this & Grenada. We're going to go on with the invasion. Tonite our men are staging a landing to gather intelligence. If everything is OK, tomorrow night is D Day—well actually it will be early morning Tues. Meanwhile Gen. Kelley (Marines) is leaving for Lebanon. We all believe Iranians did this bombing just as they did with our embassy last April.

Monday, October 24

This was really a Monday. Opened with NSC brf. on Lebanon & Grenada. Lebanon gets worse as the death toll climbs. More bodies are found & more critically wounded die. Ambas. Hartman (Russia) came by. He confirms what I believe: the Soviets won't really negotiate on arms reductions until we deploy the Pershing II's & go forward with MX. He also confirms that Andropov is very much out of sight these days.

Phoned Tip & Howard Baker to express hope they'd stay firm on keeping the Marines in Lebanon—both said yes.

[Lunch with regional radio and television news editors.]

The Pres. of Togo visited. He's anti-communist & pro West. A meeting with the Joint Chiefs—they outlined the final details for our move on Grenada scheduled for 9 P.M. take off. No evidence of any moves by Cuba.

Jack Anderson came by with some ideas about ed. & the lack of history in our schools. Also an idea to give people a chance to sound off about legitimate beefs with govt. We're looking his ideas over.

So far not even a tiny leak about the Grenada move.

[Photo sessions.]

Then at 8 P.M., Tip, Jim Wright, Bob Byrd, Howard Baker, Bob Michel & all our gang met upstairs in the W.H. & we told them of the Grenada operation that would take place in the next several hours. We gave them the complete briefing. In the middle of the meeting Margaret Thatcher called. She's upset & doesn't think we should do it. I couldn't tell her it had started. This was one secret we really managed to keep.

Tuesday, October 25

At 7:30 A.M. without breakfast, met with Mrs. P.M. Charles of Dominica who came up there to join me in telling the press. Then a meeting with the cabinet room full of Congressmen. We gave them the briefing with the fact that around 5:30 A.M. our forces had landed on Grenada at 2 points & had both airports secure. Went from there to the press room—told them & presented P.M. Charles who was magnificent. We both then joined the Cabinet & told them. We are taking some casualtys but the operation is successful.

Met with Lt. Gen. Ershad, Pres. of the Council of Ministers of Bangladesh.

He's quite a man & is taking that poverty stricken land out of statism & into the world of free enterprise. I liked him & his ministers.

An NSC meeting on trade matters with Japan—preparing me for our trip. Japan is still holding stubbornly to trade practices that are unfair. I'll have to really lean on P.M. Nakasone.

Yousuf Karsh brought me his book on his photographs. In the Red Room a meeting with Paul Laxalt, Dick Lugar & Bob Packwood. Bob has evidently regretted his attack on me & wants to start obeying the 11th Commandment. I'm all for that.

Wednesday, October 26

NSC meeting with Cap who is off to Canada for the NATO ministers meeting. He's presenting a plan for our further reduction of nuclear tactical weapons in Europe. We've already taken out about 1000 without it ever being noticed.

[Greeted winners of the New York City Marathon, including a competitor born with only one leg; meeting with Senators Paul Laxalt (R-NV) and Strom Thurmond (R-SC) on administration's crime legislation; spoke to Jewish Coalition of Republicans.]

Taping session for the events I've cancelled because of Grenada. Speaking of that it looks like the mission is almost accomplished—a few hotspots left but we've taken 100's of Cubans prisoner. The Marines, the A.F. & the Army & Navy planned beautifully & executed even better. We have 6 killed, 8 missing, & 33 wounded at last count. George B. is back from Lebanon—stopped by to report on that tragic scene. Morale of the Marines is high. I'm re writing a speech for Nat. TV tomorrow night.

Thursday, October 27

N.S.C.—naturally on Grenada. Success seems to shine on us & I thank the Lord for it. He has really held us in the hollow of his hand. We cleared the schedule so I could work on the speech. I'd managed to get about half of it done last nite by staying up til 1 o'clock.

[Meeting with Governor Bob Orr (R-IN) bringing proposals on gas deregulation bill; visit from Prime Minister John Compton of Saint Lucia, in Washington at W.H. request to speak to the press regarding Grenada; Shultz reported that the other multinational-force nations remained committed to staying in Lebanon.]

Everything is going well in Grenada. We're mopping up. We discovered a Cuban base, barracks, H.Q., Warehouse full of weapons. They were really going to move in & take over. The speech must have hit a few nerves. I did my best to explain what our Marines were doing in Beirut & then explained Grenada. ABC polled 250 people before the speech, the majority were against us. They polled the same people right after the speech & there had been a complete turn around.

1000's of phone calls & wires from all over the country flooded us more than on any speech or issue since we've been here—10 to 1 in our favor.

Friday, October 28

NSC—mainly on Grenada. Our forces have been wonderful & most effective. We've captured about 700 Cubans, most of the Grenada military has faded back into the population leaving weapons & uniforms behind. We've brought a great many Americans out, particularly our medical students from the St. Georges U. Medical School—they kissed the ground when they reached the U.S.

We had a preliminary briefing on the upcoming trip to Japan & Korea. Then a budget review. We have only achieved ½ the savings we tried for & now our own agencies are asking for more money than we'd proposed. I think it's time for a lecture.

[Left for Camp David. **Saturday,** October 29: *lunched with Marines and Seabees; took ride.]*

Sunday, October 30

Watched the Sunday talk shows—subject Lebanon & Grenada. The press is trying to give this the Vietnam treatment but I don't think the people will buy it. They are still whining because we didn't take them on a guided tour the 1st day we were on Grenada. No mention of the fact that we've flown 180 of them onto the Island today. Just got a call—we've captured Austin—top villain on Grenada.

Monday, October 31

N.S.C.—Subjects Lebanon & Grenada. The attempt at reconciliation in Lebanon began in Geneva. Grenada is almost cleaned up—a little sniping here & there. We're trying to get Don Rumsfeld to take Buds place in Lebanon.

[Met with steel executives; signed appropriations bill for Health and Human Services and Department of Labor; report from Senator Pete Domenici (R-NM) on his trip to El Salvador and Nicaragua, describing Nicaragua as the villain; social meeting with officers of UPI; videotapings; Maureen arrived for visit.]

Tuesday, November 1

Day began with G.O.P. Cong. leadership—a full Cabinet room. Last night the Repub. Sen. very irresponsibly refused to pass an increase in the debt ceiling which is necessary if we're to borrow & keep the govt. running. After we gave them all a rundown on Lebanon & Grenada we took up the budget & the necessary legislation. I sounded off & told them I'd veto every d--n thing they sent down unless they gave us a clean debt ceiling bill. That ended the meeting.

[NSC meeting and a brief on upcoming Japan trip; signing ceremony with Mrs. Reagan for National Drug Abuse Week.]

Stu Spencer came by with a monster book (like an L.A. phone book)—it is the campaign book for if & when I declare. I'm supposed to read it.

Cab. meeting on budget. I told them to go back & trim their '85 budget requests some more.

[Chat with Tom Vail, publisher of Cleveland Plain Dealer; ambassadorial formalities; visit from four-year-old boy with rare disease, let him sit behind the desk in the Oval Office.]

Our Marines have set sail for Lebanon—from Grenada. On the way they stopped at a tiny island north of Grenada & cleaned out another warehouse filled with weapons & ammunition.

Wednesday, November 2

N.S.C.—Grenada is declared at peace. Hostilities officially declared over. The Marines are on their way to Lebanon & a number of Army Units are being withdrawn. On Jamaica, P.M. Seaga has ordered several Soviet diplomats ousted for planning to murder a govt. official there.

Formal signing ceremony of bill naming Martin Luther King Jr. a nat. holiday. Mrs. King & the children were present as well as sizeable crowd. It went very well.

[Desk work; met new members of press corps, Jim Hildreth of U.S. News & World Report, and Charlotte Saikowski of Christian Science Monitor.]

Up to the gym—a massage & work on speeches for Japan & Korea.

Thursday, November 3

Don Rumsfeld came in—he will be our new envoy to Lebanon. That's quite a sacrifice for him to make—a leave from his job as Pres. of Searle Corporation.

Another briefing on the Asia trip.

Met with Conservative Repub. Sens. who are playing games trying to come up with amendments to the debt ceiling bill. We must have that bill or govt. checks will start bouncing.

[Further briefing from Lee Morgan and Dave Packard on Asia; James Roosevelt and others described plan to restore the former presidential yacht Potomac and open it as a museum; met with congressmen behind Radio Marti bill; met with winner of Hoover engineering medal; signed proclamation regarding diabetes; greeted people working regularly through industries organized for the blind, also leaders of International Baptist Movement; spoke to reunion of 1980 campaign.]

Friday, November 4

An early start for Camp Lejeune in N. Carolina: A memorial svc. for the

Marine dead of Lebanon—the suicide bombing. It was a dreary day with constant rain which somehow seemed appropriate. All the ceremonies were outdoors rain or no rain. It was a moving service & as hard as anything we've ever done. At the end, taps got to both of us.

The only indoor part was a receiving line meeting the families of the deceased. They were so wonderful, sometimes widows or mothers would just put their arms around me, their head on my chest & quietly cry. One little boy, 8 or 9, politely handed me a manila folder saying it was something he'd written about his father. Later when I could read it I found it was a poem entitled "Loneliness."

We helicoptered back to Cherry Point & there I addressed a crowd of Marines & families. Before leaving Lejeune I spoke briefly to the families I'd met in the line. That was ad-lib & the Lord was with me—the right words came. We flew back to Wash.—a few meetings & then to Camp David where it was snowing.

Saturday, November 5–Sunday, November 6

It snowed most of Sat. but the snow melted as it fell. Spent day on the briefing books for the trip. I think they tell me more than I need to know. A movie Sat. night that was d--n near pornographic. It was embarrassing.

Sunday we flew out right after lunch. I think Security had some worries about a plot to rocket the M-1.

[Phoned parents of soldier severely wounded in Grenada; met with Prime Minister Edward Seaga of Jamaica, "a staunch ally in the Grenada action."]

Monday, November 7

[Staff and NSC meetings; assignment of troops to Grenada rebuilding effort; ceremony for Grenada soldiers and students rescued by them, noted, "Ten yrs. ago kids of the students age hated anyone in a uniform—remember Kent State? It was heartwarming, indeed thrilling to see these young people clasp these men in uniform to their hearts"; taped interviews for Japanese media.]

A long luncheon meeting re contingency plans in event Iran tries to close Straits of Hormuz. All agree we can not let the Straits be closed. Then we touched on a touchy one. We believe we have a fix on a headquarters of the radical Iranian Shiites who blew up our Marines. We can take out the target with an air strike & no risk to civilians. We'll meet 7 A.M. tomorrow as to whether we order the strike now or while I'm away.

Long briefing on trip—my mind can't hold all they are trying to pour into it.

[Met with Philadelphia Urban Education Foundation; videotapings; packed for Asia trip.]

Tuesday, November 8–Wednesday, November 9

Off to Asia. A farewell ceremony in the East Rm., then Marine 1 to Andrews A.F. Base. Began the day however with a short meeting re a possible air strike in Beirut against those who murdered our Marines. Decided we don't have enough intelligence info as yet.

On my way, 7½ hours to Anchorage Alaska. There is a 4 hr. time change so it was only 11:30 A.M. but at this time of year it seemed like evening. We met with military & civilians 8,000 of them in a hangar. We were warmly received by the most enthusiastic crowd.

On our way over the Pacific another 7½ hours to Tokyo. Before we left, U.S. learned that Geo. B. had cast a tie breaking vote in the Sen. to kill an effort to delete Binary (chemical) weapons from Defense bill. Later in flight got word Sen. passed Defense appropriation 86 to 6. We tried the new plan to minimize jet lag. We sacked out for 6 hours—didn't sleep all of them. Then got up & had a brunch on what would have been lunch time in Tokyo. Arrived there at 2:15. Helicoptered to Akasaka Palace (their guest house) a magnificent French style castle. Hustled into my clam hammer morning suit. The Emperor arrived for formal greeting, then we went to Imperial Palace for visit. Back to Akasaka for meetings with P.M. Nakasone. Day finally ended & had dinner in room, a massage & to bed.

It was a funny sensation. Having crossed the international date line it was now Wed. & yet it was the same day we had left Washington.

Thursday, November 10

[Watched riding exhibition at Meiji Shrine with Crown Prince Naruhito and Princess Nori.]

We then went to the P.M. residence for meetings & a luncheon. Mid-afternoon got back to Akasaka for a nap then up & into White tie & tails. I met with a group of Japanese who have raised $1 mil. for the R.R. scholarship fund at Eureka College. Then stopped by an organization of Japanese & Am. businessmen.

[State dinner.]

Friday, November 11

Addressed the Japanese Diet—1st U.S. Pres. ever to do so. Very well received. P.M. said no other head of state had ever received such applause. Then a reception for Diet leaders. Back to Akasaka for reception for Japanese & Am. press. Then helicoptered across city to the outskirts & landed at a school where we were greeted by 100's of school children waving Am. flags. From there we motored up into the wooded hills to Nakasone's hideaway—a tiny, typical Japanese home where we sat on the floor & had a real Japanese lunch.

Back at Akasaka the Emperor came over to officially say goodbye.

Our team had a short meeting to compare notes & see if we'd left anything undone regarding the problems we'd come to discuss. Then a meeting with our very fine Ambas. to the Philippines.

[Taped Saturday radiocast; interviews with Asian media; met with U.S. servicemen and women.]

The trip was a success. The Japanese had provided the utmost in security—22,000 men.

Saturday, November 12

A last goodbye with P.M. Nakasone & his very nice wife. We're very fond to them. Then it was Marine 1 & Air Force 1 & off to Seoul, Korea—a 2 hour & 5 minute flight.

[Formal greeting at Kimpo Air Base.]

President Chun & I then went in one car, Nancy & Mrs. C. in another. Security here was overwhelming because of all the threats against me by the N. Koreans. The drive into town was fantastic—more than a million Koreans lining both sides of the streets waving Korean & Am. flags. Signs that said they loved us & seeing their faces you knew they meant it.

Ambas. & Ceny Walker had moved out of the embassy residence so we could have it.

Almost immediately we were on our way to address the General Assembly. There too we were well received. Then back to Blue House for meeting with the Pres. & a luncheon—really Korean. More meetings after lunch. Just a short time for a nap & change into tuxedo then back to Blue House for a State Dinner. On way stopped at Embassy to meet with a group of leading Koreans & Americans. Could only stay a few minutes—we were delayed because one of the bomb sniffing dogs acted up about an elevator shaft. Other dogs were brought in & didn't act as if there was a problem. On to the State Dinner—a magnificent setting. My partner, Mrs. Chun. She is a most delightful person. She & Nancy buddied up like 2 school chums. Afterward Korean entertainment & it was magic, the grace & beauty of the performers was like nothing we've ever seen. I wish I could bring them to America.

Sunday, November 13

My day at the D.M.Z. We helicoptered into Camp Liberty Bell for an open air church service. We were less than a mile from N. Korean guns. The choir was made up of little Korean orphan girls. Our G.I.'s support & maintain the orphanage.

After service we went through the gate into the actual zone—a white flag flying on our car. We went up onto a 500 foot promontory to Post Guard Collier. I met a patrol just going out. The zone is patrolled night & day. We could hear the North Korean loud speakers spewing their propaganda.

After touring the bunker etc. we returned to Camp Liberty Bell where I addressed the troops. Then into the Mess to lunch with them. I've never been so proud. These men are the closest to an enemy of any Am. forces but their morale, their esprit de corps is unbelievable.

[Drove to Korean base to watch karate demonstration.]

Back to a worried Nancy for a little nap. I learned later that U.S. & Korean artillery was on alert to fire a barrage if the N.K.'s tried anything funny. Helicopter gun ships were standing by.

At 6 P.M. went back to Blue House for final meeting between our 2 teams. Then back to the house for dinner & a massage.

Monday, November 14

[Met with American businessmen.]

Then over to Blue House to pick up Pres. & Mrs. Chun. There is no doubt this was a most successful trip. The ride to the airport was again through crowds of more than 1 mil. people. Their happiness & affection was humbling.

Nancy on Sunday had met 2 small Korean children both requiring open heart surgery. She met them through a wonderful woman, Harriett Hodges, who has found & delivered for the "save a life movement" more than 600 such children. Nancy was bringing them to Am. on A.F.1.

[Flew home, a seventeen-hour trip.]

At 2 P.M. a meeting re the same Beirut problem we'd discussed before the trip. We have some additional intelligence but still not enough to order a strike.

The trip is over but I believe it was a 10 strike for the U.S. Both P.M. Nakasone & Pres. Chun have become fast personal friends & I return their friendship.

Tuesday, November 15

Back to the routine. Staff meeting—then G.O.P. leadership for a report by us on trip & then to hear from them on agenda. They are trying to adjourn on Friday but have a h--l of a full plate including passing an extension of the debt ceiling. Without that we face fiscal disaster. Sen. Bill Armstrong is leading the charge & while he's sincere, he just doesn't know what the h--l he's doing.

[Telephoned Republican Governors Conference.]

Clarence Pendleton Chairman of Civil Rts. Commission came in re the Congress's so called compromise to keep me from making new appointments. The boys are playing games but I think I can snooker them. Haircut, then meeting Sens. who are battling to get Tuition Tax Credit—Upstairs to get some mail done.

Wednesday, November 16

The Israelis burned Naba Chit known to be a camp of the Iranian Shiites believed responsible for the car bomb attacks. That was one of the targets we were looking at but didn't feel we had enough information yet.

Cab. Council meeting on commercial entry into the Space effort. Question should the govt. department in charge of this be Dept. of Transportation or Commerce. Both wanted it. I came down on the side of transportation.

Rcvd. word Brazil had accepted the I.M.F. terms. That is good news for the world financial markets.

I've OK'd some shifting of our Marines in Beirut. I believe the Lebanese can take over the airport while we move South to an area once controlled by Israel.

Met with Geo. S. about establishing a pipe line outside the bureaucracy for direct contact with Soviets.

N.S.P.G. meeting. We've contacted French about a joint operation in Beirut re the car bombings.

Italy Parliament voted—landslide for deployment of I.N.F.'s. F.C.C. withdrew for 6 months its regulation that would have allowed networks to syndicate shows. Good news. Closed day saying a few words at farewell reception for Jim Watt.

Thursday, November 17

N.S.C.—Surprise call from France—they were going ahead without us & bombing our other target in Lebanon. They took it out completely.

Lunched with V.P. Got a call on secure phone—Bill Casey wanted me to call Ted Stevens chairing a conf. comm. of Demos. trying to kill off covert aid in Nicaragua. I told Stevens to fight to the death. Got word late in day he'd stood his ground & we won.

[Met with newly elected president of the National Council of Catholic Bishops; swearing-in of Susan Phillips as chairman of Commodity Futures Trading Commission; Cabinet Council meeting on Human Resources, noted, "Had a slide presentation of how govt. redistribution of earnings programs etc. have undermined the family. This must be reversed. It's what we came here to do."]

Personnel time—some new appointments—a fine group. The leaders of the organization to research & find a cure for Alzheimer's disease plus Rita Hayworth's daughter Yasmin presented me with a plaque for helping to publicize their cause.

[Greeted nine-year-old girl who had written letter about the presidency; met Frederick von Hayek, one of the president's most admired economists; visit from editor of Phoenix Gazette; *videotapings; called father of soldier killed in Grenada.]*

Friday, November 18

[Staff and NSC meetings; farewell for retiring ambassador to Ireland; met with Sena-

tor Laxalt (R-NV) on appropriations bill much higher than administration wanted; met
with leader of organization of attorneys general, who supported crime bill.]

George Shultz & I had a talk mainly about setting up a little in house group
of experts on the Soviet U. to help us in setting up some channels. I feel the
Soviets are so defense minded, so paranoid about being attacked that without
being in any way soft on them we ought to tell them no one here has any intention of doing anything like that. What the h--l have they got that anyone would
want. George is going on ABC right after its big Nuclear bomb film Sunday
night. We know it's "anti-nuke" propaganda but we're going to take it over &
say it shows why we must keep on doing what we're doing.

A most sobering experience with Cap W. & Gen. Vessey in the situation
room—a briefing on our complete plan in the event of a nuclear attack. The
Chiefs have been working on it for 2 yrs. in reply to my request in October,
1981.

[Received word of confirmation of Bill Clark as secretary of Interior, passage by House
of International Monetary Fund (IMF) bill and the debt ceiling.]

Saturday, November 19

Slept in & it was great. Did radio broadcast. Lunched in the Solarium—1st
time in months. Bud McFarlane called on secure phone, has checked with all
concerned & consensus is that in view of French bombing in Lebanon we carefully catalogue potential targets & be ready for immediate retaliation in event of
another attack on our forces. I agreed but asked that we maintain intelligence
efforts to see if we can forestall another attempt.

Sunday, November 20

Nancy left this morning for Phoenix. I won't see her til Wed. at the ranch.
I'm lonesome already. I even watched an Errol Flynn movie—"Robin Hood"
this afternoon.

Bud called—the Israelis took out targets (air attacks) on Beirut—Damascus
Road. One was a choke point which could shut off Syrian supplies & ammo to
the guerillas. One plane shot down but pilot parachuted & was recovered. I have
no complaints with this. Rumor also going around that Assad (Pres. of Syria) is
dead. No confirmation.

[Monday, November 21: meeting with President Gaafar Nimeiri of the Sudan with
discussion of Qaddafi; Thanksgiving turkey presentation; met with President Spyros Kypri-
anou of Cyprus; ambassadorial formalities; complained of "a lonely evening & not a very
good night for sleep."]

Tuesday, November 22

[Memorial mass for John F. Kennedy; met with Israeli president Chaim Herzog.]

Bill Smith came by (top secret). It looks like F.B.I. wants to "lie detector" 3 of our people in their investigation of the leak some time ago of a top secret NSC meeting.

Don Regan came by for another top secret item. We're going to change our currency. Copying machine including new ones due on the market in '86 can duplicate existing paper money so realistically it's almost impossible to tell the difference. Already we estimate some where near $100 mil. of counterfeit in circulation.

[Visit from Cardinal Casaroli, Vatican secretary of state; awarded medal to female long-distance pilot; visit by Andy Williams to launch Christmas Seal drive; saw cowboy who had been badly injured in rodeo.]

Met with Don R., Jim B. & Paul Volcker. Paul believes he has to limit money supply & slow down economic recovery to reduce inflation. It's true inflation is going up some these past few months but we don't believe we should slow down the recovery.

[Reception for party donors; still lonesome. **Wednesday, November 23:** *flew to California and up to the ranch by helicopter; ranch work in afternoon.* **Thursday, November 24:** *[. . .] Bess, Patti, and Dennis staying at ranch.* **Friday, November 25:** *strong winds, no riding.* **Saturday, November 26:** *ranch work.* **Sunday, November 27:** *returned to W.H.]*

Monday, November 28

N.S.C. briefing for P.M. Shamir (Israel) visit. This 1st meeting was a stage setter for tomorrow's longer meeting. I let him know we wanted to strengthen our relations with the moderate Arab states so we could help bring peace to the area. We'll get into the nuts & bolts tomorrow.

[Bill-signing ceremony for appropriations bill for Departments of State, Justice, Commerce, and for Judiciary; gathering at W.H. of hundreds of self-sustaining handicapped people; signed proclamation designating 1983 to 1992 the Decade of the Handicapped, noting, "Our goal will be rehabilitation & opportunity for all the handicapped"; videotapings.]

Spent most of afternoon signing letters of condolence to next of kin of the 239 Marines.

Tuesday, November 29

Most of morning spent briefing for & meeting with P.M. Shamir. We met in a small session, 1st in the Oval office, then a larger meeting in the Cabinet Room & finally lunch in the St. Dining Rm. He & his team of course have spent 2 days in meetings at State, the Pentagon etc. I think things are well on track & a lot of suspicion etc. has been washed away on both sides.

I had some useful desk time then met with our own team about a controversial Dairy bill. I thought I leaned toward veto but after hearing from a bi-partisan group of Reps. & Sens. I decided to sign. A veto would have left us with the present Dairy program which is a Turkey. I believe the new one may be a step toward getting government eventually out of the Dairy business. Maureen arrived in time for dinner. She is really doing a job for the Nat. Committee.

Wednesday, November 30

N.S.C. for much of morning including a visit by Ambas. Prince Bandar (Saudi Arabia). He brought a message from King Fahd re Lebanon & the chances of wooing Syria away from the Soviets. I'm afraid his plan involves us separating ourselves somewhat from Israel. No can do.

Most of N.S.C. had to do with our going forward with research leading to a defense weapon against nuclear missiles. Some 50 scientists were persuaded to look at the problem after my March 23rd ('83) declaration. They started as skeptics & have wound up enthusiasts. We'll proceed.

Lunch was a Budget strategy meeting. It was downbeat. Dave S. projected some pessimistic projections based on what we could expert Cong. to do. I said the h--l with that. We can't go on with business as usual—business being us asking for spending cuts & being satisfied with less than half what we asked for. We can't go on with $200 bil. a year deficits ad infinitem.

Pvt. meeting with George Shultz. He brought me up to date with or on his meetings with P.M. Shamir. The P.M. doesn't think our defense people are very cordial to Israel but believes I am a trustworthy friend.

Cabinet Council meeting again on budget. I'm afraid I got on a soap box about taking our case to the people—this is a crisis situation.

[Signing ceremony for revenue-sharing bill.]

Thursday, December 1

Most of NSC spent on getting ready for visit by Pres. Gemayal of Lebanon. But also on whether to step up our artillery fire on Druze batteries lobbing shells in the direction of our Marines. So far we haven't done anything. We're a divided group. I happen to believe taking out a few batteries might give them pause to think. Joint Chiefs believe it might drastically alter our mission & lead to major increases in troops for Lebanon.

Met with Gemayal & his Cabinet. We emphasized need to work harder for expanding govt. to include some dissident groups. He made a good case that Syria has a block on the main factions.

Cabinet Council on "space" & where we go. The issue is whether to move on a program for a permanent manned space station. I'm for it as I think most everyone is but the question is funding such a new course in face of our deficits.

Dave Stockman who opposed my signing the dairy bill came in to brief me on how the bill came to be & how may Reps. & Sens. broke their word on concessions they'd promised in return for the bill. I was never told any of this while it was going on. We'll have to have a different procedure so this can't happen again.

[Attended PBS's "Young Talent" program at Shiloh Baptist Church, Leontyne Price hosting.]

Friday, December 2

A hectic ¾ of a day with contentious staff & Cabinet members on both sides of a couple of issues. One with NSC had to do with attacking Syrian targets around Beirut in response to shelling of our Marines. Some wanted to do this whether they were responsible for the shelling or not. I came down on the side of only responding if we knew our target was the source unless that target was in a populated area—then take a nearby target if it was of the same org. such as Druze or PLO etc. Fire at Syrians only if they had fired at us.

Another problem came up with regard to shutting off some still exiting exports to Libya. The exports were things the Libyans can get virtually anywhere so the action wouldn't hurt them at all. I said we should see if we could persuade all our friends to join the boycott—otherwise we'd sell.

[Appearance with high school students on C-Span program; left for Camp David.]

Saturday, December 3–Sunday, December 4

Grey cold day—did radio broadcast & we took a walk. Then it was back at the fireside where I caught up on reading & some mail. That evening received call from MacFarlane that the Syrians had launched an anti-aircraft & ground to air missile attack against our unarmed reconnaissance planes during one of their routine sweeps over Beirut. Permission from me was needed for a return air strike against the guilty batteries. I'd already received a call on this from Cap in Paris. I gave the order. Sunday morning got a call—we had taken out a communications center, some batteries & an ammo dump. Two of our planes (24) had been shot down. One pilot parachuted in safe territory & has been recovered. The other 2 in the 2nd plane parachuted in hostile zone—we've heard that one was machine gunned but we've also heard they are both prisoners. We're trying to get a confirmation & will open negotiations for their return.

Back at the W.H. made statement to press—am now going to Rockville Jewish community Center for a Hannakah ceremony. Then back to a reception honoring the 5 honorees of the J.F.K. Center For The Performing Arts—Frank Sinatra, Jimmy Stewart, Eliza Kazan, Katherine Dunham & Virgil Thomson. A posse of our Hollywood friends will be here at the W.H. for the reception.

It turned out to be a wonderful evening & a great show.

Monday, December 5

N.S.C. in the morning—subject Lebanon. We took out 11 anti-aircraft & missile launching sites, a radar installation & an ammo dump. Syria says they will hold our captured pilot (they're returning the body of the other) until we all leave Lebanon. Our press & TV are hostile to the point of being pro Syrian.

[Satellite teleconference with astronauts in the shuttle.]

A Budget appeal meeting on Nasa's request. I think we're OK there & can still start to plan a space station.

We now have a U.S. attorney in each of the 94 districts—I mean one appointed by us. They are meeting in Wash. I went over to the W.H. after lunch & spoke to them on our fight against crime & they're a part of that.

[Budget-review meeting; ambassadorial formalities.]

Met with Geo. S. who is off to meetings in Europe tomorrow. We see pretty much eye to eye on our problems in Lebanon & with the little Red brothers in Moscow. We have a back channel contact, we hope can get some common sense discussion going with them.

[Tuesday, December 6: NSC meeting, no new reports from Lebanon; met Premier Kennedy Alphonse Simmonds of St. Kitts-Nevis; budget meeting regarding EPA; interviewed for People *magazine; met with Armenian American leaders; dinner honoring General Jimmy Doolittle; telephoned family of Navy flier killed in Lebanon.* **Wednesday, December 7:** *visit by King Birendra of Nepal; appeared with volunteer group, Citizens for America; meeting on Veterans Administration budget; talked about tuition tax credits with leaders of organization of church and private schools; new poll indicates improvement; state dinner, entertainment by Ferrante and Teicher.]*

Thursday, December 8

The Soviets have walked out of the "Start" talks but not so definitely as in the INF talks. This is regular time for holiday break & they didn't say they wouldn't be back. They just said they were unable at this time to set a date for their return.

Our dead Navy pilot is being returned to us by the Syrians. We still don't know cause of death. After a couple of routine meetings & lunch with Geo. B. I left for Indianapolis. Addressed the Nat. Forum on Excellence in Ed. About 2000 teachers, students, state legislators, Govs., School Board members etc. Was well received although I'm sure the few from N.E.A. weren't happy. They were on record as saying that if I didn't come with a pledge of more money for Ed. the meeting would be a "sham." Well I didn't come with any pledge—to the contrary I told them Fed. money was not the answer. I was given a very warm reception.

[Photo session with state leaders of party; returned to Washington; spoke briefly at American Enterprise Institute (AEI) dinner; telephone Bobby Allison, winner of auto racing's Winston Cup.]

Friday, December 9

Officially signed off on position I took a few days ago regarding sanctions on Libya. Bill Smith came by to tell me the F.B.I. investigation into leak from Nat. Security meeting could not claim evidence of a criminal act. At same time, F.B.I. targeted several they believed should take polygraph tests. I've decided that's up to them—I will not order it. I know such an order from me would be leaked & it would kick up a fuss not warranted by the deed.

[Signed Human Rights Day proclamation.]

A sobering briefing on Soviet offensive power & plans for a protracted nuclear war. I wish some of our pacifist loud talkers could have access to this information. A preliminary budget review on Defense budget. Cap has pulled it down from $321 bil. to $305 bil. I have a hunch it will come out around $295 but we should ask for $305 because the Dems. will cut whatever we come in with.

Off to Camp David.

Saturday, December 10–Sunday, December 11

Sat. a beautiful & not too cold a day. Phoned King Hussein. Call paid off. He was very upset with me over the press accounts of our talks with Israelis & he's due Sun. A.M. on Brinkley's show. We had a good talk & it was apparent on that Sun. show.

[Returned to see W.H. decorated for Christmas; attended Christmas show hosted by Andy Williams.]

Monday, December 12

[Report that the last of the combat troops were leaving Grenada; flew to New York for annual meeting of Congressional Medal of Honor recipients.]

Back to W.H.—met with Tony O'Brien & Anthony Aguilar (the John Wayne of Mexico). Tony had his family, his lovely wife & 2 fine sons. He does an annual 5 mo. tour of the U.S.—Rodeos etc., appears before 3½ million Hispanics. He means to use those appearances to hail me as his friend.

Tuesday, December 13

Most of the day was taken up with Budget meetings. It's that time of year. Another car bombing this time on our Embassy in Kuwait & 5 other places on the same day. Some of our employees—all Kuwaitis were killed. The same Iranian holy war group are taking credit. George S. will be back tomorrow. He's been on quite a trip.

[Budget meetings; preliminary meeting on State of the Union Address, possible announcement of candidacy; signed proclamation for Drunk and Drugged Driving Awareness Week; meeting with Secretary of Commerce Malcolm Baldrige, U.S. Trade Representative Bill Brock, and others on textile imports; videotapings; Christmas reception for members of Congress.]

Wednesday, December 14

The New Jersey finally did it! Our Recon planes over Lebanon were fired on again—this time in the area patrolled by the N.J. The 16 inch guns fired 11 rounds. All of Beirut thought there was an earthquake. No reports on results yet.

[Press availability; in separate appearance, took questions from gathering of journalists from the Gannett news organization; received positive report from Shultz on recent trip to Europe and Middle East; meeting with Senator Baker over concerns that Republican senators will continue to bolt on budget matters; interviewed by U.S. News & World Report; *farewell party for Ken Duberstein, who stepped down as Assistant to the President for Legislative Affairs; Christmas reception for press; put a sign on Mrs. Reagan reading, "I have laryngitis—Merry Christmas."]*

Thursday, December 15

P.M. Pierre Trudeau arrived. He's been travelling the world trying to arouse interest in a 5 big nation summit to try & reduce nuclear weapons. He hasn't been able to get much interest. Actually it isn't a sound idea but still we support his arousing interest in other nations, all of them to talk of eliminating such weapons. I think he went away with some added ammunition.

Lunch with V.P. He's done a great job in Central Am., especially in El Salvador inspiring more action on human rights.

An end of year interview with Time Inc. They are doing a feature on me & Andropov & Soviet-US. relations. I did my best to make it plain—we are the side that's trying to bring about arms reductions & peace.

[Met with Senator John Danforth (R-MS) on his way to Africa; awarded Congressional Gold Medal to choral leader Fred Waring; met with Rich Colino of Intellstat; met with Vietnam veteran Rich Eilert.]

Capt. Grace Hopper—Navy—77 yrs. old sworn in as Commodore. Editors of Ladies Home Journal with a bound copy of their 100th yrs. edition.

Over to the W.H. for the annual Xmas Tree lighting. 7 yr. old Amy Benham—with terminal illness had made public her dream to help me light the tree. She was a sweet little girl—she pushed the button.

Another handshaking night—2nd half of Press Corps—about 600 hands. Today the New Jersey fired only its 5 inch guns at Syrian artillery. We intercepted a message between 2 units of Syrians—saying "We don't want the N.J. to fire anymore."

Friday, December 16

Hopefully there is a cease fire again in Lebanon brought on we believe by these 16 inch guns. Don Rumsfeld is meeting with the Syrians & for the 1st time there has been an admission that Pres. Assad had a heart attack.

Over to the O.M.B. to meet about 125 people including the board of directors of the newly created Nat. Endowment for Democracy. A true mix, Lane Kirkland, a business rep., both party chairmen etc. This is outcome of my proposal to British Parliament in 1982.

[Cabinet lunch on textile trade inequities; met with Shultz regarding START (no details of discussion); budget appeal meeting; dentist appointment. **Saturday, December 17:** *radiocast from Oval Office; desk work; reception for military families.* **Sunday, December 18:** *slept late; watched panel shows; holiday reception for staff.]*

Monday, December 19

Mermie & Winter arrived—a very cold day. A Budget meeting. I think Dave S. tells us more than we need to know for budget decisions.

[Gave awards to employees in Executive branch; photo session; rehearsed for press conference; attended Christmas party given by Fred Fielding, counsel to the president; holiday reception for Secret Service.]

Tuesday, December 20

An in & out day crowned by a Press Conf. We had an N.S.P.G. meeting on study of whether & how the Soviets are violating treaties & agreements like the A.B.M. treaty. There isn't any question but that they are & yet some of those treaties are so ambiguous you might not make a case that would stand up in court. I'm afraid in the days of Détente—our negotiators were so anxious to bring home some kind of treaty that it didn't have to be good.

[Lunch with American Security Council; rehearsed for press conference, then staged the conference.]

Wednesday, December 21

Two budget meetings—coming to wrap up time. We aren't making the dent in spending we should but the biggest bundle of spending is Soc. Security, medicare, pensions etc. We do have a proposal for medicare that will result in savings; however, it calls for higher share costs for the recipients in short illnesses but makes up for it with catastrophic (no limit) insurance. Still I'll bet we get pilloried for proposing it.

[Lunch with Shultz and Vice President Bush to discuss upcoming Japanese election; meeting alone with Shultz, agreed "we must do something to get some movement in the peace process without waiting for settlement of the Lebanon problem"; desk work; Christmas visits and reception for W.H. senior staff.]

Thursday, December 22

An NSPG meeting at the OEOB bldg. having to do with plans to keep the Persian Gulf open if Iran should try to close it. All are agreed it must be kept open to shipping.

Dick Wirthlin came by office with new poll results—our figures are up in everything but the Lebanon situation.

Jeanne Kirkpatrick reported on recess of Gen. Assembly U.N. She's weary of the U.N. & I can't blame her but she wants to do whatever will best help in my re-election. We are going to walk out of Unesco. We've tried for 3 yrs. to re-direct it into proper channels with no success so we'll pick up our marbles & go home.

Did a Xmas msg. for Hispanic Am. Radio network. Actually did part of it in Spanish.

[Staff appointments; received Bibles from various groups; greeted March of Dimes poster child; invitation from American Legion to speak at national convention; interviewed by Le Figaro *French newspaper.]*

Friday, December 23

A really easy day. NSC—Cap getting ready to release study by commission investigating Beirut massacre of our Marines (241). They are going to charge there was negligence on part of officers regarding safety precautions. I'm worried about the effect of this on families that lost loved ones. Another briefing in situation room on Soviet nuclear arms & the almost impossibility of verifying whether they are cheating or not. Did year end interview with 4 wire services. Had a haircut & upstairs.

Saturday, December 24

Tis the day before Christmas & all thru the house—yes there is a bustle. Ron & Doria arrive from N.Y. for lunch. Patti & her friend Paul Grilley will be in late. Christmas Eve dinner at the Wicks is beginning to be a tradition.

One cloud in the sky. I'll keep to myself—the threat world wide by the Iranian fanatics to loose terror on everything American.

The Wicks dinner was as always a warm, wonderful time with long time friends. I played Santa Claus—another part of the tradition but my 1st time to play the role.

Sunday, December 25

Mid morning Ron & Doria, Patti & Paul (her friend), Nancy & I started the gift opening. It took until noon. A good time was had by all. Tonite the Wicks will be here for dinner. Wash. is having it's coldest Christmas ever.

The Bishop who was supposed to have been murdered by Contras (Miskito Indians) according to the Sandanista government has crossed into Honduras with 1040 Indians he rescued from the govt. forces who were the only ones trying to kill him & them.

Phoned the Bishop & a wonderful lady in N.Y. state who is celebrating Christmas & her 113th birthday.

Monday, December 26

Ron & Doria departed for N.Y. Patti & Paul are visiting the Smithsonian & will head West tonite. We leave tomorrow for Calif.—a total 6 day trip—due back here a week from today.

It's bright & cold out, possibly an all time record low.

Tuesday, December 27

9:00 A.M. & off to Calif.—a cold gray day in Wash.—a nice almost 70 degree day in L.A. at the Century Plaza. One change—we helicoptered right to the Hotel parking lot. Just one of the evidence that S.S. is taking the new terrorism seriously.

Wednesday, December 28

Taped my Sat. radio broadcast. Met with Roy Miller about our wills. Then a session with Mike Abrams. He agrees the gym routine has definitely paid off for me. I'm worried about Nancy. A deep cough continues beyond when I believe it should have dried up. Mermie came by for a short visit. We still have no break in the Mike situation. We must find an answer to that.

Thursday, December 29

Early afternoon off to Palm Springs. A quiet dinner with Lee & Walter & early to bed.

Friday, December 30–Sunday, January 1, 1984

The usual wonderful time with good friends. Golf in the day time & the Wilson, Jorgenson dinner at El Dorado Fri. The big New Year party at the Annenbergs Sat. nite. It was a wonderful, warm affair. A little excitement early in the evening when a local, somewhat mentally disturbed woman used Joan Clark's name to gain entry. She was taken away but then someone said she had a package of some kind when she came in but not when she left. A sniffer dog was brought in but—must have been a false alarm according to the dog. Sun. a smaller dinner party—informal & fun.

February 5, 1981—President Ronald Reagan and Mrs. Reagan greeting the Reverend Billy Graham at the National Prayer Breakfast in Washington, D.C.

February 10, 1981—President Reagan being filmed working in the Oval Office for an NBC program, *A Day in the Life of the President.*

February 16, 1981—With House Speaker Thomas "Tip" O'Neill in the residence of the White House before their first dinner.

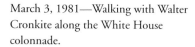

March 3, 1981—Walking with Walter Cronkite along the White House colonnade.

March 25, 1981—President Reagan and James Baker travel on Marine One to the White House after horseback riding at Quantico Marine Corps Base.

April 15, 1981—In one of the first official photos taken at the White House after the assassination attempt, the president is shown here working in his residence study on an upcoming speech to Congress.

April 24, 1981—A senior staff meeting with *(from left)* Mike Deaver, Jim Baker, and Ed Meese on the president's first official day back in the Oval Office after recovering from the assassination attempt.

June 3, 1981—With Mrs. Reagan and Frank Sinatra in the Oval Office, receiving a baseball on behalf of the Multiple Sclerosis Society.

July 14, 1981—Greeting his youngest son, Ronald Prescott Reagan, in the Oval Office prior to a reception for Dr. Loyal Davis, Mrs. Reagan's father.

August 19, 1981—At a dinner at the Bloomingdale residence in Beverly Hills, California, with Charlton Heston, Lydia Heston, Ricardo Montalbán, and Georgina Montalbán.

September 25, 1981—President Reagan and Justice Sandra Day O'Connor outside the Supreme Court after her swearing-in ceremony as the first female justice of the U.S. Supreme Court.

October 6, 1981—Walking with Mrs. Reagan through the entry hall into the Cross Hall at the White House following the statement about the assassination of President Sadat of Egypt.

October 8, 1981—Speaking at the White House prior to the departure of Presidents Ford, Nixon, and Carter for the funeral of Egyptian president Anwar Sadat.

November 22, 1981—In one of their favorite photos, the president and Mrs. Reagan pose on the White House grounds during a shoot for *People* magazine.

November 24, 1981— Horseback riding with the first lady at Rancho del Cielo during Thanksgiving break.

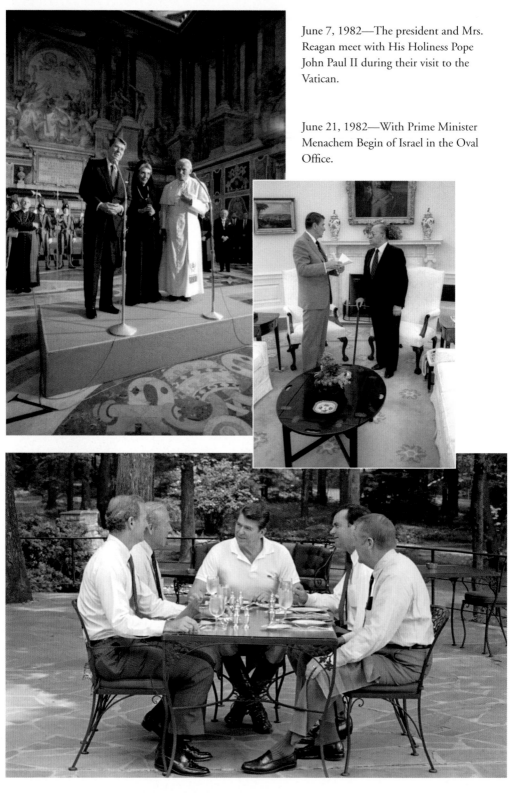

June 7, 1982—The president and Mrs. Reagan meet with His Holiness Pope John Paul II during their visit to the Vatican.

June 21, 1982—With Prime Minister Menachem Begin of Israel in the Oval Office.

June 26, 1982—A working lunch meeting with new secretary of state–designate George Shultz at Camp David, Maryland, with James Baker, Ed Meese, and Bill Clark.

August 15, 1982—Speaking to Mrs. Reagan in Arizona from Camp David, Maryland. (She was very upset over the failing health of her father.)

September 1, 1982—Working at Rancho del Cielo in Santa Barbara during a visit to California.

August 15, 1982—A question-and-answer session during a working luncheon for Republican members of Congress at Camp David, Maryland.

September 4, 1982—Delivering his weekly radio address to the nation from Rancho del Cielo.

September 28, 1982—Enjoying a rare piece of solitude during a private lunch on the colonnade outside the Oval Office.

November 4, 1982—Waving to the United States Marine Band, known as "The President's Own," in the Cross Hall of the White House.

March 4, 1983—The president and Mrs. Reagan celebrating their thirty-first wedding anniversary with Queen Elizabeth II onboard Her Majesty's yacht, *Britannia*.

June 28, 1983—Taking questions from reporters during a press conference in the East Room of the White House.

July 2, 1983—With his youngest daughter, Patti Davis, at Mrs. Reagan's birthday party at Rancho del Cielo.

September 24, 1983—President Reagan maintained a regular exercise routine. He is shown here working out for a photo in *Parade* magazine at the U.S. Secret Service gym in the Old Executive Office Building next to the White House.

November 13, 1983—Speaking to the troops at the DMZ during his trip to the Republic of Korea.

February 14, 1984—Meeting with King Hussein of Jordan (*left*) and President Mohamed Hosni Mubarak of Egypt in the Blue Room of the White House.

February 23, 1984—Working in the Oval Office at the White House.

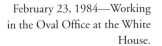

April 28, 1984—The president and Mrs. Reagan visit the Great Wall during their trip to the People's Republic of China and are shown here receiving their official tourist certificates.

June 6, 1984—Greeting U.S. Army Ranger veterans after the ceremony commemorating the fortieth anniversary of D-Day at Pointe du Hoc in Normandy, France.

June 6, 1984—In one of his most famous speeches, President Reagan addressed the world on the fortieth anniversary of the D-Day landings at Pointe du Hoc.

June 21, 1984—Shaking hands with participants in the National YMCA 1984 Youth Governors' Conference in the Rose Garden of the White House.

June 26, 1984—President Reagan and Elizabeth Dawn Alford, Muscular Dystrophy poster child for the state of Georgia, meet in the Oval Office during her visit to Washington, D.C.

CHAPTER 4

—

1984

Monday, January 2

Back to Wash. Nancy went to L.A. & then to Phoenix so I'm here alone. I'm watching the Rose Bowl—U.C.L.A. 45 Ill. 9—so far.

Tuesday, January 3

The day started before daylight with a phone ringing. Bud McFarlane called to tell me Jesse Jackson was bringing Lt. Goodman home from Beirut. I'm sure Jesse had some less than noble incentives in making that trip but I prayed every day he'd be successful. I didn't see or talk to him at all before he left. I felt that if I did he would appear to the Syrians as an emissary of mine & that would be the kiss of death. As just a citizen (along with a dozen other clergymen) it was far more possible that he might be able to get our man released. And he did. I talked to him on the phone & the Lt. They'll be home tonight.

[National Security Planning Group (NSPG) meeting on Marines in Lebanon; only twelve minutes for lunch at desk; budget meeting; cabinet meeting on health matters and violence in schools; met with Senators Howard Baker Jr. (R-TN), Bob Dole (R-KS), and Pete Domenici (R-NM) on budget; Mrs. Reagan expected home in the evening.]

Wednesday, January 4

Mermie's birthday. Big day at the office, too. Jesse Jackson, his family, Lt. Goodman (released by Syrians) & his family came to Oval Office. Jesse's debriefing on his hours spent with President Assad was most interesting. He believes (& so do I) that Assad could be sending a message. J.J. said he talked at length about wanting to establish a relationship with me.

We all trooped out to the Rose Garden for photos etc. It was a good day.

[Met with congressmen on the education budget, positions not far apart; budget meeting on possible alternative tax structures ("None of them looked good to me"); met with Shultz and McFarlane on indication that Soviets would be amenable to means of direct communication.]

Thursday, January 5

Usual staff & N.S.C. briefings. In the latter rcvd. word (and fingers are crossed) that a solution may be in sight for Lebanon.

[Briefing on upcoming visit of Chinese premier Zhao Ziyang; met with agricultural leaders; lunch with Vice President Bush; Cabinet Council meeting on improvements in government management, including reducing publications by half; met Mrs. America pageant winner; photo session; videotapings, including one for Cary Grant's eightieth birthday; noted record trading day on Wall Street, small drop in unemployment.]

Tomorrow Don Regan & I do battle with our team members who want to raise taxes because of the deficit. He & I want to reduce spending instead. Just informed this meeting will be on Mon.

Friday, January 6

Usual staff meetings plus NSC. We're still sweating out the Lebanon deal. Met with speechwriters re the Soviet speech. We want it to be a level headed approach to peace to reassure the eggheads & our European friends I don't plan to blow up the world.

Met with Geo. Shultz. The Soviets are interested in a private back door channel.

[National Security Planning Group (NSPG) meeting on visit by Premier Zhou Ziyang of China; went to Camp David; watched Cary Grant movie, Gunga Din *(1939).* **Saturday, January 7:** *radiocast on crime in schools; desk work, including Soviet speech; watched* To Be or Not to Be *(the 1983 Anne Bancroft version); commented, "It was fun."* **Sunday, January 8:** *report of Marine killed in Beirut; returned to W.H.; watched Super Bowl.]*

Monday, January 9

Called Dick Nixon to say Happy B.D. On way to office had photo with Paul Laxalt & Frank Fahrenkopf—for "News Week" I think. Staff meetings & then a drop by for business leaders who have joined together & are going to Grenada to see how the private sector can help Grenada's ec. grow through private investment etc.

Budget luncheon meeting. We're really divided. Don Regan & I want to battle down to the wire for more spending cuts as an answer to the deficits. Stockman & Feldstein plus others want a tax increase, I think they are wrong as h--l.

An NSC meeting—this one on how to handle report to Congress on Soviet violations of weapons treaties—which are numerous. We're going to low key it in the report but deal directly with the Soviets on what do they intend to do about them.

[Met with state legislators; greeted Miss U.S.A. and Miss Universe; ambassadorial formalities; reception for executive exchange program between government and private sector.]

Tuesday, January 10

Premier Zhao (P.R.C.) arrived. We had a formal S. Lawn ceremony with full military honors. The day was freezing cold which held off a rain that didn't fall till around noon. Zhao is a likeable fellow & a very capable one. China communism makes room for some pvt. enterprise etc. In our Oval Office meeting he took off on Taiwan being the block to full relationship & even hinted that if peaceful negotiations didn't resolve the problem he might have to use other means. I told him we wanted a resolution of the problem but any use of force would change our relationship beyond repair. We got along fine from then on.

Sens. Tower & Warner came by to report on their middle east trip. Both went, inclined to believe we should bug out of Beirut. Both came back convinced the Marines should stay & that withdrawal would be a disaster. This is going to be of great help to us with the Cong.

[Senator Pete Wilson (R-CA) visited to campaign for government purchase of F-20 Northrop fighter planes; state dinner, music by Isaac Stern.]

Wednesday, January 11

Snow on the ground—very pretty & very cold. An NSC brief on Lebanon—a glitch in what looked like progress toward a settlement. Jumblat balked.

Met with Kissinger's Central Am. Commission. They presented their report. It's darn good & we'll push for its adoption. In fact it really bears out what we've been pushing all along. George S. & I met on a variety of things—we're still eye to eye.

Met with my Ec. Policy Board. Being economists they are not in agreement except that we all know we have to go after the deficits. Some of them would turn to taxes. I still say no & the majority of them agree.

Ended the day. No I didn't—just remembered a meeting with Ann Armstrongs commission—Foreign Intel. Advisory Board. They reported on their overview of our intelligence activities. Had some suggestions which we'll follow. (Just received a cable—one of our Helicopter pilots in the Honduras maneuvers was shot & killed by fire from across the Nicaraguan border.)

Thursday, January 12

[Looked forward to seeing a stage production of The Hasty Heart *(Reagan starred in the 1949 movie version) in the evening; breakfast at 7:30 with Premier Zhao Ziyang and respective cabinets; signed trade and technology agreement; appeared before the Small Business Legislative Council; NSC briefing, on-going investigation of helicopter downing, report that it was over Honduras at the time, concludes, "In other words, Nicaragua was shooting at our men while they were in neutral air space & even after they landed on Honduran soil."]*

Bill Wilson & Archbishop Pio Laghi came by. His Eminence will be Papal Nuncio now that we have diplomatic relations. Lunch with the V.P. & then a meeting with Cab. Council on Ec. affairs. This was on monetary policy & of budget loans & loan guarantees. We should make this off budget spending—on budget so the people know how extravagant government really is.

A brief drop-in on our business task force that is going to Japan to pursue agreements we started on our visit. Then upstairs;—my guardian angel (Nancy) had called Mike D. about the 2½ hours of scheduled activities & he postponed them. Only thing left is "The Hasty Heart." I've never seen a one of the Broadway hits that I did on screen.

Friday, January 13

An NSC meeting on our approach to the Soviets re arms negotiations. We've notified them of our report to Congress on their violation of various treaties & agreements. They do just plain cheat. Then lunch arranged by Mermie with about 100 Repub. Women office holders from all over the country. We gave them a full day of briefings by Cabinet members etc. I spoke & did Q&A. They were a great group & went out of their way to convince me there was no gender gap.

[Meeting then with Shultz on his way to Stockholm; personnel meeting; videotapings; went to Camp David. **Saturday, January 14–Sunday, January 15:** *sat by the fire and worked hard on a draft of the State of the Union; returned to the W.H.]*

Monday, January 16

Staff & NSC meetings but the day really began in the East room at 10:00 A.M. when I went live on TV worldwide with address on Soviet-U.S. relations. The press, especially TV is now trying to explain the speech as pol. etc. The speech was carefully crafted by all of us to counter Soviet propaganda that we are not sincere in wanting arms reductions or peace. It was low key & held the door open to the Soviets if they mean what they say about having peace to walk in.

Then a meeting on the St. of the Union address. I worked the whole weekend on a re-write of a draft which represented the input of every agency. Now my re-write will go back into the mill.

[Briefing on issues at lunch; met with Grace Commission with report containing twenty-five hundred recommendations on improving efficiency in government.]

An interview for year-end round-up in Wash. Post or rather a 3rd anniversary of our admin. Lou Cannon, Hoffman & Juan Williams—3 journalists who usually kick my brains out.

The Q's were OK but I'm not sure how honestly they'll treat the answers.

[Visit from Gerald Carmen, former head of General Services Adminisration (GSA), then headed for Geneva; met with Arab American group; waiting for word on reason for disappearance of U.S. corporal in West Germany.]

Tuesday, January 17

NSC meeting—actually it was to say goodbye to Ambassador Valenzuela of Chile. He delivered a nice message from President Pinochet. Then 2nd guest was Mrs. Massey. She is a regular visitor to Russia. She's dedicated to previewing some of old Russia's art & is helping with an art museum. Widely acquainted among Russians in art & cultural activities—has no trust with the govt. types. She was most complimentary about my speech.

Lunch was a round table meeting with a group concerned about the family in America—an interesting time & a lot of interesting & diverse view points.

Met with Paul Nitze. He, too believes we must not yield to Soviet pressure to make a new offer to get the Soviets to resume the I.N.F. talks. That would be rewarding them for walking out.

Cabinet Council meeting on Natural Resources & Environment. A lot of decisions to be made on Western Water policies, toxic waste dumps, wetlands, preserving offshore sandbars & doing something about acid rain. This last calls for more research. I also approved lifting regulations restricting use of methanol as a fuel for cars.

Then Nancy & I went over to the Hay Adams hotel for a reception—"Citizens For The Republic." This is the group formed with my leftover 1976 campaign funds. About 85 people all stalwart supporters—very pleasant half hour.

Wednesday, January 18

I think I've been doing wrong in these diaries for 3 yrs. I've made them a log book of the days schedule & those schedules are all in the archives. I guess I should be noting other things so I'll start now. Prime Minister Mahatlin of Malaysia was our visitor today with a number of his visitors. He is also a Doctor. We had a good talk although there aren't any great problems between our two countries. He is a free enterpriser & is well along on bringing democracy & capital investment to his country.

Attorney Gen. Bill Smith came in—he wants to bow out after 3 yrs. & take an active role in the coming campaign which he can't do as A.G.

Sen. John Danforth has returned from his African trip. He brought over some slides of the starving people principally in Mozambique. He plead for immediate help for these people. The govt. of M. had thrown in its lot with the Soviets. Now the Soviets have failed them. I told our gang to get underway & we'll ride to the rescue.

Stu Spencer came to the house for dinner—we talked politics—what else? I'm glad he's with the campaign.

Thursday, January 19

[Prime Minister Rudolphus Lubbers of the Netherlands visited, concerned about Parliament reaction to deployment of nuclear missiles.]

I'm appointing Ed Meese to succeed Bill S. as A.G. This is his life long dream—his main interest is & has always been law enforcement. My decision was taken well by Mike & Jim. Bill will announce on Monday. I hope we don't have any leaks.

[Considered himself honored by acceptance into a "Religious Heritage Hall of Fame"; received gift from Kiwanis; photo session; budget meeting, found loopholes to close in tax code; appeared before private group, headed by George Allen, advocating physical fitness; farewell party for David Gergen, who served as director of communications.]

Friday, January 20

George S. is back. He told me it was a pleasure to represent us overseas with me here. I'm embarrassed even writing this but I feel good at the same time. There is a respect for the U.S. abroad that wasn't there a few years ago.

[Photo session; gathering for third anniversary of inaugural.]

Back to a meeting with several Gov's. regarding Acid Rain. It's a tough problem. Do we know enough for sure that mid west plants burning sulphur coal is the cause. The solution is terribly expensive—too much so if we're wrong about the cause, we're upping our research effort. Then off to Camp David—cold & snowy.

[Saturday, January 21–Sunday, January 22: sat by the fire, worked on speeches, watched Super Bowl.]

Monday, January 23

Routine visits in A.M. plus meeting with all the leaders of the annual march for Life—70,000 in Wash. against abortion. They gathered down on the Elipse at noon. I went up on the W.H. balcony & waved to them. I could hear the roar that went up. The leaders & I had a good meeting. They are more united & determined than ever.

An issues briefing lunch & later an NSC meeting—Don Rumsfeld, George S, Bud McFarlane & myself later joined by Cap W., Jack Vessey & Bill Casey—all on Lebanon. We're going to study a possible move of the Marines to the ships off shore but an Army force on shore to train the Lebanese Army in anti-terrorist tactics.

Did a taping session & then off to opening of the rebuilt Nat. Theatre. The show a great performance of "42nd St." They don't write them like that anymore.

[Tuesday, January 24: NSC meeting with General Vernon Walters, commented that he is "a roving Ambas. & what a job he does. He mends fences that might be leaning a little & makes the strong ones stronger"; lunch with Republican senators, made pitch for unity; new poll shows high approval rate; worked on upcoming speeches.]

Wednesday, January 25

Met with joint leadership of Cong. to brief them on Lebanon & Central

Am. George S. did that then as discussion opened Tip sounded off in a very partisan manner. So much so that Sen. Bob Byrd tried to moderate things. He's off & away—Tip that is on "bring the Marines home." I almost let go the controls but I didn't. There is no doubt the Dems. are going to ride this thru the campaign.

I later called Tip, Bob & our 2 Sen.—Baker & Bob Michel & propositioned them about naming some reps. to meet with on people about a bipartisan approach to the deficits. Tip ranted on that all we had to do was tax the rich more & cut the fat out of the mil. budget. He knows that is crap but it again is a tip on what their campaign will be.

Dropped in for a minute on the TV anchor men & women who were being briefed on tonite's St. of the Union address. I cannot conjure up 1 iota of respect for just about all of them.

[Met with Weinberger on defense budget; delivered State of the Union Address, considered it a success.]

Thursday, January 26

We took up the business of Beirut again & came up with a plan for redeployment of the Marines but only after sending in Army training units who specialize in anti-terrorist measures. Don R. is going out tonite because Pres. Gemayal must be willing to ask for this change.

[Photo session with tennis player Evonne Goolagong; flew to Atlanta, addressed free-enterprise meeting and then the Republican Leadership Southern Regional Conference; returned to Washington; attended ball for anniversary of inauguration.]

Back to Wash. & at 10 P.M. over to meet another 1000 supporters having a Ball to celebrate the 3rd anniversary of the admin.

Friday, January 27

Started the day with Repub. members of the House—breakfast in the East Room. Their Q's. were mainly on Lebanon & a few on the deficit. I can't give a reading on whether they'll stand with us on Lebanon. I'm sure some won't but I gave a little lecture on why we can't bug out. We're trying to get the Israeli lobby which is very effective in the Cong. to go to work on how much Israel has to lose if Congress forces a withdrawal of our troops.

Campaign time is coming closer even though I have not actually said the words to anyone (except Maureen & Nancy) that I'll run. This morning took photo with Paul Laxalt, Ed Rollins & some of our staff for the R.N.C.

Foreign Minister Abe of Japan stopped by—he's been meeting with Geo. B. & our Cabinet members. My job was to make it official that if they backed away from the agreements we came to in Tokyo on trade & defense, we would not be able to restrain the Cong. from adopting protectionist measures.

In the afternoon met with delegation of Repub. Mayors & elected Municipal Employees. They were most supportive but due to some fluke a female Mayor (Binghamton, N.Y.) a Dem. was invited. I met her—she seemed very nice—I didn't know at the time she was the interloper. She then ran for the press & did all she could to torpedo us.

The rest of the afternoon was spent on 2 interviews, one with News Week, the other with Time. Both are for Monday so the Q's. were all on the supposition that the night before (Sunday) I had announced my candidacy. I told them I understand the supposition & would answer accordingly but would not tell them what my decision was going to be.

Saturday, January 28

Already a call from Mike—the 2 editors (News Week & Time) must have my statement by midnight for their stories. They swear secrecy. I have to give in but I don't trust them for 1 min. Now it's over to the Oval O. for the radio broadcast—possibly the last since next week I'll be a cand. Spent rest of day on desk work & polishing my remarks for Alfalfa. The dinner was as always a lot of laughs & really good companionship.

My few jokes were well received & then I left early to get back here to the announcement speech & bed.

Sunday, January 29

[Quiet day; practiced announcement speech; Maureen and Dennis visiting; "It's hard to believe there would be anyone who isn't sure I'm going to run—but I still haven't said so outside the family"; telephoned widow of Marine killed in Beirut.]

Tonite I went on Nat. TV for 4 minutes & 41 seconds to announce I was a candidate & would seek re-election. The response has been terrific—calls, wires etc. running 10 to 1 in our favor.

[Monday, January 30: peppered with questions by press in Rose Garden; photo session; meeting with Shultz and Edward Rowny on START talks (no details recorded); briefing for editors; addressed the Broadcast Evangelicals, "They couldn't have been more supportive—actually I was delivering a sermon & meant every word of it"; interview with Dave Hartman for Good Morning America; *took delivery of new armored limousine.* Tuesday, January 31: *meeting in Oval Office "hassling over Beirut"; flew to Chicago to address cement, gravel, and aggregate groups; interview with* Chicago Tribune en route; *photo session with labor leaders.]*

Wednesday, February 1

Started day with a Cabinet meeting to give them the word about helping in the campaign—now that I'm a candidate. Then it was a visitor day. Pres. Spilzak of Yugoslavia was here for a meeting & lunch. He's a personable & reasonable man. I picked his brains about the Soviet U. He was an Ambas. there for a time. He believes that coupled with their expansionist philosophy they are also insecure & genuinely frightened of us. He also believes that if we opened them up a bit their leading citizens would get braver about proposing change in their system. I'm going to pursue this.

[Signed a proclamation for National Tourism Week.]

S.S. got a tip [. . .] that a terrorist act was pointed at the Presidents daughter. I'm inclined to believe they might mean Maureen because she is so visible in her pol. work. She has S.S. protection. On the other hand, Patti screamed & complained so much we took the S.S. detail away at her request. Now S.S. went to her & asked if she would accept it for no more than a week until they could get this informant out of Lebanon & check the story. She said yes. But today's the 4th day & she's screaming again about invasion of her privacy & last night she abused the agents terribly. I said take them away from her so she's again without protection. Insanity is hereditary—you catch it from your kids.

Thursday, February 2

[Attended Annual National Prayer Breakfast, looked forward to it each year.]

Back to W.H. & and a meeting with Bill Casey re some hush operations in Central Am.

Up to the Capitol for lunch with all the Repub. Sens. & Rep's. It was a good session. An incipient revolution on the part of some of our Repubs. has been nipped in the bud.

Back to the W.H. and an interview with 3 Wall St. Journal editors.

[Visit from president of Tuskegee Institute, "came to present me with a souvenir but really for me to give them a grant of $9 mil. for an engineering (aeronautical) & space center named after Gen. Chappie James"; announced recognition of Black History Month; received Boy Scout annual report; gift of model of USS New Jersey; meeting with Export Import Bank; photo session. Friday, February 3: met youth group sponsored by Hearst newspapers; with leader of Austrian Conservative Party; with Vice President Kurt Furgler of Switzerland; with congressional leadership on legislation based on Kissinger Commission report on Central America; editor of local paper near Dixon, Illinois; lunch arranged by Maureen with Republican women elected officials; met Prime Minister Eugenia Charles of Domenica, stating U.S. should stay longer in Grenada; went to Camp David.]

Saturday, February 4

Called a woman in Peoria Il. who had wired after the St. of the U. Her complaint was over freedom of choice. She was referring to abortion & she called herself an ex Repub. who wouldn't vote for me. I was going to write her & then just on a hunch I phoned. It took a little doing to convince her it was really me. We had a nice talk & I was right that her problem was abortion. I made my pitch that there were 2 people's rights involved in abortion—the mother's & the unborn child. She promised to give that some deep thought. We had a nice visit. She's a 51 year old divorcee working for less than $10,000 a year—has a 17 year old son ready for college & a married daughter. I think I made a friend.

[*Sunday, February 5: woke up to fresh snow; returned to W.H.; opened gifts, seventy-third birthday the next day; reiterated upcoming schedule.*]

Monday, February 6–Sunday, February 12

Well it was a happy birthday & also a busy one. We flew to Rockford Ill. where we were met by Congresswoman Lynn Martin & Gov. Jim Thompson. Then by helicopter to Dixon where brother Moon met us & we drove to the old house on Hennipen Avenue. It was quite an experience. They've fixed the house up very nicely & furnished it reasonably close to what we'd had in 1920. The big surprise was how much smaller the rooms seemed compared to our memories. The 3 of us had lunch in the old dining room & a birthday cake had been provided.

After lunch our motorcade went downtown to the parade reviewing stand—8 degrees below zero. After the parade went to the high school gym where I spoke to about 3000 Dixonians. It was a truly nostalgic & heartwarming day.

[*Helicoptered to Eureka; delivered first in a series of lectures sponsored by* Time; *Mrs. Reagan went to Phoenix; flew to Las Vegas for address before convention of high school principals, received ovation upon calling for prayer in school; party fund-raiser; flew to ranch in California and met Mrs. Reagan there; rode every morning, ranch work in the afternoons; commented, "But Nancy is right—Presidents don't have vacations—they just have a change of scenery"; conference calls on Beirut, where Marines were being moved out to ships.* **Sunday, February 12:** *Mrs. Reagan flew to Seattle for anti-drug appearance; returned to Washington, commented, "I'm back in the W.H. wishing to h--l I was back in Calif."*]

Monday, February 13

A hush hush cable from Geo B. re Lebanon. Margaret T. feels strongly we should tell Gemayel to abrogate the May 17 agreement with Israel. Geo. wanted to know if he could tell her we wouldn't object. That's all we can say—we won't oppose but we won't urge either.

King Hussein of Jordan came in. Tomorrow I meet with him & Mubarak (Pres. Egypt). His Majesty is a solid citizen. We had a good discussion. If we have any chance of lasting peace in the M. E. it will be because of his willingness to get into the peace negotiations.

[*Interview with journalists from Knight Ridder; stopped at Soviet Embassy to write a brief condolence in the book for Yuri Andropov, who died February 9; desk work at W.H.* **Tuesday, February 14:** *Mrs. Reagan headed for St. Louis; meeting with Egyptian presi-*

dent Hosni Mubarak, who explained need for financial help, wanted to make progress on peace initiative without waiting for accord in Lebanon; King Hussein arrived; lunch with both leaders; discussed plans for peace negotiations; in farewell remarks, President Mubarak suddenly demanded Israeli withdrawal from Lebanon and that Arafat be included in future plans, Reagan commented, "We'll be busy for awhile patching things up with Israel. It was really a shock because for hours we discussed the problems of recognizing Arafat etc. "; telephoned Olympic medalists in skiing; cabinet meeting on laws that discriminate against women; Pat Boone arrived with Easter Seal poster child, commented, "Who are those who would have us believe such children shouldn't be allowed to live"; met with Neil Kinnock, Labour Party leader from Britain.]

Wednesday, February 15

A few more days like these last few & I'll change my mind about running. Opened with Geoffrey Sperling's annual breakfast in our St. dining room. About 50 assorted members of the press & one hour of Q's. (I ate before I went downstairs). Most Q's. on Beirut. Over to the East Room for Crime Prevention Week ceremony. Rcv'd. report of United Way Fund drive. More money was given than ever in our history.

George S. came in. He gave me a T-shirt from Grenada—crossed flags & a line "Thank you America for saving Grenada." He says they are selling like hot cakes. It seems Dick Stone (our roving Ambas. in Latin Am.) & Tony Motley can't get along—Dick is resigning. We'll try to find another spot for him—he's a good man. They both are.

Meeting with Paul Volcker. I think we have him on course with money increase matching growth so as to keep the recovery going without a return to inflation.

[Attended fund-raiser arranged by Maureen in celebration of Susan B. Anthony Day; **Thursday, February 16:** *staff meetings; meeting of Emergency Committee for American Trade (against protectionist trade bills); lunched with Vice President Bush and Bud McFarlane, "a good gabfest about Soviets, arms reductions etc. & how to get moving again on 'START' talks"; Cabinet Council meeting on trade, effort to stop tariff bill on wine; visit from Simon Wiesenthal; honored by* Muscle Training Magazine *as best physically conditioned president in history; inducted as a member of the National Press Club; photo session and videotapings.]*

Friday, February 17

Dropped by a briefing for press correspondents re the budget & other things mainly having to do with issues on women's problems—most of the scribblers were women. Before that, however, I went to the cabinet room to greet 3 from "Human Events"—Alan Ryskind, Stan Evans & Tom Winter. H. E. had done a feature indicating I was a bigger spend thrift than Jimmy Carter. Dave S. &

Ed M. had invited them over to get the real figures—which refute their story very definitely.

At 11 A.M. down to the situation room for a briefing on a possible covert operation (C.I.A.) in El Salvador involving a C-130 gun ship which we'd provide for strikes against Rebels. I'm afraid since it would require Congressional Committee support we'd blow our whole aid program to El Salvador. But I was tempted. It was the kind of mission that could break the rebels back.

[Lunch arranged by Maureen for Republican women office holders; ambassadorial formalities; greeted royal family of Monaco, staying as guests.]

Saturday, February 18

A beautiful day with a touch of Spring. Not so beautiful in Beirut. It looks very much as if everything is falling apart. Some Christians are exploring partition—a Christian state including W. Beirut & the area immediately to the North. Syria indicates it won't accept the Saudi proposal which was probably Gemayal's last chance at putting together a coalition. Behind all the villany is Syria.

The big gala fundraiser for the Princess Grace Foundation was a great success. There were some moments. When I returned to the table after proposing a toast to Grace I found Princess Caroline, her daughter, quietly crying.

*[Sunday, February 19: Rainiers left; quiet day at home; desk work. **Monday, February 20:** flew to Iowa for campaign, good reception, called it "an adrenalin shot"; returned to Washington.]*

Tuesday, February 21

Met with Repub. Cong. leadership. To my surprise—back from their 10 day recess they were all cooled down about Lebanon. When they left town they thought Lebanon was the biggest issue on hand. When I suggested doing a TV speech to explain it they said no, let it lay.

[Received an award for peace efforts; issues lunch; report from Commission on Security and Economic Assistance; two-hour practice session for press conference; Mrs. Reagan in New York.]

Wednesday, February 22

George S. & I met & discussed mainly the Soviets & how we should react to Chernenko's mild sounding talk with George B. I have a gut feeling I'd like to talk to him about our problems man to man & see if I could convince him there would be a material benefit to the Soviets if they'd join the family of nations etc.

We don't want to appear anxious which would tempt them to play games & possibly snub us. I have our team considering an invitation to him to be my guest at the opening of the Olympics—July in L.A. Then he & I could have a session together in which we could start the ball rolling for outright summit on arms reductions, human rights, trade etc. We'll see.

[Rehearsed for press conference; Mrs. Reagan returned; staged conference.]

Thursday, February 23

One of those days—when one thing ran into another—still it was pleasant. Met with Nat. leaders re School Discipline. I assured them I'd keep hammering on the subject. Went over to E.O.B. & spoke briefly to about 200 representatives of our Asian Americans. They are good people & sound Americans. Back to the Oval office & lunch with about 40 publishers of leading national magazines. Strange—some of their mags. kick my head off in the current issues but these publishers are all gung ho for me & what we're doing. I wish they'd talk to their writers.

Dick Wirthlin came in with latest pollings. I can only say I wish the election was tomorrow.

[Cabinet Council meeting on management, ways to modernize and increase efficiency; made a member of the Purple Heart organization; visit from West Point cadets; life member of the Variety Club; received lithograph from Olaf Wieghorst; met the head of Independent business group.]

Lew Lehrman came by with some disturbing info about Paul Volcker & his strong belief that we should slow our business growth. Paul is obsessed with a fear of returning inflation & according to Lew would seek to use the Fed. to slow our recovery. Finally home & mother.

Friday, February 24

[Visit from Prime Minister Robert Muldoon of New Zealand.]

An hour of NSC. Our Marines will be out by Sunday. We are all agreed our training units with the L.A.F. should stay where they are on shore.

Camp David—a beautiful afternoon. I got some backlog of mail answered.

Saturday, February 25–Sunday, February 26

Sat. rain in the morning & gray & windy in the P.M. More desk work. Called R.N. [Richard Nixon]—told him about letter rcv'd. from Chernenko. I think he's made it easier for us to approach him about a meeting. Sunday bright & beautiful. Bud MacFarlane called—our 16 inch guns silenced batteries in Syria occupied Lebanon. They had loosed heavy fire against one of our recon planes (no damage to plane). Back to W.H. for dinner tonite with Gov's. of the 50 states. The dinner was very pleasant & everyone seemed to have a good time. Bob Newhart entertained & was very funny.

Monday, February 27

A short night & now a long day—but it was one of those days when everything seemed to go right. N.S.C. briefing was on Chernenko's letter. We're agreed we are going to make our plans for response with George B., George S., Bud, Cap & me—no bureaucracy.

[Appearance before governors; met with secretary of Agriculture on finding a way to use surplus food to help starving people in Africa; lunch with Vice President Bush regarding USSR and politics, commented, "He's carried the ball in N.H."; National Security Planning Group (NSPG) meeting in which Casey reported on worldwide covert actions; interviewed by military journalists; ambassadorial formality; photo sessions, including newly retired baseball player Carl Yastrzemski; videotapings. **Tuesday, February 28:** *visit by President Rudolf Kirchschlager of Austria, commented, "I think we made him & his Ministers understand our situation in Central Am."; cabinet meeting on containing "Beltway Bandits," advisory and consulting groups gaining unnecessary government contracts; greeted actress Shelley Long; state dinner, entertainment by Mel Tormé and Peter Nero.]*

Wednesday, February 29

A short day thank Heaven because it was a short night. N.S.C. was about Jeanne Kirkpatrick's need for guidance in the U.N. Security Council on a Lebanon resolution. She's been battling the Soviets who want a resolution calling for the M.N.F. & the Israelis to leave Lebanon but no word about Syria leaving. I think we came together on some wording to correct that. Otherwise it would be—veto.

[Budget meeting on position regarding spending cuts; met with Trent Lott and supportive members of Congress; addressed National Alliance of Senior Citizens; meeting with Dutch foreign minister Hans van den Broek to press for stronger stand on Intermediate-Range Nuclear Forces (INF) deployment; meeting with Shultz on answer to Chernenko's letter; greeted Olympic team.]

Thursday, March 1

N.S.C. brief—said goodbye (I hope temporarily) to Dick Stone. He's been doing a great job as a roving Ambas. in Central Am. Friction developed—a personality clash with another good man in State Dept. Dick has resigned. I hope we can find another spot for him.

[Addressed the American Legion's Women's Auxiliary.]

Had lunch with Bud, Mike D. & Mrs. Massie—just back from Russia. She's a remarkable woman with some great insight on the Russians. She reinforced my gut feeling that it's time for me to personally meet with Chernenko.

[Cabinet Council on Economic Affairs, favored proposal for lower minimum wage for teenagers; rejected congressional job plan for youth; visit from leader of VFW; new members of W.H. press corps; school prayer group; "a womans org. with more members

than N.O.W."; *athletes in favor of prayer in school; Latin American journalists; leaders of Vietnam Veterans Leadership Program.]*

Friday, March 2

Wasn't the longest day but it got to be the busiest. It got busy beginning with the N.S.P.G. meeting in the situation room. This had to do with several pieces of legis. we're putting together to enable us to have a program for anti-terrorism.

[Lunch arranged by Maureen with female Republican elected officials; meeting with Shultz on personnel changes at State; Weinberger report on his meeting with Prime Minister Thatcher, learned "She's forgiven us completely for Grenada."]

Then into the Treaty Room for a top level & secret meeting with Ambas. Hartman (Moscow), Bill Casey, Bud McF., Geo. B., Mike & Jim & Gen. Vessey. Subject was a plan to move into communications with the Soviets. I'm convinced the time has come for me to meet with Chernenko along about July. We're going to start with some ministerial level meetings on a number of substantive matters that have been on ice since the KAL 700 shoot down.

[Senator Howard Baker (R-TN) and other senators discuss upcoming vote on school prayer amendment; dinner with the Conservative Political Action Committee. **Saturday, March 3**: *radiocast; quiet day with desk work.]*

Sunday, March 4

Our 32nd Wedding Anniversary. We spent it at home. The wife of the Pres. of Israel & the Emperor of Japan helped me with my gift buying. Under the law gifts from people like that are not ours but we are allowed to buy them if we want. Two such gifts—a gold & diamond pin & a beautiful jewel box I knew were much admired by Nancy so I bought them for Nancy.

More homework & then a delightful evening with the Laxalts & Wicks—just the 6 of us. It was a happy anniversary.

Monday, March 5

A typical Mon.—no breathing room & a stack of memos plus things to sign—which I didn't get to until later afternoon. Helmut Kohl arrived (W. Ger. Chancellor). We had a good meeting thru lunch. He confirmed my belief that Soviets are motivated, at least in part by insecurity & a suspicion that we & our allies mean them harm. They still preserve the tank traps & barbed wire that show how close the Germans got to Moscow before they were stopped. He too thinks I should meet Chernenko.

[Met with Senators Howard Baker (R-TN) and Paul Laxalt (R-NV) regarding negotiating with Democrats over a program for reducing deficit; addressed National League of Cities; brief meeting with officers of U.S. League of Savings Institutions, commented,

"They are very supportive of us"; photo session with tennis player John Newcombe; video-tapings; reception for campaign advance workers.]

Tuesday, March 6
This modern age! Spent half a day in the office then by mid afternoon I was addressing the Nat. Assn. of Evangelicals in Columbus Ohio, then at Dinner in N.Y., a Repub. $1000 a ticket fundraiser & went to bed in the W.H.

Before flying to Columbus, met with Sens. Bill Cohen & Joe Biden. They've been to Russia & are all wrapped up in "Arms Reductions." I suspect that at least one of them (J.B.) doesn't believe I'm sincere about wanting them. We had an N.S.C. meeting & then I was off on A.F.1. The Evangelicals were most warm & cordial. I had 4 standing ovations. The group in N.Y. was also receptive. I met Nancy in N.Y. & we came back to Wash. together.

Wednesday, March 7
We had a meeting with our Deficit Downpayment Negotiating group. I'm afraid I blew my top at one point. It seemed to me they are willing to let the Dems. run with the ball because they don't think we can stop them. I told them before we do that, it's time for us to agree on our position & then let me take it to the people (TV) & smoke them out. The way we're going we're not exerting any leadership. Besides—our position is on the side of the Angels & leaves the Dems. the wrong side of the debate.

[Spoke briefly to state agricultural directors; lunch with editors of business magazines, regarded as "very supportive of our programs etc."]

George S. & I met. Our plans about the Soviet Union are going forward. He's giving Ambas. Dobrynin my letter for delivery to Chernenko.

Then upstairs for afternoon of homework.

Thursday, March 8
Charlie Price came by our N.S.C. briefing with some input about England's cautious attitude about the Persian Gulf if Iran should try to close the Straits of Hormuz. No real problem—they'll line up with us to keep it open.

Met with Repub. Sen. leaders on the budget. I'm convinced we must take the lead & wait no more on the Democrats for a bi-partisan approach to the deficit. They think they have a campaign issue. They've been in charge of the Cong. for 45 years & for 45 years they've run up deficits. Now they want to hang the present deficit on us & yet they've only given us half the spending cuts we asked for. We are all agreed now on the tactic but haven't reached agreement on the amount we propose for Defense spending. Some—like Hatfield & Domenici want to cut it too much. We'll meet tomorrow to settle it. We took up another item, an emergency appropriation to keep El Salvador's

Army running until the budget is passed—they are about out of ammunition.

Hatfield is chmn. of the appropriations committee. He engineered the defeat (by 1 vote) of amending it to a sure thing bill on energy. We'll try again next week but I'm frustrated as h--l. With some of our friends we don't need enemies.

[Report in Cabinet Council on future demographics of U.S.; photo sessions; met young hero of a fire; signed proclamation for Red Cross; reception for smaller newspapers, commented, "Unlike the Wash. Post & others they are very supportive of our admin." **Friday, March 9:** *cancellation of NSC meeting to prepare for visit by President François Mitterrand of France; brief meeting with Senate leadership on defense budget; signing of transportation grant bill; discussion with Shultz, who reported that Anatoly Dobrynin, Soviet ambassador to the U.S., was "very interested in getting some talks going on cultural exchange, consultants in N.Y. & Kiev etc."; went to Camp David.* **Saturday, March 10–Sunday, March 11:** *deep snow; desk work; returned to W.H.; Maureen and Dennis house guests.* **Monday, March 12:** *met with Republican House leadership on budget; visited local elementary school; another lunch organized by Maureen with a group of women Republican elected officials; interview with reporters from southeastern states; photo sessions; Mrs. Reagan was in Arizona.* **Tuesday, March 13:** *spoke to United Jewish Appeal at Hilton, used entrance that was the scene of assassination attempt in 1981, security had constructed a housing around it; met Mr. and Mrs. John Walsh, parents of kidnapped boy, who were starting a new program for missing children; swearing-in for Maureen Corcoran as General Counsel for Department of Education; ambassadorial formalities.]*

Wednesday, March 14

P.M. Soares of Portugal was our visitor. He's most impressive. He is a Socialist but he's seeking private investment in Portugal's industry & is as rabid an anti-communist as you can find. He's a great supporter of our country & the West. We had good & fruitful meetings. Then we had another meeting with the Sen. leadership on the deficit. I got a little worked up & made a speech. I think we have a deal. They are working on the figures & we'll know tomorrow.

Then upstairs to make phone calls rallying support for the school prayer amendment & doing mail.

Thursday, March 15

[Met with Republican leadership in Congress, formed agreement on proposed deficit package.]

Jeanne Kirkpatrick came in to tell me we lost a Poland Human Rights bill in the U.N. Security Council—2 votes against us—Soviet U. & Jordan. King Hussein has turned on us & blasted me personally in a N.Y. Times interview. I don't know that he's left any room for coming back & he's already flirting with the Soviet Union. Jeanne is also worried about what the Congress is doing to us on El Salvador & Nicaragua & so am I. Dick Walters came in—what a man he is. He has averaged 1000 miles travel a day for the past year. [. . .]

[Spoke to Puerto Rican American leaders; appearances at meeting of National Alliance of Businessmen and at National Federation of Press Women; photo sessions; announcement of GOP plan for deficit reduction; Mrs. Reagan returned from West ("Nancy is home—hooray!").]

Friday, March 16–Sunday, March 18

Phoned Bishop John J. O'Connor who is to replace Cardinal Cooke. This will be a good relationship—Pope John Paul has chosen well.

[Met with unidentified senators on school prayer amendment, uncertain of effect.]

P.M. Fitzgerald of Ireland was our visitor. He's a fine man. I think we gave him some different insights on Central Am. He's very brave & outspoken about the terrorism in N. Ireland. We held a St. Patrick's Day lunch which was great fun with many old friends such as Maureen O'Hara, Geraldine Fitzgerald, Merv Griffin, Eva Gabor & Ginger Rogers on hand.

Geo. Shultz & I then huddled on Angola & the Middle East. I had a photo with our Ambas. & wife to Lebanon (the Reginald Bartholomews). They should both have medals for bravery. Our embassy has been under artillery fire off & on for months. They never flinched. Then it was off to Camp David—with a brief case full of homework. Weather was fair but chilly—most of snow gone. Now we're back at the W.H. Mubarak has asked for A.W.A.C.'s which we've provided in response to an unprovoked bombing of Sudan by Libya.

[Monday, March 19: spoke to representatives of small business; met with Cuban Americans and asked for support in launching Radio Marti; issues briefing at lunch; meeting with Joint Chiefs of Staff for the seventh time, positive report on military readiness; met with senators pressing them on school prayer amendment; dentist visit for cleaning and replacement of filling.]

Tuesday, March 20

[Breakfast with farm families.]

In the afternoon back over to the East Room for a signing ceremony of the Shipping Act of 1984. On to the E.O.B. for an NSC meeting on contingency planning for what we'll do if Iran should seek to close the Persian Gulf causing a worldwide oil shortage. One thing we won't do is go down the road of govt. allocations etc. We'll trust to the mkt. for that. We'll have to help our allies, we are less dependent on imported oil than they are.

Finished the day with an interview by reps. of 6 midwest papers. I don't think they drew blood. During the day I called Congressman Ireland in a Fla. Hospital with pneumonia. He is the 4th Dem. Rep. to turn Republican since we've been here. We lost the school prayer amendment in the Senate. We had a majority but needed a ⅔ majority. The sad thing is about 15 Sens. were convinced the amendment was a mandate that schools would have to have prayer. Lowell Weicker was the head ringmaster against us as he is on everything we want. He's a pompous, no good, fathead.

Wednesday, March 21

Over to Capitol hill to talk to our Repub. Senators & Res. The subject was our $150 bil. down payment on the deficit. First I talked to the Senators in the original old Senate Chamber, then did a Q&A. It went pretty well & I answered Lowell Weicker's Q without telling him what a schmuck he is. Later over to the House side—but not Q&A. All in all everything went well.

[Met with Shultz on (unidentified) potential progress in Angola and ongoing effort to resolve problems in Nambia and South Africa; new poll numbers seemed positive.]

(Then upstairs for some desk work.) We have house guests—Jimmy & Gloria Stewart, Bets Bloomingdale, Ted Graber, & Maureen. Got word that a carrier of ours in the Sea of Japan had a scraping collision with a Soviet sub. Then came a call that Gromyko has launched a charge that we're responsible for a mine off Nicaragua that damaged a Soviet tanker & they claim injured some crewmen. I have approved refusing to accept their note. Our evening with old friends was great.

Thursday, March 22

A chill windy day but still an outdoor full dress ceremony for Pres. Mitterrand (France). Everything went well. We had an hours meeting in the Cabinet Room. I probed him about the Soviet U. His views were most informative & confirmed some thoughts I've been having. Geo. B. & I had our usual Thurs. lunch. Then I met with the speechwriters re a planned nat. address on foreign policy.

Sec. Bill Clark checked in with a report on progress we're making in calming the normal hysteria of our more extreme environmentalists.

[Delegation of Costa Ricans bring petitions of good will signed by three hundred thousand of their countrymen; state dinner.]

Friday, March 23

[Breakfast with President François Mitterrand, discussed trade and agreed against "a return to protectionist policies"; Mitterrand left; addressed government managers; appearance with Tau Kappa Epsilon (TKE) fraternity alumni.]

George Shultz, Bud & I met for a strategy session on where we go with the Soviets. I think they are going to be cold & stiff-necked for awhile. But we must not become supplicants. We'll try to get agreement on a few lesser matters.

[Videotapings, including a message for a tribute to Lucille Ball. **Saturday, March 24:** *radiocast; helicoptered to Monticello, toured Jefferson's home, "a place of surpassing beauty"; returned to Washington; addressed the Gridiron dinner.* **Sunday, March 25:** *quiet day; Maureen visiting; cleaned out part of desk ("I have many more that need it").]*

Monday, March 26

Why are Mondays all so busy. Called Paul Volcker whose Fed. board meets tomorrow to urge him not to go for any drastic action. Let this recovery continue. Frankly I don't know what he will do. An NSC briefing & then a Cab. Council meeting on—whoops—that came along in the afternoon. It was on energy. We have an energy policy & it's working. We need to get rid of excessive regulations to cut the building time on nuclear reactors now 12 to 14 yrs. down to 5 to 7 as it is in other countries.

Now back to the A.M. I met briefly with the Att. Gen's. from our 50 states. We have a good working relation with them & they all sound supportive of what we're doing with regard to crime.

At noon Nancy & I hosted the Medal of Freedom lunch. We gave 14 medals—some to old friends like Jimmy Cagney & Tennessee Ernie Ford. Also to Howard Baker & Mrs. Jackie Robinson for her deceased husband. She's quite a lady. The news headline one though was the son of Whittaker Chambers accepting for his father.

[Appearance at meeting of Council of 100, black leadership group.]

The delegation that went to El Salvador as observers of the election—part Dem. part Repub. members of Congress plus some citizen reps—came to the Cab. Room to report. They were greatly impressed by what they'd seen. The TV news was a false picture in view of what they'd seen—people who would risk their lives to vote. Story after story confirmed that the people of El Salvador want Democracy & we should help them.

[Visit from leaders of National Cattlemen's Association; greeted contributors to Senate campaign fund.]

Tuesday, March 27

Don Regan is back from China. Apparently the P.R.C. is moving more & more toward free enterprise. They are now opening up to outright capitalist investment by foreign businesses & industries. Farm communes are gone & individuals are allowed long term land leases & the right to sell produce in the market.

[Addressed Insurance Agents of America; visit by ten Hispanic American holders of the Congressional Medal of Honor, designing stamp; signed bill recognizing African American soldiers during Revolutionary War; NSC meeting on ways to counter Soviet propaganda implying "that they are the peace seekers when they are the ones who walked out of the arms

talks," believed U.S. campaign should begin with reply to Chernenko's letter; interviewed by Family Weekly; *off-the-record cocktail party for small number of journalists.]*

Wednesday, March 28

[Met with Harry Shlaudeman, taking over for Richard Stone as Special Envoy to Central America.]

Had an interview (2 reporters) N.Y. Times. They've interviewed Mondale & Hart & Jackson—now it's my turn. I enjoyed it. It gave me a chance to set the record straight on a few things.

[Interviewed by Phoenix television regarding Barry Goldwater.]

Geo. S. & Bud & I met—more talk about the Soviets & how we should handle the situation. Later our Ambas. Art Hartman came by. He's truly a fine Ambas. It was good to have a chance to pick his brains. He says they aren't going to let Chernenko become a top man & Gromyko is going to be sure everything goes thru him.

Went up to my study for rest of afternoon & some desk work.

Thursday, March 29

Met with Ambas. Pickering (El Salvador). He's a fine rep. of the U.S. He confirmed all that our observers had said about the El Salvador election. He had some shrewd insights to offer on the entire situation.

N.S.C. meeting re a chemical warfare treaty. Problem is the near impossibility to verify whether treaty is being observed. I settled finally that we should table a treaty but with the proviso that verification must be negotiated first. If there is no agreement on that—then no treaty.

[Visit from Congressman Andy Ireland (D-FL), who was switching parties; interviewed on health by USA Today; *meeting with presidents of Traditionally Black Colleges and Universities; visits by group from American Cancer Society; by Hungarian American group; executives of Smith & Wesson, with gift of a pistol; photo session, videotapings; off-the-record cocktail with newsmen; visit after dinner by the Reverend and Mrs. Billy Graham; report of good economic news.* **Friday, March 30:** *National Security Planning Group (NSPG) meeting on contingency plans in case of disruption of oil flow from Middle East; lunch with Commission on Women's Business Ownership; visit from Mr. and Mrs. Mel Blanc; discussion of Soviet situation with Shultz and McFarlane (no details recorded in diary); greeted thirty labor union leaders pledging support; left for Camp David.* **Saturday, March 31–Sunday, April 1:** *quiet days; walks for fresh air; returned to Washington; annual meeting of Trilateral Commission.* **Monday, April 2:** *ceremony announcing Cherry Blossom Time; greeted Eisenhower Exchange Fellows; helicoptered to Baltimore for baseball season opener; returned to Washington; attended Paul Laxalt's "Lamb Fry": "The delicacy of the evening (an old Basque custom) lamb gonads. They made for lots of humor in the toasts but they're not my favorite food."]*

Tuesday, April 3

Met with G.O.P. leadership—House & Sen. on 2 subjects; Our $150 bil. down payment on the deficit & why we had to stay united & our supplemental appropriations to help El Salvador. I hope we convinced them. There are military units facing the rebels with ammunition limited to the clip they have in their gun.

[NSC meeting with U.S. ambassador to Sweden; visited National Security Telecommunications Advisory Committee; couple of hours in rehearsal for press conference; Mrs. Reagan at dinner receiving award for anti-drug campaign.]

This afternoon Bill Smith came in to tell me he'd stay on as Attorney Gen. until this whole mess with Ed M. is cleared up & we can go on. This solves a real problem for us.

Wednesday, April 4

N.S.C.—staff time & finally a Cab. Council meeting regarding the street people, particularly those who have been released from Mental Hospitals & have no homes. We are embarking on a program of partnership with various pvt. org's. & groups. Had a brief meeting with Reps. of the Catholic Health Assn. Then lunch with Don Regan. He's almost certain that the Fed. is on the right course & our recovery is going to continue. He says tax simplification is not going to be easy.

I made a couple of calls to the President of Columbia & Venezuela to thank them for the Argentine rescue mission. I'll call the rest tomorrow. Geo. S. & I met to discuss our progress with the Soviets. We're quietly making some headway.

[Rehearsal for the press conference; then staged press conference; considered it went well. **Thursday, April 5:** *flew to New York City; spoke at Women in Business Conference; visited day care center financed by corporations; dinner with state organization of Catholic-school parents, escorted by Archbishop John J. O'Connor; returned to W.H.* **Friday, April 6:** *made "major foreign policy speech" at Center for Strategic and International Studies (no details recorded); met University of Southern California's championship NCAA women's basketball team; looked at plans for convention hall in Dallas; lunch with experts on China; attended meeting of National Conference of Lieutenant Governors; telephoned leaders of Latin American countries that cooperated with financial bailout of Argentina.* **Saturday, April 7:** *radiocast; greeted Georgetown championship NCAA men's basketball team; desk work; evening watching television.* **Sunday, April 8:** *quiet day; reception for restored Ford's Theatre.]*

Monday, April 9

Like all Mondays—a busy day with my desk buried in paper. The usual staff & NSC meetings then a signing ceremony for the 27th Law Day U.S.A. proclamation. A goodly crowd on hand.

The Mayor of Berlin dropped by. He's a personable young man & we had a nice visit. He presented me with a plate commemorating the 35th anniversary of the Berlin airlift.

[Issues lunch; ceremony for National Teacher of the Year; report by Brent Scowcroft on his trip to Russia, noted, "He believes the Soviet cold shoulder is due in part at least to their not wanting to help me get re-elected"; signed proclamation for Parkinson's Disease Awareness Week; informal gathering with small number of W.H. reporters; attended dinner of the Chowder and Marching Club, Presidents Nixon and Ford also present.]

Tuesday, April 10

[Visit by President Jorge Blanco of the Dominican Republic, meetings on international affairs, including El Salvador, Nicaragua, and economic affairs; state dinner later in the day; signing ceremonies for Fair Housing Proclamation and Wheat Act, providing death benefits for government employees killed by terrorists; photo sessions.]

The whole day was however under a cloud. Barry G. started it off with a letter to Bill Casey which he made public. He's raising h--l as chairman of the Intelligence comm. because of the harbor mining in Nicaragua—says he was never briefed. He was briefed on March 8 & 13. There is a rebellion which will probably lead to their shutting aid off to the Nicaraguan Contras—which will bring joy to the Soviets & Cubans.

Wednesday, April 11

Off to Kansas City to visit a Ford plant & speak to 2000 workers. I was warmly received in spite of the best efforts of the Auto workers union to see I wouldn't be. They are doing a remarkable job of making a worker management team at that plant which probably drives the Union leaders up the wall.

[Flew to Dallas for an overnight. **Thursday, April 12:** *visited construction site in Dallas; attended seminar with Builders Magazine; returned to Washington; reception for Republican fund-raisers; Maureen visiting; Patti announced engagement plans; Mrs. Reagan in Detroit.]*

Friday, April 13

Bad luck day here but a big holiday—the beginning of New Year in Thailand. P.M. Pren pd. his 2nd visit to the U.S. We have a fine relationship with him & with his country. Our meetings were productive & we are able to offer some help—M48 tanks to begin with.

George S. & Bud & I met to discuss how we fund military needs of El Salvador with Congress going on recess & no action on appropriation we asked for. El Salvador is scraping the bottom of the barrel on ammunition & medical supplies.

[Ambassadorial formalities; received awards from Adenauer Foundation and the All-

American Collegiate Golf Association; met new Hispanic appointees; photo with departing head of physical fitness program; spoke to large gathering of Baptist fundamentalists.]

Then we went on to White House Correspondents Dinner. I was picked on a little but think I might have tapped them with some Gridiron type one liners.

[Saturday, April 14: radiocast on Central America; lunch with China scholars and diplomats in preparation for upcoming trip. Sunday, April 15: quiet day; report of two observers killed in Namibia.]

Monday, April 16

A press ceremony in the Cab. room seeing George B. off for Geneva to present our treaty banning chemical weapons. An NSC working lunch with a movie prepared for our China trip showing all the places where we'll be. It was well done & most informative.

Dave S. briefed us on the budget situation. It's clear we'll have to go to work to rebuild govt. structure along different lines if we're still here next year in order to eliminate deficits.

[Met Chinese students from U.S. colleges; interviewed by Chinese journalists; video-tapings.]

Tuesday, April 17

Nancy left for Phoenix. Discussion in NSC about doing a speech on Central Am. nationwide. Geo. S. & Bud leaned toward trying to do it—like tomorrow nite. We'll do it when I return from China. Had lunch with about 100 leaders of Hispanic groups. They're having a kind of conference & our people are briefing them. I spoke & was well received even though many of those organizations are Dem.

[NSC briefing on China, especially leaders and their dispositions.]

A Cab. Council meeting on Agri. About 2 to 4% of our farmers will probably go under this year. Much of that is due to poor business judgment, over-borrowing based on inflated land values. The inflation bubble has burst & that plus high interest rates is doing them in.

[Met with editors of farm papers; visit by president of Eureka College; gave medal to widow of soldier killed in the Sinai; signed proclamation for Military Spouses Day; photo sessions; visit by Art and Lois Linkletter.]

To the East Room for a reception—the Reagan-Bush finance committee. They were an enthusiastic group. In the receiving line I was surprised that at least ⅔ of them urged me to stand firm on Central Am.

Wednesday, April 18

A busy time tying up loose ends for getaway day on the trip to China plus a complicated packing chore.

Said a farewell to Larry Eagleburger who is leaving State Dept. after 23 yrs. to go into private life. We'll miss him—he's a fine foreign service officer.

Kadafi must be insane—a demonstration by Libyan exiles in front of the Libyan Embassy in London was fired on by someone in the Embassy, 11 wounded & 1 Eng. Police Woman killed. The Eng. police have the embassy surrounded now but can't storm it under international law. Kadaffi has surrounded the British Embassy in Libya with mil. forces & that is a state of siege. Our intelligence has learned that Kadafi ordered the embassy in London to attack the demonstrators & start a wave of fire throughout Eng. He charges Eng. is in cahoots with the demonstrators.

Had an interesting lunch with 2 Cardinals, several Arch Bishops, Bishops & Monsignor. It was a mix all the way from stalwarts like Cardinal Krol & Arch Bishop O'Connor to Bishops who are part of the peace movement. I think I got a few things off my chest—but politely. Back to packing.

Thursday, April 19

Marine 1 & A.F.1 now to Seattle, then to the ranch on the start to China. I'm writing this actually on Wed. nite.

* I've just been notified that our Congressional delegation in Central Am. survived the crash of an army (U.S.) helicopter near the Honduran, El Salvador border. The craft was downed by ground fire from the Salvador side—which would mean Guerillas. Sens. Lawton Chiles & Bennett Johnston, staff & wife of Ambas. Negropante all apparently alright & on their way to Palmerola Air Base in Honduras.

[**Thursday, April 19**: *met with Weyerhaeuser executives in Tacoma, Washington; spoke with dockworkers; flew to ranch in California, met Mrs. Reagan there.* **Friday, April 20–Saturday, April 21**: *two days of riding and ranch work.* **April 22—Easter Sunday:** *flew to Honolulu; attended Easter Service at the Episcopalian Cathedral, commented, "The Bishop got in some digs in his sermon at our defense buildup. It turns out he's a marcher in peace & anti-nuke parades."* **Monday, April 23**: *went swimming.*]

Tuesday, April 24

Off to Guam—a farewell ceremony as we left Hickam field. Barry Goldwater & a few other Sens. were there on their way back to Wash. after a trip to Taiwan. Barry is upset about my trip & can't hide it. He seems to think I'm selling out our friends on Taiwan. He should know better. I've made it very plain to the leaders of the P.R.C. that we will not forsake old friends in order to make new ones.

---

* *The asterisk is in Reagan's original.*

[Eight-hour flight to Guam, arrived on **Wednesday, April 25;** *large reception at airport.]*

Thursday, April 26

We left Agana Naval Air Station at 10 A.M. enroute to Beijing. I had scheduled our meals to correspond to China time as part of our anti jet lag program. We arrived at Capital Airport at 2:05 P.M. China time which was 1:05 A.M. Wash. time. We drove directly to the East Court of the Great Hall of the People. We were greeted by Pres. Li & Mrs. Li. The national anthems were played, there was a 21-gun salute—the 1st ever given to an American Pres. Then we reviewed the troops & were greeted by at least 100 costumed children.

We then had a 30 min. meeting with the Pres. after which we were driving to the Diaoyutai guest house—our home for our stay in Beijing.

About 7 P.M. we took a 5 min. drive to the Yang Yuan Hall (another bldg. in the guest house compound) for the dinner hosted by Pres. Li. The Pres. is an affable person as is his wife. Here was our 1st go at a 12 course Chinese dinner. We heeded Dick Nixons advice & didn't ask what things were—we just swallowed them. There were a few items I managed to stir around on my plate & leave. We both did well with our chopsticks.

Friday, April 27

A breakfast meeting with my gang at our Villa. We kept a noisy tape going all through the meal to nullify any hidden microphones. We later learned there were such. Indeed Dave Fisher unscrewed the plate on his light switch & removed one for a souvenir. Later 5 were found in our quarters.

On to the Great Hall of the People for a 90 min. meeting with Premier Zhao. It was a good meeting. We mainly discussed the global situation & I scored some points on what our goals were. I was careful to make it clear that we were not seeking an alliance, that we approved of China's non-aligned status but that as friends & Pacific Basin neighbors we could contribute to Peace & Stability in the basin.

Back to the guest house for lunch & then back to the Hall where in the auditorium I addressed a meeting of Chinese leaders—bus. Acad. etc. This was the speech re-played on TV but with some of my lines about the Soviet U.—religion & free economy deleted. Then back to another 90 min. meeting with Zhao. We get along very well. I like him & I think he reciprocates. This time we got into bi-lateral matters having to do with trade, investment etc.

Following this meeting I had a 1 hr. meeting with Gen. Sec. Hu—the party leader.

He's a feisty little man & more doctrinaire about his ideology than anyone I met. He lectured me about removing our troops from S. Korea. I gave it right back to him that there was no way we'd do that, if North Korea wants better relations let them stop digging tunnels under the D.M.Z. etc. I didn't consider this meeting very important but I invited him to visit the U.S. He jumped at it. I have in mind he might learn something by seeing the outside world.

That night was the big welcoming banquet at the Great Hall. Somehow this 12 course dinner went down easier. There were the usual toasts & then some interesting musical entertainment.

Saturday, April 28

Another breakfast meeting because this was Big Casino day—my meeting with Chairman Deng. Nancy went with me for the informal opening. Deng—who has a sense of humor invited her to come back to China without me.

Then in our meeting he really waded in critical of our mid-east policy, our treatment of the developing nations etc. & our disarmament failure. He touched a nerve—when it was my turn I corrected him with facts & figures & I meant it. Funny thing happened—he warmed up although he did bring up Taiwan (the only one who did). I told him that was their problem to be worked out—but it must be worked out peacefully. We broke for lunch (he was host) & though it was scheduled to be a working lunch it became a pleasant social event instead.

After lunch we picked up Nancy & were off to the Great Wall. All the way we waved our arms off at the crowds lining the streets to see us & even in the villages after we got out of the country.

The Wall has an amazing effect even though you've seen photos & movies of it. There is a feeling I can't describe when you stand on it & see it disappear over the mountains in both directions.

This evening we hosted a dinner for a mixed group of Americans & Chinese plus Premier Zhao etc. at the Great Wall Hotel which is an American, Chinese partnership venture. The dinner was American—roast turkey.

Sunday, April 29

This was our tourist day. We flew in A.F.1 to Xian the ancient capital of China. Then an hour & a half drive to the tomb of the 1st Emperor. This is the great excavation scene where they have uncovered 800 life size Terra Cotta figures of soldiers standing in ranks complete with horse & chariots guarding the Emperor's tomb. They know there are more than 7000 they haven't uncovered yet. It is an unforgettable experience. This plus the drive past villages surrounded by endless wheat fields dotted here & there with burial mounds & relics of Chinas ancient past—made for a day we'll long remember.

Before we left they had set up an example of a free market. We wanted to see the real thing because it's part of their departure from communism—

individuals selling their wares in the open market for profit. The Chinese said no—for security reasons so they set up a temporary one at the museum locale. Nancy bought some decorative items for our Christmas tree. There was an incident—her bill came to 5 Yuan ($2.50). The poor gal selling didn't have any change. I'm aware there is no tipping in China but she was so embarrassed & looking to others for help. I said "keep it," & we moved on. She caught up with us & gave me the change she'd gotten from someone for the 10 Yuan note I'd given her. Our TV press made a big thing of it—that I had committed a blunder & tried to tip her.

Then back to Beijing & our last night at Diaoyutai.

Monday, April 30
To the Great Hall for signing of a nuclear power agreement & other protocols agreed on in our talks. A Champagne toast & the official farewell by Chmn. Zhao & Pres. Li. Then to A.F.1 & on to Shanghai.

We landed at Honggiao Airport at 12:30. A Big receiving line headed by Mayor Wang. Nancy & I separated. They had a schedule for her & I went to the Shanghai-Foxboro co. It's a high tech industry & another example of the modernization taking place in China—a partnership between Foxboro of Mass. & China. After a tour of the plant & some ceremonies we went on to the Jing Jiang guest house.

At 2:35 I left for Fudan U. I was met by a tiny Chinese woman who is Pres. of the U. but who is also a graduate of Smith & M.I.T. First we had a tea with faculty members—half of whom had gone to school in America & most of whom spoke English.

Then over to a classroom where I did Q&A with students—all of whom spoke English. I wish I could have stayed with them longer. Their questions were very much about America & American students.

Then to the Auditorium for an address carried live on Shanghai T.V. The audience was a mix of faculty & students. Most understand Eng. & for those who didn't they had a script of my speech in Chinese.

I was well received & frankly it was a darn good speech—the students ate it up. Then Pres. Xie presented me with a set of books of ancient Chinese maps by one of the Prof.'s.

That evening the Mayor hosted a dinner in the Shanghai Exhibition Hall. It was an enjoyable evening & the best Chinese food of the trip—for an American that is. There was entertainment that was enjoyable & the evening ended.

After breakfast met with the consulate staff then went to the Rainbow bridge Commune. The Deputy Commune leader briefed us on the commune—now called a township. Then we toured the township stopping at a child care center where 4 & 5 year olds entertained us with dance & song. They are the cutest children & so full of joy. Then we visited a private home—a young couple, their little boy & the father & mother of the husband. He (the husband) had built the house himself. He told us how much income they had & how long it took him to save for the house etc. We went out in a field & met some women working there.

Under modernization they get to keep & sell in the open mkt. everything they can produce over & above the govt's. quota. That ended our visit to China. We boarded A.F.1 & headed out over the Pacific toward Fairbanks Alaska. We departed at 10:35 A.M., flew for about 8½ hours, crossing the dateline again & landed in Fairbanks at 3:20 A.M. Alaska time. Believe me that can confuse you. We drove to a new home belonging to Sen. & Mrs. Murkowski who haven't lived in it yet & went to bed until about 11 A.M. but it was still May 1 & still Tuesday. We had juice & coffee & headed for the U. of Alaska for a welcoming ceremony—about 3000 people. The Sen. introduced me & I made a short speech. Some children presented us with gifts—native handcraft.

Then to the William Wood Student Center for lunch & more remarks. Back to the Murkowski home at 2:30 P.M. & into Pajamas. We saw the midnight sun—it was still up at 11 P.M. We finally got to sleep but it was strange getting away from China time.

Wednesday, May 2

We went with about 10,000 others to meet the Pope who was stopping for re-fueling on his way to South Korea. We both spoke briefly to the crowd then went into the airport for a meeting. I briefed him on our China trip. He is anxious to enter into talks with the P.R.C. They haven't established diplomatic relations because the Vatican has relations with Taiwan.

Then he saw us off to Wash. where we landed about 10 P.M.

Thursday, May 3

Into the office at 10 A.M.—I'm still under a China spell & nothing here seems real yet. Had a photo with the editor of Far Eastern Ec. Review. Then Coach Eddie Robinson of Grambling U. came by with the U. President & the coaches family. He's quite a guy & his teams have won 300 games (F.B.) but more important he sees that his players go on to make something of themselves. P.M. we went to Camp David.

Friday, May 4–Sunday, May 6

Camp David. I spent most of the time in the same chair catching up on homework. Now we're back in the W.H. & tonite having a dinner honoring Dillon Ripley for his 20 yrs. of service with the Smithsonian. It was a very pleasant affair in the Blue Room.

Monday, May 7

An easy day but a bad night. I couldn't get to sleep until 2 A.M. then I awoke at 6 A.M. Found out most of our China trip party had much the same kind of night.

[Staff and NSC meetings; Senator Paul Laxalt (R-NV) arrived with group from Virginia interested in forming independent campaign support group; lunch for volunteers.]

Met with German Foreign Minister Genscher who is off to Moscow. Wanted to touch base with us first. He gave a slant that Soviet intransigence might be due to their fear of our power (based on W.W. II memories) & our link with Germany.

A taping session then a half hour with Holmes Tuttle re Ed Meese. I think he'd like to have me push him overboard. I can't do that.

Tuesday, May 8

[Breakfast with Democratic congressmen inclined to support MX; meeting of bipartisan congressional leaders with briefing on China and Central America; support is tenuous at best; short meeting with International Olympic Committee President Juan Antonio Samaranch on Soviet statement withdrawing its team from the Olympics.]

We had a lunch in the St. Dining Room observing the 100th B.D. of Pres. Harry Truman. His daughter Margaret was there. She's quite a lady & I like her very much.

Later in the day spoke on Central Am. to the Council on the Americas meeting at the State Dept. Ended day with a brief meeting with Lyn N. & Frank Donatelli. They are worried about our confidence among Repubs. So am I.

Wednesday, May 9

[Ceremony on behalf of Small Business Association.]

Then Pres. Masire of Botswana arrived. His landlocked Co. is suffering from 3 yrs. of drought. He has adamantly stood for Free enterprise & democracy in his land. Geo. S. & I met briefly about the Soviets. He's convinced the freeze including the Olympic boycott is Gromyko's doing. I think he's right.

The observer team I sent to El Salvador came back to report. They are 100% enthusiastic in their support of the election—it's honesty etc. & the need for us to help in Central Am.

Two Dem. Congressmen came to see me, Marvin Leath & Beryl Anthony. They are solidly in my corner re the Chemical Warfare bill to indicate if Soviets don't accept our treaty to eliminate such weapons—we'll go ahead & produce them as a deterrent. From them I went before a group of Repub. Congressmen who are reluctant about this action. I hope I moved them.

Then upstairs to wood shed for my speech to the nation. Went on at 8 P.M. explaining the Central American situation. By 9:30—915 to 151 positive phone calls had come in.

[**Thursday, May 10**: *spoke to small group about MX; met with Japanese official who brought a message from Prime Minister Nakasone (no details recorded); Senator John Tower (R-TX) pessimistic on passage of chemical weapon bill in Senate; presented Enrico Fermi Award to scientists; addressed National Realtors Association, reiterated support for deductibility of mortgage interest; reception for donors; then fund-raising dinner set record, drawing in $5 million.*]

Friday, May 11

Met with Ambas. Hartman (Moscow). He believes there is friction in the Polit Bureau & Gromyko is much of our problem. He doesn't feel I could have any success in appealing to the Soviets to come to the Olympics.

Dropped by a meeting of Several columnists re the ed. situation. Then met with members of the Commission on Excellence in Ed. They have performed a miracle. One year ago they issued their report & now all over the country on their own, parents, teachers, local school boards & state legislatures have started a wave of reforms to eliminate mediocrity in ed.

[*Honored high school students for excellence; awarded medal posthumously to fighter Joe Louis; met Mrs. America Pageant winner; made life member of Italian-American War Veterans.*]

Then a Cabinet Room meeting with conservative leaders re Central Am. They are all pledged to help. George S. & I met with Bud M. It was mainly a report by George on his meetings with Soviet reps.—Ambassador Dobrynin etc. They are utterly stonewalling us. Off to Camp David & a beautiful afternoon.

[**Saturday, May 12**: *rode, but commented, "My Giminish was fighting his head the whole way—I don't know what was bugging him."* **Sunday, May 13**: *returned to W.H., stopped for Mother's Day luncheon at Catholic-run home for the aged; Maureen and Dennis visiting W.H.; learned that W.H. usher had died suddenly.*]

Monday, May 14

Met with "Sherpa" team who have been preparing (with their colleagues from Europe & Japan) the Ec. Summit meeting in London. Apparently P.M. Thatcher wants to use the informal format we used in Williamsburg.

A Ceremony on the S. Lawn to honor young Michael Jackson who is the sensation of the pop music world—believed to have earned $120 mil. last year. He is giving proceeds from one of his biggest selling records to the campaign against drunk driving. He is totally opposed to Drugs & Alcohol & is using his popularity to influence young people against them. I was surprised at how shy he is.

Had a good meeting with Mexican American business & civic leaders from the border area of Texas. They have some good ideas on changes needed in Mexican law to encourage investment etc. in Mexico to provide employment. They want me to take these things up with Pres. de la Madrid tomorrow.

[Farewell reception for Deputy Counsellor Jim Jenkins, leaving to go into business for himself; participated in ceremony to send torch on its way to L.A.]

Tuesday, May 15

Arrival ceremony for Pres. de la Madrid. Weather turned out near perfect. He was visibly impressed. After reception he & I talked for nearly an hour on Central Am. I think our end goals are the same but they have trouble seeing Castro & the Sandinistas as we see them. I think I made a dent. [. . .]

Then I went to Dan Ruge's office for an electrocardiogram. It's part of my annual physical. I didn't bother to ask him about it so I guess it's alright.

Cab. Council meeting on family affairs. Margaret Heckler opened meeting with prayer. We are doing a lot of study on how we can help strengthen the family. Some ideas require legislation, some things H.H.S. can do by regulation. One important thing is to re-establish adoption instead of abortion for the literally mil's. of people who are waiting for a chance to adopt.

Spent most of afternoon meeting with Congress men & women on the MX situation. Tip O'Neill has mounted an all out campaign to kill the MX.

The State dinner was a huge success. Gloria Loring entertained beautifully. A lot of old friends on hand plus NBC's Mr. Brokaw who recently dinged me good in an interview. He has voiced regret over that.

[Wednesday, May 16: staff meetings; continued meetings with congressmen on MX; appearance at gathering of editors of publications for senior citizens; lunch with American Retail Federation; met with Shultz, discussed possibility that a massacre took place near Sidon in Lebanon [. . .]; calls to congressmen regarding the MX; appeared at White House news photographers dinner.]

Thursday, May 17

Brkfst. with Pres. de la Madrid & our 2 teams. I think he still does not see Castro & the Sandinistas the way we do but all in all the meetings have been productive. We both appeared before the reps. of the interparliamentary Union.

[Brief meeting with representatives of the Black Mayors Council; launch of annual summer youth jobs program; visit from former football player and actor Roosevelt "Rosey" Grier and unidentified members of Yankees baseball team; telephoned Regan in Paris regarding Lewis Lehrman's low estimate of 1984 deficit, noted that "Don thinks it won't be that good but he says it is going to be less than everyone thinks."]

The day got busy. Dick Wirthlin came by with good news & not so good. Poll shows an increased concern by people about foreign affairs. An N.S.P.G. meeting re the trouble in the Persian Gulf. Both Iraq & Iran have been attacking oil tankers. The Gulf states hesitate to ask for help fearing Iran can stir up internal troubles for each of them. It's a typical Mid East problem.

Cab. Council met on some legal problems such as too many govt. programs are assuming police powers.

[Shot campaign film for congressmen; filmmaker A. C. Lyles visited to receive award; greeted a beauty queen from Texas; off-the-record gathering with members of the press corps.]

Friday, May 18–Sunday, May 20

Passed the physical with flags flying. I'm grateful to God for the health he has blessed me with.

Nancy met me at the hospital & we helicoptered to Camp D. Weather was beautiful & 80 degrees for the weekend. We rode & swam. Phone was a little busy—back & forth with Bud & Cap & Jim. It had to do with Iran attacking oil tankers in the Gulf—Persian. We're keeping in touch with the gulf states & our allies. I doubt the Arabs will ask for our help. They are afraid that Iran can kick up radicals in their own countries. Uneasy lies the head that wears a crown.

[Returned to Washington; reception for ethnic Republican groups; anticipated visit the following day by President José Napoleón Duarte of El Salvador.]

Monday, May 21

Summer has come all at once—almost 90 & humid. Pres. Elect Duarte arrived. He is a Notre Dame graduate. He told me I was his hero—as a freshman he went thru the Notre Dame indoctrination which includes seeing the film in which I played the Gipper. I think he'll do alright before Cong. tomorrow. He's outspoken against both the extremists on the Right & the Guerillas. He'll also speak of the need for Am. help.

[Spoke to Cuban American leaders, described them as "gung ho for our side."]

A touching moment with tiny Amy Smith. She is the Arthritis Poster Child—7 yrs. old & Arthritic since age 2. When are we going to find the answer to this terrible curse.

[Ceremony launching coin sales for the Olympics.]

Then a most enjoyable lunch with Prince Phillip, the Wrights & the Princes aide up in the Sun room. The Prince expressed a typical British attitude regarding the Olympics, namely that it is overdone & should get back to just youngsters having fun.

Back at the Rose Garden after lunch I signed our legislation designed to help us come down on child pornography.

Then a couple of hours practice for tomorrow night's press conference.

Tuesday, May 22

N.S.C. meeting re our security measure at nuclear installations & nuclear weapon storage here & abroad. It was a good report but I'm astounded at the precautions we must take anticipating every imaginable thing that could happen.

Walter Annenberg came by—he too has received the Eagle award. A lengthy taping session for a variety of meetings all the way from a Repub. convention in Ind. to a statement on the D-Day anniversary for France—6 in all. Then a prenews conf. brief & finally at 8 P.M.—the News Conf. Nancy isn't here—be back from N.Y. tomorrow. Everyone said the Conf. was the best I've done. Many Q's. on Persian Gulf, Soviet-U.S. relations & Central Am. Thank Heaven Pres. Elect Duarte is here—he really did a selling job on Cong. & the press.

Wednesday, May 23

Fred Fielding came in—he's worried because I turned our Pvt. Enterprise people loose to raise some money for Eureka College. They had to have $400,000 in 2 weeks or they'd be shut down. I think we saved the day. Fred's pitch is that money was raised by invoking my name. Frankly I see no reason to worry. I'd have asked them to help any other college out that was forced with such a crisis.

[Presented people successful in increasing exports; meeting with Senator Bob Dole (R-KS) about gaining support for enterprise zones bill; meeting with Shultz about getting the Soviets "to bend their backs a little."]

Howard Cosell did a 5 minute interview with me on Sports. Finished up in Dr. Ruge's office getting rundown on my physical. I'm in good shape. Told him I was worried about Nancy—she seems so lacking in energy. He's worried too. He thinks she's just not eating enough.

Thursday, May 24

[Met first woman to graduate first in class at Annapolis; spoke to CIA employees; lunch with Vice President Bush.]

Jay Hair head of "Wildlife Fed." came in with Bill Clark & Bill Ruckelshaus. Maybe we're making a little headway with the environmentalists who declared war on me about 3 yrs. ago. I think Jay wants to be fair & heard some things today he hadn't understood before.

A Cabinet meeting to hear a report on how we're doing on implementing the Grace Commission recommendations. We're doing fine & have some already in action & others ready for action by Cong.—about 25% of them.

[Met Multiple Sclerosis Father and Mother of the Year; honored as life member of AMVETS; met star City College of New York student, once a refugee from Vietnam; filmed more campaign ads; dinner at the home of Clare Boothe Luce.]

Friday, May 25

The usual staff meetings. In one of these I approved 400 Stinger missiles for Saudi Arabia—immediate delivery. This was under my emergency powers. I'll ask Cong. for more.

A briefing on the European trip & a meeting with Geo. Shultz. Then Nancy & I went up on Capitol Hill for the Unknown Soldier ceremony. He will lie in state in the rotunda til the Mon. funeral at Arlington. It's impossible to describe our thoughts when you stand before that casket. Who was he? A son, a husband & father. The answers known only to God.

Afterward we left immediately for Camp David.

[Saturday, **May 26:** *went swimming and riding.* **Sunday, May 27:** *riding.* **Monday, May 28:** *left Camp David at 11 am; attended funeral for Unknown Soldier at Arlington National Cemetery.* **Tuesday, May 29:** *two-hour briefing for upcoming European trip. Flew to Colorado Springs to tour Olympic training facility; fund-raiser hosted by Joe Coors; meeting with campaign workers; stayed the night at the Broadmoor.]*

Wednesday, May 30

Up early & out to Air force Academy for commencement. There was a crowd of about 35,000 for the biggest graduating class in Academy history—993 & I shook hands with each & every one. It took hours. I was amazed & pleased by the number who expressed faith & prayer in the few words they had time to speak to me. The General (Scott) told me several hundred cadets can be found every day at 6:30 A.M. Catholic & Protestant services.

Back to the W.H. & hosted a dinner for the foreign ministers of the 16 NATO countries. There is no question but that alliance is stronger than it has ever been.

Thursday, May 31

Last day before Ireland & it was hectic as h--l. At 9:00 A.M. met F. Ministers again this time in the Cabinet Room. Had some time for discussion of E.W. matters then a farewell in the Rose Garden.

[Interview with Hugh Sidey on D-Day '44; John Herrington, deputy assistant to the president, brought a list of appointees for approval; ambassadorial formalities; received book of letters collected by a radio station on Central American policy, noted, "One was from the sister of Fidel Castro who says he is a communist & seeks a world communist state"; met 4-H leaders; videotapings; interviews with foreign journalists.]

A haircut—a last minute flood of paperwork & upstairs to pack for Europe.

[Friday, **June 1:** *flew to Ireland; formal reception.]*

Saturday, June 2

We helicoptered to Galway where the Mayor made me an honorary citizen. Then the U. gave me an honorary degree. It was Galways 500th anniversary.

Midway in my speech it began to rain & hail & I told them I was cutting the speech to what they'd heard & we thanked them & said that's all. Back to Ashford Castle where we were met by the Mayo county council.

[Radiocast; met with Americans with factories in Ireland.]

Sunday, June 3

This was the day. We helicoptered to Ballyporeen in Co. Tipperary. This was the home of my great grandfather who left there for America. After all the greeting ceremonies I went with Father Condon to see old Father Murphy keeper of the baptismal records. There I saw the hand written entry of Michael Reagans baptism in 1829. We then crossed the street to the church where the baptism had taken place.

We walked thru the town shaking hands with as many people as we could on our way to the Pub that has been named for me. They are building a Community Center in my name. There I was presented the family tree as researched by Burkes Peerage. Our family line going back to Brian Boru has us related by way of Mary Queen of Scots to every Royal family in Europe. I'm a 6th cousin of Queen Elizabeth of Eng.

I addressed the crowd in the street from a platform. There were entertainers—dancers & musicians. Then it was off to Dublin to Deerfield Park, our Embassy where we were met by the Lord Mayor. We called on Pres. Hillery & Mrs. Hillery. There was a tree planting ceremony. Then we were off to a state dinner at Dublin Castle. Finally back to the embassy for the night.

Monday, June 4

To the office of P.M. Fitzgerald for a private meeting. Then Nancy arrived & we went to the meeting of the Dail (Parliament). I addressed that body— the speech was televised. There was an interruption at my introduction. Three members—all Left Wingers—protested my presence & walked out. Other than that I was well received & took advantage of the opportunity to explain our Central Am. policy which is not understood in Ireland. I also called for the Soviets to join us in an effort for peace. The speech I learned was carried in Europe & caused quite a stir in my favor.

[Lunch at the embassy; helicoptered to airport; formal farewell; flew to London; official greeting by Sir Geoffrey Howe; motorcade to Kensington Palace and met by Prime Minister Margaret Thatcher; reviewed the Coldstream Guards; returned to U.S. Embassy for tea.]

Tuesday, June 5

Lunch at Buckingham Palace with the Queen & Prince Philip in their private quarters. It was a warm & pleasant visit. They are both nice & she is an outstanding human being.

Back to the embassy for a few hours & then to 10 Downing Street for a one on one—then a working dinner with 5 of her people & 5 of mine—Sec. Shultz, Sec. Regan, Amb. Price, Jim Baker & Bud McFarlane.

Wednesday, June 6

The 40th anniversary of the landing on D. Day. We helicoptered across the channel. First stop was Pointe du Hoc where 40 yrs. ago our Rangers—225 of them climbed the 100 foot sheer cliff to establish a position. Only 90 were still combatable by the 2nd day. We met 62 who had returned for this anniversary. I addressed them & the large crowd. It was an emotional experience for everyone. Nancy & I went into the massive concrete pill box from which the Germans had 1st seen at daybreak the 5000 ships in the invasion fleet.

Walter Cronkite did a 5 min. TV interview with me—then we flew to Omaha beach. This was the heart breaker—row on row of white marble crosses (& stars of David) more than 9000 of them. We have a picture of one—the grave of our Ann's brother, we're giving it to her. Pres. of France, Mitterrand arrived.

Together we placed wreaths at the monument then I spoke. My speech contained many quotes from a letter I'd received a few days before the trip from a young lady whose father landed on D. Day. All her life she had heard his stories of what the day had meant to him. A few years ago he died of Cancer. He'd always said one day he would return to Omaha Beach. She promised she'd do it for him. We made it possible for the family to be there. I had difficulty getting through my speech. From there it was on to Utah beach. This was the biggest affair. Pres. Mitterrand made the only speech but on the beach were mil. formations of Ours, the French, English, Norway, Belgium, Netherlands & Canada. All the heads of state of those countries were there. The crowd was tremendous in size.

British & American naval forces were off shore. One of ours was the aircraft carrier Eisenhower. On the way back our helicopter circled her while I addressed the 5000 crew members by radio. They were drawn up on the desk in a formation that spelled Ike.

Thursday, June 7

This was a busy one. Beginning at 11:30 A.M. I had a bilateral meeting with P.M. Nakasone of Japan. Then a working lunch with our gang in preparation for more bilaterals & the Summit itself. At 4:15 it was P.M. Craxi of Italy. At 5:15 Chancellor Kohl of W. Germany. At 6 P.M., Pres. Mitterrand (France). A quick shower & clean up & then to St. James palace for reception with P.M. Thatcher & all the others. The usual press coverage & photo ops. & then a working dinner. The main subject was East-West relations & the Persian Gulf.

Friday, June 8

An 8:30 working breakfast with our gang. Then to Lancaster house for 1st real summit session. A working lunch then back to a plenary session with ministers of finance & foreign affairs—in our case, Sec. Regan & Sec. Shultz.

Margaret handled the meetings brilliantly. As we expected, P.M. Trudeau of Canada & Mitterrand were usually with a different viewpoint on some subjects—lot of it nit-picking. We overcame them with some compromise word changes.

Back at Winfield. I taped my Sat. radio program.

Then it was on to the Nat. Portrait gallery for a working dinner & a tour of the gallery.

Saturday, June 9

Another briefing breakfast 8:30 A.M. Then back to Lancaster house. The morning meeting went long—more protests by Pierre & Francois. We didn't get to lunch until 2 P.M. There was blood on the floor—but not ours. As usual, Margaret, Helmut, Yasu & I stayed together & for the most part prevailed. It was a good summit & we did make progress on trade matters, East-West, plans if the Iran-Iraq war creates an oil crisis & agreements on 3rd world matters & handling the tremendous international debt.

We were further delayed in getting to Guild Hall for the finale—the presentation of statements to the press, by 40,000 anti-nuke demonstrators who fouled up traffic. Before leaving we displayed a model of our proposed space station & I extended an invitation to all of them to join us.

Into black tie & on to Buckingham for the dinner given by the Queen. It was a very nice affair & an experience to be in that historic palace dining with the Royal family & others. I was between the Queen & the Queen Mother who is a delightful person. Across the table was Nancy between Prince Phillip & Prince Charles.

And so the Summit ends.

Sunday, June 10

A mini press conference in the garden at Winfield then photos with household staff & Eng. security people. Then out on the terrace to thank & farewell the embassy staff & their familys—about 1500 people. With that Charlie & Carole joined us in helicoptering to Heathrow where we took off for home. It is now late afternoon & we are in the White House. George B. & Barbara met us at Andrews. Maureen is here & we're going to run some tapes of the way our TV covered the trip.

[**Monday, June 11:** *quiet day; telephoning to senators on MX bill; desk work, never left residence.* **Tuesday, June 12:** *not yet on Washington time, woke up early; met with*

Sen. Howard Baker and Rep. Bob Michel on compromise budget; meeting with Republican leadership in Congress, reported on Summit, briefed them on Central America; meeting with Defense Minister from P.R.C.; signed order allowing U.S. to sell defensive weapons to China; met National Council of Farm Cooperatives; concluded, "A haircut & now we're on our way over to George Bushes to drink a birthday toast."]

Wednesday, June 13

After staff & N.S.C. meetings—into a meeting with the Sens. & Rep's. who had joined with me to draw up a "down payment" on the deficit. A triad of domestic spending cuts, defense cuts & tax revenues added by way of loophole closings—no hike in rates. The Sen. has passed it reasonably close to what we agreed to. The House version is reasonably close on tax revenues but more than doubles the Defense cuts & makes very little reduction on domestic spending. I told them to fight to the death. I'll veto the tax cut if it comes to me unaccompanied by spending cuts.

[Spoke at ceremony to open Missing Childrens Center, advocate John Walsh also spoke; received gift on behalf of 101st Airborne; greeted Boston Celtics, NBA champs; Maureen visiting, bestowed Father's Day gift early.]

Thursday, June 14

A Cabinet Council meeting on Natural Resources—specifically off shore oil. Congress is talking of legislation which would give the states a veto over whether we could lease to oil companies off shore even though we are outside of state jurisdiction. They would also share in revenues. I've authorized the Sec. of Interior to do some quiet bargaining with Congress to see just what we are up against. I'm all for states rights but this is ridiculous. Lunch with the V.P. & then a busy afternoon. Photos with the top brass of the Reserve Officers Assn. Also with the top man of the Elks Club. A newly appointed (by me) judge & his family came by—Bud Brewster. And Sen. Inouye presented me with a poster celebrating Hawaii's 25th anniversary.

Another taping session for 4 various events & then a meeting with Geo. S. & Bud. We dug into the subject of a meeting with Chernenko. I have a gut feeling we should do this. His reply to my letter is in hand & it lends support to my idea that while we go on believing, & with some good reason, that the Soviets are plotting against us & mean us harm, maybe they are scared of us & think we are a threat. I'd like to go face to face & explore this with them. I presented a White House flag to our Olympic team for the opening day parade. Then for a couple of hours in the theatre we briefed for the press conference. It is over & was successful. Everyone is calling it the best ever. My favorite moment was on a question about debating Walter Mondale & that Jimmy Carter had said I'd hide in a cave rather than debate. My answer, "There he goes again."

[Friday, June 15: *photo session; met leaders of Amway; went to Camp David, "A swim, a sunburn, movie & to bed."*]

Saturday, June 16

Read in Baltimore Sun of a young lady, Cynthia Nevers in Portland Maine who ran across the street—cleared a chain link fence cutting her hand (40 stitches) & caught a baby that crawled off the edge of a 2nd floor balcony. Thru the A.P. I tracked her down & talked to her on the phone. She sounds very nice. I'm trying to find out about her hospital bill because she is evidently poor. Rode in the P.M.—It was cloudy all day.

[Sunday, June 17: *returned to W.H.; left for Long Island for opening ceremonies of the International Games for the Disabled; returned to W.H.*]

Monday, June 18

Pres. Jayewardene of Sri Lanka arrived. We had the usual mil. ceremony. This time, however, I went out before his arrival & thanked the troops. The day was hot & muggy—11 people passed out including some of our m. during the ceremony. We had good meetings—one on one in the Oval office & then the plenary in the Cabinet Room. At the end of that one we went out on the S. Lawn where I was presented an 18 month old Elephant for the Nat. Zoo. She is a cute little thing—only 34 inches high.

The Pres. is a good man. He believes in pvt. enterprise & we're working to help bring in investment to Sri Lanka. He's faced too with a Northern Ireland kind of situation with a minority of Tamils wanting to break away & form a new country. Their tactics are terrorism.

Then in the evening a State Dinner—his toast was very moving & eloquent & he ad-libbed. It was an enjoyable evening & Frank Sinatra sang.

[Events before the state dinner: *met Teenaged Republicans; talked with Barry Goldwater, concerned about "the extent to which Congress is interfering with the Presidents Const. duty to run foreign policy"; ambassadorial formalities; talk with Weinberger, advising against negotiating a treaty against space weapons with Soviets, since U.S. was making progress on defense against nuclear missiles; dentist appointment, filling replaced.*]

Tuesday, June 19

[Met with congressional leadership on deficit and Central America; spoke at dedication of new National Geographic building.]

Jeanne Kirkpatrick came by to report on her Asian trip. She says the Chinese were greatly impressed by our visit there. Like me, she believes P.M. Lee Koan Yew of Singapore is truly a great statesman.

Then out to the S. Lawn to recognize the 20th anniversary of the Presidential Scholars Program. The crowd had been sitting on those metal folding chairs for a long time & had a long time to go yet. I couldn't resist—when I faced them I said maybe I could suggest something a little more comfortable & I took off my coat. I must have made a lot of friends. Some very pleased men followed suit.

[Signing ceremony for four environmental bills, designating land for wilderness; medal for departing military aide; visit by former members of Kitchen Cabinet in California governorship days (Henry Salvatori, Jack Wrather, Ed Mills, Stu Spencer, Joe Coors); taping with every Republican candidate for Congress; dinner with son-in-law Dennis Revell; Mrs. Reagan in Las Vegas, to be gone for a week. **Wednesday, June 20:** *flew to New Jersey to attend rally against drunk driving; flew to Hartford, Connecticut, to address convention of sheriffs; returned to W.H; Maureen visiting.* **Thursday, June 21:** *wrote of "usual meetings in a A.M., NSC etc."; met with young people from YMCA; report from National Productivity Advisory Committee; lunch with Vice President Bush; latest poll figures higher than ever; meeting with Brian Mulroney, leader of the Canadian Progressive Conservative Party, commented, "He's an Irishman thru & thru"; spoke to National Association of Broadcasters; interviewed by Tom Jarriel of 20/20 on ABC; photo sessions; fish-fry dinner for members of Congress.]*

Friday, June 22

Mayor Padilla of San Juan, Puerto Rico stopped by. He's a stalwart supporter & now President of the U.S. Mayor's Assn. He played a large part in getting that Dem. dominated group to pass resolutions supporting Line Item Veto & a number of other measures.

Then over to the Pentagon. [. . .] It was a briefing on where we are going weapon wise, communications & intelligence gathering. I can only say I left for the Oval Office filled with optimism, pride & a sense of safety.

In the Rose Garden after lunch I met with representatives of the International Youth Year Commission. Then a meeting with Sec. Shultz, mainly on the Soviet situation. No break through but further evidence that they aren't quite sure which way they want to go. George Romney came by, he is heading up a part of our Pvt. Sector Initiative called "Volunteer." He's interested in possibly a special medal for outstanding volunteers. I'm rather inclined to think maybe they should be formally included in the presentation of Medals of Freedom. Did a portrait session with Mike Evans & then off to Camp David. Got there in time for a swim.

Saturday–Sunday, June 23–24

Last night we ran "Bedtime for Bonzo"—no one had ever seen it. With Nancy gone I had guests—the Laxalts & the Deavers plus a friend, Leslie Leach of Amandas. We had a birthday cake for Amanda—her 14th. Everyone enjoyed the picture. Sat. dawned grey & temperature in the 60's. It was a no swimming weekend. A lot of tennis was played by Mike & the Laxalts. Paul & I rode in the afternoon then he went back to tennis. I joined Caroline on the lawn hitting golf balls. After dinner we ran "Star Trek III." It wasn't too good. Sun. a down pour of rain & back to the W.H. Maureen has arrived. Nancy not due to return until Tues.—I'm ready.

Monday, June 25

[Audiotaping for event; spoke to black appointees; issues briefing at lunch; spoke to Agriculture Communicators Congress.]

We had a brisk N.S.P.G. meeting re our negotiations with Nicaragua. The idea is to keep the Contradora countries from sitting back thinking we would do it all.

A long meeting in the Cabinet Room with what is called the Reagan-Bush Advisory Council. These were the people who have been in the forefront of my support for years past. They came together from all over the country.

Did some taping with a list of House & Senate cands. Mermie & I dined & about coffee time mommie came home, thank Heaven.

Tuesday, June 26

I forgot yesterday to note that I called Pres. Mitterrand about his trip to Moscow. Very interesting. He said Chernenko gives evidence of not being well & doesn't say a word without a script in front of him. He believes the Polit Bureau is kind of a collective in charge.

Did a photo with Gen. Vessey for a magazine story on him. Then an N.S.C. meeting on how to respond to the Soviet challenge to negotiate on limiting militarizing space. The problem is they are ahead of us in that dept. & want to freeze us into inferiority. I think we've worked out a plan that will "head them off at Eagle gap." Then I hurried over to the East Room to recognize a Senior citizen group who have a Nat. program called Volunteer. After lunch in the Rose Garden I presented a Medal of Freedom to Helen—Scoop Jackson's widow. Her son & daughter were with her, a large group of Congressmen, Cabinet, Supreme Court members etc. Then some quick visits in the Oval Office—1st Fuzzy Zoeller, winner of the Nat. Open golf tourney. Then Bob Anderson of the mag, Runner's World—he's doing an interview on me.

[Met poster girl for Cystic Fibrosis; also former quarterback great Otto Graham; parents of missing child; videotapings.]

Wednesday, June 27

Staff & NSC meetings. Jesse Jackson off on his own diplomatic mission to Cuba, Nicaragua & other points South has tried to reach me by phone. That is a call I'm not taking. We have put Under Sec. of State Armacost on the phone at this end. J.J. has us on some pretty thin ice with his adventuring. Commander Eric Liu (W.H. Staff Dr.), his wife & baby boy came by for a pic. He's being assigned to Bethesda Naval hosp. He's a good man.

Met with 3 Gov's. from Midwest on farm problems. Between Tornadoes & Floods, agriculture is having a rough time. But worse than nature is what reduced inflation has done. Land prices went out of sight during the high inflation of the 70's & 1980. Farmers borrowed using their land as collateral. Now that sanity has returned to the market they are faced with big credit problems.

After lunch Geo. S. came in to report on our Ambassador to Central America & his meeting in Nicaragua.

[Addressed Conference on U.S./Soviet exchange; also National Association of Minority Contractors; meeting with Eureka boosters on college's financial problems.]

Thursday, June 28

One of our diplomats came in this morning. Bob Seton headed up an effort I asked for to find out how we could better help get food to 3rd world countries. He's done a great job & brought back a plan that makes sense.

We had a Cabinet meeting on meeting our goal of reducing Federal employees by 75,000. We're about 3000 short. We also took up the matter of employee morale & how to increase productivity.

[Lunch with Vice President Bush; meeting regarding Jesse Jackson, commented, "Later we learned he was on his way home with the 40 odd prisoners Castro is allowing him to bring home"; taped a 4th-of-July message; photo session; met Asthma poster child; National Guard generals; president of Rotary; head of Community College Association; Miss U.S.A.]

Tonite dinner at Geo. Will's. A very pleasant evening. Alistair Cooke & several other very nice & interesting people—8 of us in all.

Friday, June 29

Jesse Helms came by to present me with a banner made for the N. Carolina 400th birthday celebration—which we can't attend. There was an N.S.P.G. meeting in the situation room. First order of business a report by Ambas. Shlaudeman on his meetings in Central Am. His talk with the Nicaraguan went no where. They obviously aren't about to give in on anything. He (Harry) is on his way back down there. Last part of the meeting had to do with the Soviets & their offer to meet with us on one subject only—"the demilitarization of space." We've replied we'll talk to them on space including nuclear weapons which after all go up into space. Hosted a lunch (State Dining Room) for a group of elected Repub. Women—State Legislators, City Councilwomen, Co. Supervisors etc. A good meeting. They were here for a day of briefings. Met with George Shultz & Bud & came to an agreement on our statement to the Soviets. Some visitor picture taking & off to Camp David. Clouds closing in but managed to get in an afternoon dip.

Saturday, June 30

Nancy guested with me on my Radio 5 min. For the 4th weekend in a row it was raining so no ride, no swim—just indoors doing homework & reading. Called R. Nixon to bring him up to date on Europe, Jesse Jackson, et al. Jesse landed Friday evening with his Cuban & Am. prisoners released by Castro. Jesse also stopped in Nicaragua & probably castigated the U.S. for picking on that poor little Dem. govt.

Sunday, July 1

Weather breaking but had to leave early. We are hosting the Diplomats supposedly out on the S. Lawn at a B.B.Q. If the weather double crosses us we'll have to move indoors.—Guess what? It looked so threatening & was raining just across the river at Nat. Airport that after the receiving line we moved indoors. Then it didn't rain one drop. Anatoly Dobrynin (Soviet Ambas.) was at my table along with Geo. Shultz. Anatoly wanted to talk about our situation—the Russians wanting us to meet in Sept. to talk about weapons in space & our reply that we'd like to discuss this and nuclear weapons etc. which they have refused to do. We didn't settle anything but I got a few things off my chest.

Monday, July 2

Began the day 9 A.M. in the Rose Garden to "Hail To The Chief." The Army trumpeters are celebrating their 25th anniversary. They've also written a "tribute" to "A New Beginning"—my line in my inaugural address. We had the usual staff times but this one attended by Geo. Shultz—our 1st chance to compare notes about Anatoly. We're telling the Soviets we'll be in Vienna in Sept. waiting for them—our terms.

[Visit by oldtimer ballplayers, unidentified by name.]

This afternoon I taped 6 interviews with anchor persons on 6 TV stations in Fla. & Texas. They'll be played later this week when I'm in their towns. About 4 of them talked only about the Simpson-Mazolli immigration bill. I hoped I calmed their fears. Walter A. sent me a copy of an article in a Nat. Catholic paper by a retired N.Y. Times obituary writer. He takes off on me as a reincarnation of Hitler. If I am he should be 1st into the gas chamber.

Tuesday, July 3

[Ceremony honoring fiftieth anniversary of the Duck Stamp.]

Lunch in the Roosevelt Rm. with leaders of several prominent environmental org's. They are up in arms over my appointment of Ann Burford to chair the Advisory Committee to Sec. of Commerce on "Oceans & Air." They had to admit our environmental record is great but want me to cancel Ann because their people will take it as a symbol that we are anti-environment. I stayed cool. But I told them Ann had done nothing wrong—she had been railroaded & I owed her something like this to restore her. They were deaf to my suggestion that if they used their publications to tell how good our record is (which they admit) there won't be a fuss over Ann. I spoke to ears that wouldn't hear. They were arrogant & unreasonable. That was it for the day—I'll get back to cleaning out my desk.

Wednesday, July 4

We're both off—in different directions—Nancy heading W. to S.F. & I'm going South. At 10 A.M. from A.F.1 I started the Daytona "Firecracker 400 Stock Car Races." We arrived there while there were still about 35 laps to go. They call them stock cars but they do about 200 miles per hour. By that stage of the race it takes a while to figure out what you should be watching for. There are cars all around the track. Finally I found the 2 who were fighting it out for 1st & then it got exciting. The winner is engaged in a program among drivers in behalf of Reagan-Bush. After the race there was a picnic in the infield—after the 100,000 fans had gone home.

[Spoke at Spirit of America Festival in Decatur, Alabama; flew to Huntsville; talked to Mrs. Reagan. **Thursday, July 5:** *party reception; flew to Michigan; toured GM facilities in Warren with Chairman Roger Smith; flew to San Antonio, Texas; talked to Mrs. Reagan back at W.H.* **Friday, July 6:** *spoke to Texas Bar Association; returned to Washington; went to Camp David; celebrated Mrs. Reagan's birthday quietly.* **Saturday, July 7:** *riding and swimming.]*

Sunday, July 8

Another beautiful day. Spent a couple of hours after lunch filming for a documentary to be shown at the Repub. Convention. And back to the W.H.

Monday, July 9

It was really a Monday—the schedule was thick as a phone book. The usual staff & N.S.C. meetings. We have to ask Cap. W. to change the speech he was going to give the Am. Legion on the 11th. The one he has opens up too many cracks for the Demos. to distort & paint me as wanting to start a war.

We had a Cabinet Council on Natural Resources. We have a darn good record on environment which the environmentalists ignore in their bigotry. At noon a lunch (W.H.) with journalists & electronic press from around the country. They are a different breed from the W.H. Press corps.

[Presented a medal posthumously to diplomat killed by terrorists in Namibia in April;

continued filming documentary for convention; brief meeting with former Japanese consul in San Francisco; ambassadorial formalities; appeared before American Coalition for Traditional Values, called it, "A good group who are doing much to spark the spiritual revival in our land."]

Tuesday, July 10

A brief half hour in the office—Staff & N.S.C. 15 mins. each. Then Marine 1 & off to Chesapeake Bay. First stop the Wildlife Refuge (14,000 acres). A briefing, then a town. Saw 2 bald eagles among other things. Then on to an island where the oyster & crab boats headquarter. They are fearful of the growing pollution of the bay. We're working with 3 states, Va., Maryland & Pa. on a clean-up program. Visited some of the boats & crews. Then to the volunteer fire house for lunch cooked by the women's auxiliary of the vol. firemen. It was a good time among real people. The ever present W.H. Press Corps kept trying to squeeze in questions that had nothing to do with the day's activities. Arrived back at the W.H. in time for a signing ceremony commemorating the 30th anniversary of the Food for Peace program. Then another hour of filming for the convention documentary. A pvt. meeting with Dick Wirthlin—re a book involving him & me by an Eng. author. It is a sleazy hatchet job. He is sending by way of a lawyer a threat to sue for libel.

Another pvt. drink with a half dozen of the press corps—just an off the record get-together. The TV press, evening news did a trashing job on the Bay visit tagging it as purely a campaign event.

Wednesday, July 11

A steamy hot day. Met with "Pfiab"—the President's Foreign Intel. Advisory Board. Anne Armstrong chairs it & it's really a distinguished group of people. They report we have no accurate way to determine the cost of & peace of Soviet mil. build-up. I've asked that they study how we can measure the extent to which Soviet arms bldg. reduces the civilian standard of living. Went over to the E.O.B. & met with Lew Lehrman & Jack Humes's Citizens for Am. Shook hands & was photographed with each of the 240 there. Then a luncheon meeting with our Economic Advisory Board. Goodbye to Martin Feldstein who goes back to Harvard. Went over to Roosevelt Island in the Potomac—spoke to the Environmental Quality people & signed their report. It was really hot there. Dropped by the Situation Room for a birthday party for Bud McFarlane & John Poindexter. A meeting with John Tower on the Defense Budget. He's doing a magnificent job. We're going to miss him. The Demos. in the Conference are trying to reduce a $299 Bil. Sen. version by several bil's of $ & wrap it in some restrictive language. Our decision is to hang tough. Came upstairs early & signed off for the day.

Thursday, July 12

Off (9 A.M.) for Bowling Green Ky. to visit Mammoth Cave & address the Nat. Campers & Hikers Assn. Convention which they call a Campvention. It was a great day—9 A.M. to 4 P.M. Seeing the cave for the 1st time was an experience made more enjoyable by the high morale of the Park Rangers. Our bil. $ refurbishing of the parks was needed & has done a great deal for them. Of course the press had one thing on their minds & it wasn't a cave—it was Mondale's announcement that he had chosen Mrs. Ferraro as his V.P. choice.

[Addressed crowd of twenty thousand on the environment; returned to W.H. **Friday, July 13:** *photo session with American Farm Bureau Federation, loyal supporters; met West German minister of defense Manfred Wörner; lunch with Republican women officeholders; brief meeting with NASDAQ board; photo session; conference with Kenneth Khachigian, writing acceptance speech for convention, commented, "I remember when I did all such things myself. No way now—no time"; went to Camp David.* **Saturday, July 14–Sunday, July 15:** *political strategist Stu Spencer weekend houseguest; noted press coverage of Congresswoman Geraldine Ferraro (D-NY); returned to W.H.]*

Monday, July 16

A hot but easy Mon. Called the Coach of the Phil. Stars who won the U.S.F.L. championship last night—23 to 3 over Ariz. Then in the East Room had a signing ceremony of Captive Nations Week proclamation. I addressed about 200 people mainly refugees & veterans of the Holocaust. A Priest told me to ignore Cong. Woman's charge that I was less than a good Christian. He said I was as good a one as there is anywhere. I was quite touched. We had an issues briefing lunch. We're going to have trouble with the House Demos. re the Defense budget. D--n them we've gone as far as we can go with those appeasement, isolationist crackpots.

Spent some time cleaning up mail etc. then came upstairs.

Tuesday, July 17

Last nite the Dem. Convention opened—we didn't watch but this A.M. I read the speeches of Jimmy Carter & Gov. Cuomo. I couldn't believe the outright falsehoods & pure demagoguery of both. But I won't reply—yet.

A signing ceremony in the Rose Garden—on a bill to induce states to raise the drinking law to age 21. Right now 23 states have it, 27 don't.

We had a 90 min. Cabinet session—60 on the ec. & 30 on commercialization of Space. Dave Stockman gave 6 year projections & again I think he's too pessimistic as always. We're doing better fight now than he projects for this year. Dick Wirthlin figures on the campaign was next on the agenda. The gap between Mondale & me has closed somewhat but we expected that. Home & Mother.

Wednesday, July 18

A tough decision I have to make after a Budget review board meeting. Do we continue with our Land Sat. program even though it means a large subsidy while commercial interests only see a few mil. $ of commercial use? On the other hand are there undiscovered commercial uses yet to be found? France & Japan are both showing signs of getting into this business.

[Lunch with Vice President Bush; farewell to China Trade Commission, leaving for Beijing; addressed two members of Central America outreach program and then one thousand members of businessmen working with Grace Commission; dinner with Mr. and Mrs. Don Regan and Mr. and Mrs. Charles Wicks.]

Thursday, July 19

Off at 9 A.M. for the U.S.[University of South] Carolina where the heads of state of the Caribbean nations are meeting. I joined the meeting & then spoke to them at lunch. We have developed a good rapport with them & are trying to help them develop economics. By 2 P.M. we were on our way back to Wash. It was a very successful trip. Back in the Oval office had a photo with Terry Misfeldt of the Jaycees publication, "Future." I'm doing an interview for them. Then photos with 2 winners of Sport & Physical Ed. Awards. They were 2 very lovely ladies with fine husbands. Gen. Polk of the Horse Cavalry Assn. came in & presented a print of a fine painting of the Dragons charging in the Mexican War.

Three leaders of the Anti-Defamation League (B'nai B'rith) came in for a visit. They are pleased with our positions & our relationship with Israel.

Finally a speech meeting on what we're going to do in next week's campaign swing. Tonite we watched (finally) the Dem. Convention & listened to V.P. candidate Ferraro whose speech was pure boiler plate with no facts at all. Then it was Mondale's turn & his speech was pure demagoguery. He promised to do things—the very things we're doing & sounded as if he were going to do it so as to cure an ec. crisis. He did revert to type once—he pledged that he'd raise taxes—on the rich to reduce the deficit. He was introduced by millionaire—son of wealth Sen. Ted Kennedy who assails me as the friend of the rich.

Friday, July 20

On the S. Lawn—recognition of POW/MIA Day. Many former P.O.W.'s & the familys of many of the M.I.A.'s. A moving experience. Taped a 4 min. TV campaign piece & in the P.M. spoke to 100 young ladies (High school age)—the Girls Nation group. They endorsed me for Pres. again. Geo. Shultz came by to report on his trip to S.E. Asia. He says we're doing fine there. Then to the East Room for a ceremony honoring the Apollo Astronauts—commemorating the 15th anniversary of the Lunar landing. Off to Camp David with Dennis & Maureen.

[Saturday, July 21: rode and swam. **Sunday, July 22:** *watched panel shows; returned to W.H.]*

Monday, July 23

Pres. Duarte—El Salvador came by on his way home from Europe. I really believe he's going to solve some of the problems down there & he certainly can help us with Congress & getting appropriations passed for assistance to Central Am. This week it was "Boys Nation" in the Rose Garden. I was overwhelmed by their unabashed patriotism & love for country. These are high school kids & so unlike an earlier generation. I've had a lump in my throat all day.

[Lunch with off-the-record discussion for seven columnists, George Will, Jack Kilpatrick, Hugh Sidey, Ted White, Nicholas Thimmesch, Ben Wattenberg, and Morton Kondracke; visit from widow of appointee.]

A lengthy taping session then a practice for tomorrow nites press conf. Went upstairs for a meeting with Don Regan. We are threatened by a collapse of Continental bank & a giant Savings & Loan in Calif. Then Bud brought Cap, George S. & Gen. Vessey in re the answer to the Soviets demand for a meeting on ASAT. We're holding out for talks also on reducing nuclear weapons.

Tonite Geo. B. & Barbara for dinner. A nice evening.

Tuesday, July 24

Manuel Frago, head of Spain's opposition party came by. I wish his party was in power as against the Socialists. Did some more taping for the convention film. Then a lengthy meeting with our Cong. leadership—they expect trouble on our Central Am. policies & on foreign aid in general, also on the defense budget. Sen. Baker brought a friend by for a photo. The friend was presenting me with some sculpture but which by way of me will be in the Cherokee nat. park museum in Tenn. Toby mugs (Eng.) have made a limited edition of a mug of me. It—No. 1 was presented to me & No. 2 to Jim Brady. Proceeds will go to the Jim Brady foundation. A string of photos with several Congressmen then a reception for the Future Farmers of Am. More practice for the press conference—then at 8 P.M. the conf. It went well & I got in some licks about the Dem. convention demagoguery—read that—falsehoods.

Wednesday, July 25–Thursday, July 26—Trip

Left at 9:35 A.M. for Austin Texas. Sen. John Tower, former Gov's. Clements, Connally & Shivers, & V.P. George B. joined me there for outdoor rally before 27,000 people. I believe most were Dems. but they behaved like the most enthusiastic Repubs. I've ever seen. Cong. Phil Gramm there also. Back stage Bill Clements gave me a diatribe on how our campaign in Texas was a failure.

I believe he's unhappy about Sen. Tower being Chairman & I believe he thinks Tower is slated to be Sec. of Defense & that's what Bill wants to be.

[Met with Texans for Reagan-Bush; flew to Atlanta; spoke to meeting of Southern Republican leaders.]

Thurs. A.M. departed for the Cumberland Mall where a crowd almost as big as the one in Austin held an outdoor rally in suffocating heat but they were just as enthused as the Texans. In both talks I laid into the D. Convention as not being representative of the rank & file Demos.

Then it was on to N.J. In Elizabeth to a large outdoor rally again heavily Democratic—I was interrupted by cheers about every line. Then the 20 yr. Dem. Mayor endorsed me for re-election. After that we helicoptered to Hoboken where Frank Sinatra joined us. He was baptized in St. Anne's Church where they were having the 74th St. Anne Celebration. The crowd was largely female & Italian but they let me know they were for me. In all the appearances in the 2 days there were signs—"Women for Reagan"—plus E.R.A. signs "Elect Reagan Again." Then back to the W.H. Only 2 days but it seemed like a week we got so much done.

Friday, July 27

A really packed up schedule. Pres. Elect Barletta of Panama came by. He's a great guy—a grad. of the U. of Chi. Ec. school. An NSC meeting where there was some differences between Cap & Geo. S. over our attempt to negotiate with Nicaragua. I think we got it straightened out. Had lunch with the network TV anchors. I think maybe I might have softened them up a bit. An interview with Lee Edwards—"Conservative Digest." Meeting with Geo. S. (more Nicaragua). Dropped by a meeting of the Americans Society. This is a high powered group of business & industrial people chaired by David Rockefeller who are trying to help out in Latin Am.

[Photo session; interviews with local television stations; ambassadorial formalities; made member of the Non-Commissioned Officers Association; visit from third-cousin; received honors from Consumer Digest *and the American Hardware Association; personnel session.]*

Saturday, July 28

Lv. for Calif. to open the Olympics then off to the ranch. Met Nancy—then addressed our Am. team at Olympic Village & on to the Coliseum. The opening ceremonies were magnificent. My 2 sentences to open the games I thought were written in reverse. The applause line—"hereby declare open etc." was 1st followed by a second line about the 23rd Olympiad. I asked permission to change the order & did so. The press having a copy of the lines as written are gleefully tagging me with senility & inability to learn my lines.

Had dinner on Marine 1 enroute to the ranch. Then from July 28 until late

Sunday afternoon of Aug. 12 it was ranch time. I believe this is the longest un-interrupted time we've ever spent there. The weather was terrific except for one foggy day but even that day we had our regular morning ride & in the afternoon worked pruning the woods along the entrance road.

There were of course the usual Presidential chores—things to sign, daily intelligence briefings, phone calls to the Nat. Gov's meeting, taping interviews, lunch with George Bush, then with Stu Spencer & Jim Baker. Phone call to Football Hall of Fame ceremony, another to young people working to decorate Convention Hall in Dallas. An interesting lunch with the Papal Ambas. to the U.S. & of course my Sat. radio tapings. On one of them I gave the press an opening to display their irresponsibility which they did. Doing a voice level with no thought that anyone other than the few people in the room would hear I ad libbed jokingly something about the Soviets. The networks had a line open & recorded it and of course made it public—hence an international incident.

[Ranch life and chores.]

The press is doing everything they can to suggest however I say it that I don't mean "no taxes." They find every excuse to say I'm really hedging. Well d--n it there will be no tax increase on my watch & Mondale is stuck with his campaign promise to raise the income tax.

On Aug. 12th we came down to the Century Plaza—dinner in the suite. On Mon. the 13th we met all the U.S. medal winners & were made honorary members of the team. Then I lunched with Bud & George S. & we looked at the Soviets from several directions. I approved asking Gromyko to the W.H. if he comes as he usually does to N.Y. for the U.N. General Assembly opening.

I have a feeling we'll get nowhere with arms reductions while they are suspicious of our motives as we are of theirs. I believe we need a meeting to see if we can't make them understand we have no designs on them but think they have designs on us. If we could once clear the air maybe reducing arms wouldn't look so impossible to them.

Then out to Bel Aire Hotel—oops after lunch with F. Minister Andreotti of Italy. He brought me a book by Quadaffi & said that gentleman expressed a de-sire to have better relations with us. Now to the hotel for a wedding rehearsal.

Tues. Aug. 14 was the wedding. It was a lovely ceremony & our daughter Patti is now Mrs. Paul Grilley.

Wednesday, August 15
A.F.1 to Wash. & here we are.

Thursday, August 16
A little stiff getting back into the routine. The usual staff & NSC meetings then a session with some very nice ladies each of whom heads up a women's

Jewish organization. They are concerned that the 1985 Woman's conference in Nairobi may slide over into some anti-Israel resolutions. I assured them if it did we'd walk out. They were reassured. A Cabinet lunch—Don Regan reported on the economy—& it's good. George S. gave a sum up on International affairs since we've been here & that's d--n good. Dave Stockman gave O.M.B.'s projection on budgets & deficits & the deficits are projected to be on a downward slide.

[Signed bill on Child Support Enforcement; report from Council on Integrity and Efficiency, which claimed to have saved $37 billion in federal spending; named National Guard Man of the Year; met Epilepsy poster child; visit from his Eureka football coach's grandniece; taped ad for Senator Jesse Helms (R-NC).]

Friday, August 17

A Lunch in the St. Dining Room with about 100 Polish Am. leaders. The spirit of freedom & independence is very much alive in these people. They've been fighting for freedom for 200 yrs. I told them we would hold the "powers" including the Soviets to the Yalta Agreement that all the captive nations had a right to free elections.

A meeting with George S. & Bud—we are agreed I will invite Gromyko to come to Wash.

A meeting with Pete Domenici (Sen.). He has had a documented report put together disproving the Dem. charges that we've been unfair. Then we talked copper. Our mines are being hurt by imports but our copper fabricates an industry 5 times as big as mining wants the imports. It's a sticky problem.

In interview with Donnie Radcliffe about Nancy & up to the living quarters.

Saturday, August 18

A warm pleasant day by the pool with no sudden calls to disturb the peace.

[Sunday, August 19: flew to Missouri for the state fair in Sedalia; heard problems of farmers with high interest rates and decline in land values; addressed a crowd; returned to Washington.]

Monday, August 20

[Flew to Cincinnati for tour of Proctor & Gamble's new headquarters, under construction; spoke in Fountain Square.]

On the way back to the airport—stopped the whole motorcade to say hello to a bunch of hard hats working on freeway construction. I believe they are for me—it was a most pleasant 20 min's. Two of them were attractive young girls—one running a giant bulldozer. On to Decatur IL. A tour of Archer Daniels Midland. They are experimenting with Hydro phonic farming. On 4 acres under glass they produce 4000 pounds of lettuce every day in Hydro phonic trays. All employees in that project are "handicapped."

Then to Milliken U. for agri. seminar—same routine as in Mo. then back to Wash.—Watched Conv. on TV. Jeanne Kirkpatrick was excellent in a Foreign Policy address & Katherine Ortega was good as the keynote speaker. The enthusiasm of the delegates was genuine.

[**Tuesday, August 21:** *desk work, taped TV ads for Senate candidates; other videotapings; report from Dick Wirthlin that lead over Mondale had shrunk to eight points.*]

Wednesday, August 22

Depart for Convention in Dallas. Still working on acceptance speech—it's too long. And I didn't get it much shorter but there were 81 interruptions for applause & demonstrations so it ran 51 mins.

Upon arrival in Dallas I had a meeting with Jerry Ford. He gave a fine speech Tues. & most supportive of me. Has pledged to go all out in the campaign. A little later there was a rally in the Hotel Atrium. It was jammed & all around the balconies up 14 floors were ringed with people. George B. & I spoke & were very well received. Then Tom Landry, Danny White (QB) of the Cowboys & one time great, Roger Staubach presented us each with D. Cowboys jerseys—number 84.

We had an early dinner—Nancy went to the convention hall where there was a tribute to her & the running of a film. George B. & I watched the proceedings on TV. At the end she turned her back on the crowd to look up at the screen behind her & there was George & me (live TV). She waved & I waved back. Then came the nomination speeches. Mine was preceded by an 18 minute film which was very effective. When the Roll of States was called Nancy & Barbara joined me & George. At the end of the vote we met the committee which came to officially declare us as nominees.

Thursday, August 23

This was a busy one. A prayer breakfast—17,000 people—a magnificent 2200 voice choir. George introduced me & I spoke of the place of religion in a Democracy. It was a truly inspirational meeting. Went back to interviews with George Skelton & Hugh Sidey. Then to address an Hispanic lunch then a fundraising lunch—didn't eat at either one. Met with Ron Walker & Mike McManus re the evening event. They had put together a TV tape to act out what I'd be doing & it was funny as h--l. Then free time until I headed for the hall to do my acceptance speech. After the speech & demonstration—Ray Charles sang "America The Beautiful." We had a reception for about 160—many of them are old friends. We were both pooped & glad to get to bed.

Friday, August 24

Dropped by & spoke to the new Nat. Repub. Committee. Then on to Chi. where I addressed the Convention of Veterans of Foreign Wars. Gov. Jim Thompson, Sen. Percy, Congresspersons Hyde & Lynn Martin accompanied us. The convention had ended but they stayed over to allow me to speak. I am the 1st President they have ever endorsed.

[Signed bill establishing a National Heritage Corridor for Illinois and Michigan; went to Camp David. **Saturday, August 25:** *radiocast; ride; surprised Mrs. Reagan by screening one of her movies,* Night Into Morning *(1951) with Ray Milland.* **Sunday, August 26:** *returned to W.H.; picnic for female appointees, entertainment by Juice Newton; Vice President Bush learned by telephone that his daughter had a baby son.* **Monday, August 27:** *staff meetings; desk work; visited junior high school cited for excellence; addressed gathering of educators.]*

Tuesday, August 28

It's wonderful with Congress away. I signed some bills & some photos. We had an N.S.C. meeting on Poland & agreed on a policy of lifting sanctions in stages based on the Polish govt's. real adherence to an amnesty program etc.

[Awarded medals to two youthful heroes; ambassadorial formalities.]

Wednesday, August 29

Things are looking better in Grenada. The moderates have at last come together which could close the door on the former P.M. who is trying again. The matter of Gromyko & whether I can meet with him is iffy—we've had no response to our suggestion that he visit Wash. after his U.N. speech.

Another day of desk time—vetoed a bill increasing by 38% the funding for public broadcasting. Sen. Barry G. had written asking that I sign. I called him—he was nice as could be & had no complaint because I vetoed.

[Ambassadorial formalities; photo for Fortune *magazine; reception for campaign staff; met with Shultz and McFarlane to discuss Gromyko situation further.* **Thursday, August 30:** *went to Greenbelt, Maryland, to tour Goddard Space Flight Center; lunch with Vice President Bush, talked about campaign; shuttle took off; noted that Mondale cast the only vote against the shuttle program in Senate; Cabinet Council heard report on anti-drug campaign, commented, "It has been a great success—we've intercepted drugs with a street value in excess of $13 bil."; photo session.]*

Friday, August 31

An N.S.P.G. meeting on 2 subjects—Pakistan & the Middle East. The Soviets have been getting bolder in their & the Afghan border raids aimed at the refugee camps in Pakistan. At the same time India is building up militarily on the Pak. border. Their concern is whether Pakistan is developing nuclear weapons. Pres. Zia looks us in the eye & says he isn't but he'll expect help from us if either threat materializes. I favor sending some high level emissaries to both give Zia some additional defense weapons but lean on him to prove once & for all he isn't making a bomb. As for Russia & the whole middle East, I believe we should make it plain by our presence in the Persian gulf & all that the Soviets might be risking war with us if they get grabby & tried moving in.

Met with Katherine McDonald—widow of the Congressman who died on this day 1 year ago in the KAL-007 shoot down of that passenger plane.

George S., Bud & I met on Israels ec. problems—they are horrendous due to their extended mil. deployment. We want to help but they'll have to demonstrate some willingness to help themselves. Peres will be P.M. for 2½ yrs. now & Shamir will be Foreign Minister.

[Addressed Catholic Golden Age Society Chapter presidents.]

Saturday, September 1

Nancy left at 10 A.M. for Phoenix. This has to be the loneliest place in the world when she's gone. Did my radio show & then a phone call to the Astronauts in space. They've been tremendously successful—launching 3 satellites. Those are ensured for $90 mil. each & the private enterprise ones like the one for A.T.T. pay us $10 mil. for the launch.

Spent the day shuffling papers.

Sunday, September 2

Get away day—off at 2:45 for Anaheim & Nancy. Uneventful trip. Phoned Jerry Lewis—Muscular Dystrophy telethon. Then when I called the number about my pledge—couldn't convince the operators I was who I said I was. Not sure they ever did believe me.

*[*Monday, September 3: *met with party leaders; recalled large rally of Marine Corps families at airport on landing the previous day; political rally; flew to Santa Clara County for another rally; flew to Salt Lake City; Mrs. Reagan to L.A.* Tuesday, September 4: *meetings with "Reagan-Bush" leaders; then Mormon Church leadership; addressed American Legion Convention, commented, "Was well received, particularly when I gave my own version of religion's place in public life. That has been very distorted since the Prayer Breakfast in Dallas"; flew to Chicago.]*

Wednesday, September 5

Started the day with "Reagan-Bush" leaders. Did Q&A. Then into the "Choosing a Future" conference sponsored by Ec. Club of Chi. (I'm 1st Pres. to ever address them) & Ameritrust Bank of Cleveland. Speech was followed by a Q&A session. Almost 2,000 people. The Lord was watching over me. Q's. on deficit etc. & I had the answers.

Departed immediately for Wash. & here I am in the W.H. Nancy is in Oregon.

George S. & Bud came by. It's just between us for now but I am going to meet with Gromyko. Sept. 28 is the day.

Thursday, September 6

A full day—it always is when I've been away for a few days. The big event in the A.M. was an address to the Nat. B'nai B'rith Convention. I know I was following Mondale who it had been rumored was going to do a job on me. I was not going to make a pol. speech but talk about matters of interest to them— Israel etc. I was right & he was wrong. Reports are that he was coolly but politely rcvd. I have to say my own reception was overwhelming—interrupted 18 times by applause. Come to the evening news—TV-ABC-CBS-NBC & you would never have known it. All 3 did a slick hatchet job.

Lunch with V.P. then met with 4 of our Senators from farm states to hear their reports on farm problems which are considerable. A Cabinet Council on Copper & Steel. One decision made—we cannot help the Am. Copper Mining industry with protectionist measures. We can offer help to unemployed min- ers—relocation, job training etc.

[Met bowling officials; Statler Brothers; editor of London Times; *audio- and video- tapings.* **Friday, September 7**: *meeting with Secretary-General Emile van Lennep, retiring from Organisation for Economic Co-operation and Development; appearance at briefing of journalists who cover space; remarks to gathering of female executives; meeting with Shultz and McFarlane ("on possible Gromyko visit"); went to Camp David.* **Saturday, Septem- ber 8**: *radiocast; ride.]*

Sunday, September 9

Phoned Alf Landon on his 97th B.D. Nancy came off the mountain by car & I helicoptered to Doylestown, Pa. to attend the Festival of Our Lady of Cze- stochowa (The Black Madonna). Gov. Thornburgh & Cardinal Krol met me. I saw the Shrine, put a wreath on the monument to Paderewski & then addressed 40,000 Polish Americans. Wonderfully received & magnificently introduced by the Cardinal. Then it was home & Mother. Maureen is here. A couple of hours helicoptering really leaves one tired—the vibration.

Monday, September 10

A good meeting with Black Clergy—Rev. T. J. Jemison, Pres. of the Nat. Baptist Convention U.S.A. I really had a chance to explain the record (Civil Rts. et al) & the hatchet job that has been done on me with regard to Blacks. We are going to have further meetings. This may be a breakthrough.

[Issues lunch; signing of Hispanic Heritage Week proclamation; received rose from

little daughter of a congressman; briefing on airport matters; launched federal charity campaigns; desk work; noted that "Mondale revealed his budget balancing program—it's smoke & mirrors & a fat tax increase."]

Tuesday, September 11

Met with Congressional leadership about defense budget etc. I've never seen them so Gung Ho & high in their optimism about my winning in Nov. They've just returned from being home campaigning. Then a Rose Garden ceremony to award Gold Medal posthumously to Hubert Humphrey. His widow received it & Hubert Jr. spoke—he's a Repub. I don't know what was with Pres. Carter—this medal was voted by Congress in '79. Did a 10 minute press availability in Press room—most questions were on meeting with Gromyko—which I announced—only 2 on Mondale's plan to reduce the deficit. I thought that would be the big issue & I was loaded for them.

Met very pleasantly with Lord Carrington who is new Sec. Gen. of NATO. I think he's going to do well.

(An N.S.P.G. meeting) [. . .] (Another issue has to do with possible shipment of fighter jets from Bulgaria to Nicaragua [. . .]. Third item also unresolved as yet—how to help El Salvador intercept arms to rebels from Nicaragua.)

Had a good meeting with farm state Senators & Reps. Was able to announce we're upping the Soviet grain agreement by 10 mil. metric tons. A later meeting with Sens. & Reps. from Steel states wasn't so happy—they are really lobbying for protection Quotas on imported steel. It's a decision I have to make in the next 2 weeks. Home for the evening.

[Wednesday, September 12: flew to Buffalo and Endicott, New York, for campaign appearances, commented, "I'm amazed at the enthusiasm & the high percentage of young people at all these gatherings"; returned to W.H. Thursday, September 13: flew to Tennessee to campaign; at Grand Ole Opry, singer Roy Acuff called for president's reelection; returned to W.H.]

Friday, September 14

Staff meetings & desk work in the morn. A ceremony in the Rose Garden to award 8 individuals (Hispanic Americans) for their great contributions to excellence in ed.

Then a taping session—one pol. for Cong. Jim Martin—running for Gov. of N.C., a tribute to Charles Wick, a plug for Texas C. of C., one for Eagles Forum & one for Future of Small Town Am. Then upstairs & more desk work. An easy day. Some of the staff work had to do with campaign. It's hard to keep from slugging at Mondale. His charges are beginning to be outright lies yet he keeps dropping in the polls so I guess the high road is best.

Saturday, September 15

Radio broadcast & some audio tapings in the A.M. for campaign commercials. Big event of the day was the dinner for the Nat. Italian-Am. Foundation at the Wash. Hilton. Speakers were Ferraro, Bush, Mondale & me.

A little Hanky Panky! Arriving at the hotel we were told Ferraro & Mondale were late because her dress was late in getting to the hotel. We & the Bush's sat in a holding room because we were told that no one by protocol could be announced or introduced after our entrance. A room full of a few thousand people were in there sitting on their hands & waiting for dinner. I finally told our advance people to tell them what was holding things up & that I was waiving protocol & we were coming in. I don't know whether I have the right to waive protocol or not. Anyway, we went in to a happy & tumultuous welcome. Then M. & F. came on about 4 minutes after we'd entered. Someone told us they'd been standing in the wings. I think they were playing games. If so—we won for while they were applauded there was also a chorus of Boo's. Then a 2nd hitch. Protocol demanded that both of them speak first, then George & I. We suddenly heard the Toastmaster Jack Valenti putting George on before Mondale. Again as I say none of it paid off—we carried the evening.

Sunday, September 16

A really easy day at home. We miss Camp David. Bud M. called. It seems that a self propelled barge up off Alaska blundered into Russian waters. The Soviet Coast Guard captured them, put their crew (5) in a Soviet hotel. This was last Wed. They didn't notify us until Fri. night. But the St. Dept. didn't get around to telling us until last night. We have a little shaking up to do over in Foggy Bottom.

Monday, September 17

Met with Japan-U.S. Advisory Comm. (our half) they've done a great job & our relations with Japan are getting better & better. Steering Comm. of Dems. for Reagan-Bush came in—they are a great bunch & demonstrate that a shift is taking place in what the 2 parties stand for.

A meeting with speech writers—it's time to shift gears a little in our campaign speeches.

[Meeting with Dillon Ripley, retiring secretary of the Smithsonian; visit with Ukrainian Americans; met Olympic medalists in target shooting.]

End day with an off the record cocktail round with Dave Broder, Pat Buchanan, Fred Barnes, Emmett Tyrell, Warren Brookes, John Fogarty & Eliz. Drew. Nancy is out on another 3 state swing visiting drug care centers.

Tuesday, September 18

This was quite a day. By 9:45 I was out in the Rose Garden thanking &

greeting our U.S. Committee for Pacific E. Cooperation. This is really another example of private initiative. Then back in the Oval Office for an NSC briefing preparatory to a N.S.P.G. meeting in the Situation Room. It was a free wheeling discussion of how we might get the Soviets back into arms reduction negotiations. I'm hopeful my upcoming session with Gromyko might get something going. I intend to open up the whole matter of why we don't trust them. Maybe if we can ease the mutual suspicion—arms talks can move better.

After lunch I met in the Cab. Room with the Exec. Comm. of the Farm Bureau Fed. plus a lot of Cong. men & Senators. I announced a credit plan to help on a case by case basis those farmers who are going broke under high loans & higher int. rates.

Then Jesse Helms brought 4 N. Carolina farm leaders into the Oval (tobacco & cotton). I think they are pleased with what we're doing. Finally a Cabinet Council meeting on the Steel problem. The gang was really divided between doing nothing & maybe doing too much. I came down with a decision that doesn't violate our "free trade" position but still offers some help. Photos with Field & Stream magazine & the Nat. Boy's Club winner. Then an hour in the dentist chair—cleaning.

[Wednesday, September 19: briefing (topic not identified in diary); television interviews regarding steel decision, commented, "The press with the exception of the N.Y. Times has played it as if we aren't going to do anything to help the industry"; photo session; flew to Connecticut and New Jersey for rallies; back to W.H.; Maureen and Dennis visiting.]

Thursday, September 20

Awakened at 5:40 A.M. by Bud. A suicide bombing of our embassy annex in East Beirut. A van crossed street barricades but did not enter embassy compound. It exploded in the street but the explosion wrecked the bldg. So far we know 2 of our servicemen, 1 Navy, 1 Army were killed as were a number of civilian employees both Am. & Lebanese. Our Amb. was injured but walked on his own to the hospital. I called him from Marine 1 on my way to A.F.1. He's quite a guy. This was my day to go to Iowa & Mich. so we made the trip. First was an airport rally at Cedar Rapids. Then we went to a farm (a wonderful family). I did Q&A with about 75 farmers. Jack Block was with me. From there we went to Fairfax for a picnic lunch. Every place we met with the warmest reception. The Des Moines Register poll has me 58 to 35 over Mondale.

[Flew to Grand Rapids, Michigan; appearances with President Ford; rally; returned to W.H. Friday, September 21: meeting with the Japanese minister of finance, hinted strongly that Japan could do more to raise the value of the yen; ambassadorial formalities; signed proclamation for National Drug Abuse Education and Prevention Week; meeting with Ambassador to USSR Arthur Hartman; latest poll shows campaign lead holding; noted that "George S., Bud & I had our usual Friday meeting," subject was Israel; met with

*Airmen and -women of the Year; videotapings; meeting with Bill Smith on the fact that the
State Department was handling its own litigation, commented that was "a job that should
belong to Justice. Al Haig started that"; telephoned parents of servicemen killed in Beirut;
visited seven-year-old African American boy who wrote a letter to the W.H.]*

Saturday, September 22

Radio program & then to the Situation Room for some discussion of the
Beirut Bombing of our embassy. It is virtually impossible to retaliate without
risking killing many innocent people. I've told George S. to let Syria know that
we are convinced this couldn't happen without their tacit approval & we d--n
well will keep this in mind if it continues to happen.

Spent most of day on the Mon. U.N. speech. It's a ticklish one. I don't want
to sound as if I'm going soft on Russia but I don't want to kill off the Gromyko
meeting before it takes place.

Sunday, September 23

Off to N.Y. & a busy 2 days at the U.N. Went right to work with a lunch for
Sec. Gen. of U.N. & top echelon. Then a series of bi-laterals with Pres. Alfonsin
of Argentina, Mobutu of Zaire (a darn good leader & friend of the U.S.) & P.M.
Willoch of Norway. A brief meeting with Cardinal Sin of the Philippines. He's
determined Marcos should step down. Finally the 6:30 reception—we shook
about 400 hands. I met Gromyko—we kept it cordial. He reminded me we met
in California while I was Gov.

Monday, September 24

Usual courtesy calls at U.N. then addressed the Gen. Assembly. I'm the only
Pres. to ever do this 3 times. As usual they sat on their hands although they did
interrupt once with applause. Gromyko & the Soviet reps. were front row center
right below the mike. I tried to catch their eyes several times on particular points
affecting them. They were looking through me & their expressions never changed.
One last meeting with prince Schanouk & P.M. Son Sann of Cambodia—teamed
up to try & oust the Vietnamese & return Cambodia to freedom.

Lunch with Dick Nixon & Ed Muskie—arranged by Howard Baker. They
have an idea—a good one—of putting a bi-partisan group of citizens together
to support our disarmament proposals after the election. Back to W.H. & the
Oval office. A bill signing in Rose Garden then a proclamation in East Room
declaring an Historical Black College & University week. Presidents of these
inst's. were on hand. It was a good session.

Tuesday, September 25

This was really a full day but a good one. Early on I addressed the World

Bank & International Monetary Fund Joint annual meeting. Had a rough time with the teleprompter. One reflector was all screwed up & I had a hard time trying not to get glued to the single screen.

Back to the Oval & the meeting with new P.M. Mulroney of Canada. He has just won a tremendous victory. He's a super fellow. We got along fine & will continue to do so. He's all for continued meetings. As part of our meeting we met with several of our astronauts plus a Canadian scientist who will be making his 2nd shuttle flight, & Kathy Sullivan who will be the 1st woman to walk in space. Henry K. came by with more insight re Gromyko. I found we are tracking very close on the approach to take. The meat of the day was an NSC meeting in which I got a briefing on Soviet espionage within our embassy in Moscow. I was overwhelmed by the evidence of high tech alterations in our typewriters which have delivered our top most papers & messages to the Soviets.

Later a meeting with Bonnie Consola—the lady born without arms. She is a charming, tiny woman totally at peace with herself. She is going to work in Army Procurement. Finished the day with the Supreme Ct. It is traditional for them to report in when they return from summer recess. Tomorrow off to Ohio & Wis.

Wednesday, September 26

Depart for Bowling Green U. in Ohio. Q&A with 4500 students. Arrived on campus & there were about 4,000 or 5,000 students at the helicopter landing who couldn't get in to the auditorium. There certainly is a new generation on hand. The crowd both in & out were wildly enthusiastic & supportive. I thought I was at a Repub. convention. Then on to Canton & the Timken Co. Faircust steel facility. It is still being built & is a high tech mill. I addressed a couple of thousand workers who were also supportive. Mondale was speaking to their union leaders in Cleveland who were endorsing him in the name of the Union. I'll take the rank & file.

[Marveled at computerization in new steel plant.]

Flew on to Milwaukee to the annual "Oktoberfest." A crowd of about 10,000 at this annual recognition of Milwaukee & Wisconsin's ethnic heritages. Again a good reception and (then back to W.H. where at 9 P.M. had a meeting with C.I.A., Defense, State, et al. We have intelligence regarding a barracks in the Bekaa Valley in Lebanon which could be the origin of the terrorists who blew up our embassy. I still was not satisfied that we had enough to go on to risk an air strike. We're getting closer to being able to tie these attacks to the govt. of Iran. I'd like (when sure) to tell them any more attacks & we'll retaliate at the source.)

[**Thursday, September 27**: *meeting with President Fernando Belaúnde Terry of Peru, discussion of economic problems; Regan brought economic data; taped commercial;*

briefing for next day's meeting with Andrei Gromyko; meeting with Representative Bob Michel (R-IL) regarding help for congressional candidates; with Foreign Minister Esmat Abdel Meguid of Egypt, with message from President Hosni Mubarak; noted that Jordan has signed treaty with Egypt; photo session and taping of commercials for candidates.]

Friday, September 28

The big day—Andrei Gromyko. Meeting held in Oval office. Five waves of photographers—1st time that many. I opened with my monologue & made the point that perhaps both of us felt the other was a threat then explained by the record we had more reason to feel that way than they did. His opener was about 30 min's. then we went into dialogue. I had taken notes on his pitch & rebutted with fact & figure a number of his points. I kept emphasizing that we were the two nations that could destroy or save the world. I figured they nurse a grudge that we don't respect them as a super-power. All in all—3 hrs. including lunch were I believe well spent. Everyone at our end thinks he's going home with a pretty clear view of where we stand.

[Spent afternoon negotiating position on Simpson-Maggli immigration bill; bill-signing ceremonies; drove to Camp David; studied for Mondale debate en route. **Saturday, September 29–Sunday, September 30:** *quiet days by fire; prepared for debate; returned to W.H.]*

Monday, October 1–Wednesday, October 3

Off to Detroit, Gulfport & Biloxi Miss., Corpus Christi, Brownsville & Houston Texas & back to W.H. In Detroit I addressed 1548 newly sworn in citizens of our country. It was a moving ceremony. Then in the same bldg. addressed (7000) a luncheon of the Ec. Club & the Women's Ec. Club. Without naming him I laid into Mondale—1st time I've done that. Was very well received. Then it was on to Biloxi Miss. The Grand Jury handed down an indictment against Ray Donovan. The whole thing seems to smell a bit. Of course the press is drooling.

[Rally in Mississippi. **Tuesday, October 2:** *flew to Texas to campaign.* **Wednesday, October 3:** *returned to W.H.; bill-signing ceremonies; accepted endorsement of the VFW; meeting in Situation Room, noted, "We still don't have quite enough info to OK an aerial strike against a terrorist center in Lebanon. It's frustrating"; preparation for debate on Sunday; taped campaign ads.]*

Thursday, October 4

Opened the day with briefing & rehearsal for debate. Then an East Room meeting with all our Congressmen & cands for Congress. It's a campaign type of kickoff for them. Two yrs. ago we did it on the capitol steps. This time Tip O'Neill refused to allow it. It was a good affair & I think they go home all revved up.

Two practice sessions for debate with Dave Stockman playing Mondale—
Mondale should be so good. Viewed film of Mondale's "primary" debates.

Friday, October 5

A brief NSC meeting—further intelligence that the Hezbollah terrorists are
gearing up for more action—spell that murders. Then 2 session of debate rehearsal, a Cabinet meeting & Rose Garden ceremony honoring a dozen mil. &
civilian defense employees who have played roles in uncovering waste & fraud.
And off to Camp David where we saw Jessica Lange's movie, "Country" which
was a blatant propaganda message against our Agri. programs.

Saturday, October 6–Sunday, October 7

Sat. another rehearsal—the whole crew at Camp D. An afternoon ride &
lots of cramming. Back to the W.H. Sunday morning & off to Louisville—the
debate at 2:45 P.M. Well the debate took place & I have to say I lost. I guess I'd
crammed so hard on facts & figures in view of the absolutely dishonest things
he's been saying in the campaign, I guess I flattened out. Anyway I didn't feel
good about myself. And yet he was never able to rebut any of the facts I presented & kept repeating things that are absolute falsehoods. But the press has
been calling him the winner for 2 days now.

Monday, October 8

We left Louisville not feeling too bad. There was a rally at the hotel last
night—1000's of people who had all seen the debate & they thought I'd won.
But now on to Charlotte, N.C. Sen. Helms & several candidates & Congressmen were on hand for a rally of more than 50,000 people. They had been there
for hours but they were wildly enthusiastic.

*[Appearance in Baltimore; returned to W.H.; videotapings; reception for campaign
steering committee.]*

Tuesday, October 9

A brief meeting with Bill Smith re the Immigration bill now in conference
committee. Decided we could not accept a couple of amendments of Dem.
origin—so bill will probably die. Too bad, too because we've lost control of
our borders. Did a couple of bill signings in the office—one on Fire Prevention
Week & the other on some child abuse legislation.

Then it was time for P.M. Peres & F.M. Shamir of Israel. We had a good
meeting & on through a working lunch. He has a tremendous ec. problem &
we are trying to be of help. I think he is the most flexible & reasonable Israeli
P.M. that I've known since I've been in office.

A Cabinet Council meeting. One part had to do with a vast reduction in the agencies regulating financial inst's. There is great duplication. I approved the action. The rest was an ec. report by Niskanen which we couldn't finish because I had a signing ceremony in the Rose Garden of the Older Americans Act of 1985. Then an interview with U.S. News & World Report. A reception for the leaders of the Nat. Fraternal Congress of America.

Wednesday, October 10

[Flew to Michigan to visit Catholic high school and to attend rally.]

Both the luncheon which featured the ethnic Am. groups was a pol. rally itself. Then the big one at the college. I'm feeling humbled by the affection these crowds show. Over & over I hear them shouting that they love me. Back to the W.H. The press is beating the drums about Mondale winning the debate—but the polls still have me out in front as a candidate.

Thursday, October 11

[Staff and NSC meetings; bill-signing ceremony; gathering of minority entrepreneurs.]

Lunch with George Bush—his office. He's a little uptight—tonite he debates Ferraro in Phil. Then an afternoon of taping—1st with about 2 dozen Congres. candidates & then campaign commercials. Tony Quinn & his wife stopped by. He presented me with 2 sculptures of his doing—one bronze, one marble. He's d--n good & his art work is getting great acclaim.

[Friday, October 12: flew to Ohio for whistlestop tour by train; noted, "I called Margaret Thatcher from the train re the I.R.A. bombing at Brighton which almost got her"; returned to Washington and then on to Camp David. Saturday, October 13: riding. Sunday, October 14: called Queen Elizabeth, wrapping up a vacation in Wyoming. Monday, October 15: flew to Alabama for campaign appearances; returned to W.H.]

Tuesday, October 16

[Flew to Illinois to campaign.]

Second stop was at Bolingbrook High School—1700 students. I spoke & then took several questions. They asked good ones which allowed me to refute Mondale's charges re Social Security, Arms limitations & Environment. It was a stimulating affair. Last stop was at Du Page College—a political rally & again an overwhelming crowd very much on our side. Gov. Jim Thompson & Sen. C. Percy were on hand. It was a good day. Back in Wash. had to resolve an issue whether to sign or veto a bill letting western states lumber co's. off the hook for contracts made to buy timber from the govt. at prices based on the Carter inflation rates. They were going broke with prices down. I signed. I believe it was best for the ec. recovery & also that govt. had to share the blame for letting the ec. get out of hand as it did in 1979 & 1980.

Wednesday, October 17

Arch. Arietta of Costa Rica came by. He's a very fine man who is totally aware of the evil of the Sandinistas. From then on the subject was debate—some interruptions of course. Out on the South Lawn a ceremony recognizing the Young Astronauts program. Little Drew Barrymore—the child in "E.T." was one of the children. She's a nice little person. Then a signing of a consumer bill that makes it possible to recall dangerous toys as adult products can be recalled. Over to E.O.B. for a debate practice session. It went pretty well. Since the 1st debate I'm a little edgy about letting myself get bogged down in specifics. I did better today. Then a double session of TV tapings; some spots for candidates & then some lengthy visuals for a Navy League Dinner, a Boy Scout award to Maude Chasen, The Nat. Elec. Contractors, Intenational Police Chiefs etc.

Thursday, October 18

[Rehearsed for debate; met Portuguese Olympic marathon winner.]

Then it was off to N.Y. for the Al Smith dinner. Henry Kissinger came by the hotel, we had a good session. He feels my firmness about the Soviets has worked & that if I'm elected they'll want to get together on arms talks.

Sonny & Leah Werblin came by—he is Toastmaster for the dinner. Then Arch. Bishop O'Connor came to escort us to the reception. It is a tradition of the affair that there be a receiving line. We must have shaken more than 1000 hands (1 hour, 15 min.). Mondale sent regrets which was taken as quite an affront by the people of N.Y. Nancy stayed over & I came back to the W.H.

Friday, October 19

A bill signing—anti-terrorism legislation. Two hours of debate briefing. Then NSC meeting re the terrorist targets in Lebanon. We have one in the Bekaa Valley we are convinced is a training center. My decision, in view of the threats hanging over us for the next several days (the anniversary of the Marine holocaust) is to be ready with an air strike if they attempt another attack. To strike now & then to have them attack would look like it was a reprisal for our raid.

A long afternoon of working on a closing statement for the debate.

Saturday, October 20

The radio program & then I can sum the day up in one sentence. I've been working my tail off to master the 4 min. closing statement I want to make in the debate tomorrow night.

Sunday, October 21–Wednesday, October 24

Off to K.C. for debate Number 2—on Foreign Policy. The consensus seems to be that I won although some want to call it a tie. A rally before the debate was a little like the Homecoming bonfire before the big game. I felt fine—certainly different than I felt in Louisville.

The next morning we parted company—Nancy off on a campaign trip by herself & me off to Calif. Ron & Doria had surprised me by showing up in K. C. for the debate. I was quite touched. They rode A.F.1 back to Calif. with us. My 1st stop in Calif. was at Palmdale for the view of a B.1. bomber. Some 10,000 U.A.W. members in a giant hangar & as enthused as the G.O.P convention. Then onto San Diego for an open air rally. Again a great turnout of young people. Next an airport rally at Medford, Oregon—18,000 people. We went on to Portland, Oregon for overnight. The day was a great one & there were more to follow. Tues. morning the rally was at Portland U. A small group of hecklers gave me a chance for a little repartee which went over big with the students & others. We left immediately for Seattle & another giant rally. Then on to Columbus Ohio for the night. Wed. morning a brief talk to a local broadcasters convention. Then to Ohio St. U. for a giant rally. The O.S.U. students were on fire. Another small heckler group only added to the fun. By this time I was so in love with young Americans I was all choked up. For lunch we went to the Teke house (T.K.E.), my fraternity. These have been the greatest four days I can remember. Mondale dug up a letter I wrote to Nixon in 1960 & had tried to say I likened him to Marx & Hitler. Now the press is after me because of a little band of clerics who charge that I'm basing my policies on the expectation of Armageddon. Mr. Mondale is desperate.

[Ceremony for medical students rescued on Grenada; Mrs. Reagan expected imminently. **Thursday, October 25:** *briefing for supporters; honored in Sportscasters Hall of Fame and in poll of high school students; photo session and taping of campaign commercials; interviewed by Knight Ridder reporters and then by* Newsweek; *dinner with Mr. and Mrs. Charles Wicks and Charles Price.* **Friday, October 26:** *flew to New York for speech at Jewish Temple, commented, "I believe I settled for them the issue of separation of church & state"; helicoptered to campaign rallies in Connecticut and New Jersey; returned to Washington and went to Camp David with Mrs. Reagan.]*

Saturday, October 27–Sunday, October 28
A beautiful summer like week-end & back to W.H.

Monday, October 29
Off to Lancaster & Media, Pa. & Parkersburg, W.Va. A good day! Lancaster rally was at Millersville College in the gym. As always with the students—joined by several thousand adults it was a great reception. Then we helicoptered—after having had lunch at 11:30 A.M. at the college, on to Media, Pa. for an outdoor rally—again a great reception. A.F.1 to Parkersburg, W.Va. & a rally in the H.S.

gym—an even bigger crowd heard the speech by loud speaker outdoors. Back to the W.H.

Tuesday, October 30

Kind of hard to change tempo & get back to the office routine. I phoned a little speech to the U.S. League of Savings institutions convention. Then a couple of taped phone calls for the get out the vote drive. Met with a group of Dems. for Reagan leaders including such individuals as Willy Mays, Don Newcomb, Rosy Grier, et al. A bill signing in the Rose Garden—the omnibus trade bill. Spent much of the afternoon in an interview by Wm. R. Hearst & about 20 of his editors from all over the country. Another one with 3 from Time mag. Then an NSC meeting on where we go from here on Nicaragua & finally upstairs. Nancy due home from her 2 day trip West any minute.

Wednesday, October 31

Don Regan came by—he believes the economy is in a lull & the Fed. holds the key. The Fed. has pulled the string on money supply & it's down to ½ the rate of growth in the ec. Don is trying to goose Paul V. into a little action before he brings on more of a lull.

To the India Embassy to express condolences of the death of P.M. Indira Ghandi. Then over to Reagan-Bush H.Q. for a tour & to thank the volunteers. They were an inspiration—an attractive gang of mainly young people—heart & soul into the campaign. Back at the Rose Garden the 10 surviving Cong. Medal of Honor—(Hispanic Am.) medalists on hand for a ceremony unveiling the very beautiful stamp commemorating the heroism of Hispanics. A routine session with John H. approving appointments. A meeting with Geo. S. & Bud—subject negotiations, whether arms or Latin Am. or what our overall policy should be. I declared that no more agreements just to get an agreement. They would have to accomplish our goal. In the oval office I video-taped the half hour speech to be played on election eve. It went well. Then I taped the Sat. radio speech cause we'll be on the campaign trail. A haircut—a pile of paperwork & home & Mother. The elec. polls are holding up but I'm still edgy—what happens if our people look at the polls & decide their vote isn't necessary?

Thursday, November 1

Off to Boston, Rochester, N.Y., & Wayne Co., Michigan.

This was last campaign swing before election. An outdoor rally in front of Boston City Hall (foreign territory for me). Hopes were high we could elect Ray Shamie to the Sen. but it wasn't to be. Then on to Rochester, N.Y. for an indoor rally & on to Wayne Co., Michigan & a Shopping Center (outdoor) rally. It had been raining on the crowd for 3 hours but when I suggested I cut my remarks down the crowd screamed no—so we all got wetter. Then to the hotel for dinner & overnight.

[Friday, November 2: hopscotched by plane to rallies in Saginaw, Michigan; Cleveland, Ohio; Springfield, Illinois; and Little Rock, Arkansas. **Saturday, November 3:** *flew to Des Moines and then Winterset, Iowa, noted, "After the rally I visited the John Wayne birthplace. Two of his daughters were on hand"; flew to Milwaukee, Wisconsin, and met Mrs. Reagan; campaign rallies.]*

Sunday, November 4

We penciled in a stop in Minn. for a press conference in Mondale's home state. Even with no advance notice there was a big crowd at the airport.

Then resumed the schedule in St. Louis Mo. in the biggest rally of all at the St. Louis Arch. Back on the plane & on to Chi. where we met up with the Bushes at a rally in the Rosemont Horizons Arena. George introduced me then I excused him because he & Barbara were due at a rally in Houston, Texas. At 4:40 we were back on the plane & headed for Sacramento California & the Red Lion Inn. Our suite was quite luxurious including having the bed on a raised platform. At 3:30 A.M. Nancy took a header off that d--n platform when she got up to get an extra blanket. She got an egg sized lump on the left side of her head. We tried ice water packs [. . .].

Monday, November 5

Nancy needed help walking. I was scared—she was really wobbly. The doctor said it was to be expected & that there was no serious or lasting damage. We went over to the Capitol where Gov. Deukmejian unveiled a picture of me & a plaque naming the Cabinet Room the Ronald Reagan Room. Then out to the West Steps to address a huge crowd where twice before I'd been sworn in as Gov. We went on to L.A. for a rally in the Pierce College F.B. stadium—about 35,000 people. Back on the plane & on to San Diego to close out the campaign where I'd closed the '80 campaign. A huge crowd. Then back to L.A. & the Century Plaza—Nancy still unsteady.

Tuesday, November 6

The great day! We helicoptered to Solvang to turn in our absentee ballots— a beautiful day. We flew right over the ranch. Back at the hotel I had interviews with Lou Cannon, Newsweek Mag. & Time. That evening our traditional dinner with our friends at the Jorgensons. It was really great to be with our friends. While there Mondale called to concede. There were also calls from P.M. Mulroney of Canada, King Fahd of Saudi Arabia & Pres. Pia of Pakistan. Finally the get-together with all our workers at the Hotel & the evening was over.

Wednesday, November 7

Well 49 states, 59% of the vote & 525 electoral votes. A short press conf. The press is now trying to prove it wasn't a landslide or should I say a mandate?

Then to the ranch on a beautiful day.

Thursday, November 8–Saturday, November 10

Beautiful weather every day. One night 3 inches of badly needed rain but clear & bright by morning. I rode all 3 days—Nancy didn't but by Sun. she was OK except for soreness where the bump was. Barney, Dennis & I spent the afternoons cutting up a couple of downed Oak trees & splitting the wood. I bought a wine closet & we hauled wine down from the barn. We found a few other chores that needed doing so all in all it was a happy time.

Sunday, November 11

Left the ranch at 7:55 A.M. (we'll be back on Sat.). Back to the W.H. & on directly to the Vietnam Vet's Memorial where I briefly addressed 100,000 people for the acceptance of the Memorial statue. It was quite an event & I hope it finally makes up for the way the Vietnam Returnees were treated when they came home.

Tuesday, November 13

[Visit by Grand Duke and Duchess of Luxembourg; formal ceremony and state dinner with entertainment by Twyla Tharp.]

During the day we had a full cabinet meeting & I laid down the charge— we must reduce spending. By the time we got through Congress had passed $114 Bil. increase for '85. I've told our people not to spend it.

Wednesday, November 14

An interesting lunch with Dr. Richard Leakey the anthropologist & the Nat. Geographic board. A good time was had by all. The Dr.—son of the Leakeys who've done so much work on the origin of man, is carrying on & lives in Kenya. A long meeting with Sec. Shultz. We have trouble. Cap & Bill Casey have views contrary to George's on S. Am., the Middle East & our arms negotiations. It's so out of hand George sounds like he wants out. I can't let that happen. Actually George is carrying out my policy. I'm going to meet with Cap & Bill & lay it out to them. Won't be fun but has to be done. A more pleasant task—meeting then with task force on legal equity for women. They've made great progress & accomplished a great deal. Had to sit for a sculptor who is doing the inaugural medal—took about 40 minutes. Home & Mother.

[Telephoned young man with terminal cancer.]

Thursday, November 15

At 8:20 A.M. called the crew of the Shuttle. Signal had worked out a connection with Houston Space Center so that I saw the crew on TV while I was talking to them.

Thanks to Bill Buckley I had a meeting with an editor from Doubleday who brought suggestions on how I should keep records like this for an eventual book or books on my terms in office. He was a fine man & most informative—his name—Sam Vaughn. We had a 2 hour Cabinet meeting on the matter of our ec. program. The meeting of 2 days ago was on the front page of the papers yesterday. I wish we could learn who is leaking to the press. We had sworn everyone to secrecy & did so again today. It's a d--n uncomfortable feeling & inhibits all of us from voicing opinions. We're facing a battle to get the spending cuts we must have but I intend to take our case to the people. Cap W. came in re some defense problems. I didn't take up the Sec. St. problem with him—pending a session with the V.P. who has some input on that matter.

[Visitor at reception became ill.]

Friday, November 16

We're all in an uproar over the continued leaks. Yesterday's Cabinet meeting was on the front P. of the Wash. Post. Don Regan was so mad he sent me a letter of resignation. I refused to accept it & before noon he was cooled down. From now on we dispense with the back benchers—only Cabinet members will attend meetings.

Then we had a session with Carol Dinkins of Justice & Shultz of St. Justice is pursuing a case—criminal charges of anti-trust violations with England's Laker airline. P.M. Thatcher has really dug in her heels. George thinks our relations with U.K. are more important than the case. Hearing both sides I came down on the side of foreign relations—case closed. In N.S.P.G. what could have been an in-house battle—again involving State Dept. & CIA was settled peacefully in favor of CIA. [. . .] Tomorrow morning I'm meeting with Cap W. & Bill Casey to iron out (if I can) some difficulties involving George S.

Saturday, November 17

Off to Calif. Nancy parted from me at Andrews A.F. Base. She went to Phoenix to be with her Mother. I arrived at the ranch in late afternoon. Barney & Dennis had started laying out the string lines for the irrigation system we're putting in beside the house. I got to putting our wine in the new wine closet I'd bought on the last trip.

[Sunday, November 18: missed Mrs. Reagan; rode; ranch work.]

Monday, November 19–Sunday, November 25

The whole week was wonderful as usual except that Nancy picked up a

Official Presidential Schedule for November 15, 1984

8:30 am (5 min)	Phone Call to Crew of Space Shuttle "Discovery" (Fuller)	Residence
9:00 am (30 min)	Staff Time (Baker, Meese, Deaver)	Oval Office
9:30 am (15 min)	National Security Briefing (McFarlane)	Oval Office
9:45 am (15 min)	Senior Staff Time	Oval Office
10:00 am (2 hrs)	Personal Staff Time	Oval Office
12:00 m (60 min)	Lunch	Oval Office
1:00 pm (30 min)	Meeting with Sam Vaughn Principal Editor of Doubleday (Darman)	Oval Office
1:30 pm (2 hrs)	Cabinet Time (Fuller)	Cabinet Room
3:30 pm (45 min)	Personal Staff Time	Oval Office
4:15 pm (30 min)	Personnel Time (Herrington)	Oval Office
4:45 pm (15 mins)	Administrative Time 1. Photo with Clarence Brown, Commander of the American legion (Whittlesey) 2. Photo with Tih-Wu-Weng, Chairman of the Board, World Journal (Meese) 3. Receive Report from the Presidential Advisory Committee on Small and Minority Business (Fuller)	Oval Office
5:00 pm (20 mins)	Reception for the American Security Council's Coalition for Peace through Strength (McFarlane/Rosebush)	Residence

Official Presidential Schedule for November 16, 1984

9:00 am (30 min)	<u>Staff Time</u> (Baker/Meese/Deaver)	Oval Office
9:30 am (15 min)	<u>National Security Briefing</u> (McFarlane)	Oval Office
9:45 am (15 mins)	<u>Meeting with Peter McPherson</u> (McFarlane)	Oval Office
10:00 am (60 mins)	<u>Personal Staff Time</u>	Oval Office
11:00 am (60 mins)	<u>NSPG Meeting</u> (McFarlane)	Situation Room
12:00 m (75 mins)	<u>Lunch & Personal Staff Time</u>	Oval Office
1:15 pm (10 mins)	<u>Presentation of Thanksgiving Turkey</u> (Whittlessey)	Rose Garden
1:30 pm (30 mins)	<u>Meeting with Secretary Shultz</u>	Oval Office
2:00pm (30 mins)	Meeting with Foreign Minister <u>Yaqub Khan of Pakistan</u>* (McFarlane)	Oval Office

---

* *In copy pasted in the diary, the 2:00 pm meeting was circled with a handwritten note from President Reagan stating, "This meeting was well worthwhile. It had to do with Pakistan's nuclear energy program. They've been reluctant to allow inspection to confirm their claim they are not building a weapon. I think the matter is settled."*

cold. The routine became morning rides & work on laying the pipes, etc. in the afternoons. Thursday—Thanksgiving, Patti & Paul, Ron & Doria, Maureen & Dennis & Moon & Bess came up for our annual turkey. Patti & Paul stayed over & went back on Friday. We set off the annual brush pile on Friday.

Word came of the firefight in Korea. A Soviet tourist ducked across the line in the demilitarized zone seeking sanctuary. The N. Koreans opened fire & the Am. & S. Koreans returned fire. One of our men was wounded—not seriously, a S. Korean was killed & the North Koreans lost 2. The Soviets must be serious about their decision to meet with us on Jan., 7 & 8. We feared they might get exercised about the defection & the shooting. They haven't mentioned it. Sat., the 24th was our one bad day—cold, foggy & raining. Well it gave me a chance to catch up on some homework. Sunday was bright & beautiful & the flight back to Wash. was uneventful. One other sour note on Thanksgiving had to do with Mike R. He blew up at something on the TV news based on an interview Nancy had given. He called me & when I tried to straighten him out he screamed at me about having been adopted & hung up on me. [. . .]

Monday, November 26

Back to the regular routine. George S. came by to tell me I should tell Lucky Roosevelt (Protocol Officer) I want her to stay. I'm not out to change anyone so if they don't tell me they want to leave, I first assume they are staying. Anyway I had her in & gave her the word & he was right—she bubbled over with joy.

Deputy P.M. of Iraq Tariq Aziz came in—today we re-established diplomatic relations which Iraq had broken off 17 yrs. ago. We maybe helped in our peace efforts because Iraq is pretty cool toward Syria—the bad boy of the Middle East.

For 2 hours we had our 1st briefing on the Treasury Dept's. tax plan. I'm worried. I'll hold my fire until I understand it better but as of now it sounds a little complicated & not as simple as I'd hoped.

To top the day off I called Mike R. We talked for half an hour & I'm more than ever convinced that he has a real emotional problem that is making him paranoid.

Tuesday, November 27

[Visit from Prime Minister Ratu Sir Kamisese Mara of Fiji.]

Howard Baker came by in a sort of farewell ceremony. He's keeping his options open regarding a Presidential try in '88. He'd like to keep a foot in govt. so I'm appointing him to P.F.I.A.B. He checked me out as to whether I'd be neutral in '88. I must say this will be a tough one for me. I've always believed the party should choose the nominee but when the time comes I'm afraid my heart will be with George B. if he makes the run. An interview with 5 editors & writers from "Wash.Times." I think it went OK—at least they are friendlies.

Tonite dinner for Frank Sinatra. The Wicks, Laxalts & Shultz's on hand. A good time was had by all. We should do this more often.

Wednesday, November 28

Called John Stennis (Sen.). He's at Walter Reed Hospt. to have a leg taken off at the hip—Cancer. Met with the Cab. working group on bud. cuts. Dave Stockman presented a case to bring the deficit down to 4% of G.N.P.—then 3% by '88—2%. It is a really draconian cut including elimination of scores of govt. programs entirely. Frankly it's time to reverse the course of the last 50 yrs., but I can't help but be suspicious that Dave made it so tough hoping we'd turn instead to a tax increase. Well I won't. Met with Geo. S. re the upcoming arms reduction talks. We agree that since Chernenko has talked as I have of total elimination of nuclear weapons that should be our goal in the negotiations.

Had a visit with representatives of the Alzheimer's Disease Foundation. Yasmin—daughter of Rita Hayworth (a victim) was with them. She is a fine young lady totally dedicated to the cause. Went upstairs to do homework & select a pile of books to give the Anacostia High School Library.

[Attended Senate Republican Unity dinner.]

Thursday, November 29

Met with new Sen. Leadership & Leadership of House (Repubs.) to discuss our attack on the deficit—spending cuts & tax reform. Consensus is to try & put it all in a single pckg & strive for a "freeze" on budget—making next yr.'s look like we are holding the line & repeating this yrs. Problem is defense. We can't hold that down to no increase without endangering our arms reduction talks.

In P.M. our Cabinet Core group resumed meeting on same subject. We'll be at that again tomorrow. We're making tough decisions on freezing, reducing or canceling about 130 programs. Later on I presented Congressional gold medal to Leo Ryan's mother. He was killed in the Jones Cult Affair in Latin Am. Met also with Capt. & 2 seamen who rescued 85 Vietnamese "boat people." They rcvd the International Nansen medals. Finished with some tapings. Another call from Mike. He is a really disturbed young man. I've contacted his Minister & believe maybe we can get through to Mike. [. . .] Mermie is here for overnight.

Friday, November 30

Another meeting with the Budget Core group to find areas where we can cut. It's a laborious process going line by line & weighing the effect. Things got sparked up a bit when Cap W. was called out several times for phone calls. It seems that a civilian oceanographic vessel under lease to the Navy had become disabled & drifted into Cuban waters—about 9 miles off shore. We had a Coast Guard cutter not too far away & called the Cuban govt. for permission to have it go into Cuban waters & tow our ship out. The Cubans refused & said they were sending a fast Naval vessel to take our ship in tow. We very much feared a hostage situation. Actually the law of the sea permits any vessel to go into national waters on a rescue mission. I ordered our Coast Guard vessel to go in. Just in case—we ordered the Carrier Nimitz in the Virgin Islands to move on Cuba & put planes in the air. Evidently the Cuban ship with a crew obviously manning it's stern gun, put a line on the disabled ship. Somehow that line got cut & entangled the Cuban's prop. Our cutter towed the ship safely into American waters. End of incident. Not knowing all the details the press has treated it as "much ado about nothing." Frankly I think our Navy & Coast Guard deserve a "Well Done."

Helmut Kohl arrived & we had a good meeting & pleasant lunch. He's a good friend & solid ally. He's concerned about V.E. day coming in May & a repeat of Normandy D-Day with the Germans ignored. They suffer a great guilt complex over the Nazi period. I'm suggesting including them this time & making the occasion one of celebrating when the hatred stopped & peace & friendship began which has continued for 40 yrs.

An N.S.P.G. meeting about the forthcoming arms talks with the Soviets. I made it plain there must be no granting of concessions (one sided) to try & soften up the Soviets. Off to Camp David. Had to cancel the movie—some kind of main fuse blew & shut off the power.

Saturday, December 1–Sunday, December 2

Sat. a quiet day of catching up on reading. Sunday back to the W.H. early. This afternoon the reception for the Kennedy Center honorees—Lena Horne, Danny Kay, Gian Carlo Menotti, Arthur Miller & Isaac Stern. Then to the K. Center for the program. The Shultz's & the Roger Smiths (G.M.) in our box. The show got a little political this time—film biographies of Lena dwelt on her participation in the Civil Rights demonstrations. For Arthur Miller it was his persecution by Sen. McCarthy, etc. The show closed with a production number by Roberta Peters—opera star & a cast of hundreds dedicating the finale to me as they sang "Let There Be Peace." I made it a point to be the 1st to stand & applaud when the last note was sung.

Monday, December 3

Asst. Sec. Crocker who has been doing a great job in Africa shuttling between S. Africa & Angola came in to report. It looks like progress is being made in a most difficult situation.

[Budget meeting; looked at new poll numbers; met singer Lou Rawls regarding show

to raise money for the United Negro College Fund; bipartisan congressional delegation reported on trip to famine-affected areas of Ethiopia; visit from Roy Brewer; met new members of W.H. press corps.]

Bud McFarlane. He's as irreplaceable as any one could be. He's an 80 hour a week man & I keep trying to get him to spend more time with his family. He feels he can only do the job on his present schedule. Result—he feels he must leave out of fairness to his wife. I hope we can work something out.

Tuesday, December 4

A visit by Pres. Lusinchi of Venezuela—a bright cold day with a formal ceremony on the So. Lawn. Almost a diplomatic Faux Pas. I slipped into my topcoat. When time came for my opening remarks I took my notes out of my pocket & horror of horrors it was a speech welcoming "His Royal Highness" the Duke of Luxemburg. I couldn't imagine how they could have given me the wrong speech. I started to ad lib & then remembered I'd worn my top coat last on the day of the Grand Duke's arrival. Today's speech was in my suit coat pocket. I made a switch & got away with it pretty well. In the afternoon Pearl Bailey & others of the Am. Lung Assn. came by to present me with the first page of Christmas seals. She's a wonderful lady & a long time friend. She had a scarf for Nancy made up of a silk reproduction of the sheet of stamps. Then she unveiled a robe for me made out of 8 of those scarves.

A long arduous Budget Core meeting—this time on Defense. No final decision but I know we can squeeze some out so as to make our claim of a freeze valid. We're up to $33 Bil. of the $45 we need.

[Visited Senator John Stennis (D-MS) in the hospital.]

Tonite the State dinner for the Pres. of Venezuela. Andy Williams is entertaining. It was as usual a nice, warm occasion. Pres. Lusinchi was carried away by the fact that I told him & the others about the speech mix-up.

[Wednesday, December 5: met freshmen Republican congressmen; Attorney General Bill Smith reported on finding a single building for his staff; cabinet meeting on budget; National Security Planning Group (NSPG) meeting to discuss approach to Soviets in upcoming meeting; signing ceremony for Food Aid to Africa program; donated four suits to charity.]

Thursday, December 6

Had the Repub. Leadership in for the whole load on our budget cuts. They took it pretty well although they expressed a desire to study some of the cuts & maybe they'd have a suggestion or two. We explained this was a working paper & that our Cabinet was reviewing it also & that we wanted their suggestions. I told them to please not go out & tell the press our proposal couldn't get passed: that I would take our case to the people & they certainly had made it plain they wanted us to do what we've been doing. Good meeting.

[Met Doug Flutie, winner of Heisman Trophy in college football, received honorary Heisman Trophy; met with Glenn Campbell, director of the Hoover Institution, and Martin "Marty" Anderson, fellow there, regarding location of presidential library at Stanford, commented, "I know I'm Pres. but I can't get used to hearing that $90 mil. will be raised to build the Library & Museum"; briefing on international steel situation; visit from March of Dimes poster child, received from her a ten-week-old sheepdog, noted "There's nothing so warm as a cuddly puppy."]

Taped a tribute to Joe Coors—did an interview with Human Events. Like a cloud over everything is the hi-jacked plane in Iran. It looks like they've murdered 2 Americans—officials in our A.I.D. program. Iran is if anything an irresponsible participant on the side of the Terrorists.

Friday, December 7

Bishop Tutu of S. Africa came in. I'm sure he is sincere in his belief that we should turn our back on S.A. & take actions such as sanctions to bring about a change in race relations. He is naïve. We've made considerable progress with quiet diplomacy. There are S. Africans who want an end to Apartheid & I think they understand what we are doing. American owned firms in S.A. treat their employees as they would in Am. This has meant a tremendous improvement for thousands & thousands of S.A. Blacks. There have been other improvements but there is still a long way to go. The Bishop seems unaware, even though he himself is Black, that part of the problem is tribal not racial. If apartheid ended now there still would be civil strife between the Black tribes.

We're facing a closer to home problem which has to do with the upcoming talks with the Soviets re arms reductions. We're convinced they want above all to negotiate away our right to seek defensive weapon against ballistic missiles. They fear our technology. I believe such a defense could render nuclear weapons obsolete & thus we could rid the world of that threat. Question is will they use that to break off the talks & blame us?

[Ceremony for Columbia *astronauts; regular meeting with Shultz; left for Camp David with new puppy in hand.]*

Saturday, December 8–Sunday, December 9

The usual restful weekend—except for watching over an un-housebroken puppy. Now we're back to the Christmas decorated W.H. & it is magnificent. Bless Ted Graber—the decorations & the tree are unbelievable.

Tonight the Christmas entertainment at Const. Hall.

Monday, December 10

Over to the E.O.B., Room 450 twice today. Once for the Pvt. Sector Initiative group. We're starting a new program to honor businesses that do outstanding charity & public service work. Each yr. a number of them will receive "C" flags—akin to the wartime "E" flags. C is for commitment. The 2nd time was a talk & signing ceremony proclaiming this "Human Rights Week." It will be interesting to see if the press (they covered the event) will carry my strong words against S. African Apartheid. They've covered every charge that I haven't criticized S. Africa.

A Red Cross delegation including Chuck Heston reported in on their trip to Ethiopia. Chuck who is a very fine sketch artist had sketched an 11 yr. old boy—a famine victim who only weighed 16 pounds. I need to remind our people the Red Cross can use donations directed to the famine in Africa.

[Devoted hours to budget meeting focusing in Defense Department; National Security Planning Group (NSPG) meeting, defining strategy for talks with Soviets.]

Tuesday, December 11

[Visits by Charles Douglas–Home, editor of the London Times, *and by President Seyni Kountche of Niger.]*

We have confirmed the Iranian storming of the hi-jacked plane & the rescue of the hostages. Our 2 citizens are on their way home so I can say I'm suspicious of Iran & I suspect the rescue was staged in complicity with the terrorists.

Jeanne Kirkpatrick came by to give me her letter of resignation. We'll keep it quiet until after the inaugural & she'll leave no earlier than March 1st. Four years of the U.N. is longer than any of her predecessors have taken. We want to keep her in the admin.—she's done a brilliant job at the U.N. We talked about her successor & found her ideas & mine were in tune. I'm looking at several things for her but didn't mention what they were until I've sorted them out.

Tonite the Congress Ball here at the residence. We drop in—start the dancing—then mingle—always edging toward the elevator. It was a great party—about 500 present. This had always been a tradition but Muffy Brandon—when she was here—shut it off in our 1st yr. Rex S. brought it to our attention & properly so.

Wednesday, December 12

George Bush took the watch early this morning & went to Andrews for the arrival of the bodies of our 2 men murdered in Iran—the hijacking.

[Addressed gathering of publishers of Gannett news organization; meeting with the International Private Enterprise Task Force; planted a sugar maple tree on the W.H. lawn; lunch discussion on defense budget.]

Met with Barry Goldwater—he is stubbornly holding out against the MX. He could well be the kiss of death.

Bill Dickinson, R. (Ala.) also came by. We need to meet with him more often on defense matters.

[Met with CEO's on volunteer committee for savings bond drive; press party. **Thursday, December 13:** *cleaned out desk, noted, "I didn't throw away half of what I should"; annual economic briefing by Don Regan, some positive news; called William J. Schroeder, recipient of artificial heart; photo session.]*

Then it was time to light the Xmas tree. I had Nancy throw the switch & the tree way down on the Elipse came to life. We were surrounded by small children up here in front of the W.H.

[Reception for press corps.]

Friday, December 14

Dropped in on Inaugural Directors meeting. Their job is unbelievably complex & tedious but they volunteer & do it in great style. Then out to the driveway to be presented our Inaugural License plate—No. 1.

We had a most unsatisfactory N.S.P.G. meeting with Charley Wick present. He is not the reason it's unsatisfactory. We're ready to go with "Radio Marti" our station broadcasting truth to Cuba—part of our Information Program. Cuba, however, threatens retaliation; not just jamming our program but jamming Am. radio stations all the way to the Mid-west. They are actually completing the transmitters to do this. We can join all of theirs but at great cost & only after several months. It will take time to set up a system. If we retreat we lose face which can hurt us in all of Latin Am. If we go forward we could knock many of our commercial stations off the air. What to do? Right now I don't know.

Billy Graham for dinner tonight. Very interesting—his experiences in the Soviet U. & S. Africa. He questions the stature of Bishop Tutu. He feels the Soviets are experiencing a wave of religious revival, particularly on the part of young Russians.

*[***Saturday, December 15:*** *desk work, answered letters from constituents; reception for military families.* **Sunday, December 16:** *answered more letters; attended reception for staffers, added, "We also sneaked down to the ground floor to visit 'Lucky' [new puppy]. We'll take her to Camp D. Friday."]*

Monday, December 17

[Further discussion on defense budget; presented awards to civil servants for efficiency.]

We had an N.S.P.G. meeting again on our negotiating posture in the upcoming meeting with Gromyko & the arms talks. I believe the Soviets have agreed to the talks only to head off our research on a strategic defense against nuc. wpns. I stand firm we cannot retreat on that no matter what they offer.

The international champion Barbershop Quartet came by for a photo & sang a song (my kind of music).

Greeted incoming White House Fellows then dropped by 3 Xmas parties at

E.O.B.—N.S.C., Speechwriters & Personnel. Last go around at 6:15, the W.H. reception for the S.S. Had a meeting—Arthur Burns—who plugged for retaining the Ec. Advisory Board. Dave Stockman had suggested elimination. I think Arthur's right. Jack Block came by to tell me his plan to turn Agriculture back to the free market—I buy that.

[Tuesday, December 18: *attended farewell party for Ted Bell, secretary of Education, and Christmas party at the counsel's office; meeting with the Joint Chiefs regarding U.S. military strength versus that of the Soviets, commented, "More & more I'm thinking the Soviets are preparing to walk out on the talks if we won't give up research on a strategic defense system. I hope I'm wrong"; private dinner party with the McFarlanes, General Brent Scowcroft, Nancy Reynolds, and the Laxalts.]*

Wednesday, December 19

A brief meeting with Gen. Ed Rowny re the upcoming Arms Talks (Start) [START]. He's a good man. We've decided he'll try to take up the talks as a resumption of the talks they broke off a year ago. He'll do that hoping to head off a "start all over again" tactic by the Soviets.

A Cab. meeting—Joe Wright reported on where we are in management reforms. The potential for savings is tremendous. To do a number of things we'll have to get Congress to repeal legislation that literally ties our hands as the executive branch. Dick Whalen came by—he's really crusading that Geo. S. has been taken over by the St. Dept. soft liners & that he's planning a purge of all our appointees. I've already taken the matter up with George & his account refutes the charges being made.

Met with Peter Grace. He has a good idea about giving some of his task force some continuing charge so they can help mobilize public opinion in support of our budget cuts. Upstairs early—I'm getting spoiled by these short days while Congress is away.

Thursday, December 20

[*Budget meetings on HUD, Agriculture, and Small Business Administration.*]

Today I asked John Tower to be Amb. to Germany—Arthur Burns is retiring. I think John is taken with the idea but he hasn't given an answer.

I have a real problem to solve. George Shultz's plan to do some shuffling of Ambassadors & Asst. Secs. at State has raised quite a holler. I trust George but there are indications that the Bureaucracy may be leading him astray. For example, he's told an Ambas. wants to quit. Now I'm told the Ambas. was told by the bureaucrats that I wanted his resignation etc. George gets his info. secondhand—not directly from the person being dumped. Tomorrow I have to show him my evidence.

[*Photo sessions and Christmas reception for W.H. staffers.*]

Friday, December 21

Well this was a meeting day with George S. By the time of the meeting I had rcv'd. evidence countering the word that had been given me to a certain extent. Part of the problem was one ambas. who is being removed for cause who has gone all over town spreading false charges. Geo. & I had our meeting & everything is squared away. We had an NSC meeting regarding our trade & defense matters with Japan—all preparing for my Jan. 2 meeting with P.M. Nakasone. We've made progress but still have a way to go particularly with regard to our entry into their markets.

[Received giant Christmas card from Johnstown, Pennsylvania; drove to Camp David.]

Saturday, December 22–Sunday, December 23

Sat. dawned clear & bright which was fine because P.M. Margaret Thatcher was coming in for a visit. I met her in a golf cart & took her to Aspen where she & I had a brief visit in which I got a report on her visit with Gorbachev of the Soviet U. In an amazing coincidence I learned she had said virtually the same things to him I had said to Gromyko. In addition, she had made it clear there was no way the Soviet U. could split Eng. away from the U.S.

Then we joined the others—Ambassadors, Shultz, MacFarlane, Bush, et al at Laurel for a plenary meeting & working lunch. Main topic was our Strategic Defense Research ("Starwars") I believe was eased some concerns she had. Then she was on her way to Eng.

Sunday was also a bright, beautiful day—like Spring & we returned to the W.H.

Monday, December 24

[Telephoned enlisted personnel in foreign posts; Mrs. Reagan's brother, Dick, and his family visiting for Christmas.]

Senator Hatch is after me to grant clemency to the Rev. Moon. I've explored this & find I just can't. I have, however, taken action to see if I can grant him a furlough over New Years. It seems that day is the holiest in that religion.

Tonite dinner at the Wicks—getting to be an annual event & a happy one. Nancy played Santa—it took about 3 long distance calls from Ron to get her to do it. She was great.

Tuesday, December 25

Merry Christmas & it was. Gift opening, a walk on the S. Lawn with Lucky & then cast of last night's party here at the W.H. for an early evening dinner. A beautiful book on Iran from the Shahbann & a poignant letter. She can only spend about 3 months a year here in the U.S. where her children are going to

school because of a law or legis. of some kind. I'm going to try to get her a diplomatic passport. That would, at least up it to 6 months.

Wednesday, December 26

Cold weather but bright sun & no snow. A brief meeting with Bud &
John—NSC. John became a Vice Admiral & I presented him with his stars.
An easy & quiet morning—cleared up a few items. Turned over to Mike D. the
matter of a passport for Her Majesty the Shahbann Farah Pahlavi. Taped a Rose
Parade greeting & upstairs to pack for trip to Palm Springs.

Nancy's Brother & family returned to Phil.

Thursday, December 27—Wednesday, January 2, 1985

Depart for Palm Springs—well L.A. for a couple of days first. On Fri. will
meet with Mike & Colleen to see our granddaughter for the 1st time. Had time
for a trip to Dr. House. My hearing has suffered no loss since last year. The
other good news was our family meeting. I think we passed a watershed & the
wounds are healed.

From Dec. 29 (Fri.) to Jan. 2nd the beautiful weather was beyond belief.
This New Year party at Lee & Walter Annenberg's house is a tradition. It's also
become my once a yr. golf game. With all the socializing, had time for meeting
with Bud, George S. & Cap pinning down approach George will take with Gromyko in Geneva Jan. 7 & 8. Then Jan., 2nd it was back on A.F.1 & the White
House—with happy memories of another 4 day reunion with old friends.

Almost forgot—on Jan. 2nd met with P.M. Nakasone of Japan. It was a
good meeting & I believe further progress was made re our trade deficit with
Japan. The P.M. is a true friend.

Chapter 5

—

1985

Thursday, January 3

Vacation is over—Congress is in session & I'm back in the Oval O. Biggest part of day was spent hearing appeals from Justice & Interior on proposed budget cuts. Managed a compromise between O.M.B. & Justice. The other problem was whether to cancel the civilian dept. having to do with water problems—irrigation, etc. & give it all to the Corps. of Engineers. Teddy R. was the 1st who wanted to end the duplication between these 2. Then Herbert Hoover tried. I've told our people to sit down & work out a plan for doing this but not in connection with our budget problem.

[Ambassadorial formalities; photo sessions; call from King Juan Carlos of Spain to say hello.]

Friday, January 4–Sunday, January 6

Started the day off with the first meeting with the leadership of the 99th Congress. We spent our time on arms control. Bud & George S. held forth on the Geneva meeting—they departed this evening. I found support even among Demos. for our Strategic Defense Initiative.

A budget appeals meeting re transportation. Dave was holding out for canceling Amtrak. The case for getting Govt. out of the RR business has no opponents. Libby Dole however does have a good argument against doing it right now. She has 3 live customers for Conrail & RR labor is helping lobby Cong. to allow the sale. She fears (and so do I) that if we suddenly announced the end of Amtrak, labor might reconsider. I haven't given an opinion yet but I favor closing the Conrail sale 1st—then getting at the other.

[Budget meeting with Secretary Margaret Heckler of Health and Human Services (HHS); Shultz said farewell before Geneva trip; went to Camp David. **Saturday, January 5:** *reported weather was clear.* **Sunday, January 6:** *returned to W.H.]*

Monday, January 7

[Scheduled budget meetings with State Department and Veterans Affairs canceled; Shultz and McFarlane have nothing to report yet from Geneva, commented, "I'll try to remember 'no news' may be good news."]

Did a couple of tapes—one for the Inaugural & one for a ground breaking in Palm Springs for Bob Hope Cultural Center. Met with 10 rather distinguished men—some former Cabinet members of both Dem. & Repub. admin's. They represent about 200 citizens who want to start a drive to change the Presidential term to a single one for 6 yrs. I don't really know whether I approve that. I do agree with another idea—that Congressmen should serve for 4 rather than 2 yrs. Did a B.B.C. radio interview for a program about P.M. Thatcher. I'm a great admirer of hers.

Biggest event today was a meeting with Don Regan & then with Jim Baker. They want to trade jobs. I've agreed. As soon as Jim is confirmed by the Sen. he'll be Sec. of Treasury & Don will come to the W.H. as Chief of Staff. I think it will resolve a lot of problems.

Tuesday, January 8

Word from Geneva continues to be good. George B. & I were presented with the new Inaugural medals. This morning I went to the Press Room with Don Regan & Jim Baker & announced they were exchanging jobs. The press was really astounded. They thought I was coming in to talk about Geneva or something of the kind. This was one story that didn't leak.

I was in the family theatre briefing for tomorrow nite's press conf. when I was called upstairs to take a call from George S. on the secure phone. The meetings in Geneva are over & the Soviets have agreed to enter negotiations on nuclear weapons, etc. Within the month a time & place will be agreed upon. Did a brief interview with 2 men from the Dallas Morning News—a friendly paper. Then a reception in the East room celebrating "Human Events" 40th anniversary.

[Mrs. Reagan in NYC; president lonely in W.H.]

Wednesday, January 9

George S. is back & things are better than I'd thought & I thought they were pretty good. Lunched with the V.P. & called Les Aspin, new chairman of the House Mil. Affairs Comm. I have a feeling maybe we're going to get along pretty well.

[Presided over press conference.]

Thursday, January 10

Good returns continue to come in on the press conf. but more on the Baker, Regan job switch—we scored a 10. Today announced Don Hodel would go from Energy Sec. to Sec. of the Interior. John Herrington moves from Personnel to Energy & Dick Darman will be Baker's deputy Sec. at Treasury.

Cab. meeting was on reorganizations—mainly to meld Trade Rep. & Dept.

of Commerce. The briefing made it evident this is very complicated & right now I don't know which side I'm coming down on. Did a big session on tapings & dropped in on a farewell party for Bill Ruckleshaus.

Friday, January 11

[Report of three U.S. soldiers killed in base accident in Germany.]

Somehow things here had a staff "snafu." Faith Whittlesey is returning to Switzerland as Ambas. but no one told me that John Lodge, present Ambas. was more or less summarily fired & I learned of it through a news leak in a Swiss paper. He of course is hurt & upset. I didn't know, nor did George S. that this action had been taken.

[Light workload; photo sessions; meeting with Shultz; interview with People *magazine; left for Camp David.]*

Saturday, January 12–Sunday, January 13

Got a call on Sat. that David Murdock's wife Gabrielle 43 yrs. old died Sat. afternoon of Cancer. Nancy & I reached him by phone about 10 P.M. our time. What does one say? When we got back to Wash. the snow was completely gone & the weather quite mild. Mermie is here.

Monday, January 14

Not bad for a Monday—climate good too, temperature up & bright sun.

Patti called about her visit here for the inaugural. She'll get in Sun. & be on hand for the big doings on Mon. Paul can't come. She says they can't afford for him to give up a couple of day's pay.

Visit by P.M. Martens & Foreign Minister Tinderman of Belgium. Big issue was whether Belgium will hold off deploying cruise missiles as their part of the '79 agreement with NATO. They put up quite a pitch for us to approve delay. I know they both favor deployment but don't have the backing of their Parliament. I stood firm on their agreement to deploy. I'm sure this is what, in reality they wanted me to do. Now they can tell their parliament they tried & I said no.

[Visits from Glamour *magazine's annual honorees; talked with Kay R. Whitmore, president of Eastman Kodak. The president noted, "He's been most helpful to us"; Paul Gottlieb, publisher of Michael K. Evans's book* People and Power: Portraits from the Federal Village; *LaVell Edwards, coach of Brigham Young football team.]*

Then after dinner it was off to open at Corcoran Gallery Mike Evans's exhibit. Maureen went with us. It was a case of saying a few words to a crowd of more than 1,000 & then declaring the exhibit open & home to bed.

*[*Tuesday, January 15: *met with Council for a Black Economic Agenda, with leaders of a wide array of business groups; cabinet meeting heard report on consolidating govern-*

ment departments; ambassadorial formalities; discussion of trade imbalance with CEO of
Caterpillar; appearance at reception for Hoover Institution.]

Wednesday, January 16

N.S.C. briefing—I made a decision to equip several planes with equipment capable of joining Cuban radio & TV. We may never use them—I hope not. But we intend to start Radio Marti—broadcasting truth to Cuba. We intend to offer Castro a channel upon which he is free to broadcast to our people. But we'll also tell him that if he jams our radio & (as he has threatened) interferes with our commercial stations we'll black out Cuban TV & radio. We must be prepared to carry that out instantly.

Greeted & thanked our entire Geneva (Soviet talks) delegation. Then I dropped by a meeting of the "Committee of The Next Agenda." Surprisingly this was a group representative of a number of Inst's. from Brookings to Hoover etc. They have come up with an agenda for our 2nd term that is pretty basic to what I want to do. Then over to E.O.B. to speak to Nat. Assn. of Republican Mayors who are here for the Nat. Conf. of Mayors.

And upstairs early to work on Inaugural address which I think I've finally finished.

Thursday, January 17

Sent Ambas. Goodby on his way to Stockholm for the continuing European peace talks. He says, strangely enough the Eastern Block Nations are getting bolder & bolder about siding with the West & disagreeing with the Soviets.

[Interviewed by two journalists; lunch with Vice President Bush; uneventful cabinet
meeting, except "Dave S. telling us we will come in with about $50 Bil. in cuts for the '86
budget"; met Nobel Laureate in Chemistry, then new members of W.H. press corps; recep-
tion for Citizens for America.]

Mike, Colleen, Cameron & Ashley arrived—here for the inaugural.

Friday, January 18

Called Pvt. Initiative to tell them about a young Black man interviewed on TV last nite in hospital. He rescued a 10 yr. old boy from a man holding him with a 12 inch knife at his throat. Thomas Locks—the rescuer, 25 yrs. old is an unemployed landscaper. He will have a job when he gets out of the hospital with a nursery in Alexandria. I've just called him & the man who has the job for him.

(N.S.P.G. meeting—settled an action we'll take immediately against Iran if the Hisballah terrorists carry out their threat to punish as spies the 5 Americans they are holding in Lebanon. It's time to serve notice we won't hold still for their barbarism.)

Met with George S.—we have a problem with General Ed Rowny (retired). We've named the 3 chief negotiators in the arms talks but we want Ed—who headed up the last negotiations to stay here as a special advisor to me & George. He sees this as a demotion. I met with him & did my best to convince it was nothing of the kind—that we need him & his expertise right here when these talks begin again. I'm not sure I convinced him.

Mermie has arrived & Dennis. Also Dick Davis's family. Tonite the 1st of the inaugural events. A show & fireworks down on the ellipse. It was wonderful. There were 100's of young people—volunteers from all over the country. I puddled up at their obvious patriotism & enthusiasm.

Saturday, January 19

This morning went down in the Rose Garden with Cameron & Ashley & we built a snowman. I'll bet that's the 1st snowman ever to inhabit the R.G.

A wonderful Luncheon hosted by the Annenbergs & Tuttles. It was at Blair house & was like a replay of New Years—all our friends plus some Cabinet & staff (who are also friends). Back at the W.H. spoke briefly to the Inaugural Trust—the people who raise the money to fund the events.

Tonite the big Inaugural Gala—produced by Frank Sinatra. It was a smash—a cast of 52 plus the Naval Academy chorus.

Sunday, January 20

Up for the traditional prayer service at the Nat. Cathedral. Billy Graham preached the sermon & it was as always inspirational. Back to the W.H. where George & I both took our oaths as the Const. requires. Then all our friends & family were here for a buffet. Later in the afternoon entertained the cast of last nights gala & some others. Late in day members of Inaugural Committee came to me about canceling tomorrow's parade & moving the ceremony into the Rotunda of the Capitol. The weather with wind chill factor is predicted to be more than 20 below zero. At the temp. exposed skin surface is subject to frostbite in about 15 mins. There is no way we should inflict this risk on all the people who would have to be out in the cold for hours.

King Hussein called with Congratulations. The Super bowl was won by S.F. & I flipped the coin on live TV.

Monday, January 21

The big but very cold day—below zero early morning without factoring in the wind chill. We made the right decision about canceling the parade & moving the ceremony into the rotunda. First stop was St. John's Church. Learned that Congressman Gillis Long (LA.) died last night in his sleep—heart failure. Back to the W.H. for a family photo then on to the Capitol with Tip O'Neill &

Mac Mathias as escorts. About 1000 people in the rotunda. When the ceremony & my inaugural address were over we moved into Statuary Hall for the Congressional lunch. Back to the W.H. & we helicoptered out to Capitol Center where the thousands of paraders from 50 states were gathered all in their uniforms & regalia. A tremendous job had been done of putting things together—just overnight. We & the Bushes & Ron Walker (Chmn.) were on what would have been the final float in the parade. Those marvelous young people really had an enthusiastic time. The All American band put together of bands from all over the co. provided music & a good time was had by all.

Back at the W.H. we met the 50 state parade marshals. At lunch Tip told me privately he was very conscious of the fact that I had received 59% of the vote.

Then 11 Inaugural balls to attend & we did them all—11 of them. At almost all we were supposed to (and did) dance about a minute. Except for 1 orchestra they all played very slow music. I figure they were being nice—or thought they were because of my age.

Tuesday, January 22

Back to work. A late start 10 A.M.—Phoned & addressed about 71,000 Right to Life people down on the Ellipse. Then met with our arms negotiators. Told them we wanted a good agreement or no agreement. Then a meeting with top Congressional leadership of both parties. There seemed to be a better relationship all around—several remarked about it to me afterward.

Met with "Right to Life Leaders." Thanks to modern science (ultra-sound) they have filmed inside the womb a baby being aborted. The Dr. responsible had performed 10,000 abortions—this turned him around. He is now anti-abortion activist. I scheduled a showing for tonite. In the meantime had a reception for about 400 of the people—volunteers who served on the inaugural committee. The movie (28 minutes) was most impressive & how anyone could deny that the fetus is a living human being is beyond me.

[**Wednesday, January 23**: *Baker nomination going smoothly through Congress; report of National Guard plane lost on training mission in Honduras; visit from Miss America; interviewed by Helen Thomas of UPI and Mike Putzel of AP; lunch with directors of inaugural; met with team planning summit in Bonn, West Germany, for May; meeting with Shultz and McFarlane.* **Thursday, January 24**: *visit from Italian defense minister, former prime minister Giovanni Spadolini; addressed meeting of Latin American legislators; lunch with Vice President Bush, who was planning trip to Africa and Brazil; meeting with Malcolm Baldrige on consolidating Department of Commerce and Office of the U.S. Trade Representative, commented, "I know I have to come up with an answer soon but I've got to listen to some others—this subject really has 2 sides"; Cabinet Council meeting to hear report on making U.S. industry more competitive; greeted Special Olympics competitor, then karate instructor of son Ron; videotapings.]*

Friday, January 25

Don Regan & I met to talk about the transfer—him to Chf. of Staff. We are in agreement about some personnel changes & he wants to alter the chain of command a little to simplify it. I agree. He's also completely in agreement that we don't make any change with regard to Jim Brady. I addressed the Exec. Forum at Constitution Hall. This is an annual affair attended by all W.H. top staff & Cabinet. They were really Gung Ho—the execs from all the depts. George S. & Bud & I met to talk about the Middle East. The problem is to get the moderate Arabs moving toward peace. Their problem is fear of Syria. Just recently a top P.L.O. person openly spoke of peace—he's dead now.

No Camp D. this week so I went upstairs early & did some housekeeping. Tomorrow the radio program starts up again.

[Saturday, January 26: radiocast; live radio interview followed; answered mail from constituents; Alfalfa Club dinner. Sunday, January 27: continued answering letters; visit from Cliff White, 1968 campaign manager and conservative activist, encouraging administration to concentrate on major issues.]

Monday, January 28

Set the fellows in motion about a news story: an El Salvador young lady who entered the country illegally but who now has an Am. husband & baby is slated for deportation. I want it stopped. We shouldn't be breaking up families.

A succession of meetings—first with national trade assn. leaders. They are ready to turn their constituents on in support of our Budget reductions. An issues briefing lunch—nothing out of the ordinary. Then Repub. members of Sen. Finance Comm. & Reps. members of House Ways & Means. This meeting was on the tax simplification plan. It was worthwhile. Finally a meeting with Trent Lotts Whip organization. This was a good one mainly on budget reductions. I phoned Berke Breathed—cartoonist who does Bloom County. He obviously thought I was calling to bitch about something. I called to thank him for the Sunday strip where he had Nancy in the strip looking lovely. He's sending me the original.

Tuesday, January 29

A long NSC session to decide what degree of violation of SALT II provisions by the Soviets we would report to the Congress. I don't think there is any question there have been violations but could we prove it in court. Most of our reports will refer to them as almost positive violations.

Foreign Minister from Hungary was in briefing. Hungary is a target for friendly efforts to woo them a little away from the Soviets. They don't seem too unwilling. Had a good meeting with 3 controversial members of the Civil Rts. Commission—my appointees Linda Chavez, Morris Abrams, & Clarence

Pendleton—an Hispanic, a Jew & a Black in that order. All 3 are solid as a rock & while totally dedicated to civil rights they are opposed to quotas which actually bring on reverse discrimination.

A long Cabinet meeting—with a fine report on our success in improving govt. management but a long fruitless discussion of the farm debt problem. A lot of farmers are going to go belly up & there really isn't anything we can do about it. A taping session & tonite a dinner for the new freshman class in Congress & their wives & husbands. It was a nice affair & I think most if not all went away with a leaning toward bi-partisanship.

Wednesday, January 30

F.M. Rabin of Israel came in. I was able to tell him we are asking Congress for $1.8 bil. in Foreign Mil. Sales for his country—a sizeable increase. I tried to impress on him why we feel we must sell weapons to the moderate Arab States if we are to ever bring them around to making permanent peace with Israel. It isn't an easy sell even when I tell them we'll never let Israel lose its qualitative edge.

Did a Q&A with a bi-partisan roomful of Sen's. on the Arms talks, M.X. etc. I hope we're making them realize we can't unilaterally disarm & hope to persuade the Soviets to enter into arms reductions talks.

Jeanne Kirkpatrick came in—this was D-day. She didn't feel she could do the job I wanted her to take & wants to return to academia. She has a lifetime endowed chair at Georgetown U. Sorry to see her go & the conservatives who worry that I'll go soft will lose a lot of sleep.

[Thursday, January 31: Annual Prayer Breakfast; met with House Armed Services Committee; lunch with Vice President Bush; considered plan for two-year trial of line-item veto; decided against Shultz's advice to maintain U.N. ambassador as a cabinet-level position; received photo album of whistlestop tour of Ohio; report that Edwin Meese would be approved by Senate Judiciary Committee as attorney general; called Michael Jackson about an appearance at World Youth gathering in Jamaica.]

Friday, February 1

[Visit by President-Elect Tancredo Neves of Brazil.]

An interesting lunch with "The Futurists." This is a group of men & women who are engaged in various kinds of projections for business & industries. Some are academicians. I'm happy to say we weren't too far apart on our ideas about where we are going & what we should be doing. We agree that many of our social programs were started without thought as to where they might be taking us & how they could make worse the problem they were intended to solve. For example did the "war on poverty" actually increase poverty because it removed any need or incentive for people to become independent. Off to Camp D.—had to drive again—low clouds, rain & icing conditions.

[Saturday, February 2: radiocast; caught up on reading. **Sunday, February 3:** *sunny day.]*

Monday, February 4

Met with So. African Chief of Zulus, Gatsha Buthelezi. He's a very impressive man—well educated & while dedicated to ending apartheid in S. Africa still is well balanced & knows it will take time. I'd quoted him in my own speeches before I ever met him.

[Presented copies of budget congressional leaders of both parties; addressed other congressmen; met with private sector leaders; spoke at convention of National Religious Broadcasters.]

Tuesday, February 5

New French Foreign Minister Dumas came by. He's an improvement over Chaysson. It was just a get acquainted meeting.

I told our people I wanted to use the renewed Sat. Radio talk to educate the people on such things as who really is responsible for the deficit & the budget, etc. We'll start doing that Sat. The Sen. Judiciary Committee approved Ed Meese for A.G. & the Sen. will vote on it right after the recess. All Repubs. & 2 Dems. gave him a 12 to 6 win.

A 90 minute NSC meeting on the upcoming visit by Saudi Arabia's King Fahd. This will have its touchy moments.

[Visit by leaders of new industries; cabinet meeting on administration economic report; call from King Juan Carlos regarding planned visit to Spain in May.]

Wednesday, February 6

Happy birthday to me. It was a nice day & I am touched by some of the greetings from staff groups with everyone signing the cards.

A meeting with Congressional leadership (Repub.). Everyone got into the act giving a kind of preview of what I'll be saying tonite. Bob Michel then presented me with a cake. We have a little strain. Yesterday appearing before a committee Dave Stockman blasted the mil. as being more devoted to preserving their pensions than they are to Nat. Security. He'd been badgered for 3 hours by some like Sen. Metzenbaum & sort of exploded emotionally but still should never have sounded off as he did.

Later our people were backgrounding the TV anchor people at a lunch. I stopped by & received a chorus of "Happy B.D."

[Photo session with son of old friend; visit from former Vietnamese refugee graduating from West Point, expressed intention to introduce her during State of the Union Address that evening.]

Well—it's done & over with. The speech went over fine & the Cadet &

Mother Hale of Harlem received a great ovation. Calls to W.H. came in 10 to 1 favorable.

Thursday, February 7

Ambas. Bosworth—our man in Manila came by & reported on the situation there. Marcos is still the best man on the horizon to straighten out the problems except his ill health may prevent that. His kidneys are the problem & he puts in less than 2 hours a day & Imelda keeps people from seeing him.

[Visit from Clara McBride "Mother" Hale, of Harlem's Hale House; swearing-in for new cabinet members.]

Rest of morning a lunch with Bob Hawke—P.M. of Australia. He's a good man. We're both upset about New Zealand—refusing to allow our destroyer to use the port. Has to do with N.Z.'s ban on ships powered by or carrying anything nuclear. We've cancelled the Anzus war games.

A colorless interview by the Wall Street Journal & then signed off on Maureen to go to Nairobi next July to the women's conference.

Friday, February 8

We had a meeting on putting together a program to help our Central Am. neighbors develop energy supplies. Right now they are dependent on imported oil which impoverishes them & holds back the possibility of ec. development. There is a potential in that area of hydro & geothermal, etc.

Mayor Buck Rinehart of Columbus Ohio came in. What a breath of fresh air he is. With other Mayors belaboring us as to how our budget cuts will put them in a hole he has launched a local program to help us cut. One feature is a gift catalogue given to his citizens which contains everything the city has to buy—from light bulbs to fire engines & asks his citizens to give gifts of these things to the city. The take so far is almost $5 mil.

[Hosted seventy-fifth-anniversary lunch for the Boy Scouts; meeting with Shultz on variety of minor matters; went to Camp David. **Saturday, February 9–Sunday, February 10:** *one brief walk; desk work; returned to W.H.; anticipating state visit by King Fahd of Saudi Arabia.]*

Monday, February 11

The big day, King Fahd arrived—the weather was unbelievably nice, clear sky & sunshine. Everything went well in the morning meetings. In afternoon an interview—me & the N.Y. Times. I think I did well even though they (4 of them) were pitching curves. The State dinner was a success. Everyone seemed to have fun. His Majesty is a really shy man & I'm convinced he felt he had been befriended. I invited him to come to the W.H. for breakfast.

Tuesday, February 12

Brkst. was a big plus. We talked of our mid east peace plan. He is agreeable to backing King Hussein & I think he sees merit in our proposals. He's also going to increase the funding he is secretly giving the Contras in Nicaragua.

[Luncheon for scientists from various disciplines.]

Cabinet Council took up matter of what our alternatives are in cleaning up toxic waste dumps. Some would have us use the Superfund to do any & everything with regard to environment. That could become a giant boondoggle. I favor a fund of about $1.1 Bil. a year strictly for the toxic waste dumps. Presented a citizens medal to John Rogers—youngest man ever to serve on W.H. exec. staff. He's done a fantastic job on refurbishing the E.O.B. & the W.H. among other accomplishments. Dick Wirthlin came by with new figures—they hold up well although there has been some slippage in areas where media & opponents have been propagandizing.

Just saw a fund-raising letter signed by Tip O'Neill for Dem. Cong. Committee. It is the most vicious pack of lies I've ever seen. It's aimed at Sr. Citizens & has me out to destroy Medicare & Soc. Security. We can't let him get by with this. Tomorrow off to Calif.

[Wednesday, February 13: left for California; interviewed by Santa Barbara newspaper en route; met Mrs. Reagan at airport and went to the ranch.]

Thursday, February 14 (Valentine's Day)–Sunday, February 17

[Weather was hot and clear; much ranch work.]

There were a few duties—phone call to King Fahd. Cap. W. says King told him our meeting was best he's ever had with a Head of State. Newsman Levin, one of hostages held in Lebanon by terrorists escaped & was taken in by [. . .]. He's home in U.S.A. now. I called Pres. Assad (Syria) to thank him. He says he's trying to arrange return of the other 4. We'll see. Now we're back in Wash. with a holiday (Presidents Day) then back to work.

Monday, February 18

[Telephoned young cancer survivor.]

I made another call on my own to a young lady in Reseda, Calif. I'd read a news story about her grandfather, 84 yrs. old—his mind getting a little dim but he believes in our admin. Every time he got one of those computerized fund raising letters he contributed until he'd given his total savings of $4200 away. Then crying he called his granddaughter because he couldn't give me more. I asked these magic working W.H. operators if they could get this young lady on the phone. All they had to go by was her name. They got her. I told her I wanted to talk to her grandfather to thank him but then to tell him about computer mail & that he must throw any more letters in the waste basket. He's just been put in a nursing home & she didn't know the phone number. She'll get it & send it to me.

[**Tuesday, February 19**: *met with staff; presented National Technology Awards; NSC meeting to prepare for upcoming Canadian trip; visit from Marines who landed on Iwo Jima in 1945.*]

Wednesday, February 20

Margaret Thatcher day. A good visit & meeting covering all the subjects from currency to arms control. We do have a close alliance on this the 200th anniversary of diplomatic relations between the U.K. & the U.S.

Before she arrived I saw a secret film of one of our cruise missiles from its launch by Sub. from beneath the Pacific [. . .].

Then a pre-press conference briefing & some more wood-shedding until time to leave for the British Embassy to dinner with the P.M. It was a very nice evening & only emphasized how firm is the friendship between our countries & with the P.M.

Thursday, February 21

[*Spoke briefly to leadership at National Conference of State Legislators; lunch with Vice President Bush.*]

The press conf. went well although I thought the questions were rather pedestrian & didn't make for much news. All this time a blackmail attempt was being made in the Sen. by Democrats trying to link Ed Meese's confirmation to a costly bailout for the farmers. Bob Dole stood firm in the face of their filibuster.

Friday, February 22

Morocco's foreign minister came by for an early morning visit. Mission was to determine if our friendship for King Hassan was still on firm footing. I believe I convinced him it is. Truth is I do like the King.

Later I did about 10 minutes of Q&A with 27 editorial writers from all over the country. Some revealed some deep biases, particularly regarding the Nicaragua situation.

Then an interview with "Business Week" & a reception in the Roosevelt Rm. for Bill Smith. Ed's confirmation still hanging fire. I'll take 2 versions of my radio broadcast to Camp David—one for if he's confirmed before noon Sat. & the other if he isn't.

[*Photo sessions, military groups and little girl who brought a puppy named Lucky as a gift.*]

Saturday, February 23–Sunday, February 24

Did the "no-confirmation" script. The weather was like Spring. We lunched outdoors—took a walk & then came the phone call—Ed was confirmed 63 to 31. Our side had won & Bob Dole has proved his right to be Leader of the Senate. Weather still great & back to the W.H. for dinner hosting the Nat. Gov.'s Conference. It was a nice dinner even though some of the Governors have been a little partisan—sniping at us mainly over the farm issue. They want more money at the same time they complain about the deficit.

[Monday, February 25: meeting with governors, main concern was distressed farmers; follow-up with Grace Commission on implementation of recommendations; interviewed by Jack Germond and Jules Witcover who were writing a book on the 1984 campaign; photo session with beauty pageant winners.]

Tuesday, February 26

A 9:30 meeting with our Repub. Congressional leadership. They are determined we are going to have to cut deeply into the Defense budget or we won't get support for our budget cuts. I'm being stubborn as h--l & intend to stay that way. N.S.C. briefing was a report indicating that Assad of Syria seems to be making an effort to get our 4 kidnap victims back from the terrorists—the Hisballah.

[Positive news on administration influence on education from Secretary of Education William Bennett and David Gardner, head of Commission on Excellence in Education; photo session for National Wildlife magazine; met with Senate Armed Services and Appropriations Committees on MX missile; Regan reported on divorce scandal involving member of the FCC; telephoned young crime-stopping hero in Cleveland.]

Wednesday, February 27

[Annual breakfast with the Senate freshman class of 1980; presented National Medals of Science.]

After lunch Jerry Parr & his family came in for a photo. He's retiring from the Secret Svc. Jerry is the one who threw me in the Limo on March 31st, 1981 when Hinckley was shooting at me.

Geo. Shultz came in with Bud. He was reporting on what our man is reporting from S. Africa about the Namibia & Angola situation. Angola has made an offer about sending the Cuban troops home but it's not a good one. We're proposing a counter which also has the approval of Savimbi—head of the UNITA faction which is in revolution against the Angolan govt. He's a good man & has offered a plan for peaceful settlement in Angola. We're also stepping up our help to Mozambique: Some of our Congressmen & Sen's. are upset by this since that country has been in the Soviet camp. What they don't know is that Mozambique wants out of the Soviet connection.

[Met with representatives of Eureka College; quiet afternoon; noted that Deaver was back from a planning trip to West Germany.]

Thursday, February 28

Met with Generals Gorman & Galvin. Gorman has been our Commander of the Southern region in Panama. I awarded him the highest medal we have for the retired. He's leaving the service. Galvin will replace him. Gorman is extremely well informed about Central America & I must say his assessment of the situation, Nicaragua, El Salvador, Honduras, et al was rather optimistic.

[Addressed the National Association of Independent Schools; met with Senator Don Nickles (R-OK), Representative Mickey Edwards (R-OK), and several independent oil drillers regarding proposed changes to tax structure.]

Next in was Gov. Terry Bradshaw [Branstad] of Iowa—re the farm problem. Only 10.4% of the 2.2 mil. farmers are involved in the financial & credit crunch. But they are the kind of farmers you find in Iowa, Nebraska, Illinois etc. So you have a farm ec. that's 90% sound nationwide with the crisis all bunched up in one area. The Demos. are demagoging all over the place smelling an issue for '86. They howl for multi billion dollar bailouts at the same time they demand I do something about the deficits.

[Photo session, including West Point football team.]

A drop by for a reception for Bill Clark & his family. He's going back to the ranch.

Friday, March 1

Just learned we're putting together a program for me to lean on individual & small groups of Congress re our problems—budget, MX, etc. Looks like a lot of breakfast drop bys & cocktail hour meetings. Sen. Pete Domenici is working on an idea of getting the Sen. to send a drastic budget resolution over to the House with cuts way beyond ours to force the Demos. to negotiate.

The Nicaragua problem continues to fester. The most sophisticated, high priced lobby effort in years—Nicaragua sponsored is taking place on the hill & a number of members are buying their falsehoods which means they accept the lobbyists words over ours including our intelligence services, the state dept. etc.

Met briefly with several representatives of Arab-Am. groups. They were most supportive of us during the campaign. We had a brief talk about the middle east.

[Spoke briefly to key appointees; spent afternoon in the residence; Conservative Political Action Conference dinner, noted, "It was a good affair—in the 1st place it ran 15 mins. ahead of schedule which has to be a 1st of some kind." **Saturday, March 2:** *answered constituent letters; dinner with the Regans and Doles.* **Sunday, March 3:** *finished answering letters.]*

Monday, March 4

Our 33rd Anniversary. Other than that it was another Monday morning. Why do they always seem different than other days?

Met with the new Sec. Gen. of OECD—Jean-Claude Paye. It was a brief but pleasant meeting. He is all for urging European members of OECD to take steps to free up their economics, etc. so as to catch up with our ec. recovery.

We had an N.S.C. meeting with our Arms Talks Leaders looking at various options for how we wanted to deal with the Soviets. It's a very complicated business. I urged one decision on them—that we open the talks with a concession—surprise! Since they have publicly stated they want to see nuclear weapons eliminated entirely, I told our people to open by saying we would accept their goal.

Nancy came to the oval office for lunch & we cut an anniversary cake & had a few of the immediate staff share in it. That was the extent of our celebration except that at dinner we opened a bottle of Chateau Margeaux 1911.

Right after lunch I addressed the N.A.C.O.—Nat. Assn. of County Officials. I wasn't sure how I'd be received since they've taken positions opposing some of our budget cuts & that was what I talked to them about. But they were very cordial.

Fred Fielding, Don Regan & Mike D. came in to see me about the Arabian Horses that King Fahd wanted to give me. I had stated I couldn't accept them as a gift—due to our stupid regulations. As it stands they are now in Prince Bandar's (Ambas.) name & he has asked Bill Clark to take care of them for him. Now what happens 4 yrs. from now is anyone's guess.

[Interviewed by Newsweek *on MX missile.]*

Had Sens. Dave Boren & Sam Nunn over for cocktails & to talk about the MX. I believe we'll have their support. In fact they talked of how wrong it was for Congress to interfere with a President in Foreign affairs & how both parties must come together at the water's edge.

Tuesday, March 5

[Congressional breakfast with questions on MX; visit from Prime Minister Bettino Craxi of Italy.]

Pat Brown—former Gov. of Calif. I defeated in '66 came in. He's now 80 & there were no vestiges of the bitterness he used to feel & display. This was day for receiving Ambas's. About 6 of them including one from New Zealand. That could have been a little touchy but it wasn't. Then over to the W.H. for another quiet cocktail do with 5 of our Congress people—more MX. Well only 3 showed up. One of them, Arlen Spector is holding out—his reason, that MX is vulnerable if put in Minute Man silos. All in all though I think we did some good. Nancy is in N.Y. for a dinner for Princess Margaret. It's lonesome here.

Wednesday, March 6

Up early—I don't sleep well when I'm alone. Another brkfst.—this time for

House Demos. Usual Q&A on MX. Imperceptibility there seems to be a tide running a bit our way. Over in the Cabinet Room met with bi-partisan leadership both Houses. Had our 3 chief Arms Negotiators on hand. They spoke briefly & then opened for discussion. A press fuss over House members going to Geneva for opening of talks, made it seem like we were complaining that it was just another junket. I called Jim Wright to tell him—not true that we believed their going would show unity to the Soviets. So at this meeting I asked him to go.

Had lunch with Sen. John Stennis. He's a fine old gentleman & while a life long Dem. he really stands by us on things like MX & the Defense budget.

[Spoke to citizen's group on MX; tests for annual checkup; vetoed farm spending bill; cocktails with congressmen with discussion of MX; Mrs. Reagan home.]

Thursday, March 7

[Congressional breakfast; cabinet affairs briefing on Treasury tax proposal; spoke to National Newspaper Association.]

Lunch with Tip O'Neill, Don Regan & Max F. Tip surprised me—He won't make an issue of MX but will not personally vote for it. He says it's a matter of conscience; having the MX he says will provoke a Russian Nuclear attack. He can't respond when asked how we can remain defenseless & let the Soviets have thousands of missiles aimed at us.

We had a Cabinet Council meeting on management & admin. We're working out a plan to sell off excess properties the govt. owns.

Big event was meeting with Polit bureau [Politburo] member (Soviet) Shcherbitskiy. He had Ambas. Dobrynin & a couple of others with him. I had George S., Bud, Don Regan & a couple of others with me. He & I went round & round. His was the usual diatribe that we are the destabilizing force, threatening them. It was almost a repeat of the Gromyko debate except that we got right down to arguing. I think he'll go home knowing that we are ready for negotiations but we d--n well aren't going to let our guard down or hold still while they continue to build up their offensive forces.

[Photo sessions; cocktail party with congressmen, commented, "I made the usual MX pitch & they seemed pretty affirmative"; met widow of drug enforcement agent killed in Mexico.]

Friday, March 8

A large breakfast with members of Sen. & House teams who are going to Geneva for opening of arms talks plus our negotiators. There seemed to be a feeling of unity even including Sen. Ted Kennedy. Then over in the Roosevelt room we had the formal send-off.

[Photo session with young winners of VFW honors.]

Then it was off to Bethesda Naval Hospital for my annual check-up. I'm so healthy I had a hard time not acting smug.

[Report that ranch manager in California had died suddenly; went to Camp David. **Saturday, March 9–Sunday, March 10:** *reading and walking; returned to W.H.]*

Monday, March 11

Awakened at 4 A.M. to be told Chernenko is dead. My mind turned to whether I should attend the funeral. My gut instinct said no. Got to the office at 9. George S. had some arguments that I should—he lost. I don't think his heart was really in it. George B. is in Geneva—he'll go & George S. will join him leaving tonight.

[Photo session with young entrepreneurs; lunch with regional journalists; spoke to organization of conservative state legislators, the American Legislative Exchange.]

Word has been received that Gorbachev has been named head man in the Soviet.

Carolyn & John Kennedy came to see me—they want help on fundraising for the endowment for the J.F.K. Library.

A Haircut & then over to the Soviet embassy to sign the grief book—this is my 3rd such trip.

[Dinner in honor of Queen Sirikit of Thailand, commented, "She's a delightful, shy & lovely lady who does all kinds of good work—helping the poor, the refugees etc."]

Tuesday, March 12

Today was Mubarak Day (Pres. of Egypt). The usual meetings, 1st in Oval O. then Cabinet room & finally lunch at the W.H. He has some financial problems—mainly about $500 Mil. a year in interest to us for past military purchases. He needs help & we'll do what we can but of course we have our own problems. Most of time spent on peace talks between Jordan & Israel. The sticking point is with the Palestinians who insist on representing themselves in the negotiations. I think we've settled for his idea that the U.S. meets with Jordan & the Palestinians & then (our idea) we repeat with Israel present. Then of course the negotiations are directly between Hussein of Jordan & Israel.

[Met with Representatives Olympia Snowe (R-ME) and John McKernan (R-ME) regarding subsidized imports from Canada; hosted dinner for Egyptian president Hosni Mubarak, noted that no business was discussed, talk was mostly of show business stories; Mrs. Reagan on trip west.]

Wednesday, March 13

A tough meeting with Mac Baldrige. He's pushing his idea of a consolidation of the International Trade Rep. & the Dept. of Commerce. I felt I had to say no even though I lean that way. Don R. is moving to re do our Cabinet Councils & I want to see how Mac's plan looks when that is finished.

Henry Kissinger came by for lunch along with Don R. & Bud. Naturally we talked a lot about the Soviet situation. He agreed I was right not to go to the funeral. He believes my attitude has put us in the best position we've ever been in to make progress with them.

[Short speech to American Business Conference; desk work done; cocktails with several Republican senators, termed it "Arm twisting for MX."]

Thursday, March 14

Ryan came by with a plan for distributing surplus left over from Inaugural. I don't think anyone has ever done [what] we're doing with that $3½ mil. One mil. will be given to the fund in the Treasury for reducing the Nat. debt. Then the rest is doled out to scores of worthy programs from one end of the country to the other.

[Addressed Magazine Publishers Association; desk work in the afternoon; greeted youthful hero from Cleveland mentioned in entry of February 26, also W.H. Marine guards; videotapings; cocktail party to press for MX with Democratic senators.]

Maureen is back from Vienna with astounding tales of the 3-day meeting preparing for the World Conf. on Women in Nairobi in July. Even here the Soviets were obstructionists.

Friday, March 15

[Saw film in preparation for upcoming meeting with President Raúl Alfonsín of Argentina; NSC meeting regarding trip to Canada; Irish Ambassador brought gifts for St. Patrick's Day.]

Dick Wirthlin came in with the usual mix of good news & here & there a little bad. Our communications on Nicaragua have been a failure, 90% of the people know it is a communist country but almost as many don't want us to give the Contras $14 mil. for weapons. I have to believe it is the old Vietnam syndrome. They are afraid we're going to get involved with troops, etc.

Nancy did well—an approval rating of 83% & rated higher than the last 7 1st ladies. She'll be home soon—in fact here she is! Thank you Lord.

Saturday, March 16

No Camp David this week—we're off to Canada tomorrow. Did my radio bit on the forthcoming trip. Ted Graber is here. Mermie & Dennis left for Calif. yesterday. I spent the day wood shedding for the meetings. Sometimes I think they try to tell me more than I need to know.

It's less than 3 weeks to Calif. & the ranch & I'm ready. The press is once again doing their psycho job on the new Soviet boss, Gorbachev & whether I should see him or not.

I've issued him an invitation so the ball is in his court. In the meanwhile the negotiations on arms reductions are underway in Geneva.

Sunday, March 17

[Flew to Canada for meeting with Prime Minister Brian Mulroney; formal reception.]

At the hotel he & I had a meeting, one-on-one & then a press appearance where we made statements re acid rain. He told me of his meeting with Gorbachev in which I was the subject of conversation. I must say he is truly a friend—he went to bat for me with our Soviet friend in no uncertain way.

[Dinner party and entertainment.]

Monday, March 18

A morning of meetings then a lunch where I made a speech on Am. Canadian relationship—well received. On to the Citadel for the signing of several agreements, treaties, etc. & to the airport & home. I have to believe U.S.—Canadian relations have never been better & certainly not at the leader level. Brian M. & I have really established a warm personal friendship. He will be a welcome new comer to the Bonn Summit. Back home. Funny, we were only gone overnight—about 30 hours total but arriving home seems like we were gone 3 days.

Tuesday, March 19

[Briefing for visit by President Raúl Alfonsín of Argentina, then greeted him on South Lawn.]

In our meeting he brought up a variety of issues. I reassured him that we weren't planning mil. intervention in Nicaragua. He voiced concerns about the Chilean situation.

The violence there is communist caused. I agreed with him on that. He has great financial problems & is seeking a large "bridging loan." I don't know what we can do to help him on that. All in all though it was a good meeting. Sen. Mac Mathias came to the office. I was able to clear up a question he had on the MX & he'll be with us. Then over to the Capitol for lunch with the Sen. Repub. Policy group. Think I did some good on MX in a Q&A session. Back at the Oval O. a series of meetings with Congressmen on MX. Think we made some gains. Then the phone call came—M.X. passed in the Senate 55 to 45. Ten Demos. voted yes, 8 Repubs. voted no. The State dinner was an enjoyable evening. Doug Flutie the Heisman trophy winner was at my table. He's a fine young man. With a $7 mil. contract in his pocket he's going back to school to get his degree. Pete Fountain entertained. It was a wonderful evening of Bourbon St. jazz.

Wednesday, March 20

[Decided to make William "Bill" Brock secretary of Labor and announced it to the press; meetings with congressmen pressing for passage of MX bill.]

A group of Rabbi's—Am. Rabbinical Reps. here on a meeting to promote freedom of Jews in the Soviet to Emigrate, came by & presented me with a book on, among other things, the Passover. I told them how much I shared their concern & asked them to understand why my efforts had to be quiet but the efforts are there. George S. came by with Bud & we talked about "Quiet Diplomacy"—the need to lean on the Soviets but to do so one on one—not in the papers. Over to the family theatre for briefing—press conf. tomorrow night.

Thursday, March 21
Was shown the photo recently taken by the bastards who are holding our kidnap victims in Lebanon. Heartbreaking, there is no question but that they are being badly treated. The day was filled with meetings—Congressmen (mainly democratic) on the MX. I'm worried—just rcvd. word it looks like we're behind in the House. Some wanted to argue that I should spring for a tax just dedicated to defense spending. Briefed for press conference & then at 8 P.M. the press. The conference went well although I think their questions were kind of dull.

Friday, March 22
[Report on armed services from the Joint Chiefs, considered to be in prime shape.]
An almost 2 hour lunch in the White House with our Repub. Sen. leadership. The subject was the Senates budget committee version of the budget vs. ours. They want deeper cuts in defense but they use them to add back cuts we want to make in domestic spending. They also want to suggest a freeze on the Soc. Security "Colas." Finally I raised a little h--l. To go to the Dem. House & have Repubs. suggesting a cut in S.S. is to give the Dems. the issue they want for '86. I told them I could give in on S.S. Colas but it had to come from the Dems. On defense I reminded them if they went in with our bottom line—the Demos. are going to reduce whatever we bring in so leave a little padding to protect our bottom line. Also if we are going to give in on further defense cuts—then it's trade off that it must be in addition to our domestic cuts—it can't be a replacement for them.

George S. & I had a short meeting about getting more active in the Middle East talks. I'm all for it.

[Signed proclamation on skin cancer prevention; greeted Camp Fire Girls, noted, "I was made an honorary Camp Fire Girl—& I didn't have to go to Sweden for an operation"; photo sessions with appointees; interview for program on Gridiron Club; dinner with Edmund Morris, biographer of Theodore Roosevelt. **Saturday, March 23:** *met with Special Olympians; settled on budget compromise with Regan, Weinberger, Shultz, Stockman, and several senators; attended Gridiron dinner.]*

Sunday, March 24

A busy day at home—phone calls re MX & homework. Sixty Minutes had segment featuring the Carters—I got a pretty good going over.

Monday, March 25

Awakened at 7 A.M. by Bud M. An American Major based in East Germany was shot by a Soviet guard in what has to be called murder. The Major was one of 14 U.S. military in East G.—the Soviets have 14 in W. Germany. Our man was only doing what both sides are permitted—indeed are supposed to do. The sentry opened fire—they kept the Sgt. pinned down so he couldn't get to the Major (37 yrs. old) & administer 1st aid. The Soviet medics arrived one hour later. The Major had died.

[Annual press breakfast; noted that Max Kampelman, arriving from talks in Geneva, spent the day lobbying congressmen on the MX; meeting with more than one hundred Central American leaders, commented, "They can really lift your morale"; launched effort to allow lower minimum wages for teens in summer jobs; more campaigning on behalf of MX.]

Tuesday, March 26

Again started the day lobbying Congressmen on MX. Then an NSPG meeting re Afghanistan. We are really delivering help to the Afghans who are fighting the Soviets. [. . .] I have approved keeping this aid at its present high level.

Then presented the gold Congressional medal posthumously to Harry Truman by way of his daughter Margaret. Harry was the last Dem. President I campaigned for.

Had lunch in the office with the Donovan family—his wife & children plus a fiancé of one of his sons. I pray that N.Y. court will clear him but I fear the less than honorable intentions of N.Y.'s courts & prosecutors. I have every confidence in Ray's innocence.*

Then other than hours of phoning any number of Congressmen on MX we had a reception in the East room for our Victory 84 Committee. Then up to watch the House MX debate & vote on TV. We won by 6 votes. Right does triumph. Tomorrow I believe they will vote the appropriation for the missiles & that will be it. We lost 24 Repubs. but got the votes of 61 Demos.

Wednesday, March 27

A happy atmosphere—what with yesterday's vote on MX. Have learned Tip

---

* *Ray Donovan resigned as secretary of Labor on March 15. He was charged with larceny and fraud in connection with a New York construction project. He was acquitted in 1987.*

is asking Demos to vote against aid to the Contras as a farewell gift to him since he's retiring in 86.

P.R.C. Ambass. Zhang Wenjin came by to say goodbye. He's on his way back to China & retirement. We've had a good relationship with him.

[Meeting with Jeane Kirkpatrick, who expressed herself willing to come back if the administration needed her help again; cabinet meeting on anti-drug actions; lunch with technology inventors; economic briefing, a lull in recovery detected; meeting with Shultz who suggested former Senator Nicholas Brady (R-NJ) as replacement for departing deputy secretary of state Ken Dam, president approved; signing ceremony; photo session; received new magazine, Commerce in Space; telephoned gratitude to Democrats who voted for MX.]

Thursday, March 28

Off to N.Y.—breakfast on the plane. I opened the N.Y. Stock Exchange—a brief speech & rang the bell at 10 A.M. I was amazed at the reception. They were cheering like it was a political rally. Then upstairs to talk to & answer Q.'s from the 650 or so board members. I was told I'm the only sitting Pres. who has ever visited the place.

Then it was off to St. Johns U. & again I was a first. What a morale booster that was—1000's of students telling me they loved me—cheering every other word. On the way back to Kennedy Airport Max F. got Don Regan, who was with me, on our car phone to give us the count on the House vote for MX (the all important 2nd vote). We slowed the convoy down so we could have the final vote before we met the press at plane side. With about 1 min. to spare—word came—we won 217 to 210. I greeted the press with the news.

Tip O'Neill had told me he would vote against it but that was a matter of conscience & that he would not try to persuade anyone else to vote with him. There are witnesses to his statement. But right down to the wire he twisted arms, threatened punishment of the 61 Dems. who went with us—in short he was playing pure partisan politics all the way.

[Friday, March 29: lunch with the National Space Club; said farewell to Vice President Bush's chief-of-staff; went to Camp David. Saturday, March 30: ride canceled due to weather; homework and reading. Sunday, March 31: foggy day.]

Monday, April 1

Late to office—Cap called—upset by Soviet Ambassador's appearance on TV last nite after visit to St. Dept. He was smiling & jovial while he was talking about the murder of our Major Nicholson. Frankly I saw it & I was too.

Met with Pres. Niemeri of Sudan—he's O.K. I think we're going to be able to help him. Quadaffi's giving him a bad time. Issues lunch—nothing of world shaking note. Lyn N. came by, he's been representing us in Europe re our walkout of UNESCO. Needed a little guidance—do we want back in, etc. I explained he was simply to observe whether any effort is being made to clean up their act. If so then we'll see whether we want in. No effort is being made & it looks like Eng., Japan & some others may follow us out. My favorite paper??!! The Wash. Post came by for an interview—Cannon & Hoffman plus an editor. I don't think I gave them anything to run with.

[Photo session; received gifts, including "a magnificent painting of me"; met family with children suffering from Ushers syndrome; farewell for Jeane Kirkpatrick; videotapings. **Tuesday, April 2:** *visit from Prime Minister Turgut Özal of Turkey, noted, "We got along fine—better than I could get along with the P.M. of Greece"; addressed the leaders of Associated General Contractors; celebrated presidential advisor John McCloy's ninetieth birthday, president of West Germany and mayor of Berlin present; ambassadorial formalities; reception for Presidential Trust Fund; dinner with Republican Eagles.]*

Wednesday, April 3

[Met with Republican congressional leadership on the budget.]

Lord Carrington (NATO) came by. He's done a good job of tightening up the Alliance. He's concerned somewhat about our S.D.I. & I believe it's because of the possibility that without our nuclear shield the Soviets will pose a threat because of their superiority in Conventional weapons.

In N.S.C. we're putting together an idea for trying to frame our spending request to Congress for the Contras in Nicaragua in connection with peace proposals. I don't believe the Dem. House will vote us the money just on a straight up or down basis. Tip O'Neill & Bob Michel came in—they are talking a bi-partisan Cong. group to the Soviet U. Just wanted a last min. briefing & our blessing. Gave them both. Geo. S. stayed on after & told me of a possible meeting I could have with King Hussein whose son is graduating from Brown U. Then a meeting with activities on subject of missing children. I've asked them to come up with a partnership—everyone joining in & to bring in a plan for same within 3 weeks.

Thursday, April 4

Ambas. Mike Mansfield (Japan) is home on leave. He came by & is as distressed as I am over some of the Reps. & Sens. who are hysterically ranting about a trade war with Japan. The truth is we've made substantial gains in getting Japan to open her markets. A big factor in this is the best P.M. Japan has ever had—Yasu Nakasone. I hope he'll get some press on what he has to say—whoops I mean Ambas. Mansfield. The new Nat. Champs—Villanova's Basketball Team came to the Rose Garden. I wound up with a jacket.

Then Pres. Betancurt of Colombia arrived—we had a good meeting. He's done a great job of cracking down on the drug trade. We told him of our peace plan for Nicaragua. He's enthusiastic about it & is going to stop off in Nicaragua tonite on his way home to meet with Ortega.

At 3 P.M. I went before the press & outlined the plan to them. At 4 P.M. Don Regan went before them & told of our budget plan. Yes we finally worked it out with the Sen. Earlier Don had told me our Sens. were mad because I wouldn't agree to a freeze on the Soc. Security "Cola" [COLA]. My beef is that the Dems. should be made to propose this. If we do it they'll base the '86 campaign on us being the enemies of Soc. Security. I told Don I'd agree to a compromise—a 2% raise each yr. for 3 yrs. regardless of the level of inflation. They bought it.

[Visit by Jack Anderson to report on progress of Young Astronauts program; received Oscar in recognition of the administration's Arts program; met with premier of Bermuda; presented Cancer Foundation award; met with Nicholas Ashford of the Times *of London; photo session.]*

After a haircut I met briefly with 3 leaders of the Contras—Cruz, Robelo & Carello. All these men were leaders against Somoza & helped overthrow him. Then the Sandinistas double crossed them & stole the revolution for Cuba & the Soviets. They were here to be briefed on our peace proposal. Tip O'Neill & his cohorts are already bad-mouthing the idea. Indeed Tip sounds irrational.

Tomorrow on to Calif.

Friday, April 5–Sunday, April 14

[Coffee with senators and staff members to express thanks for work on budget compromise; flew to ranch in California. **Saturday, April 6:** *too foggy for outdoor activities.]*

On Sunday (Easter) the day was beautiful as was every day including Sun. the 14th—get away day.

Ron & Doria & C.Z. Wick came up for the afternoon & dinner on Easter. Bud McFarlane also spent part of day filling Ron in on the Soviet U. Ron has a mag. assignment to do a story on May Day in Moscow.

The rest of the vacation was as usual—beautiful weather, a horseback ride every morning & wood cutting & brush hauling with Barney & Dennis in the afternoons. There were some new additions—a day when Roy Miller & Fred Fielding came up with our Inc. Tax forms. [. . .] Then on 2 different days we interviewed applicants for Lee's job—hired one Courtney Trissler. He was Henry Mudd's foreman back when we had neighboring ranches in the Malibu mountains.

During most of the week—the press had a field day assailing me because I'd accepted Helmut Kohl's invitation to visit a German mil. cemetery during our visit to Bonn. I had turned down a not official invite from a W. German politician to visit Dachau in his district. All of this was portrayed as being willing to honor former Nazi's but trying to forget the Holocaust. Helmut had in mind observing the end of WW II anniversary as the end of hatred & the beginning of friendship & peace that has lasted 40 yrs.

I have repeatedly said we must never forget the Holocaust & remember it so it will never happen again. But some of our Jewish friends are now on the warpath. There is no way I'll back down & run for cover. However Helmut is upset & thinks this may become such an uproar it will color the whole ec. summit. He may change the program—we'll wait & see.

I still think we were right. Yes the German soldiers were the enemy & part of the whole Nazi hate era. But we won & we killed those soldiers. What is wrong with saying "let's never be enemies again?" Would Helmut be wrong if he visited Arlington Cemetery on one of his U.S. visits?

Today Walter Annenberg called—his news wire had picked up a Pravda story lacing into me about their Jewish slight in my not going to Dachau. I want to respond to Pravda & point out that today 40 yrs. ago after the Holocaust, the Soviets are the only ones officially practicing anti-Semitism.

We're back in Wash. & struggling to re-adjust after 8 wonderful ranch days.

Monday, April 15

This was not only 1st day back—it was Monday & a hectic schedule. First off—Pres. Monge of Costa Rica came in. He has generously flown up here to talk to members of Congress & lobby for our Nicaragua proposal. Then I met with Brezenski, Jeanne K. & Schlesinger who are supporting our plan & came in to tell the press so.

[Went to Ringling Bros. and Barnum & Bailey Circus, commented, "I hadn't seen a circus in years. I'm impressed—it really is the greatest show on earth."]

We had a Cabinet meeting on the budget & the Nicaragua plan. It was to bring everyone up to date & enlist them to lobby Congress. In the middle of everything a cable arrived from Helmut Kohl & Mike D. took off for Germany. Helmut may very well have solved our problem re the Holocaust. The invite I turned down about a visit to Dachau was a private thing. Helmut is making it official. He'll invite me to visit the camp as well as the cemetery. I can accept both now that it's official.

Back to the schedule—a meeting with Cong. Hamilton, Chmn. of the House Intelligence Committee. He feels we can't get our Nicaragua plan passed. Actually he's for it. He made a suggestion about a change. I don't know whether we can do it. Had a room full of lawyers including Chief Justice Burger for the signing of a proclamation making May 1—U.S.A. Law Day. Then a Mr. Wymbs came in & left a manuscript. He's chairman of the foundation for my boyhood home in Dixon. He's also writing a biography & wants me to check it for accuracy.

[Received letter signed by 146 representatives pledging support for stand against tax

increases; cocktail party for W.H. barber; reception for Citizens for the Republic and a reception for a Nicaraguan fund-raiser for refugees.]

Tuesday, April 16

A meeting with our G.O.P. Congressional Leadership. Subjects were the Nicaragua Peace Plan & the Budget. There was quite a feeling of unanimity.

Then a meeting with Pres. Diouf of Senegal—his 2nd visit. He's a 6'10" impressive man who is doing a good job & believes Pvt. enterprise is the answer to Africa's problems. After lunch with V.P., went to E.O.B. to speak to Conference on Religious Liberty. Prior to that we went at my German problem again. The press has the bit in their teeth & are stirring up as much trouble as they can. At the close of my speech I made a statement acknowledging that we had been confused about the Dachau suggestion—that it was part of the official itinerary & that I was going to visit the Bitberg German Cemetery & a Concentration Camp. By the nightfall the TV press was distorting that statement.

[Addressed key interest groups mainly about the budget; presented Congressional Gold Medal to Danny Thomas; dentist appointment; cocktail-party meeting with five Democratic senators on Nicaragua, noted, "I think they'll be helpful & they gave good suggestions about possible points of flexibility on our plan that could get Dem. votes." **Wednesday, April 17***: met with President Chadli Benjedid of Algeria, found areas of agreement; met with Shultz; then met with the House Intelligence Committee on Nicaragua peace plan, all agreed to act in a nonpartisan manner; videotapings; individual meetings with congressmen regarding Nicaragua; state dinner, entertainment by American Ballet Theatre.]*

Thursday, April 18

Another backbreaker. The press continues to chew away on the German trip & my supposed insensitivity in visiting a W.W. II Germany mil. cemetery in spite of the fact I'm going to visit a Concentration camp. They are really sucking blood & finding every person of Jewish faith they can who will denounce me.

Made a call to the Space Shuttle & talked to Sen. Jake Garn. Then a very interesting meeting, a group of European leaders such as Winston Churchill III plus former P.M. of Australia Malcolm Frazer. They have signed a petition to Congress asking that Congress support our Nicaragua plan.

Most of the day was taken up then with meetings with members of House & Sen.—both parties & some one on one sessions—all having to do with Nicaragua.

[Lunch with news editors from all over the country, commented, "They ask better Q's. than the White House Press Corps"; presented award to Teacher of the Year; received Harry Truman Award from the U.S. National Guard Association; photo session; entertained about thirty-five senators and pressed for budget plan.]

The evening TV news was again filled with my sinning against humanity by going thru (In May) with the visit to the German Mil. cemetery.

Friday, April 19

A brief signing ceremony opened the day then we got back to my "Dreyfus" case—the trip to a German cemetery. I told our people Don et al there was no way I could back away in the face of the criticism which grows more shrill as the press continues to clamor. Mike D. is back & said Kohl was going to phone me. Our Ambas. Art Burns met several hours with Kohl. Our people want me to suggest a Nat. German war memorial as a substitute for the mil. cemetery. I said only if it presented no problem for Kohl.

The call came while we were meeting. Helmut told me the Camp would be Bergen-Belsen not Dachau. Then he told me my remarks about the dead soldiers being the victims of Nazism as the Jews in the Holocaust were had been well received in Germany. He was emphatic that to cancel the cemetery now would be a disaster in his country & an insult to the German people. I told him I would not cancel.

Then we brought in Elie Wiesel—survivor of the Holocaust & several others who were on hand for the Jewish Heritage Week Ceremony in which I was presenting Elie with the Congressional Gold medal. I explained the situation to them & must say—made some gains even if later Elie in his prepared remarks implored me not to visit the cemetery. We've invited Elie to accompany me on the trip. He's said yes except that he won't be present at the cemetery.

Later in day met our Ambas. to the Soviet U. Art Hartman. He's a d--n good man. He confirms what I believe that Gorbachev will be as tough as any of their leaders. If he wasn't a confirmed ideologue he never would have been chosen by the Polit beaureu [Politburo].

While I was on the phone to Kohl—the V.P. was in the room with our gang hearing my end of the call. He wrote me this note.*

Re Kohl Phone Call

Mr. President,

> *I was very __proud__ of your stand.*
> *If I can help absorb some heat—*
> *send me into battle—It's not*
> *easy, but you are __right__!!*

George

---

* *The note from Vice President Bush is written on a White House notepad stationery and was taped into the diary by the president.*

[Signed proclamation on Victims of Crime; left for Camp David.]

Saturday, April 20–Sunday, April 21

Weather wonderful—in the 80's—swam both days & rode on Sat. Sun. an early return. Nancy went on to Calif.—just for overnight to see a house. That comes under the heading of looking ahead. Mermie here for dinner.

Monday, April 22

Got a report from Don & Bud of the weekend meetings of Repub. & Dem. Senators trying to find a consensus position on the Nicaragua vote—$14 mil.—negotiations—cease fire etc. We had the Dem. plan—it differed in a few important points from ours. Most important—it wanted us to negotiate with the Sandinista govt. We feel negotiations should be between the S's. & the Contras. At 10:30 we met with them. I listed all the points of their plan we could agree with then I had to leave to meet with Ec. Advisory Board. The Sens. & our people stayed at it for 6 hours—no agreement.

At noon the annual Vol. Action awards—lunch in the East Room. Back then to meetings with Congressmen & W. on Nicaragua. The meetings went well & I think I answered some of their worries. It's apparent though that the lack of support on the part of the people due to the drum beat of propaganda "a la Vietnam" is influencing some of them. Finally some desk time & then a reception for Repub. Congressional Leadership Council—supporters of the Repub. Congressional Campaign Committee.

The uproar about my trip to Germany & the Bitberg cemetery was cover stuff in Newsweek & Time. They just won't stop. Well I'm not going to cancel anything no matter how much the bastards scream.

Tuesday, April 23

[Met with House Republicans to press for support on Nicaragua; addressed national Association of Realtors; brief visit from new president of European Community, French writer Jacques Delois; ceremony for Peace Corps honorees; visit from Western painter Sherry Beadle; Arts and Humanities luncheon; meeting with Democratic congressmen; later received word that Nicaragua plan passed Senate, commented, "Now I'll sleep well"; cabinet meeting on tax plan, decided against dropping tax-deductible status of charitable contributions.]

Wednesday, April 24

Every day seems to begin with latest press—muck raking over whether I should or shouldn't go to the Bitberg cemetery in Germany. Well d--n their hides I think it is morally right to go & I'm going. I dropped by Nancy's get together—18 first ladies from that many countries even from far off Japan all brought together by the drug problem. Nancy's idea. Then back to some Congressional phone calls trying to get votes for Bob Michel's bill in the House to further our Nicaraguan peace plan. Tip has engineered a partisan campaign to hand me a defeat—never mind if it helps make another Cuba on the American mainland. Geo. Shultz came by to get my blessing on a speech he wants to make tomorrow on Nicaragua—calling a spade a spade. I gave it gladly.

In every spare moment I was working on my TV speech on the budget.

[Cabinet Council meeting on tax-reform plan; swearing-in ceremony for ambassador; delivered televised speech on budget; attended Senator Paul Laxalt's "Lamb Fry" after dinner; noted that the House voted down aid to Nicaraguan rebels, despite efforts of Representative Bob Michel (R-IL).]

Thursday, April 25

[Mrs. Reagan left for Atlanta on anti-drug campaign; met with a couple of senators on budget; ceremony recognizing youth volunteers; briefing on upcoming economic summit in Bonn; received award from National Troopers Coalition; interviewed by foreign press; visited by Asthma poster child; signed fair-housing bill; photo session; videotapings.]

I called Pres. of Honduras. After news of Congress vote last night his military intercepted ammunition headed for the Contras & made noises about driving them out of their camps in Honduras. The Pres. was outraged, swore we must all continue helping the Contras & said he would order the Gen. (Logig) to deliver the ammo. I told him we weren't licked yet.

Friday, April 26

My call to Pres. of Honduras worked—Bud reported the orders have gone out & the military is moving supplies to the Contras & has restored guns to one outfit they had disarmed.

[Recognized volunteer senior citizens group, RSVP, with actor Harry Morgan as honorary chair; met Navajo boy and his grandmother for making heroic rescue of fire victims.]

Then it was Pres. Chun of S. Korea for a meeting & lunch. Mrs. Chun & Nancy had some time together. We have pleasant memories of our visit to S. K. in November '83. Everything went well. A meeting with Geo. S. then a private meeting with 2 gentlemen—long time friends who are going to take over our blind trust.

Sen. Metzenbaum along with others of his kind such as Cranston got a non binding resolution passed asking Germany to let me out of the Bitberg cemetery visit. Unfortunately, some of our Repubs. went along. Well I don't want out. I think I'm doing what is morally right.

Saturday, April 27

A day of reading my eyes out—briefing materials for trip. The press is still chewing on the Bitberg visit. I'll jut keep on praying.

The White House Press Corps dinner—Mort Sahl entertains. He's funny as always. He also pledged his all out support for what I'm doing. I had some good material & was well received. Turned our clocks forward so it was a short night.

Sunday, April 28
More homework—a nice day & lunch on the Truman Balcony. Called Jerry Ford to thank him for his words re Bitberg. I'm worried about Nancy. She's uptight about the situation & nothing I say can wind her down. I'll pray about that too.

Monday, April 29
[Signing ceremony for announcement of National Partnership for Child Safety; appeared at National Chamber of Commerce convention; briefing on summit at lunch; television interview with foreign journalists; swearing-in for William "Bill" Brock as secretary of Labor; haircut.]
Dick Wirthlin did a weekend survey—my approval rating went up to 62%. Bitberg was approved 49 to 47 but he made the Q. one to lean toward negative—"do you think he should honor the Nazi's, etc."
Tomorrow's the day.

Tuesday, April 30
Now it's really getaway day. I'm scratching at drafts of some of the 14 speeches I'll be giving in Europe. We had a Cab. Council meeting—another on the tax simplification plan. We're making some progress. Then a short meeting with our 3 head negotiators who are home on recess from the Geneva arms talks. Nothing much to report—they said the Soviets in this 1st round seemed to be positioning themselves for some kind of propaganda efforts. Speculation is that it will be Gorbachev's speech at the U.N. in Sept. If so that means the next session beginning June 1st won't see much progress.

I hope I'm not being too optimistic but it seems there are a few signs that the Bitberg issue may be turning. Cap just came back from a ceremony installing the new military Vicar, a Catholic Priest or Bishop named Ryan. He told Cap I was a "Christian," a good man & doing the right thing about the visit to the cemetery.

Dave Murdock came by—he's laying off 1000's of textile workers because of the foreign competition. I don't know what we can do & still hold our position on free trade. We already have dozens & dozens of quotas in effect limiting imports. An interview with Hugh Sidey—he's a fair & common sense man.

[Made statement for the press and then left for West Germany; formal reception; settled in at Schloss Gymnich, a castle used as a guest house by the German government; strolling in garden; quiet dinner.]

Thursday, May 2

The day started with a working breakfast with all our staff then at 9:35 Nancy joined us. We helicoptered into Bonn & motored to Villa Hammerschmidt—home of Pres. Von Weizsaecker. Met by him & his wife plus a greeting party of our embassy people & German officials. It was a formal ceremony (the Presidents functions are largely ceremonial). He & I inspected the troops etc. A photo opp. for the press then Nancy left us for her trip to Rome—an audience with the Pope & a schedule that will keep her away 'til May 4.

The Pres. & I (plus staffs) had a half hour meeting then we were off to the Fed. Chancellery to meet Chancellor Helmut Kohl. He & I had a 45 minute private meeting. I assured him I was not upset by the press furor over my scheduled wreath laying at Bitberg military cemetery. He said I had won the heart of Germany by standing firm on this.

We went into a plenary session & got into discussions about the Summit for another 45 minutes. Then on to our Deputy Chief of Missions residence (they are a very nice couple). We had a working lunch preparing for my one on one visits. At 2:15 P.M. Nakasone (Japan) arrived. We had a half hour meeting discussing mainly trade matters. He's supportive of our desire to have a Gen. round of trade talks in early '86 to head off "protectionism."

Next visitor was Pres. Mitterrand of France. He's still pressing for a formal monetary conference to deal with exchange rates etc. We've had our ministers & a task force discussing & studying financial matters for 2 yrs. since the Williamsburg Summit. Their report is due in June. I proposed that we wait for that report & then see what further is needed. He's not satisfied with that & wants to tie a monetary meeting to any trade talks. I briefed him also on our Strategic Defense Initiative (S.D.I.) & sensed a reluctance on his part as to any participation in this.

Then it was back to Schloss Gymnich where I had a meeting with our solid ally Margaret Thatcher. This was a half hour—mainly spent in discussing the upcoming meeting.

Back to the helicopter & off to Schloss Augustusberg to meet Helmut—attend a reception—which was jammed—no reception line, just mix & mingle.

Back in a motorcade & off to Schloss Falkenlust for the Summit dinner meeting. During the course of dinner I told them how successful Nancy's anti-drug session with 17 other 1st ladies had been & by the time I finished they picked up on it & we voted to move together on a program of cooperation in dealing with the drug problem. Then it was back to Schloss Gymnich for the night.

Friday, May 3

The Summit really begins. Let me interject that in all our motoring the streets are lined with people clapping, waving, cheering—all I'm sure to let me know they don't agree with the continuing press sniping about the upcoming visit to Bitberg. This continued throughout our entire stay in Germany.

The meeting was going pretty well with consensus on most of the points that would wind up in the final statement. Then we came to the matter of an early '86 round of trade talks. Pres. Mitterrand expressed his own opposition to protectionism but absolutely refused to agree to an '86 meeting or re-opening of trade talks to further reduce or eliminate protectionist measures that presently exist. We're all guilty of some. His big hang up is the fact that France subsidizes it's agric. so they can compete in export trade at lower than world mkt prices. Couple that with his upcoming election in '86 & you have the story.

We tried everything, wrote & re-wrote the clause. The debate grew heated & then they took on the United States as being an interloper in European affairs, etc. The battle went on way past lunch hour. We finally settled for wording that bluntly said most of us felt there should be such a meeting & our ministers would meet in July to lay plans. Lunch came at about 2:30.

Our dinner that night was heads of state only. Instead of business we got into story telling. Brian Mulroney (Canada) started it & I got on with some & a good time was had by all—(all except Mitterrand).

Saturday, May 4

The usual working breakfast then back to the Chancellery for the plenary session. This was the windup. After lunch we went to the Bundestag for Helmut's reading of the statement. Our press treated the lack of a specific date in '86 for trade talks as a repudiation of me by the summit. First of all I had never sought a specific date—nor had anyone else. We simply asked for a meeting sometime in early '86 & all heads of state plus the Pres. of the European Community were for it.

Later I had a brief one on one with P.M. Craxi of Italy. He's very interested in the Middle East problem. I briefed him on where I thought we were on that.

Nancy arrived about 5:30 & that night we went to a dinner for heads of state, wives plus a few others at home of Pres. Von Weizsaecker. A pleasant evening.

Sunday, May 5

Dawns the day the world has been hearing about for weeks. By 9 A.M. we were on our way to Konrad Adenauers grave site with the Kohls—our wives put flowers on the grave. The press had only been given an hours notice on this. We didn't want them claiming we were doing it to soften the criticism on Bitberg.

From there we helicoptered to Bergen-Belsen Concentration Camp. This was an emotional experience. We went through the small museum with the enlarged photos of the horrors there. Then we walked past the mounds planted with Heather, each being a mass grave for 5000 or more of the people—largely Jews but also many Christians, a number of Catholic Priests & Gypsys who had been slaughtered there or who were just starved to death. Here I made the speech I hoped would refute the phony charges that had been made. I declared we must not forget & we must pledge, "never again."

Before the day was out there were reports that my talk had been effective. It was carried live on German TV & elsewhere in Europe.

Next stop later in the afternoon was Bitberg. Here the people were jamming the streets—most friendly but some demonstrators. We went to the cemetery & met Gen. Ridgeway—91 yrs. old—last surviving top W.W. II leader in Am. & Gen. Steinhoff, a German General who had been shot down in flames & whose face had been rebuilt by a American Army Dr. at wars end. Kohl & I & the Gen's. walked thru the tiny cemetery & then at a monument there the Gen's. placed wreaths. The German "taps" was played & then in a truly dramatic moment the 2 Generals clasped hands. There had been no leak to the press that the Generals would be there.

Then we motored to the Air Base where both German & American units are based. There were several thousand people—families of the military plus a number of citizens of Bitberg—the Mayor, City Council et al. The German mi. band played our Nat. Anthem—then the Am. band played theirs. My speech was sort of a sequel to the one at Belson. It was enthusiastically received & our people thought it turned the issue around. I felt very good. I was told later that during my speech—the 2 Generals sat—holding each other's hands. General Ridgeway & his wife returned to Bonn with us on A.F.1.

Back in Schloss Gymnich we got done up for the White Tie State dinner at Schloss Augustusberg. There was a small reception then into dinner. After dinner there was a half hour entertainment—Chamber music.

Well this was the day—everyone—well not everyone but much of the press had predicted would be a disaster. Dick Wirthlin did some before the day & after polling. Before the trip 49% said I should go to Bitberg, 47% said no—4% undecided. After the trip 59% said I should, 38% said no & 13% were undecided.

I always felt it was the morally right thing to do.

Monday, May 6

A 10 A.M. farewell ceremony complete with troops, national anthems & all at Villa Hammerschmidt with the Pres. & his wife. Then the Kohls joined us & we departed for Hambach castle. We made this trip in Marine One & landed at Nevstadt. There was some ceremony & then we made the drive up the mountain to the castle. Again there were lines of happy cheering people except for one little group of dissenters. When we went by they turned around, bent over & showed us their bare bottoms. There were about a dozen—boys & girls.

At the castle there were about 10,000 teenagers gathered. I'd been told that German youth were a little down—tired of being told of the Nazi horror & not sure there was a future for them. Accordingly I addressed them in an effort to show them the pride they should have in what their country has achieved in these 40 yrs. & how they have freedom & the chance to make their dreams come true. They really responded. I had a chance to meet a small group of them afterward. They want to participate in an exchange program with us. I'm going to help.

We helicoptered to our Ramstein Air Base—the C.O. of our air forces in Europe, Gen. Donnely, Jr. & other officers met us. We could only wave to the turnout of base personnel & families. It was goodbye to the Kohls—they were quite emotional. Helmut swore undying friendship. Then we boarded A.F.1 & on to Madrid. Had lunch on the way & landed at 4:15—Barajas Airport. King Juan Carlos & Queen Sofia (We have known them for some time) met us. More ceremony—review of troops, then a march by—the Nat. Anthems & of course 21 gun salute. Then we met the official party—Pres. Felipe Gonzalez & other officials. I've already met Gonzalez in Wash. The King & I rode in one car, Nancy & the Queen in another to Pardo Palace. Again the streets were lined with cheering people. There had been a hostile demonstration before we arrived. It was as if the people were trying to tell us not to feel upset by that. At Pardo another ceremony with troop review, guns & anthems. Then the King & Queen departed—Pardo was to be our residence. At 6 P.M. I had a meeting with Gonzalez who is both Pres. & P.M. We got along fine. I gave him the whole load on Nicaragua & think he'll be no pushover for Ortega who is visiting him after his visit to Moscow. By the time our meeting was over we were Felipe & Ron.

At 8:50 we motored to the Zarzuela Palace—the Royal home for dinner with their majesties & their daughters the Princesses Elena & Cristina. There was an exchange of gifts—one of his to me was a most interesting saddle. It will dress up our tack room.

They have refused to live in the Royal Palace preferring this smaller, home-like place. The evening was a family dinner & very pleasant although we did manage to get in some talk about Central Am. & Spain's place in NATO.

Tuesday, May 7

Nancy off with Queen Sofia for a lunch & a visit to an art school where Nancy was persuaded to join the students in dancing the Flamenco. She made every paper in the world & a lot of magazines.

I made a morning speech on free enterprise to about 350 business leaders—some Americans who head up Am. business in Spain. The King accompanied me. Then I was into a meeting with Gonzalez & members of the Cabinet & Parliament.

The day was topped off with a State Dinner—White tie again. The dinner was in the official Palace which is used for such occasions. All in all it was a good day & a fine relationship exists. The people in the streets were really friendly & enthused.

Wednesday, May 8

The King & Queen came by to take us to the airport. There was a farewell ceremony—review of troops—nat. anthems, 21 guns—the works. Then it was goodbye & we were off to Strasbourg to address the European Parliament. Let me just say, the monarchy in Spain like an anchor to windward is an important factor in keeping Spain on a democratic course & the King is dedicated to that purpose. He & I have a solid relationship & his friendship for the U.S. is sincere & lasting.

We arrived at the Strasbourg airport about 12:15—met by our Ambas. Van Galbreath plus a large delegation headed by Pierre Pflaimlin—Parliament Pres. The usual ceremonies—review of the French Troops, etc. Nancy was then taken for a tour of historic old Strasbourg. I went to a lunch with the greeters & members of Parliament, signed the Golden Book of Strasbourg with the Mayor & then on to the Parliament. I was aware that 38 members out of the 434 had voted that I shouldn't be allowed to speak, so was not surprised when I was greeted with something of a demonstration. I am the 1st Am. Pres. to ever address the European Parliament. The political coloration of the demonstrators was obvious. They reacted to any criticism of the Soviets—held up signs about Nicaragua, etc. I felt it necessary to direct a few comments their way which brought ovations from the majority. My theme was "Freedom works," & I recognized the near miracle that the Parliament represents.

At the beginning there was a breakdown in the teleprompter & I had a momentary problem picking up my place in the script. After a time the TP began working.

After the speech had a brief meeting with Pres. Ahrens of the Parliamentary Assembly, Sec. Gen. of the Council of Europe, Oreja & Chairman of the Council of Europe Comm. of Ministers Deputies.

Then Nancy joined us & it was off to Lisbon—a 3 hour flight across France, the Pyrenees & Spain. I was struck again at how much the Spanish & Portuguese countryside looks like Calif.

We were met by Pres. Eanes & wife. Motored into town (this was my 2nd meeting with the Pres.—he had been to Wash. on a State visit). We had a chance to visit on the drive—Nancy & his wife were in another car.

In town a huge, warm welcoming crowd & the usual military review—except this time part were horse cavalry. Then we went into the ancient Jeronimos Monastery for a wreath laying ceremony assisted by mi. cadets. A meeting with Lisbon's Mayor & his wife—signed guest book & exchanged gifts. Then to Belein Palace escorted by the cavalry. Another exchange of gifts. Pres. Eanes & our people had a plenary meeting. He held forth most of time about NATO & need for additional help in mil. spending. We have a solid partnership what with our bases on the Azores, etc.

Then we departed for the Queluz Palace & the luxury of dinner in bed & good night's sleep.

Thursday, May 9

A busy day. Briefing before leaving Queluz at 10 A.M. for Sao Bento & a meeting with P.M. Soares (we too have met previously in U.S.). We met privately for about 15 mins.—some photos, etc. then downstairs to a plenary. We covered a number of subjects with a great deal of agreement on all. He & I then went into garden & delivered statements to the press. Incidentally there had been some demonstrations about me & a lot of graffiti but I noticed it usually linked me with Soares. The dissidents didn't like either of us.

At noon he took me over to the Assembly where I met the officers of that body plus reps. of the 7 parliamentary groups of counsel of the Parliament. There was a gift exchange & a guest book signing. Then into the Assembly Chamber. Pres. Amaral made an opening speech & introduced me, at which point a group to my left—both physically & philosophically got up & walked out. The great majority was warm in its response to my remarks.

Nancy joined me & after a switch of cars at Queluz we drove to Sintra Palace. We were in a Ford Grenada because the limos couldn't get up to the palace because of the narrow road & short turns. It's an old palace started by the Moors on a steep mountain side. It was a 2 P.M. lunch but very pleasant. Back to Queluz for 10 mins. then a meeting with Dr. Lucas Pires—Pres. of the Center Social Dem. party—he is a conservative. The govt. is basically Socialist but is moderate regarding private enterprise.

We had a few hours rest then before we had to get into Black tie. First downstairs a press reception for all the assembled press, foreign & American. From there it was on to the St. Dinner at Ajuda Palace. We & the Eanes's did a receiving line before dinner. The dinner was held in a magnificent room—three long tables the length of the room & a speaker's table across the room at the head of the 3 tables. He & I did brief toast remarks then back to Queluz.

Friday, May 10—Going home day

I did a briefing & then an outdoor press conference in the garden. Then

Nancy & I & our party joined the Eanes's on the grounds at the Horse Pavilion for our exhibition of their Lusitanians Horses. This was Haute Ecole—dressage on the order of the Lipizzaner in Vienna & very good.

From there direct to the airport—a mil. review, this time of Marines. I was struck by the fact that I was a head taller than almost all the men in the ranks.

Ceremony over—it was farewells & onto A.F.1 & out over the Atlantic—a 7 hour, 40 min. trip to Wash. D.C. & a beautiful 80 degree day. A nice welcome home on the S. Lawn. We're here for overnight & then to Camp David.

[Saturday, May 11: radiocast; left for Camp David; ran news coverage of trip. **Sunday, May 12:** *ran more news coverage, commented, "The media is doing its best to suggest the trip was a failure."* **Monday, May 13:** *relaxed by pool; returned to W.H.; met with Regan on tax plan; haircut.]*

Tuesday, May 14

Still waking up on Europe time—early. An early meeting with a bipartisan group of Congress leaders—minus Tip O'Neill & Jim Wright. We reported on the trip which was a different story than they've been getting in the press who are determined to picture a very successful trip as a failure. Bad cess to them! Then an N.S.C. meeting to talk about stock piling strategic materials. We've had a study going which found the 1979 plan a shambles.

[Issues briefing lunch; cabinet meeting to discuss budget plan; cabinet affairs briefing on administration's tax-simplification plan; Maureen visiting. **Wednesday, May 15:** *woke up early; meeting on tax proposal, some progress made; met with trustees of Reagan Library and Museum regarding plans for buildings at Stanford University; commented, "I can't get used to the idea that mil's. of dollars are going to be spent to house my papers etc."; met with Congressman Bill Dickinson (R-AL) about a plan to present Defense Department affairs to the public; greeted women's NCAA basketball champions from Old Dominion; said farewell to reassigned military aide; received an award from Senior Citizens Council; met a Michigan county executive who changed to Republican Party; videotapings; went upstairs, noted, "Have to stay hidden—Nancy has some guests for tea."]*

Thursday, May 16

Bill Casey came in to explain about the Wash. Post story that broke this week about the March car bombing in Lebanon in March that killed so many people. The story was that the CIA had trained the bombers in anti-terrorist tactics & this was the result. CIA tried to tell them the story wasn't true but the Post went ahead with it. Well it wasn't true. [. . .]

Pres. Duarte of El Salvador came by—We had a good but short meeting. He's totally supportive of what we're doing in Nicaragua.

[Lunch with Vice President Bush; met with four oil men with counterproposal on suggested changes in energy taxation.]

An ec. policy meeting with the council. Much time spent on textiles & our protective quotas. We're getting a lot of heat that we must do something to curb imports of textiles. Frankly they are crying wolf & they aren't hurt that bad. They close a plant because it's old & no longer profitable—then blame the closing on imports. There are 28,000 mills in the U.S.

[Attended party reception and dinner, commented that it was the "biggest Repub. fund raiser in the history of the Party."]

Friday, May 17

Met with Jack Kemp—he wants the top rate in our tax proposal to be 30% rather than 35. We'd have to eliminate some pretty major deductions to make up for the revenue loss.

Then an NSPG meeting re Radio Marti. We've been ready with that station for some time. It is to broadcast to Cuba & finally bring some honest news to the Cuban people. Now some of Castros flunkies have gone public laughing at us because we haven't gone on the air. Our problem has been that Cuba has the ability to join not only Radio Marti but radio & TV stations halfway across our nation. We need the ability to retaliate & knock his TV & Radio off the air. Chairman Addabbo in the Congress has refused to approve the funds for the needed equipment even though Congress passed the legislation that established Radio Marti. Monday is Cubas freedom day. I've ordered us to start broadcasting on Monday May 20. If Castro retaliates we may have to shut down but make it temporary & demand that Congress move. After lunch I spoke to the Convention of Nat. Repub. Heritage Groups Council. I learned I was the 1st Pres. to even address their convention. They put 42 ethnic groups together to support Repubs. & me in the last election.

I met with Sec. Shultz. He's tired & wants to bow out before the summer is over. I told him I had never envisioned being here without him but didn't have the heart to lean on him if he really wants to go. I'm afraid he really wants to go.

Cong. Sam Hall—Dem. from Texas came by to say farewell. I've appointed him a judge in E. Texas.—Off to Camp David. We did a TV taping in the early evening—part of the NBC documentary airing in June.

[Saturday, May 18: took a two-mile hike; taping for NBC documentary. Sunday, May 19: weather clear; returned to W.H.]

Monday, May 20

Cuban Freedom day & today Radio Marti went on the air. Castro responded immediately by canceling the immigration agreement by which we were returning to him the criminals & mental patients he dumped on us in the Mariel boatlift & he was sending us pol. prisoners he's been holding. We'll have to wait

& see if he [. . .] that mil. watt jammer he has which could knock out commercial radio & TV in half our country.

[Launched program for summer jobs for teenagers; received mixed news in economic briefing; met with three congressmen with plan for better public relations in explaining administration projects; attended farewell party for Mike Deaver, leaving to work in private sector.]

Tuesday, May 21

Met with Repub. Leadership—House & Sen. I got a little ticked off & told them I was tired of foreign policy by a committee of 535. Our discussion was on the budget & aid to the Contras—I'm not sure we can't get the latter since Ortega went to Moscow.

Pres. (of Honduras) Suazo for a meeting & lunch in the W.H. It was a good meeting. They are not flinching from the threat of Nicaragua & they intend to help the Contras. I think we tightened the friendship.

[Photo session; final briefing on tax plan, called it "truly tax reform"; addressed the Council of the Americas; met with Ed Zorinsky (D-NE), considered him a staunch ally.]

Wednesday, May 22

[Photo session with Scottish Rites group; flew to Annapolis to address graduation at naval academy.]

Back to the W.H. for a late lunch then met with George S. re a meeting with Gorbachev. George will be seeing Gromyko before long—at Helsinki. I told him to suggest mid November here in Wash. & if they insist on a neutral locale—make it Geneva. We're going to offer Wash. with a commitment that a subsequent meeting would be in Moscow.

[Ambassadorial formalities; photo session; taped radio broadcast.]

Bud MacFarlane is up on Hill negotiating with Sen. on a bill by Nunn who wants to cap MX at 40 missiles & only 12 this year. We want 21 & 50 (we'll try for the 2nd 50 down the road). Bud thinks he has Nunn willing to go for 50 & possibly 18 this year. I okayed that.

Thursday, May 23

Always there is trouble—for hours yesterday Bud M. was on the hill trying to persuade Sam Nunn & a few of his fellow Demos.—namely Bob Byrd to change Sam's bill to limit production to 12 instead of 21 MX missiles in the coming year. No go! Finally this morning I phoned Sam & offered a compromise—he didn't say yes or no—only that he'd talk to Bob Byrd. As the day went on Bob dug in his heels. By this afternoon we got an agreement with Sam that flanked Bob in a way. Sam is usually with us on defense matters but on the MX his big gripe is the basing mode. He just doesn't want to put them in existing silos. Even our best mil. experts have said that's where they should go.

[Telephone conversations with school children.]

Dick Wirthlin's poll figures were interesting & holding up well—except for the Nicaragua issue. We have to do a job of education with the people—they just don't understand it. Lunch in the East Room to award a dozen Medals of Freedom—friends Jimmy Stewart & Frank Sinatra were among the recipients.

[Meeting with Senator Russell Long (D-LA) on employee-stock ownership aspect of tax reform; greeted National Arthritis Foundation poster girl; award-winning high school basketball player; Multiple Sclerosis Mother and Father of the Year; photo for Italian press; met with retiring ambassador to Australia; interviewed by Molly Dickinson, writing book with James Brady (later published as Thumbs Up: The Life and Courageous Comeback of White House Press Secretary Jim Brady*); met George Ward, president's first agent in Hollywood.]*

Friday, May 24

We had an N.S.C. meeting on the whole subject of Central Am. I've asked all the Cabinet members present—dept. heads, etc. to come back within the month with ideas on how we can more effectively help our Latin Am. friends upgrade their economies, create opportunities for entrepreneurship, etc. We are providing (our govt.) 61% of all the funding to Central Am. We need to find ways to get more in the line of ec. development for that money.

Signed a bill today that rectified one of Congress's major errors. We cancelled the requirement for detailed log keeping of use of Co. or business cars which were also privately used.

Had a meeting with Dith Pran whose story was told in the movie "The Killing Fields" & Dr. Haing Ngor, who played the part in the picture getting an Oscar for it—& his wife. They plead for help in restoring their country's freedom. I assured them we were & would do everything we could.

George S. came in—we have a problem with an Ambas. we appointed at the request of one of our Congressional leaders. He has become a pain in the Anus & George wants to fire him.

[Addressed convention of National Association of Manufacturers; videotapings. **Saturday, May 25:** *watched celebrity tennis tournament at W.H. (Shultz was among the players) to raise money for anti-drug efforts.]*

Sunday, May 26

A special dinner—the MacFarlane's, Shultz's & Regans with our special guests—Mr. & Mrs. Shevchenko—he is the defector who turned double agent & then came over to our side from the Soviets. It was an interesting evening & it goes without saying that our guest sang for his supper. He confirmed that the Soviet leaders do have an inferiority complex about their Super power standing—that they are a super power only in mil. power. That will be a factor in the arms control talks. He also affirmed that they do nurse a feeling that we may be a threat to them.

[**Monday, May 27:** *laid Memorial Day wreath at Arlington National Cemetery; flew to Orlando, Florida, for performance by marching bands; flew to Miami for party fund-raiser; returned to W.H.* **Tuesday, May 28:** *mentioned meeting various children in Orlando the previous day; received report of kidnapping in Beirut; cabinet meeting on tax reform; delivered televised speech on tax-reform proposal; noted that calls to W.H. afterward were favorable.*]

Wednesday, May 29

Brian Crozier—England came by. I've corresponded with him & read his writings frequently but this was the 1st meeting. He made a suggestion—we should not allow the Soviets to have so many more agents in our country than we do in theirs. I know our estimate is around 4000 here.

This was King Hussein (Jordan) day. He's doing a great & courageous thing in his efforts to bring about peace in the Middle East. We are working with him & intend to carry on. We'll face a battle in Congress but he needs "arms" & we should provide them. He's threatened by Syria because of his efforts to bring about peace with Israel. Syria is fully armed by the Soviets. Our problem is the need for the P.L.O. to participate in peace talks but their reluctance to agree that Israel has a right to exist as a Nation. Meetings went well. Back at the office a brief meeting with the widow of our agent Camerena who was murdered in line of duty in Mexico. She's a lovely little lady, without bitterness but certainly with grief. Her 3 sons weren't with her. Geo. S. came by [. . .]. We've agreed.

[*Addressed group of citizens on tax reform.*]

Cleaned up desk & went upstairs. Had to stay under cover. Nancy has the Queen for tea.

[**Thursday, May 30:** *flew to Williamsburg, Virginia, for speech on tax reform; flew to Oshkosh, Wisconsin, for same purpose, noted it was the "home of Oshkosh B'Gosh overalls. I wore them as a kid"; flew back to Washington for dinner in honor of Clare Boothe Luce.* **Friday, May 31:** *met with Madame Prime Minister Milka Planinc of Yugoslavia; visit from Senator Howard Baker (R-TN), on his way to Japan; flew to Malvern, Pennsylvania, to deliver speech on tax reform; met with entrepreneurs; helicoptered directly to Camp David.*]

Saturday, June 1

My radio broadcast was on the tax reform. Then in the afternoon—a beautiful day—we had a horseback ride followed by a swim. I phoned a Mrs. Gerney—woman who was trapped under an overturned derrick in N.Y. (her legs crushed) for 6 hours. She's captured everyone's hearts with her courage & faith & I told her that when I reached her in Belleview. This weekend has us under a possible problem. Kuwait, it is rumored, may execute several terrorists convicted of murders in the embassy bombings in Kuwait. There is a rumor that the terrorists would execute some of our Americans they have kidnapped in Lebanon. If all that happens I'm in favor of an air strike against a target in Lebanon that we know is a terrorist base.

[Sunday, June 2: *relaxed by pool, swam; returned to W.H.; reception and dinner for Ford's Theatre.*]

Monday, June 3

An NSC meeting dealt with what to do about the Salt II un-ratified treaty which expires in Dec. So far the Soviets & us have operated on an agreement that even though the treaty had not been officially adopted we'd both conduct our arms policies within the treaties restraints. We've done so but they have cheated a number of times. Now we are coming to a time when we'll have to dismantle Poseidon subs as we add Tridents to our Navy. Question—do we do that even though they are cheating or do we announce that we'll no longer abide by the agreement? I've had 5 options presented to me & now it's up to me & frankly I don't have the answer yet. Dr. David Owens, former labor foreign minister in Eng.—now head of a new party that's coming on like gang busters in old "blighty" was last on my schedule this afternoon. He put in a pitch for continuing to observe the restraint. He brought up the subject without knowing where we are or that we've been meeting on the subject. He voiced one of our concerns—what do our allies in NATO want us to do. Well the hot potato is mine to handle.

[*Issues-briefing lunch; interviewed by columnist James J. Kilpatrick.*]

A photo session for a new official photo—my very unfavorite thing to do.

Tuesday, June 4

[*Meeting with GOP congressional leadership to report on tax reform, Nicaragua, and chemical weapons.*]

Sen. Jake Garn brought the Shuttle Crew he flew with by for a picture complete with wives & one husband. Cap came in re the resignation of Gen. Vessey as Chmn. of the Joint Chiefs. I'd named Admiral Watkins to replace him but Cap says he prefers to remain as Chief of Naval Operations. We have another Admiral I may name.

[*Attended Domestic Policy Council meeting, chaired by Attorney General Edwin Meese with reports on further management reforms.*]

Then I dropped by a meeting of several CEO's of Corp. Am. who are here to plug our tax reform before Cong. An N.S.C. briefing on what to do about the Salt II treaty provisions we & the Soviets are pledged to observe but on which they are cheating. Come Sept. under our agreement we'd have to dismantle a

Poseidon missile Submarine. I was presented 5 options with no consensus on any of them. Our allies want us to continue observing Salt II. Our friends here at home think we look silly doing that in the face of Soviet cheating. I have to have an answer before Monday.

[Dinner party at the home of columnist George Will. **Wednesday, June 5:** *flew to Oklahoma City to speak on tax reform and attended party fund-raiser; flew to Atlanta, with speech on taxes and fund-raiser.]*

Thursday, June 6

[Visited high school that turned itself around in ten years, called it "an exhilarating experience."]

Then off to Alabama to do a fundraiser for Sen. Denton. This too set a new record in $ raised in the state. This was a lunch for about 4000. Jerry returned to Wash. with us so he could be on hand to vote on the aid to Contras bill. On the ride back I made my decision on the Salt II matter. We will continue to practice restraint on the building of Nuclear weapons. That restraint will keep us generally within the frame work of the Salt II but only commensurate with the Soviets observance of the Salt II restraints & for only as long as the Soviets abide by Salt II restraints.

Friday, June 7

More talk about Salt II & legislative strategy re the situation in Nicaragua. We carried the Sen. on a bill to provide $38 mil. in help to the Contras—non lethal. We headed off attempted amendments from our friends among others which would have endangered passage.

Called Don Devine who withdrew his name from the Sen. confirmation process. This was another lynching. He did just what I wanted him to do in his appointed position & he'll be greatly missed. He saved the govt. $6 Bil.—something his Senate critics have never done.*

[Had lunch with Vice President Bush; appeared before group of editors; left for Camp David. **Saturday, June 8–Sunday, June 9:** *relaxed by pool; received NSC intelligence reports on upcoming visit by Prime Minister Rajiv Gandhi of India and position of allies on SALT II; returned to W.H. and watched a classified film on Prime Minister Gandhi, commented, "These films are good preparation for his visit. They give you a sense of having met him before."]*

Monday, June 10

Pres. Barletta of Panama in for a short visit. He needs Ec. help & I don't

---

* *Donald Devine had been nominated for a second term as director of the Office of Management and Budget. He was criticized in Congress for the extent of his cuts in the federal workforce.*

know how much leeway we have. He's supportive of our position on Nicaragua & believes we can't use a carrot unless we have a stick to back it up. Today was the day we told Congress & the world what we intended doing about Salt II. Apparently my decision was right—at least I'm being called a Statesman by both the left & right.

This morning an unusual & exhilarating experience. I shook hands with & addressed some 250 local & state office holders who in recent months have switched from the Dem. Party to Repub. Bulk are from the South. Geo. Shultz came in—just back from the NATO Ministers meeting. It went well. George saw Helene V.D. in London & told her our feeling that having married a native Austrian she should resign as our Ambas. It hit her pretty hard as I was sure it would but she took it well. She's been a darn good Ambas. & is very well thought of by the Austrians. I don't think this will take effect until the end of the year.

[Greeted NBA champion L.A. Lakers; interviewed by Reader's Digest.*]*

Tuesday, June 11

Had a group of Dem. & Repub. Congressmen in the Roosevelt Room for a session on why they should support the bi-partisan bill before the House to give aid to the Nicaraguan freedom fighters. I think we did some good.

Lunch was with 5 top space scientists. It was fascinating. Space truly is the last frontier & some of the developments there in astronomy etc. are like science fiction except they are real. I learned that our shuttle capacity is such we could orbit 300 people.

Later in my office one of the guests Dr. Edward Teller reported on where we are on our Defense research for a way to halt nuclear missiles. The bad news is that our Congressional advocates of lower defense spending are cutting our research funds at a critical moment that will be very hurtful to the program.

We have house guests, Nancy's brother & wife, Betsy Bloomingdale & Maureen—a very pleasant dinner. They are all here for tomorrow nights State Dinner—P.M. Gandhi of India.

Wednesday, June 12

Today was supposed to be a rainy one & I've been praying since last night. So this morning at 10 A.M. the outdoor mil. ceremony & welcome to the P.M. of India took place in the sunshine on the S. Lawn. Less than an hour after the ceremony, it rained.

In our meetings, I think we got along fine & that was his purpose in being here. He made it plain to me that while India wants to maintain the friendly relations over the years with the Soviet U., at the same time however he says India does not want the Soviet U. to have a foothold anywhere in S. Asia.

Later this afternoon I met with the Sen. Steering Committee. This is a solid group of conservative Sens. who generally support the Admin. We talked strategy of getting things like budget cuts, the tax reform etc. through the Cong. After this meeting I had an emergency meeting—NSC & Don Regan. We can't get the arms package for King Hussein (Jordan) thru Congress. The Am. Jewish lobby which consistently opposes arms to Arab states. In this case though Hussein is our greatest hope for an Arab-Israeli peace. I have to ask Hussein to let me delay our approach to Cong. about the arms. I can however on my own slip him about 60 Stinger missiles. Good news—Congress passed our aid to Contras bill.

[State dinner for Indian prime minister Rajiv Gandhi. **Thursday, June 13:** *flew to New Jersey for speech on tax reform; returned to W.H.; photo session; interviewed by James Buckley for Radio Free Europe; ceremony honoring General James Doolittle; videotapings.* **Friday, June 14:** *awakened with news of plane hijacked in Middle East, mostly Americans onboard; gave awards to corporations for volunteerism; received positive poll report; went to Fort McHenry with Mrs. Reagan for Flag Day observance, noted, "At last minute my actor's instinct told me to skip the part of an otherwise patriotic speech that dealt with tax reform. I was right"; went to Camp David and continued to receive reports through the night on the hijacking.* **Saturday, June 15:** *following hijacking situation closely.* **Sunday, June 16:** *decided to cancel scheduled appearance and returned to W.H.]*

Monday, June 17
Day started with what's getting to be my regular early call from Bud. Nobbi Barri has taken up the hostages from the hi-jackers & has them someplace in W. Beirut. His price is release of the 760 Shiites held by Israel. Israel is probably saying they will—but the U.S. at the highest level of govt. must ask them to do it. This of course means that we,—not they, would be violating our policy of not negotiating with terrorists. To do so of course—negotiate with terrorists—is to encourage more terrorism. I suggested that if Israel said to Barri—we were going to release these detainees anyway, we'll expedite it if you let the Americans go—no one would be giving in to terrorists. After all Barri took the hostages away from the hijackers.

I just told Bud "why don't we ask Israel by saying you kidnapped & are holding 760 hostages—we ask you to release them in order to free Barri's hostages." That way we wouldn't be dealing with terrorists—just asking that hostages be freed by both sides.

[Meeting with Economic Policy Council on our trade disputes with the European Commission; met with bipartisan congressional group on need to upgrade chemical warfare stock with binary weapons; announced appointment of David Packard to chair commission to study Department of Defense, commented, "It's the only way I can see to let the people really know how much we've improved things."]

Then a little practice for tomorrow night's press conference. I'm lonesome—Nancy is in N.Y. for tonite & tomorrow nite.

Tuesday, June 18

Well there were good reasons when we talked it over why some of my ideas about Berri & the Israelis wouldn't work. Today, however, Berri did release 3 of the hostages. One was a Greek citizen with an American girlfriend & the 3rd an 18 yr. old American.

[Meetings with CEO of Dart Kraft; with leader of inner-city organization and with family group, all pledged support for tax reform.]

Then it was time to get ready for our sessions with Pres. Borguiba of Tunisia. They were good sessions & he is really a friend of America. Tunisia is the only Moslem or Arab country that practices Monogamy & gives women equal rights. The Pres. is also a declared enemy of Quadaffi. He is 85 yrs. old, in bad health [. . .].

[Staged press conference, considered it a success. **Wednesday, June 19:** *received view of pilot on hijacked plane; photo session with Donald Peterson, head of Ford; flew to Indiana for appearances on tax reform; returned to W.H.; saw NASA film on space shuttle.]*

Thursday, June 20

Awakened by a 6 A.M. phone call. In San Salvador guerillas opened fire on an outdoor cafe killing 15 & wounding 13. Of the 15, 2 were Am. business men & 4 off duty Marines from our embassy guard. They were in civilian clothes & unarmed. We called a Security Planning Session to take up this & our hostages in Beirut. We finally settled on availing ourselves of the generous proposal by Algeria to ask that the hostages be turned over to them—a face saver for Nabbih Berri. Algeria wants us to find out what Israel would do with the 766 Shiites if there were no hostages left in Lebanon. We would then give the answer to that question to the Algerians who would use it in dealing with Nabbih. In the meantime the Israelis are not being helpful. They have gone public with the statement that they would release their prisoners if we asked them to. Well we can't do that because then we would be rewarding the terrorists & encouraging more terrorism.

[Greeted winners of high school awards.]

Another happy time in the Rose Garden—presenting the Medal of Freedom to Mother Teresa.

[Photo session with president of international committee of the Red Cross; legislator from Arizona; Ray Charles, with representatives of disability organization; presented commendation to Willard Scott for the Today *show for encouraging volunteerism.* **Friday, June 21:** *flew to Dallas; met families of hijacking hostages; noted, "We are proceeding with a plan to have the Algerians intercede & ask that the hostages be turned over to them. In turn they want to be able to tell Berri that once the hostages are free the Israelis plan to continue*

returning the Shiites they hold. We were able to get a statement approximating that F.M.
Rabin (Israel) by going public with a statement that the U.S. should ask them to release the
Shiites loused things up by establishing a linkage we insist does not exist"; speech to Lions
Club convention; flew home and on to Camp David.]

Saturday, June 22–Sunday, June 23

On Sat. we dressed up & flew back to Andrews to meet the families of our 4 Marines who were murdered in El Salvador. This was an emotional trial—meeting their families but I believe the impressive Marine ceremony had to ease some of the pain of those whose loved ones were not combat casualties but victims of a vicious massacre.

[Received report of Air India air explosion; mentioned that on Saturday, Vice President Bush, National Security Advisor McFarlane, and Assistant Treasury Secretary Craig Roberts lunched at Camp David and talked about the hijacking; Shultz on his way to speak to allies about anti-terrorism measures.]

Monday, June 24

Well my heart is broken—we all agreed that in view of the hostage situation we should cancel our trip to the ranch. I can do everything there I could do here but the perception of me vacationing while our citizens are held in durance vile is something I can't afford—so no trip.

Met with several Governors Repub. & Dem. regarding tax reform. Most were supportive. Gov Babbit of Arizona had a couple of questions—I think we answered them & that he'll go along. After an issues briefing lunch we called an NSPG meeting re the Beirut situation. Bill Casey feels we must come up with a fig leaf for Berri or releasing the hostages would cause his assassination by the fanatics. Bill suggested a spokesman approaching him about offering Israel an assurance of safety in S. Lebanon for which they would free the 700+ Shiites they are holding. At the same time I've urged that we approach Assad of Syria to go to Berri & tell him he can be a hero by releasing our people or he can be stubborn & we will begin some actions such as closing down the Beirut airport, closing Lebanon's harbors etc. until he releases our people. This is all being staffed out now. I can agree to the need for a fig leaf but there also has to be a threat of action (non-military).

A meeting with Armand Hammer re the Bethesda research on Cancer which appears to possibly be a real breakthrough. He recommends more money to enlarge the trials—more Cancer victims treated, etc. He also has had a strictly 1 on 1 with Gorbachev & feels I should go to Moscow for a meeting. He's convinced "Gorby" is a different type than past Soviet leaders & that we can get along. I'm too cynical to believe that.

[Photo session with champion gymnasts from the University of Utah.]

Tonight Nancy & I went to Teddy Kennedy's home for a big fundraiser in a tent—purpose—money for the J.F.K. Presidential Library. John & Caroline had invited me some time ago. The whole family was most warm in their thanks. I wonder if Teddy will still be able to blast me as he has?

Tuesday, June 25

Met with Ambas. His Royal Highness Prince Bandar of Saudi Arabia. His uncle, King Fahd had written me a letter about a message that someone had given him that his purchase of mil. weapons from us was contingent on his publicly supporting King Hussein's middle east peace effort. King Fahd was understandably upset. I told the Prince to tell him there was no such linkage. There is however a plea from us; we have a better chance of getting Congressional approval of the sale—which isn't easy, if we can point to the King's participation in the peace effort. Then a meeting with Fred Fielding about another problem in Dixon Il. Bill Thompson one of the movers & shakers in the restoration of our boyhood home is off on his own as if he is an appointed ambas. for me. He reproduces autographed photos I've given him & his wife Jean & sells them etc. His wife is upset by what seems to be his obsession with me. So am I.

[Lunch with congressional leadership of both Houses.]

Then an NSPG meeting on Beirut—I've called for action now—beginning with the closing of the Beirut airport which would cut off a lot of revenue for Berri's Arnal militia.

[Photo session with golfer Andy North.]

Wednesday, June 26

Dropped in on Brkfst. downstairs for House Ways & Means Comm. Repubs. Another pitch for tax reform—a few had a sticking point re non-deduction of local & state taxes. We gave them some hard to rebut figures. I think we made some points. Then NSC & newest hostage development. A light at the end of the tunnel. Berri (Assam's doing) has offered to let the hostages move to a Western Co. Embassy in Beirut or Syria—supposedly to remain until Israel frees the Shiites it's holding. We prefer Syria & I have no intentions of letting them be held there. The Israelis are already planning to begin returning their prisoners. We want no linkage between what they're doing & the release of the hostages. Berri released one today because of a heart condition. We are really optimistic. I've just learned that Berri owns a couple of mrkts & some oil stations here in the U.S. We might consider that a pressure point. A meeting in the P.M. with Repub. members of committees (Sen.) who confirm our appointees. We've had trouble with some of them going over to the Dem. side & mistreating our candidates. I put in a pitch for sticking together & showing that Repubs. have the capability of governing.

In East room a pleasant gathering—141 finalists in the competition to choose the Am. Teacher who would be 1st teacher to go into space in the shuttle come January. Back to my desk for a load of mail & then home. Maureen is here but leaving tomorrow. Nancy is in L.A. after a visit with her mother in Phoenix. As usual this place feels empty as h--l.

[*Thursday, June* 27: *word of slight movement on hostage situation; report that "Quadafi talking to Iran & Syria about a joint terrorist war against us"; spoke to assemblage of local officials on tax reform; meeting with Senator Alan Dixon (D-IL) with plan for tax amnesty; received commemorative edition of* Reader's Digest; *greeted race driver and supporter Richard Petty; gave medal to Camp David military commander.*]

Friday, June 28

Yesterday we lost in the Judiciary Committee. Brad Reynolds nomination (by me) to be the No. 3 man at Justice was rejected. They even refused to pass it out to the floor with a no pass recommendation because of their fear the whole Sen. would do what they were unwilling to do—approve him. & they couldn't have done what they did without the help of 2 Repubs. Sens. Spector & Mathias. Well there are 2 Sens. I won't have to help campaign.

[*Flew to Chicago; visited school with local officials; met families of hostages and of kidnap victim; speech on tax reform; returned to Washington; National Security Planning Group (NSPG) meeting; received indication that hostages might be moved to Damascus the following day; noted, "We also launched a plan to strike by air a guerilla base connected with the murderers of our 4 Marines."*]

Saturday, June 29

We would be riding at the ranch if it were not for the hostage situation. We learned the hostages minus 4 had been moved by Red Cross bus to a bldg. a mile S.E. of the Beirut airport. I was ready with a statement for the press to be read at 9 A.M. Having breakfast in bed we turned on TV—there were our hostages still in Beirut not in Damascus—apparently neither Mr. Barri or Assam could spring the missing 4 from the bastardly Hisbollah. The hours passed & it became apparent there would be no further movement today.

Now we pin our hopes on tomorrow. Cap & Don called & I went down to the Oval office where I'd be doing my radio broadcast. [. . .]

[*Watched* Rambo: First Blood Part II *(1985) with guests.*]

Sunday, June 30

No early calls—but later in the morning word came that the hostages were gong to leave in a Red Cross motorcade for Damascus. It was a long ride—we then were told that celebrations in small villages along their route was delaying them.

About a quarter to 3 our time they arrived at the Sheraton hotel in Damascus.

[Dinner at the Shultz home.]

Monday, July 1

Awoke with knowledge our people were in Wiesbaden at our air base there. As far as we know now they & some of their families who flew to Germany to meet them will arrive here tomorrow afternoon. Nancy & I plan to meet their plane. An N.S.P.G. meeting to plan strategy now. We are limited because the 7 kidnap victims are still being held. I phoned Assam to thank him & to call upon him to work for their release. He's the only one who has any possible influence on the fanatics who did the kidnapping. He got a little feisty & suggested I was threatening to attack Lebanon. I told him nothing of the kind but we were going to do everything we could to bring the murderers of our young man to justice.

A full Cabinet meeting to report on the whole episode. Upstairs early—with Congress gone these are an easy few days.

Tuesday, July 2

NSC briefing had to do with Gorbachev's latest movement, the naming of Gromyko as Pres. of the U.S.S.R. It is a ceremonial job & possibly it was given primarily to get a new face in the Foreign Ministry.

[Met with representatives of civic and business groups concerned with continuing spending cuts; greeted Disabled Veteran of the Year; also president of Rotary International, telephoned YMCA meeting.]

Then Nancy & I went to Arlington Cemetery to put flowers on the grave of Robbie Stethem—the hostage murder victim. His sister was there. Then we helicoptered to Andrews Air Base to see the returning hostages & their families. It was nice homecoming ceremony & a heartwarming one.

Wednesday, July 3

We're all agreed the new Soviet Foreign Minister is there to hold the fort for Gorbachev. We also decided—now that the meeting with him has been announced—that we should do nothing to raise public expectations. I said we must paint with a broad brush & not give the press specifics as to our agenda.

Then we had a frustrating N.S.P.G. meeting re the 7 kidnap victims & the matter of Lebanon generally. Some feel we must retaliate. I feel to do so would definitely risk the lives of the 7. We are going to proceed to enlist other nations in closing down the Beirut airport. We are also going to maintain a posture that if another hijack takes place & the plane is headed for Beirut landing—we'll get there 1st & bomb h--l out of the runway. We are also discussing a plan to take care of the Libya situation—including bombing some known terrorist training centers.

We know the identity of the 2 hijackers who murdered Robbie Stethem. The problem is how do we get to them for trial in the U.S. All in all it's d--n frustrating even though we are overjoyed at our success in getting the hostages back.

That's about it for the day.

Thursday, July 4

[Lunch on the balcony, saw people celebrating the holiday on the mall; cleaned out desk; took call from Robbie Stethem's father; dinner party to watch fireworks included, as noted, "the Regans, Wicks, MacFarlanes, the George Wills, Oatsie, Bill Webster, Pam Wick & her boyfriend." **Friday, July 5:** *left for Camp David.* **Saturday, July 6:** *rode; contacted stewardess from hijacked plane.* **Sunday, July 7:** *swam and relaxed by the pool; returned to W.H.* **Monday, July 8:** *spoke to American Bar Association convention about using international law to combat terrorism; had lunch with Regan, McFarlane, and Vice President Bush; budget briefing; received new poll numbers, job rating up since hijacking crisis.]*

Tuesday, July 9

Met with Repub. Cong. leadership—subjects bud. deficit, tax plan & legislative agenda. I made a pitch that we heal our difs. & prove that Repubs. are able to govern. Another regional press luncheon—took Q's. I'm always impressed with the difs. between these press & media people from outside the beltway & the W.H. press corps. These people ask Q's. that are legitimate requests for news not traps.

[Signed bill on coins for Statue of Liberty fund-raising; videotapings; met with leaders of both parties on budget impasse, commented, "Made some headway but Tip was his usual pol. self."]

Wednesday, July 10

Got word my brother is home after an operation for Cancer of the Colon & gall bladder out.

[Stopped at breakfast for House Ways and Means Democrats, regarding tax reform; met with Vernon "Dick" Walters, the new U.S ambassador to the UN.]

Then a meeting with Bud. Conf. Committee. Things got a little feisty & I sounded off on Govt. doing a h--l of a lot of things it had no business doing & gave examples. Meeting ran long then it was over to a W.H. lunch with U.N. people & women Maureen is taking to Nairobi for U.N. Women's Conf.

*[***Thursday, July 11:*** *Cabinet Council meeting on Senator Alan Simpson's (R-WY) immigration bill, intended to support it with addition of certain amendments; lunch with*

Regan to block out schedule for coming months [. . .]; photo sessions; noted that Justice Department issued warrants for three hijackers and the murderers of Robert Stetham; noted procedure scheduled for the following day for removal of a colon polyp.]

Friday, July 12

[*Mentioned previous day's meeting with Senators Strom Thurmond (R-SC), Paul Laxalt (R-NV), and Orrin Hatch (R-UT) about a balanced-budget amendment; received good economic report; had cake in honor of McFarlane's birthday; noted Senator Grassley (D-IA) was leveling attacks on farm and defense issues.*]

I've had 80 ounces of Golightly fluid since 9 A.M.—it's now noon & I am drained.

Fri. afternoon & on to Bethesda Naval Hospital. We were supposed to go on to Camp D. Saturday morning but plans changed. They had removed the small polyp from my colon & examined the entire colon. Far up in the colon they found a large flat polyp—the kind that with time can become cancerous. So we held a strategy meeting. Did I go home & then have to go through the whole preparatory business again or do I go back in & get it over with. That of course was the decision we made.

I left the world—Sodium Pentothal, & awoke hours later somewhat confused. I had an incision that ran up past my naval to my chest. I was laced with tubes & very much a patient in for a stay.

Saturday, July 13–Monday, July 15

Nancy decided to decorate the place & brought pictures from home. They did make things better. But I was in a lot of discomfort which continued over through Sun. July 14. The polyp was not cancerous but they are checking around the Lymph nodes & we won't know about them until Mon. Sun. night was a bad one—I didn't sleep well.

Mon. July 15—Awoke to the usual things I'm supposed to do to keep my lungs clear etc. Nancy came with more pictures & brought me up to date on messages, flowers etc. This morning though I went back to sleep & did pretty well until noon. It turns out there were Cancer cells in the tissue they're checking—but they swear they got it all. It means however, I'll have to have annual checks for the next 5 years.

Tuesday, July 16

Mon. night was miserable—I kept waking up & felt I'd had no good sleep at all. Did some walking around. Mon. nite was better & Tues. the 16th was a good day. Don Regan comes each day & we do what chores are necessary. Nancy brought more pictures. The walls are now pretty well covered with large framed, colored photos of us at Camp D., the Ranch etc. They really help. I've taken some short walks down the hall. Nancy visited the children's ward & gave them some balloons that were sent & flowers. The children responded with home made cards. Nancy visited one 13 yr. old boy & his heartbeat went from 70 to 130.

Tues. night was the best for sleeping—all the way thru til 5 A.M.

Wednesday, July 17

A big day—they took the tube that ran through my nose & into my stomach out. It was the suction pump that kept my insides clean. I continue to have some bodily functions & I now only have one tube in my arm through which I'm fed. Nancy wasn't here—she's out on a carrier talking about drugs to the sailors. Geo. Bush, Don Regan, Larry S. & Craig F. came & we had a pretty good meeting on the budget. Then miracle of miracles I had my 1st food by mouth. Only a couple of spoonfuls of broth, Jello, some water & ½ cup of tea. I'm waiting to see what happens. By evening I repeated that menu with good results. The Doctor says maybe he'll take the feeding tube out tomorrow.

Some strange soundings are coming from the Iranians. Bud M. will be here tomorrow to talk about it. It could be a breakthrough on getting our 7 kidnap victims back. Evidently the Iranian economy is disintegrating fast under the strain of war.

Thursday, July 18

A good night's sleep until about 5:15 A.M. when I had a bathroom call. What a morning—the Dr. (Oller) took the metal clips out of my incision— what an improvement that made. Then I had a light breakfast & he told me he'd take the feeding tube out before lunch & he did. Such a feeling of freedom—no more tubes harnessing me to the machinery.

Nancy & I had lunch. I find I can only eat a few mouthfuls. A lot of cards & messages to look at. Then around 4:30 P.M. a wave out the window to the press & down to X-ray.

Bud came by—it seems 2 members of the Iranian govt. want to establish talks with us. I'm sending Bud to meet with them in a neutral country. Gorbachev has passed the word he'd like to establish a private channel of communications. We tried to get such a thing with his predecessors & couldn't make it. I gave the word to proceed.

I watched a TV Press roundtable on their handling of my "illness." I detected an effort on the part of some—particularly Helen Thomas to use the term "The President has Cancer." My Dr.'s said use of the present tense is a misstatement. "The President had Cancer—it has been removed."

Friday, July 19

Woke up to a breakfast of Juice, Fruit, Poached egg on toast, Bacon & toast

& honey. I feel 1000% better. Good meetings with Don R.—Bud M. & then George S. joined us to report on his literally round the world trip. He says our Assean allies are much more aware of the Soviet threat than they were a short time ago. Our relationship with Australia is solid as a rock. The 2nd term of the arms talks in Geneva has ended with little or nothing to show for it.

Nancy arrived bringing joy as always plus an announcement that I'm going home tomorrow. Had soup & a hamburger for lunch. Slept most of the afternoon.

Saturday, July 20

Last nite my routine was a hike to the bathroom about every 2 hours. Still I managed to get back to sleep between times. But here it is—getaway day. Yes I'm pleased to be going home but I've developed a real affection for the people who've been caring so well & so cheerfully for me. They've been just great—all of them.

I showered, got dressed, Nancy arrived & I did my radio broadcast from the hospital then it was goodbye time. I walked out of the hospital to meet rank after rank of Navy personnel & the band playing Anchors Away. I was deeply moved. Then it was Marine One back to the S. Lawn where more than 1000 staff & family, the Cabinet etc. were waiting with music, balloons & wonderful signs—many invoking the name of the "Gipper." This too was moving—I'm a very lucky man. I hope I can be worthy of such affection.

Upstairs—did some walking then sacked out for a nap. Dr. Smith came up about 5:45—blood pressure was 120 over 80 & temperature 97 degrees.

Sunday, July 21

Had a good night & really feel physical improvement. First time I've cleaned my plate—breakfast & lunch. Dr. Smith came up & checked my vital statistics—Temp 98.4—Blood pressure 132 over 80—pulse 66. Quiet morning, watched Sun. talk shows. A lot of them were devoted to whether I'll get some attention from Congress out of sympathy as against, "I've lost momentum & am a lame duck." One suggestion was that I'd not finish out the term.

Patti phoned. Speaking of phones—I have calls in for 4 Sen's.; Durenberger, Spector, Heflin & Bentsen. I'm asking them for a parliamentary vote to end Hatfield's filibuster against the Line item veto bill. Spector [Specter] said yes, Heflin says he's leaning our way, Bentsen is a no & I've yet to hear from Durenberger. All in all a good & restful day.

Monday, July 22

"Day by day, in every way." Don & Geo. B. came by at 9:30 after the Dr. had checked my vital signs & determined I was still alive. We talked "deficit" &

made it clear I was desirous of mixing in the fight. Then Bud & John P. came in for a discussion of our strategy with the upcoming Summit. A haircut & manicure—lunch & after a nap (Nancy's orders). Then some hall walking, home work & reading. Ate a good dinner—probably the biggest yet.

[Tuesday, July 23: visit from President Li Xiannian of China arrives; formal greeting and then meeting, noted, "Taiwan was 1st on his agenda & I repeated our position—it's a Chinese problem for them to settle peacefully"; meeting ending at 11 am; rested in the afternoon; state dinner in evening.]

Wednesday, July 24

[Meeting with House and Senate leadership budget imperative; lunched on Truman Balcony.]

Rcvd. top secret word on our Walker spy case.* We have to assume that since 1966 they have completely compromised our Navy communications system & our codes among other things.

Called Rock Hudson in a Paris Hospital where press said he had inoperable cancer. We never knew him too well but did know him & I thought under the circumstances I might be a reassurance. Now I learn from TV there is question as to his illness & rumors he is there for treatment of AIDS. Usual homework, some reading & so to bed.

Thursday, July 25

Another big day—had my vitals checked then climbed into a Sincere blue suit & went to the Oval O. & a cabinet meeting for reports on the success of our visit by Pres. Li, the budget, the farm bill, the tax reform, & the Justice dept. brought us up to date on the effort to prosecute the TWA hi-jackers.

[Interviewed by Hugh Sidey; changed into robe and pajamas; noted storm hit Washington; Mrs. Reagan returned from trip to Ohio.]

Friday, July 26

Staff meetings here in W.H. & then an NSC meeting in the old treaty room. Subject S. Africa. We've quietly influenced the S.A. G. to a number of changes benefitting Blacks. Now our Congress yielding to demonstrations, etc. is debating legislation to impose sanctions on S.A. Govt. We're opposed. It isn't a solution to the problem of apartheid & it will hurt the very Blacks we're trying to help. I think we're all agreed on continuing & even stepping up our present program but resist-

---

* *John Anthony Walker Jr., a Navy employee, was the main character in a spy ring that began selling military secrets to the USSR in 1968; he was arrested in May 1985 and received a life sentence.*

ing the other. We're off to Camp David. From there I'll call Margaret Thatcher at George Bushes house—sorry we can't get together while she's here.

Saturday, July 27–Sunday, July 28
[Mentioned telephone call to British prime minister Margaret Thatcher on Friday; walked in pool for exercise.]

Sunday A.M. called P.M. Nakasone of Japan. He is sending an emissary (very hush hush) to Iran to put the squeeze on for return of our 7 kidnap visitors in Lebanon. Japan has considerable clout with Iran because of trade. I thanked him & told him how important it was to us. The day was beautiful—more pool walking. Then back to the W.H. for a meeting with Mike D. & Stu Spencer. Both were advising me that I must grab hold & push the budget plan through before Congress breaks for the Aug. recess. There's nothing I'd rather do & I intend to try it but it's no easy task. The Demos. are a major part of the problem but our Repub. Sens. are a good part of it also.

Monday, July 29
Maureen arrived victorious from Nairobi.

[Met with Regan on ways to press budget bill through Congress, telephoned Senator Bob Dole (R-KS), who agreed to plan for no tax increase; met with Admiral Bill Crowe, choice for chairman of the Joint Chiefs of Staff.]

Don R. called—Dole was upset even though he was peaches & cream to me on the phone. Pete could hardly conceal his anger Don said when they met. He feels that he stuck his neck out—pitching for a tax increase & then I undercut him. Well he never consulted me before he went public with his tax idea & he's known for 2 yrs. or more of my opposition to a tax increase.

[Regan trying to arrange lunch with Senator Pete Domenici (R-NM) and Representative William H. Gray (D-PA).]

Tuesday, July 30
A message from Somalia that Maureen's visit to head of state was the best high level meeting they've ever experienced.

[Met with Republican leadership of House and Senate, then Regan, Senator Pete Domenici (R-NM), and Representative William H. Gray (D-PA) regarding budget compromise.]

About noon I stopped by the Dr.'s office. Nancy's dermatologist cut a little bump off my nose. I think it was aggravated by the tape holding the tube in my nose at the hospital but it's been evident for quite awhile although not as prominent as it became.

An early dinner—Nancy was to replace me at the Boy Scout Jamboree. Mermie & I will watch T.V.

Wednesday, July 31

Usual staff & NSC meetings after a photo session with Mermie—her supposedly reporting to me on the UN Women's Conf. in Nairobi.

[Noted that budget meetings continue; met Senator Arlen Specter (R-PA) about an idea to allow courts to handle trade disputes; staff meeting; met with Senator Paul Laxalt (R-NV), who said he didn't want to run for reelection in 1986, commented, "I told him we really needed him & can't afford to have the Demos. pick up the seat. Of course we'd miss him like H--l. We've traveled a long road together"; attended reception for departing budget director David Stockman.]

George Shultz called from Helsinki on "safe phone." He reported an interesting 3 hr. meeting with Shevardnadze—new Soviet Foreign Minister. Before I could reply we lost the connection. I hope he doesn't think I hung up on him.

Thursday, August 1

A lengthy session with Ec. Policy Council on Farm Bills & the Agric. situation. The coming year appears to be worse than what we've just gone through. Half of farming is outside all govt. agri. programs & it's having no ec. problems. Govt. has led to the present situation for the other half. Having caused the problem, govt. has to help bail out the victims now & then work to return farming to its free mrkt.

[Addressed members of the religious press; lunch with Vice President Bush; Regan brought news of possible budget compromise; cocktails with Senator Bob Dole (R-KS) and Regan to smooth over contretemps; commented that press was erroneously calling bump removed from nose cancerous.]

Friday, August 2

NSPG meeting—updating our situation re security of nuclear weapons—here & in Europe etc. Progress has been & still is being made.

Met with Geo. S. & Bud. I proposed Mike D. as a possible back door personal channel to Gorbachev. Mike will be making trips to Europe now & then in connection with his business & he certainly would have credibility as a personal emissary of mine. 12:15 off to Camp D. After all the fuss about my nose & our efforts to tell them the truth I learned at Camp D. that a biopsy had been taken & that pimple on my nose was a Basal Cell Carcinoma—same thing Nancy had a yr. or so ago on her upper lip. They are very common, also not dangerous & are the result of too much sun. First I have to give up popcorn & now sunbathing. I've been tan virtually year round all my life. I guess we'll have to square ourselves with the press & correct the statements we've been giving out.

[Saturday, August 3: relaxed by pool, wearing a hat. Sunday, August 4: quiet morning; returned to W.H. Monday, August 5: briefed for short press conference; lunched alone; presided at press conference; went upstairs for a nap but had to work instead.]

Tuesday, August 6

Baseball went on strike—they are nuts.

Staff & NSC meetings—I'm back to going to work at 9 A.M. now. Rumors have it that 5 of our 7 hostages are going to be released. We have no confirmation whatsoever.

Cap W. & Gen. Vessey came over for a slide presentation & a new plan for responding to an attack. There are 4 different responses for different levels of attack.

Sen. Byrd came in—he's taking a bi-partisan group of Sen's. to Moscow & anticipates a meeting with Gorbachev. I gave him a letter to the Chairman & we discussed topics he could take up.

[Meeting with executive board of Knights of Columbus.]

Geo. S. came & we planned a group to start working on the Nov. meeting with Gorbachev. He's agreed to using Mike Deaver as a back door messenger when I need to get something directly to Gorbachev.

A haircut & home.

Wednesday, August 7

Don R. and I discussed a couple of staff changes. Max F. who has been doing a good (temporary) job with the Cong. relations people must now give it up for health reasons. He'll return to being Consul General in Bermuda. Bea Oglesby will take over. Ed Rollins wants to go into the Pol. management business & make some money. We'll move his assistant up.

At 11 A.M. we had an NSPG meeting on coming up with a new directive & new program for counter espionage. It will involve replacing a lot of Russian employees in our Moscow embassy with Americans & getting a reduction in number of Soviets who are in our country on one pretext or another but who are in reality KGB.

[Economic Council meeting on countering trade-protection legislation, commented, "Our trade policy must be—free trade"; answered mail, went upstairs early.]

Thursday, August 8

Awakened at 6 A.M. by an NSC call. A car bomb at our air base in W. Germany killed an Am. solider & one other believed to be his wife, 17 were wounded, 14 of them Americans. D--n terrorists—h--l is too good for them.

[Cabinet meeting on budget matters.]

A signing ceremony for Foreign Assistance Authorization bill. The first one I've had since 1981. This is one where I wanted more money than they allowed. They just won't recognize this is part of our nat. security. Security assistance for example to Turkey—a Turkish soldier only costs $6,000 a year. If we have to replace him with an American it's $90,000.

[Lunch with Vice President Bush; relatively bright economic briefing; session to con-sider appointments; noted Mrs. Reagan away until Saturday. **Friday, August 9:** *quarterly meeting with Joint Chiefs, positive report on readiness.]*

George S. gave me an update on the Middle East peace process. We're hav-ing problems with some of the Palestinians King Hussein wants us to meet with—several are unacceptable to the Israelis.

[Meeting with Economic Policy Council on dominance of imported shoes, commented, "Options include tariff increase, quotas or rejecting the I.T.C. ruling & doing nothing. On simple Ec. & Nat. Interest this last is what I should do"; autographed photos; went upstairs.]

Saturday, August 10

The place is still empty but won't be this afternoon. Radio B.C. was in observance of V.J. day. Busy packing for Calif. *!!* Nancy is home! The place feels different already.

Sunday, August 11–Sunday, September 1

This was our long awaited ranch vacation. We arrived there on Sunday Aug. 11 to beautiful weather. In fact we had such weather (a little on the warm side our last week there) for the entire trip except for one day. That one day was the delayed birthday party for Nancy—Sat. the 17th of Aug. We had fog right down to the ground. Fortunately there was a tent for the 90 guests plus portable heaters. All our family except Moon & Bess & Nancy's brother Dick were there—both absent by reason of distance. Just about going home time the fog blew away & all those who were at the ranch for the 1st time got to see the beauty of the place.

[Did light chores and took walks.]

Then on Tues. the 20th of Aug. we went into Los Angeles & the Century Plaza hotel. That night a family dinner, only Maureen & Dennis absent—they were back in Sacramento. The rest of the kids & our grandchildren were all there & it was a warm, family get together. Nancy had all kinds of errands the next couple of days. I holed up in the suite—had a haircut & a visit from Dr. Bookman for an update on my allergies. But both Wed. & Thurs. evenings were great fun. Wed. was a gathering of 14 at Tom Jones's place—all old friends. Thurs. night was a drop by & short talk to a Repub. fundraiser at the Century Plaza. We didn't stay for dinner but went on to Betsy Bloomingdale's for another old friend's dinner. Again it was wonderfully warm & pleasant. Incidentally the fund raiser was to raise money for a Repub. State H.Q's. bldg. It will be named the Ronald Reagan bldg.

Fri. the 23rd.—back to the ranch by lunch time. Sat. was 6 weeks since my operation & we had our 1st ride—a short one. That was also the day I received a "secret" phone call from Bud MacFarlane. It seems a man high up in the Iranian govt. believes he can deliver all or part of the 7 Am. kidnap victims in Lebanon sometime in early Sept. They will be delivered to a point on the beach north of Tripoli & we'll take them off to our 6th fleet. I had some decisions to make about a few points—but they were easy to make. Now we wait. From that Sat. through Sun. the 1st of Sept. the days were pleasantly all the same except for Wed. the 28th when we went down to Santa Barbara for the late afternoon Press party.

[*Ranch work.* **Monday, September 2—Labor Day:** *flew east; spoke on tax reform in Independence, Missouri; returned to W.H.*]

Tuesday, September 3

At the office at 9 A.M.—meeting with Don R. & Geo. B.—Somehow this seems like the longest time we've ever been away (it was by only a few days) it's a little strange getting back into harness.

[*NSC meeting included Suzanne Massie, authority on Soviet Union.*]

Had a frustrating meeting with N.S.P.G. on our relationship with Egypt. Pres. Mubarak believes Quadhavi of Libya is a threat to all the N. African countries & mainly Egypt. We've been meeting with the Pres. on how we can assist him if Egypt is attacked. Do we all go out & join in a war & would our people support such a move? There is no doubt Egypt & the Suez canal are very important to our security. Still it's a complex problem. For now we'll keep on studying contingency plans with Pres. Moubarak.

A full Cabinet meeting for an update by Max F. on the legislative agenda, Joe Wright on the budget & George S. on international relations. There is no question but my veto pen is going to be needed in the coming months. The budget resolution passed by Cong. claims about $17 bill. in savings that just aren't there. Some desk work & a haircut & home to my 1st gym workout in more than 7 weeks.

Wednesday, September 4

Staff & NSC meetings concerned with whether to talk about S. Africa on Sat.'s. Radio talk—I'm for it. Next subject was a call I'm to get from King Hussein at 2 P.M. He wants a meeting with Ambas. Murphy, himself & some Palestinians. We can't go for that because 5 of the Palestinians are connected with P.L.O. The Israelis would climb the wall. When call did come all His Majesty talked about was my health. We were on an open line. We'll send our Ambas. a message for the King for personal delivery.

Had lunch with Don Regan—we talked about what we want to accomplish in these next 3 yrs. One thing for sure is control of govt. spending, a real arms reduction deal with the Soviets & a real relationship with our Latin neighbors.

A meeting with Geo. S.—he leans toward my signing the Sanctions bill against S. Africa. I've been sure I'd veto but he says the bill isn't all that harsh. I'll wait & see.

Nancy is in N.Y. for the day & night. Ted Graber & Maureen are here.

[Thursday, September 5: brief talk to federal marshals; flew to Raleigh, North Carolina, for speeches to students at North Carolina State University, received many ovations; returned to W.H.; vetoed South Africa sanction bill, with executive order supporting some parts of it; received good employment report from Regan. **Friday, September 6:** *photo session with new ambassador to West Germany, Richard R. Burt; noted that another nominee was withdrawn due to expressions of racial prejudice early in his career; photo with Federal Reserve Board member Martha Seger; meeting with Wirthlin on possible backlash if vetoes were to be overridden; lunch with female Republican state legislators; videotapings; left for Camp David.]*

Saturday, September 7–Sunday, September 8

A hot day Sat. mostly by the pool & in it. My radiocast was on the subject of protectionism—hoping we could cool down some of the hotheads on the hill. I announced a few actions we are taking where there are violations of fair trade on the part of three of our trading partners. Late in the day had a call from P.M. of Canada Brian Mulroney. Brian wanted me to know the Canadian govt. was voting not to join us in the S.D.I. research but that he was voicing approval of our doing it & that any private researchers, businesses & institutions in Canada were free to join us. All in all I think there is no problem.

Sat. night we ran "Hellcats of the Navy" starring N.D.R. & R.R. It was fun.

[Watched panel shows on Sunday, commented, "most of which were painting me as being in a deep hole & showing no leadership"; returned to W.H.; had conference call with Regan and McFarlane regarding congressional response to veto of South Africa bill.]

Monday, September 9

Saw our Ambas. off on his return to S. Africa. Then at 10:30 I went before the press & read a statement about the exec. order I would sign listing things we were going to do with regard to Apartheid in S. Africa. Many were things included in the Cong. bill calling for sanctions. I explained these were things I would agree to but eliminated parts of the bill I did not favor & that I would veto the bill if it came to my desk. This wouldn't have been necessary if I had line item veto.

At 11 A.M. I welcomed the 18 Generals commanding their country's forces in NATO. They are here visiting our mil. bases. An issues lunch & then a meeting with Bill Casey. An amazing number of K.G.B. agents are defecting these days. We have a serious problem with the Philippines—a communist takeover is a distinct possibility.

Met briefly with leaders of the Soviet Jewry movement. Then an ec. policy Council on trade matters. I've approved a trade policy as part of our effort to head off the flood of protectionist bills now upon the hill.

A very enjoyable interview for the 1300 college & U. radio stations conducted by 3 students from 3 different campuses—2 girls 1 male. They were fine young people & I enjoyed myself immensely. In the evening called Gov. Alf Landon on his 98th Birthday.

Tuesday, September 10

Arrival ceremony for P.M. of Denmark, Schlueter. The usual mil. formation—held in considerable heat. We had a good meeting. He's a conservative with a parliament of the opposition party (sounds familiar). He's managed to make some ec. progress including tax reform.

In PM met with Tax Reform Coalition a group of C.E.O.'s, heads of organizations etc. who are supporting our tax reform. Then Sen. Byrd & the Sen's. in his Moscow junket came by to report on their meeting with Gorbachev. Really not much new—he's clever, articulate etc. They think he may be willing to make some changes.

Later with our staff I made a decision we would not trade away our program of research—S.D.I. for a promise of Soviet reduction in nuclear arms.

State dinner. Mrs. Woll—sister of Sgt. Jack Elwood Wagner on hand. Invited because her brother was shot down off Danish coast (age 19) in W.W.II. The people of a small village found his body & for 40 yrs. a couple now in their 80's have been taking care of his grave. When they no longer can their village has ruled it will take over the task. For 40 yrs. the people there have stated he gave his life for them.

During dinner got word that Pres. Duarte's daughter was kidnapped in El Salvador by terrorists who machine gunned & killed two security men who were with her.

Wednesday, September 11

[Met Prince Sultan bin Salman of Saudi Arabia, who flew as astronaut in shuttle program; meeting with Republican leadership in Congress; noted cooler weather.]

Over to the E.O.B. for a Q&A with the Specialty Press—no news-making Q's. Met with Geo. Shultz about Summit. I sense he & Bud feel that "Defense" is going to be uncooperative & not want to settle anything with the Soviets. I can't quite agree on that. One thing I do know is I won't trade our S.D.I. off for some Soviet offer of weapon reductions.

[Photo session with Elizabeth Dole to launch charity fund drive; interviewed by James Dobson, conservative Christian broadcaster; photo sessions with departing appointees; fitted for new bulletproof armor.]

[**Thursday, September 12:** *flew to Tampa for speech to senior citizens and meeting with state legislators; noted British coup in Soviet intelligence, bringing a KGB official to the Western side; reception for Republican Eagles.* **Friday, September 13:** *received annual report from Intelligence Oversight Board, called it "good"; attended a lunch hosted by Maureen for Republican female politicians; met with Shultz on upcoming summit; interviewed by Hispanic broadcasters; videotapings; left for Camp David.*]

Saturday, September 14–Sunday, September 15

Finally a horseback ride—beautiful weather. A lot of homework to do—briefings for an ABC interview & a Q&A lunch with regional press plus more on Summit.

Sun. a call from Bud M. on the secure phone. Rev. Weir the Presbyterian Minister has been delivered to our embassy in Beirut & is now aboard the U.S.S. Nimitz. We're trying to hold it secret because of the other 6 kidnap victims. An unverified source says they will be delivered in 48 hours. Everything is top secret but suddenly on the TV talk shows they quoted a Reuters story that an anonymous call had reported Weirs rescue. We of course are stonewalling. Back to the W.H.

Monday, September 16

Dropped in on meeting of Nation's top business leaders who have plants in S. Africa. Urged them to tell the Am. people what that means in employment & quality of life for Blacks in S.A.

Our G.O.P. Hispanic Assembly met in Cabinet room. Had a nice session with them. They were largely responsible for upping Hispanic support from the 28% we had in the '80 election to 48% in '84.

[*Attended luncheon for regional press; interviewed by Peter Jennings and Ted Koppel of ABC for special on World War II.*]

We're setting up a plan to get Weir out of Europe & back here to his family. They know he's free now & also they must keep it very secret.

Tuesday, September 17

Rev. Weir & his family are at a "safe house" here in our country. His family was a little hard to handle. They insisted on going to a hotel but we managed to move them when he arrived. So far the secret is holding & they are all together. We've been told by the mystery man in Beirut the others (hostages) will follow.

[*Met with David Packard's commission on Defense Department; ambassadorial formalities; met with Chief Justice Warren Burger on upcoming bicentennial of the Constitution; meeting with congressional leaders on farm legislation; practiced for press conference; staged press conference and considered the response favorable.*]

Wednesday, September 18

[Greeted representatives of the Future Farmers of America; flew to Concord, New Hampshire, for rally on tax reform.]

Back to Wash. & a meeting with speech writers—discussed importance of putting facts & figures in speeches. Then a meeting with Bob Packwood—chairman of Sen. Committee that will have to do with tax reform. I stressed importance of Sen. voting on tax reform this year if House sent a bill over before years end.

[Hosted barbecue for members of Congress on W.H. lawn, entertainment by country music star Mickey Gilley.]

Thursday, September 19

A busy one—a State visitor—Pres. Machel of Mozambique. Turned out to be quite a guy & I believe he really intends to be "non-aligned" instead of a Soviet patsy. We got along fine.

Cabinet meeting dealt with bi-lingual ed., per diem pay for travel, Aircraft safety & Synthetic fuels. This last one is unfinished & we have a split on it. The ed. one is a problem I believe should be solved by getting Uncle Sam out of it & turning it back to the states.

[Photo session for New York Times Magazine; personnel meeting; visited delegation of National People's Congress of the People's Republic of China; more photo sessions; interviewed by Pat Robertson; videotapings; meeting with Senators Bob Dole (R-KS) and Bob Michel (R-IL) regarding tax reform; fitted for new hearing aid.]

Evening news brought word of a gigantic earthquake that caused incalculable damage in Mexico. First TV shots were horrible.

Rev. Weir (former kidnap victim) all over TV. His wife also suggesting we had nothing to do with his rescue.

Friday, September 20

[Met with leaders of farm and agribusiness groups; NSC meeting on upcoming summit.]

Then off to Bethesda hospital for routine (1st) post operative check up. Last time I had this kind of pre Camp David schedule, I didn't get to Camp D. but had an operation instead. Not this time. The result was a unanimous decision by all the Dr.'s that I have had a 100% complete recovery.

At Camp D. I had a good conversation (phone) with Sen. Byrd. He's most supportive of our S.D.I. & wanted me to know he had taped a TV talk show making that point.

The news was full of the horror of Mexico.

Saturday, September 21–Sunday, September 22

Another quake hit Mexico City, the death toll is climbing into the 1000's.

Late Sat. night I got through to Pres. De La Madrid by phone. We're mustering all the help we can. Nancy is going to go to Calif. by way of Mexico City to emphasize our concern & cooperation.

Back to the W.H. Nancy packing & I found a load of homework that kept me busy all afternoon. Sam Donaldson on the Brinkley Show quoted me 3 x's as saying in my press Conf. that the Soviets have a 3 to 1 edge on us in every kind of weapon. No one disputed him. I dug out the transcript & Mark Weinberg is calling him to tell him he's off base. First of all I could never have said that because I know it isn't so. What I did say (and is true) is that they "lead us in virtually every type of offensive weapon."

Monday, September 23

Day opened with Nancy leaving for Calif. by way of Mexico City. Then a letter was hand delivered. It was from Sam Donaldson apologizing for his misstatement on Sunday TV show & promising to clear the record on next Sunday's Brinkley Show. That's enough to make this a red letter day.

[Addressed trade leaders and met new appointees to export council; issues lunch; met with President Hosni Mubarak of Egypt, discussed advancing peace talks, noted, "Hosni told me privately that Arafat told him he could deliver our 6 hostages to our embassy. I don't know whether he has in mind kidnaping them from their present kidnappers or whether he has a connection"; photo sessions; met NFL alumni players; Vietnam veterans group; attended Inner Circle party reception. **Tuesday, September 24:** *flew to Knoxville for University of Tennessee seminar on public-private partnership; addressed rally on tax reform; returned to Washington.]*

Wednesday, September 25

[Ambassadorial formalities; received new poll numbers showing steady approval rating.]

Most important meeting was with Cap on nominee to be Sec. of Air Force (Verne Orr is retiring). Cap & Bud are on opposite sides on this one—divided between 2 candidates. I have to come down on Cap's side on this one. It isn't going to be easy telling Bud, he's all up in arms on this—threatens to quit. A phone call from Margaret Thatcher on a communications system for military. Eng. & France are both trying to sell us one. I hope we can come down on the side of Eng.

Spent afternoon up in W.H. Study doing mail.

Thursday, September 26

An N.S.C. briefing for my visit tomorrow with Soviet Foreign Minister Shevardnadze. I'm getting d--n sick of cramming like a school kid. Sometimes they tell me more than I need to know. George Bush & I had lunch together then I had visitors, Napoleon McCallum & his parents. He's a fine young cadet at the U.S. Naval academy. He sat out last yr.—1st Navy footballer to ever Red Shirt. He'd broken his leg. Now he's a Sr. & so far this season has averaged carrying the ball 130 yards per game. He's quite a fine young Black man & will be a credit to the Navy.

[Gave certificates to new W.H. Fellows; met with Domestic Policy Council on disbanding of Synfuel board; met with McFarlane, noted, "squared him away on the Air Force Sec. nomination. It will be Caps man, Rourke"; videotapings; Mrs. Reagan's return possibly to be delayed by weather.]

Friday, September 27

Woke up to a surprise—the twin doors that open onto the living room from the bedroom were wide open (and they open in). Apparently when "Gloria" blew through Wash. before dawn it did that.

A brief meeting with P.M. Gonzalez of Spain then into a jam session on the upcoming Shevardnadze meeting. He arrived at 10 A.M.—a 2 hr. meeting, then I had 10 min's. alone with him & then lunch (St. Dining Room) until 1:30. He's a personable fellow but we had our differences. My goal was to send him back to Gorbachev with a message that I really meant "arms reductions" & I wasn't interested in any détente nonsense. For the 1st time they talked of real verification procedures.

After lunch George S., Bud & I met preparing now for King Hussein's visit Monday.

Afternoon, hurricane Gloria blew away, the sky is blue, the sun is shining & Nancy will be home at 6:40. It's an answer to a prayer & I mean it. Gloria shifted course a little & the threatened disaster melted away. There was some coastal damage but no deaths, few if any injuries & all's well with the world.

[Saturday, September 28: desk work. Sunday, September 29: more desk work; lunch on the Truman Balcony; noted that press was carrying unconfirmed report of ransom deal for kidnap victims.]

Monday, September 30

Our friend King Hussein arrived—we had a good visit. There are still a few glitches in trying to put an international meeting together that would lead to direct negotiations between Jordan & Israel but I don't think they are insurmountable.

[Attended farewell ceremony for General John Vessey, retiring chairman of the Joint Chiefs of Staff; dinner party in honor of royal family of Lichtenstein, entertainment by Burt Bacharach.]

Tuesday, October 1

Breakfast with Sen. Repubs.—got a few things off my chest & took some

Q's. from a few of the dissidents—mostly on the trade imbalance & some need for protectionism. I let them know I'm against it.

Then the swearing in of Admiral Crowe, Jr. who replaces Jack Vessey. An NSC meeting—a report on S.D.I.—it really is showing promise. There have been some remarkable breakthroughs. A brief talk in the East Room to about 200 of the Tax Reform Action Coalition.

Margaret Heckler & I then went before the press. I'd received a call that she'd accepted the Ambassador post. I denounced as false the rumors & charges against her & introduced her as the new Ambas. Had trouble then as press wanted to change subject to the Israeli bombing of a building in Tunis said to be Arafat's P.L.O. HQ. Tunis says it was a hotel with innocent women & children victims. It's also suspected they used Am. F-15's which is against our agreement that weapons provided by the U.S. would only be used in self defense. It's a sticky one. We then went on to Cabinet meeting on budget & deficit. The Dem. House budget just isn't going to meet the deficit crisis. I'm going to have to do some vetoes & I can't wait (Lrnd. Israeli raid did hit P.L.O. H.Q.).

[*Addressed educators group; meeting with president of the Sons of Italy; reception for returning Supreme Court justices.* **Wednesday, October 2:** *meeting with congressional Republicans on foreign relations and balanced budget proposals; NSC meeting with Vernon Walters; visit from envoy from South Korea regarding trade problems; addressed convention for Interpol; videotapings; desk work; attended reception for Citizens for America.* **Thursday, October 3:** *flew to Cincinnati, visited Proctor & Gamble soap factory; press conference; addressed business leaders; returned to W.H.]*

Friday, October 4

Appeared before press with a bipartisan group from Cong. to announce our support for the Gramm-Rudman plan to balance budget over next 5 years. Their plan is virtually identical to our own plan I'd suggested to our people & which we had not yet announced.

[*National Security Planning Group (NSPG) meeting regarding the Philippines; asked Senator Paul Laxalt (R-NV) to be emissary to President Ferdinand Marcos; flew to Parsippany, New Jersey, for party fund-raiser; went to Camp David.* **Saturday, October 5–Sunday, October 6:** *rode horseback; telephoned college football coach Eddie Robinson; returned to W.H.; awaited news of Graham-Rudman bill, delayed by filibuster by Senator Robert Byrd (D-WV).]*

Monday, October 7

Meeting with Lord Carrington, NATO Sec. General. He was positive & optimistic about our relationship with our European allies. In other words Gorbachev propaganda hasn't succeeded in splitting the NATO nations from us.

Our Nat. Repub. leadership was in the East Room & I went over for short
speech. The filibuster is still blocking our efforts to get the deficit plan passed or
the extension of the debt ceiling. In a couple of days the govt. will be faced with
closing down unless we break the log jam. Sen. Byrd is the villain. He also is personally blocking almost 100 of my appointees from Sen. confirmation—some
of them for many months past. His beef is that I named 7 appointees while
Congress was in recess. That's my constitutional right. He disputes this & wants
me to concede his right to prior consultation.

We had an NSPG meeting to have a briefing on the Soviet Unions progress
in defensive weapons against nuclear missiles. They are raising h--l about our
research & they've been at it for 20 yrs. & we're just starting.

[Received word of terrorist hijacking of Achille Lauro ocean liner in Mediterranean
Sea. **Tuesday, October 8:** meeting with Prime Minister Lee Kwan Yew of Singapore, called
him "an old friend" and "truly one of the world's great statesmen"; presided at swearing-in
ceremony for James Miller, director of the Office of Management and Budget; NSC meeting regarding hijacked ship, received unconfirmed word of murders of Americans; noted,
"We are planning a naval operation employing the 'Seals six.' It's a high risk effort but we
feel justified."]

Wednesday, October 9

[Photo session with the New York Times; spoke to National Security Telecommunications Advisory Commission; attended fund-raiser in Virginia; lunch with Shultz and
McFarlane to discuss upcoming Geneva summit.]

Word came that the Italian Liner had returned to Port Said—the hijackers
were taken by the Egyptians who turned them over to the P.L.O. who took them
out of Egypt. They were only 4 in number but then we learned they had killed
an American—a 69 yr. old man in a wheel chair. So we never had a chance to
launch our rescue attack. The hostages minus 1 are on their way home.

The Senate passed the Gramm, Rudman deficit plan 75 to 24—now it's
up to the House. The Sen. also passed a temporary debt ceiling increase. If the
House doesn't pass it before midnight we are out of money.

Thursday, October 10

[Flew to Chicago; addressed Sara Lee employees; visited high school.]

On the way back to Wash. I O.K.'d a plan to intercept an Egyptian plane
& try to force it to land at a base of ours in the Mediterranean. It carries the 4
hi-jackers (murderers). We want to turn them over to Italy for prosecution. Of
course we will not attack the plane—just signal it to turn & crowd it a bit. Our
friend—Pres. of Egypt had said the hijackers in the hands of the P.L.O. were no
longer in Egypt. They took off from a hangar while we were on our way home.

Back at the White House for another go at my nose. They didn't get all of the skin carcinoma the 1st time. Now I'll have 4 days at Camp D. to hope it won't show enough for the press to notice when I come home.

Friday, October 11

Well my nose will show—so I prefaced a statement to the press on the hijacking by calling attention to the plaster on my nose & explained what had been done.

Of course the big news was that our Navy F 14's had intercepted the Egyptian plane carrying the hi-jackers & forced them down on Sicily—the NATO base in Sigonella.

Americans as well as friends abroad are standing 6 inches taller. We're flooded with wires & calls.

This happened late Thurs. evening & there were other kinds of calls half the night—such as my call to P.M. Craxi of Italy asking that we be allowed to fly the 4 to the U.S. for prosecution here. He explained that he didn't have the authority—Italian magistrates are independent of the Govt. Well the upshot is, Italy will prosecute but we are putting in an extradition request just in case. Moubarik is offended & called our act piracy. I think he's playing to his own audience. The Egyptian people are partial to the P.L.O.

Well back to Fri. An N.S.P.G. meeting about our handing of all these matters—then a meeting with George Shultz. & off to Camp David. There were more calls there.

Saturday, October 12–Monday, October 14

The usual weekend except for heavier than usual phone traffic & the additional day (Mon. Columbus Day). We managed to get in rides Sat. & Sun. I called Mrs. Klinghoffer, widow of the man the hi-jackers murdered. I'd called her daughters before she arrived in the U.S. She & some other passengers had a stop in Rome to identify the hi-jackers. She told me I'm really hated by them that every few minutes during the ordeal on ship they were sounding off about me. Italy raised our hackles when they smuggled one of the Palestinians who had accompanied the 4 on the plane out of Italy & into Yugoslavia. We have evidence that his top rank aide to Arafat was a conspirator with the hi-jackers. We had sent evidence & a warrant to Italy.

[Returned to W.H.; small dinner party. **Tuesday, October 15:** *flew to Boise, Idaho, for fund-raiser; was well received at rally at Boise State; flew to Milwaukee, Wisconsin, for another fund-raiser; returned to W.H.]*

Wednesday, October 16

The body washed up on Syrian shore is that of Mr. Klinghofer, complete with 2 bullet holes which lays to rest the lies told by the P.L.O. that he died of a heart attack. A cable informed me Defense Minister & party chief Spadolini in Italy pulled the rug out from under P.M. Craxi by taking his party out of the coalition. Craxi has been on thin ice & I guess this was just taking advantage of the high jacking. Susumo Nikaido of Japan came by. He's been here on a trade mission & I think he's going back convinced that the protectionism threat on the hill is real. He'll go back promoting more effort by Japan to open up markets.

Then former Gov. of Vermont, Dick Snelling came in. We've been urging him to run for the Sen. against Pat Leahy. Today he said yes to us.

[Meeting with Shultz regarding upcoming summit; addressed a gathering of Jewish philanthropists; new poll showed rising job approval; stitches removed from nose.]

After dinner went by the farewell dinner for Sen. Russell Long. I said a few words about him—he is really a good guy.

Thursday, October 17

Bob Michel brought a Japanese Labor leader over for a pic. before I went to the office. Then it was the usual staff meeting, some desk time & a short visit some Repub. Governors. Terry Brandstead wants a separate meeting about the plight of Iowa farmers. He says there have been 3 farmer suicides in his state. I feel as bad as he does but what can we do? We are spending more on the farmers plight than any admin. in history.

[Spoke to a coalition of ethnic and fraternal groups that support administration's tax-reform plan.]

Then a meeting with P.M. Peres of Israel. I hope he remains P.M. He's a statesman & a fine man. We talked the peace process—our desire to sell Jordan some arms. He has trouble with that one—mainly because of opposition in the Knesset.

[Photo sessions with old friends from Dixon, Epilepsy poster child; dropped by Republican governors' dinner.]

Friday, October 18

A huddle on the speech to the U.N. next week. Some wanted it more harsh toward the Soviets than I think it should be. I won. NSC meeting—wide disagreement on whether to make a new presentation on the M.B.F.R. talks in Vienna. They've been going on for 10 yrs. Kohl & Thatcher want a new proposal—D.O.D. opposes. I'm inclined to go with K & T. For one thing they hang their proposal on a strict, intrusive verification procedure. If the USSR doesn't agree—no reduction in forces. If they do agree it will be the 1st time ever.

[Met with CEO's to solicit help with Congress on deficit plan.]

The Egyptian Ambas. came by with a lengthy letter from Mubarak. Pres. M. is pleading for understanding but still charging us with humiliating him, etc.

The Ambas. almost in a whisper said—"put yourself in our place." I said "that should be mutual."

[Spoke about deficit plan to supportive groups; met with Mrs. Reagan and presidential library board members to choose an architect; met with Shultz about summit.]

Saturday, October 19

No Camp David—a dinner tonite—Nat. Italian-American Foundation honoring Frank Sinatra. It was a fun evening. The Italian Ambas., it was very obvious, wanted to make sure I was not upset or angry toward Italy. I publicly assured him in my remarks to the dinner that friendship between Italy & the U.S. was unshakeable.

[Sunday, October 20: Mrs. Reagan left for NYC, won't see her until Wednesday.]

Monday, October 21

This is or was an easy Monday. I think they are easing up because of the 3 days in N.Y. this week. Anyway I went from a brief session with Anne Armstrong about PFIAB's finding [. . .] (which we must do something about) to a brief meeting with Mr. Jerzy Milewski—Polish labor leader. He thanked me for the help the U.S. has been to Solidarity & the people of Poland. Then a brief speech at the E.O.B. to about 100 of our U.S. Attorneys—my appointees. Our issues briefing lunch was more or less routine. I had written a 4 page essay on my views of what our approach to the summit should be & read it to the group. Then an NSC meeting on the coming U.N. anniversary in N.Y. It has been decided that I will not sit with the U.S. delegation at the 2 hr. ceremony. I will already have addressed the Gen. Assembly that morning. Our intelligence suggests several of us have probably been fingered for possible assassination attempts. Some 5000 UN badges have been given out & the UN refuses to magnetometer the attendees or look at briefcases, etc. Security doesn't feel I should attend & I'm not going to 2nd guess them.

[Photo sessions with Miss America and the Times of India.*]*

Tuesday, October 22

Started the day with a meeting of Repub. Cong. leadership. I hope we did some good with them on the Gramm-Rudman bill & Tax reform. Then an NSC meeting—Paul Laxalt reported on his visit with Pres. Marcos of the Philippines. His report as to the President's health & alertness completely contrary to widespread press reports that he is failing & not mentally alert. I suspect an element in the St. Dept. bureaucracy is anti Marcos & helps the false reporting along.

Then an N.S.P.G. meeting on how we reply to the Soviets arms proposals. We're still working on that. My own idea is that we undermine their propaganda plan by offering a counter proposal which stresses our acceptance of some of their figures—such as a 50% cut in weapons & a total of 6,000 war heads etc. Those are pretty much like what we've already proposed.

[Met with Prime Minister Herbert Blaize of Grenada; greeted archaeologist Kamoya Kimeu; met with House Appropriations Committee and expressed concern about Congress reducing funding for SDI; met with Republican congressmen about 1986 election.]

Wednesday, October 23

[Flew to NYC for the fortieth anniversary of the U.N.; briefed by Deputy Secretary of State John Whitehead who had met with Egyptian president Hosni Mubarak and Italian prime minister Bettino Craxi, noted, "I think that flurry is all over—neither of those 2 want a rift with us"; had lunch at U.N. with other heads of state; met Mrs. Reagan at the Waldorf-Astoria.]

Back at the hotel I had meeting with Pres. Zia of Pakistan, P.M. Gandhi of India & old friend Margaret Thatcher of U.K. At 7 P.M. downstairs for a reception—heads of state & spouses. After the receiving line a few mini bi laterals as I made my way through the crowd. Oretega of Nicaragua wanted one but I said "no." Then upstairs for private dinner & beddy bye.

Thursday, October 24

This was the big day. Nancy & I went over to the U.N. I addressed the Gen. Assembly & a few thousand U.N. guests. I had to wear my iron undershirt. [. . .] a sizeable group of reps. carry guns. The U.N. refuses to allow any magnetometering or checking of briefcases. My speech went over extremely well. In fact veterans at the U.N. said no western speaker had ever gotten such a warm applause. It was broadcast live & we all agreed the crowds on the street had been affected by it—they were cheering like for a Super bowl.

Nancy & I parted company. I went over to Walters office in the U.N. Mission Building. There I had a good meeting with P.M. Craxi. We really cleared the air re the hi-jacking. They now have the info we had on Abbas & are trying to get him back. Then a working lunch with Geo. S., Don R., Bud & Ambas. Walters. After lunch I met with 5 of our Ec. Summit partners—Margaret T., Brian Mulroney of Canada, Helmut K., Yasu Nakasone (Japan) & Craxi. It was a good 90 min. session—all on the upcoming meeting with Gorbachev. They were all most supportive & jumped for joy when I volunteered to go to Brussels (NATO) & brief them after the summit. Back to the hotel & a meeting with F.M. of the Soviets, Shevardnadze. It was as usual a polite exchange with him saying nothing of importance.

He was followed by Brian Mulroney of Canada—this was a good meeting. He accepted my invitation to an official visit to Wash. this Spring. We're making progress on the Acid rain problem.

A brief cocktail time—then a dinner of just the 6 of us who had met in the afternoon. They told me I would go to Geneva with their trust, their love & their prayers.

[Friday, October 25: met individually with Prime Minister Yasuhiro Nakasone of Japan and Chancellor Helmut Kohl of West Germany, then the family of Leon Klinghoffer, victim in the Achille Lauro *hijacking; flew to Camp David, noted that the case of a Soviet seaman trying to defect in New Orleans had been bungled and the sailor returned to his ship.* **Saturday, October 26:** *radiocast; rode; watched the World Series.]*

Sunday, October 27

No change on the sailor. I have issued orders that we take him off the ship by force if necessary so that he can be questioned away from the Soviets as to what he wants to do. Back to the W.H.

Monday, October 28

The story of the Soviet Seaman goes on all day & will go on to Tues. morning. Our people were on the Soviet ship to make sure they didn't sail away. This afternoon word came through Ambas. Dobrynin that we could take the sailor to a nearby Coast Guard Cutter for interrogation. He was moved but shortly after arriving on the Cutter he became nauseated. The sea was quite choppy & the diagnosis was sea sickness. He went up on deck for an hour. We asked permission to move him ashore to the Naval hospital. That's where he is now & it is planned that he be questioned tomorrow.

[Attended domestic-policy meeting with report on the Grace Commission's 2,478 recommendations, 83 percent of which had been implemented; met with groups supporting the Grace Commission's work; issues lunch; presided at ceremony proclaiming "Year of Liberty," in reference to the centennial of the Statue of Liberty; met with families of hostages held in Lebanon, noted, "We've tried to keep them informed of all we're doing to try to free their loved ones."]

Tuesday, October 29

A long d--n day. Finale to our Soviet sailor episode. Finally today our team of experienced experts questioned him at great length. He now says he fell off the ship. When asked why he jumped off the boat taking him back to the ship he says he has no memory of that. Now he declares he wants to see his mother & father again—so he's been turned over to the Soviets.

[National Security Planning Group (NSPG) meeting on strategy for Soviet arms reduction proposal, believed a compromise was near; spoke to Protestant lay leaders and clergy on youth problems, also briefed them on tax reform; interviewed by BBC; received gifts from visitors, including a painting of himself on horseback, candlesticks made by a former Hungarian freedom fighter, quilt made by schoolchildren; greeted professional imitator and signed an antique fan

autographed by nearly every president since Grant; attended party fund-raiser and reception for
the program the McLaughlin Report. **Wednesday, October 30:** *meeting with Bill Clark on*
recent visit with President Chiang Ching-kuo of Taiwan; meeting on arms negotiation, noted,
"I gave a 'go ahead' on our arms proposal in reply to the Soviets Counter proposal"; celebrated
anniversary of longtime W.H. employee; greeted thirty-seven elected officials, former Democrats
who switched to GOP; met with Shultz and McFarlane on their plans for a trip to Moscow; saw
exhibit at National Gallery of treasures on loan from English homes.]

Thursday, October 31

Pres. Duarte (El Salvador) & his daughter—the kidnap victim came by. She
is a charming young lady. He has a few speaking dates while here & he is lacing
into the Sandinistas. The World Series Champs the K.C. Royals came for a Rose
Garden ceremony. I wound up with a cap, a bat & a warm up jacket.

After lunch—during which I phoned King Hussein to encourage him about
our arms deal—I did an hours briefing for an interview with 4 Soviet journalists—from Tass, Novosti, Pravda & Izvestia. I wonder if they'll print my answers
as I gave them? If not I have a tape which U.S.I.A. can use to expose them.

A short session with Ambas. Charles Price. I'm afraid our friend Margaret is
in some political trouble. The U.K. economy just isn't picking up and, of course,
they jump on her.

I sneaked upstairs for a meeting with Chairman Rostenkowski of the House
Ways & Means Comm. He's really on the edge. He's been working like a dog to
get our Tax Reform thru Committee. He's being opposed by everyone who has
been bought by a special interest & that includes 9 Repubs. He wanted to assure
me he was really trying & wanted to get it for me. I assured him I knew what he
was doing & was grateful.

[Send-off for ambassador; greeted head of American Legion; videotapings.]

Friday, November 1

Well the Goblins didn't get us but we have another Soviet defector case. In
Kabul—our embassy—there is a deserter from the Soviet military asking asylum. Our embassy is surrounded by Afghan & Soviet military. We have refused
to turn him over to them. Our problem is that we have no way to get him out of
the country. The minute he steps out of the embassy he can be seized. He's only
safe as long as he is in U.S. territory which is what the embassy is.

Met with Cap W. The bad news is that we have no choice but to go with
the French communications system for our military. They are competing with
the English. I wanted the Eng. to win but the cost difference is such we cannot
justify awarding them the contract.

Then it was off to Bethesda Hosp. for another check up before going to
Camp D. Results were the same—A OK—100%.

Saturday, November 2–Sunday, November 3

A good ride under gray & threatening skies. Nancy didn't go, her cold is still hanging on. Our defector in Kabul can't make up his mind. He's 19 yrs. old. The Soviet Ambas. visited him in our embassy & gave him a fatherly pitch as to how he could go back to Russia—no punishment etc. Now the lad wants to see him again. That will take place about 11 P.M. Sunday our time. We in turn have offered him asylum here in the U.S. (on my orders).

Over the weekend I called Nixon & Ford to get any suggestions they might have on the Summit. Dick had a h--l of a good idea on the arms negotiations. We probably won't have them settled by the time the Summit ends. His suggestion is that we state what we have agreed on, that we will continue negotiating on the other points & as a token of our resolve to achieve results we each take 1000 missiles out of the silos & store them for a set time. If we can't come to a reduction agreement we put them back in the silos. Back to the W.H.

Monday, November 4

Mondays seem to be coming around more often. This one is gray, wet & dismal. Our Soviet lad in Kabul made it final today. He left our embassy with the Soviet Ambas. who promised he could go home & not be punished. On top of that our defector from the K.G.B. held a press conf. here in Wash. to announce he was returning to Russia. Something smells fishy. Gov. Candidate Durretti (Va.) came by for a photo & a Q&A on the lawn with the press. He sounded genuinely optimistic even though the polls don't look good. Well tomorrow is the day. Issues briefing lunch. In the middle I got a secure phone call from George S. & Bud in Helsinki. They'd just had 8 hours of meeting with Shevardnaze with little or nothing accomplished. Tomorrow they meet with Gorbachev.

[Meeting with Economic Council on farm matters.]

A good meeting—Geoff Swaebe, our Ambas. to Belgium. He's really a solid citizen & I'm glad he's willing to stay on. No good news from Cong. If we don't have an extension of the debt ceiling by the 15th we will have to sell gold or default on bonds. D--n their hides (the Cong.), we'll default for the 1st time in our history. Something has to wake those d--n prima donnas up.

Tuesday, November 5

N.S.C. meeting was a movie. We saw a demonstration of our new Bomber, one of the greatest advances in aircraft in years & years. It is of course most hush hush—I should call it what it is—a fighter bomber.

Geo. S. called from Moscow on scramble phone—7 more hours of talks—4 of them with Gorbachev. Apparently not much progress. Gorbachev is adamant we must cave in our S.D.I.—well this will be a case of an irresistible force meeting an unmovable object. Met with Edmund Morris who is going to do my official biography. I'm pleased—his book on Teddy Roosevelt was wonderful. Of course I can't charge up San Juan Hill. Had an Ec. briefing—our recovery is continuing—or by now I should say our expansion & growth is progressing at a slow but steady rate & on employment we're doing extremely well. A higher percentage of the potential work force (all between 16 & 65) is employed than ever in our history.

[Ambassadorial formalities; Mrs. Reagan in NYC.]

Wednesday, November 6

Briefing not the way to start the day—what with news of the games Cong. is playing with regard to the debt ceiling, deficit & tax reform. And yes that goes for Repubs. as well as Dems.

[Met with Republican leaders; briefing for interview with wire services; had interview; lunch with Republicans and Democrats raising money for presidential library; videotapings]

Then George S. & Bud came upstairs with Don R. & George B. to report on their Gorbachev meeting. It seems Mr. G. is filled with a lot of false info about the U.S. & believes it all. For example, Americans hate the Russians because our arms manufacturers stir them up with propaganda so they can keep selling us weapons.

Nancy & Maureen arrived.

Thursday, November 7

This one got to be a busy day. The usual staff & NSC meetings then into the Cabinet room to meet 40 or 50 heads of all manner of Nat. Org's.—Vet's. groups, K. of C. etc. all in a coalition to support S.D.I. Then former Gov. Otis Bowen of Ind. came in—I've named him Sec. of H.H.S. to succeed Margaret Heckler. Next meeting was with Sen. Dan Evans who is fighting like h--l to halt the textile protection bills. He wanted to be reassured that I'm with him & I am.

[Met with Russian scholars on upcoming summit; interviewed by U.S. News & World Report; gave Presidential Medal of Freedom to Cold War diplomat Paul Nitze and nuclear strategist Albert Wohlstetter and his wife, Roberta.]

Back to the P.M. Met a lady Mrs. Barbara Newington who has spent 10's of 1000's of her own money taking out ads to build support for things like our Nicaraguan policy.

Ed Meese came in to brief us on the Medved affair—he is the sailor who jumped ship (a Russian ship) in New Orleans to be a defector. Then he recanted & is back on shipboard. Ed gave us the detailed story of how it all happened. Now Jesse Helms has gotten the Agri. Committee of the Sen. to subpoena him—a staff member is on his way to N.O. I don't know how this would work & think it's pretty silly. The guy was very positive that he wanted to go back to Russia.

A ceremony for swearing in of Ed Hickey who is leaving our Mil. Office to be Chmn. of the Fed. Trade [Maritime] Commission. Upstairs Lyn N. came by briefly—wanted to be reassured that I wouldn't roll over & settle for something at the Summit just to come home with an agreement. I reassured him.

Friday, November 8

A meeting with our Repub. Cong. leadership. Geo. S. & Bud reported on their Moscow trip & we had a discussion about the Summit & the part our Defense Bud. plays in that. Jack Kemp kicked up a fuss when he challenged the St. Dept. about not supporting $27 mil. aid to Savimbi in Angola. Geo. replied that our objection was to Cong. making the aid overt. We want a covert operation for real help. Our problem is Cong. interference in what should be exec. office management of international diplomacy. Things got hot for awhile. Then we got around to the extension of the debt limit. If we don't have that by the 15th we will have to default on our bonds for the 1st time in our history. I let them know if the fun & games continued—trying to hang unacceptable amendments on the debt bill I wouldn't be blackmailed into signing even if my veto brought a default.

Bob Dole started talking about a nice clean bill with no amendments.

[NSC meeting on case of sailor, Miroslav Medved, going back to USSR, noted, "We're convinced his Mother & Father have been threatened if he doesn't"; lunch with religious leaders regarding human rights and the summit; met with Shultz and McFarlane on summit; signed proclamation for Alzheimer's Disease Foundation; photo session for Time *and* Newsweek; *accepted unexpected visit from two students introduced on University of Tennessee trip.]*

Saturday, November 9

The big day. Prince Charles & Diana arrived at 11 A.M. for coffee. They are very nice & it was an enjoyable 40 min. or so. Then they had to move on & I had to hightail it to U.S.I.A. where my 5 min. Saturday Radio broadcast became, for just this once, a 10 min. Radio & TV show to the Soviet U. & a number of other countries in 9 languages—an estimated audience of 120 mil. people. I wonder if Mr. G. listened. The dinner for the Prince & Diana was a great success. It had not a hint of "State Dinner" flavor. Everyone seemed to have a wonderful time. Ted G. had decorated the W.H. more beautifully than we've ever seen it. Leontyne Price sang wonderfully & then everyone danced like at a Junior Prom.

[Sunday, November 10: noted that the ship carrying Miroslav Medved departed

New Orleans; lunch with friends on the sun porch. **Monday, November 11:** *attended Veteran's Day ceremony at Arlington National Cemetery.]*

Tuesday, November 12th

Met with Cong. Ldrship—Dem. & Repub. Things got a little tense re the deficit & the need to extend the debt ceiling. I told them there wasn't a state in the union that didn't have a better budget process than the Fed. Govt. We submitted a bud. on Feb. 5th—some weeks ago the Cong. finally passed a bud. resolution which I agreed to—but here it is Nov. & they are fighting over appropriations entirely at odds with their own bud.

Later an N.S.P.G. meeting on aid to Savimbi in Angola. [. . .] Had a TV interview with 5 foreign press—5 countries—U.K., Switzerland, Italy, France & darn if I can remember number 5. A meeting with Rep. Strangeland—he's concerned about the farm situation but aren't we all.

[Latest poll results showed 65 percent job approval.]

Wednesday, November 13th

[Cabinet meeting on plans for Geneva, discussion of 1987 budget; NSC meeting on details for Geneva, report from Weinberger on Soviet treaty violations; met with Red Cross leaders, concerned about financial shortfall; bill signing.]

Bill Casey brought in 3 of his experts on the Soviet U. Their presentations on the people of Russia were great & confirmed things I had heard from unconfirmed sources. The Soviet U. is an ec. basket case & among other things there is a rapidly spreading turn by the people to religion.

Thursday, November 14

Met with our fine Ambas. to the Philippines Steve Bosworth. He believes Pres. Marcos is underestimating the danger from the Communist insurgents— not so much as to mil. threat but their progress in winning converts. It's a touchy mess. An N.S.P.G. meeting to talk about arms control & how we present (& what) to Mr. G. in Geneva.

Called Jimmy Carter to ask any input he might have on the meeting. He was remarkably cordial.

Met with our 3 arms negotiators who have just received the 3rd session. They are aware that the Soviets have gone farther in the sense of actually proposing numbers but still the Soviets shade things in their favor.

[Delivered televised speech on summit.]

Friday, November 15

At 9 A.M. Wash. time P.M.s Thatcher & Fitzgerald (Ireland) signed an agreement on bringing peace to Northern Ireland. Tip O'Neill came down &

we were photographed together endorsing their action & making statements of support.

An NSC meeting was a run through by George S. of the Geneva day to day Summit schedule. Then Geo. & Bud & I had our usual Fri. meeting—nothing very important. Now we know what will happen to our re-defecting KGB agent Yurchenko; He's doing appearances in Russia repeating how we drugged him.

[Received letter of support on SDI from thirty-nine senators; videotapings; packing. **Saturday, November 16:** *flew to Geneva.]*

Sunday, November 17

I awoke at 9:15 after a pretty good night's sleep. Nancy slept till around 10, then we had breakfast. It's a gray day but the view from our windows over the formal lawn & gardens & on to Lake Geneva is lovely. A visit to where we'll hold the Tues. meeting & where I hope to get Gorbachev aside for a one on one. There is a pool house down on the Lake shore complete with fireplace. I'll try to talk him into a walk. The press is excited about the leak of a letter to me from Cap on why I shouldn't be trapped into endorsing continued observance of Salt II. It is a great distortion & is not as the press would have it an in house battle. I agree with Cap & wanted his factual accounting in writing.

Ron came for dinner—a nice time for the 3 of us.

Monday, November 18

Both of us had a wakeful night. The day was colder & darker. Don R. came for me at 10:55 then we walked across to the Pometta House for a meeting plus a working lunch—Lord I hope I'm ready & not overtrained. After lunch we picked up Nancy & went to Le Reposoir for welcome ceremony with Pres. Furgler. Usual Nat. Anthems & review of troops. Then a plenary meeting with him & his top people. We have a top relationship—very warm & friendly & few if any problems. They were happy with my position on S.D.I. & Protectionism. Back to our quarters & met the owners—the Pometta family—very nice. They rent this house to the Aga's son. Nancy surprised me with a masseur coming in at 5 P.M. I fell asleep about 4 x while he was massaging.

Tuesday, November 19

This was the day. Mr. G. & I met. We were scheduled for 15 min. of pvt. one on one—we did an hour which excited the h--l out of the Press. Then we joined the plenary meeting. I gave him the floor 1st & he did a pitch about us not trusting them, etc. We should have no pre-conditions set before any agreement about better relations. "Our ruling class (munitions makers) keep our people upset at the Soviets so they can sell more weapons." He also took off on the Heritage Foundation & think tanks that do the same. He said we had declared zones of special interest around the world but attacked the USSR when they did the same thing. We must recognize right of people to a revolution. Well finally it was my turn & I took them all the way through the history of Soviet aggression, etc. We broke for lunch but I assured him he'd have the floor to rebut me after lunch. Our gang told me I'd done good. In the P.M. session he had quite a prepared thing that had us suspicious without cause, etc. Again I rebutted with some pretty solid examples—WW II then refusal to let Am. warplanes use Soviet fields etc. When I finished I suggested he & I leave the group & do another one on one. We walked down to a pool house on the lake shore. Eddy had a fire going & we did about 2 hours on S.D.I. He's adamant but so am I. I scored one we've worried about;—that the meetings should be on an ongoing basis. He accepted my invite to U.S. next year & I'm invited to U.S.S.R. in '87. That in itself could make the meeting a success—Tonite to their place for dinner. And what a dinner—they must be influenced by the Orientals. Course after course & for half of them I thought each one had to be the entrée. Finally dessert & by this time it was time to go home & that's what you did because the host & hostess pushed back their chairs & escorted us to the front door. When you have dinner with the Russians—dinner is the full evening's entertainment.

Wednesday, November 20

The last day of the Summit & this time Mr. G. was host. We went to the Soviet mission & he took me into a small room with interpreters. This was my chance to have at human rights. I explained that I wasn't telling him how to run his country—I was asking for his help; that I had a better chance of getting support at home for things we'd agreed to if he would ease some of the restrictions on emigration, etc. I told him I'd never mention what he was doing out loud but he'd find that I could better meet some of his requests for trade, etc. He argued back sort of indicating that he thought they treated their people better than we did ours. He quoted statements made by some of the feminist extremists to prove we were unkind to women. I fought back—only time will tell if I made any headway.

In the plenary I took off on arms control then he fired back about S.D.I. creating an arms race in space & the stuff really hit the fan. He was really belligerent & d--n it I stood firm. That took us til lunch. In the P.M. session I tried out a written proposal for a joint statement. Upshot was we cut short the meeting & our teams went at the problem of a joint statement. He & I & the interpreters went into a small room & wound up telling stories. We were there 'til 5:30 then the teams came in with a number of things agreed upon & several we didn't. We broke up to leave them still at it so he & I could get ready for the reception at the Swiss Presidents home. Then they were here for dinner. It was a pleasant evening & a small informal dinner. Over coffee some of our aides (both his & mine) came over to tell us they were having trouble on the joint statement—his people were trying to withdraw some things they had already agreed to. There was some brisk language & at 5 A.M. the statement ended up the way we'd wanted it! And I think it was because Mr. G. told his guys to quit what they were doing.

Thursday, November 21

[Formal farewell ceremony; spoke briefly, as did General Secretary Mikhail Gorbachev; noted that Shultz and Minister of Foreign Affairs Eduard Shevardnadze signed agreements; stopped en route to America in Belgium and delivered report to NATO, commented, "The air of success in the meetings is widespread—but probably not with our cynical press"; arrived in Washington and slept; addressed joint session of Congress.]

I haven't gotten such a reception since I was shot. The gallerys were full & the members wouldn't stop clapping & cheering. So home to bed at what is about 4 A.M. by Get up time.

Friday, November 22

A good night's sleep & into the office at 11:15 A.M. They are being kind to me. Almost instantly I was rushed over to the family theatre to take Q's. from all the leading TV commentators & some top columnists. It was all on the summit of course & designed to keep the upsurge going. My approval rating on handling the summit was 81%.

Back to the office for a brief NSC. Subject was our hostages in Beirut. We have an undercover thing going by way of an Iranian which could get them sprung momentarily.

Geo. B. & I had lunch. He's been to Iowa to do a fundraiser for Gov. Brandstead who has been kicking my head in on the farm situation. He also met with some farm leaders. Sen. Grassley has really mobilized the farmers against me. I have compassion for them but there is a limit to what we can do to help them in this situation & we're doing all we can.

George Shultz came in after the Cabinet meeting. The meeting was for both of us to report on the Geneva trip. George wants to leave govt. I told him he could stay as long as I was here but I couldn't try to talk him out of his decision which is based on the fact he's just burned out.

[Met with James Baker regarding tax bill; visit from Michael Deaver.]

Saturday, November 23

Awoke to the 1st sunshine we've seen since before leaving for Geneva. Did my radio stint on the Geneva meetings. After lunch a session of sorting out background papers on the trip. Some I will burn—some I'll let Bud Mac-Farlane burn. Tomorrow Nancy leaves for the coast—engagements in S.F. &

L.A. I leave Tues. I don't look forward to 3 days without her. She'll get to the ranch Wed. We're still sweating out our undercover effort to get hostages out of Beirut.

[Sunday, November 24: *received ongoing reports of hijacked Egyptian airliner, ending with a firefight.* Monday, November 25: *further reports on "savagery" onboard the hijacked airliner. Attended farewell ceremony for Secretary of the Air Force Verne Orr; new poll shows high approval for dealings with Soviets; received annual live turkey; visit from John Fling, humanitarian award winner, and Pearl Bailey, honorary chair of American Lung Association; videotapings; dinner with Maureen.*]

Tuesday, November 26

Get away day. At 9 A.M. we lifted off the S. Lawn—we being Lucky & me. Lucky is on her way to her new home in Calif.—Rancho del Cielo. A.F.1 was a new experience for her & never did quiet down. Of course I sweated the hours out worrying about a possible biological problem. She made it all the way without a boo boo & we went down the steps at Point Mugu to be greeted by the top Naval Brass & their wives plus a good sized crowd plus the press with cameras clicking. Then in front of all—Lucky proved she was a healthy dog. As one Navy Captain said—"better there than on the helicopter." A call from Pres. de la Madrid of Mexico, some brush cutting with Barney & Dennis and, of course some paperwork as always. Problem of the moment is the House Ways & Means version of Tax Reform. It's being studied by our team in Wash. & I'm looking at a thick stack of figures. There is a great question as to whether I can accept this. Then my first night ever to be at the ranch without Nancy.

[Wednesday, November 27: *Mrs. Reagan arrived; rode.*]

Thursday, November 28—TURKEY DAY

Rose to dense wet fog. About 9:45—called off the ride. Thurs. and the family arrived in early afternoon for our annual Thanksgiving dinner which was also early—around 5 P.M. By that time the fog had been joined by heavy rain. Ron & Doria stayed over night.

Friday, November 29

Weather unchanged—no ride. Awakened early by phone call from Bud M. It seems Moubarak wants to fly some of the Egyptian wounded & his commandos home from Malta & is concerned that Quadaffi might attack the plane. He has asked us for fighter cover. I said h--l yes. We are flying F-15's plus a tanker & a communications plane down from Europe. Operation should take place some time tonight—I think. Ron & Doria who had stayed overnight departed around 11 A.M. Called Rostenkowski to assure him we were studying the tax reform as produced by Ways & Means Committee. Some press accounts had us feuding, etc.—absolutely no truth to it. For a time it looked as though it might be clearing—the sun even came out, but it was all temporary. Before the afternoon was over the rain was pouring down & continued into the night.

Saturday, November 30

[Radiocast; weather not conducive to riding.]

Then along about noon the skies cleared, the sun came out & we had an afternoon outdoors. I painted—(a little touch up) of our Rooster weathervane then Courtney, Barney, Dennis & I pruned the big walnut tree in the yard & hauled the cuttings to the dump. A letter, "for my eyes only" came from Bud MacFarlane. He wants to resign. He says after 30 yrs. in govt. service he owes it to his family. I can't argue against that. I believe his successor should be John Poindexter—presently Buds Deputy in NSC.

Sunday, December 1

[Helicoptered to L.A. for Variety Club charity fund-raising dinner.]

Right after we arrived I saw Bud M. for a few minutes to make sure he knew he could stay as long as I'm here but I could not put up an argument about his feeling he should retire for family reasons. He assured me that was his reason so we'll part Dec. 31st.

The Variety Club Dinner was a wonderfully warm event. The room was totally Hollywood—people we've known for years—some of whom we were in pictures with. Frank Sinatra M.C.'d—Steve & Eydie sang, Dean Martin, Burt Reynolds, Chuck Heston & on & on. Variety is naming a children's hospital bldg. at U. of Neb. for me. Our family all there except for Patty & Paul (N.Y.) & Ron & Doria—he'd caught the flu.

[Monday, December 2: flew to Seattle; spoke briefly at party fund-raiser; met school-children; returned to Washington; met at airport by family with thirteen-year-old boy, ardent supporter, suffering from brain cancer.]

Tuesday, December 3

A crisp day—but finally bright sunshine all day & how I've missed it.

[NSC briefing with Shultz; meeting with Republican congressional leadership, mainly about moving tax reform bill through Congress.]

In the NSC meeting Bud brought up his retirement. John Poindexter was there & maybe Bud intended that I should name him then as his successor. I hadn't discussed it with anyone so remained silent. Later talked to George & Don—then phoned Cap in Brussels & Bill Casey. Tomorrow we'll announce I think—Bud's resignation & that John P. will replace him. Over to the East Room to address a group who have set up an org. to link technology to help the disabled. Two young people who have been helped were present—one a young girl who cannot speak & can only move her thumbs. A computer scientist has worked out a method where by with thumbs alone she can create a message on a small screen attached to her wheelchair.

A Cabinet meeting to discuss a bill before the Sen. to give the Pres. the power to shut off private credit & bank loans to the Soviet U. I came down on the side of not doing that.

[Met with Senator Paul Laxalt (R-NV) to describe summit. **Wednesday, December 4:** *spoke at high school in Maryland; met with Shultz on midlevel meetings in advance of 1986 summit; announced John Poindexter as McFarlane's successor as National Security Advisor; visit from industrialist and philanthropist Eugene Lang; met with senior citizens group.]*

Thursday, December 5

N.S.C. Briefing—probably Bud's last. Subject was our undercover effort to free our 5 hostages held by terrorists in Lebanon. It is a complex undertaking with only a few of us in on it. I won't even write in the diary what we're up to.

[Telephoned celebration of EPA's fifteenth anniversary; met with Southern congressmen hoping for support of textile bill, commented, "I listened but feel I must veto it. It is pure protectionism"; spoke to retired members of Congress; flew to NYC for thirtieth anniversary party for the National Review.*]*

Friday, December 6

Kathy went out in the morning & picked up the new member of our family—a one year old "King Charles Spaniel." Nancy had fallen in love with the Buckleys' & this one is its brother. This is a small dog & we can keep it at the White House.

Back in Wash. I was just in time for a meeting in the Cabinet room with the Joint Chiefs of Staff, OMB & others on the impact further defense cuts will have on our Nat. Security. I'm between a rock & a hard place. I can't let the mil. budget be hacked away any more by Cong. & at the same time I can't veto the Gramm, Rudman, Hollings bill—which threatens to do that. It is the 1st plan ever offered as a way to reduce the deficits. I fear a veto would set us off into an ec. tailspin as people lost confidence that anything would ever be done about deficit spending.

[Met with Shultz and Poindexter regarding midlevel pre-summit meetings; cabinet meeting on 1987 budget; videotapings.]

Saturday, December 7

Day opened with "Rex"—(our new dog) on our bed. I then had a meeting with Don R., Cap W. & Bud M., John P., Geo. Shultz & Mahan of CIA. This has to do with the complex plan which could return our 5 hostages & help some officials in Iran who want to turn that country from it's present course & on to a better relationship with us. It calls for Israel selling some weapons to Iran. As they are delivered in installments by air our hostages will be released. The weapons will go to the moderate leaders in the army who are essential if there's to be a change to a more stable govt. We then sell Israel replacements for the delivered weapons. None of this is a gift—the Iranians pay cash for the weapons—so does Israel.

Sunday, December 8

A quiet day with our guests—then at 5:15 the reception for 350 guests connected with the Kennedy Center awards. Everything went well. The show was the best of these events since we've been here. The only sad note was the absence of one of the honorees—Irene Dunne. Irene is in Wash. but in George Washington Hosp. with the flu.

Monday, December 9

Bud is back from London but not in the office yet. His meeting with the Iranians did not achieve its purpose which was to persuade them to free our hostages 1st & then we'd supply the weapons. Their top man said he believed if he took that proposal to the terrorists they would kill our people.

We had an issue briefing lunch & I'm calling reluctant Repub. Reps. trying to persuade them to vote for the tax reform even though it isn't what we want. But we need it to pass it so the Sen. can take it up & hopefully correct it.

Dick Wirthlin came by with new polling. No question but that curbing the deficit is the number one issue on the people's mind.

[Photo sessions; met Mrs. America pageant winner; haircut.]

Tuesday, December 10

[Meeting with congressional leadership to discuss Gramm-Rudman-Hollings budget bill, commented, "Not as good as we would have wanted but still a bill I'll have to sign."]

Some of the time we talked tax reform. I've done a lot of telephoning all day & still don't know whether it will pass. Bud M. back from Eng. & his meeting with the Iranian "go between" who turns out to be a devious character. Our plan regarding the hostages is a "no go."

[Bill signing for military appropriation; signed proclamation for Human Rights Day; ambassadorial formalities; photo session; attended annual congressional W.H. dance.]

Wednesday, December 11th

Several Rabbis of the Lubavitch came in to present me with a silver "Menorah" the Hebrew symbol of Hanukah. They have a giant one lighted in Lafayette Park. Then a Budget review meeting—wait til our Cabinet members see what we have for them—there will be blood on the floor. We're going to try for all the cuts we couldn't get last year.

[Photo session with widow of William Loeb, publisher of the New Hampshire Union Leader.*]*

Then a Cabinet room session with 52 top business leaders who are a Volunteer Committee to sell U.S. Savings Bonds. I had 3 meetings with Repub. Cong. men & women—about 20 in all, soliciting their votes for the tax reform. But up on the hill our Repub. Reps. led a Repub. attack that prevented the bill from coming to the floor. I saw them gloating on TV. Tip O'Neill on TV said they had "humiliated the man who led them to victory" (me) & d--n it they had.

[Greeted independent business people in favor of tax reform; after dinner, participated in holiday receiving line. **Thursday, December 12:** *staff and NSC meetings; presented awards to executives who saved the government money; lunched with Vice President Bush, noted, "of course talked about what a foolish thing the (or some of) the Repub. Congressmen had done"; cabinet meeting regarding proposed budget; met with Republican Eagles about inaugural anniversary fund-raiser; greeted March of Dimes child ambassador; met officials of National Federation of Republican Women, noted, "I took advantage of the visit to urge them to pitch for observance of the 11th Commandment in the '86 elections"; photo sessions; went outside with Mrs. Reagan and Rex to light the official Christmas tree; participated in holiday receiving line; telephoned Kentucky commander of 101st Airborne, which lost 248 soldiers in noncombat plane crash in the morning.]*

Friday, December 13th

John Poindexter (NSC) reported in on his return from short but busy trip to El Salvador, Panama, Guatemala, Costa Rica & Honduras. It was a successful trip & tied up some loose ends.

Our "Dreyfus" case with Cong. is on a shaky hold. Repubs. have offered up enough votes to carry the day if Tip will make a slight change in the rule under which the bill was blocked. Last word is that Tip is meeting with Rostenkowski. Mon. I'm going up to the hill to meet with the Repub. Conf.

[Photo sessions; desk work.]

Saturday, December 14

A bright but crisp day. Did my radio show on 2 subjects—Tax reform & Nicaragua. in the 2nd half I referred to Ortega as a dictator in designer glasses. In N.Y. at the U.N. opening he & his wife bought $3500 worth of designer glass frames.

During the day a meeting with Deaver & Glenn Campbell to see an architect's idea for the Presidential library, then a greeting to all the S.S. & familys & the W.H. military. The day winds up with a dinner for the Foundation Bd. of Governors for the Library.

The architect's presentation was outstanding. His concept is a bldg. in the Calif. mission style—very beautiful.

Sunday, December 15

At 11 A.M. went down to the East room & spoke briefly to the Foundation board of Gov's. Then at 2 P.M. Trent Lott came by to see me. He was one of the leaders (Repub. Whip) who surprised & disappointed me when he scuttled—or helped scuttle the Tax reform bill. As it turns out he'd like to be helpful but doesn't think we could get the help in the Senate to improve the Ways & Means bill. He says Chmn. Packwood (Sen. Committee) has told him he won't or can't make the changes. Trent suggests we let the bill hang & create a bi-partisan task force to work something out.

[Holiday reception and filming of Christmas show broadcast by NBC.]

Monday, December 16

This morning we flew to Ft. Campbell, Ky. home of the Screaming Eagles—the 101st Airborne. In a hangar there we were part of the memorial service for the 248 soldiers who died in the crash at Gander. I said my few words & then Nancy & I met each & every grieving "loved one," wives, mothers & fathers, brothers & sisters & children. It was a heartbreaking time. They were all wonderful—we'll never forget them.

Back in Wash. I went up to the Capitol & met with all our Repub. Reps. It was a straight talk session with all sides having their say. I told them I was asking them to vote for a tax bill that I would veto if it reached my desk as is. If they'll pass it my idea is that we get an improved version in the Sen. This will bring on a conference & here the Sen. will have my pledge that I won't sign the Dem. version.

Back to the W.H. I addressed a group working against drunk driving & then a meeting (brief) with wonderful Mother Teresa. Finished the day with a private dinner at the Richard Helms'. The Deavers & the Rex Harrisons.

Meanwhile I've been told we have 50 Repub. Reps. who will vote for the "rule" vote on tax reform. Tip has said he'll make a change to get the bill before the House if we get 50 Repub. votes.

Tuesday, December 17

An hour of meetings with 3 different groups of Congressmen. Apparently my meeting with the Repub. Conf. yesterday had an impact. The Cong. will vote on the rule & that depends on us being able to deliver 50 Repub. votes. Then they will take up the actual tax bill—if the rule has passed.

We had an issues lunch in the cabinet room but our minds were on what was happening on the hill. The end of the afternoon was a farewell reception for

Bud M. & his family. The news began to come in—the rule had passed & we delivered 70 Repub. votes. Then came the bill & it passed by voice vote.

Appropriately the evening was very pleasant. We had the Baldridges & the Sideys for an early dinner & then to Kennedy Center—all of us for the Claudette Colbert & Rex Harrison play.

Wednesday, December 18

Everybody's spirits up & no one calling me a lame duck since yesterday's vote. Cap came by for the NSC briefing. Things are going fairly well for Defense budget items.

[Signing ceremony for the Gramm-Rudman-Hollings bill; lunch with leading medical researchers on new developments; received new poll numbers, noted, "my ratings are up quite high."]

Last night I vetoed the Trade Bill which was a protectionist attempt for textiles, shoes & copper. I called Strom Thurmond—author. He took it better than I thought he would.

[Met with Shultz on his return from Eastern bloc trip; holiday reception for Secret Service agents.]

Thursday, December 19

A lot of talk last night about signing a bill before midnight & then getting the office phone calls telling me Congress had adjourned. I signed the bill alright at about 10:30 p.m. in bed. Then we waited for the 2 phone calls. When you are expecting a phone call you don't go to sleep. We turned the lights out around 1 A.M. but we weren't doing too well. Then the phone rang at 15 min. to 2. It was the operator telling us there wouldn't be a call because they didn't want to wake us.

We had a budget appeal meeting—Dept. of Defense offering a compromise figure in response to the O.M.B. proposal. I haven't given an answer yet but I definitely lean to the compromise.

Then there was a joint meeting of the Ec. policy & Domestic policy Councils & again I held my fire.

[Received list of appointee recommendations; photo for American Legion *magazine; greeted Park Mounted Police; photo sessions; videotapings; staff dinner party, entertainment by Peter Duchin Orchestra.]*

Friday, December 20

A flap has blown up about our plan to utilize Polygraph tests in trying to uncover espionage. In Europe Sec. Shultz was badgered by press who played it up as a widespread compulsory program involving 100,000 employees including Cabinet members. This of course is untrue & a distortion of an NSPG directive to come up with an anti-espionage program. I think we got it settled down now that I've had a chance to straighten it out with George who is a non-believer in lie detector tests.

[Send-off for Margaret Heckler as ambassador to Ireland; shared in birthday cake for Regan; met with NASA officials on budget appeal, commented, "Once again I'm going to settle for a compromise on funding of the Space Station that won't delay it's progress or crimp our research too much"; swearing-in of Dr. Otis R. Bowen as secretary of Health and Human Services; received Christmas card from town of Westminster, Maryland; went to Camp David.]

Saturday, December 21

Did radio show, called Walter Annenberg to ask him to serve on exec. committee for our library foundation. [. . .] Then a call to Casey to stroke him a little over the polygraph flap. Mostly it was a day in front of the fire.

[Sunday, December 22: noted cold temperatures; returned to W.H.]

Monday, December 23

Mac Baldridge came by with his report on the meeting with Gorbachev. It was somewhat similar to mine. G. on human rights gave him the same pitch I got that the basic human right was everyone's right to a job & in the Soviet U. everyone is given a job. Of course he doesn't also add that they can't choose a job—they take the one the govt. tells them to.

[Signing ceremony for farm bill; interviewed by local broadcasters with Secretary of Agriculture John Block; photo sessions.]

Tuesday, December 24—Xmas Eve

[Made telephone calls to service personnel around the world.]

We've spent most of day opening gifts so the tree will be clear tomorrow for Dick's family & Maureen & Dennis. Tonight the annual affair—dinner at the Wicks. It's Mary Jane's turn to be Santa. It was as always a most enjoyable time—all the Wicks & Doug's bride-to-be. Nancy has been the marriage broker on this one. Then Nancy's brother & his family & Maureen & Dennis. A good time was had by all.

Wednesday, December 25—Merry Xmas

The day began at 11 A.M. with gift opening & closed with a 5 P.M. Xmas dinner. All of last nights group plus the Deavers & our special guest Claudette Colbert, also Nancy Reynolds & her son. This, too was a fun time & Claudette was wonderful as always.

Thursday, December 26

Some homework & signing to do & some thank you calls. Don R. gave me news that Japan teetering between European Airbuses & Boeing 747's came down on the side of Boeing for 25 planes. It's another indication they want to maintain a solid relationship & better reciprocal trade.

Everyone (our guests) have departed so it's Nancy, Rex & me.

Friday, December 27

Sport clothes & all—off to Calif.

On Sat. the 28th did my radio cast at the Century Plaza where we were staying & taped my New Years greeting to the Soviet people. This is a big 1st. Gorbachev is doing one for us Americans—it's never been done before. There'll be no editing & we each provide our own translator.

Ron & Doria came by for dinner. In the afternoon I was measured for new hearing aids that have the ability to muffle background noise when I'm at a noisy shindig.

Sunday, December 29–Friday, January 3, 1986

Won't try to break this down by days—it was just the usual wonderful time at the Annenberg's for our New Years. We arrived in the afternoon—no other guests on hand as yet. Monday some were on hand and we got in 9 holes of golf. As usual I had some enjoyable shots but many more of the other kind. After all it was my 1st round since last year. That night was the Wilson, Jorgensen party at El Dorado Country Club. A good time—mainly with the old friends we'll be with the rest of the trip.

Ended 1985 with 18 holes of golf—shot the 1st 9 in 45. Then the New Years party and as always it was great. New Years Day—we watched the Rose Bowl—U.C.L.A. vs. Iowa. And a smaller dinner & songfest with mainly the other House guests. Of course hanging over all of this was a cloud we tried to ignore temporarily—the matter of Khadafy & his connection with the massacres at Rome & Vienna airports. We all feel we must do something yet there are problems including thousands of Americans living & working in the mad clown's country.

Thursday Jan. 2nd we left the Springs & returned to L.A. There my new hearing aids were delivered & they work. I'm the experimental animal—these are the only 2 in operation.

CHAPTER 6

—

1986

Friday, January 3

Nancy off to Phoenix & I departed for our meeting with Pres. De la Madrid of Mexico at Mexicali. It was a nice ceremony & fruitful meetings. Then after lunch on to Wash. In the meetings I tried to sell him the idea of opening his country up to outside investors. My pitch was that Baja could be one of the worlds great resort areas. He's handicapped by Spanish tradition. Constitutionally an outside resort operator would not be allowed to own the beachfront land for a hotel etc. Arrived home around 10 P.M.—Nancy had gotten here around 7.

[Saturday, January 4: radiocast; desk work; briefing for Tuesday press conference. **Sunday, January 5:** *quiet day; watched football; read.* **Monday, January 6:** *staff meeting including Regan and Vice President Bush; NSC briefing; visit from president of Mormon Church; desk work; National Security Planning Group (NSPG) meeting on Qaddafi and how to handle him and terrorism, commented, "I think we're all agreed we must do something in view of the massacres in the airports at Rome & Vienna"; issues lunch without major business due to congressional recess; haircut; desk work.]*

Tuesday, January 7

(It seems that this was a day spent on 2 issues)—one was Quadafi (Libya) (and the other our 5 hostages in Lebanon.)

First on Libya's top clown, after quite a session I finally came down on the side of an exec. order bringing Ams. & Am. business home from Libya & canceling relations—trade, etc. with them. At the same time we beef up the 6th fleet in the Med. Sea. If Mr. Q. decides not to push another terrorist act—O.K. we've been successful with our implied threat. If on the other hand he takes this for weakness & does loose another one, we will have targets in mind & instantly respond with a h--l of a punch. At tonite's press conf. I announce the ex. order.

The other issue is a highly secret convoluted process that sees Israel freeing some 20 Hizballahs who aren't really guilty of any blood letting. At the same time they sell Iran some "Tow" anti-tank weapons. We in turn sell Israel replacements & the Hisballah free our 5 hostages. Iran also pledges there will be no more kidnappings. We sit quietly by & never reveal how we got them back.

[Prepared for press conference and then presided over it in the evening.]

Wednesday, January 8

Called Dick Nixon to wish him happy birthday tomorrow. He endorses what we're doing with Quadhafi—called it ominous silence. We found out he is making noises like taking Libyan funds out of American banks ($240 mi.) so we've frozen those assets. He'll probably do the same with ours but we were afraid he was going to do that anyway.

[Received report on acid rain from Drew Lewis and Canadian Bill Davis; interviewed by independent television bureau chiefs, commented, "They asked good sound questions not zingers trying to catch me—as the W.H. press corps does."]

Geo. S. came by for a kind of year end round up re the Philippines, Middle East, Nicaragua & Soviet U. He's seeing Dobrynin this P.M. & wants to settle the next summit date as late June—our suggestion. They've been talking Sept. Then a busy taping session which included a tribute to Irving Berlin—now in his 90's. Then upstairs for rest of day & some desk work.

Thursday, January 9

[Staff and NSC meetings; learned unemployment decreased; received report from Canada-U.S. Permanent Joint Board on Defence on minor dispute over sovereignty of Northwest Passage; met with Asian American voters, concluded that more appointments must be found for Asian Americans.]

Met with 2 Congressmen, Bill Hendon & Bob Smith. Bill is way out yonder on the issue of M.I.A.'s. He claims Mil. Intelligence has all sorts of proof that Vietnam is holding live prisoners but the bureaucracy is covering it up from even the Sec. of Defense etc. After they'd gone I turned George Bush loose on it—to check out some of his stories—one of which involved me in a meeting—which never took place. Geo. reported back after lunch. I'm afraid Bill is off his rocker—even Bob (who said nothing in our meeting) told Geo. he can't go along with Bill on his rantings.

[Bought cookies from champion sales agent of Girl Scouts; greeted new members of W.H. press corps; farewell meeting with retiring political editor of Time.]

Friday, January 10

Saw Foreign Minister Abe to Japan. Last time he was in the Oval O. I had told him Congress was ready to go for protectionist legislation I might not be able to stop mainly because of their auto sales. He returned to Japan & a short time later Japan announced its <u>voluntary</u> limit on export to U.S. of their autos. Things are better now but let him know there was still a protectionist sentiment in Congress. NSC meeting about whether to take on Congress for mil. aid to the Contras in Nicaragua. Everyone had his say & I then announced they wouldn't have to wait for my decision. The answer is "<u>yes</u>" we must come to their aid. This does not mean however use of Am. troops.

[Interviewed by five European journalists; meeting with Shultz, called it "our usual Friday round up"; went to Camp David. **Saturday, January 11–Sunday, January 12:** *quiet weekend; returned to W.H.;* **Monday, January 13:** *received report of U.S. freighter stopped and soon freed by the Iranians in the Persian Gulf; greeted* Glamour *magazine's honorees; lunch with Citizens for America; meeting with Council for a Black Economic Agenda; desk work.]*

Tuesday, January 14

A St. visitor—Pres. Febres Cordero of Ecuador. Of all the Lain Am. leaders he is the closest to us in philosophy & politically. He's a self made businessman who filled his cabinet with people like himself. Our outdoor ceremony was in the cold—28 degrees plus a wind & falling snow. It was really cold—7 spectators including one Marine passed out from the cold.

Our meeting went well.

[Presentation of a painting by representatives of National Geographic; *report on organized crime by Judge Irving R. Kaufman, commented, "It is shocking the extent to which organized crime has moved in & taken control of a number of unions"; took several tests at doctor's office; state dinner, entertainment by Jessye Norman.]*

Wednesday, January 15

As usual after a St. Dinner it was a short night & an early morning. I could have slept longer. Had a meeting with Deputy Sec. of St. John Whitehead who is off on a 10 day, 9 nation trip to persuade our allies they should join us in clamping down on Kadhafi. The press has a story today about the variety of spellings they are using of his name. I notice I use the last one I've read so this book has it a dozen ways.

[Spoke at Washington's Martin Luther King Jr. High School.]

In the afternoon an interview with Trudy Feldman for the Jewish Press. I'm not sure I pleased her—she seemed to want me to ramble on about what my coming birthday meant to me. I was expecting more direct & specific questions.

A bill signing ceremony for a bill to have states provide sites for disposing of radioactive waste. Then a long meeting with Geo. Shultz & John Poindexter on our Khadafi problem, also our response to a letter from Gorbachev who surprisingly is calling for an arms reduction plan which will rid the world of nuclear weapons by yr. 2000. Of course he has a couple of zingers in these which we'll have to work around. But at the very least it is a h--l of a propaganda move. We'd be hard put to explain how we could turn it down. Then we moved on to the Middle East. We're sending Crocker over & we have to get the peace talks between Jordans King Hussein & Israel's P.M. off dead center & moving again.

A lengthy taping session, a rub down & days end.

Thursday, January 16

Staff meetings & then a meeting with our team just back from meeting with the N. Vietnamese Govt. in Hanoi on the matter of our still missing P.O.W.'s. Progress was made but no admissions about possible live P.O.W.'s being held. There is agreement on joint excavations of crash site, etc. The lady who heads the M.I.A.-P.O.W. family organization gave further evidence that Cong. man Bill Hendon is absolutely wrong in his wild charges that our govt. has info about the P.O.W.'s & is covering it up.

[Meeting with Child Safety Partnership; lunch with Vice President Bush; awarded medals to young people for bravery and service; cabinet meeting on the 1987 budget, noted, "urged all cabinet sec's. to take to the speaking circuit to inform the people about our need to cut spending"; met with Chief Justice Warren Burger on the need for further funding of the celebration of the Constitution in 1987; approved list of appointees; presented medal to civil rights activist Roy Wilkins's widow; gave award to the Air Force Academy football team.]

Photos with 2 of our Ambassadors & familys.

Then an hours meeting with Coretta King (widow). I believe I managed to erase any lingering ideas that I am a bigot.

Friday, January 17

[Met with Ciriaco De Mita, leader of Italy's Christian Democratic Party; brief talk before to American Legislative Exchange Council, conservative state legislators group.]

Only thing waiting was N.S.C. wanting decisions on our effort to get our 5 hostages out of Lebanon. Involves selling TOW anti-tank missiles to Iran. I gave a go ahead. Then office time & the swallowing of 8 or 9 glasses of Golightly & frequent walks to the little room in the hall. All part of preparation for physical at Bethesda Naval Hospital. It went well—X-rays of lung then the colonoscopy—found 3 tiny polyps, size of a half a grain of rice—all benign. Then for 2 hours I drank a quart or more of barium & waited 2 hours for it to coat my innards for the cat scan. It took 100 photos of slices of my innards about 1 milimeter [millimeter] apart—another hour & a half. My day was made when the Dr. (a woman) on the catscan said my innards looked like those of a man 20 yrs. younger than me. & so off to Camp David.

[Saturday, January 18: took two hikes. Sunday, January 19: noted foggy weather; sat by fire; watched The Color Purple (1985), called it "a magnificent job of picture making." Monday, January 20: left early due to weather; rough trip back to Washington in helicopter; haircut; dinner with Maureen.]

Tuesday, January 21

It seems like Monday but then yesterday seemed like Sunday. Geo. Bush reported on his day in Atlanta—the various programs celebrating Martin Luther King, Jr.'s birthday. Bishop TUTU of S. Africa took advantage of the day to kick me & our admin. around.

Then at 9:30 A.M. met with Repub. leadership of House & Senate. We briefed them on '87 budget also on ideas for covert help to [. . .] the Contras in Nicaragua. Then we discussed tax-reform & I made it clear I wanted it.

Later I met with Bill Casey who is leaving for Saudi Arabia. I gave him word I've received that there may be some hostile actions on our Southern border & that some of the illegal Mexican entrants are actually from the Middle East.

Dick Wirthlin came by with some amazing poll results. My general approval rating is at 74%.

[Received United Way report; desk work; annual Republican Eagles gathering.]

Wednesday, January 22

Brkfst. in the State dining room for almost all the Repub. Senators. It went very well & I think we're off to a good start. Then a meeting with Japanese Finance Minister Takeshita (emphasize the 2nd syllable).

[Telephoned Right to Life group congregated on the Ellipse; lunched with group from Public Sector Initiatives.]

Geo. Shultz came by to talk about a possible meeting with Pres. Botha of S. Africa. Having him here will present some touchy problems. We'll talk about this some more.

[Met with leaders of Right to Life movement, pledged support for them; Baker and Regan came by to discuss tax reform.]

I think the ghost of Abe Lincoln is stirring around upstairs where we live. Rex sets up a holler & goes barking down the great hall for all the world as if he's barking at someone. Finally I accompanied him all the way to Lincoln's bedroom. There he balked at going into the room.

[Thursday, January 23: breakfast with Republican senators, heated discussion with those in favor of tax increase; staff meeting; briefed one hundred business leaders on effort to induce Congress to approve sale of Conrail; meeting with Economic Policy Advisory Board, with no consensus resulting; cabinet meeting to brief them on State of the Union Address; meeting with twenty-five CEO's on tax reform; greeted youthful winners of Vocational Olympics; also Unser family of car racers; address RNC; worked out in gym; dinner at the apartment of writer Arnaud de Borchgrave, noted, "At about 7 P.M. our time our 6th fleet began maneuvers right of Kadaffi's [Qaddafi's] beach."]

Friday, January 24

Another brkfst. drop by—this time the Sens. who were elected with me in 1980—a good give & take. Over to the Oval O. for staff & NSC meetings & then 5 Afghan children were brought in. They were mere babys—but all victims of Soviet bombings. One little girl with her face virtually destroyed. Three with one arm each & one with only one leg. I'd like to send the photos to Gen. Sec. Gorbachev.

[Met with Senator Packwood, on plan to work out tax reform plan; dropped by a meeting of Republican mayors; then briefed gathering of supporters on budget; meeting with Shultz; briefed group of CEO's on budget.]

Saturday, January 25

The radio cast then upstairs for an afternoon at my desk—catching up. Put in a phone call to Margaret Thatcher. She's being rousted around by the Parliament & her integrity challenged—the 1st time ever that has happened. I told her I just thought she needed to hear a friendly voice. She was most appreciative. We had a few pleasant words & I really think I left her feeling better.

[Attended Alfalfa Club dinner.]

Sunday, January 26

A lot of homework & then an NBC interview as part of the pre-Super Bowl program. Then the game & the Bears did it 46 to 10. Two phone calls—an easy one to the Bears locker room—a tough one to the losers. I must say, however, Mr. Sullivan, owner of the Patriots & the Coach Berry were splendid in defeat & ended up thanking me for what I was doing.

Monday, January 27

At NSC briefing—with snow falling outdoors, Sen. Ted Kennedy came by—he's off to Moscow Wed. & wanted to check in to make sure he wouldn't say anything there that would be at cross purposes with us.

[Briefing for 180 citizens on budget; issues briefing lunch on budget and tax reform; meeting with twenty business executives positioned to help lobby Congress; finished work early, commented, "I think they are taking it easy on me because of the "State of the Union" tomorrow nite."]

Tuesday, January 28

A day we'll remember for the rest of our lives. Started off with a staff meeting & then a session with the Cong. leadership of both parties. Had a go around with Tip—think I came out pretty good. Then Sen. Murkowski brought in a family of 4 just recently united here in Am. An American husband, his Vietnamese wife & their 2 children who had been held in Vietnam.

Then I was getting a briefing for a meeting I was to have with network anchors—an advance on the St. of the Union address scheduled for tonight. In came Poindexter & the V.P. with the news the shuttle Challenger had blown up on takeoff. We all then headed for a TV & saw the explosion re-played. From then on there was only subject—the death of the 6 crew & 1 passenger—Mrs. McAuliffe, the teacher who had won the right to make the flight. There is no way to describe our shock & horror. We cancelled—I should say postponed the St. of the Union address til next week. Cong. closed down for the day. Nancy's brother Dick, Patty & son Geoff were already on their way here for the speech—also Maureen was here. Well they'll all be back next week.

Wednesday, January 29

Shades of Déjà vu—a meeting with the Sherpas who are preparing for the Tokyo summit. They wanted to be sure I'd approve "terrorism" as a subject. My answer? H--l yes. Fred Fielding came by to tell me he wanted to return to civilian life. As always I said yes & thanked him for all he's done. Actually he stayed longer than he'd planned to. Now he has to make some money. A memorial service has been set up for Friday in Houston for the 7 heroes who lost their lives yesterday. Incidentally, phone calls & wires poured in with nice words about my 5 min. speech yesterday on TV. Now this afternoon I had a tougher job—phone calls to the families of the 7. All of them asked that we continue the space program—that's what their loved ones would have wanted.

[Thursday, January 30: *dropped by breakfast for Republican congressmen, discussion of budget; short meeting with Angolan anti-Soviet rebel Jonas Savimbi; lunch with Vice President Bush; met with supporters who paid for television campaign in support of administration; brief meeting with Bob Tuttle; videotapings; appearance at Conservative Political Action dinner.*]

Friday, January 31

At 8:45 on way to Andrews. A.F.1 taking us to Houston for memorial service to Astronauts. Met with families at NASA Center—an emotional time. Then out to join some 14,000 people which included all the employees, family etc. of the entire space center. Nancy & I sat between Mrs. Scobee—wife of leader of the Challenger crew & Mrs. Smith, widow of one of the crew. It was a hard time for all the families & all we could do was hug them & try to hold back our tears. When it was over we flew to Andrews & then helicoptered to Camp D. It was really winter there—in the 20's & covered with snow.

[Saturday, February 1: *took walk in snow; read by the fireside.* Sunday, February 2: *noted clear weather, returned to W.H.*]

Monday, February 3

Staff meeting & NSC as usual. This time I had an issue I wanted looked into. Last nite on "60 min's." they had a segment on homeless welfare recipients in N.Y. being put up in hotels. In one case a woman & three children in a 10 x 12 room for which the govt. was paying $2000 a month. They were blaming it on the Federal govt. I thought I knew the answer but wanted it checked out. I was right—that was a practice of N.Y. City not us. Another question had to do with Scharansky. We have a deal to get him out of Russia. Last nite & this morning it was all over the news. I feared the publicity might queer the deal. Turns out the leak was from Moscow.

[Addressed several hundred appointees with gratitude.]

Then it was N.S.P.G. time in the situation room re Gorbachev's proposal to eliminate nuclear arms. Some wanted to tag it a publicity stunt. I said no. Let's say we share their overall goals & now want to work out the details. If it is a publicity stunt this will be revealed by them. I also propose that we announce we are going forward with SDI but if research reveals a defense against missiles is possible we'll work out how it can be used to protect the whole world not just us.

[Issues lunch; announced commission to investigate shuttle tragedy; received annual Boy Scout report; visit from Director of the U.S. Commission on Civil Rights Linda Chavez, leaving to enter political race in Maryland; addressed another group of CEO's on subject of budget; haircut.]

Tuesday, February 4

Some of our very conservative friends & strangely enough some liberals are getting worked up about Chevron Oil having wells in Angola & paying a royalty to the "Bad guys" the govt. of Angola while we're helping Savimbi who is trying to oust the bad guys. The situation is not as simple as they make it out to be. Chevron owns the oil wells—indeed discovered them. If they shut off the royalties—Angola will expropriate the wells, get Soviets & Cubans to help run them & they'll keep all the money. Better to pay the royaltys.

[Interviewed by print media editors.]

I phoned Widow McNair—her husband the Astronaut had done a musical tape recorded in space. It was something to do with loving our fellow man. She has wanted it played as a message from him. I was able to tell her it was part of the memorial service this morning at the Air & Space Museum. Some desk work & then upstairs to wait for the St. of the U. speech. Dr. Dick & family plus Maureen will be here for it.

Wednesday, February 5

A busy day. Short meeting with Don R. then the presentation & signing of the Budget for each of the Cong. leaders—both parties. This was followed by a Cong Repub. leadership meeting for a briefing on the budget & presentation of printed copies of each. It was a good meeting & my pitch that we had to stick together or we'd be hung out to dry was well received. Last night's speech must have been better than I thought it was. The public reaction is 95% favorable &

we're getting 3 x the response we've had previously. Many of the Cong. leaders went out of their way to praise it.

[A short but pleasant meeting with Johannes Rau, leader of West German Social Democratic party; desk work; addressed Treasury employees, then H.H.S. employees, noted "While at H.H.S. we had a meeting of the Domestic council on the need to study how we can reorganize our welfare & social programs so they can succeed in making people independent of welfare."]

Thursday, February 6

At 7:30 on our way to the annual Prayer Brkfst.—about 3,000 in attendance. Billy Graham called me a Pastor to the Nation. I must have said something effective—Nancy was teary when I sat down.

[Meeting with French Ambassador; signing ceremonies; addressed Executive Forum; noted, "A big surprise when I finished—Nancy & Maureen to my complete surprise wheeled a giant cake out on the stage."]

After a lunch with the V.P. & some present opening Dick Wirthlin came by with latest polling re the State of the Union. All in all pretty good.

Then it was personnel time followed by Roy Brewer, then Edmund Morris for some biography background & finally upstairs laden with homework & gifts. After dinner a lot of gift opening. I don't want to seem ungrateful—the gifts were beautiful but I feel guilty. I wish they'd give money to a good cause in my name.

One of our unpublic plans is underway. Sometime after midnight a mil. plane of ours will spirit Baby Doc Duvalier & his coterie of family & friends out of Haiti & on to France. We have another deal underway to get Sahkarov out of the Soviet U. but the date on that one is Feb. 11.

[Friday, February 7: greeted high school students sponsored by Hearst Foundation; staff and NSC meetings; met briefly with Foreign Minister of India, noted of Duvalier, "Our Haitian fugitive is out over the Atlantic"; visited high school in Virginia; luncheon in honor of Walter Annenberg; gave medal to longtime W.H. employee; meetings with Shultz and then Casey; large birthday party at W.H. Saturday, February 8: birthday lunch with friends, noted "I called it the 3rd day of my birthday." Sunday, February 9: Maureen and Dennis continue visit.]

Monday, February 10

Dave Packard & 3 of his commission plus Cap W. came by to give me a verbal preview of the report they'll be presenting to me re the Defense Dept. As I had hoped when I created the Commission they will report all the fine things Cap has achieved over there which will refute the drumbeat of propaganda that has most of the country believing the Dept. is an extravagant spendthrift. There will be recommendations for certain organizational changes but even here they are things Cap would like to see done. Another regional press luncheon & as always a good time. They are all so different from the blasé, adversarial Wash. press.

Interview with Lou Cannon & Hoffman—Wash. Post. Heaven only knows how that will turn out. Then I spent 2 hrs. in the family theatre rehearsing for tomorrow nite's press conf.

At 6:30 Nancy & I go to the St. Dept. & surprise Mermie who is honored guest at a reception. She was surprised & especially so because we brought Dennis with us. She thought he'd gone back to Sacramento & she'd been upset about that.

Tuesday, February 11

1st news of the day "Scharansky freed by the Soviets." After years of imprisonment he was made part of a spy swap & allowed to rejoin his wife. We flew him to our base at Frankfort & an Israeli plane flew to Tel Aviv. Later in day I received a call from P.M. Peres & Scharansky thanking us. I told them Kohl of W. Germany played a big part in putting this together.

Sen. Lugar & Cong.man Jack Murtha who headed up our observer team in Manila (the election) came in to give me a preliminary report. While they had no hard court room evidence they said the appearance of fraud in the Phil. election was overwhelming. They won't issue an official report, however until the vote counting stops which probably won't be for several days.

Here on TV we've seen day after day mobs apparently hating Am. burn me in effigy, the flag etc. According to Dick & Jack this was never more than a small group. The Filipinos love Americans they said & opened their arms to our people & hailed them as protectors.

We had an ec. council meeting on the need to help our neighbors in the Caribbean. We have a plan to increase their exports to the U.S. by allowing them to export to us clothing without limit—if it is made from fabric woven & cut in the U.S.

Next item was re the sugar quota which again is hurting these same countries. I agreed to a quota change at the expense of some of the better off countries.

Last item was to try for change in the farm bill to prevent us from subsidizing grain sales to the Soviet U. Believe it or not Cong. will give us trouble on this.

[Photo session with heart transplant recipient; attended farewell reception for Jack Block, secretary of Agriculture; meeting with Shultz on reconciling warring factions in Manila, decided to send Philip Habib as envoy.]

More practice for the Press Conf at 8 tonite. It went O.K. Most of questions were on Philippines & frankly they were all so busy trying to trap me into a headline maker they didn't ask very good Q's. Phone calls first half hour were 57 to 25 in my favor.

Wednesday, February 12–Sunday, February 15

[Flew to St. Louis for party fund-raiser; flew on to California.]

Arriving at Point Mugu we were told we could only "helo" to the Santa Barbara Airport—the ranch was closed in. We drove up—the 1st time in more than 6 yrs. All we can say about the few days there is that 3 Pacific storms nose to tail came through. We never could see the ranch around us—just fog, rain & wind. On one day a let up between showers allowed Barney, Dennis & me to cut up a fallen oak limb for fire wood & haul 2 loads of the brush to the dump. Lucky hasn't forgotten us—we have muddy prints (paws) on our clothes to prove it.

There were more birthday & valentine gifts to open—phone calls to make & our new tractor to see.

There was also homework—a statement on the Philippine election—stolen by Marcos. I ordered Admiral Poindexter to contact our friends in Japan, Korea & Singapore about what course we might follow together. Then a call came re the Shuttle tragedy. Apparently the representative of the company that makes the rocket boosters refused to sign the certificate for launch & someone in Florida got 2 Nasa officials to sign off without telling them of the negative report on the booster.

[Returned to Washington; Mrs. Reagan went to Texas until Wednesday, noted "Rex & I are lonesome." **Sunday, February 16:** *desk work.]*

Monday, February 17

A holiday so Rex & I are roommates again. Finished up the homework—including a 7 page handwritten letter to Gen. Sec. Gorbachev.

Only a haircut & a lot of desk work.

Tuesday, February 18

Back to work. Met with GOP Congressional leadership—subject Nicaragua. We're starting our campaign to get approval from Cong. for Covert aid to the "Contras." It isn't going to be easy. The Sandinistas have a disinformation program that has fooled a lot of people—some want to be fooled.

[NSC meeting regarding upcoming visit by Prime Minister Mulroney of Canada, commented, "He's had some bad times & we will send him home with some good news for his people"; lunched off-the-record with journalists Jack Kilpatrick, Bob Novak, and Arnaud de Borchgrave; photo with new appointee; visited doctor as part of preparations for Grenada trip; long meeting with Domestic Policy Council regarding "Report on Acid Rain"; met with members of Congress; ambassadorial formalities.]

Upstairs for another lonely night—just me & Rex. Nancy will be home tomorrow.

Wednesday, February 19

Watched two short films—one on my unfavorite actor, Quadafy & one on how C.I.A. agents are made. [. . .]

[Met with Governor Martha Collins (D-KY) to discuss mining regulation.]

Dick Wirthlin came in with more polling figures. Mainly the news was good but it also revealed where the drum beat of press propaganda has it's effect such as on defense spending & such. A great majority believe we are now stronger militarily then the Soviets: also no support for our helping the freedom fighters in Nicaragua.

George S. came by with some advance reports by Phil Habib in Manila. The situation with Marcos has worsened. The question really is whether he can continue to govern.

[Addressed gathering of industrial leaders to elicit support on budget; Mrs. Reagan returned home. **Thursday, February 20:** *visited Grenada to commemorate war and meet with regional leaders.]*

Friday, February 21

Most important part of day—addressed the House Repub. Conf at the Capitol Hill Club. It went well. I took Q's. which revealed some of what troubled them. I think their greatest prob. is that they see us dealing with the Repub. leadership of the Sen. & the Dem. leadership of the House. Well there is some truth to this but it's because the Dems. are the majority in the House & we can't get anything done if they don't go along. However, I think we could have done more to communicate with our Repubs. & go to the Dem. leadership with Repub. support already in line.

Had to motor to Camp David.

Saturday, February 22–Sunday, February 23

It started snowing Sat. afternoon & by Sunday morning we had several inches of new snow. I got some homework done but also had conf. calls re the Philippine situation. On Sun. morning I approved a letter from me to Marcos begging him not to resort to force. His Defense Minister & acting Chief of the Army have resigned & taken some troops with them. At one point Marcos ordered troops & tanks to go to where the rebels were based but 100's of 1000's of civilians blocked them & the tanks turned back.

Back in Wash. I met with Cap, John, George & Don in the situation room—well the room was full with reps. from St., Defense, Sec. Treasury, V.P., etc. & of course Phil Habib—just back. It was a long meeting with no disagreement but lots of frustration. Pres. Marcos is stubborn & refuses to admit he can no longer govern. I made the point that a message from me must appeal to him on the grounds that if there is violence I'll be helpless to continue support for the Philippines. We must not try to lay down the law. All we can do now is send the message by way of NSC staffer Sigure [Sigur] who is in Japan & pray.

[Attended Governors Conference dinner, entertainment by Rich Little.]

Monday, February 24

The day started at 5:30 A.M. with a call from John P. & Don R. The situation in the Philippines is deteriorating. The Marcos family & the Vers left the palace & went to the airport. Then Gen. Ver apparently talked them out of leaving. Back in the Palace they went on TV. The Pres. & the Gen. They got in an argument. The Gen. wanted to launch an attack on the military that has gone over to the anti-Marcos people. The Pres. said no. Well all of this ended sleep for me.

In the office at 9—the staff meeting & NSC were on the same subject. I was approving statements for delivery to the Pres.—pleading for no violence.

Later Gov. Lamar Alexander came by—he's a fine Gov. & a true gentleman. If Ambas. Mike Mansfield (Japan) retires, Lamar would like the post—he's no longer Gov. as of next year.

[Addressed the governors at W.H.; issues lunch; dropped by gathering of O.M.B. employees who work on budget; interviewed by associate of Jack Anderson; met with new Deputy Secretary of Interior for Indian Affairs, noted "He's a Cherokee Chief but looks like a lawyer from the East."]

Then a call from Nancy—what to say to Imelda Marcos who was calling her? At same time I'm told Paul Laxalt, Geo. S., John Poindexter & Don R. were coming in about Paul's call to Marcos. We've agreed that he should be told I'm recommending he step down & we'll take the lead in negotiating his safety & offering him sanctuary in the U.S. He says he wants to live out his life in the Philippines. Well we'll try to negotiate that. Wound up the day in the dentist's chair—time for inspection—I passed.

Tuesday, February 25

The call this morning was at 6:45. Pres. Marcos & his family & close circle I was told are in our Clark Air base. We don't know yet his destination but he's said he wants to stay in the Philippines. He has a home in Northern Luzon. In the office I was met by George S. & the V.P., Cap, John, Don etc. We are ordering our Ambas. & others to contact Aquino to see if we could persuade her to accept his staying in the Islands with a promise of security. As the day went on we learned she wasn't going to do that. He incidentally is quite ill & is bed ridden at Clark. By evening we learned his party had left by medi-vac plane for

Guam. He was carried to the plane on a stretcher. It was a day of bill signings &
a luncheon with exec. committee of our library fund.

The big deal here was an hour interview with Barbara Walters—then another with Nancy on the 3rd floor. There will be more tomorrow. It's for a
special to precede the Academy Awards.

*[Report during meeting with Domestic Policy Council indicates progress in manage-
ment.]*

Wednesday, February 26

Pres. Marcos & his party departed Guam & are now at Hickam Field Hawaii. They will stay at the base possibly 72 hrs. then he will go to his home on
Diamond Head—Yes, he has one there, too. The deal at Hickam is because of
the large Filipino pop.—there might be demonstrations. We're going to provide
S.S. protection for a limited time. So—no civil war & we've proceeded to recognize the new Philippine govt.

Had a group of Congressmen in for a pitch on aid to Nicaragua. Then met
with Board of Dir's. of Nat. Peace Institute. After lunch a meeting with Geo.
S. We're having some trouble with the lobby (Jewish Cong.) pro Israel about
our A.W.A.C.'s sale to Saudi Arabia. The Saudis are the stabilizing force in the
Persian Gulf & we're determined to lend them mil. support such as sale of arms.
The lobby opposes on the grounds of possible renewal of the Arab-Israeli war.
We've pledged to Israel we'll never let them be outgunned.

Geo. is going to lean hard on the Lobby.

Some more filming for Barbara Walters special & then upstairs—Well
there was one more Cong. briefing on Nicaragua, this time Sen's. Dem. &
Repub.

Then 8 P.M. my TV address from the Oval office—on Defense. We got
more calls & wires than on any other speech on defense & the favorable ran
91.4%. ABC put a Soviet commentator on the air to reply to my speech.

Thursday, February 27

A meeting with bi-partisan Reps. & Senators. They were warm in their
praise for our handling of the Philippine crisis. Then we moved onto defense &
Nicaragua. I threw my best pitch. We'll have to wait to see if we scored.

*[Meeting with President Paul Biya of Cameroon, found common ground on free-
enterprise approach; visit with Casey regarding selection of new deputy; met with Council
of Economic Advisors; received list of new appointees; Regan reported record closing for
Dow Jones Industrial Average and split in Federal Reserve Board attitude to economy;
greeted Easter Seal poster child, then Hubbard family, broadcasting entrepreneurs, then
Congressman James Santini, former Democrat, switching to GOP; reception for contributors to conservative causes.]*

Friday, February 28

Here it is Friday & the last day of Feb. The sun is finally shining & the snow is melting. Best of all we'll finish the day at Camp David. This morning more word about the possibility of getting our hostages out of Lebanon. This has been a long tragic time for the families. We are supposed to know by next Thurs.

[Received Packard Commission report on Defense Dept.]

A meeting with Geo. S.—we'll have to work out a group to attend Pres. Aquino's formal inauguration and he expressed optimism about getting Cong. to O.K. our aid to the Nicaraguan freedom fighters. A photo with Susan & Bee Oglesby—he's going to work for Stu Spencer. A taping session & away we go to Camp D.

[Saturday, March 1–Sunday, March 2: relaxed and did some homework; returned to W.H.]

Monday, March 3

A briefing by NSC—not all is well with Marcos. They are still at Hickam Field. It seems he does not own a home on Diamond Head & now suits are being filed by a number of Filipino Americans & the Central bank of the Philippines. So far the military has not allowed the process servers on the base. Paul L. called—he'd had a call from Marcos who asked for help with regard to papers. They have no passports so travel could become a problem.

Met with Russell Rourke our new Sec. of the Air Force. An issues lunch—plans discussed for getting Congress to O.K. aid to Nicaragua. Then 3 UNO leaders Calero, Robelo & Cruz came to see me. They are here to help sell Congress. We dropped in on a meeting of private citizens who have been raising pvt. funds to help the Contras.

[Received report from Kaufman Commission on organized crime; attended dinner in honor of Senator Laxalt.]

Tuesday, March 4—"Our 34th Anniversary!"

A light day. The usual staff meetings & a sum up on Marcos & his people who fled to Hawaii. We are appealing to Aquino to provide passports. She doesn't want him in the Philippines, so it would serve her purposes to make it possible for him to travel throughout the world.

[Met with Bennett and two hundred educators; photo session with Congressman; Cabinet meeting with briefings by Shultz and Weinberger on international affairs and defense; spent an hour briefing for Sperling breakfast the following morning.]

Wednesday, March 5

Downstairs early for a last min. report of any happenings that might have taken place overnite—then into the St. Dining room for the Sperling breakfast. It went well & the Q's. were better than they are at a press conf.

Met with Dem. & Repub. Committee chairmen & leaders whose committees will be voting on Nicaraguan aid. I asked if they could (failing support) pass the issue on to the floor without a recommendation—we'll have to wait & see. Geo. S. came in for a meeting. He's ready to battle all the way for the aid. We all came together on another matter—arms sales to Saudi Arabia. Geo. thinks he has headed off a lobbying effort by the Jewish Congress & we have word the Israeli govt. will not actively oppose us.

Over to EOB to address about 150 Presidents of major Jewish organizations—Subject Nicaragua. Very well received. Met with our speechwriters on the speech on Nicaragua. We agree on 15 min. length. Some tapings for different groups—all on Nicaragua We're making a full court press. Then wound up day with a meeting—Sen. & House Repub. leaders—this time on the '86 election.

Thursday, March 6

Gave Don a job to see how we can get citizenship for Doria's dad who has lived in Am. for 60 yrs. married to an American & father to 2 Am. offspring.

V.P. & Admiral who directed commission on terrorism presented me with the commissions report.

A good meeting with Trent Lott's Whip organization. The subject was the Nicaragua arms bill. I think they are revved up to really push it.

Geo. B. & I met briefly with one of the group who are raising money & putting commercials on TV promoting our program of support for the Nicaragua freedom fighters. He left with me a slick paper magazine—part of the Sandinista disinformation program. It's a propaganda journal making the Sandinistas sound like a mix between Geo. W. & Abe L. The freedom fighters of course are all Hitlers. Then I came upon the "subscription" envelope & there in black & white it said contributions & money for the magazine was tax deductible—we're looking into that. The mag. H.Q. is Berkeley, Calif.

[Photo sessions; reception for National Newspaper Association; attended party fundraiser.]

Friday, March 7

Charles Wick reported in on his trip to Russia on the matter of cultural & youth exchanges. There must be some spin off of the Summit. He had meetings with the very top echelon of Soviet leaders & apparently they are willing to cooperate on exchanges. A meeting with 3 top Contra leaders (Nicaragua) Robelo, Calero & Cruz. They are here to help us persuade Congress to help the freedom fighters. They are an answer to the left wing propaganda that the freedom fighters are led by former Somoza honchos. Two of these 3 were imprisoned by Somaza & the other one is a defector from the Sandinistas.

Had my regular meeting with Geo. S. & then off to Camp D.

[Saturday, March 8—Sunday, March 9: noted cold weather; returned to W.H.]

Monday, March 10—A beautiful sunshiny day

Old friend Jack Hume dropped by with a proposal to get tax reform off dead center—but I think it's already moving.

Over to E.O.B. for a pitch on Nicaragua to an overflow crowd of leaders of a number of supporting organizations. Then after our usual issues briefing lunch, a meeting with Stu Spencer to hear his views on prospects in the coming election. He told me I'm not a lame duck & the 3 important things we must do in these 3 yrs. are tax reform, deal with the deficit & make progress toward peace.

[Swearing-in ceremony for new secretary of Agriculture Dick Lang; visit by Baker regarding progress by Senator Packwood on tax reform; met with Senator Daniel Evans (R-WA) on Nicaragua, noted, "I think I moved him somewhat"; delivered pre-dinner speech at fund-raiser honoring congressional leadership; received advance report that two bodies of Challenger *astronauts had been found off Florida.]*

Tuesday, March 11

A day that should never happen again. By the end of the 1st hour I was behind schedule. It was one meeting after another, most of them with Reps. & Sens. some in groups, a few singles. The subject was Nicaragua & why they should vote for aid to the Contras. On one meeting it was however the immigration bill & another was tax reform.

[Lunch with regional press; Economic Policy Council meeting, approved moving $700 million to loan program for farmers; meeting with Meese and Baldrige; ambassadorial formalities; off-the-record cocktail party with journalists; made birthday call to Representative Michel.]

Wednesday, March 12

Saw Phil Habib off to Central Am. 1st stop Pres. Duarte—El Salvador, then the other countries. Mission to try & persuade the Sandinistas in Nicaragua to negotiate with the Contras.

Met with 4 Dem. Congressmen—all from the S. My pitch was to support us on the aid to the Contras. I don't know whether I have them or not. They want me to get their constituents revved up. It's true that the public has been steered away from this cause & has a kind of "Where's Nicaragua" feeling.

An interview with 4 journalists from the Baltimore Sun. Regular Wed. meeting with George S.—a number of items including Marcos. Last night Marcos called Paul Laxalt. He wants out of Honolulu & wants Mexico, Panama or Spain & Air force transportation. We're working on it & are trying Panama first.

[Presented National Medals of Science and Technology; photo session with volunteers on Savings Bond drive; videotapings; dinner with twenty-three major donors to Presidential Library [. . .].]

Thursday, March 13

[Greeted top high school basketball players from around the country.]

Then more meetings morning & afternoon. Most of them were here for the Nicaragua pitch but one who was in on the farm bill.

The high spot was trip over to the St. Dept. to see a display of weapons the Sandinistas delivered to guerillas in other Latin Am. countries. There were speeches by 3 men, 2 of them had turned away after being on the wrong side. One had been a communist rebel in El Salvador, another a member of the Sandinista govt. & one was a victim of the Sandinistas, imprisoned & tortured. Their stories were impressive. I wish they could appear on T.V. but I'm afraid T.V. won't do such a thing.

[Made honorary president of Olympic committee; met Young Republicans leadership, then new president of the National Press Club, then Secret Service champion pistol team; ambassadorial formalities; met with Democratic Southern senators, commented, "We had a good session on Nicaragua & they seem to be so much with us I asked them to work on some of our Repub. Sens"; met with Representative Henson Moore (R-LA.) on sharing federal offshore oil revenues with states.]

Friday, March 14

Another solid group of citizens over at E.O.B.—Jeanne Kirkpatrick & I both talked to them about Nicaragua. If they have their way we'll get the help we're asking for. Then an N.S.P.G. meeting about next week's Naval operation off Libya. Our ships & planes will now & then cross over the line Quadaffy says marks his territorial waters. It doesn't—the area is still international water open to navigation. He has threatened to take action. I have approved—indeed ordered that we respond to any hostile move on his part with appropriate action. My regular meeting with George S. He's going to the Olaf Paline funeral in Swedin. While there he's going to see if Gonzalez (Spain) will take Marcos.

[Interviewed by Hispanic TV network, mainly on the subject of Nicaragua.]

Saturday, March 15–Sunday, March 16—Camp David

We drove up yesterday—Camp D. was socked in—fog & rain. Sat. was better. Made a phone call to about 10,000 Hispanics in Miami who were holding a rally in support of the Nicaraguan Contras. It was well received. Then I called Phil Habib—just back from Central America. We do have the support of the leaders in Costa Rica, Guatemala, Honduras & El Salvador. At question is how far they'll go publicly in making that known.

Today (Sunday) a nice day & back to the W.H. Tonite 8 P.M. on TV to the Nation—subject—aid to the Contras.

Monday, March 17—St. Patrick's Day

Phil Habib came in with a more complete report on his trip to Central America. There is no question but that Central Am. wants us to continue helping the Contras.

[Gathering for St. Patrick's Day with Prime Minister Fitzgerald of Ireland and Ambassador Heckler; positive economic briefing; more St. Patrick's Day festivity at issues lunch.]

Later in the afternoon Danny Kay [Kaye] came by with a letter from the P.M. of Denmark. Then Al & Cynthia Schwabacher dropped in.

[Interviewed by Hugh Sidey; haircut; received gift of an aquarium from the Aga Khan; attended Boston College dinner for Tip O'Neill.]

Tuesday, March 18

P.M. & Mila Mulroney arrived. The sun came out & we had the usual S. Lawn ceremony with the troops & all. Then a good meeting. I hope we can work out some things in the area of trade that will benefit him. Right now pol. opponents are trying to portray him as an American puppet. After lunch I met with 14 Dem. Congressmen & 1 woman who are possibles on Contra aid. I hope we have them but no way to know. Then a meeting with Dole & Lugar who believe we have the votes in the Sen. A final session with Don R. & Pat B. to polish up the Gridiron speech for Sat. nite.

7:15 the evening begins—the State Dinner for the P.M. etc. It (again) was a most enjoyable evening. I give Nancy credit. I've been to state dinners before I had this job & they were stuffy, impersonal & dull. Our state dinners are enjoyable, everyone has fun & the guest list is always an interesting mix.

Wednesday, March 19

The pot is coming to boil on Contra Aid. Tip is putting the squeeze on his gang as if this is a personal test of loyalty to him. We've countered with a proposal that I'll hold still for a time without using the money while we try again to negotiate providing they'll pass the bill. A visit (brief) from the outgoing Foreign Minister & wife of Costa Rica. Then a meeting with the leadership of the U.S. Sister Cities program. It's been going on for 30 yrs.—They made me honorary president. Some more Congressional phone calls & then P.M. Mulroney came by & we had lunch in my study followed by a signing of a 5 yr. extension of the

Norad agreement between our 2 countries. Then I announced our approval of the Acid rain report by Drew Lewis (USA) & Bill Davis (Canada). Saw the P.M. off—he's greatly pleased by the whole trip.

A meeting with George Shultz & upstairs for the rest of the day. I have a huge bowl of gold fish given me by Aly Khan whose home we occupied in Geneva. This was a reward for taking care of his son's aquarium while we were in occupancy. I'm sending them over to Kathy for her daughter Sherrie [Shelley].

Thursday, March 20

At about 5 P.M.—it was the 1st day of Spring. That's about all I can say for it. Spent a large part of the day on the phone to Congressmen—they were voting on the aid to Contra bill. We lost—16 Repubs. deserted us—if 8 of them had voted for us we'd have won. Tip's trick worked—he promised any number of his team if they'd vote no today they could vote on the bill again on April 15th & vote yes. We think he plans to tie the bill to a lot of big spending bills & that I'll have to sign it. H--l of a way to run a country.

Geo. B. & I lunched together then we went back to our phoning. I had some taping & bill signing, a hasty N.S.P.G. meeting to plan strategy on the Contras problem.

[Approved new appointees; met breast cancer survivor; photo session with pianist Vladimir Horowitz, going to USSR in cultural exchange; greeted boxer Marvin Hagler; photos with retiring employees; dinner with Dennis and Maureen.]

Friday, March 21

Earliest news—Panama Pres. reversed himself & now says Panama won't take Marcos. We're still working on it.

Brief meeting with Ambas. Hartman (Russia). He says Gorbachev hasn't tied everything down his own way—according to recent Soviet Party Congress.

[Interview with New York Times *reporters; had off-the-record lunch with Meg Greenfield, David Brinkley, and editor of* New Republic *(Michael Kinsley); meeting with Secretary-General of U.N. Javier Perez de Cuellar; brief session with several hundred people representing groups supporting administration on Contras; finished writing speech for Gridiron Club; had dinner with Billy Graham; received report during dinner that "Tip O'Neill has mustered a gang of his favorite toadies in the House to hold a press conference demanding that I cancel today's nuclear test (underground) in New Mexico. I said a flat no."]*

Saturday, March 22

The test (nuclear) went off as scheduled—after a 30 min. delay due to high winds.

It was Gridiron dinner night. As usual the longest evening of anything we do—but fun. They put the skits & speeches between each course which makes dessert on at about 11 P.M. Mine is always the last speech & thank heaven it was well received & I got a lot of laughs.

Sunday, March 23

This evening our navy maneuvers off Libya begin. Will he (Qadaffy) open fire or not?

Monday, March 24

[Report by Abrams, back from trip to the region, that leaders of El Salvador, Honduras and Guatemala endorse administration plan to aid Contras; meeting with senators on up-coming vote on constitutional amendment mandating a balanced budget; videotapings.]

But now the real news was that Quadafy fired Sam [SAM] missiles at our planes in the Mediterranean war games & sent a couple of missile ships out toward our naval vessels. Score—no hits on any of our craft but 2 Libyan ships sunk & our air missiles took out the radar without which the Sams [SAMS] can't be fired. Qudafi claims he shot down 3 of our planes—he lies.

On top of this the Sandinistas crossed the border of Honduras & are 15 miles inside that country attacking Contra encampments. The Pres. of H. has asked us for transportation of Honduran mil. to the scene. We have helicopters there in numbers enough to do it. I've said yes. I guess I'll hear how that came out tomorrow.

Tuesday, March 25

Well it's tomorrow & Honduras has asked for more than transportation—some weapons. Under our law concerning an emergency situation in a friendly country I can give such aid so I've done so.

The score now in Libya is 4 of their Navy vessels sunk & one possible—we can't confirm. Reaction is practically unanimous that we're doing the right thing.

Scheduled to see Sen. Wendel Ford on "Const." vote. He cancelled. Said—he's with us so doesn't have to have a meeting. We had an NSPG meeting on whether to continue observing the Salt II restraints in view of Soviet violations. State Dept. put up an argument to continue doing so. Others including Cap want to give it up. I'm inclined to vote for replacing that informal agreement with our arms reduction proposal now in Geneva. Tell the Soviets we can have a real reduction in weapons or an arms race but we're not going to sit by & watch them keep on fudging.

After lunch an Ec. Policy Council meeting. I've OK'd telling European Community we won't stand by for unfair trade practices brought on by Spain & Portugal's entry into the E.C. Other subject was Davis-Bacon Act & should we cancel or what. I can't reverse myself & cancel but I've ordered that entry level contracts must be $100,000 & up. Right now they are $2,000. Well $2000 in 1931 when this started would be $100,000 now so we're just keeping even.

Made some phone calls, got a sneeze shot & I'm home.

Wednesday, March 26

[Ambassadorial formality; no news from Libyan situation; NSC meeting on deterioration of machine tool industry in face of imports; had lunch with astronomers.]

Dick Wirthlin's report illustrates how well the Sandinista disinformation network has done its job. Our people do not support what we're trying to do in Nicaragua.

Fred Fielding & family came by. He's returning to private practice. Then a photo with Mr. & Mrs. Ralph Harris (Reuters)—he's being transferred. A haircut, a visit with Cap for his report on Soviet mil. power & upstairs. Tomorrow Calif. here I come. Nancy went out a day early to visit her mother in Phoenix.

Thursday, March 27

[Flew to New Orleans for party fund-raiser; noted, "I learned that a man had appeared demanding to meet me & S.S. told him it wasn't possible. When he finally turned away one of our S.S. saw a strange bulge in his coat & grabbed him—he was wearing a gun"; flew on to California.]

At Pt. Mugu—landed in fog—Nancy there to meet me. We took off for the ranch in a smaller helicopter than Marine 1. Turns out this is an economy measure in response to Gramm-Rudman. It's O.K. with me.

[Discouraging weather forecast.]

Friday, March 28–Sunday, April 6

[Inconsistent weather; rode every day; ranch work.]

During our stay got a night time call re the bombing of the Disco in W. Berlin where 50 or so of our servicemen were wounded & killed. Evidence is adding up that the villain was Kadaffy although that hypocrite went on T.V. to say "it was a terrorist act against innocent civilians & he wouldn't do such things."

Roy Miller came up one day with our income tax forms. We really need tax reform! Final day, Ron & Doria came up—that was our rainy day ride.

[Returned to W.H.]

Monday, April 7

Staff time spent largely on terrorism & the latest info. Learned only 1 of our mil.—a soldier named Ford was killed. Then Bill Casey, Geo. S., Bill Taft +

Admiral Crowe etc. came in for meeting. Our intelligence is pretty final that this bombing was the work of Kadaffy. We have definite info. on other plans—some of which we've been able to abort. We discussed targets for retaliation. I'm holding out for mil. targets to avoid civilian casualties because we believe a large part of Libya would like to get rid of the Col.

[Report from two congressmen returning from meeting with Gorbachev, considered nothing new; issues meeting; helicoptered to Baltimore for baseball opening day; returned to W.H. for presentation to military officer leaving presidential service; received check-up and typhoid shot in anticipation of trip to Bali; dinner with friends.]

Tuesday, April 8

The 15 min. meeting with Ambas. (now Sec. of Politbureau) Dobrynin went 45 min's. My feeling is the Summit will take place—if not in June or July—sometime after the election.

[Briefing on upcoming Summit in Tokyo.]

Dick Wirthlin in with latest polling & all my ratings are up—70% in approval etc. However we don't have a majority support for aid to Nicaragua. I think it's the word "aid." People just don't understand.

[Prepared for press conference; attended reception for Advertising Council.]

Wednesday, April 9

A full—in fact 2 full NSC meetings planning targets for retaliation against Kadaffy. Our evidence is complete that he was behind the Disco bombing in W. Berlin that killed an Am. Sgt. & wounded 50 G.I.'s. We have 5 specified military targets in mind.

[Meeting with Republican congressional leadership regarding Contra aid and budget; practice for press conference; appeared before American Society of Newspaper Editors; lunch with Shultz; discussed his meeting with Dobrynin and upcoming meeting with Shevardnadze; haircut.]

Press Conf. at 8 P.M.—went well. Almost all Q's were on Kadaffy. The TV news is doing everything but telling Kadaffy we're going to hit him & when.

Thursday, April 10

Another session with Admiral Crowe on potential Libyan targets. I think it will be Mon. night. I've sent a long message to P.M. Thatcher explaining in generalities what we're up to. She has replied with a long message pledging support but expressing concern about possible civilian casualties. That's our concern also.

[Meeting on domestic policy on drug policy, suggested making a film for schools showing ugly scenes of long time addicts.]

Then a big & good meeting with about 15 Dem. Reps. on the Nicaragua-Contra situation. I think we have a chance to get the Contra aid voted this time. But Tip has opened up & my hunch was correct—he's going to tie it to a big spending bill which I just can't sign. Our job now is to get rank & file Dem's. to oppose this.

[Photo sessions and greeting visitors.]

A high spot was meeting 4 yr. old Candi Thomas Kojella. Three yrs. ago when Candi was 16 months old she was dying. I asked for a donor to give a liver to her—did it on T.V. A wonderful couple whose baby had just died came through. And here was Candi—a cute, beautiful little girl. Some more visitors including Japanese journalists then a one hour taping session & upstairs.

Friday, April 11

Anne Armstrong & Howard Baker brought in the "Pfiab" [PFIAB] report on the Yurchenko defector [Miroslav Medved] & re defector. He was definitely not a plant—he did just change his mind.

Doug Morrow came by, he's gung ho to make the shuttle trip into space even though he knows it will be delayed now. A briefing for our Nakasone visit—he'll be with us Sun. at Camp David. Then a meeting with Geo. Shultz re our coming Libya operation & our Soviet Summit. He's going to go all out with Shevardnadze to pin it down for July.

[Meeting with Baker on international currency arrangement; left for Camp David. **Saturday, April 12:** *radiocast; ride; phone calls to family of FBI agents killed in the line of duty.* **Sunday, April 13:** *visit from Prime Minister Nakasone of Japan, confided plan for attack against Libyan targets.]*

Monday, April 14

Our 1st meetings (staff) had to do with what happens tonite. For one thing I can't go to Paul Laxalt's Basque B.B.Q. At 7 P.M. our time, A.F. F-111's from an Eng. base & planes from our 6th fleet carriers will hit mil. targets in Libya. It wouldn't be seemly for me to be out socializing. Some time later I'll go on TV to tell the people the details.

[Awarded the National Security medal to retired General Lincoln Faurer; final meeting with Nakasone, considered that a warm relationship existed; luncheon for donors to Presidential Library; spoke to Association of General Contractors; brief meeting with Ambassador to Britain; had secret meeting with congressional leaders to divulge plan for Libyan attack.]

Back to the Oval O. to present the Nat. Tchr. of the Year to Guy Doud a most impressive young teacher from Minn. Then upstairs to wait for air time & first results from Libya.

Well the attack took place right on the nose 7 P.M. our time—about 11

MIN's. over the target areas. Preliminary report—all planes withdrawn but 2 of our F-111's are reported. Maybe it's only radio failure—maybe they are down we don't know as of this time. One thing seems sure—ours was a success.

Tuesday, April 15

An easy day but I'm told tomorrow will make up for it. Of course the entire atmosphere has to do with Libya. Public support is massive. The morning was taken up with study & analysis of the attack. We fear one bomb was off target & caused civilian casualties. At the same time it's possible it wasn't our bomb but one of their anti aircraft missiles that came down & exploded. One of our aircraft is missing, an F-111 with crew of 2.

[Made brief speech to American Business Conference; interviewed by Edmund Morris for upcoming biography; spent evening with Mrs. Reagan in N.Y.C.]

Wednesday, April 16

I learned that in the last two 24 hr. periods the W.H. has taken 126,000 phone calls—15,000 couldn't get through—then 160,000 & 16,000 couldn't get through. They were more than 70% favorable. The sad note is we have to assume the 2 missing airmen are dead. Evidence indicates their plane was shot down just off shore after dropping its bombs.

An NSC meeting—this time subject was Salt II & what to do about Soviet violations. The Soviets have called off the May 15 meeting with Shevardnadze. But that was after a meeting with W. German Foreign Minister Genscher.

I made a couple of phone calls to Reps. re the Contra aid. The Dems. as we expected are playing tricks & games. Over to E.O.B. faced about 140 lawyers & signed a proclamation making May 1 Law day.

[Greeting men's and women's NCAA basketball champions.]

A meeting with Geo. S. & John P.—we're getting reports there may be a revolution (which we would welcome) kicking up in the streets of Tripoli. No one has seen or heard from Quadhafy—rumors are flying like crazy. The Libyan A.F. strafed the barracks of Quadafys special guard.

Back to the mundane—a meeting with Sen's. Dole & Domenici about the budget. Cleared the air a little but not too much progress. At least they know I won't stand still for a tax increase. I phoned the 2 widows of the lost airmen—each had a young son. Dinner tonite with Dennis & Mermie. Nancy will be here tomorrow.

Whoa!—Evening news—reports Quadhafy has made an appearance & is in charge. Tripoli radio is calling on Arabs to kill Americans.

Thursday, April 17

April showers—all day. A meeting with a few Congressmen who want me to declare we'll no longer abide by the terms of the Salt II un-ratified Treaty. Frankly I'm ready. The Soviets are guilty of about 30 odd violations & show no signs of making any redress.

[Friendly meeting with Prime Minister Robert Hawke of Australia.]

Dick Walters reported in on his trip to Europe to talk to our allies about Libya. P.M. Thatcher as always was solidly behind us. I'm surprised that P.M. Chirac of France was violently against us. Mitterrand, Craxi & Kohl are easing up—Kohl & Craxi indicated if it happens again we could fly over their countries.

[Photo session; approved appointees; spoke to Republican Congressional Leadership Council; attended dinner for W.H. correspondents.]

Friday, April 18

A brief session in the office for NSC briefing etc. Learned one of the 3 murdered in Beirut is one of the 6 hostages we've been trying to get freed for almost 2 yrs. Unconfirmed reports had it that Kadhafy had been trying to buy them from the kidnappers so he could use them against us.

[Flew to N.Y.C. for party fund-raiser; went to Camp David; watched movie Winning Team *(1952) starring Ronald Reagan.* **Saturday, April 19:** *rode.* **Sunday, April 20:** *returned to W.H.]*

Monday, April 21

Prime rate dropped from 9 to 8½ & the Yen is the highest with regard to $ it's ever been—171 & a fraction. An interview with 4 wire service operators. Most questions were on Libya plus how it tied in to the coming Summit. I managed to get some things said I've wanted to say—i.e. where does Khadafy get over moaning about our bombs killing a little girl? What about the baby blown through the side of a TWA plane at 15,000 feet. What about the 11 yr. old girl mowed down in cold blood at the Rome airport. Mr. K. called that one a "noble deed."

[Issues lunch; noted that Congress was working on a spending bill that would be vetoed; briefing by farm experts, commented, "The simple truth is the whole world is subsidizing farming & farmers are overproducing," photo session for TV Guide; *videotapings; work-out.]*

Tuesday, April 22

More briefing on Tokyo. I know I'll have to study the briefing material closer to the time. I can't retain all that we talk about for 2 or 3 weeks. An interview by press reps. from each of the Summit countries. It went well—more about Libya etc. Then a meeting with several Congressmen from Oil states. They had a program of things we can do to help the oil industry which is hard hit by the nose diving price of oil. I was able to tell them we are already prepared to do most of what they are asking. They left on cloud nine & I issued a public statement. Later in the day 2 Dem. Govs. of oil states came by. We did the same routine. In between times I was phoning Sens. to persuade them to vote for a "fast track" on trade talks with Canada. This is a Finance Comm. vote & we're behind. Sen. Bill Livingston of Col.[LA] refuses to go with us because he says I don't pay enough attention to the Sens.—& they can only get my attention by throwing a monkey wrench in the works. I got a little mad!??!

[Visit from loyal California politicians; addressed Heritage Foundation dinner to honor Joseph and Holly Coors.]

Wednesday, April 23

[Met with Senator Bob Packwood (R-OR) and other members of Senate Finance Committee, most against administration plan for a free-trade zone arrangement with Canada.]

On another front it's been decided I should phone Pres. Acquino (Philippines) before I leave on trip & before I phone Marcos. Frankly, I'm not happy about it.

[Addressed International Forum of the Chamber of Commerce; photo session with foreign journalists; press availability with columnists and commentators; desk work. **Thursday, April 24:** *Mrs. Reagan on her way to Phoenix before joint trip to Asia; meeting with Bob Dole and other senators who voted to make negotiations for a free-trade zone with Canada a priority; briefing for Tokyo summit; economic session; photos with Japanese press; made appearances to thank volunteers in government; telephoned daughter-in-law, noted that she "was nervous about Cameron & Ashley what with Khadafy threatening my family. I think I gave her some reassurance."* **Friday, April 25:** *flew to L.A. on the way to Asia; reception for contributors to presidential library; taped radiocast; haircut; met Mrs. Reagan.]*

Saturday, April 26

We were airborne bound for Honolulu at about 7:30 A.M. their time. 5½ hours later landed at Hickam Field. A big reception—mainly mil. & their familys plus a big group of kids—members of "Just Say No" Club. Greeted by Gov. & top brass. Then on our way to Hemmeter home which had been made available to us. On the way surprised everyone with a side trip to a hospital. We visited a 13 yr. old Filipino boy who is a terminal Leukemia case. We had helped get a visa for his grandmother to lv. the Philippines & come to be with him—she had raised him.

Hemmeter home is right on the Ocean—a lot of sailboard action here. Admiral Hay—Pacific Commander came to brief me on our mil. setup in the Pacific. Later I phoned Pres. Marcos & Imelda—Nancy talked to her for an hour. They have not given up on returning to the Philippines & to the office which he claims is still his.

[*Sunday, April 27: took walk on the beach; desk work.* **Monday, April 28–Tuesday, April 29:** *flew to Bali with stop in Guam.*]

Wednesday, April 30

This was to be a day off like in Hawaii. It was interrupted by a briefing on the nuclear plant emergency in Chernobyl Russia. As usual the Russians won't put out any facts but it is evident that a radioactive cloud is spreading beyond the Soviet border. Had a walk on the beach & surprised at how early the sun went down. Then remembered—So. of the Equator—it's Fall not Spring & the days are getting shorter.

Thursday, May 1

This is the busy day. First a briefing then a bilateral with Pres. Soharto at the Putri Bali hotel. He's done quite a job—ending at 600% inflation rate, making his country self sufficient in rice production. They had been World's largest importer of rice. He's anti-Libya & P.L.O.

The meeting was really a plenary with both our teams on hand but he took me into an adjoining room for a one on one. He told me how he has promoted a nat. language among his 165 mil. people on 13,600 islands. The islands have phone connections by way of a satellite. They were to send up a 2nd one but the Challenger tragedy has interfered with that. He was gratified by my position on protectionism. We rejoined the others for the last 15 min. of meeting.

At 2:30 met with V.P. Lowell of Philippines. He brought a request from Pres. Aquino for $10 Bil. at 2 Bil. a yr. for 5 yrs. I told him we had a few problems of our own.

Then we joined the full meeting of the Asean foreign ministers. They have done a good job of making their economies sound—G.N.P. growth has been 6.8% a year for the last several.

That night a dinner by the Sohartos [Suhartos]—wardrobe—Batique shirts—a gift from the Pres.

Friday, May 2

Day started with a message from Wash.—the budget passed out by Sen. does not do what it should. I'll wait & see what happens in Conf.

This is the day Nancy splits off & goes to Kuala Lumpur & Thailand & I go to Tokyo. My trip was 7½ hours. It was interesting though because much of it was over the islands including the Philippines. Met at the Haneda Airport by Japanese Chf. of Protocol & Mike & Maureen Mansfield (Ambas.). Another fabulous suite at Hotel Okura—actually a penthouse.

Saturday, May 3

Taped my radio show—had a briefing session & then off to a bilateral with P.M. Nakasone at our embassy. He has problems as I do with Cong. I'm fighting off protectionism & he is too except his people oppose him as he attempts to eliminate protectionism which dominates their ec. Next meeting was with P.M. Craxi of Italy. I think we've found an answer to his not belonging—or asked to belong to the Group of 5. Then I spoke briefly to the Am. C. of C. We apparently have chambers in a number of Asian states & their reps. had come from a variety of clubs from a number of S.E. Asian states.

Dinner alone in my suite. I'll be glad when Nancy joins me. CNN has an English language channel in Japan. I ate dinner watching "The A Team" & "Hart to Hart." Then word came to me that we'd had our 3rd calamity in the Space program. A Delta missile carrying a $57 mil. weather satellite had to be blown up when it malfunctioned. You can't help but wonder about sabotage.

Sunday, May 4

Morning off—then working lunch at Embassy. Then a bilateral with Chancellor Kohl (W. Germany). We planned a little strategy to get some things done at the summit.

Then a trip to the Akasaka Palace for formal arrival. Met by Yasu—we inspected the troops etc. Learned later terrorists had launched 5 homemade missiles that missed the Palace by a country mile. Then back to embassy for a bi-lateral with P.M. Thatcher (U.K.). A good meeting as always. Then off to Hotel Otani for big reception—probably 300 people. Then to Yasu's residence for dinner—just heads of state. We got into plans for unified action against terrorism & Libya in particular. It grew late.

Monday, May 5

Usual early briefing then on to meeting at Akasaka Palace. We were ordered to drive slowly because Protocol called for me as Sr. member of summit to arrive last but Pres. Mitterrand was up to his old tricks of trying to be last. I finally said to H--l with it, let's go to the meeting. The meeting was long & arduous over wording of paper on terrorism & our statement of taking Russia to task for being secretive about the nuclear accident when there was a radioactive threat to other countries. We finally got everything we wanted including using word Libya in the terrorism document.

A group photo in the Palace grounds—then lunch at a Japanese tea house— shoes off & chopsticks, not forks. Back to afternoon meeting & lengthy hassle over G-5. We fought for inclusion of Italy & Canada & won again. Back to hotel & Nancy arrived. But I had to leave again for a working dinner.

Tuesday, May 6

The big & final day. Morning session spent on Ec. statement Yasu would deliver to press. A lot of nitpicking but in the main everything worked out. I have never seen an Ec. summit where there was so much real unity.

A pleasant lunch at the Okani then back to Embassy for bilateral with Mitterrand & Chirac—it went well. Finally the press session & Yasu's report. Back to the Hotel & into a tux for dinner at the Imperial Palace as guests of the Emperor & entire royal family. It was truly an Imperial affair with a table easily 150 ft. long with 2 shorter tables across—one at each end of long table. My dinner partners were the Crown Prince & his younger brother.

Wednesday, May 7

A press conf. at 10 A.M.—most Q's. about summit & terrorism. Then a farewell thank you to Embassy staff & on to Haneda Airport for 12:15 take off for Alaska & home—14 hours of flying & an hour for refueling. Crossing the dateline makes for our interesting situation. We leave Tokyo on a Wed. noon & arrive at the W.H. lawn at a little after 1 P.M. on Wed. In between is the refueling—an entire night & sunrise. It can be confusing. You fly into darkness & through it into daylight again. We managed to sack out for a few hours sleep & then got our meals on to Wash. time so that dinner tonite in Wash. will be at the regular time following breakfast & lunch at 8 A.M. & noon Wash. time.

A crowd of several thousand plus Geo. Bush & the Cabinet were awaiting us. Maureen is here & now dinner time approaches.

Thursday, May 8–Sunday, May 11—"Mother's Day"

[Haircut; cabinet meeting to report on trip; went to Camp David; rode and swam in heated pool.]

On Sun. I called Kay Graham about a story the Post has been sitting on but threatened to run. It refers to a really top secret achievement of ours with regard to our ability to intercept USSR communications. She promised to take the matter up with Ben Bradlee. She doesn't want to order him—says that wrecks an editor's influence on the writers etc. Sun. morning has gone by without the story appearing. I hope we're in the clear.

Sleeping is coming hard to both Nancy & myself—we're having trouble re-adjusting to Wash. time—13 hours different from Japan.

[Monday, May 12: swearing-in ceremony for James Fletcher as director of NASA, commented, "I think morale will be restored there among the people"; caught up on desk work; luncheon for Medal of Freedom recipients Earl Henry "Red" Blaik, Vermont Royster, Dr. Albert Sabin, General Matthew Ridgway, Barry Goldwater, Helen Hayes MacArthur, and Walter Annenberg; photo session with schoolchildren for Parade; videotapings; dinner party with friends the Annenbergs, the Wicks, and Ted Graber.]

Tuesday, May 13

[Met with Republican congressional leadership, delivered report on Tokyo summit, discussed budget and tax reform; learned that friend had been diagnosed with Lou Gehrig's disease; addressed high school students.]

Another Congressional meeting—this time with some Sen's. who voted against me on the sale of arms to Saudi Arabia. Now I'm trying to persuade them to uphold my veto of the thing they voted for. Jewish groups lobbied them hard & some are politically scared.

Met with Anatoly Scharansky. It was fascinating to hear the story of his imprisonment by the Soviets. I learned that I'm a hero in the Soviet Gulag. The prisoners read the attacks on me in Tass & Pravda & learn what I'm saying about the Soviets and they like me.

[Met with gathering for Responsible Government for America. **Wednesday, May 14:** *staff and NSC meetings; met with Senator Jesse Helms (R-NC), discussed why he voted against arms sales to Saudi Arabia; greeted Vietnam veteran who overcame disability; spoke to Tax Reform Action Committee; meeting with Shultz on negotiating with Soviets for reduction of land-based nuclear missiles; desk work.]*

Thursday, May 15

A brief meeting with the Vice Premier of China—Yao Yilin. He'll be here for 3 days—meetings scheduled with George S. & George B. & Sec. Baldridge. Our relationship with China seems to be on solid ground & they continue to swing more & more to a free mkt ec.

[Lunch with Vice President Bush; made appearance at luncheon for former press secretaries; signed proclamation of Transportation Week.]

A meeting with Sen. Gordon Humphrey of N.H. He's been opposed to us on Saudi arms. I think we answered his legitimate Q's. on a few points & maybe we'll have his support.

Then an NSC meeting on getting NASA back in operation of launching satellites. It's a tough problem & a costly one. We arrived at no answer.

[New poll shows approval rating of 75 percent; photo session with departing Secret Service agents; received word that House passed a budget bill; stopped by dinner for W.H. photographers.]

Friday, May 16

At staff meetings I took up matter of 5 Soviet soldiers (all 20 or 21 yrs. old) who deserted in Afghanistan & came over to the Freedom Fighters. They want to come to America. All or each of them wrote me. Their letters came to me by way of a lady from "Freedom House." I'm determined we should give them sanctuary. I'm told the problem is Pakistan through which they'd have to come to reach the U.S. Well I want to try. We had an N.S.P.G. meeting about how to break the deadlock on aid to the Contras in Nicaragua. Phil Habib reported on the meetings he's held in Central Am. re the Contra Dora effort to get a treaty signed. Our Cong. is afraid the Sandinistas will sign & won't mean it & our aid would be wasted. I think the other way—the only way the Sandinistas can be brought around is if the Contras represent a threat.

My meeting with Sec. Shultz was wide ranging. He's planning a trip to the Middle East with some ideas of getting the Arab-Israeli peace talks on track. I agree. Saw 5 Ambas's. & their familys off to their stations & called it a day. Nobody home when I got here—Rex is being operated on.

[Saturday, May 17: had lunch with members of the Honor Guard; radiocast; Rex returned in good condition. **Sunday, May 18:** *Mrs. Reagan went to Boston; watched panel shows; homework; Mrs. Reagan returned; Maureen visiting.]*

Monday, May 19

[Met with Foreign Minister Jean-Bernard Raimond of France; scheduled meeting with Jewish leaders postponed.]

In the Rose Garden a signing of a proclamation on trade week, then presentation of E & E star awards to about 20 businesses. It was a beautiful day. We issued a statement critical of S. Africa for sending armed forces into 3 neighboring countries. I have sympathy & understanding for the complexity of S. Africa's problem but there is no justification for acts of this kind. An issues briefing lunch—1st in some time. Bill Wilson came in—suggested resignation from his Ambassadorial Post (The Vatican). I was relieved & told him so. The Capitol Hill Lynch mob has been gearing up to go after him as they have so many others. He has made some mistakes—trying to stay on Board of Penzoil, visiting a high official of Libya, etc. I didn't want to see him smeared by the little gang that has done this to so many in my admin.

Another bill signing—this time S-49, the McClure-Volkmer bill that undoes so much of the 1968 Gun law which was a lemon.

[Photo sessions.]

Tuesday, May 20

Suzanne Massie came by for a half hour visit—Nancy joined me. It should have been an hour at least. She is the greatest student I know of the Russian people. She's convinced the Russians are going through a spiritual revival & are completely tuned out on Communism.

Nancy stayed with me for the signing of a proclamation naming "Just Say No" week. It's catching on as a real anti drug program for youngsters.

Prince Bandar back from Saudi Arabia told us the King has withdrawn the request to buy "Stingers." We've passed the word to the Sen. & I think picked up some votes to support my veto.

[Met with Senator Russell Long (D-LA) regarding Saudi arms deal; expressed thanks to Senate Finance Committee for work on tax reform plan, commented, "A few of them started suggesting amendments. Most felt we should leave well enough alone. I agree."]

After lunch & some desk time Paul Laxalt brought Jack Dreyfus by. Jack has spent $50 mil. of his own money on a foundation trying to get the drug Dilantin (used for Epileptics) recognized as useful in dozens of other ailments. He has no financial interest—it's purely a belief that we are neglecting a great beneficial source of relief. The villain in the case is the Fed. Drug Admin. & they are a villain. Then a meeting with a dozen Jewish leaders on the Saudi arms sale with mixed results.

[Wednesday, May 21: staff meetings; spoke to high school students; Senator Chic Hecht (R-NV) changed vote to side with administration on Saudi arms resolution; met with Shultz, on his way to Middle East; signed proclamation on risk of skin cancer; photo session with party contributors; large GOP fund-raising dinner.]

Thursday, May 22

After the usual meetings I went over to the E.O.B. & addressed the Am. Retail Fed. This group is a good supporter of our program & policies. I pushed tax reform this A.M.

Lunch in my new patio & then some desk work plus a drop by—Bill Casey on the problem of leaks. Then Jerry Carmen with his resignation as our U.N. Ambas. in Geneva. He'll be back with us in another job I think. He's done a great job in everything we've asked him to do.

A photo with the speechwriters & staff. They're a great bunch. Ben Elliott—head man is leaving. We're thinking of giving the job to Jim Brady whose Dr. says it would be good for him to have more to do.

[Presented medal to longtime W.H. employee; greeted football coach Eddie Robinson of Grambling University; said farewell to military aide.]

Then a lengthy taping session & upstairs. Two phone calls—one to the Queen of Eng. who is visiting her racing stable in Ky. And the other to an 11 yr. old Vietnamese girl who won her state's (Hawaii) contest for best essay on "What the Statue of Liberty means to me."

Friday, May 23

Last night I learned I'd been invited to stand in the "Hands Across America" Sunday at 3 P.M. I also learned the line would go right past the W.H. Security of course had ruled out my participating. Upon learning the line up was past the W.H., Nancy, Mermie, Dennis & I got the idea of having the line come through the W. gate & out the E. gate. Then we'd ask our people—staff—W.H. employ- ees, etc. to volunteer & that would take care of the security problem. Mermie & Dennis volunteered to go door to door in the E.O.B. getting volunteers. I took the idea to Don at the staff meeting & the plan is on it's way & announced to the press already. Each vol. is supposed to give $10.00. We told them it was already taken care of. Don R. & I split the bill for about 225—the number it will take from gate to gate.

N.S.P.G. meeting was cancelled due to absence of 2 members. So I lunched on the patio, then stayed there for meeting with George S. Subject was S. Africa. We're going to send their mil. attaché home just to emphasize our feeling about their raids into 3 neighboring countries.

Dick Wirthlin came in with some astounding figures. My approval rating is still 74%. Finally I did a P.B.S. interview with long time friend Bill Rusher. I think it was a good one. It will be pieced into a 90 min. special. Then upstairs for a short day.

[Saturday, May 24: *attended celebrity tennis tournament on W.H. grounds.* **Sunday, May 25:** *participated in Hands Across America day.* **Monday, May 26—Memorial Day:** *ceremony at Arlington National Cemetery; waited for word on hostage situation in Lebanon.*]

Tuesday, May 27

Well let's start by saying we still don't know whether our hostages will be freed. Bud's call revealed that 2 of the Iranians who had involved us were on the phony side. However through them Bud was put in touch with a rep. from the P.M.'s office. Outrageous demands were made by the Hisballahs [Hezbollah] such as Israel must leave the Gulan [Golan] Heights & So. Lebanon. Kuwait must free the convicted murderers they've tried & imprisoned etc. Bud said no dice so they got back to the original price—sale of some weaponry—now we'll know possibly in the next 48 hours. Next event was arrival of Pres. Azcona of Honduras. We had very satisfactory meetings with him & reps. of his Congress & Cabinet capped by a luncheon meeting. He's a solid citizen & totally approving of our aid to the Contras. Then a drop by of a meeting of top heads of all phases Fed. law enforcement.

Then an hour & a half meeting with Chief Justice Burger who is also heading up our commission to plan 1987 observance of the 200th B.D. of the Const. His main mission was to tell me he wanted—well—he didn't want to but will resign from the Court at the end of this session in July. We'll keep it very hushed up 'til closer to the date then try & announce it plus his successor & if it's a present member of the ct. who his replacement will be.

Then a reception for 175 members of Citizens for the Republic. This is the org. I created with left over campaign money in 1976.

And waiting for us after the reception were Don & John. Another call from Bud in Iran. Again they tried to exact some outrageous terms—delivery of the weapons & spare parts before release of the hostages. Bud told them deal was off. They backed down & said we had a deal—but they'd have to get through to the Hisballah in Beirut. Bud thinks Iran—conscious of the Soviet forces on their border & their own lack of competence want a long term relationship with us & this could be what's behind their negotiations. Now we wait some more. The deal is the plane carrying the material takes off from Tel Aviv. If at the end of 3 hours we have not received the hostages we signal the plane to turn back.

Wednesday, May 28

And that's just what we did—signal the plane to turn back after over ½ hour. It seems the rug merchants said the Hisballah would only agree to 2 hostages. Bud told them to shove it, went to the airport & left for Tel Aviv.

This was a heartbreaking disappointment for all of us.

[Lunch with twenty cartoonists, commented, "The laughter was continuous."]

Then Nancy & I taped a little segment for a comedy bit on Ron to be played at a big ABC shindig.

And that wound up the day.

Thursday, May 29

A meeting with the 2 new Chiefs of A.F. & Navy. Everything has been settled peacefully. They are good men.

[Addressed the National Association of Manufacturers.]

George B. & I had lunch on the new patio & then a meeting with Ed Meese on the coming retirement of Sup. Ct. Chief Justice Burger. We're holding this close until we can also name his replacement. Right now I lean to Rehnquist. A photo with Al Singer's Cabinet affairs staff. Then personnel time with Bob Tuttle followed by a lengthy Admin. time.

[Presented Harmon Foundation trophies to Senator Goldwater and NASA test pilot; received report from Boys Clubs of America; met with Public Diplomacy Advisory Board; greeted Arthritis poster boy, escorted by Victoria Principal; met retired NYC policeman running for Congress.]

Nancy & I dropped by the dinner of the Vets of O.S.S. & I rcv'd. the Bill Donovan award. Then we went on to the Kennedy Center where with the Helms & Regans we saw Chuck Heston in the "Caine Mutiny." It was a great evening & the show was magnificent.

[Friday, May 30: presented posthumous medal for Captain Joseph Rockefort, WWII cryptographer; greeted finalists of the National Spelling Bee; addressed business gathering on tax reform; went to Camp David. Saturday, May 31: rode and swam. Sunday, June 1: some homework; returned to W.H.]

Monday, June 2

[Noted hot weather; staff and NSC meetings; attended luncheon for volunteer action group led by George Romney.]

Then I went to the Map Room & did 7 tapings for various events. From there to the Situation room for an N.S.P.G. meeting on how to get a handle on leaks—particularly those revealing classified info. that threatens our Nat. Security. I'm holding out for going to the Nat. Publisher's Board & persuading them to adopt a code of ethics by which they'll come to us with tips or leaks they get which could endanger our Nat. Security. No decision as yet. A haircut & upstairs to get ready for dinner at George Stevens, Jr.'s home. The Hestons, Andy Williams & John Guares [Guare]—the playwright who won 4 Tonys last night will be there. Well it turned out to be a most enjoyable evening. Pierre Salinger's French wife was there—very charming & approving of our country.

Tuesday, June 3

[Positive meeting with Republican congressional leadership.]

A brief NSC meeting—in which we talked about yesterday's N.S.P.G. meeting on leaks & how our account of the entire meeting was on the front page of today's Wash. Post.

After lunch a meeting with Cap which is going to become a regular thing. A quick trip over to the E.O.B. to say a "well done" to our W.H. Personnel Liaison team. They round up the people we appoint to jobs & they do great work. Then a lengthy meeting with Geo. S. & a discussion about the Soviets & what they are trying to do. They certainly are sending mixed signals. A sneeze shot & upstairs for rest of day.

Sent a check ($500) off to Mrs. Sheila Petersen who has created a program—run out of her kitchen to help familys with a child requiring more medical care at home than the family can provide.

[Wednesday, June 4: visited Parris Island, South Carolina, basic training center for the Marines; attended fund-raiser in Greensboro, North Carolina; returned to W.H.]

Thursday, June 5

9 A.M. breakfast in St. Dining room with about 75 of our 100 Sen's. Dem. & Repub. It was a good session & I think reduced some acid in the partisanship. They are even talking about a clean sweep—100 votes for tax reform.

At the office later saw our Ambas. Don Lowitz off to Geneva—he is in charge of our efforts to get an agreement on chemical warfare.

George B. & I lunched on the patio then he took off for the Hill to preside over the Sen. which was taking up the matter of overriding my veto of their resolution to prevent the Saudi arms sale. About 2:30 when the dust settled my veto was upheld 34–66. For me that's a big win.

[Photo session with staff; greeted Asthma and Allergy poster child; went to Camp David.]

Friday, June 6

A meeting with Sec. of Agri. Dick Lyng. Because of the law—we are spending $24 Bil. this yr. on the farm problem. One farmer is getting $12 mil. A number are getting a mil. or more. He confirmed what I've been saying—those parts of agri. not in the govt. programs are having no problems at all. We have to find an answer to this.

[Short National Security Planning Group (NSPG) meeting regarding tactics in dealing with the USSR; lunch with Mrs. Reagan and Suzanne Massie, Soviet authority, commented, "She's convinced the govt. there is having real problems with Gorbachev & the old guard at odds. I believe this is true"; spoke to supports from all over the country; signed bill reforming government employees; pension plan; met with Robert Tuttle on personnel matters; went to Camp David; swam. **Saturday, June 7:** *swam.* **Sunday, June 8:** *returned to W.H.; attended annual fund-raiser for Ford's Theatre.]*

Monday, June 9

Framed a letter to new Austrian Pres. Kurt Waldheim but avoided using the word "congratulations." The press would have run with that because of the charges against him of having a Nazi background.

King Hussein visited—Bless him—he's still working to bring about peace negotiations between the Arab states & Israel. He now agrees Arafat & his P.L.O. faction cannot be counted on to help.

The usual issues briefing lunch & then a report or I should say <u>the</u> report of the Rogers commission on the Challenger tragedy. They did a great job for 120 days at no cost to the govt.

A meeting with Ed M. re the replacement on Supreme Ct. for Chief Justice Burger. Only 4 of us are involved in this which must remain quiet until we're ready to announce. Curiously enough each of the 4 of us came up with the same choice for Chief (Rehnquist) & the same choice of a replacement then for Rehnquist.

[Interviewed by Radio Marti; photo session for Management Magazine; *dentist appointment.]*

Finally over to the Mayflower Hotel to the CSIS conf. I made a strong pitch for aid to the Contras & was well received. Back to the W.H. for gym & dinner.

Tuesday, June 10

9:30 meeting with Repub. Leadership—A general discussion of legislative strategy. Heavy emphasis on "Aid to the Contras." Bob Michel fears Tip is going to try & put all the appropriation bills in one big costly package & since I don't have line item veto they figure on getting away with budget busting spending. Later in the morning I went over to the E.O.B. & addressed an enthusiastic

group of supporters of Tax Reform. About a dozen or more Sens. Dem. & Repub. were on the platform with me.

*[Had lunch with state leaders of Project '88 supporters group; visit from victim of
hijacking in Malta.]*

A session in the Family Theatre to prepare for tomorrow night's press conference. Then to the Dr.'s office for some of the exam (post operative). Blood
pressure 120 over 76, an electrocardiogram etc. Then home & Mommie—a
quiet evening.

Wednesday, June 11th

Staff briefings—this time with Sec. Shultz on hand. Some discussion of S.
Africa & the legislation some in Cong. are cooking up to have us take extreme
actions like taking Am. investment out of S.A.—Sanctions etc. Of course then
we'd have no contact or ability to help rid them of apartheid. Besides Am. investment is only 1% of S.A. investment.

At 11:15 went to the Air & Space museum to talk to 600 kids from all
over the U.S. who are in the "Young Astronauts" program—totally funded by
the pvt. sector. Back to the W.H. for lunch & then a meeting with 16 Repub.
Reps. & 16 Dems. all supporters of Aid to the Contras—a good meeting &
really bipartisan.

[Rehearsal for press conference; staged press conference, considered it went well.]

Thursday, June 12th

More talk about the Soviets plus discussion about budget. It's now in conference & our leader Sen. Domenici like a broken record keeps harping on a tax
increase—my word to him which he refuses to take, is no tax increase & I'll veto
it if it ever comes to me.

I had a surprise visitor. Donna Ardwin & her husband. She is the granddaughter of my mother's sister Jenny so is a 2nd cousin. She's a nice, attractive
young lady. My Aunt Jenny was a favorite of mine so I was glad to meet her.
We received the Packard Commission report. I was glad to learn it calls for less
Congressional micro management of defense.

Then a lunch with Fred Barnes, Morton Kondracke & Paul Harvey plus a
few of our fellows. It was fun & I think maybe Fred Barnes might have discovered his image of me is wrong.

N.S.P.G. meeting. It was on what to do about Soviets & their proposals for
some kind of arms deal. Are they for real or just trying for propaganda.

I finally proposed that we offer a reduction of I.C.B.M.'s & at same time
an agreement that both of us should continue research on a possible defense
against them. If & when such research should indicate such a defense weapon
is possible both of us observe "tests" & we agree that deployment must follow

elimination of all I.C.B.M.'s & then the defense be made available to all. I think our team bought it.

At 3 P.M. met with Justice Rehnquist—he is our choice to become Chief Justice & he said yes.

[Greeted high school students going on exchange program to USSR; met mother of two Marines killed in Korean War; photo sessions with retiring military officers; signed proclamation for Baltic Freedom Day; spoke for an hour with biographer Edmund Morris, concluded, "Home to Rex—Nancy is in N.Y. until tomorrow. She's speaking [to] the Foreign Affairs Council on drugs."]

Friday, June 13th

[Presented medals to federal employees; visited by Senator Jim Abdnor (R-SD), running for reelection.]

Met with Jim Fletcher & told him I wanted the Rogers Commission Rpt. implemented & that I be kept informed of it's progress. I think he'll restore NASA to its high standing.

Mother Teresa dropped by for a brief visit & to tell me she prayed for me every day. She's a most remarkable little woman.

*[Participated in regional press luncheon; met with Shultz on Soviet situation (no details in diary); ambassadorial formalities; greeted two high school students formally considered to be at risk; went to Camp David; swam. **Saturday, June 14:** rode and swam. **Sunday, June 15:** returned to W.H.]*

Monday, June 16

More S. Africa in morning briefing. Later, over to E.O.B. to speak to 200 reps. of various groups who are all gung ho for our effort to help the Contras in Nicaragua. Then a meeting with 4 Afghan leaders who head up something called the Alliance. They would like recognition as a sort of govt. in exile. We can't do that but did let them know we'd keep on helping the resistance. The usual Mon. issues lunch then an interview for 4th of July issue of "People Magazine." Frankly I thought the Q's. were kind of flat & of little interest. I wonder what the story will look like. A good meeting with Cap W.—he's all out for my idea of offering a treaty to the Soviets in which we offer to involve them in any testing of S.D.I.—if research develops that our anti-nuc. missile defense is practical. Then we share the system if all agree to eliminate I.C.B.M.'s.

A meeting with Judge Scalia—he will be my appointee to Sup. Ct. as Rehnquist moves up to Chief. We'll announce the whole pckg. tomorrow.

Then a photo with the new committee to create a museum on the battle of Normandy. It will be housed in the old German H.Q. bunker.

[Photo sessions with visitors brought in by congressmen; haircut.]

Tuesday, June 17

[*Met with President Julio María Sanguinetti of Uruguay.*]

This afternoon we astonished the press with a news break of a perfectly kept secret. I took Chf. Justice Burger, Justice Rehnquist & Judge Scalia into the press room to announce Burger's resignation, Rehnquist's nomination to succeed him & Scalia's nomination to take Rehnquist's place on the court.

A meeting with John P. & Don R. on my proposal for a proposed treaty with the Soviets to share the S.D.I. if & when it is a reality & to then do away with I.C.B.M.'s. John seems to want to sneak up on it in phases, etc. I'm holding out for a simple plan—yes or no.

[*State dinner with entertainment by Dave Brubeck.*]

Wednesday, June 18

Our U.H.O. friends Salero, Rebolo & Cruz (Nicaragua) were in for a meeting. They are here to lobby Cong. on aid to the freedom fighters.

Then a meeting with Bishop Dario Castillon of Colombia. He's supportive of what we are trying to do in Central Am. I think he was relieved to hear though that we have no plans for armed intervention. He believes that at least 840 of the 900 bishops in Latin America are supportive of what we are trying to do.

[*Gathering for flag awards.*]

A photo with officers of the Nat. Homebuilders Assn. Then a meeting with Sec. Shultz. Two subjects—Soviet's & S. Africa. He believes it is one turn to move with a proposal to Gorbachev. On S.A. the problem is once again harsh language in the Cong. which may force me to veto something.

[*Videotapings.* **Thursday, June 19:** *morning briefings; report from Mrs. Reagan that Director of Drug Abuse Policy Office Carlton Turner was feeling ignored by Department of Justice and was ready to quit, the president asked Regan to intervene with Meese; congressional meeting on Contra aid obviated by vote call; lunch with Vice President Bush; photo sessions with staff and candidates; flew to New Jersey for speech to high school students and parents; commented, "I came away feeling ten ft. tall. I like people"; returned to W.H.; drank liquids in anticipation of following day's medical checkup.*]

Friday, June 20

The check up after a couple of hours in the office & then off to Camp D. It was a long spell in the Hospital, winding up with the Cat scan. Everything was A-OK & the Dr. on the cat scan said my insides were 25 yrs. younger than my age.

[**Saturday, June 21–Sunday, June 22:** *walks and swimming; returned to W.H.*]

Monday, June 23

A half hour interview with 3 members of L.A. Times. It was wide ranging & I hope I didn't give them any possible distortion material. Then the issues briefing lunch. We made a decision that I would ask to appear at noon tomorrow before the House of Reps. to make a speech in support of aid to the Contras. Don called Tip & reached him on a golf course. Tip refused to let me speak to the House. I'm going to rub his nose in this one. We are sending a copy of the speech to every rep.

[Interviewed by Time *magazine; presentation of awards to high school students; photographed for French magazine; report from Senator Paul Laxalt (R-NV) that Ferdinand Marcos wanted the president to know that he did not instigate demonstrations in Philippines.]*

A meeting with Clarence Thomas my man on the Equal Opp. board. I've nominated him for another term. He's done a h--l of a good job. The Lynch gang on the hill is after him. Then received 9 Ambas's. The 9th was Dobrynin of the Soviet U. He brought a letter from Gorbachev. We had quite a conversation I'm not going to form a quick opinion but he sure is different from the old timers I've met. Nancy is in Calif. This place feels empty.

Tuesday, June 24

I'm supposed to be writing this at the ranch but here I am in the lonely old W.H. without Nancy.

When I got to the office Don R. told me I just couldn't go with the Contra vote hanging over us & Tip's refusal to let me address the House. Well I knew he was right but I was d--n mad. There was the matter of my appearance tonite in Las Vegas—a fundraiser for our cand. for Senate Cong. man Santini. Well we got them to postpone it until tomorrow night. At 12 noon I went on TV with the speech I would have given to the House. I spent the day phoning & meeting with House members trying to line up votes—with some success. Oh, I met the new champ Ray Floyd & his wife. He just won the U.S. Open. He gave me a new putter. I did a photo for Time mag. & Here I am upstairs with the trip scheduled for tomorrow.

Wednesday, June 25

Dick Wirthlin's poll—all things considered the figures are good. My overall approval dropped from 73 to 70 though.

[Made calls to members of Congress on behalf of Contra aid; flew to Las Vegas for fund-raiser; received news that Contra aid bill passed; went to ranch.]

Thursday, June 26–Monday, June 30

For the 1st time this year the weather was beautiful each & every day. We rode every morning, then Barney, Dennis & I cut up some downed trees. On Sun. we learned of Sen. East's suicide—a sad time for all of us. Now it is Mon. night & we are back at the White House.

[Tuesday, July 1: received briefing on good progress with SDI, commented that "Our stumbling block is the d--n Congress who keep cutting the funding which pushes delays of one or two yrs. every time they have their way"; photographed for Life *magazine; desk work; ambassadorial formalities; haircut; attended reception for diplomatic corps.]*

Wednesday, July 2

Bill Casey & his team briefed us on plans for Contra aid including training of Contra officers etc. They have really worked out a fine operation which if it succeeds will or can bring about Sandinista troop defections.

Had an interview with 2 gals from "U.S.A. Today." I think it went well. Q's. were on S. Africa, Soviets etc. George B. & I had our usual lunch then I met with Geo. S. just back from the Philippines & several other countries. He had quite an agenda for us on stepping up pressure on S. Africa. I agreed with much of it but told him I thought we should offer a plan to Pres. Botha that could offer something other than just turning the 5 mil. Whites over to a govt. by 26 mil. Blacks. I suggested something like Switzerland's "Canton" type of govt. He agreed. Then a taping session which included a toast to Lew & Edie's 50th wedding anniversary (July 14) at Universal Studios in Calif.

[Thursday, July 3: briefing for visit with President François Mitterrand of France; taped radiocast; joined Nancy's brother and his family to helicopter to Tarrytown, New York, staying at Rockefeller mansion at Pocantico Hills; joined President Mitterrand in New York for unveiling of refurbished Statue of Liberty, commented, "Upon finishing my remarks I pushed the button that lighted up Miss Liberty. It was spectacular"; returned to Pocantico.]

Friday, July 4

Left Pocantico at 8 A.M. Landed Marine 1 on Stern of battleship Iowa. We sailed slowly down the river & bay passing warships of 30 odd countries & several of our own warships who fired 21 gun salutes & of course on all sides were the 40,000 private boats loaded with people. Then we landed on Governors Island for the parade of Tall ships in company of Pres. M. & Madame M. Shortly after noon Nancy & Mrs. M. went to a luncheon at one of the Admiral's home on the island. Pres. M. & I joined Cabinet members of mine & several of his ministers at a working lunch. Pres. M. is going back to a meeting with Gen. Sec. Gorbachev. Our meeting had to do with East-West relations. I tried to give the Pres. all the info I could about our dealings with the Gen. Sec. on arms reductions so that he would know our side if he was subjected to any propaganda. It was a good meeting although Pres. M. can be very unpredictable.

[Dinner at Pocantico; celebrated evening on USS John F. Kennedy. **Saturday, July 5:** *toured Rockefeller mansion; returned to W.H.]*

Sunday, July 6

Happy Birthday Nancy! A morning phone call from John P. The V.P. candidate on the Marcos election is holed up in a Manila hotel surrounded by several thousand of his demonstrators. Had himself sworn in as acting Pres. of Philippines per orders of Marcos. Acquino is on Mindano—500 miles away. Don't feel civil war will result. Army remains loyal to Acquino. Nothing for us to get involved in.

Quiet day—then to the Wicks for dinner. Just the 4 of us I think. Still waiting for news on Nakasone & the Japanese election. Dinner turned out to be at the Jockey Club. We slipped out of the W.H.—no flags or police escort or press. It was nice for a change.

Monday, July 7

A waiter leaked the word about last night & the press is horrified that I went someplace without them knowing. They are now threatening to stake out all the gates to the W.H.

[Desk work; photo session with Antonin Scalia; short staff and NSC meetings.]

Tonite a dinner at John & Ann McLaughlin's with the Charlton Hestons & the Jack Welch's—he's the new & sensationally successful head of G.E. It was a lot of fun. We did something unusual—we slipped out of the W.H.—no we didn't. This time we went in the usual procession.

Tuesday, July 8

[Noted Congress in recess, commented, "It's so nice with Cong. gone"; economic briefing with Baker and Treasury Undersecretary Beryl Sprinkle and staff, report that economy is sluggish, projections have been revised to the downside; interviewed by John Casserly for biography of Barry Goldwater (later published as Goldwater*); lunch with Vice President Bush; interviewed by New York* Daily News.*]*

A meeting with Strom Thurmond who feels it's time to have a Sup. Ct. Justice from the South. His name is Wilkins & he's a good man. I think maybe it's time. Strom also thinks Justice Powell is soon going to resign.

[Cocktail hour with W.H. press.]

Back home & lonesome—Nancy is in N.Y. tonite.

Wednesday, July 9

Another short day—upstairs by 2:30 to await Nancy's return. A lengthy meeting re the appt. of Bob Brown as Ambas. to S. Africa. I had some doubts. He's Black. My 2 concerns were would S.A. be receptive & would run into the Uncle Tom syndrome among their Blacks & our own Black community. Bob is a remarkable person, extremely well qualified to be a middleman in S. Africa's situation. Now all that remains is for us to quietly probe Pres. Botha's attitude.

We also moved on my reply to Gorbachev's arms proposal. I phoned Pres. De La Madrid (Mexico) to personally invite him here Aug. 13 for our annual meeting. He accepts.

Al Haig came in for an unscheduled meeting—to tell me he plans to run for the Presidency in '88—has been all over the country testing the water.

[Met with Shultz, who decided to cancel Middle East trip, summit with Soviets looks definite; photo session with staff and scout troop leader active in the president's Hollywood days; awaited Mrs. Reagan's return. **Thursday, July 10:** *noted that federal discount rate had been dropped; flew to Alabama for speech on tax reform, noted, "The welcome was so warm from the moment we left the plane until we re-boarded for home that once or twice I choked up"; returned home.]*

Friday, July 11

A call from R. Nixon. He's going to the Soviet U.—a strictly pvt. meeting. Some of our people thought I should try to talk him out of it. I didn't. Visit came out of a meeting with Dobrynin whom he's known for 20 yrs. His only scheduled meeting is with Dobrynin. We both feel he'll probably end up with others possibly, including the Gen. Sec. Mr. G. I have great respect for his international talent & think he might come up with some help. An NSC meeting—briefing for visit by P.M. of Pakistan. [. . .] I'm going to lean on him a bit. Under our law I have to certify every year to Cong. that they have no bomb or aid to Pakistan stops. After lunch I signed off on some appointments including Bob Brown as Ambas. to S. Africa. Then a drop by to the E.O.B. for a meeting of our Exec. Exchange Commission. This is our program where Pvt. execs. spend a year in a govt. job & govt. execs. take their place out in the private sector.

[Met with Shultz, concerns over lack of progress scheduling Soviet summit; went to Camp David; watched movie. **Saturday, July 12:** *rode, swam, watched movie.* **Sunday, July 13:** *returned to W.H.; dinner with friends.* **Monday, July 14:** *staff meetings were routine; report by Fletcher that NASA will launch new shuttle no sooner than the first quarter of 1988; presented awards for the arts; campaign photos with Senator James Broyhill (R-NC); meeting with Weinberger about lack of provision for raise in military salaries; videotapings; haircut; Maureen and Dennis at W.H. for dinner.]*

Tuesday, July 15

[Met with Republican congressional leadership regarding upcoming campaign season; presented National Security Medal.]

Then an N.S.P.G. meeting on the subject of Nuc. testing. For the 1st time I was briefed on the nature of our testing. It isn't some part of buildup for the new weapons. It is necessary to learn when existing weapons deteriorate & also how radio activity from an explosion can disable things like communications. At my request Cap is going to put together a statement that can be declassified to allow us to tell the people—the real nature of testing. I think the public believes the testing has to do with a nuclear arms buildup.

Paula Hawkins came by—she's promoting legislation to do away with the 3% cost of living trigger for Social Security recipients. Just give them an annual increase based on the inflation rate. I surprised her when I told her I was all for it & she could tell the world that.

[Met with teenaged Republicans from all over the country; attended reception for party contributors.]

Wednesday, July 16

This was a visitor's day—P.M. Junejo of Pakistan. He has been P.M. 16 mo's. & has brought about the end of martial law & a start back to Democracy. We had the usual mil. ceremony. Then he & I met alone in the Oval O. for a short meeting. Our problem has to do with evidence that Pakistan is on the way to having a nuc. bomb in spite of a pledge to us that they wouldn't. I told him of how I had to regularly certify to Cong. that they weren't doing that in order to continue getting help from them with regard to Afghanistan. I believed him when he said he had no knowledge of that but he would get into it & pledged to me they would keep their agreement. Then we went into the plenary meeting in the Cabinet Room. I'm quite impressed with him. Then tonite was the State Dinner. His daughter is with him instead of his wife.

Thursday, July 17

A brkfst. meeting (8:30 A.M.) with Sen. Packwood & Rep. Rostenkowski re the conf. on tax reform. I'm reasonably optimistic. Rosty asked if I'd not pub-licly ding them when there were rumors & reports on supposed items. I agreed. Then we had a big NSC meeting on my letter to Gorbachev. Cap & George S. differ on some elements—such as what to say about the ABM Treaty. On this I'm closer to Cap. I want to propose a new treaty for what we do if & when SDI research looks like we have a practical system. Make the treaty now & it eliminates, if & when, any problem with ABM.

[Spoke to trade groups on protection issue; had lunch with Vice President Bush; met with Shultz, mainly regarding South Africa, agreed president should deliver speech on the subject; approved new appointees; photo session with departing W.H. employees.]

Got a call that the Sen. had ratified the extradition treaty with Eng. 87 to 10. This will stop the U.S. from being a shelter for Irish terrorists. I called Margaret Thatcher—tracked her down at a dinner party. She's delighted.

Friday, July 18

Well we finally came up with a letter to Gorbachev that I can sign. In fact it's a good one & it should open the door to some real arms negotiations if he is really interested. Sir Geoffrey Howe—Eng. Foreign Minister came by. He's off to S. Africa to see Botha. We had a good talk & I suggested some points for his talk which he bought. After lunch a pleasant get together in the Rose Garden with the "Girl's Nation"—these are high school girls—sponsored by the Am. Legion Auxiliary. Lorraine Wagner & her family came by. She goes back to 1943 when she was a teenage member of my fan club.

The press has taken a crack at Don Regan charging that he said Am. Women should be against S. African sanctions because it would cut off their diamond supply. We have a tape of the off the record backgrounder in which he was supposed to have said it & he said nothing of the kind. Cong.man George O'Brien (a good man) died today of Cancer. No Camp D. this week.

Saturday, July 19

No Camp D.—Nancy packing for the royal wedding in London. She leaves at 9 A.M. tomorrow. I did my radio speech. I'm really upset. We have to hold off on Bob Brown as Ambas. to S. Africa. There is a grand jury hearing in S. Carolina about some college laundering hot money from Nigeria. At one time Bob's p.r. firm handled Nigeria. He'll be involved in the hearing. Frankly I don't believe he'd ever be involved in anything shady.

Sunday, July 20

Well Mommie is on her way to Eng. & Rex & I are alone. The Bushes took pity on one of us (me) & invited me over for lunch.

Monday, July 21

[Signed proclamation for Captive Nations Week; had lunch with contributors to presidential library.]

Later Bob Dole, Nancy Kassebaum & Dick Lugar came down to talk about our S. Africa policy. I made it plain I could not support punitive sanctions. Tomorrow I make a speech outlining our policy on S. Africa.

[Photo sessions with Indianapolis 500 race winner, and others; videotapings.]

Tuesday, July 22

[Met Future Farmers of America.]

Then at 2 P.M. did my speech on S. Africa in the East Room before an audience of Diplomats, Congressmen & reps. of all kinds of groups involved in foreign Relations. It was live on CNN TV also. Of course when it was over & Cong. Grey at 4:00 P.M. broadcast the official Dem. reply—(pure demagoguery) & the media sought out every enemy they could & put them on TV I was a colossal failure. I don't think I was. Dick Nixon called & thought it was masterful. Met with Morris Abrams who is leaving our Civil Svc. commission to go with a Jewish org. as C.E.O. He's done a great job. Then Dr. Thomas Paine came in to present me with the report by the Nat. Commission on Space. Finally a reception for the Congressmen & their spouses who voted for Contra Aid—this was bipartisan at its best.

[Wednesday, July 23: *flew to Dallas, Texas, for campaign rally; flew to Miami, Florida, to campaign, commented, "We motored to the Hotel Inter-Continental—Jeb Bush met us & we went to the lobby where I spoke to several thousand very enthusiastic Repubs";* made further appearances. **Thursday, July 24:** *met nine-year-old cancer victim; flew to South Carolina for schedule of campaign appearances; returned to W.H.; appeared at reception for National Republican Senatorial Committee Trust, party donors; awaited return of Mrs. Reagan from wedding of Prince Charles and Lady Diana in London.*]

Friday, July 25

French Nat. Assembly Pres. Jacques Chalan-Delmas in for a photo & brief meeting. He's a great friend of P.M. Chirac. He believes the P.M. & the Pres. are trying very hard to work together. He's strong for privatizing the business & industry Mitterrand nationalized when he came into office. Some progress has been made. A Domestic Policy Council meeting on drugs. It had to do with going nationwide to mount a national anti-drug campaign on every facet of the trade. One plan for starters is for Nancy & me to share in a TV speech. There was much praise for Nancy on what she's accomplished. Geo. S. said her work with other 1st ladies has had a worldwide impact.

[*Had lunch with Vice President Bush; greeted participants in Boys Nation; had regular meeting with Shultz, noted, "He was reluctant but I think I convinced him to send a team to S.A. to follow up on my speech"; photo session.*]

Saturday, July 26–Sunday, July 27

Nancy taking it easy—she picked up some kind of throat virus in England. Weather was supposed to be rain & thunder storms. Actually after some morning showers Sat. it turned out alright. Sat. good word one of our hostages—Father Jenco was released in Lebanon & turned over to the Syrians. Now on Sunday he's in W. Germany on his way home. The Hisballah sent a video tape out with him on which one of the remaining hostages—Jacobson dressed me & our govt. down for, as he put it, not lifting a finger to try & get their freedom.

This release of Jenco is a delayed step in a plan we've been working on for months. It gives us hope the rest of the plan will take place. We'd about given up on this.

We're back in the W.H.

Monday, July 28

[*Staff meetings; met with Weinberger, who was concerned about move in Congress to make military fringe benefits subject to income tax.*]

Then a meeting in the Roosevelt Room re a part of farm bill that would have us sell subsidized grain to the Soviet U. The conflict is between subsidizing agri. to make our farm produce more exportable & unfairly competing with countries like Australia or seeing our surpluses grow plus favoring the Soviet U. & China. Of course it is contrary to our efforts to persuade the allies to wipe out protectionism. I have to make a decision & soon. Frankly I don't have one.

[Issues lunch; presented Medal of Freedom to Vladimir Horowitz; haircut; met with Jim Broyhill about move in Congress to raise cigarette tax, the president pledged to fight it.]

Then a stop at the Dr.'s—I've been passing blood in my urine since last night. He thinks it's an inflamed prostate.

Talked on phone to Father Jenco in Germany. I'm looking forward to meeting with him.

Tuesday, July 29

Sen. Jake Garn came in to see me about the 4th Shuttle to replace the Challenger. The problem is money & O.M.B. has proposed that we delay till 1988. This will set back our plans for a Space Station etc. The alternative is to put up a half Bil. in the '77 ['87] budget & start then. That's what Jake wants. From him I went in to an NSC meeting on the same subject. Defense, State, Transportation, Justice & NASA go for the '87 start. The problem is now mine & it's a tough one. I want to do the '87 bit myself but we face a $220 Bil. deficit in '87 because Congress won't give us the spending cuts we've asked for.

[Had lunch with Regan; spoke to W.H. and Capitol Hill interns; received steady poll numbers.]

Wednesday, July 30

Our staff meetings were a little more meaty than usual what with S. Africa, Congress stirring about sanctions etc. To add to that the Sen. has opened the hearing on my appt. of Rehnquist to be Chief Justice. The 1st meeting was last evening—highlighted by vitriolic attacks on TV by Sens. Kennedy, Metzenbaum & Biden. They really are a lynch mob.

At 11:30 A.M. I addressed the leaders of a number of org's. ranging from Kiwanis to Knights of Columbus & Girl Scouts & a dozen others. The subject was their involvement in a national effort to wipe out the drug problem. At lunch I read a letter & a speech sent to me by Buthelezi the Zulu Chief in S. Africa. Both were the most statesmanlike works I've seen in a long time. I've never heard or read the case against sanctions expounded better.

Geo. S. came by and I gave them to him to read. I wish everyone could see them. We will probably name our present Ambas. to Liberia to go to S. Africa. He's a remarkable man & self-made. He's also Black.

Thursday, July 31

 Dropped by a breakfast in the St. Dining Room—about 50 or 60 members of Cong.—both parties. They'd heard from Cap, Geo. S. etc. I spoke to them about not undercutting our negotiations with the Soviets by slashing the defense budget—then I took Q's.

 [Desk work; had lunch with group from Vote America Youth Initiative; met with Lyn Nofziger who was pressing for L.A. as site of 1988 GOP convention, concurred; photos and film clips with congressional candidates; visits from Governor James Rhodes (R-OH), disfigured Vietnam veteran, and others.]

Friday, August 1

 [Met with members of Congress supportive of veto of trade bill; visit from bicyclist Greg LeMond; photographed for Success *magazine; interviewed by* Newsweek.*]*

 Lunch on the patio then a meeting with the Joint Chiefs of Staff. They have gone along willingly & on some matters enthusiastically with instituting changes recommended by the Packard Commission, but they brought to my attention some changes (organizational) being proposed by Congress that would be counter productive to say the least. George Shultz dropped by for a little more talk about S. Africa & then came the high spot of the day. The arrival of Father Martin Jenco—just released by the Hisballahs in Beirut after being a hostage 19 months. His family were all with him & it was an emotional experience. There was a hint of his possibly approaching the Stockholm syndrome*—or maybe it was just his holy forgiveness of even those who had so mistreated him.

 Then it was off to Camp David where we ran a video tape of a new film— well at least none of us had ever heard of it. It was "The Last Days of Patton." Geo. Scott again played Patton & it was wonderful.

 *[***Saturday, August 2:*** *Mrs. Reagan still under the weather; rode and swam.* **Sunday, August 3:** *watched panel shows, main topic was congressional opinion against Chief Justice nominee Rehnquist; returned to W.H.; Mrs. Reagan planning trip to NYC for several days [. . .].]*

Monday, August 4

 A Cabinet meeting on the campaign we are launching on the drug situation. We are all agreed—on matters such as offering healing & treatment for users but stiff penalties for pushers.

 [Addressed three hundred appointees; met with congressional leadership on drug campaign; made press announcement about it.]

---

* *Stockholm syndrome is a mental response to abduction or even abuse in which the victim feels loyalty only to his or her tormentor.*

A meeting with Doug Morrow, he's here for a Space Council meeting. He says morale at NASA is shot & urges announcement that we go forward with the 4th Shuttle. Then I saw 3 Ambassadors & their familys off & went home.

Tuesday, August 5

This was a day to test the soul. At 9:30 I met again with the Repub. Leadership of House & Sen. The subjects were several of the issues before all of us—starting with my veto of the textile bill & went all the way to the Contras. We began to get behind schedule. Then a short meeting with Sec. Shultz. He's thinking of retiring. Later a meeting with Cap—just to report on what the mil. has accomplished on drug control—a 67% reduction.

After lunch a meeting with Barry Goldwater & Sam Nunn. They presented a report on what the proposed cuts in defense budget would do. They would wipe everything we've accomplished out.

[Met with Domestic Policy Council about finishing Petroleum Strategic Reserve; telecast speech to Knights of Columbus convention; taped campaign ads; cocktail party with reporters, main topic of conversation was South Africa.]

Wednesday, August 6

Started the day again with a Congressional meeting on leadership plus some members with concerns about S.D.I. They've been reading things in the press that disturbed them. I had to make them understand the press doesn't know what it's talking about.

Then we had an N.S.C. meeting. Geo. Shultz brought Ambas. Crocker in who has been talking S. Africa with our allies. Most of our allies share our reluctance to impose punitive sanctions.

[Met with Weinberger regarding defense budget.]

After lunch I dropped by the Roosevelt Rm. to greet some ball players—the Texas Rangers who are in the pennant race in the American League.

Then it was on over to E.O.B. to help brief a group of citizens who are active in supporting S.D.I. Back at the W.H. I left for the Hyatt Regency Hotel to address the Nat. Conf. on Alcohol & Drug Abuse. Found out I'm the 1st Pres. to ever do that. By the time I was being introduced I was given word that my veto of the infamous trade bill had been sustained—276 votes to override but 149 (we needed 142) to sustain. I announced that to open my speech. I'd been calling Congressmen for 2 days asking for help. I guess it paid off. Came back to the W.H. Don Regan came upstairs with me & we talked about George S. & his retiring. I'm afraid he may be doing it thinking I'm a little cooled off. That isn't true. Don's going to try & find out if he really wants to return to pvt. life or if something is bugging him. As far as I'm concerned he can stay as long as I'm here. Mermie is coming back from NEW ORLEANS this evening & tomorrow Nancy will be home—hooray!

Thursday, August 7

[Briefings on South Africa, considered that President Pieter Willem "P. W." Botha may be growing more flexible.]

Then it was desk time until lunch with V.P. Well one more thing before lunch—Ed Meese & Peter W. (W.H. Counsel) came in. The lynch mob (Sen. Judiciary Comm.) wants again some confidential papers from Justice Dept dating back to when Scalia was a Deputy Attorney Genl. I should be able to claim exec. privilege but if I do, they'll just hold Scalia hostage. I therefore have agreed to waive my right in this case.

[Met with gubernatorial nominee from Michigan; had Economic Policy Council meeting on NASA, noted. "I came out with what I always wanted—we'll start the new shuttle in '87 not '88 and we'll shift unmanned satellite launches to a pvt commercial business."]

Ended the day with a bunch of photos & acceptance from American Hellenic Education Progressive Association (AHEPA) of their Socrates Award—their highest honor. Up stairs & waited for Mommy to come home & come home she did—Rex got to her 1st but that's because he can outrun me.

Friday, August 8

Fridays are getting to be my favorite day—even when like this one I won't be going to Camp D.

Main thing this morning was an interview with Fortune mag. There were 3 of them and they were doing a feature on leadership. I think they're trying to suggest that I can give tips to the business & corporate leaders who read Fortune. I told them I thought I was imitating the successful business leaders. I think it was a good session, now we'll see how they write it up.

[Positive economic briefing; greeted twelve-year-old African American musician.]

Now I'm waiting to see the Dr's who are going to give me a bladder exam tomorrow. A team is in from Mayo—friends of Nancy's brother. Of course the press has me at death's door & I never felt better in my life. This really is a check up I should do every several years since 20 years (almost) ago Burton Smith operated & removed a couple of bladder stones. There has been no recurrence so far.

*[*Saturday, August 9*: radiocast in defense of William Rehnquist's and Antonin Scalia's Supreme Court nominations; medical examination, with results entirely favorable; watched TV in the evening. *Sunday, August 10*: homework.]*

Monday, August 11

An NSC meeting—briefing for Wed. meeting with Pres. de La Madrid of Mexico. We're getting good cooperation on the drug problem but their economic woes are very serious. I'm going to approach him on things he could do to open Mexico to outside investment and more entrepreneurship.

Then an issues briefing lunch. They're usually pretty pleasant & this one was too.

I met later with Steve Symms, Chic Hecht, and Elizabeth Dole. The two Cong. men want to return speed laws to the states & so do I. Liz is worried that lifting the Fed 55 mile limit may lead to an increase in traffic fatalities. I held out on the side of turning power back to the states—it was none of Wash's business in the 1st place. I think Liz left a little upset—but the 2 Reps [senators] were happy and so am I.

A meeting had been scheduled with the editor of *La Prema*—the paper in Nicaragua the Sandinistas have closed down. His family persuaded him he might not be allowed back in Nicaragua or if he was, he might not live long.

[Rehearsed for press conference the following day; haircut; videotapings. Tuesday, August 12: flew to Springfield, Illinois, for campaign appearance at fair; flew to Chicago; party fund-raiser; returned to W.H. Wednesday, August 13: briefings and then meeting with President Miguel de la Madrid Hurtado of Mexico, covered Nicaragua, drugs, and Mexico's economic problems; met with Peter Ueberroth regarding cost of drug program; photo session with congressmen and guests, including Miss Teenage America.]

Thursday, August 14

An NSPG meeting on our continued attitude toward Qadhafi. He's got some problems & I'm pleased to say a lot of his people are unhappy with him. He's been pretty quiet but if he lets go with another terrorist act we're ready to respond immediately. Geo. B. & I had a pleasant lunch on the patio. Then it was a meeting with Cong. Boulter of Texas. He had 2 oilmen with him. We discussed the problems of the industry as a result of the price collapse. We could find ourselves so dependent on OPEC oil that it would be a Nat. Security problem.

[Meeting with Shultz who decided against retiring; with Bolivian cabinet ministers regarding cooperation on drug problem; photo sessions with departing staff members, W.H. Fellows, and reporters.]

Up to change into slacks for the BBQ. All of the Congress & their families—1st time their children have ever been invited to an affair in the W.H. It was a fun night & everyone seemed to enjoy themselves. I made a decision—next year there will be no printed programs. I think I autographed every d—m one.

[Friday, August 15: NSC meeting with briefing on upcoming visit by President José Sarney Costa of Brazil; addressed leaders of Conference on Small Business; taped radio-cast, commented, "subject—the Houses irresponsible witling [whittling] away at Defense. They're doing everything but give the soldiers close order drill by legislative action"; photo sessions with W.H. Fellows; packed for trip to L.A.]

Saturday, August 16

A stop at the Drs. they dropped some kind of acid on 2 tiny spots on my face that they thought possibly become the kind of Basel Cell Carcinoma I had on my nose. No big deal—I have a couple of pink spots (tiny) and that's it.

Then it was Marine 1—A.F.1 and Calif here I come. I got to the ranch ahead of Nancy & had time to get in blue jeans & meet her when she helicoptered in. The ranch is beautiful as always and Dennis & Barney have already picked out a couple of Oak trees—nearly dead that we can spend a few afternoons on.

[Met family of Vietnam MIA. **Sunday, August 17***: ride and ranch work.* **Monday, August 18–Tuesday, August 19***: ranch work; helping Mrs. Reagan get used to a new horse.]*

Wednesday, August 20

A morning ride & back to our Madrone thinning. I'm really getting some exercise & I feel it. Things seem quiet on the Presidential side—so far I'm just a rancher. I called Bill Rusher about some of the items in the *National Review* that echo the errors in the press about our position on arms control & SDI. He was pleased to get the inside info. Weather still hot.

*[***Thursday, August 21***: more ranch work; called* Washington Post *columnist David Broder to correct recent column charging that he did little to reduce deficit.]*

Friday, August 22

It was Madrone thinning again in the A.M. Caterers were here putting up a tent for tomorrow's delayed Birthday party for Nancy given by the Wilsons & Jorgensens. After lunch they (the Wilsons & Jorgensens) arrived. Bill & Earl & I went for a ride. There was some late afternoon paper work, a decision memo I haven't signed yet. Cap W. is concerned that if I pursue an interim reduction of the Intermediate Range Missiles without an agreement to total elimination at a time certain we'll never get our zero zero goal. On the other hand others—John P. etc. think I should go ahead & take our chances on zero. Frankly Cap has a point. I'll confess I've been talking (in my letter to Gorbachev) of an interim agreement. In my fuzzy way I guess I was seeing it as a 2 step agreement not 1 agreement & then try for another.

*[***Saturday, August 23***: hosted birthday party, noted, "Even a little excitement when someone left the back yard gate open & 2 of your horses joined the party."* **Sunday, August 24***: received news of death of NSC employee from cancer, telephoned widow; ranch work, prepared barn for new floor.* **Monday, August 25***: rode; cleared trail; telephoned party fund-raiser in Las Vegas.]*

Tuesday, August 26

Into Los Angeles for a few days of seeing friends etc. We left the house at 10:40 A.M. and arrived at the suite at 11:45. Coming down the coast in the helicopter I watched for landmarks I remembered and was a little upset when I could locate them & then couldn't remember their names—Topanga Canyon for example. After lunch a haircut & manicure—both overdue. Then George Scharfenberger one of the trustees of our blind trust & a lawyer came by—mainly to talk about our wills in view of the changes in inheritance taxes. My 1981 tax change eliminated such tax on surviving spouses. Then Mike Abrams to check on how we're doing with his workouts. He was quite pleased with me.

[Mrs. Reagan out on errands; signed bill on funds for territories. **Wednesday, August 27:** *Mrs. Reagan out much of the day; signed bills; underwent ear test and hearing was deemed to have improved; received new aids; dinner at Bloomingdales' home.* **Thursday, August 28:** *desk work all morning; reception for presidential library fund-raisers; dinner with friends.* **Friday, August 29:** *Mrs. Reagan worked on domestic chores; had lunch with speechwriter Ken Khachigian regarding September 14 telecast on drug problem; returned to ranch.* **Saturday, August 30:** *rode; interviewed by C-SPAN and AP; met with local residents hoping presidential library would be located in Santa Barbara vicinity, commented, "They didn't know it was already set for Stanford"; received word of U.S. News & World Report journalist arrested in USSR and charged with spying; dinner with friends, watched TV in evening.* **Sunday, August 31:** *rode; pruned trees.]*

Monday, September 1—Labor Day

For the 1st time this trip fog rolled in. We changed our planned ride and rode up Bald Mountain from where we could look down on the fog. The whole ride was in sunshine. Another afternoon of tree pruning—same location. Ron was on "Good Morning America" subbing for David Hartman. He handled himself very well. My only criticism was his neck tie. I'll have to talk to him about solid colors.

Tuesday, September 2

Another morning of fog but we managed to find sunshine. A sunny afternoon & we took down a dead Oak tree down by Pa. Ave. Finished the job. Nancy cut ride short because "No Strings" was stumbling. Dr. says he can only be ridden on the flat & not every day.

Wednesday, September 3

Nancy didn't ride. Dr. was testing "No Strings"—problem is age & a little arthritis. Rest of us rode & then in P.M. Dennis, Barney & I went down to Snake Lake and cut up a giant Oak that had tipped over. It was on a steep slope & hard going but we got it done.

The Soviets are holding American journalist (U.S. News & World Report) charging him with being a spy. It is of course a frame up & the 4th time they've done it. Each time we have arrested one of their KGB agents they have done this. The last time before this one was in '78. Each time before they grabbed an American businessman. Then they try to arrange a prisoner exchange.

Thursday, September 4

Had our ride, but 1st I called Geo. S. re our man Daniloff in the Soviet Union. I asked his opinion of my thought that perhaps I should communicate directly with Gorbachev & tell him Daniloff was not working for our government. At about 5 P.M. I signed such a message.

[Splitting timber; ranch worker had finger tip crushed.]

9 P.M. a call from John P. A Pam Am plane with 300 + people on board bound from Bombay to Frankfort landed in Karachi. Three or 4 men in some kind of military uniform hijacked it but the crew got away so the plane can't be flown. Hijackers are demanding that a crew be provided to fly them to Cyprus.

My next call was for an OK to send our trained experts in hijack situations to Pakistan. They got it.

Friday, September 5

By now we know there are around 380 or 90 people on the plane & possibly 80 or more are Americans—one a naturalized citizen of Asian birth has been shot & thrown out on the tarmac.

We went for our next to last ride this trip. About half way our Secret Service agents got a radio call for us to pull up—the motorized group trailing us would catch up—they had an emergency call for me on secure phone. I learned that the Pakistanis had captured the 4 hijackers and there had been shooting & a great many casualties—no figures as yet.

Later in day got more details—roughly 18 dead & 100 or more wounded. 2 of the hijackers were killed, 2 captured & there is a possible 5th. Ron & Doria arrived in afternoon & stayed for dinner. It was a nice visit.

[Saturday, September 6: rode; cut wood; watched video on President José Sarney Costa of Brazil; radiocast.]

Sunday, September 7

Left the ranch at 11:45, as usual it was a beautiful day. Had lunch at the hotel & spent the P.M. with some homework. Word came the Soviets were going to officially charge Daniloff with espionage. Gorbachev response to my letter was arrogant & rejected my statement that Daniloff was no spy. I'm mad as h--l. Had a conference call with Geo. S., John P., Don Regan. Decision was to wait until Tues. in Washington where we could explore our course of action. This whole thing follows the pattern. We catch a spy as we have this time &

the Soviets grab an American—any American & frame him so they can then demand a trade of prisoners.

[*Attended party fund-raiser.* **Monday, September 8:** *flew to Denver, Colorado, for fund-raiser; flew to Washington; homework en route, commented, "Then the W.H. & an enthusiastic welcome from Rex."*]

Tuesday, September 9

A meeting—Geo. S., John P., Don R. & myself re the Daniloff case. We are going to try to get him released to our Ambas. pending trial. We'll offer the same here with their spy. If its possible we'll do something of an exchange but only if they'll release some dissidents like Sakharov etc. Once we have him back I propose we kick a half hundred of the U.N. KGB agents out of the country so there can't be a repeat of this hostage taking.

[*Met with congressional leadership of both parties, discussed upcoming agenda, including drug crusade; met with Portuguese prime minister Aníbal Cavaco Silva; cabinet meeting to update Daniloff situation; met widow of NSC employee taken by cancer; haircut; telephoned Alf Landon on his ninety-ninth birthday; videotapings; telephoned retirement party for Senator Barry Goldwater (R-AZ); caught up on backlog of homework.*]

Wednesday, September 10

Daniloff has been charged with espionage. This was Brazil day. The usual ceremony—parade of troops, 21 gun salute et al. In our pvt meeting I told him we were called the Colossus of the North but Brazil was the Colossus of the South and together we could be a great force for good. He liked that. I think we struck a chord and can become very close as allies. He's doing a great job democratizing Brazil.

[*Greeted honorees of Boys Club of America; met with Maryland senatorial candidate; dropped by reception for members of Boston Red Sox baseball team; state dinner with entertainment by Paul Anka; late word from Shultz that Soviets would trade Daniloff for man arrested as spy in U.S., commented, "I told him to do it. I think it's very important to get our guy out of their jail & away from that 4 hours a day interrogation."*]

Thursday, September 11

It was a short night & a long day. The usual meeting with Don R. & the V.P. then the NSC briefing. John P. was hesitant about the Daniloff deal without us holding out for more. Since it's only a move pending the trials I reiterated my stand on getting him out of jail. Our Repub. Chrmn. Frank Fahrenkopf came by, he wants Mermie to become his assistant but it needs me to say the same thing—which I did.

[*Addressed volunteers from U.S. as program of public/private partnership is being copied overseas; had lunch with Vice President Bush; lengthy cabinet meeting on proposed legislation for anti-drug crusade; photo sessions with refugee from Cuban prison, departing staff, and Republican National Committee magazine.]*

Friday, September 12

The usual staff meeting. We've adopted a new policy for meetings anymore—only principals participate—we have to stop this d—m leaking. After a brief meeting with Bob Tuttle to OK appointments, and NSPG Meeting re Daniloff. We have agreed to turn Zakharov over to the Soviet Ambas. pending trial & they will deliver Daniloff to our Ambas. This does not mean a trade—this will not do. Their man is a spy caught red handed & Daniloff is a hostage. After lunch I spoke to 272 principals of elementary schools in the Rose Garden. They were picked as the top in our Nation. I was able to open by announcing that Daniloff was in the U.S. Embassy in Moscow. They cheered the news.

[Weekly meeting with Shultz, mainly about South Africa; met with representative of Telephone Pioneers; ambassadorial formalities; photographed for Philippine newspapers; went to Camp David; noted, "Called Dick Nixon re Daniloff matter—he approves what we are doing." **Saturday, September 13:** *radiocast; rode; practiced speech to be delivered jointly with Mrs. Reagan; watched movie.]*

Sunday, September 14

Watched the usual talk shows. Nancy was on "Meet the Press" and did very well on the subject of drugs. Most of the shows dealt with the Daniloff matter and a number of the media & press involved launched a campaign that I had blinked & softened—giving in to Soviet demands. That's a lot of crap & they don't know what they are talking about.

Than back to the W.H. where our quarters are all torn up with the TV crew readying for the broadcast. A rehearsal at 5 & on the air at 8. Everything went well and the broadcast was a success.

Monday, September 15

A meeting to talk about my speech next week at the U.N. General Assembly. I've suggested opening the question of whether the U.N. is living up to its charter.

[Had lunch with fund-raisers for Presidential Library Foundation.]

Then it was Shimon Peres time—P.M. of Israel. I admire him very much and am sorry the pol. rotation agreement will see him replaced with P.M. Shamir. Of course he will be Foreign Minister then. That will help some. He's done a great job seeing the way toward peace in the Middle East.

A hastily called meeting with John P. & Don R. to agree on our approach to Gorbachev if Daniloff is not turned over to us at once. We are going to tell about 25 KGB members on their U.N. staff to get out of the country.

[Ambassadorial formalities; dinner with Maureen and Dennis.]

Tuesday, September 16

A bad night—my hay fever. Nancy on her way to N.Y. until late tomorrow afternoon. I had a 9:30 meeting with the GOP leadership—both houses. We were talking a little strategy on the things on the legislative plate. I think they know now I'll veto those wild hair things that have been coming out of the House. They also know I think the budget process is minor league. Then we had an NSC briefing on tomorrow's visit by Pres. Aquino of the Philippines. I can't wait to tell her what I know about communism & why she must take them on—not try to take them as allies.

[Addressed gathering of Hispanic citizens; taped campaign ads; received positive poll figures; chose successor to Lew Lehrman as head of Citizens for America.]

A meeting with Cap W. & then an unscheduled visit by Geo. S. on the Daniloff case. He is going to notify the Soviets tonight that we know the evidence they claim to have & we won't buy it—Daniloff is not a spy. Then tomorrow we tell them we are shipping 25 KGB agents on their U.N. staff back to Russia.

Wednesday, September 17

The press is obsessed with the Daniloff affair & determined to paint all of us as caving in to the Soviets which they of course say is the worst way to deal with them. The simple truth is we've offered no deal and are playing hard ball all the way.

Today I met with Pres. Aquino of the Philippines. She & I met 1st in the Oval Office while our teams were in the Cabinet Room where we were supposed to join them. I'd been told she was really up tight about seeing me possibly because of my friendship with Marcos. We had all agreed that we wanted to assure her of our support for what she's doing. In our private meeting I took up the subject of her problem with the communists & related my experience with them in Hollywood 1946. She really relaxed & we ended up spending the whole hour in the Oval O. & finally went into the Cabinet Room only to get everyone moving to the W.H. for lunch. Lunch was a lot of fun for everyone & I think she has gone home reassured & very happy.

[Met with Chief Justice Warren Burger on upcoming anniversary of Constitution; interviewed by Edmund Morris for biography; farewell party for longtime aide Jack Svahn.]

Then an NSC meeting getting ready for the Soviet Foreign Minister's visit & how we treat the Daniloff problem with him. We've notified the Soviet U we're sending 25 of their U.N. staff home—all are KGB agents.

Thursday, September 18

Off to New Orleans & Montgomery Alabama. In N.O. first stop was an outdoor rally for our Senate Candidate Herman Moore. 6,000 were expected,

40,000 came. It was an enthusiastic crowd and all of us were roundly cheered. Then into the city to a lunch fund raiser for him. Afterward the usual reception & photos with a few hundred top donors. Back to the plane & on to Maxwell Air Base at Montgomery. All the mil. & families were out to see us but all we could do was wave. Then into the Civic Center—6,000 on hand for Sen. Jeremiah Denton. This crowd was on it's feet a half dozen times cheering. Prior to facing them we did another handshake & photo session with about 200 people. Back to Air Force 1 & everyone pleased. During the day we were receiving bulletins about the Daniloff case. Gorbachev has gone on TV to declare our man is definitely a spy. I have told him in writing twice he is not. Shevardnadze has arrived to meet with Geo. S. Whether I see him or not is up in the air. Gen. Sec. of the U.N. stated our action ordering 25 U.N. Soviet staff out of the country is against U.N. Charter. He'd better be careful if we cut off U.N. allowance they'd be out of business.

[Returned to W.H.; received word that Senate confirmed Supreme Court nominees.]

Friday, September 19

George S. brought F.M. Shevardnadze (Sov. U.) over to the Oval Office to deliver Gorbachev's letter to me. Then he discovered he'd left the letter with his own team. But he had a good set of notes on what it contains so he did a 20 min. speech on it. The Gen. Sec. wants a meeting between him and me in London or Iceland—I opt for Iceland. This would be preparatory to a Summit. I'm agreeable to that but made it plain we wanted Daniloff returned to us before anything took place. I let the F.M. know I was angry & that I resented their charges that Daniloff was a spy after I had personally given my word that he wasn't. I gave him a little run down on the difference between our 2 systems & told him they couldn't understand the importance we place on the individual because they don't have such a feeling. I enjoyed being angry.

[Addressed presidents of 101 Traditionally Black Colleges and Universities; bill signing; telephoned Hispanic Chamber of Commerce convention; went to Camp David. **Saturday, September 20:** *radiocast; ride; calls from Poindexter on developments with USSR, commented, "Nothing sensational—the ball is in their court now."* **Sunday, September 21:** *watched panel shows; returned to W.H.; hosted reception for donors to Blair House renovation.]*

Monday, September 22

On this 1st day of Fall went to N.Y. to address the U.N. Gen. Assembly. Left South Lawn at 9 A.M. Marine 1. Then from Andrews to Newark A.F.1. The usual crowds in N.Y.—met by Mayor Koch who told me everything I was doing was right. At U.N. a brief meeting with King Juan Carlos & Queen Sophia of Spain. Then a meeting with Gen. Sec. & Pres. of Gen. Assembly.

It was a frank discussion of our problems with the Soviets as well as the regional hot spots, Afghanistan, Angola, Ethiopia, Nicaragua etc. I had expected some delegations to walk out but no one did. In fact I got a pretty good hand. Lunch at the Am. Mission with Dick Walters, George Shultz, Don Regan then a meeting with our European & Japanese Allies—the Foreign Ministers—add Canada & Australia. We briefed them on the Shevardnadze meetings. Then I met briefly with Lord Carrington—NATO head and we were on our way back to Wash.—arrived at 3 P.M. Did some TV footage with our Repub. Cand. for Gov.—Wyoming & Alaska.

[Tuesday, September 23: meeting on House defense cuts; had lunch with Mrs. Reagan and Soviet authority Suzanne Massie; spoke to organizations supporting tax reform; haircut; filmed campaign ads; dinner for Reagan Scholarship at Eureka College, commented, "I was given a 1932 class ring which I couldn't afford in 1932."]

Wednesday, September 24

George S. & Shevardnadze at the U.N. still dickering about Daniloff. It's getting more apparent that it's the Soviets who are blinking. We're getting closer. The Soviets don't call him a spy anymore—they refer to him as "the American citizen."

[Flew to Detroit, Michigan, and Omaha, Nebraska, for party fund-raisers; returned to W.H.]

Thursday, September 25

Nancy off to Pittsburg early in the A.M. She'll be back this afternoon. For me it's a long & busy day ahead. Day started late for me—10:30 A.M.— meeting with Don R. & then with John P., Geo. S. had called from N.Y. to counsel with us about Daniloff. He has had (Geo. I mean) with Shevardnadze & the deal cooking is Daniloff free—Zakharov free in exchange for Orlov + others if possible. I think we'll have to settle for Orlov but I recommended only if Orlov comes here as Z. leaves. The Soviets want Z. first & then Orlov about 15 days later. Of course we hold fast that the 25 KGB's leave the U.N. & go home.

After lunch I met briefly with Committee on Future of Western Community. This is a group of our people & rep's of Japan, France & several other European Countries—U.K. etc. The idea is a program including U. of S. Carolina to help spread democracy world wide just as Communists are doing. Then I went to the Marriot Hotel to address the National Fraternal Congress on it's 100th B.D. This is an org. of more than 100 fraternal like clubs organized to do charitable work. They gave away over $300 mil. last year.

[Met with George Will to counter impression lent in his columns of U.S. caving in on the Daniloff case; signed pledge for federal fund charity drive; met American Legion

commander and postmaster general; also David Carmen, new director of policy and com-
munications of Citizens for America; videotapings; received word that Congress passed tax
reform bill, commented, "Cloud 9."]

Friday, September 26

High spot was the swearing in of Chf. Justice Rehnquist & Associate Justice Scalia in the East Room. After lunch a meeting with Geo. S., Cap W., &
Bill Casey plus our W.H. people Don R., John P. etc. It was a sum up of where
we stand in the negotiations between Geo. & Shevardnadze. The difference
between us is their desire to make it look like a trade for Daniloff & their spy
Zakharov. We'll trade Zakharove but for Soviet dissidents. We settled on some
bottom line points beyond which we won't budge.

[Helicoptered to Fort Meade to open new National Security Agency buildings; went to
Camp David; swam. **Saturday, September 27:** *radiocast; rode; received news that Senate*
passed tax reform bill, commented, "Finally after all the years—tax reform—2 rates 15%
& 28% & 34% for Corp's." **Sunday, September 28:** *returned to W.H.; Mrs. Reagan left*
for NYC; telephoned fund-raiser.]

Monday, September 29

Didn't sleep at all well last night—I need my roommate. Into the office for
a brief meeting before taking off for some politicking. George S. has won the
day. Mr. & Mrs. Daniloff will be on their way home before the morning is over.
That will be announced. Then tomorrow George will announce that Zakharov
will be found guilty & sent to Russia on probation, so long has he never returns
to the U.S. and Orlov & his wife will be freed to leave Soviet Union in one
week. Then I'll announce a meeting with Gorbachev in Iceland October 10,
11, & 12.

[Flew to Kansas City, Missouri, for political rally, then on to Sioux Falls, South Da-
kota, for another campaign appearance.]

There was a hostile demonstration outside the arena by the Farmers Union.
Two leaders had publicly challenged me to meet them & hear their story. I
agreed & they rode out to the airport in the limo with Don R. & me. To blame
us for the farm problems is a little ridiculous. We will provide more money for
farmers $26 bil. this year than Carter provided in all his 4 years put together. By
the time we reached the airport they were on our side.

So back to Wash. & my room mate.

Tuesday, September 30

A hectic day. Arrived at the office about 10:05 & was rushed into the press
room to announce Iceland meeting Oct. 11 & 12. Geo S. had just done the
announcement about Zakharov & Orlov. Already it's plain the press is going to

declare I gave in & the trade was Daniloff for Zakharov. By the end of the day the network anchors were laying into me for having given up. I addressed the annual meeting of the I.M.F & World Bank members at the Sheraton. Then a meeting in the Cabinet Room for members of Cong & rep's of the oil industry. I expressed support for a program to help in this period of real hard times for that industry.

[Noted that Nicholas Daniloff had arrived in Washington; meeting with Weinberger; bill signings; greeted woman he taught to swim years before; appeared at gathering of Republican Eagles.]

Wednesday, October 1

Off to the Carter Library dedication in Atlanta.

It was quite a day. I had to speak to the large crowd and said some very nice things about J.C. which was appreciated by the crowd. Jimmy said I was very generous. We had a tour of the Library & got a lot of ideas from it for our own.

Back to the W.H. and a visit by the Daniloff family. Needless to say they were happy, but so were we, and grateful to the Lord for making it possible to save him.

Thursday, October 2

[Called senators and met with Senator John Stennis (D-MS) seeking support for veto of bill enforcing sanctions against South Africa.]

Then a short time with a bunch of columnists including some who regularly kick my brains out. Subjects were of course the Daniloff and a new one—a story in the Post that we were lying to the press in a disinformation campaign against Khadaffi. Lunch with the V.P. was pleasant—then a meeting with P.F.I.A.B. getting their report on our defector, Ex-CIA agent Howard—now in Russia. There were also recommendations about CIA cooperation with FBI & CIA's employment policies. A meeting with Pres. & Board of Trustees Chairman of Eureka College. The old college is doing pretty well. Coach Mac was 92 yrs. old yesterday.

[Greeted visitors; videotapings; received word that Senate overrode veto.]

Friday, October 3

Most of staff time taken up with assault by press based on distortions of the Daniloff case & dishonest charges that we had sought to use the press with falsehoods about Khadaffi. And of course hailing my defeat in the veto override.

Then an NSC meeting which was much ado about little. Really a discussion of problems in areas Cap W. is going to visit—India, Pakistan, China etc.

[Ambassadorial formalities; visit from head of Salvation Army; visited doctor about persistent cough, told it was a result of allergy attack; went to Camp David.]

Saturday, October 4–Sunday, October 5

Had a call that a Soviet Nuc. Sub. is on fire several hundred miles No. of Bermuda. Russian ships are on hand—no danger of nuclear accident. Message was from Gorbachev. Seems Chernobyl had an effect.

[Radiocast; rain precluded riding; read.]

Sun. was cloudy day. We came down early because of Horowitz concert in the East Room. About 200 hand picked guests. It was wonderful and everyone enjoyed it greatly.

Monday, October 6

The Soviet Sub—sank, the crew was saved. Nancy came down & we kicked off Nat. Drug Abuse Ed. & Prevention Week—in the Rose Garden. The news is that Mrs. Gorbachev is coming to Iceland. Our news is that Mrs. Reagan isn't. We had a working lunch about the Iceland meeting. Consensus was that Gorby is trying to make it a one topic agenda—arms control. We think we should get into Human Rights, Afghanistan, etc.

[Greeted eight-year-old who used Heimlich maneuver to save woman; met with Shultz on Iceland (no details recorded in diary); haircut; desk work; received message from Democratic Party leader Robert Strauss about speech at Carter Library, commented, "He says I should be outlawed from getting before a crowd—it isn't fair to people like him who go out & speak. He liked what I said about Carter."]

Tuesday, October 7

An 8:30 meeting with bipartisan leaders of Congress to tell them I wanted them to drop their attempts to impose restrictions on me regarding arms when I'm on my way to try for arms agreements with Soviet U. I think I scored.

[Met with Director of the Office of Management and Budget Jim Miller about impending budget deadline; signed proclamation making rose national flower; photos with candidates; NSC meeting, noted subject was "arms control & how to handle it with Gorbachev"; met Soviet citizen released from Gulag, joined meeting of human rights leaders; received list from Senator Chic Hecht (R-NV) of twelve hundred Jews seeking right to emigrate from USSR; spoke to Republican governors.]

Wednesday, October 8

Into the office at 9 for usual staff meetings then out to Marine 1 for flight to Andrew's A.F. base. On the way to the chopper I stopped and told the press the Dem. Leadership in the House was holding up the budget process tying amendments having to do with National Security & the bill. They would in effect be doing Gorbachev's work for him.

[Flew to Raleigh, North Carolina, and Atlanta, Georgia, for campaign appearances.]

This time my speech seemed to go better then ever. On way home learned that Bernard Kalb was resigning at State Department claiming it was in protest over our trying to use the press by spreading disinformation (lies) in the press. He exempted George Shultz from any blame thus dropping it on me. Naturally the press was screaming Q's at me. I simply but forcefully replied "we haven't lied to anyone." I'm getting suspicious—here I am 12 hours away from leaving for Iceland and a half dozen attempts are doing much to render me impotent & helpless in the face of Gorbachev. Well I'll just sleep on it.

[**Thursday, October 9:** *flew to Iceland.* **Friday, October 10:** *homework.]*

Saturday, October 11

A.M. a briefing session then a 5 minute drive to the meeting place—a waterfront home. I was host for the 1st session. Gorbachev & I met 1st with interpreters & note takers. Then he proposed we bring in Geo. S. & Shevardnadze. That's the way it went for all the meetings. We got into Human Rts, Regional things & bipartisan agreements on our exchange programs etc. I told him I couldn't go home if I didn't bring up why they reneged on their commitment to buy 6 million tons of grain. He claimed lower oil prices—they didn't have the money.

Then it was plain they wanted to get to arms control—so we did.

In the afternoon we had at it and looked like some progress as he went along with willingness to reduce nuc. weapons.

At the end of long day Geo. S. suggested we take all the notes & give them to our teams to put together so we could see what had been agreed & where were sticking points. They worked until 2 A.M.

Sunday, October 12

Final day & it turned into an all day one even though we'd been scheduled to fly out in early afternoon. Our team had given us an agreement to eliminate entirely all nuc. devices over a 10 yr. period. We would research & develop SDI during 10 yrs. then deploy & I offered to share with Soviets the system. Then began the showdown. He wanted language that would have killed SDI. The price was high but I wouldn't sell & that's how the day ended. All our people thought I'd done exactly right. I'd pledged I wouldn't give away SDI & I didn't but that meant no deal on any of the arms reductions. I was mad—he tried to act jovial but I acted mad & it showed. Well the ball is in his court and I'm convinced he'll come around when he sees how the world is reacting. On way out I addressed our mil. forces & families at Air Base. They were enthused & cheered my decision.

[**Monday, October 13—Columbus Day:** *worked on speech regarding Iceland summit, delivered it on national TV; poll indicated positive response.* **Tuesday, October 14:** *continuing good feedback on summit and speech; applauded on entrance to meeting with congressional leadership; lunch with TV anchormen and columnists; taped campaign commercials; dinner with Monaco's Prince Rainier and his daughters.* **Wednesday, October 15:** *growing sentiment according to polls that electorate approved of Iceland results; flew to Baltimore, Maryland; spoke to high school history class; attended fund-raising luncheon; returned to W.H.]*

Thursday, October 16

Al Simpson came by to see if he had my support. After 5 yrs. of trying (during which I've been on his side) the House finally passed his immigration bill. They have one or two amendments we could do without but even if the Sen. in conf. cannot get them out, I'll sign. It's high time we regained control of our borders & his bill will do this.

[*Swearing-in ceremony for new Science Advisor; usual Thursday lunch with Vice President Bush; farewell ceremony for ten students leaving for USSR; cabinet meeting with report on Iceland; briefings on legislation and political campaigns; reception for new book on the history of the White House by William Seale.* **Friday, October 17:** *flew to Grand Forks, North Dakota, for campaign appearances and meetings with leaders of farm groups to hear suggestions; returned east and went to Camp David, watched* Kings Row *(1942), commented, "I'd forgotten how really good that pic. was."]*

Saturday, October 18–Sunday, October 19

Signed a number of bills plus our absentee ballots. One bill was "Continuing Resolution"—3 weeks late so govt. can stay in business—6 yrs & I've still never been given a budget. Have a team working on a new budget process proposal.

A horse back ride & then today home early because at 7:15 we entertain a lot of corp. C.E.O.'s who are helping to fund the Reagan Library. It was a really nice party. A real group of top C.E.O.'s who have donated—some of them as much as $1 mil.

Monday, October 20

A meeting to discuss how to respond to the Soviets kicking 5 of our diplomatic people out of Russia. Four are from the Soviet Embassy & 1 from the Leningrad Consulate. This is their reply to our sending 25 of their KGB types home from the U.N. We had announced we were going to reduce their staff at the U.N. which is greater than the next 2 nations put together. Now they have hinted at further action if we reply in kind. Well we're going to reply with 4 from their embassy & 1 from the S.F. [San Francisco] Consulate are going to be ordered out. In addition we're going to reduce their staff to the size of ours in Moscow—that will be maybe as many as 80 or so.

[Lunch with sportscasters in honor of centennial of Sporting News; *desk time; haircut; videotapings.]*

Tuesday, October 21

Well the weather did it again for us. At the 10 A.M. arrival ceremony for Chancellor (West Germany) Kohl, the sun was out, the sky blue, & the temp. about 70 degrees. Then we went to the Cabinet Room for a meeting—his ministers & my team. It was a good meeting & he's very supportive of our position at Iceland.

Before his arrival we had a session on our own about our election & whether to do anything in response to the Soviets expelling 4 of our embassy personnel & 1 from our consulate in Leningrad before the Nov. 4. We settled that—about 55 of their embassy people here & in S.F. are going to on their way home by Nov. 1.

After the Oval Office meeting with Kohl I was free to do piled up desk work, which I did until about 4 P.M. at which time upstairs. A long evening lay ahead of us—a State Dinner & all. The dinner was a great success—Joel Grey entertained & brought the house down. Jerry Lewis was on hand & told me my phone call to his Muscular Dystrophy Telethon brought in $2 million.

Wednesday, October 22

The Soviets have responded to our actions by sending 5 more of our embassy & consulate people home & declaring the huge staff of Russian employees in our Moscow embassy can no longer work there. We still have the edge if they're going to play that way—they still have much greater staff here & at the U.N. than we do.

[Ceremonial signing of tax reform bill; meeting with Shultz and Poindexter; spoke to black preachers group, commented, "No press so I could really get down to the issue of my frustration about the image makers painting me as a racist. I think I succeeded in changing a lot of minds. I feel real good about it"; latest poll numbers indicate increases.]

An awful lot of desk work then up to my lonely quarters. Nancy is off to N.Y. until Fri.

*[***Thursday, October 23:*** flew to Wisconsin, Missouri, and Oklahoma for campaign appearances. ***Friday, October 24:*** attended campaign rallies; returned to W.H. ***Saturday, October 25:*** radiocast; telephoned Morality in Media Prayer Breakfast in Chicago; had flu shot; consultation about hearing aids; quiet afternoon and evening.]*

Sunday, October 26

Reading & homework at 5 P.M. downstairs for the 1st of season's "In Performance at the White House" for Public Broadcasting. Old friend Kitty Carlyle, Marvin Hamlish, Sara Vaughans, "The Manhattan Transfer" (a quartet) George Merritt & Priscilla Baskenville. Nancy did a chorus. It was a tribute to George Gershwin.

Monday, October 27

An NSPG meeting on the Iceland arms proposals. The Joint Chiefs wanted reassurances that were aware of the imbalance with Soviets in conventional arms & how that would be aggravated by reduction in nuclear weapons. We were able to assure them we were very much aware & that this matter would have to be negotiated with the Soviets in any nuclear arms reduction negotiations. Signed a bill for freshman Colorado Congressmen Mike Strang having to do with water conveyance in National Forests. Then over to the East Room for a big signing of the Drug Bill. Some Olympic athletes were on hand & some kids—members of the "Just Say No" Club. Charles Price—Carol & their son & daughter came by—they're also coming to dinner tonite for Angus & Princess Alexandra. Just a small, private dinner. Cap W. & John Poindexter for more talk re the arms negotiations. We have a problem with Cong. & its cuts in the defense budget. Conventional arms are more expensive then missiles. If we have to rev up that part of the mil. the Cong. is going to have to recognize it & raise the ante. I feel however the Soviets if faced with an arms race would have to negotiate—they can't squeeze their people any more to try & stay even with us. A long taping session & then upstairs for dinner.

I'll host it alone for a while—Nancy's hairdresser was late.

[Tuesday, October 28; flew to Columbus, Georgia; signed new veterans bill; made campaign appearances; flew to Birmingham, Alabama, and North Carolina for more appearances; returned to W.H. in time for dinner. Wednesday, October 29: flew to Evansville, Indiana, and Rapid City, South Dakota, for campaign rallies, commented, "the routine for all the stops on this trip—crowds along the highway & streets—multiple signs of approval & of course always a little cluster of hostiles along the way. Sometimes their beef was aid to Contras—sometimes SDI & now then a splinter group or some other subject but also a lot asking that I run again in '88 which of course I can't do"; flew on to Colorado Springs, Colorado.]

Thursday, October 30

Early met with a most remarkable woman—Mrs. Jean Sutherland. Her husband is a hostage in Lebanon. She is very strong person and wanted me to know she is involved in a move to free her husband—could not tell me the details as I could not with regard to our ongoing efforts. As she put it she & her husband have put their lives on the line.

Then I went downstairs & signed a bill—the Wild Scenic Rivers Act—HR 4350—very important to Colorado. For one thing it marks the 1st Colorado stream to be classed as a wild river & preserved as such.

[Received enthusiastic reception at campaign rally; flew to Reno, Nevada, commented, "Again wild enthusiasm"; flew to Spokane, Washington. **Friday, October 31:** *campaigned for local candidates; flew to Twin Falls, Idaho, for appearance; flew to L.A. and met Mrs. Reagan at Century Plaza Hotel.* **Saturday, November 1:** *radiocast; attended party fund-raiser in Anaheim, California; helicoptered to ranch, noted, "Puttered around in the afternoon."]*

Sunday, November 2

The most beautiful Calif. day we've seen in a long time. A horse back ride in the morning and then an afternoon of pruning some massive Oak trees. Only one day of it but it was wonderful.

*[***Monday, November 3:** *campaign appearances in Las Vegas, Nevada, and Anaheim, California, noted mix of young and old in attendance.* **Tuesday, November 4:** *returned to W.H. to await election results.]*

Wednesday, November 5

Morning off. About 1 P.M. over to office. We now know we lost the Sen. but gained a number of Gov's. In fact, Republican Gov's now govern 51% of the American people. At 1:45 went over to E.O.B. and spoke to about 150 of our team. Message was telecast. I called for going to bat to get the rest of our program by taking our case to the people.

Thursday, November 6

Back on a regular schedule. Staff meetings at 9 A.M. Nancy is off to N.Y. to receive an award. She'll be home tomorrow.

[Signed immigration bill; addressed delegation leaving for Paris to introduce public/ private initiative program; had lunch with Vice President Bush; cabinet meeting to review election results and plan future; photo session with officers of United Stations Radio Network; all the while telephoned winners and losers in election.]

Friday, November 7

Usual meetings—discussion of how to handle press who are off on a wild story built on unfounded story originating in Beirut that we bought hostage Jacobsen's freedom with weapons to Iran. We've tried "no comment." I've proposed & our message will be "we can't & won't answer any Q's on this subject because to do so will endanger the lives of those we are trying to help." We had a meeting with Cap W. & Admiral Crowe (Chrmn of the Joint Chiefs of Staff). This was on forming a plan for continuing the modernization of nuc. weapons at same time we are talking eliminating of such weapons. Our own people are going to challenge our doing this—while Russians will stiffen their backs if we show signs of not doing it. I've charged the chiefs to come up with a plan that meets this double problem.

[Economic briefing indicates positive outlook.]

Then our returned Hostage arrived with his family—David Jacobsen & sons & daughters. His statement to press was same line I've advocated—in short—"they should shut up." It was an emotional & heartwarming meeting. He's been a prisoner 17 months. Now we are off to Camp D.

Saturday. November 8–Sunday, November 9

Lousy weather so no riding. Did get in a couple of brisk walks between showers. The Sat. night & Sun. morning talk shows continued to hammer on the hostage & Iran arms story giving credence to every rumor & supposed leak (unidentified official). They can do great harm with their irresponsible drum beating. Now the Dem's (House & Senate) have joined the chorus. Don R. called & we've decided to meet Mon. morning to get a handle on this.

Monday, November 10

A bright, pretty day & back to the W.H. At 11:30 a meeting in the Oval Office—Don R., Geo. S., Geo. B., Cap W., Bill C., Ed M., John P., & 2 of his staff. Subject the press storm charging that we are negotiating with terrorists kidnappers for the release of hostages using sale of arms as ransom. Also that we are violating our own law about arms sales to Iran. They quote as gospel every un-named source plus such authorities as a Danish sailor who claims to have served on a ship carrying arms from Israel to Iran, etc. etc. etc. I ordered a statement to effect we were <u>not</u> dealing in ransom etc., but that we would not respond to charges or Q's that could endanger hostage lives or lives of people we were using to make contact with the terrorists.

[Attended ceremony marking 211th birthday of the Marine Corps; photo session with congressman; lengthy videotaping session; long dental appointment for replacement of worn filling.]

Tuesday, November 11

A holiday. Morning press largely ignoring our statement re the hostages & Iran & continuing their false stories. Saw Ron on ABC's "Good Morning America" swimming in Fla. with Dolphins. He really does well. Yesterday it was sky diving & tomorrow a cattle drive.

Wednesday, November 12

I'm glad I taught Ron to ride when he was just a lad. He looked pretty good on the cattle drive.

This whole irresponsible press bilge about hostages & Iran has gotten totally out of hand. The media looks like it's trying to create another Watergate. I laid down the law in the morning meetings—I want to go public personally & tell the people the truth. We're trying to arrange it for tomorrow.

[Ambassadorial formalities; greeted New York Mets, World Series winners.]

A lunch in my study with Lehrer of TV show, King of C-SPAN, Rollie Evans of "Evans & Novak," plus Don Regan, Pat Buchanan & a couple of staff members. It was an off the record session & came off very well. A meeting with George Shultz. He'll be a team player but he was never happy about our Iran policy. Then to the Situation Room where we briefed the Cong. Leadership. (Bob Michel couldn't come so Lawton Chiles replaced him) Bob Dole, Bob Byrd, & Jim Wright. We gave them the whole load on hostages & Iran & explaining why we couldn't go public with some of the info—it would actually endanger some lives, including the hostages. I'm betting Bob Byrd will double cross us. He's out to undo one—President Reagan.

Then upstairs & a quiet evening.

Thursday, November 13

1st order of business—I will go on TV at 8 P.M. tonite and reply to the ridiculous falsehoods the media has been spawning for the last 10 days. Had a photo taken with Richard Godwin—new fellow at Defense in charge of buying etc. He comes to us from Bechtel Corp. Usual lunch with V.P. he had some disturbing news about Mexican Pres. De la Madrid. The Pres. is upset & thinks I've cooled off on him. I'll have to get into this & see what has him upset.

[Met with U.S. ambassadors on anti-drug crusade; worked on speech most of afternoon.]

At 4:30 a big admin. list. The people from the Lung Assn., with the annual Christmas seals. The White House Historical Assn. Photo with Epilepsy poster child—a pretty little girl. The Harper family—friends of my brother & so it went. Then at 5 P.M. spoke to Citizens for America in the East Room & took Q's. Then came 8 P.M. & I did a 12 minute TV speech on the Iran incident & gave the facts to refute the firestorm the press is raising based entirely on unsubstantiated rumors & out right inventions. Several thousand phone calls came in & about 2/3 were favorable.

Friday, November 14

A meeting with a Cabinet Room full of religious leaders covering every denomination—Protestant, Catholic, & Hebrew. They were gathered to support Ed Meese's study on child pornography. They made it plain they supported me & were praying for me.

[Interviewed with Mrs. Reagan on Camp David; went to Camp David with Charles Price, ambassador to Great Britain.]

Saturday, November 15

Margaret Thatcher arrived. I met the helicopter in a golf cart & brought her back to Aspen where we had a good one-on-one re our Iceland meetings & what we are trying to achieve in arms reductions. She had some legitimate concerns. I was able to reassure her. Then we went down to Laurel where I did the radio cast then lunch—a working lunch with her sec. & Ambas. in attendance plus Don R., Geo. S., John P., & some W.H. staff. We covered the Iran setup etc. She & the others left. Later in Wash. she did a press conf. & went to bat for us. Most helpful.

[Sunday, November 16: returned to W.H.; desk work; telephoned daughter of farmer who committed suicide on losing property to bankruptcy.]

Monday, November 17

Told Sec. of Agri. about the farmer & to see if there was anything we could do for the widow.

The press continues to harp on the Iran situation to the point of writing & broadcasting pure fiction.

A Domestic Policy Council meeting to receive the first result of the study's I'd asked for. This one was on Federalism & how the Nat. govt. has violated the Const. in making our states admin. districts of the Fed. Govt. rather than sovereign states. I want this reversed.

[Issues-briefing lunch; signing ceremony on water projects bill; meeting with President Raúl Alfonsín of Argentina; haircut; meeting with officials of the Alzheimer's Disease Foundation (no details recorded in diary).]

Tuesday, November 18

Our usual meetings were again taken up with what the press continues to do about Iran & the hostages. A meeting with N.H. Gov. John Sununu. He is head of the G.O.P. Gov's & Vice Chairman of the Nat. Gov's Assn. He's a good man & I approached him about helping us with restoring Federalism, the rights of States that have been taken over by the Fed. govt. Then an NSC meeting about Chile & how we can persuade Pinochet to move toward a democratic form of govt. We're agreed we must try. After lunch, Edmund Morris (my biographer) came in for a session. I brought him up to date on Reykjavik. Then a session in the family theater preparing for tomorrow nights press conference. A taping session & then over to the Hilton to address the Ethics & Public Policy Center dinner. Bill Buckley is getting an award. It was a brief stop for me but it was a very nice event & I was well received.

Wednesday, November 19

The staff time & NSC briefing were taken up with our own "Dreyfus" case—the coming press conf. and the lynch mob attitudes of the press. Day didn't start until 10:00 A.M.

After lunch a meeting with Geo. S. I think we're going to have some top level meeting with me, the 2 George's, Cap, Bill C., and John P. I'm suspicious that some of George's people have been busy making him feel that he's being left out of the inner circle.

Then it was back to the family theatre for more practice for the press conf. Went upstairs, exercised & did more wood shedding for tonite. Nancy came home from New York about 5 P.M. so there's a different feel at the W.H.—It's been a barn for about 36 hours.

Now "press conference"—They were out for blood—every Q. had a sharp barb. Our gang seems to feel "I done good."

Thursday, November 20

A meeting with Bob Michel R. Minority leader & Jim Wright D. Majority Leader & probably the next Speaker of the House. I queried them on what they saw as the agenda for the coming session. I didn't hear any smashing ideas of importance.

Then a meeting with our Ec. Policy Advisory Bd. This is once a year thing. They are all top nationally known economists & it's surprising how many things they can disagree on.

Lunch with V.P.—Don Regan told me Geo. S. was about to lay down an ultimatum—either I fire John P. or Geo. S. quits. I don't like ultimatums.

Dick Wirthlin shows this Iran drum beat has lowered my ratings with the people. He's going to do some polling now after my press conf. Surprisingly the telephone call since last night were 84% positive & the telegrams 88%.

[Photo sessions with staff members.]

Then it was admin. time—photos with Pierre Pflimlin—Pres. of the European Parliament & Field Marshal Abu Ghazala of Egypt. Received a print of painting that will be the Conservation Stamp for this year. A couple of S.S. agents departed for other assignments. Then upstairs a touchy meeting with Geo. S. & Don R. George is very upset about the Iran affair & I fear he may be getting ready to say "Either someone else is fired or I quit." I've called a Mon. afternoon meeting of him & Don—me and Cap W., Bill Casey, John P. and the V.P. to get everything about Iran effort out on the table.

[Attended annual Senate dinner.]

Friday, November 21

Late start 10 A.M. Meeting with Sen's Dole & Byrd to hear their ideas of upcoming agenda. Byrd <u>now</u> Majority Leader is going to keep Sen. in Wash. much more than has been the case. Some of the Sen's are going to be unhappy.

We had an NSC briefing—it seems there is a thing having to do with Israel & and some Hawk Missiles in the Iran mix that has to be straightened out. Ed Meese assured us again that I'm in the clear legally on what we were doing. Some of the Congress are swiping about the delay in telling them what was going on.

[Had short meeting with General Henri Namphy, acting president of Haiti; met with Weinberger, who wanted to make a supplemental request for more defense money; ambassadorial formalities; went to Camp David; watched The Hasty Heart *(1949).]*

Saturday, November 22–Sunday, November 23

Sat. was cold & clear. Usual radio broadcast—on Turkey day this time. A horse back ride in the afternoon—probably the last at Camp D. until Spring. We all got pretty chilled. The evening TV talk shows Agronsky & Co. & McLaughlin were laying into me on the Iranian affair.

Sun.—bright & cold—spent morning watching the Sun. talk shows, Meet the Press, David Brinkley & again I was being lynched. This time Nancy was brought into it. Sam Donaldson referred to Nancy as the "Smiling Mamba." The Mamba is a large very poisonous snake. After lunch back to Wash.

Monday, November 24

Met with Zulu Chief Buthelezi of S. Africa. I admire him greatly. I told him I was not a fan of Bishop Tutu. [. . .]

An issues briefing lunch. I called for a script for Saturday's radio [speech] on the doomsayers who said our ec. plan wouldn't work & it did. Now they are predicting a recession by Thanksgiving. But that's only a few days away & no recession.

Big thing of the day was 2 hour meeting in the Situation Room on the Iran affair. George S. is still stubborn that we shouldn't have sold the arms to Iran—I gave him an argument. All in all we got everything out on the table. After meeting Ed M. & Don R. told me of a smoking gun. On one of the arms shipments the Iranians paid Israel a higher purchase price than we were getting. The Israelis put the difference in a secret bank account. Then our Col. North (NSC) gave the money to the "Contras." This was a violation of the law against giving the Contras money without an authorization by Congress. North didn't tell me about this. Worst of all John Poindexter found out about it & didn't tell me. This may call for resignations.

[Ambassadorial formalities.]

Tuesday, November 25

John P. came in this morning & announced he was leaving the NSC & returning to the Navy. I told him I wouldn't refuse his resignation but regretted it.

I explained that I know the press would crucify him if he stayed & he didn't deserve that. What it was all about was that Ed Meese learned that several months ago the Israelis delivered some of our arms to Iran but exacted a higher price then we had asked. They sent us our price then past the balance in a Swiss bank account belonging to the Contras—their way of helping the Contras at a time when Congress was refusing aid to the Contras. John resigned because he had gotten wind of this game but didn't look into it or tell me. In the old Navy tradition he accepted the responsibility as Captain of the ship. We broke the story—I told the press what we'd learned. This headed them off from finding out about it & accusing us of a cover up. I've asked Ed Meese to continue digging in case there is anything we missed & I'm appointing a commission to review the whole matter of how NSC Staff works. Ed Meese stayed with the press & took their Q's. They were like a circle of sharks.

Lunch was at the W.H. with returning Justices of the Supreme Court. It was a fun time. Then an NSC meeting to see how we'd handle the roll out of the 131st B2 bomber equipped for nuclear cruise missiles. It puts us 1 plane above the restraints of SALT II which the Soviets & us had agreed to observe even though the treaty had never been ratified. The Soviets have regularly violated the agreement. My decision is to h--l with them we roll out the plane. Upstairs to the lonely W.H. <u>Mommie</u> left for the West today. I join her tomorrow.

Wednesday, November 26

Off to Calif. 11:30—1st a further report on the Iran case—nothing new to add. Then into Rose Garden for annual receipt of the Thanksgiving Turkey. We don't kill this one—he's a 58 lbs. Tom who has been here before. A meeting with George S., I think he's feeling some guilt—he's sworn fealty until I leave the job.

[Flew to California; met Mrs. Reagan at Point Mugu; greeted schoolchildren who won awards for fitness; went to ranch, noted, "I got into jeans & joined Barney & Dennis in cleaning up a lot of brush around the entrance gate."]

Thursday, November 27

Had a good horseback ride—then back to the House to wait for our Turkey Day guests. Ron & Doria arrived early, then later Moon & Bess, Maureen & Dennis. A pleasant evening & we all stuffed ourselves. During day had a nice call from Margaret Thatcher.

*[***Friday, November 28:*** *rode; worked on beautifying entrance by pruning, noted, "Press is still shooting at us. Now they claim some one shredded documents."* **Saturday, November 29:** *rode; received call from Ambassador Charles Price, noted, "with some suggestions re the Iran mess"; completed pruning of entrance.]*

Sunday, November 30

Departed for Wash. 8:35 A.M.—Don R. and some staff meeting on the plane re the mess. We'll be meeting with Ed Meese tomorrow morning. This evening at the W.H. Mike Deaver brought his attorney over to talk about the situation. They feel I should get an attorney (criminal practice) and have him get affidavits from any who were involved in the Iran gambit that I was not informed of what was going on. Their idea is that someone might try to involve me to protect his own hide.

Monday, December 1

Ed Meese on hand for 9 A.M. meeting. We've decided to name Frank Carlucci as Nat. Security Advisor. Then I've told Ed Meese to ask for an Independent Council to do an investigation & to see if we can get a Joint Committee of Congress to also do one & I'll announce all these steps hopefully on Wed. My truthfulness seems to have become the issue.

At 11 A.M. I introduced to the press our review board I've appointed to look at NSC procedures & recommend any changes needed. They are John Tower, Ed Muskie, & Brent Scowcroft.

An issue briefing lunch—then Dick Wirthlin—his polls weren't happy making. A sample—71% of the people like & think I'm a nice fellow. But 60% don't think I'm telling the truth.

A Budget review meeting—we're going to have to sweat $50 Billion out of the '87 budget.

Met with Carlucci—he's on board & will take the job.

Dinner with Mommie, Maureen & Ted Graber who's back here to do the Xmas decorations.

Tuesday, December 2

Things moved fast today because announcement day instead of tomorrow. GOP House & Sen. Leadership came down. We met in the Cabinet Room. Again I told them the whole story about Iran and answered their Q's & we agreed to meet again tomorrow. Then I went on TV and announced Carlucci's appointment. Also that Ed Meese at my urging was going to ask the Court to appoint an Independent Counsel—we used to call them a Prosecutor. I assured our TV viewers we would cooperate with the Congress etc. I think we cooled some of the savage beasts.

Later in the day Dick Wirthlin called to say my nose dives in the polls had turned around & I was up from 3 to 5 points since Sun.

We had a Domestic Policy meeting on the report on "The Family." We're trying to fix so well indented welfare programs don't become an incentive to break up family's or induce teenage girls to become pregnant.

Later we had a budget meeting—there will be some unhappy Cabinet officers because we are cutting back their budget requests. We must reduce the deficit. Wound up the day with a taping session. Stock Mrkt went up 40 points.

Wednesday, December 3

Met with same group of GOP Sen's & Rep's met with yesterday plus Jack Kemp. They are still undecided as to what they can do with a Cong. Committee to investigate the Iran fuss. Bob Dole thinks Bob Byrd want to wait until Cong. meets in January. & then a dozen committees will hammer away well into summer with the 88 campaign in mind. They are satisfied with what I've done & feel my TV speech yesterday has checked the assault by the press.

[Cabinet meeting on budget reductions; spoke to female entrepreneurs.]

Long meeting with Geo. S. He'll soon be off on some foreign trips & he's totally behind me now on my actions in the Iran flap.

Went home early—met Xmas tree donors on way. A lot of homework to do. George B. making a speech today—a good one totally on the Iran affair.

Thursday, December 4

Usual meetings—mostly on latest stories—the Iran mess. Then a briefing for meeting with Pres. Arias of Costa Rica. And then a meeting—usual format—15 minutes in Oval O. with his Foreign Minister & Geo. S. There is a great bond between our 2 countries. Costa Rica was the 1st true democracy in the Latin American country's & to this day has no army. It was 4 yrs ago today that I visited Costa Rica.

After Oval O. we moved to the Cabinet room for plenary meeting with our staffs. They are have some ec. problems & we're trying to be of help. Pres. Arias has been most outspoken leader in Latin America against the Communist govt. of Nicaragua. He'll be meeting with different departments tomorrow.

Met with Frank Carlucci for a photo & welcome aboard. He's already chosen his Deputy—a Black General in our army who is believed to be on his way to being Army Chief of Staff.

Bob Dole came in. He has worked it out with Byrd to not have a Special Session called. Sen. will create 1 special committee to hold hearings on Iran.

Time with Bob Tuttle to OK some appointees. Press Sec. Larry Speakes came in to submit his resignation. He has a great job—public relations on Wall St. with the big Wall St. finance co. He was followed by Carlton Turner head of our anti-drug program. He's resigning to have more time with his family—2 young children—feels he has the program (drugs) well started.

[Met with congresswoman who failed in try at Senate.]

Upstairs for a meeting Mike Deaver, Bob Strauss & Bill Rogers. They stayed 'til dinner time while I picked their brains on the Iran mess.

Friday, December 5

Met with GOP Cong. Leadership. Quite a feisty meeting. Big disagreement between them on whether I should call for a Special Session of the Congress. Hasn't been done since Truman in 1948. Then it became known as the Do-nothing Cong. All kinds of advice as to who should be fired etc. Also that I should bring in an outsider as my personal council. I finally brought peace by reading a tribute to Strom Thurmond who was there. It was his 84th Birthday. We sang <u>Happy Birthday</u> to him & presented him with a cake. Lunch with Geo. B. I think he would like to see Don R. go. I'm not of that mind.

The Soviet Trade Minister who is in town trying to build up trade with the Soviet Union. I told him how much easier that would be to do if the Soviet Union would change its emigration policies & its approach to human rights.

A brief meeting with Sec. Shultz. He has agreed to change Ambas's in the Soviet U.

Then a meeting with top 4 in Cong. Bob Michel & Jim Wright from House and Bob Dole & Bob Byrd from the Senate. They have agreed to 1 committee from each house to look into Iran matter & I believe they are united against me calling a Special Session of the Congress. I've already decided not to do that.

Then it was off to Camp D. where temp. was 30 degrees.

Saturday, December 6–Sunday, December 7

Radio script was on Iran. I admitted there were mistakes in the implementing of policy but not in the policy itself.

[Looked over exercise building constructed on site for use by Seabees.]

Sun. we got an early start for home because of reception at 5:20 in W.H. for Arts Honorees at the Kennedy Center. They were Lucille Ball, Ray Charles, Hume Cronyn & Jessica Tandy, Yehudi Menuhin, & Antony Tudor. The usual receiving line (350) then presentation of honorees in East Room. Ted & our people have finished the Xmas decorations and the W.H. is a fairy land—indeed the theme is "Mother Goose."

<u>Almost forgot had a call late Sat. nite</u>—conf. call with Geo. S., Admiral Crowe, & Al Keel. I OK'ed providing U.S. helicopter transport to 400 Honduran troops to point near Nicaraguan border. The Sandinistas has crossed the border & attacked a Honduran base—some casualties. I gave an immediate go ahead & slept like a baby.

Sun. nite- The annual Kennedy Awards. A reception in the East Room, as always. This time the guests were Lucille Ball, Ray Charles, Hume Cronyn & Jessica Tandy, Yehudi Menuhin, & Antony Tudor. Upstairs for a hasty dinner then over to the Kennedy Center for a very fine entertainment.

Monday, December 8

Staff meetings taken up with Iran & the Honduran matter which has been carried out. At noon the usual issues briefing lunch. Our Cabinet has asked for restoration of virtually every cut in the proposed budget—not in their entirety but enough to drastically raise the deficit. I think I'll hold out for "no give in."

[Gave Distinguished Executive Awards; desk work; went to dentist for cleaning; Maureen and Dennis staying for the whole week; anticipated visit by President Joseph Mobutu of Zaire the following day. **Tuesday, December 9**: *met with Mobutu regarding Zaire's economic problems; dialogue with Republican governors; attended Republican Eagles dinner; then congressional ball, commented, "I was surprised how many of the Repub went out of their way to express support for me in the Iran mess."]*

Wednesday, December 10

Short meeting with newly elected Congressmen 27 Dem—24 Repub. Took a few Q's. One young Democrat asked a "have stopped beating your wife" Q. What would I do for the poor & hungry etc. the inference being that we were <u>not</u> doing anything for them. His name—Joseph Kennedy.

[Spoke to gathering on human rights; signed proclamation declaring Human Rights Day; met with Natan Sharansky and assured him of administration attention to cause of Soviet Jews; met with U.S. Savings Bond Committee; signed proclamation for United Way Centennial; desk work; attended Christmas party for press.]

Thursday, December 11

At 9:30 this AM down to Situation Rm for a meeting re the request of the Dutch for mil. transport of about 700 of there Marines to Suriname to take over the govt. of the Brutal Dictator who is endangering & taking the lives of the people there including about 6,000 Dutch citizens. Cap W. & Admiral Crowe were hesitant about giving an OK to Geo. S. who is in Europe meeting with the Dutch NATO rep's. I felt there was no way we could just say no. We agreed there are very real problems in simply saying yes without knowing the tactical problem of a mil. operation. Do we just fly them in & land on the airstrip in Suriname which might be surrounded by hostile military? Well result is a military team of ours is on it's way to Holland to meet with their top brass.

[Had lunch with Vice President Bush; photo session for staff members, also with National Republican Bulletin, leaders of Future Farmers of America; budget review meeting; lit Christmas tree on Ellipse; second press Christmas party. **Friday, December 12**: *noted military team in Holland cooperating on Suriname situation; photo session with Heisman Trophy winner Vinny Testaverde; Domestic Policy Council meeting with reports on reforming welfare program; addressed conservative group, American Legislative Exchange Council; noted new poll results indicate rise, commented, "Maybe the people are getting bored with the Lynch mob."* **Saturday, December 13**: *radiocast; stopped in on party for security personnel.]*

Sunday, December 14

A call to Margaret Thatcher to respond to her warm handwritten letter re our "Irangate." George Shultz called about a short talk. So he & Al Keel arrived at 11 A.M. Geo. had several things on his mind—most important was our arms proposal re the Reykjavik talks. We'll meet on that this week to discuss possible changes to get the Soviets on board. One idea is to go for the 5 yr. 50% cut on ICBM's & then negotiate out further action beyond 1991.

A little before noon met with Mike Deaver's attorney Jack Miller. He is going to find out from Poindexter's lawyer (if he can) if John & Ollie North are taking the 5th thinking they are protecting me. If the answers are right it's possible I might ask the Cong. committees to offer them immunity so they can tell the whole story.

Did my (or our) appearances before exec. branch staff & families downstairs. Then back up to a phone call from Barbara Walters. She is sending a lot of material obtained from a top Iranian figure re the money exchange. It sounds fantastic.

[Attended performance of Christmas in Washington; dinner at home.]

Monday, December 15

I don't like Mondays. George S. came in at 9:30 A.M. I've talked him out of firing Ambas Kelly in Lebanon. At 9:45 our Quad Commission came in with the report on top echelon salaries. The recommended big raises for the top level appointees, the Cong. & judges. The Commission was appointed by the Cong., the Chief Justice of the Supreme Court & me. I know there will be the usual screams but we are losing judges by the dozen who can not afford to stay on the bench & and are returning to pvt law practice. The same is true of Cab. members & others who have sacrificed 6 fig. incomes in pvt. life to lend a hand in govt.

Our Domestic Policy council met on the subject of catastrophic illness insurance. Dr. Bowen (HHS) had brought forth a plan aimed at protecting the elderly which I think has promise. I'm concerned though as to whether we can find something for those working stiffs.

Issues briefing lunch—looks like I'm going to be busy as h--l in January.

[Signed proclamation for Drunk and Drugged Driving Awareness Week.]

Met with Cap W. on plan for having 50 MX missiles on railroad cars. He & I agree we'd like to cancel the "Midget" missile Cong. forced on us a requirement for getting the MX.

Ed Meese came by for a few minutes. I'm releasing a statement tomorrow that I want Cong. to grant Poindexter & North immunity if they'll come forward & testify. Finally a haircut & upstairs.

Tuesday, December 16

An NSC meeting on the Midget Missile. I'm afraid we're stuck with a $40 billion program the military doesn't want & I don't want. Congress tied the small missile (mobile) to the 2nd 50 MX missiles. We couldn't have the MX's unless we agreed to go forward with the Midget.

Right after lunch a GOP Leadership meeting—House & Senate. It was a good discussion of how we all saw the State of the Union Address & the agenda for the coming year. Don Regan not present he was testifying before the Durenberger Sen. Committee. It turned out he took their Q's for 5 hours. This meeting was followed by a full Cabinet meeting to discuss the budget & the State of the Union Address & of course some talk of Iran. They all felt I should not—volunteer to appear before any of the committees.

[Made appearance at meeting chaired by T. Boone Pickens, of large donors for 1988 congressional campaigns; videotapings; reception for Secret Service.]

Early in the day I had Larry Speakes read a statement to the press asking Cong. to grant immunity to John P. & Ollie North so they can testify & get this d—m thing over with. By late afternoon I learned the Dem's say no. They want to keep this going for a year if they can.

Wednesday, December 17

Cap W. came by for a meeting in the Situation Room. It was an update on SDI and a proposal that we deploy partially about 1993—not the finished system but 2 stages that will give us a partial defense but more important valuable information for completing the system. There are some problems but I'm inclined to go forward with it.

After lunch an Ec. Council meeting on the farm problem & on a job training plan for jobless who are that way because imports or industrial changes had eliminated their jobs. We're talking about some $400,000,000. One suggestion was a .2% import fee to pay for it. I'm opposed. This would begin a breakdown of our ban on a tax increase.

George S. came into the Oval Office for a regular visit. He touched on several items including trips he feels he must make & state visits here by P.M. Shamir (Israel) & Pres. Mubarak of Egypt.

It was a light day & I gave the Dr a pint of blood to hold for my operation January 4th.

Late afternoon Stu Spencer dropped by with Mike Deaver. They are good friends & honestly want to help me but I can't agree with their recommendation—that the answer to my Iran problem is to fire my people—top staff & even Cabinet.

Thursday, December 18

Marine Reservists came by for annual "Toys For Tots" drive. They brought

7 Tots with them. Then to the Situation Room for a briefing & decision on replying to a covert sounding that we should contact Dobrynin re a possible meeting (secret) of someone on our side about a possible arms deal. Lunch with the V.P. then a meeting with our people on Welfare reform. This was a group of individuals including Governor Tom Kean on how to reform Welfare to make it less of a disincentive. It should be designed to get people off Welfare.

[Economic Policy Council meeting on increasing competitiveness; greeted delegation from Yeshiva University and received honorary degree; attended reception for speech-writing department.]

Word on Bill Casey—several hours of surgery & it looks like his brain tumor is malignant.

Tonite 8:15—Staff party here at W.H.—receiving line & entertainment. A magic night—a magician who baffled even the most sophisticated among us. I confess I have no idea how he did what he did.

Friday, December 19

Met with Senior advisors in Situation Room—re a third party message that Gorbachev (possibly) but Dobrynin definitely want a secret rep. of ours to come to Moscow. We argued about it but between Xmas & New Years Paul Nitze & Perle will go & return.

Met with Joint Chiefs of Staff—our quarterly meeting. They have concerns about our policy of striving for zero—zero on ballistic missiles. I believe I reassured them of my belief that conventional weapons must be involved or we'll hand the Soviets a margin of superiority.

[Met with Senator Lloyd Bentsen (D-TX) on trade legislation; regular meeting with Shultz; photo with president of Veterans of Foreign Wars.]

Then a hush, hush meeting in the Map Room with Senator Durenberger—Chairman of Intel. Committee investigating Iran mess. I think they are about ready to go public & their finding will be that I didn't cover anything up.

Then off to Camp David—Thank Heaven.

[Saturday, December 20: radiocast; photo session with members of Camp David military force; homework.]

Sunday, December 21

Finished homework & back to W.H. Called Barbara Walters about info she provided—re an interview with a principal (Iranian) in the Iran situation. It's believed a hit squad is tagging him. Thanks to him I know a little bit more about the money deal in the situation—but still not who did what etc.

Monday, December 22

Have to begin thinking of possible Dir. for CIA. The prognosis on Bill

Casey is not too good. Will now have to have radiation in addition to chemo-therapy. If we must—our U.N. Ambas. Vernon Walters might be a very good choice.

[Paid holiday visit to W.H. switchboard operators; signed proclamation for Nation-al Day of Prayer (May 7, 1987); issues-briefing lunch on possible plan for Catastrophic Illness insurance, commented, "It isn't an easy idea. Heads of private insurance have been of little help"; received silver menorah from American Friends of the Lubavitch, a Jewish religious order; photo session with employees of motor pool; received pictures from Charlie Wick of Ronald Reagan Museum in Rome, Italy; interviewed by biographer Edmund Morris.]

Tuesday, December 23

Well Fred Fielding can't do the job we asked him about—checking through all the documents having to do with the Iran case. He is on a case for a client. We think we'll get our Ambas. of NATO who is coming home anyway.

[Photo session with departing congressman; addressed Small Business Conference, noted, "I had a few lines about Iran. It'll be interesting to see what the press does with it"; watched on TV the return of the around-the-world flight of the Voyager.]

In the afternoon a Domestic Policy Council meeting. Subject was Secre-tary Otis Bowen's plan for catastrophic illness insurance. It's my decision to make—I've got a lot of studying to do.

We had another smaller meeting on the "Quad Commission's" report on executive salary raise at top levels. There is widespread belief that politically I should reject it because of the sizable increase for only the top 3,000 or so of govt. including members of Cong. & Judges. Frankly I think we should have the raise—present salaries are so out of line with the private sector.

Finished day with taping session—including taping my Sat. radio show because I'll be on way to Cali. Saturday. Really ended the day when after din-ner we opened our Xmas presents. We go out to dinner at the Wicks Xmas Eve—that's become traditional. We have people in for dinner Xmas nite so our tradition—the 2 of us—do package opening 2 nites before Xmas.

Wednesday, December 24

Not a very Christmassy weather—drab & raining. Don R. called 9 A.M.—Ambas. Abshire has agreed to do the counseling job in W.H. re Iran. That's good news.

At 11:15 A.M. I went to the Oval Office made Xmas phone calls—one for each branch of the service to our enlisted men & women selected by the par-ticular service. They ranged all around the world—a couple were in bases over the date line so it was already Xmas day.

3 P.M. Dick & family arrived. Learned that out in Cali. 3 of the Wick

kids plus Ron & Doria are going to Mid-night Mass tonite. I couldn't be happier—I've worried for years about Ron deciding he's an agnostic.

The Wicks party was—as always a wonderful, warm part of the Holiday. Xmas Eve with & at the Wicks has become our institution—a tradition for all of us. Highlight this time was a homemade video tape featuring Ron & Doria & the Wick kids who are in California. They all did a humorous skit with Ron as MC. Dick was this years Santa Claus.

Thursday, December 25

Merry Christmas! It was a happy day—gift exchanging with the family—phone calls from everyone but Patti & Paul, Moon & Bess. Early evening Deavers, Wicks & Nancy Reynolds & son Mike joined us for dinner & that was a happy event. Early to bed.

Friday, December 26

Saw Dick, Patti, Geoff & Ann plus Ann's boyfriend Jon on their way. Jon & Ann got engaged while they were here. I made a call to NATO Ambas. Dave Abshire who is going to be our NSC advisor. Then downstairs for a haircut. Back up to packing for our trip to Palm Springs which begins tomorrow morning.

Saturday, December 27

Wash. in the A.M. & beddy bye in Calif. We're off for our traditional visit to the Annenbergs' for New Years. Tonite we'll dine with Ron & Doria at the Century Plaza. Received word this A.M. the Soviets have said "no" to an exchange of TV New Years greetings. I was to have taped a New Years greeting to the Soviet people & Gorbachev would have done the same of Americans. They've said "Nyet."

I'm wrong about Ron & Doria—that's tomorrow nite.

Sunday, December 28

A quiet day at the hotel, then a nice evening with Ron & Doria for dinner. Earlier he had come over for a meeting with Maureen & Dennis—subject again was Iran & why don't I display anger etc.—fire someone. I think I'm doing it right.

Monday, December 29

Late morning went downstairs & presented Service Medals to crew of the *Voyager* Richard Rutan, Jeana Yeager & Burt Rutan who designed & built the plane. A fair sized audience—many of them ground crew etc. Then back upstairs to get some allergy tests—Dr. Bookman. Then I taped a New Years greeting for Voice of America—to be beamed to Russia. It was the greeting we'd wanted to exchange with Gorbachev. Found out later—Russia radio then ran it—with several points edited out.

2:40 in the afternoon went down to the helicopter & on to LAX (airport) & A.F.1 to Palm Spring. Visited with Lee & Walter at Sunnyland. Then dinner—just the 4 of us & watched the evening news. CBS ran the medal presentation & declared I did it to divert attention from Iran.

[Tuesday, December 30: *played golf, with inconsistent shots considered "as usual"; dinner with friends.*]

Wednesday, December 31

Golf again—Sec. Geo. S. & I played Charlie Price & Bill Smith. With 4 holes to go we were 3 down. I won 2 games for us & George won 2 & finished winners 1 up.

Then it was the big New Years Eve Dinner Dance at Sunnyland. A whole room full of long time friends—much nostalgia and a lot of fun. Happy New Year!

July 5, 1984—President Reagan working aboard Air Force One during a trip to Alabama.

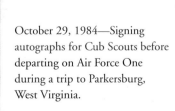

October 29, 1984—Signing autographs for Cub Scouts before departing on Air Force One during a trip to Parkersburg, West Virginia.

August 22, 1984—
President and Mrs. Reagan
with Vice President and
Mrs. Bush waving to the
delegates at the 1984
Republican National
Convention via satellite
from their hotel suite in
Dallas, Texas.

November 3, 1984—
Outside John Wayne's
home during a trip to
Iowa.

October 29, 1984—Speaking at a Reagan-Bush 1984 campaign rally at Parkersburg High School Field House in Parkersburg, West Virginia.

December 24, 1984—With the first lady and their dog Lucky, on Christmas Eve in the White House residence.

January 19, 1985—Building a snowman in the White House Rose Garden with Cameron Reagan, Michael Reagan, and Ashley Reagan.

March 17, 1985—The president and Mrs. Reagan with Prime Minister and Mrs. Brian Mulroney at a dinner at the Château Frontenac Hotel, Québec City, during the Shamrock Summit in Canada.

November 19, 1985—Greeting General Secretary Mikhail Gorbachev of the USSR at Villa Fleur d'Eau during their first meeting at the Geneva Summit.

November 26, 1985—As part of her anti-drug crusade, Mrs. Reagan attended a "Just Say No" rally with children in Oakland, California.

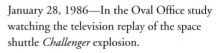

January 28, 1986—In the Oval Office study watching the television replay of the space shuttle *Challenger* explosion.

February 7, 1986—With
the first lady, Jimmy
Stewart, and Gloria Stewart
at a private birthday party
in honor of the president's
seventy-fifth birthday at the
White House.

April 8, 1986—Mrs.
Reagan and Rex, a Cavalier
King Charles spaniel, on
the South Lawn of the
White House.

April 13, 1986—President Reagan hosted Prime Minister Yasuhiro Nakasone of Japan and Secretary of State George Shultz for a meeting at Camp David, Maryland, during the prime minister's visit to the United States.

November 15, 1986—Greeting Prime Minister Margaret Thatcher of the United Kingdom before their meetings at Camp David, Maryland.

July 4, 1986—The president and first lady during the ceremony observing the centennial of the Statue of Liberty, Governors Island, New York.

November 30, 1986—Riding his horse,
El Alamein, at Rancho del Cielo.

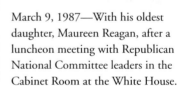

March 9, 1987—With his oldest
daughter, Maureen Reagan, after a
luncheon meeting with Republican
National Committee leaders in the
Cabinet Room at the White House.

December 30, 1986—Playing his annual
golf game at the Annenberg estate in
Rancho Mirage, California.

March 13, 1987—Flanked by Secretary of State George Shultz (*left*) and Secretary of Defense Caspar Weinberger (*right*), during a meeting in the Cabinet Room.

May 10, 1987—President Reagan participated in the *Bob Hope Salute to the United States Air Force 40th Anniversary* celebration in North Carolina.

June 12, 1987—Delivering his famous "Mr. Gorbachev, Tear Down This Wall" speech at the Brandenburg Gate in West Berlin, Germany.

October 7, 1987—Listening to a session of the Organization of American States through an interpreter's headset at the OAS Building in Washington, D.C.

September 10, 1987—Talking with Pope John Paul II at Vizcaya Museum in Miami, Florida, during the Holy Father's trip to the United States.

December 31, 1987—Dancing with Mrs. Reagan at the annual New Year's Eve party at Sunnylands, the Annenberg estate in Rancho Mirage, California.

January 25, 1988—Addressing a joint session of Congress on the state of the union. The president is holding just a portion of the national budget, which was a focus of his speech.

March 29, 1988—President Reagan listens intently to a national security briefing in the Oval Office.

March 30, 1988—On the White House colonnade during an interview for a film documentary, *The Reagan Years.*

June 20, 1988—The annual "class photo" at the G-7 Economic Summit in Toronto, Canada. *From left to right:* Jacques Delors, Ciriaco de Mita, Margaret Thatcher, President Reagan, Brian Mulroney, François Mitterrand, Helmut Kohl, Noboru Takeshita.

October 13, 1988—
Speaking to reporters
after a bill-signing
ceremony in the Rose
Garden of the White
House.

May 31, 1988—With
General Secretary
Mikhail Gorbachev
touring Red Square,
with St. Basil's
Cathedral behind,
during the Moscow
Summit, USSR.

October 24, 1988—
Speaking at a White
House ceremony on the
South Lawn to welcome
home U.S. Olympic team
members who competed in
Seoul, Korea.

November 9, 1988—
President Reagan waves to
the White House staff who
gathered in the Rose Garden
the day after the presidential
election.

November 10, 1988—
Congratulating President-
Elect George Bush in the
Oval Office just two days
after the election.

November 16, 1988—
President and Mrs. Reagan
posed with Prime Minister
Margaret Thatcher and
Denis Thatcher of the
United Kingdom at the last
state dinner hosted by the
Reagans at the White House.

December 6, 1988—Ronald Reagan alone in the Oval Office during his last two months as president of the United States.

January 20, 1989—Following the inaugural ceremonies, President Reagan salutes the troops and the nation as he boards Marine One with Mrs. Reagan at the U.S. Capitol for the trip home to California following eight years in the White House.

CHAPTER 7

—

1987

Thursday, January 1

No Golf—got up late watched some of Rose Parade—as usual it topped itself. The floats were sensational. In the afternoon after a big luncheon—watched the Rose Bowl game. Mich vs. Ariz St. Mich. Started out as if to go all the way—then came unglued as Ariz. began to convert. Ariz. St. won.

That night—a nice dinner—again our same circle of friends. During all those days at Sunny Land I was getting memos & papers requiring some work. Holland called off it's proposed assault on Suriname in which they'd asked us for transportation. Don R. called on what we're going to do about submitting commissions report recommending big pay raises for Congress, Judges & top exec's. I think we'll have to hold back or cut back on amount recommended—because of deficit. Truth is the recommended increases make sense.

[Friday, January 2: returned to Washington; was interviewed live during halftime of Fiesta Bowl football game.]

Saturday, January 3

Don R. & Peter Wallison & his assistant came up. We settled on course to take with regard to the Quad Commissions recommended pay raises. Also put together some sequence of events on the Iran affair by comparing our memories.

Then some desk work. A call from Billy Graham & also from Henry Kissinger.

Watched Redskins beat Chicago Bears.

Sunday, January 4

Off to the hospital for check up—today. Tomorrow the operation.*

Sometimes I think the check up is worse than the operation. I swallow about a gallon & a half of Golightly. Then there is the exam.

---

* *The president underwent a colonoscopy, followed by surgery to remove an obstruction from his prostate.*

Monday, January 5

Operation day. My 1st spinal. Was unprepared for the feeling of total numbness from the waist to my toes. Closed the day with a CAT-SCAN. During day got some paperwork & phone call. Report—a complete clean bill of health.

Tuesday, January 6

The day & night stretch out when you are hitched to an intravenous plus an ongoing irrigation of the bladder.

*[***Wednesday, January 7***: medical tubes removed; met with Regan and Frank Carlucci; pain eased up during day.* ***Thursday, January 8***: returned to W.H.; some improvement in condition.]*

Friday, January 9

Don R. & Frank C. both came over. Frank is taking hold of NSC in great fashion. Last nights TV news NBC featured excerpts of the Sen. Committee on Iran. The whole report was leaked to them but they used selected lines unfavorable to me. Today Abshire & Frank met with the press to straighten out the record.

We have a problem bordering on blackmail. I want Dennis Patrick to be new Director of FCC. Sen. Packwood is dead set on a young lady & threatens us with all out opposition on our agenda if we don't take his choice. I will not be blackmailed. Speaking of such things the Soviets are demanding that we not have Perle (Dept. of Defense) as member of our team in a scheduled arms meeting in Geneva. Well they are not going to dictate who we send—Perle stays. My Dr team from Mayo Clinic came by twice—at noon & later 4:30. They are pleased as all get out with my recovery.

Saturday, January 10

Don R. is talking with Bill Casey's Dr's about our idea of telling Bill we want to replace him at CIA but then when he's up to it we'll put him in a Cabinet level position at the W.H.

[Called Congressman Silvio Conte (D-MA), diagnosed with prostate cancer; doctors visited W.H. residence, said president's recovery was good; watched PBS show on postwar conservative movement. ***Sunday, January 11***: met with Dr. John Hutton (president's personal physician); said farewell to doctors from Mayo Clinic; desk work; watched football game.]*

Monday, January 12

Into office at 10 A.M.—A light schedule for the rest of the week. Met with Don R. & V.P. The press is quoting the usual unnamed W.H. source that our game plan is to sit tight for 2 years—protecting what we've already accomplished. Truth is we've got so many <u>new</u> programs we want to go after we're having trouble setting the agenda.

Then Frank came in for NSC briefing. He's really taking hold. We have a situation where Soviets are talking about establishing a covert channel—an individual to carry messages between Dobrynin & Frank???

Then 3 of our arms negotiators came in. We are naming our top one as leader of the group to meet with a Soviet team in Geneva. We're promoting Max Kampleman to Counselor to the Sec. of State to give him more status as he leads our team. Then upstairs to lunch & sit in a hot tub & back to the office at 3 P.M. Met with David Abshire who is putting all the paper together bearing on Iran mess. A haircut & upstairs to another hot tub.

[Telephoned French woman who helped U.S. airmen during World War II.]

Tuesday, January 13

Began with usual meetings, Don R. & V.P. We have a tough problem Bill Casey. It's pretty certain that he'll never be able to head up the CIA again but we don't want to say that now & set back his recovery. Maybe in a few weeks we'll talk to him about a Cabinet level post in the W.H. & go ahead with a replacement at CIA. Suggestions are possibly Sen. Baker or Sen. or .

Then Frank C. came in with a his new assistant Gen. Colin Powell. He's a good man. Frank believes Soviets refused my son a Visa to fit their propaganda that I have worsened our relations. There is also possibility it had to do with ABC's big special "Amerika." Ron's trip was to be for ABC's "Good Morning" show. We have some problems too about possible deployment early of 1st phase of SDI.

[Cabinet meeting; discussion of State of the Union Address with Regan and Ken Khachigian.]

Wednesday, January 14

[Considering Howard Baker for head of CIA.]

Then meeting with Frank C., I gave him some information—a copy of a message Ghorbanifar is sending to King Fahd on Iran & the factions waiting to succeed Khomeini. I also told him to turn NSC loose on a possible Govt. in Exile for Nicaragua which we would recognize.

[Approved construction of security tunnel connecting Oval Office to Situation Room; noted postponement of interview with journalist Trudy Feldman.]

Thursday, January 15

Don R. had a reply for the latest Wash. Post falsehood. The Post proclaims that he (Don) had made the decision on pay raises for Legislators, Judges, executives, etc. & that once again I was off on a cloud unaware of what was going on. Well the record shows that we had 8 meetings on the subject & I was presiding at all 8. Frank C. brought in our new member of the NSC group an authority on the Soviet Union with a most prestigious resume. He will be a big help.

[New poll numbers show slight drop in approval rating; delivered speech to schools on Martin Luther King Jr.; lunch with Vice President Bush; met with Shultz on his trip to Africa, some improvement in situations in Nigeria, Ivory Coast, and Angola.]

Friday, January 16

Into the office for usual meetings. Then a meeting with Sen. Packwood. We have a difference, he wants a former staffer of his, now on the F.C.C. to be chairman of F.C.C. I have decided on another member for the job. I've assured him that I'm looking for another high level job for his candidate. Watched Bud McFarlane's televised testimony before Sen. Committee. He was masterful for more than 2 hours.

[Went to Camp David; watched For Whom the Bell Tolls *(1943) for the first time.* **Saturday, January 17:** *radiocast; short walk.* **Sunday, January 18:** *short walk in rain; sat by fire.* **Monday, January 19:** *returned to W.H.]*

Tuesday, January 20

Into the office at 10 A.M.—Met with V.P. & Don R. We are considering Ed Bennet Williams for Director of CIA. He'd like to do it & will give us an answer soon. His problems are 2—Cancer & financial. As to Cancer it hinges on a possible upcoming surgery. I hope he can, he'd be a great choice.

Frank C. & Gen. Powell came in for their meeting. Frank wants to make a trip to Central America, which is O.K. with me. We discussed upcoming problem re the future of SDI. Cap wants to go beyond the ABM Treaty restraints. Apparently our progress has updated our anticipated attainment of deployment. Geo. S. is adamantly opposed. I'll be hearing from each of them tomorrow. Topped off morning with a phone call to Pres. Febres-Cordero of Ecuador congratulating him on his release from the kidnapers who had held him for about 10 hours.

[Issues meeting, suggested bipartisan commission on budget planning.]

Wednesday, January 21

[Maureen and Dennis visiting; decided to remark in State of the Union Address on John Chancellor's negative commentary the previous night.]

Frank C. & Gen. Powell had some things to say about the Soviets & disarmament and a few suggestions for my meetings with Cap W. & Geo. S. Cap came in at 11 A.M. We discussed SDI. He agreed we should pin our scientists down on progress. Some times they get a little over optimistic. [. . .]

After lunch—back to office for some desk work—then a meeting with Geo. Shultz. He brought word—Soviets are suggesting a foreign ministers meeting in Moscow in Feb. & then—based on its success—a summit here in the Spring. We also discussed our arms control strategy to try on them based mainly on getting a 50% reduction in ICBM's—period. Then we'd negotiate follow up plans.

David Abshire came by—he's really clearing away a lot of confusion on the Iran mess & turning Cong. men around who just plain didn't understand the situation.

Upstairs & deskwork.

Thursday, January 22

[Received positive economic statistics; NSC briefing on Iran-Iraq War; telephoned Right to Life marchers in Washington; made bet with Australian prime minister Robert Hawke over America's Cup race.]

Upstairs for lunch & got out my diary for 1985—to check on chronological layout of the Iran situation prepared by NSC & Dave Abshire. It sure is helping my memory. Then at 2 P.M. a briefing by Frank C. & Dave for my Monday session with Tower Commission.

[Brief meeting with Vice President Bush; continued to check diary until going to bed. **Friday, January 23:** *staff meeting with Regan and Vice President Bush; NSC meeting with brief discussion of Manuel Noriega, dictator of Panama, commented, "Corruption there is rampant & he's too fond of Castro"; had another prebriefing of meeting with Tower Commission investigating Iran-Contra covert operations; worked with policy makers and writers on State of the Union Address; went to Camp David; watched James Bond movie.* **Saturday, January 24:** *radiocast; took walk; worked on speech; received word of new kidnappings in Beirut; approved positioning of Delta Force on a British base in Cyprus; dinner with Mr. and Mrs. Ken Khachigian.]*

Sunday, January 25

An early morning message re the Delta Force. England asked us not to citing the inability to keep such a thing secret & the possible threat to hostages & Terry Waite still in Lebanon. We said yes to their objection. Weather turned bad in Wash. and we had to drive down & a rush start at 9:30 A.M.

Back in W.H. communications coming from Frank C. A nameless voice speaking for the "Organization of Oppressed People" (believed to have a link to Hezbollah) phone a Beirut radio station demanding the return of the terrorist we're trying to extradite from W. Germany also that we stop helping Iraq??? & intervening in Middle East or more specifically Persian Gulf affairs or hostage will be killed 5 P.M. our time. Next message indicated Hezbollah was holding Terry Waite as a hostage.

Monday, January 26

No word about Waite & no word about execution of a hostage. But 2 more have been seized. It's a frustrating business—you feel like lowering the boom on someone but how do you do it without getting some hostages killed? We need more intelligence on who & where.

John Tower's committee—whoops!—Commission questioned me for an hour & 15 minutes—not in an adversarial way—just trying to round out their information.

[Positive economic briefing; haircut; finished work on State of Union Address; sent message of congratulations to Helmut Kohl on reelection as chancellor of West Germany]

Tuesday, January 27

In at 9 A.M. for usual meetings—[. . .] Chiefs of Staff are working on potential targets if we need to retaliate directly against Iran.

[Meeting with Republican congressional leadership to outline speech; called Canadian prime minister Brian Mulroney, then traveling in Zimbabwe, to discuss Canadian-U.S. relations; rested in afternoon before speech.]

Walked into the warmest reception from Congress—well the Repub. Congress I've ever received at a St. of the Union appearance. Speech was interrupted 39 times by applause. Dem's were if anything more partisan than ever. When they did come to their feet they were obviously being sarcastic.

Wednesday, January 28

[Noted favorable messages on speech, more cynical comments in press.]

No word yet on whether Terry Waite is a hostage. He hasn't been seen in 8 days since he went into a house alone to discuss hostages. We are going ahead with planning mil. targets & trying to get a better intelligence set up on Lebanon.

Noon & upstairs for lunch. Back over to office for 2 P.M. meeting with George S. Most of discussion was what our position should be on ABM Treaty and SDI. I said I believe that we were correct in a broad interpretation of treaty which would leave our development of SDI within the terms of ABM. I asked George to look into some $½ million proposals regarding the embassy residence in Rome for Ambas. to Vatican. It would easily be a half mil. wasted. We should sell the house. He'll look into it. Dr. Smith & wife came in to say goodbye—they're going back to California. Then I did a live TV broadcast to all the NASA stations—anniversary of Challenger tragedy.

[Videotapings.]

Thursday, January 29

Back to a 9 A.M. start. Colin Powell handled NSC briefing, Frank C. is on a visit to Central America. At 10 A.M. met with Bi-partisan leadership of Cong. We reported to them more fully on specifics of programs I'd proposed in St. of U. Address. Then Jim Wright sounded off on his usual pitch that we were the big spenders & they were trying to balance the budget. I blew my top and heaped some honest figures on him & Bob Byrd who somewhat came to his rescue. My figures are correct & they had no real answer. I've learned that our Repub. leadership were on cloud 9 they were so pleased with my response. I'll admit I felt good.

Upstairs for lunch with Nancy & Ted & back down at 2 P.M. for a Domestic Council Meeting on the latest thing in high tech. The "Collider," that isn't it's full name but it is the successor to the Accelerator where atoms are split. The present one in Illinois is a tunnel in a 12 mile circle. This new idea calls for a 52 mile circle with results 1000 times greater. We had a briefing & my answer is going to be "go" event though it adds to our budget. I think we can get some help from our trading partners who'd like to share in it.

A short meeting with George Bush to replace our usual Thursday lunch.

Don R. went to the hospital with Ed M. to see Bill Casey. Bill told them he wanted to resign from CIA. Don told him (to his obvious joy) that when he was ready to come back he'd be my counselor—the job Ed M. had before he became Attorney General.

[Meeting with woman with possible channel to release of hostages in Beirut.]

Friday, January 30
[Meeting with Deputy National Security Advisor Colin Powell on hostage situation in relation to Germany and its hostage problems.]

This morning I received Bill Casey's resignation—also word that Ed Bennett Williams can't take the job. Surgery revealed cancer of the liver—quite extensive. I have appointed Bill's deputy—now acting Dir., as Dir. in Bill's place.

[Addressed gathering of supporters of budget cutting; farewell meeting with W.H. Spokesman Larry Speakes; discussed hostage crisis with Shultz; interviewed by biographer Edmund Morris on Hollywood career.]

A brief session with Peter Wallison & Dave Abshire—more probing of what I remember about the Iran deal.

*[*Saturday, January 31: *trip to Camp David canceled due to weather; radiocast on administration efforts in education and anti-drug crusade.]*

Sunday, February 1
At 10 A.M. the Dr's came up & hung a lot of gear on me which I must wear 'til 10 P.M. It is a 12 hour electrocardiograph recorded on a 12 hour tape. Just a check to see how the old man's ticker is doing. Colin Powell called. In Teheran the Iranians arrested an American press man (Wall Street Journal) took his

passport, accused him of being a Zionist spy & threw him in jail. He's a Roman Catholic. I'm ready to kidnap the Khomeini.

[Homework and reading.]

Monday, February 2

The Groundhog saw his shadow! Pat B. is leaving us and I've told Don R. to see if we can augment the speech writers by bringing back Landon Parvin.

[Short meeting with foreign minister of Italy Giulio Andreotti; new poll figures show slight increase in approval rate; issues lunch; greeted champion Penn State football team.]

Then a pre-meeting to be briefed for tomorrow's NSPG meeting on SDI & whether to go for the liberal interpretation of ABM Treaty which is really legal & correct to me. Doing this would help research on SDI but the Soviets & possibly some of our allies might scream.

[Videotapings.]

Tuesday, February 3

A meeting with Repub. Cong. Leadership. I pitched a plan that they stand together so that even with the Dem's out voting us we can point out to the people how different the Dems & Repubs are. I don't think they got the message. In the House today only 26 Republicans supported my Veto of the Clean Water bill.

[Received word that Wall Street Journal *reporter kidnapped in Iran would be released; greeted heroes of Maryland train accident; National Security Planning Group (NSPG) meeting on SDI and furthering research in loose interpretation of ABM Treaty, commented, "On this one I don't think Geo. S. & Cap are as far apart as we thought"; domestic-policy council meeting on Secretary of Health and Human Services Otis Ray "Doc" Bowen's plan for catastrophic-illness coverage.]*

Wednesday, February 4

After all this talk by the Dems of wanting us to work together on matters we had a bipartisan congressional leadership scheduled for 9:30 A.M. Meeting canceled—Bob Byrd called Jim Wright & told him to cancel for the House as he did for the Senate. His word was that if I wanted a meeting I could come to the Hill & see only one house at a time. He'll wait a long time for that—NSC briefing—only rumors on Seib [*Wall Street Journal* reporter] being released so far. We talked about SDI and how we handled the ABM part.

Lunch in the study then met with Sen's Hatfield & McClure—back from trip to Asia. They visited the refugees in Thailand—agree we can't take them all—they should be allowed to return to Cambodia if Vietnamese will clear out. They believe there is a desire on part of Hanoi to establish a relationship with us. Can't be done 'til they get out of Cambodia & clear up once & for all the POW-MIA problem.

[Met with Shultz on Soviet pianist Vladimir Feltsman, noted, "who got word to Nancy he would be allowed to come to America if he had invitation to play at the W.H."]

Upstairs & met with 6 individuals including FBI agents on whether Mike Deaver ever lobbied on behalf of his clients after he left the W.H. I can't remember a single instance of his ever doing that.

[Telephoned Landon Parvin about returning to the W.H. as a speechwriter. **Thursday, February 5:** *received birthday greetings from Eureka College classmates; attended Annual Prayer Breakfast; met with Prime Minister Turgut Özal of Turkey; lunch with Vice President Bush, who had to leave early to cast vote as president of the Senate; received Charles Russell painting for W.H. from Mr. and Mrs. Armand Hammer; farewell visit with retiring park mounted policeman; received unconfirmed word on escape attempt by hostage Terry Waite.]*

Friday, February 6

My Birthday! After usual meetings went over to EOB to address meeting of Private Sector Initiative supporters—prepared speech in hand. Walked into the auditorium which was filled with Cabinet & Executive Staff people & the Marine Band. The band played "76 Trombones." Nancy appeared from nowhere wheeling a giant birthday cake. The whole meeting had been a frame up to achieve a very successful surprise party.

Then back to work. A good meeting with Dave Abshire who is doing very well on his digging into the Iran matter.

[Received word of release of Wall Street Journal *reporter held hostage in Iran; received birthday gifts.]*

Ended office day with a meeting with George S. Well it didn't quite end the day—Bob Byrd & Bob Dole came in regarding ratification of Nuclear Test Ban Treaty. I had proposed that it be ratified in its present form with a proviso that I would submit a protocol for ratification continuing agreement to be negotiated regarding verification procedures. Bob Byrd was adamant that a double ratification could <u>not</u> be done. Believe it or not he settled for treaty with clause that would only be in effect after I had submitted verification clause satisfactory to <u>me</u> alone.

[Dinner with friends. **Saturday, February 7:** *radiocast; wrote thank-you notes; quiet evening.* **Sunday, February 8:** *telephoned Shultz with praise on TV appearance; wrote more thank-you notes; read background for decisions on SDI, the ABM Treaty, and catastrophic-disease program; noted that Mrs. Reagan was to leave the next day for California.* **Monday, February 9:** *noted death threat to hostages in Beirut; advised Regan to set high priority on welfare reform; addressed gathering on same subject; greeted Dennis Connor and winning sailors of America's Cup; received Boy Scout Annual Report; met with Weinberger on SDI, commented, "we must not engage the Soviets in a discussion that could lead to watering down our ability to proceed with development"; haircut; Maureen visiting; noted that no word received of any attack on hostages.]*

Tuesday, February 10

Don R. & V.P. & I talked about Bud McFarlane & going to call Johnny, his wife, & see if I can phone him. We also talked about an active schedule for me—having some special groups in for lunches etc. NSC was about SDI & the ABM Treaty—getting ready for NSPG meeting at 11 A.M. The problem is how to continue arms reduction dialog with the Soviets & at same time eliminate any compromise on SDI. I have proposed a plan to seek an agreement that we will—when & if SDI is ready for deployment—put it in hands of an international force as a defense against any & all nuclear missiles from wherever they are launched in the World.

[Spoke to American Legion members regarding defense budget; met with Dave Abshire to prepare for meeting with Tower Commission; ambassadorial formalities; telephoned Mrs. McFarlane regarding Bud, who had attempted suicide, commented about the call, "She seemed very grateful."]

Wednesday, February 11

Learned this A.M. that a tape has turned up & been revealed to Tower Commission & the Sen. investigation. It is complete fiction. It has Ollie North telling the Iranians he's held meetings with me at Camp David and that I was willing to go all out with arms deliveries in order to get our hostages back & I wanted Iran to win the war. I OK'd (according to the tape) providing Iran with strategy & intelligence to help them beat Iraq. I'm also said to have wanted Hussein of Iraq to abdicate & because of my support for Khomeini.

Later in morning me with Abshire and told him the tape was complete fiction. There have been <u>no</u> meetings with North at Camp D. He's never been there while I've been President.

Another part of the tape said I went away to pray for 2 days & brought back the Bible with my favorite verse & my signature inscribed on the fly leaf. That's the Bible John P. asked me to sign in the Oval O. and to copy the verse while it was supposed to be the favorite verse of one of the Iranians. The Bible was then delivered to Rafsanjani.

Later in the day, I met with the Tower Commission & told them it was a complete fabrication.

[Addressed gathering of private charities on welfare reform; met with Shultz, who had plan for inviting pianist Feltsman from USSR, commented, "We also discussed a pol. plan for Central America to show that we would like a pol. settlement of the Nicaraguan mess."]

Had a couple of tiny red spots on my nose frozen in case they are beginning of basal cell carcinoma. Upstairs for the night—the last one alone. Nancy gets home tomorrow.

Thursday, February 12—Lincoln's Birthday

[Waiting for response from Stu Spencer on W.H. job; NSC and NSPG meetings on

need for renewal of Middle East peace process; spoke to junior high students on Lincoln; considered list of potential appointees with Robert Holmes Tuttle; greeted Maxwell Football Club; signed proclamation making 1987 the National Year of Thanksgiving; greeted members of Ducks Unlimited; photo sessions with departing Secret Service agents; Republican Eagles dinner; Mrs. Reagan returned home. **Friday, February 13:** *staff meetings including welfare reform experts; lunch with Vice President Bush; meeting with Shultz; greeted NFL champion New York Giants; went to Camp David.* **Saturday, February 14:** *work and reading.* **Sunday, February 15:** *took walk; spoke to Michael Reagan about Easter plans.* **Monday, February 16:** *returned to W.H.; Maureen visiting.]*

Tuesday, February 17

Looks like Jack Koehler will replace Pat B. & Frank Donatelli will come aboard as Pol. Liaison. Neither have said yes—but want to talk to me & we believe their answer will be yes. An NSC briefing was mainly on the P.M. Shamir (Israel) visit tomorrow. I've suggested that after that we have Jeanne Kirkpatrick in to tell us about her visits with Gorbachev & Schavansky [meant Sakarov]. Then Cap came by for a brief meeting on SDI—we're all on the same wavelength.

[Addressed gathering of CEOs on administration competitiveness plan.]

A brief meeting with Don R. on receiving the Tower report next week. On the advice of a lot of contacts I will go on the air & make the report public the day after we receive it. Then several days later we'll have a press conference.

[Videotapings, including one in Spanish, read phonetically; noted that the need for veto was avoided with amended version of appliance-standards bill; stock market closed at record high; took light workout in gym.]

Wednesday, February 18

An 8:30 A.M. breakfast with Bi-partisan Cong. leadership. We briefed them on our competitiveness proposal—Jim Baker, Bill Brock, & Clay Yeutter. Then a discussion—all in all a pretty good tone to the meeting. Then an NSC brief—I approved an approach to the SDI hassle. Later on briefing for P.M. Shamir's meeting—then one-on-one between me & the P.M. I asked him to not lobby our Cong. against arms sales by us to certain Arab States. Israel's position is one of opposition on the grounds that a state of war exists between them & the Arab States. All in all though we solidified our relationship & the subsequent plenary meeting & working lunch went very well. He's going home quite happy. His departure statement was a glowing tribute to me & our country.

Geo. S. came by with some long range plans for me—visits to Philippines & India etc. Ended day meeting with Frank Donatelli who is taking over the Pol. Liaison post & Jack Koehler who replaces Pat B. Upstairs for rest of day—some mail & then a work out in the gym.

Thursday, February 19

A lot of desk time today—nothing very exciting in Staff meeting or NSC briefing. Some briefing on meeting coming up at 11:30 with large Polish delegations (Polish American). I signed a measure lifting Polish sanctions in answer to pleas by Pope & Lech Walesa. They were beginning to hurt the Polish people & that was never our intention. Former Polish Ambas. & his wife who defected to us in '82 were present.

Lunch with V.P.—he offered to have a talk with Don R. about retirement.

Then it was desk time 'til about 2:30 & I met with Dave Abshire & Pete Wallison re the Tower Commission report. I'm sending the commission a letter clearing up my seeming conflict in answering Q. re did I OK Israeli arms sales to Iran in a meeting with Bud McFarlane in August '85. Letter will say I just plain don't remember.

[Looked at personnel matters; greeted Father Bruce Ritter, founder of home for run-away youth; photo session with U.S. Chamber of Commerce leaders; met March of Dimes poster children; photo with ambassador to Colombia; signed proclamation for National Consumers Week; had cocktails with senators and congressmen, also columnist James Kil-patrick, who supported veto of clean-water bill.]

Upstairs for private meeting with John Herrington, Sec. of Energy. He had some info on W.H. Staff & Chief & low morale. I have a problem.

[Telephoned conservative Political Action Committee dinner.]

Friday, February 20

Learned that our W.H. press corps has learned that Jack Koehler, who I have picked to replace Pat B. at 10 years of age joined the Hitler Youth movement in Germany. They are raising a fire storm over this. They ignore that he came to this country served in our military & went from Corp. to Capt. with 14 years of active & reserve duty—including service in our mil. intelligence. He's been top man at A.P. and had a lifetime career there heading the International Bureau among other things. Well I'm not going to back down in the face of the lynch mob.

Later on NSPG meeting—subject was Central America. We've concluded we must make an all out effort to go to the public on the need to support the Contras & get rid of the Sandinista Communist govt.

[Spoke to Conservative Political Action Committee luncheon; went to Camp David; telephoned prospective participants in ad campaign for NASA.]

Saturday, February 21–Sunday, February 22

Did the radio show then a long walk with Nancy & Rex. It was a cold but beautiful day. Press reported that Don R. had told the Staff that Nancy was responsible for the appointment of Jack Koehler. That does it—I guess Monday will be the showdown day. Nancy has never met J.K. and certainly had nothing to do with his appointment.

[Sunday, February 22: watched panel shows; returned to W.H.; attended annual governors' dinner, entertainment by Sarah Vaughan.]

Monday, February 23

This was my day to deal with the Don R. business. At staff time he & I & Geo. B. talked about the upcoming Tower Commission Report & agreed to get some advisors in to round table how we should deal with it about going public. NSC time was spent with Defense Minister Spadolini of Italy. It was a short but good meeting. He too could be a future Prime Minister.

Then Don came back alone & we agreed it was best he go. He had expressed a desire to do so several months ago before the Iran trouble broke—then he stayed on because of the trouble. He'll announce his leaving in a few days after the Tower report comes in. I'll try for Drew Lewis as his replacement.

[Addressed gathering of governors, who expressed interest in proposal for welfare reform.]

Usual issues lunch with a meeting on the Tower report—Paul Laxalt, David Abshire, Dick Wirthlin, Don R., Al Kingon, Dennis Thomas, Geo. B., & me. Decided on a program of speech & press conference several days after report delivery.

[Greeted West Point football team.]

A haircut & then down to the hide away bomb proof command post for a briefing on what we do if Soviets launch a nuclear attack.

Tuesday, February 24

[Met with bipartisan House leadership regarding competitiveness legislation; NSC meeting, followed by NSPG meeting on terrorism.]

After lunch a good meeting with Jeane Kirkpatrick reporting on her meetings with Gorbachev & Sakharov to Geo. B., Geo. S., Cap W., Frank C., Don R. & me. A very useful & good session. Then I went into a meeting with the Council for a Black Ec. Agenda. Our 3rd meeting and we're making progress in improving situation of Blacks. The council has been a Private Sector Initiative effort & is doing well.

A farewell party for Pat Buchanan. Then upstairs for an hour with Ross Perot. He has laid on me a story of chicanery & corruption in our executive branch including the mil. & CIA. It's a shocker & has me asking where do I start. Of course all he told me was based on circumstantial evidence. Called Drew Lewis about taking Don's job. He's thinking hard about it.

Wednesday, February 25

Don R., Geo. B. & I talked over the Ross Perot story. He had talked to both of them as he talked to me. Well this A.M. I had talked to Ed M. I'm going to turn this over to him & our Dir. of the FBI. First however I'm going to give it all a good going over—the material Ross left with me.

[Bipartisan Senate leadership meeting; brief NSC meeting; addressed gathering of CEOs from Midwest on competitiveness project; met with the Reverend Jesse Jackson, commented, "He was off on funding for the Black colleges, student aid, S. Africa, & our coming visit with P.M. [President] Moi of Kenya. I tried to explain our position to him."]

After lunch—a farewell to Mitch Daniels—his wife & 4 daughters. He's going back to private life. Geo. S. came by with some suggestions for a speech after the Tower report comes in & another later speech on us & our relation with Soviets. Then a fine meeting with Suzanne Massie. Very interesting—she suggests maybe I should got to Moscow instead of Gorbachev coming here. Then she dropped bomb. A top Soviet official told her Gorbachev might well be killed if he came here. There is so much opposition to what he's trying to do in Russia—they could murder him here & then pin the whole thing on us. I don't find the warning at all outlandish. The KGB is capable of doing just that.

A Calif. St. Sen. (James Nielsen) (Minority Leader & his wife) came by—Herrington came by. It seems Frank Zarb (former Nixon staffer) a friend of Don R. told John he'd been talking with Don. It seems the conversations got around to Don's resignation & that Don was thinking of April & a promotion to another job—the Fed. Reserve Board. That just can't be, whether he deserves it or not he is tagged in the minds of a majority as responsible for our problems.

Thursday, February 26

At 10 A.M. received the Tower report—almost 300 pages. We had about a 45 minute brief but I won't know what's in it 'til I've read it. At 11 A.M. over to EOB with the 3 members of the Commission—hundreds of media. I made a statement, turned the meeting over to John Tower & left with all of them screaming Q's at me—Sam Donaldson the loudest. Went to TV set & watched John, Ed Muskie & Brent Scowcroft field Q's. Press obviously was seeking answers that would let them keep on their distortions & this lynch party. No I've got to read the report.

Lunch with V.P.—he met with Don R. and for 1st time the side of him he hadn't seen—the outburst of temper. Finally he snarled he'd be out of here by Monday or Tuesday. I went upstairs for a meeting with Paul Laxalt. It was a good meeting with an old friend. Paul took himself out of contention for Don's job. Said he might want to run for Pres. & didn't want to take himself out of the running by taking this job. His candidate is former Sen. Howard Baker. It's not a bad idea. He thinks Howard is looking for a graceful way of getting out of running for Pres. I'd probably take some bumps from our right wingers but I can handle that.

V.P. just came up—another meeting with Don. This time totally different. He says he'll hand in his resignation 1st thing Monday morning. My prayers have really been answered. And this answer continued. I called Ed M. to let him know I was going to ask Howard Baker to become Chief of Staff. Ed approved heartily. Howard was in Florida visiting his grandchildren. When he called back I asked him to take the job. Paul was right—he accepted immediately & is coming to Wash. tomorrow. Now I'll go on reading the "Report" 'til I fall asleep. Thank you God.

Friday, February 27

A meeting with Repub. Cong. Leadership. Whole conversation was how I was to stage a comeback, my upcoming speech etc. Ed Meese dropped by with a word about Lennis the man we'll have to extradite to the Soviet U. He's been convicted there of W.W. II crimes—he'll be executed. We have no choice under the law. We've tried to find a country that will take him & none will.

A couple of photos with visiting State legislators & then upstairs. Before the afternoon was over I'd been visited by Howard Baker & Paul Laxalt. Howard has agreed to be Chief of Staff when Don R. resigns on Monday. Things moved up—a leak about Howard got out. I received a one line letter from Don resigning as of now. He was understandably angry. So guess what the news is tonite?

Early evening John Tower came by & we discussed my next week speech about the Tower Commission Report etc.

Saturday, February 28

Strange to be in Wash. instead of Camp D. but too many things going on. It's a gray dreary day. Did radio cast in Oval Office. Kathy says there is an uplift in morale as a result of yesterdays happening. A phone call from P.M. Margaret Thatcher & Dennis to wish us well & tell us we're still liked by the people there. Ron arrived. He came from the Coast to plead with me—out of his love for me—to take forceful action & charge of the situation. I was deeply touched.

[Sunday, March 1: finished reading Tower Commission Report, commented, "Now it's up to me to say the right thing Wednesday nite on TV"; had massage.]

Monday, March 2

9 A.M. & the Chief of Staff & V.P. came in for usual meeting—only this time the Chf. was Howard Baker—his 1st day. He's going to be fine & there was a great feeling in the West Wing of improved morale. Ron asked if he could spend a day in my office to see what my job was like. So I said yes & he was in on all of the meetings etc. NSC had to do with Gorbachev's new arms proposal & the matter of Bob Gates as Dir. of CIA. The Sen is giving him a bad time & may not approve him.

[Received copies of upcoming speeches; cabinet meeting, including briefing on Tower report; latest polls show job approval rating down to 44 percent; issues-briefing lunch; meeting with Republican mayors.]

Before day ended Gates had come in & asked me to remove his name from consideration as Dir. of CIA. Now we're trying to see if we can get John Tower to take the job. We should know by tomorrow.

Nancy had to got out for an early evening drug do so Ron & Mermie & I are dining at home.

Tuesday, March 3

Ron is on his way back to Calif. We cancelled the Cong. Leadership meeting & had our own—Cap W. attending. We decided to try for Webster—Dir. of FBI to take the CIA job. I called him—he was quite emotional—torn between saying yes & having some reservations about it. Promised to call back & did so but not until 6:15 P.M.—He said yes & that made my day. We'll make an announcement right away.

This morning I went over to EOB the NSC's room & met with the entire NSC Staff (many new faces). Then back to lunch. Then an NSC meeting in Cabinet Room in preparation of my summit in Canada with P.M. Mulroney. He's in big pol. trouble & part of it is based on pol. attacks that he's beholden to me & the U.S. We're trying to find some things to bolster him. One possibility is (if we can do it) some pickup in the acid rain situation.

[Photo session with Mrs. Reagan for wedding anniversary; read statement to press regarding Mikhail Gorbachev's INF arms proposal; videotapings; attended W.H. dinner for freshman congressmen.]

Wednesday, March 4

Our Wedding anniversary. Nancy says my speech tonite is her present from me.

Usual morning meetings—Howard & the V.P. & then Frank C. & Gen. Powell. Nothing of great import. Then at 9:45 A.M. Mayor Diepgen of W. Berlin came by for a visit. I've agreed to make a speech in Berlin in June for the 750th Birthday of Berlin. Then a press opportunity with Bill Webster our new Dir. of CIA.

Nancy surprised me by joining me for lunch in the study. After lunch a brief huddle with Howard, Stu Spencer, Dick Wirthlin, & Landon Parvin about the speech & upstairs to wait for 9 P.M.

The speech was exceptionally well received & phone calls (more than any other speech) ran 93% favorable. Even the TV bone pickers who follow the speech with their commentaries said nice things about it.

Thursday, March 5

[Meetings on aftermath of speech; visit from Doug Morrow, head of U.S. Space Foundation; addressed National Newspaper Association; meeting with Senate Steering Committee, conservative supporters.]

Friday, March 6

[Noted that Vice President Bush was in Iowa; NSC meeting with update on Soviet situation.]

The NSPG meeting was on Afghanistan & Pakistan. Pakistan is necessary to our help of the Mujaheddin [. . .]. We must & will continue to give aid to the freedom fighters.

Then a working lunch with Max Kampleman & Ron Lehman & Mike Glitman. We brought them home form Geneva to talk about the Soviets & the proposal to go zero, zero, on INF forces in Europe. It looks good but we mustn't get too carried away until we see how far they'll go on verification.

[Meeting with staff to boost morale; photo session with Time *magazine; went to Camp David.* **Saturday, March 7:** *radiocast; telephoned British prime minister Margaret Thatcher with condolences on ferry accident in North Sea; telephoned widow of Senator Edward Zorinsky (D-NE).]*

Sunday, March 8

An early start home—We have a musical at the W.H. at 5 P.M.—Marvin Hamlisch, MC—Vic Damone, Liza Minnelli & Bobby Short—part of "In Performance at the White House." It was a fine show & wonderful evening.

Monday, March 9

Jack Koehler resigned this morning. I guess he & the job just didn't jell. He's going to be a consultant over at Charlie Wick's shop & couple that with a private enterprise job.

[Staff and NSC meetings; heard positive report from Shultz on recent trip; lunch with Republican National Committee Executive Board; legislative strategy meeting.]

Cap W. came by. I read him the statement I'm putting in next Sat's Radio cast about him & George S. putting to rest the criticism of them in the Tower report. He told me how our mil. has already moved equipment & men into Ecuador to give aid in the Earthquake area. I had a call from President Febres-Cordero asking for help.

A haircut & upstairs. Mermie is here.

Tuesday, March 10

A meeting with GOP Cong. Leadership. We had a session on Contra aid which Dem's are opposing. Then a session on the budget. Our opponents want us to join them in a summit. Well we've presented a budget—they haven't. We all agreed it would be silly to have a summit before both sides have presented their budgets. It was a good meeting.

[Brief NSC meeting, projected ideas for future foreign-policy speech; lunch with Vice President Bush; domestic-policy meeting regarding ways to help Prime Minister Brian Mulroney of Canada, who had been criticized for being overly friendly with the U.S.; dental appointment.]

Wednesday, March 11

I wrote a sample letter to the editor rebutting an article in *People* magazine quoting a Prof. of Pol. Science at Duke U. who took me apart on supposed misstatement of FACTS back all the way to W.W. II when I suspect he was only a child. Howard Baker & George Bush applauded my letter but thought someone else should sign it. I think they were right.

[New polls show upturn in approval rating since March 4 speech; National Security Planning Group (NSPG) meeting with Under Secretary of State for Political Affairs Michael Armacost, who was going to meet with the Soviets in Moscow, commented, "I told him to give them h--l about getting out of Afghanistan."]

A Bi-partisan Cong Leadership meeting. We briefed them on Central Am. and the negotiations in Geneva (Max Kampleman). It was a pretty good meeting. A half hour with Geo. Shultz followed the NSPG meeting which was entirely about SDI—the ABM Treaty and we were all pretty much in agreement that SDI would stay within the narrow interpretation as to research until December 31, then go into the broader interpretation. George's meeting was on his intended trip to Moscow & whether I should send a personal letter to Gorbachev re a summit this summer here in the U.S.

[Ceremonial meeting with former Supreme Court justice Warren Burger.]

A brief meeting with Ed M. on his way to a board meeting of trustees of Presidential Library. They are going to discuss whether to take it away from Stanford since there are demonstrations protesting it being there. I'm in favor of USC. Then Dick Allen brought in a wonderful little Korean gentleman C.J. Chung. He raised a mil. $ for a program in true pvt. enterprise fashion. His son—a student here interpreted. Chung took Korea from the wreckage of war and literally created its industrial system.

[Dinner party with friends at the home of Nancy Reynolds.]

Thursday, March 12

This A.M. we agreed on Ken Duberstein to be Howard's deputy. It's great having him back on board. NSC was brief & no sensational news. Later dropped in on PFIAB having regular meeting in Roosevelt room. Henry K. gave a brief sum up of meeting with Gorbachev in Moscow. I'm afraid I can't agree with one of his views. He doesn't think we should go for zero option we're negotiating with Soviets on INF. He says they have missiles <u>not counted</u>, hidden, & even on site inspection wouldn't reveal them.

[Meeting with President Daniel Arap Moi of Kenya, commented, "Kenya thanks to him is the best organized country in Africa & the most free enterprise one."]

Some desk time then admin time—photos with Ambas. Constable & her family—she's Ambassador to Kenya. A group of the "Citizens for Reagan" came by—they are working hard to line up public support for our aid to the Contras. Yesterday afternoon the House voted a moratorium which I will veto if it passes the Sen.

[Greeted departing Secret Service agents and military guards; read through new wills and trust agreements.]

Friday, March 13

[NSC briefing on situation regarding arms control with Soviets and Central American policy; spoke to gathering of leaders who support balanced budget amendment.]

Got word NBC has a leaked story & has aired it about a ship load of T72 Soviet tanks on their way here—a deep dark secret we wanted to hold. At my suggestion we arranged for Jack Welch, CEO of GE (which now owns NBC) to come to Camp D. later this P.M. with Frank C. & Gates.

[Regular visit with Shultz, noted, "wanted to outline continuing peace efforts in Middle East. I'm all for it"; ambassadorial formalities; greeted newly appointed senator from Nebraska; went to Camp David.]

Our visitors arrived. My idea was that we see if Jack Welch would be interested in getting media leaders to set up their own Code of Ethics about going public with leaks like this one about the tanks. He's definitely in favor. We told him we'd heard NBC News had another about our Mi-24 Soviet helicopter [. . .]. They all left with Jack agreeing to pursue our idea. This very evening NBC came on with the Mi-24 story

Saturday, March 14–Sunday, March 15

Radio address—I gave a sound plug to Geo. S. & Cap W. I think they deserved it. By the evening news it was being played as rebuke to the Tower Commission which it wasn't. Called Ed Rollins in hospital for operation. Received call from Henry Kissinger with some ideas about arms control—some of which I do not agree with. Sat. nite a wet snow was falling

[Sunday, March 15: returned to W.H.; dinner with the Howard Bakers and the Carluccis.]

Monday, March 16

Staff time was short with no issues to speak of. NSC brought up some strategy problems for Contra aid etc. Later in the morning Howard, Dave Abshire & our new legal counsel Culvahouse came in for a brief session on the upcoming press conference. Also they need to see my diary so they gleam any further notes on Iran problem. Otherwise the special prosecutor & or the Sen. might raise a fuss & demand to see it themselves.

[Issues-briefing lunch, noted, "no great problems & all in all a pleasant time."]

3 P.M. Howard, Dave, Culvahouse in for a session on upcoming press conf. We're all convinced it will mainly be about Iran. Then Jim Baker—re the Aug. end of term for Volker as Chrmn. for the Fed. Reserve Board. We're going to see if Alan Greenspan will take the job if Paul will step down gracefully.

[Photo sessions with guests chosen by various members of Congress. **Tuesday, March 17—St. Patrick's Day:** *NSC briefing with light load of subjects; greeted Soviet sailors rescued on freighter by U.S. Coast Guard; attended St. Patrick's reception at Irish Embassy and holiday lunch hosted by Representative Jim Wright (D-TX), noted, "By the time they were over I was talking with a brogue"; met with Howard Baker, Abshire, and lawyers to prepare for upcoming press conference; visit from Senator Larry Pressler (R-SD) with idea for recruiting religious leaders to approach pro-Sandinista Americans; videotapings.]*

Wednesday, March 18

[NSC briefing, including report by Undersecretary of State for Political Affairs Michael Armacost, on recent visit to Moscow; met bipartisan group of senators favorable to Contra aid; further preparation for press conference.]

Had a meeting with George Shultz. He presented some material on his coming trip to Moscow having to do with Human Rights. He really wants to go after observance by the Soviets of the Helsinki Pact.

He also laid out some ideas on our Central American policy—the short-comings of the Arias (Costa Rica) proposal. Hooray!!—Phone call first to tell me the Sen. voted 52 to 48 for the $40 million aid to the Contras. I'd thought I might have to veto.

Thursday, March 19

Nancy off to Atlanta, GA back this afternoon. Our Staff & NSC meeting involved with Ross Perot who wants to go on his own to Hanoi on the P.O.W. matter. He says he'll only go if they assure him he'll be able to talk to American P.O.W.'s still being held. He also wants a letter from me saying I approve his going.

[Lunch with Vice President Bush; visit from Florida supporter Tommy Thomas; two hours' rehearsal for press conference; staged conference in evening; regarded response as positive. **Friday, March 20:** *signed proclamation for Afghanistan Day; lunch with regional campaign directors; met with Shultz regarding strategy for upcoming meetings with Soviets; received word that diplomat Max Kampelman suffered mild heart attack; interviewed by*

Trudy Feldman; went to Camp David with reports on energy and Soviet arms. **Saturday, March 21:** *reading.* **Sunday, March 22:** *returned to W.H.; Mrs. Reagan left for NYC; received word that plane flown by Dean Martin's son, a National Guard pilot, was missing in rugged California terrain.]*

Monday, March 23

Howard B. reported Ross Perot who is evidently upset about something or other having to do with Vietnam & our POW's. He had told me his arrangement with Hanoi was that they had to write him that he could meet some of our POW's still being held captive if he'd go to Hanoi. Well no such message came. It was due Friday.

Colin took Frank's place at the NSC brief. Frank is in South America. Most of this meeting was taken up with report on Dean Martin Jr. He & his crewman it is known bailed out over San Gorgonia peaks but still no word of them. Weather has been interfering with the search. Nancy & I called his folks Dean & Jeanne—they of course are all torn up.

Late morning with Wm. Webster. He requested that he not be made a Cab. Member as Dir. of CIA. I'd done that with Bill C. and found out he shouldn't be in the Cabinet so we had no problem with Bill W's request.

[Issues lunch; met with Bob and Elizabeth Dole regarding highway bill and hope to sustain veto of it; photo session with editors of Conservative Digest; *National Security Planning Group (NSPG) meeting on Persian Gulf strategy, noted, "I made it plain that if Iran tried to close the Gulf—we'd reopen it"; meeting with Weinberger regarding Special Services Command.]*

Then a haircut & upstairs—Mommies home.

Tuesday, March 24

[Report that Martin and crewman did not bail out; Mrs. Reagan left for Florida.]

A GOP congressional leadership meeting. Much of the time spent on the Highway Bill & my veto of same. Bob Dole made a speech about supporting me that drew applause from his colleagues—1st time I've ever seen that.

Then to NSC briefing—info on Space Station which costs out above our estimates. Also selection of Army Chf. of Staff—Gen. Vuono & Geo. S. & Cap's differences on SDI & whether to go for the broad or the strict interpretation of ABM.

[Spoke to American Business Council; cabinet meeting, with report on management-improvement program.]

Later a meeting with our advisor group for them to recommend how to promote our agenda.

Then a session with biographer Edmund Morris.

Later a visit by Howard & Frank C.—we've found another spy—a Marine

Guard at Moscow embassy. He's confessed and has probably compromised our communications [. . .].

Wednesday, March 25

[Short meeting with Howard Baker and Vice President Bush regarding highway bill and Bank Board appointment.]

At NSC the talk was mainly on the upcoming visit of Sec. of State to Moscow. I think we're going to have some differences between Geo. S. & Cap on arms negotiations involving SDI.

[Spoke with Congressional Republican Conference; met with Archbishop of Greek Orthodox Church, noted, "His Eminence whispered that P.M. Popandreau would like an invitation to Wash."; visit from Shultz on arms deal topics to be presented to Soviets; meeting of conservatives hosted by Senator Paul Laxalt (R-NV); greeted foreign-exchange high school students. **Thursday, March 26:** *received word that Martin and crewman were confirmed dead; flew to Columbia, Missouri, appeared at schools; returned to W.H.; photo session with terminally ill boy; Maureen visiting.]*

Friday, March 27

Staff meeting was light—a few minor items. Then NSC and the subject was our spy case in Moscow—but only preparing for a later meeting. Then a matter of $10 million worth of arms from present supplies to Chad which seems to have Quadaffi on the run.

British Labor party leader Neil Kinnock along with former Defense Minister Healey & the Eng. Ambassador came by. It was a short meeting but I managed to get in a lick or two about how counter-productive "Labors" defense policy was in our dealing with the Soviets.

[Signed veto of highway bill; meetings with congressmen.]

Lunch with V.P. then a meeting with George Shultz about the coming trip. Finally down to the Situation Room. First subject was about Chirac's meeting next week (P.M. of France). Second the security matter at the Moscow Embassy. We think the 2 Marines did us great damage with regard to security [. . .]. We're busy planning how to restore things. Present was Jack Matlock our new Ambassador to Soviet U. Had a photo with his family. A few more photos during day with friends of our people—then desk work & home.

[Saturday, March 28: radiocast; spoke at Gridiron Dinner, received laughs. **Sunday, March 29:** *Maureen visiting; received word that William Casey, gravely ill, was moved to his home.]*

Monday, March 30

Six years ago today I got shot. Howard, Ken, George & I talked over a few domestic problems. Then NSC—reports we have a big disagreement between

Sec of St. & Sec. of Defense over SDI & our approach to the Soviets. I don't want it negotiated at all.

[Spoke to three thousand federal employees at Constitution Hall; meeting with R. Emmett Tyrrell Jr., founder of American Spectator magazine; honored vigilant inspector general; meetings with Senators Thad Cochran (R-MS) and Mitch McConnell (R-KY) to elicit support for highway bill veto; visit with Daniel Boorstin, retiring Librarian of Congress; video-tapings. **Tuesday, March 31:** *briefing on visit of Prime Minister Jacques Chirac of France; arrival ceremony; meetings on private initiative program and announcement of agreement on joint research on AIDS; photo with disbanding team of David Abshire, W.H. Special Counsel investigating Iran-Contra scheme; courtesy call from King of Spain; state dinner.]*

Wednesday, April 1

A staff meeting—spent most of our time talking about the Sen. vote on my veto. We don't know whether we've got the votes to sustain or not. But when 1 Dem. Terry Sanford broke ranks & voted with us—the winning vote—Bob Byrd switched his vote to us so he could demand reconsideration. You have to be a vote on winning side to call for reconsideration. All day on A.F.1 & in Phil. We've been back & forth on the phones trying to line up votes by then Byrd has been busy too. Terry Sanford stood up and announced he was changing his vote. It looks likes we're behind.

[Flew to Philadelphia; spoke at bicentennial celebration for College of Physicians; greeted political donors.]

Ride home in A.F.1 was again a series of phone calls to the Sen. I'm afraid we're going to lose.

Thursday, April 2

Staff time again on the Highway Bill. I decided to go to the Hill & make my pitch to our Repub. Sen's including the 13 who are voting to override. Some of them thought I would only look worse if I was overridden. I decided I couldn't live with myself if I didn't go.

In NSC the subject was a presentation for the Joint Chiefs delivered by Colin. They feel in our arms talks we need more than 5 yrs—6 or 7—in order to offset the advantage to the Soviets in conventional weapons.

[Had good meeting with Lord Peter Carrington, secretary general of NATO.]

Then I left the meeting & headed for Capitol Hill. I met 1st with a large group of Sen's in old Sen. Chamber. Then with the 13 in Bob Dole's office. I knew when I left I'd failed but I have no respect left for that 13. They were voting on strictly the pressure they were getting from the construction industry and they were voting <u>against</u> trying to balance the budget.

[Greeted women's NCAA basketball champions from University of Tennessee; economic policy meeting on upcoming Canada meeting; received word that veto was overrid-

den; greeted group of Hasidic Rabbis, then members of American Cancer Society escorted by Minnie Pearl.]

Friday, April 3

Stop by the Doctor for a look at my nose—clean bill of health. Staff meeting—very short. NSC—report on brkfst. Meeting with Sec's Weinberger & Shultz. They have differences regarding arms talks but nothing that can't be worked out.

At 11:35 Intel. Board came by with an interim report on Iran. On their own they were overseeing the "intelligence" situation. Then a little ceremony in the Roosevelt Room with Elizabeth Dole. She presented me with a check for $1,575,000,000.00 the purchase price we got for selling Conrail.

[Lunch with donors to the presidential library; meeting with Shultz on foreign-debt problem, specifically countries that owe so much they probably won't be able to pay it back; greeted men's NCAA basketball champions, Indiana University. **Saturday, April 4:** *read Canada briefing book; radiocast.* **Sunday, April 5:** *flew to Ottawa, Canada, formal arrival ceremony; initial meeting with Prime Minister Brian Mulroney; dinner reception.]*

Monday, April 6

Meeting 9 A.M. with John Turner, Leader of the opposition party & Sec. Shultz & Charles Bouchard, Policy Leader of opposition.

About 9:30 meeting over—left for Parliament Hill. Signed books for both Houses of Parliament then a meeting in P.M.'s office. About 10:30 a plenary meeting in Cabinet Room. It went well & I think we resolved some difficult problems having to do with Acid rain, Arctic waters sovereignty problem & defense budget.

[Working lunch with Prime Minister Brian Mulroney; addressed Parliament; returned to W.H.; Maureen and Dennis visiting.]

Tuesday, April 7

Staff Time—decided on a statement to press on Soviet Embassy spy situation. Then about 1:30 P.M. I went into Press Room with it.

Before that however there was our NSC briefing—subject there was also our Marines & their part in the situation. An NSPG meeting to discuss George S's coming trip to Moscow & offer suggestions to him re subject matter.

[Attended W.H. lunch for corporate sponsors of commemoration of the bicentennial of the Constitution; cabinet meeting with projection for future efforts; visit from second-cousin; received new poll numbers; haircut.]

Wednesday, April 8

[Staff meeting with discussion of appointment in bank program; NSC briefing on topics including security at Moscow Embassy; meeting with U.S.-Canada free-trade group.]

Then I went over to the Pentagon for my quarterly meeting with the Joint Chiefs. It was a fascinating report they had for me. Some of it reorganization plans but the real kicker a report on new weaponry. We can't match the Soviets tank for tank so we use our technology & came up with a weapon that nullifies their superior numbers. There was much top secret & brand new in aircraft.

Back to the W.H. & after lunch a meeting with Ed Meese on replacement of Bill Webster as Dir. of FBI. A possible candidate is Byron White—presently on Supreme Court. This too will be checked out while we're in Calif. Then a meeting again with business people volunteers. This time the Advisory Committee on Trade Negotiations. Again it was to report on our Canada trip.

[Ceremony for stamp honoring Dr. Harvey Cushing, the mentor of Dr. Loyal Davis, Mrs. Reagan's father; Mrs. Reagan's brother present. **Thursday, April 9:** *flew to Lafayette, Indiana, for appearance at Purdue University; flew on to L.A.* **Friday, April 10:** *spoke at L.A. World Affairs Council; Mrs. Reagan arrived in evening.* **Saturday, April 11:** *radiocast; lunch with Reagan Presidential Library Foundation; went to ranch.]*

Sunday, April 12

We rode in the morning—then Mike & Colleen brought Cameron & Ashley up for Ashley's birthday. It was a new Mike. He's writing a book & it had led to a soul searching about himself which resulted in a confession of how he had done things to all of us & that he now saw himself as he had been & what he wanted to be.

*[***Monday, April 13–Wednesday, April 15:*** *rode every morning; cut branches; homework; received call from Shultz in Moscow.* **Thursday, April 16:** *rode; Ron and Doria arrived; chainsaw work; Shultz arrived to discuss Moscow trip, commented, "There is reason to believe we may be on the path to some arms reduction."* **Friday, April 17:** *rode and cut branches.* **Saturday, April 18:** *rode without Mrs. Reagan; visited children from local Camp Ronald McDonald for Good Times; tree pruning.]*

Sunday, April 19—Easter

We went to church—1st Presbyterian in Santa Barbara. The Minister only learned 15 min. before arrival that we were coming. It was good to be in church again—we've missed going. In the afternoon our last day with the chainsaws.

*[***Monday, April 20:*** *left for Washington.]*

Tuesday, April 21

Back to routine—9 A.M. meeting with Howard, V.P. & Ken D.—talked about putting together an agenda & campaign for the next 21 months. Then an hour with Repub. Cong. Leadership. We talked budget & then Geo. S. gave a report on the Moscow trip.

At NSC the subject was Soviet missiles & General Rogers who is retiring as Chief of NATO forces. He has a nuc. missile theory a little different than mine.

Then Japanese Minister Abe came by. We had a good & friendly meeting but settled nothing about trade problem.

[Presented award to National Teacher of the Year; brief cabinet meeting.]

Ed Meese came by about new Dir. of FBI—He's going to try for former Gov. of Pa. Dick Thornburg—I hope we can get him.

[Videotapings; meeting with Representative Jim Wright (D-TX) to discuss his trip to Moscow. **Wednesday, April 22:** *usual morning meetings; lunch with Citizens for America; meeting with Shultz, mostly on Middle East; haircut; bipartisan congressional leadership meeting on arms control; attended W.H. correspondents dinner; noted call from son Michael "to thank us again for our session with him at the ranch."]*

Thursday, April 23

Staff time taken up with discussion of Cong. inroads on Presidential Power and choice I must make between Sen. Montgomery's nominee for Chmn. of Fed. Home Loan Bank Bd.—Philip Winn (He made a good pitch) and Sen. Jake Garn's candidate Don Wall. Jake due in at 3:45 to present his case.

In meantime had a nice lunch with V.P.

In morning NSC meeting subject was the PLO meeting in Algiers & the manner in which Arafat continues to dominate. Jake Garn made a good case & now it's up to me. The choice isn't easy. They are both good men.

Personnel time, a meeting with Culvahouse who has been through diary looking for items on Iran situation.

Then Admin. time—photo with former Sen. John Sherman Cooper & with the Easter Seal Poster Child & his family. She is 11 yr. old. Susie Wilcox—West Simsbury, Connecticut. Adult representative is 57 yr. old Dr. Andrew Vangelatos of Eureka, Calif. Pat Boone & Shirley also—he is Chmn. of Nat. Campaign. Susie is confined to a wheelchair.

Then up to a lonely W.H. Nancy is in N.Y.—home tomorrow.

Friday, April 24

Nancy due home today. In for the 9 A.M. day starter—the usual routine—nothing very exciting. Then NSC at 9:30 Frank reported on situation in Congress where they are hanging amendments on bills having to do with the SALT II & ABM Treaties. Not only do those impinge on my prerogatives they are most helpful to the Soviets in our ongoing negotiations. I'm ready with a veto—in fact I can't wait.

[Signing ceremony for Crime Victims Week.]

After a lunch a preliminary meeting with our team preparing for the Ec Summit in Venice. Then a meeting with Geo. S. This was on Middle East—King Hussein, P.M. Shamir & F.M. Perez. He also gave me a report on efforts to trim down

Embassy personnel. Problem is these are people serving agencies other than State Dept.—such as Coast Guard Loran Team. State doesn't have a say so over them & they resent efforts to reduce their numbers. We discussed the situation of Ambassador Barnes. Back in Carter Adm. his wife became entangled with a Romanian driver who supposedly was romancing her out of State secrets. The whole case was investigated, the records sealed, & the Barnes's reconciled. Now a defector from Romania has written a book & is telling this story. Mike Wallace apparently is going on 60 Min's this Sunday with it. It's tragic & could destroy 2 peoples lives.

[Asked Representative Dan Rostenkowski (D-IL) about making appearance in Poland; dinner party for presidential library board; left for Camp David to avoid demonstrations planned for W.H. vicinity following day. **Saturday, April 25:** *radiocast on trade; walk with Mrs. Reagan and Rex; watched movie; phoned Justice Byron White on twenty-fifth year on Supreme Court.* **Sunday, April 26:** *returned to W.H.; received word that William Casey had medical setback.* **Monday, April 27:** *addressed National Chamber of Commerce; issues lunch; signing ceremony for National Volunteer Week; reception for donors to National Federation of Republican Women, commented, "I've urged them to stage a drive for the '88 election to revive the Repub. 11th commandment"; photo session with Japanese newspaper; attended Senator Paul Laxalt's annual "Lamb's Fry."]*

Tuesday, April 28

A session on our upcoming agenda. I agreed to a general plan but with proviso that specific events be checked with me.

Then Bob Dole, Bob Michel, Trent Lott, & Al Simpson were in with our team for a session on budget & trade & how to handle the Dems who are moving on both with acts that look like veto candidates.

NSC brief was on upcoming meeting with Nakasone plus some talk on Soviets & arms reductions.

Then an NSC meeting—all on the Japan problem & the coming visit of the P.M.

After lunch Senator Gordon Humphrey came by to deliver a letter from P.M. of Pakistan—where he has been on a visit.

They are asking for an AWAC plane because of violations of their border with Afghanistan. More people have been killed by Soviet bombing & strafing in these 3 months than in all of last year.

[Oval Office interview with six correspondents; new poll shows job approval up to 55 percent; meeting with Howard Baker and Frank Carlucci regarding Soviet arms talks; Maureen visiting.]

Wednesday, April 29

Staff meeting had to do with comments on statements re Soviet arms deal by R.N. & General Rowny.

[NSC meeting regarding arms situation related to Germans, who were concerned about giving up short-range missiles; meeting with Republican congressmen about forthcoming veto of trade bill; joint economic and domestic policy meeting on increasing domestic oil production, commented, "Faced with 2 options—I came down on the side of one that should make a real difference—using the market & not a tax increase."]

George S. meeting—subject Nuclear weapons & our allies. He also had a report on Israeli relations—visit by Arens.

[Interview with biographer Edmund Morris; party fund-raising dinner.]

Thursday, April 30

Opened day with a briefing for P.M. Nakasone's visit—then the moment was at hand. Ten A.M. the S. Lawn ceremony. I never seen a bigger public turnout for one of these events. Weather was perfect. He & I had a very brief one-on-one & then into the Cabinet Room for the plenary. He's most anxious that our trade sanctions on the semi-conductor issue should be lifted before the Ec. Summit this June in Venice. Well we'd like that too but it depends on their adherence to the agreement we had on the marketing of semi-conductors. I invited him, his wife & daughter to have breakfast tomorrow morning with Nancy & me. We did that in 1983—our 1st meeting.

He was very pleased.

Had lunch on the patio with V.P. He said he'd understand if I felt I should lend a hand to Paul Laxalt now that Paul was a candidate for President. I told George I had already told Paul I'd remain absolutely neutral in the primaries.

[Considered new appointees; ceremony for Law Day; meeting with Weinberger, noted, "He is opposed to a zero zero deal for short range nuclear missiles but all for it on the long range weapons. He & I disagree on this one"; state dinner, entertainment by Henry Mancini and Johnny Mathis.]

Friday, May 1

Started day by having the Nakasone's including their daughter for breakfast—a pleasant hour. Then he went up to meet Cong. & I went to the office. Usual staff meeting had to do with the budget & the games the Dem's are playing trying to play off cuts in defense budget for a tax increase. Then there was the oil problem & again their desire for a tariff on imported oil. Some discussion on my upcoming speech to the Nat. Publishers Assn. Sunday in N.Y. The subject is going to be Central America. NSC brought an announcement that the Bank of Japan had publicly repudiated Prime Minister Nakasone on his proposal that they lower interest rates—something we'd asked him to do in our trade negotiations. Then we got into SDI and Sen. Levin's amendment which would really cripple & eventually close down the project. And a lot of discussion of short range nuclear missiles. Soviets have proposed to eliminate them but our NATO Allies aren't too happy about that. We're waiting for this discussion.

[Final meeting with Prime Minister Yasuhiro Nakasone; farewell ceremony; addressed American Legislative Exchange Council (ALEC); meeting with Shultz on short-range nuclear missiles; report from Suzanne Massie on last Soviet visit; discussion of assistance to domestic oil industry with Howard Baker and others; videotapings. **Saturday, May 2:** *radiocast; homework delivered by staff members, noted, "They've interrupted my search for books to give away"; dinner party at British Embassy.* **Sunday, May 3:** *flew to NYC, Mrs. Reagan pursued her own agenda; addressed National Newspaper Publishers Association; returned to W.H., while Mrs. Reagan not due to return for two days; Maureen visiting.]*

Monday, May 4

Staff meeting—subject budget & Dem. determination to cut defense. I've proposed that we submit supplementals but make each one a particular weapons system. See if Demo's can afford to oppose specific weapon systems.

NSC—Discussion of Ross Perot who wants a meeting with me. I said OK but he has become a loose cannon on the POW matter. I have chosen Gen. Vessey to be our sole negotiator with Hanoi.

[Signed proclamation for Asian Pacific American Heritage Week.]

An issues briefing lunch then a Domestic Policy Council meeting. Subject was AIDS. I've ruled that we appoint a commission to see if various groups dealing with AIDS—Nat. Health, Ed., etc can help find a way to combine our efforts.

[Meeting with departing W.H. Fellows; haircut; received update on William Casey, noted, "Casey is still in intensive care & is in & out of comas. It sounds last stage." **Tuesday, May 5:** *staff meeting to plan upcoming agenda; NSC meeting on Central America; signed Executive Order on W.H. Conference on Drugs; NSC briefing on upcoming visit by President Vinicio Cerezo of Guatemala, commented, "We'll have to work on him to get him in line with Honduras & El Salvador against the Sandinistas—he's known to play games"; advisor meeting on two-year agenda, decided on priorities, including budget reform and balanced budget amendment; Mrs. Reagan returned home.]*

Wednesday, May 6

Short staff meeting to make time for George S. to come in. NSC postponed also. George came in to tie up loose ends about arms deal with Soviets. I approved trying to persuade Helmut Kohl to join in offering zero zero on short range missiles.

Then Ross Perot came in on our dealings with Vietnam. I have named General Vessey as my rep. to seek info. on our POW's. Ross is convinced some 350 or so are being held in Laos. I'm trying to persuade Ross to step back & not indicate we should try normalizing relations—trade, etc. until we get the truth on our POW's.

A quick lunch & then to Marine 1 & off to York, PA the Harley-Davidson motorcycle plant. They have done a remarkable job climbing out of a slump. Japanese competition was destroying them. We invoked a 201 a temp. use of tariffs to allow them to reform to meet that competition. A year early they told us to cancel the protection. It was thrilling experience. They haven't just improved production, they have a team from the workers on the line to top management & they can out compete any one. It isn't a factory, it's a religion.

[Received word early in the day that William Casey had died; telephoned widow.]

Thursday, May 7

At Staff—we had a full plate—Secord's testimony that North had led him to believe I knew about the Iran money going to the Contras. Then the run away Bud. resolutions passed by the House & Sen. with $118 billion over 4 years tax increase. Senator Hollings trying to hold up confirmation of Webster as CIA Director. Nothing done about debt limit. I've agreed to promote Jim Hooley to Deputy Assistant to the Pres.

NSC—Report House has passed an amendment that we'd have to abide by SALT II Treaty but Soviets wouldn't. I approved a set of bargaining terms for our Arms negotiators.

[Meeting with Zbigniew Brzezinski, commented, "He was quite informative about European attitudes"; met with American Association of Editorial Cartoonists; lunch with Vice President Bush; photo session with W.H. Personnel Department, Air Force Thunderbirds, Naval Academy basketball star David Robinson, Diabetes poster child; attended reception for National Endowment for the Humanities—about forty people; noted meeting early in the day with South Korean defense minister Lee Ki Beak, in Washington for discussions with Weinberger.]

Friday, May 8

9 A.M. meeting with Sen. Repub. Leadership—Discussed our problems of getting a clean bill increasing the debt limit. Congress has a way of loading this bill with amendments knowing how impossible it is to veto it. If we don't get this passed before the 28th of this month the U.S. will be in default for 1st time in 200 yrs. This could be a disaster in the money markets world wide. We also touched on the terrible trade bill the Dem's in the House passed. They have hopes of clearing it up in Cong.

Back to the Oval O. for a short session with NSC. We talked about the restraints on my ability to deal with the Soviets built into the Budget Bill by Dem's in House.

Then a brief meeting with delegates from the Arab League. They are in the U.S. to urge the U.N. to take steps to halt the Iran-Iraq war. Iran is the hold out.

[Meeting with Republican House members, along the lines of earlier meeting with

senators; signed bill making Santa Fe Trail a historic site; met with sponsors of report by Working Seminar on the Family and American Welfare Policy; taped radiocast; donated books to Eureka College Library. **Saturday, May 9:** *flew to NYC for William Casey's funeral, noted, "Bishop called attention to what he said was Bill's inability to understand it was wrong to help the Contras. Jeanne Kirkpatrick took care of that in her Eulogy & got a big hand. First time I've ever heard at a funeral"; returned to W.H.* **Sunday, May 10:** *flew to Alabama to speak at commencement of Tuskegee University; participated in ceremony to unveil plaque citing General Daniel "Chappy" James; flew to Pope A.F.B. in North Carolina; joined Bob Hope in show; returned to Washington.]*

Monday, May 11

Staff: Ed Meese came by—he has called for a special investigation to look into his case. The Cong. is charging him with some kind of hanky panky involving a defense contractor. It's the usual phony lynch mob attack. I think he's wise to do what he's done. Then we talked about Bud McFarlane's return as a witness before the Joint Committee. None of us feel alarmed at all. Last subject was the debt limit problem. Dem's are playing game hoping to trap me into tax increase. We have to have an extension on the debt ceiling by May 28 or face default.

NSC—Discussed terrorist movement & possibility that Qaddafi is trying to set up a network in the U.S.

At 11:30 Cap W. came in for his usual meeting. He's off for Europe. One of his targets is Norway & their making available to Soviets technology for silent running submarines. We also talked about irresponsibility of Cong. with regard to the defense budget. What they are proposing will set us back years. It will also play h--l with our arms negotiations.

[Issues lunch; ambassadorial formalities; met with Republican members of Ways and Means Committee regarding debt ceiling; noted Mrs. Reagan away until Thursday.]

Tuesday, May 12

At 9 A.M. a phone call (secure) to Helmut Kohl (W.G. Chancellor). This was a session on our intermediate range nuclear missiles. I wanted him to know we weren't going to pressure him on the short range missiles. They have real concerns about being left with nuclear weapons that would explode on Germany & being left with Soviet superiority in conventional weapons. But I think he'll be cooperative.

Then it was a good meeting with the Repub. Cong. Leadership. Main problem is getting an extension on debt ceiling before May 28 when if we don't have an extension we'll be in default. Talk is of getting a 60 day extension so something can be worked out. I asked them to consider a ceiling based on Gramm, Rudman, Hollings (G.R.H.)—to carry us until budget was balanced with the ceiling each year based on G.R.H. deficit allowed each year.

Then back to an NSC meeting—subjects Nicaragua, Contra leaders are planning a Democratic government system. Then the Philippines election. It looks like Acquino will win big. On Malta there has been an election & for 1st time in years a pro U.S. Prime Minister has been elected.

[Addressed Council of the Americas; signed proclamation for Just Say No to Drugs Week; meeting with Shultz on arms control; photo sessions; visit from Rush Hill (friend from California government); met with Kuwaiti journalist; attended reception for Eureka College Reagan Scholarship Fund.]

Wednesday, May 13

[Staff meeting; noted that House voted for a sixty-day debt extension.]

NSC briefing dealt with my call to Kohl but mainly with the Persian Gulf & Iran's attacks on neutral shipping. Question is—will Kuwait sail its ships under American flag. If they will we will respond forcefully to any attacks on vessels flying our flag.

[Meeting with President Vinicio Cerezo of Guatemala, noted, "I spoke to him about need to stay aligned with Honduras & El Salvador at coming meeting of Central American Presidents"; briefing for interviews with magazine writers the following day.]

Thursday, May 14

At staff meeting received some recent figures by Wirthlin. They are encouraging. My approval rating holds at 53%. Then some discussion on the press coverage of the hearing & the way in which they have linked Iran & the Contras as if it's one incident & I'm covering something up. I've been going public for several years on aid to the Contras. Then it was NSC time. Frank reported on Cerezo visit & his other meetings while here. We seemed to have done well. Mrs. Cerezo attended a dinner on the yacht Sequoia former Presidential boat sold by Carter. I think before I leave I should get the craft back for future Presidents. A report on the Sofaer study of the ABM Treaty with regard to our SDI development. It proves the ABM Treaty does not forbid our testing of SDI. We are paying a price because of this SDI debate. Otherwise friendly members of Cong. opposed to the liberal interpretation of the treaty turn against us on other issues.

[Photo session with Philippine-U.S. business committee; briefing and then half-hour interview with writers from four national magazines; signed personal financial statement; videotapings; finished the day, as noted, "Upstairs & waiting for Nancy's return & and here she is. Me & the dog are very happy."]

Friday, May 15

A staff—quite an agenda. We looked over a list of names for AIDS Commission. List included Nancy's brother Dick. Then I learned that Jonathan

Miller who was mentioned in testimony yesterday as helping an aide of Col. North cash $1,000 in travelers checks for distribution to the Contras immediately resigned. But it seems he had already informed us he was resigning as of today several days ago before there was any such testimony. Press of course is declaring he resigned because of the testimony.

NSC—Cap W. & George S. apparently have conflicting views on proposed INF treaties with regard to the zero zero proposal vs. leaving some short range missiles in place. And there is possible disagreement about SDI and whether Cap should send report to Cong. directly bearing on progress & relation to ABM Treaty or go through State Dept. In this one Cap is correct, the law is very specific on how it's to be done.

Rcvd. word RENAMO—the insurgent group in Mozambique has kidnapped an Army Nurse. We have asked the Red Cross to provide us with some details. I want her rescued if we have to blow up the whole d—m country.

Then some discussion of Pakistan & their request for an AWAC plane.

[Interviewed by journalists from regional media; took afternoon off; quiet evening. **Saturday, May 16:** *addressed commencement for military doctors; radiocast; attended Mrs. Reagan's annual tennis tournament to raise funds for anti-drug program; reception afterward; sat with Tom Selleck; telephoned Yasmin Aga Khan about death of her mother, Rita Hayworth.]*

Sunday, May 17

A beautiful, easy day—until I got a call from Frank C. In the Persian Gulf off the coast of Bahrain our Frigate the USS Stark was hit by 2 missiles fired by an Iraq Mirage F.1. The USS Wadell went to the scene—both ships made port. One American injured. We have filed a protest. Apparently the plane did not have visual aim but fired at a radar image. More bulletins through the day & situation looking worse. Dinner with Mermie & Dennis.

Monday, May 18

Staff Time—some talk on schedule & how to announce the ship situation. Then NSC and more news on USS Stark. We're still waiting for an apology from Iraq's Pres. We've heard from Foreign Minister but that's not enough.

[Meeting with President Mário Soares of Portugal; presented awards at World Trade Ceremony; issues-briefing lunch; meeting with speechwriters requesting input on upcoming speeches; new polls indicated falling approval rating, though some categories were positive.]

Then to Situation Room for hour meeting on USS Stark. I've declared that we let everyone know I've ordered our naval vessels to open fire any craft positioning itself as a possible attacker. Saw 8 Ambas's & their families off. A haircut & up to get ready for dinner at Dick Helms. A small group Bill Webster

& Sally Thomkins, Senator & Ann Simpson, John Brass, Evangeline Bruce, Mr. & Mrs. Robbie Evans, the Hank Greenbergs, Mel Laird, & David Rockefeller. A most enjoyable evening.

Tuesday, May 19

To the office 1st Geo. Bush has called Sulzburger about an editorial which actually calls me a liar. So much for the <u>New York Times</u>. I'm waiting to hear the outcome. We decided I'll order flags half mast on the USS Stark tragedy.

[NSC meeting; Frank Carlucci brought unconfirmed report that West German chancellor Helmut Kohl was open to administration position on short-range nuclear missiles; flew to Chattanooga, Tennessee, to speak at high school commencement.]

Before leaving for Wash., I phoned Harry Carrey in the mist of his broadcast of the Chi. Cubs—Cinn [Reds] B.B. game in Chi. He has returned to being the voice of the Cubs after on absence because of a stroke. Finally home & Mother.

Wednesday, May 20

Opened day talking about news story that Navy wasn't providing transportation for the next of kin to memorial service in Jacksonville Friday. I changed that quick. Later called parents in New Jersey named in story & told them they'd be hearing from the Navy. Then a little talk with Howard & Ken about Solicitor General Fried who issued a ruling that Unions could use Union members dues for political purposes without their permission. I'm in total disagreement.

NSC—It seems Dobrynin is passing word Soviets can & will cancel missile reductions if I don't call off S.D.I.

Then a meeting with Foreign Minister Yaqub Khan of Pakistan. I got in a few words about nuclear weapons & how they must not build any. Later a talk with or mainly by R. Perle. He's brilliant & gave a great presentation about the arms reduction process & the course we should follow. Basically it was to stay tough. Then over to EOB met with & addressed a group of great teachers & principles. Bill Bennett presented me with 3rd book on "*What Works*"—title of the one—"*Schools That Work—Educating Disadvantaged Children.*" Great progress is being made all across the land in upgrading education.

Met with George S. He asked that we send someone to W. Germany to talk to Kohl about 0-0 in S.R.I.N.F. I agreed & Linhart is on his way. George moved onto L.R.I.N.F.—he's all for 0-0 there. Also discussed modernizing our Lance Missiles. Then some discussion of Middle East Peace process. And finally a problem with our designated Ambassador to Mozambique—Melissa Wells a Foreign Service Officer. Sen. Helms is blocking her approval by Sen. Committee. I want her approved. Then our meeting with Chinese delegation headed by Yang—a top level man in their govt. It went very well. Then a meeting with

Baker & Meese re the choice for head of the FBI. List is down to 3—we picked one but won't make it final 'til Ed has discussed it with Webster. Then upstairs to do my exercises before the 6 P.M. reception of 220 Repub. women holding elective office across the country. Mermie is hosting the affair. I speak & then a reception line. It was a fine time & we shook every hand.

Thursday, May 21

Word has come from Iraq Ambas. that they will pay compensation to the family's of the victims on the USS Stark. Space Shuttle set for 1st flight, June of 88. Geo. S. said there was talk of Soviet Summit here in late Sept. or early Oct. I've agreed to meet with the Independent Investigator as I did with Tower Commission. Then a briefing on tunnel that is going to run from Oval O. to the W.H. shelter. End of meeting.

[NSC meeting with report that West German chancellor Helmut Kohl will make decision on short-range intermediate-range nuclear forces (SRINF) by early July; group called RENAMO admitted to kidnapping missionaries and American nurse; awaited word on semiconductor situation with Japan; signed bill lifting regulations on natural gas; met with Chancellor Franz Vranitzky of Austria, noted, "Purpose to make it plain our banishing of Austrian President Kurt Waldheim did not mean any lessening of our friendship with Austria."]

After lunch Personnel time then NSC meeting to discuss foreign policy issues likely to come up at Ec. Summit. A meeting with Cap W. I have a problem—Geo. B. want to settle treaty with Soviets including 100 missiles in Asia & Alaska. Cap wants us to go for 0-0. Frankly that's what I want.

*[Photo session with departing appointees; meeting with Michel Rocard of France; presentation at W.H. by booksellers of several hundred books; attended reception for Vote for America group. **Friday, May 22:** flew to Jacksonville, Florida, for memorial service honoring those killed on USS Stark; returned to Washington, then on to Camp David.]*

Saturday, May 23

Mixed clouds & sunshine. Nancy had to go back to Washington to get an honorary degree from Georgetown Medical College. I lunched with the troops at Camp David then had my 1st ride in a long time. Nancy returned. I took a dip in the pool & it was Saturday night as usual. Ran a movie about Big Foot & to my surprise I was in it—a shot of me & Bonzo on a TV set.

*[**Sunday, May 24:** rode; signed condolence letters for Stark family victims; swam; watched taped show on Hollywood's one-hundredth birthday.]*

Monday, May 25—Decoration's Day

It was foggy & gray & chilly. Moved up departure time to 2 P.M. & back to the W.H. There I received word Ray Donovan & the others been found innocent of all 10 charges. Justice has truly been done. The only injustice was that the indictment was handed down in the 1st place. I've always suspected some conniving in the N.Y. prosecutor's office.

Tuesday, May 26

At staff I brought up my brother's query to me about whether he could go fishing off the Mexican coast as a guest of a friend who lives there. A couple of years ago someone here in Wash. told him no. I cleared that up—he can go fishing in Mexico. Then some discussion again about the "hearings."

NSC—Well first some discussion of how to re-open negotiations about doing away with the 100 missiles each on China border & ours in Alaska. We won't go public but bring it up in negotiations.

We have a problem with our friends in Canada—they sell natural gas across our border but they pile most of the cost of production into the price they charge us. Top Secret—We've tapped a line that gives us access to much of Libya's (Quadaffi's) communications.

[Briefing for interview with foreign journalists; gave interview.]

After lunch a brief meeting with Howard & Culvahouse & his assistant. This is going to be a weekly thing to give me legal opinions & inside on hearings etc.

[Photo with Crown Prince Felipe of Spain; presummit briefing; swearing-in ceremony for William Webster as CIA director; videotapings.]

Wednesday, May 27

In our 1st morning meeting we decided I'd sign the Farm Disaster Bill & later in the morning I did. It's somewhat controversial but it isn't a typical farm subsidy bill. It really is a disaster bill to compensate farmers who were prevented from planting or lost crops because of the weather.

Then Joe Coors name came up to my surprise. He wants a job in the West Wing. He's a great guy & obviously does not need a job. He just wants to be of help. We'd have some problems public relations wise. I decided we should put him on one of our temporary commissions (such as the AIDS group) for a few months & then find a place for him. Next subject was the Persian Gulf & what we are doing there. Then Bob Byrd's 50th Wedding Anniversary. We're having him & a few people to a lunch Friday in the study off the Oval Office. Final topic was Paul Volcher's term winding up—to re-appoint or not. Well he told Howard Baker he wants to leave. So we put a list of possible replacements together.

NSC—First subject was need for me to meet with President of Mexico—in Mexico. I've tried to meet each year with him & P.M. of Canada. This time it's Mexico's turn. We've decided I'll see him in a border city on way back to

Wash. from the ranch in August. Another less pleasant subject was evidence we have that Govt. of So. Africa is planning terrorist operations against the A.N.C. We're looking into how we can turn that off. Subject of Ambas. Straus Hupe (Turkey) came up. He's 85 & getting a little slowed down. We discussed a replacement. Last item was W. Germany & the matter of their Pershing missiles & the SRINF.

Then Ed Meese came in for a brief meeting. First item was possible extradition of terrorist Hamedi from W. Germany. They hold him for trial there but he is the murderer of one of our Navy men in the hijacking of our plane. Problem is terrorists in Beirut have kidnapped 2 W. German business men & are demanding a trade. Then came problem of new director of FBI—another prospect turned us down. Now we're going to look at a L.A. Police official—named Vernon. Last subject was a group of our Sens are demanding we appoint former Sen. Slade Gorton (Wash. defeated in 1986) to Court of Appeals. We might settle for a district judgeship if there's an opening—but he has been an opponent of everything I've tried to do.

[Television interviews with foreign journalists; met with Shultz regarding Persian Gulf; met with Prime Minister Amintore Fanfani of Italy; long phone call with Trudy Feldman, reporter for Trans-Features.]

Thursday, May 28

A very brief staff time with little to report—then at 9:15 A.M. into Cabinet Room & a meeting with the Repub. Cong. Leadership. First subject was budget & budget process the onto Persian Gulf. There were a few there who were challengers as to why we were in the gulf & why weren't our Allies there too. Pete Dominici asked why we didn't ask Japan to share the expenses. We had the answers to all those Q's. Whether we satisfied them remains to be seen.

Then NSC & more Persian Gulf. Some of the Sen's have been pushing Sen. DeConcini of Arizona for FBI Chief. Apparently he has good qualifications but their real reason is that they believe we could grab off his seat in the Sen. by way of Arizona Repub. Governor.

[Meeting with Prime Minister Wilfried Martens of Belgium; addressed annual meeting of National Association of Manufacturers; lunch with Vice President Bush with, as noted, "a discussion of the ongoing hearing on Iran-Contra & the press handling of the same"; appeared before Republican support group, Go-PAC; Domestic Policy Council meeting on AIDS; session on new appointees; photo session with departing Secret Service agents.]

Friday, May 29

Staff time turned over to NSC. A full agenda. Persian Gulf—we're submitting a report to Congress but in meantime will continue to defend ourselves &

the Gulf. I think some of Dem's agree with this although others are making a fuss. Cap W. is going to rebut press stories that we've decided to delay escorting Kuwait vessels. We're going ahead as soon as they have raised the American Flag.

Frank C. had called me last night to say it looks like Kohl is going to join us on the SRINF idea. I'm to expect a call from him Monday morning at 9 A.M. Then at 9:30 (today) we went to the NSPG meeting in the Situation Room. This was all about Gulf. Navy briefed us on plans etc. We're all agreed on what we want [to] be prepared to do. Then my meeting with Geo. S. Won't be seeing him now until Venice. Our talk was about the disarmament talks & things are looking better.

[Lunch party, including Bob Strauss, Howard Baker, Senator and Mrs. Byrd.]

My lawyers meeting again re the continuing Iran-Contra hearings. An interview with Edmund Morris—on the biography. I also did a Persian Gulf statement to the press. Into ranch clothes & off to Camp David. On way had photo with 13 year old girl who won the Nat. Spelling contest. Some 700 of the contestants were on hand to see us off as were some 22 wives of all our helicopter pilots.

Saturday, May 30

A beautiful, hot day—the usual schedule—the radio broadcast, lunch, & a horseback ride followed by a swim. The 17 yr Cicadas serenaded us for lunch. I phone Paul Laxalt about the strained relations between Frank F. & Mermie. Also phone Helen—Scoop Jackson's widow re the memorial service being held for him.

Sunday, May 31

Some homework and home early—leaving Camp D. at 1:20—threat of thunderstorms. Tonite 8 P.M. to big program on AIDS—Elizabeth Taylor in charge. I addressed group before dinner & then we leave. Well I addressed them in a tent. Well received until I mentioned routine testing for AIDS. A block of the Gay community in the tent booed me enthusiastically. All in all though I was pleased with the whole affair.

Monday, June 1

[Staff meeting, discussion of AIDS commission; NSC meeting on message from Japanese prime minister Yasuhiro Nakasone, noted, "he's desperate to have us do something about the sanctions we've imposed. I'm not sure we can do that before the Summit"; visit from Nina Shea, head of group publicizing Sandanista persecutions; met with Ambassador Perkins regarding South Africa's reluctance to adopt reforms; met student winners of essay contest on U.S. Constitution; signed bill to make G.I. Bill of Rights permanent; presum-

mit briefing; speech on summit, accepted Paul Volcker's resignation as chairman of Federal Reserve Board; phoned Alan Greenspan who accepted post.]

Tuesday, June 2

[Staff meeting, discussed gay demonstrators who heckled AIDS speech.]

NSC—Ed Meese came in for a few minutes to report on the Ministers meeting in Europe regarding terrorism. He was very upbeat & found a lot of unanimity among the Ec. Summit countries. Our need at the summit is for all of us to loudly re-affirm the decision on terrorism we agreed to last year in Tokyo. We discussed the problem of how to handle Iran's missiles—the "Silkworms." If they move them on line do we shoot or wait & fire back? Problem is if we wait 'til they fire we may lose a lot of lives—there isn't much chance of intercepting a Silkworm. I think we have to do what we've decided about hostile aircraft. If they deploy their Silkworms we have to assume they have hostile intent & defend ourselves.

[Announced appointment of Greenspan; photo session for Time *magazine; call from West German chancellor Helmut Kohl affirming support for administration zero-zero proposal on INF; visit to doctor regarding bug bite received at Camp David, no cause for concern; packed for European trip.]*

Wednesday, June 3

[Flew to Venice.]

We were met by Fanfani's and Ambas. & Mrs. Rabb. I reviewed the honor guard & then it was Marine 1 to Condulmer—an ancient palace. Very beautiful but not quite as convenient as a Hilton hotel. It was about midnight when we arrived and 1:30 A.M. for lights out. I took a Dalmaine & actually dropped off to sleep & awoke at 8:45. I sneaked out of bed & had breakfast. Nancy awoke at about 10 A.M.

Thursday, June 4

Condulmer is a 300 year old palace which has become a hotel complete with pool and 18 hole golf course. But now it is again a palace for us. The St. Dept. took it over & it's manned by W.H. Stewards. We took a walk around the gardens this A.M. After lunch it was studying for the coming meetings. We ran tapes also of last night's U.S. news & tonite a John Wayne movie—video cassette of course. Sometimes I have to say these grand old buildings are beautiful, magnificent with their marble floors, tapestries, & paintings—and inconvenient. This last word has to do with bathrooms & showers & no place to put your razor, toothbrush etc. Finished evening with old John Wayne movie.

Friday, June 5

A beautiful sun shiny day. Took a morning walk—after I did a TV speech for "Worldnet" and taped my weekly radio broadcast. We lunched by the pool and basked there (under umbrellas) for a few hours. I had brief meetings with Frank C. & Ken D. It seems Kohl & Genscher are at odds over handling the shortest range nuclear weapons but I think it's a matter that can be resolved. Word came that Paul Laxalt has cleared the way to run for President. He has settled his suit against the McKlatchy press out of court

Back in our Ancient castle—more homework—shower, hair wash & ready for dinner—followed by a Laurel & Hardy old movie.

Saturday, June 6

Departed 9 A.M. by helicopter for Marco Polo Airport—met by Venice Prefect and his wife—Ugo Trotta. They saw us off on A.F.1 for Rome. Arrived at Ciampino Air [Force Base] at 10:15 A.M.—a sizable greeting party included Papal rep's & Air Force Base Commander. Boarded Marine 1. Arrived at the Vatican at 10:40. Motored 5 minutes to Pontifical Palace. Much pomp & ceremony—with Swiss Guards and all. Nancy went on a separate tour while Bishop Monduzzi took me to the Papal library where I met the Pope. We talked for an hour—an interesting hour. I filled him in as best I could on Nicaragua & General Secretary Gorbachev. Then Nancy arrived, press photographers, our entire team & his. He recited my prepared remarks—he responded, gifts were exchanged. Then we left the Pope & went to the Sala Clementina room to meet & be greeted by the American Seminary students from No. Am. College. From there it was to the motorcade—back to the helicopter & off for Castel Porziano. There we lunched Pres. Francesco Cossiga, P.M. & Mrs. Fanfani, & 2 interpreters. It was very pleasant. Then back to the motorcade, Marine 1 to Marco Polo, A.F. & A.F.1. From there to Ciampino A.F. back into a helicopter & on to Villa Condulmer. We were greeted there by the 6th Grade students (American) from Aviano Elementary School (U.S. A.F. base) a nice bunch of kids.

Upstairs—tired & time to peel down for a quiet dinner & evening.

[Sunday, June 7: slept late (until 9:15 am); decision needed on Soviet point regarding scientists having access to nuclear testing in each other's countries; preparation for meetings.]

Monday, June 8

Get away day—breakfast at 7:45 then up & going. I left at 10 A.M. for the Cipriani Hotel on one of the small islands of Venice—our home for the next 5 days. Nancy left at 10:15 for Stockholm—another program in the anti-drug crusade. She'll rejoin me on Wed. This was a busy day. We helicoptered to an Italian Naval station then took a boat to the hotel—many official greeters along the way. Arriving at the hotel a little before 11 A.M. then to a room for briefing on bi-lateral with P.M. Fanfani & his team.

[Meeting on farming and subsidies; working lunch; briefing for meeting with Chancellor Helmut Kohl of West Germany.]

I had a phone call from P.M. Mulroney—Canada about taking some action on S. Africa. I urged him to hold off until after Margaret Thatcher's election Thursday.

Finally, my meeting with old friend Yasu Nakasone, P.M. of Japan. I was able to tell him of a partial lifting of the sanctions imposed because of the transistor dumping by some of their companies. They've made some improvement so we lifted $51 million of $300 million in tariffs. This was made public at 4 P.M.

[Dressed; participated in formal inception of summit.]

Then dinner lasted til midnight mainly because Margaret & Helmut did battle over whether to go zero on the very short range & tactical nuc. weapons. She says no & I had to differ with her although I explained it shouldn't happen until after we had negotiated on end to chemical weapons & reduced conventionals.

For a while it looked as if they were going to try to settle the whole summit in this one evening. Bed at last.

Tuesday, June 9

Up at 7 A.M.—a short night. A briefing then a meeting with Margaret Thatcher only half an hour—as usual we were on the same wave length. Then on to the morning. Most of the session taken up with farm situation. I think we've come out with a good position. A half hour break & then lunch with some of our Sherpas. Back to the meeting—a plenary this time—Sec's & Ministers of State & Sec's & Ministers of Finance. Much of the day devoted to economics. A news item came that a car bomb & several missiles had exploded outside Am. Embassy in Rome & British Embassy—no deaths or injuries. Later a report that a floating mine off Venice was detonated. That turned out to be an old water heater. Back to the hotel—homework waiting & then after shower back to a dinner meeting. Well it was really a social dinner at a large round table—30 ft. across, in a magnificent palace.

Wednesday, June 10

[Plenary meeting in morning.]

Ended meeting at 12:30 had half hour quiet time with Howard, Jim & Geo. S. I reported to them on talk I'd had with Helmut K. They intend to try Hamedi for murder of our young flyer (Navy) during hijacking. We thought they were only going to try him for a relative minor crime they'd caught him doing in Berlin. I see no point in extraditing him now. Then over to the Palazzo Ducala for our last lunch at 1:30 P.M. Then by boat to Giorgio Cini Foundation for bilateral with Mitterrand and Chirac. From there into Foresteria Ante Room where Fanfani was reading our statement to press & audience. That ended the summit & it was back to the Cipriani where Nancy was waiting—just in from Sweden. Now a quiet, private dinner in our suite.

[Thursday, June 11: *bilateral meeting with Prime Minister Brian Mulroney of Canada to make progress on free trade; staff meeting to prepare for press conference; held press conference; went to Agnelli home to meet Private Sector Initiatives Task Force; quiet evening with Mrs. Reagan.*]

Friday, June 12

[Prepared to leave; helicoptered to airport.]

At the airport we were met by Fanfani's, the Rabb's, and several others. There was the usual ceremony—Nat. Anthems, reviewing troops then goodbyes & aboard A.F.1. We had a brunch and arrived at Templehof Central Airport in West Berlin at 11:45 A.M. Met by Ambas. Rich Burt & Gahl as well as a considerable group of West German officials. A ceremony there—Nat. Anthem & review of troops—American. From there to the Berlin home of Pres. & Mrs. Von Weizsaecker accompanied by Mayor & Mrs. Diepgen. A lot of press photos then a pleasant half hour or so indoors. At 1 P.M. we drove to the Reichstag— the Kohls joined us. We viewed the "Wall" from balcony. We saw the exhibit of the Marshal Plan & Germany's rebuilding. Met several elderly ladies who had been part of the female force that cleaned bricks from the rubble & played a role in Berlin's rebuilding. Then it was on to the Brandenburg Gate where I addressed tens & tens of thousands of people—stretching as far as I could see. I got a tremendous reception—interrupted 28 times by cheers.

Then it was back to the airport and on our way to Bonn except at airport we viewed displays of Berlin Airlift. Met 3 pilots of that airlift, one the father of my Mil. Aid Steve Chealander.

Finally inside building several thousand people—mainly Am. mil. & family's etc. Again well received in a speech tailored for that audience. A birthday celebration with cake for Berlin's 750th anniversary. And then we flew to Bonn with Kohl's & Burt's on board. Left for Bonn about 4:45 P.M.—The Nat. Anthems—review of troops. There were meetings & photos with various people in Bonn Airport building. Then a bi-lateral with Kohl & finally back out for farewell statements to largely American crowd and back to A.F.1.

Departed 7:45 P.M. for Andrews A.F. Base 8 ½ hours away. I was surprised that we traveled in bright sunshine for about 8 of the 8 ½ hour flight. It didn't get dark until a little less than an hour out and yet it was after 3 A.M. back where we left. Well home at last.

[Saturday, June 13: *radiocast; massage.* Sunday, June 14: *noted that it was Flag Day; awoke early; telephoned Hugh Sidey to acknowledge supportive television appearance night before; reading.*]

Monday, June 15

1st meeting with Ken D. & the V.P.—Howard late getting back from Tennessee. Some talk about Sen. Biden—now cand. for Pres. I saw him on CNN last night speaking to the John F. Kennedy School at Harvard U. He's smooth but pure demogog—out to save Am. from the Reagan Doctrine.

[NSC meeting included ambassador to South Korea with report on student riots there.]

A meeting with bipartisan Cong. leaders—Byrd, Jim Wright, Bob Michel, Bill Armstrong etc. We talked mainly about the budget & the Dems of course took no blame for the deficit. I sort of corrected them on that with the 50 years history of deficit spending under Dem. control.

[Staff meetings; haircut; rested in afternoon; telephone Jeane Kirkpatrick in France to thank her for a supportive column; delivered nationally televised speech on economic summit; telephoned Jack Kilpatrick to thank him for positive comments on PBS show Agronsky & Co.*]*

Tuesday, June 16

[Staff meeting; approved David Ruder of Northwestern University as chairman of the SEC.]

NSC—a little difference between Cap & George about an upcoming nuc. missile test. Geo. wants to announce it as in keeping with SALT II Treaty which we've said is no longer binding. Cap wants to just go ahead & forget SALT II. On this one I lean toward Geo.—not because of the treaty but because we're on a kind of friendly basis with Soviets leading toward the arms talks. Just learned we have a new Ambas. from Israel. This has been hanging fire for 6 months—an argument between Shamir & Perez.

[Photo session with departing W.H. reporter for AP; appeared before regional journalists; had lunch at Capitol with senators, noted, "Weicker had the gall to question me—I think I did him in"; National Security Planning Group (NSPG) meeting regarding the Persian Gulf, commented, "We have some Congressmen stirring up popular sentiment against our presence there"; addressed reception for Republican Congressional Leadership Council.]

Upstairs—Nancy was entertaining Steve Lawrence and Eydie Gorme. I stopped in for a few minutes & then headed for the gym. Early to bed.

Wednesday, June 17

Staff meeting & NSC kind of ran together—much discussion of the Persian Gulf. Charges are being made that we are being frivolous in basing some naval vessels there & running unnecessary risks of casualties. I brought up the fact that our naval strategy since W.W. I has been to base elements of the fleet all over the world where our national interests are involved. We've based ships &

planes there in the Gulf & the Arabian Sea for 38 years. I've proposed explaining this to the people on Saturday's radio cast.

Some briefing for meeting with Pres. Arias of Costa Rica. He has a peace proposal he wants to make on behalf of his Central Am. neighbors & to us to the Sandinistas. We're in favor of the idea but believe his plan has some loop-holes the Sandinistas could take advantage of to get rid of the Contra threat & hold on to their power. Meeting with him was to take 15 min's. It took 45 & we went round & round. I don't know whether I sold him but our group thought I did well.

Lunch on the patio then photos with 5 of our outbound Ambassadors & their families. A meeting with Cap W.—I persuaded him to notify the Soviets of our missile test but not to mention we were doing it as a requirement of SALT II Treaty. For us SALT II does not exist.

[Greeted presidential scholars.]

Then a taping session for a half dozen events—a Happy Birthday call to Holmes Tuttle & a call to Ed Meese in Vienna. I'm leaning toward appointing Senator DeConcini (Dem) to Dir. of FBI. A lot of our Senators want this very much & truth is he'd be replaced by Repub. Sen.—I've just made the call to Ed & he really cooled me down on the Sen.—maybe that is not the way to go. Ed will be home next week & we'll talk about it.

Well a quick jump in the gym, the shower, early dinner, & to the Sheraton Hotel to do a fundraiser for Orrin Hatch. He raised $530,000 & I was home at 8:20 P.M.

Thursday, June 18

The V.P. is back in off the road and says the people out there are fed up with the Iran-Contra hearings. Howard & Ken are opposed to my talking about the Persian Gulf on Saturday. I'm still not convinced. They thought it would tempt Iran into a hostile or terrorist act. My belief is the Ayatollah already hates us with an insane passion. Word has now been received that an American press man has been kidnapped off the street in Beirut by the Hezbollah. We also have evidence [. . .] that Iran ordered the crime—our man was accompanied by the son of the Lebanon Defense Minister—he was also taken.

NSC—More of the same debate plus more info on Iran's acquiring Silk Worm Missiles—even a photo of me.

A brief session with Culvahouse checking my memory against some of the "hearing" testimony.

Dick Wirthlin came in—not with a poll but with his "pulse" taken on my speech—a couple of low points but mostly good.

[Presented Medal of Art awards.]

Then a meeting of Domestic Council. I'm faced with making a decision on instructions to our team on what we seek world wide in reduction of Fluorocarbons that are destroying the Ozone layer. Right now I don't have the answers.

[Signed agreement with Israel for Radio Free Europe and Radio Liberty installations; session to review appointees; photo sessions.]

Friday, June 19

Short meeting & a crowded morning. I gave in on the radio script & said "no" to doing it on Persian Gulf. It will be on the Budget. I also started something on Presbyterians 189th meeting in Biloxi, Miss. They've passed a resolution against us helping the Contras in any way. They sent a group of laymen & clergy to Nicaragua. They were given the Sandinista tour & came back anti-freedom-fighters. I called Rev. Donn Moomaw last nite & gave him some true pictures. He volunteered to help so I started our gang on getting together with the Presbyterians top people.

NSC was very brief—just long enough to say that if CIA was sending their Persian Gulf paper to Congress we should send a notice under the War Powers Act.

[Reviewed upcoming bills with GOP Senate leaders; briefing and then meeting with President Hissène Habré of Chad, commented, "He's quite a guy & I think we made him happy—he deserves it. He did a great job on Quadaffi"; vetoed bill to perpetuate fairness doctrines in broadcasting; went to Camp David; had swim. **Saturday, June 20:** *radiocast; had ride, just barely outran thunderstorm; telephoned president of Argentina, on vacation in San Diego.* **Sunday, June 21:** *returned to W.H.; attended gala evening at Ford's Theatre, noted, "Nancy was a hit when they took her up on stage & she danced with Baryshnikov."]*

Monday, June 22

Off on Marine 1 for Andrews & A.F.1. There was a little boy on the Lawn jumping up & down waving a paper pad & a pen. I couldn't resist. I ran over to him & signed the autograph. Found out later he had only recently lost his father. His mother they tell me was shedding tears of happiness.

[Flew to Melbourne, Florida; met with Representative Bill Nelson (D-FL) en route; toured Dictaphone; addressed Chamber of Commerce.]

Only dark spot of day was a real one—a notice that Fred Astaire had died. He was a truly wonderful man.

[Returned to W.H.]

Tuesday, June 23

More talk about our budget campaign. It was also decided I wouldn't call Bill Nelson on his interview after I left Melbourne yesterday in which he took a crack at me after being like an ally all day.

At 9:30 a meeting with Repub. Leadership of House & Sen. This too was mostly budget talk. I campaigned for our Repub. to get together & unite to make this issue a contest between the 2 parties & the people would be on our side. We finished with some talk about the Persian Gulf. We had some disagreement.

[NSC meeting and briefing; Frank Carlucci to summarize agreement denying sale of certain technology to USSR; luncheon for Medal of Freedom recipients, including the late Justin Dart and Danny Kaye; addressed convention of National Federation of Independent Businessmen, got some standing ovations; visit by major party donors; photo sessions with people brought by various congressmen, including Susan Butler, winner of Iditarod; noted Parade *magazine featuring exchange of articles with Soviet magazine.]*

Wednesday, June 24

A busy staff time with discussion about putting together a citizens committee to rouse up a public cry about budget reform.

Then discussion of a commission, 11 members, on advising us about AIDS. We're going ahead with this. Then it was re the extradition of Hamadie from Germany. They are going to try him for the murder of our sailor.

NSC—Began with Hamadie. His mother was on TV very bitter that he wouldn't be tried here & saying we hadn't done enough. (Let me correct this—it was Stethem's mother—not Hamadie's) Anyway I phoned her and I think I straightened things out. She was quite emotional & crying bitterly throughout our talk. It seems some of our lower level bureaucrats in the Justice Dept. had been sounding off about us. When the cat's away the ~~mice~~ rats will play. Ed M. is in Europe. Then we had a report on S. Korea. Pres. Chun has met with leaders of the opposition but no real settlement or agreement reached.

Then Culvahouse came in—another report on the hearings. I'm embarrassed by how many times I have to say "I don't remember." Then a NSC meeting in Cab. Room—it was report time on our security measures at our nuclear installations. I must say they are so thorough & well planned you can't see how any threat of theft or sabotage could be successful.

[Photo session with Chris Wallace; addressed exchange students; visited mail room volunteers to express gratitude, noted, "I greeted one in particular—John Martin—77 years of age who has signed 252,000 W.H. birthday cards"; waited to hear word on Lew Wasserman, who had colon operation during the day.]

Thursday, June 25

[Noted jeep accident involving Deputy National Security Advisor Colin Powell's son; discussions on upcoming campaign on budget; considered possibility of veto should current energy bill pass Senate.]

NSC—Marcos is back on agenda. We have information that he's plotting a return to the Philippines and a coup to recover control. We can't permit that. Persian Gulf—Saudi Arabia has agreed to make AWACS, coverage to us, & tankers to re-fuel our planes. S.A. also said would make their mine sweepers available.

We have learned Cubans are flying the gunship helicopters in Nicaragua.

NSC was followed by a meeting again with Culvahouse. He says North will be publicly testifying July 7.

[Presented National Medals of Technology and Science; lunch with Vice President Bush; National Security Planning group (NSPG) meeting, as noted, "to explain to all the changes we'd made in NSC and covert operations"; personnel meeting and decision to appoint Dr. Eugene Mayberry to chair commission on AIDS; accepted design for medal honoring bicentennial of Constitution; photo session with Reserve Officers Association; received copy of George C. Marshall biography from author, Forrest Pogue; met Lord McGregor, new head of Reuters; attended reception for Citizens for the Republic; noted that Mrs. Reagan was in New Orleans.]

Friday, June 26

A short but busy day, I started taking the "Go Lightly" at 7:30 A.M.—8 oz. glass & every 15 minutes thereafter for 13 glasses. All preparation for my checkup. At staff time learned Justice Powell was resigning from Supreme Court. I'll have an appointment to make. We had Frank in early for NSC. Colin has gone to Germany to see his son. Had a report on his injuries— they sound horrible. Our main discussion was over a program to head off Cong. from reversing us in the Persian Gulf. In their efforts to "ding" me they are destroying our foreign policy in the entire Middle East. They have to be stopped.

[Greeted volunteers from boyhood home in Dixon, Illinois; meeting with Shultz and Sigur, just back from Korean trip; had checkup, went to Camp David. **Saturday, June 27:** *relaxed by pool; received word that two small polyps found in exam were benign.* **Sunday, June 28:** *returned to W.H.; attended televised performance honoring Cole Porter and starring June Allyson, Patti Austin, Kaye Ballard, and Mel Tormé; telephoned Wasserman, surgery went well; received word that Paul Nitze's wife died.]*

Monday, June 29

[NSC meeting on situation with demonstrators in South Korea; spoke to political activists, loyal party supporters; met with Governors John Sununu (R-NH) and George Sinner (D-ND) to discuss progress on Federalism.]

Then meeting with our lawyers, Ed Meese & Howard on Supreme Ct. vacancy. We are together on thinking it should be Robert Bork. Another meeting with Bill Webster—He's back from Europe & reported on intelligence matters involving Allies.—Haircut & end of day.

Tuesday, June 30

Staff time (short) had to do with plans for checking some Sens about our choice for Supreme Court—Robert Bork. V.P. then told of our Cuban defector General del Pino's report to him on the Cuba situation. He says 56,000 mil. people have deserted in the last 3 yrs. A brief time with NSC prior to meeting with Cong. Leadership—both houses & bi-partisan. This meeting had to do with Persian Gulf. Sam Nunn has made this his cause & is fighting us to make us delay & especially not go through with protecting Kuwait tankers. He's as wrong as he can be. We'd lose any standing we have & appear spineless if we did as he proposes. I don't know whether we blunted the opposition or not. After meeting Culvahouse & Howard came in—there are rumors that Ollie North—tomorrow may take cracks at me in his testimony. If he does he'll be lying. Later in day Howard got a call that suggested those rumors weren't true.

[Awarded medals for volunteer work; interviewed by biographer Edmund Morris.]

Then upstairs & tonight the Kay Graham dinner (9:30 P.M.). Well it started about 9:45 & instead of getting home at 11:30 it was 12:20. But it was a spectacular party enjoyed by all. I proposed a toast to Kay after several speeches had been made.

Wednesday, July 1

Started day with Howard suggesting that on back to Wash. from ranch in Sept. we stop in Kansas & see Alf Landon—100 years old. I had proposed we send A.F.1 for him & bring him to Wash but his health wouldn't permit. We have a problem with George Alan—Chief of our Physical Fitness Commission. He is also Chairman of a Foundation & there is a conflict. Howard is going to tell him to choose 1 or the other. Then we settled on fact that I would name Bob Bork to Supreme Ct. today.

NSC—I've signed a directive on recommendations of Tower Committee. Geo. S. has a problem with it but I think it can be straightened out. Les Aspin is taking some Congressmen on a junket to the Persian Gulf. Everyone's a diplomat.

[Met with Senator Strom Thurmond (R-SC), who made recommendation of a person for Supreme Court vacancy; received report of U.N. International Conference on Drug Control; lunch with Shultz, on his way to meetings in Canada; photo session with members of cabinet; announced selection of Robert Bork for Supreme Court.]

Thursday, July 2

Staff time without the V.P. He's in Iowa. Some talk about Bork & last night's broadcast featuring all the Dem. Cand's for Pres. There was a lot of demagoguery.

NSC—Some talk about continued Iran hearings. That continued in following meeting with Culvahouse. Top Secret is call I got from Howard last night. He has learned North in his private testimony said he had kept all news of diversion of Iran funds to Contras from me & no one else had told me. We have to hold that secret til hearings resume in week or so.

[Cabinet meeting to plan launch of Economic Bill of Rights; desk work; positive economic report on employment; met eleven-year-old transcontinental pilot; reception for Human Events *radio program and Radio America.]*

Friday, July 3

At 9:45 A.M. over to Jefferson Memorial for speech to several thousand people. Main purpose of speech is to launch our "Ec. Bill of Rights" program which is going to be our main theme for rest of my term. It includes Const. Amendment to wipe out the deficit spending, right for the Pres. to line item veto, Cong. required to reveal cost & ec. effect of any new programs proposed, etc.

Well speech made, we're back at the W.H. & changed into clothes for Camp D. I don't ever remember being hotter than I was on that platform in the sun. I don't know how those people (5,000 of them) standing there could take it.

[Went to Camp David; rode and swam.]

Saturday, July 4–Sunday, July 5

Did my radio show & in the afternoon we both rode. A beautiful day. Back to the pool for a swim. Our movies Saturday & Friday were Fred Astaire pictures. Sunday another beautiful day spoiled some what by crud I heard on Brinkley's "This Week" program. Back to W.H. in afternoon.

Monday, July 6

Happy Birthday Nancy! Staff time a discussion of Judge Bork and plans of those who are creating trouble for him—Ben Hooks of NAACP for example. We'll get Bork confirmed to Supreme Ct. but it will be a battle with left wing ideologs.

NSC—Update on Persian Gulf, Korea, & Gorbachev's attempt to make changes in Soviet govt.

[Addressed Kiwanis convention; issues lunch, remarked "always a pleasant chore"; quit work early; read diary for early 1986.]

Tuesday, July 7

Staff meeting was mostly discussion of obstacles in Cong. to getting a decent budget. They are determined to cut defense back to where we were in the Carter years. They would dump a quarter of a million young men out of uniform & on the job market.

NSC was more of the same. We have to find a way to beat them.

Mel Laird & his commission came by with their report on the Embassy situation in Moscow. They say State Dept. bureaucrats have no business trying to run the construction of an Embassy. They suggest it would more properly belong in the Army Corps of Engineers.

Then word came that at last Ollie North had declared he did <u>not</u> tell me about the extra money in the Iran arms deal. Found out later that when that was announced the market went up 20 points.

[Photo session with head of Home Loan Bank; meeting with Peter Grace and Jack Anderson to discuss private group to campaign for budget policy; National Security Planning Group (NSPG) meeting about Korea.]

Ed M., Howard, & Ken D. came by re the FBI Dir. It may turn out to be Strom Thurmond's choice Judge Wilkins—maybe.

[Met with Weinberger to discuss defense budget. **Wednesday, July 8:** *flew to New Britain, Connecticut, with Secretary of Commerce Malcolm Baldrige; met briefly with widow of Stark victim; appeared at Elks Lodge; made speech at city hall; returned to W.H.]*

Thursday, July 9

Howard was in N.Y. addressing the NAACP. We spent a half hour on our situation with regard to some of the shenanigans going on up on the Hill. Every pen I look at is a veto pen to me.

[NSC meeting about effort to fly U.S. flag on Kuwaiti tankers; met with business leaders supporting Vote America effort; met with Secretary of Labor William Brock about labor legislation in Congress; report from Secretary Lyng that farm situation improving; watched some of Colonel Oliver North's hearing, noted that W.H. received overwhelmingly positive response to North; addressed reception for party contributors; went to dinner party honoring Bill Webster.]

Friday, July 10

Howard is back—so our usual staff meeting only V.P. absent. Discussed how to deal with Congress efforts to cut defense budget to ribbons. They're planning to add things as amendments to Debt limit bill which I <u>must</u> sign or create a world wide financial panic.

NSC—In arms negotiations, Soviets want us to push Kohl on eliminating 72 Pershing missile launchers. We claim we can't negotiate for a 3rd country. In other parts of the arms discussion, Soviets are trying to link them to SDI. We can never give on that.

Culvahouse came in after this meeting. Again it was to check some points in Ollie North's testimony. Incidentally the witch hunt against him has made him a national hero. Thousands of phone calls to the W.H. etc.

[Addressed Deficit Reduction Coalition gathering; National Security Planning Group (NSPG) meeting on security for nuclear installations; had regular meeting with Shultz regarding use of military craft by administration officials.]

Another matter—a signing I may do on an N.S.P. which Geo. questions. I heard him out but this I must sign.

[Received update from Packard Commission; went to eye doctor for exam. **Saturday, July 11:** *Mrs. Reagan in Maine for the day; radiocast; reading and massage; phoned Prime Minister Robert Hawke on reelection in Australia and Representative Mickey Edwards (R-OK) on birthday.]*

Sunday, July 12

A quiet day ahead—spent most of it reading. First—homework then a book about the *N.Y. Times* which explains a lot of things about the "Times" that annoy me. A Raymond Price (on the other hand) in the *Wash. Post* has a column pointing out that the only time Cong. engages in carnivals like the Iran-Contra Affair is when the Cong. is Dem. & the Pres. Republican. Cong. has been Dem. for 46 out of 50 years. Mermie arrived too late for dinner because of big storm, streets blocked by fallen trees, etc.

Monday, July 13

Met by morning papers headlining that a memo initialed by John P. proves I did know of the Iran money dispersal. It's phony as a $3 bill, but the media lynch mob is grabbing at anything. I'll control myself & not answer their shouted questions as I go for the helicopter. I'm off to Indianapolis—selling the Ec. Bill of Rights again. In the "Dip." room on the way I met a young father & mother & their little boy. He is a terminal Leukemia case but looks quite healthy. His wish was to meet me so we did. Then I asked Dottie to come over for some papers. It's her birthday and when she got here I surprised her with her 3 grown sons & the rest of the day off.

[Flew to Indiana; met with community and county officers; addressed National Association of County Officials; commented, "Then back to A.F.1 where I learned that Dick Cheney in this mornings Ollie North session blasted the press & established that there was no truth at all in those stories"; returned to Washington and quit work early. **Tuesday, July 14:** *talked with Vice President Bush about hearings; short meeting with GOP congressional leadership, discussed ways to fight Democratic initiatives; NSC briefing from Ambassador Jack Matlock on return from Moscow; desk work; meeting with President's Foreign Intelligence Advisory Board (PFIAB) on security problems at embassies; met with colonel new to W.H. military office; haircut; teeth cleaning; watched North and McFarlane at hearings.]*

Wednesday, July 15

Opened day talking about Contra Aid, the Graham, Rudman, Hollings

Bill & Cong's attempts to amend it to either force a tax increase or an across the board spending cuts—that would fall most damaging on Defense.

Then Ed Meese came in asking that I appoint a Fed. Judge in St. of Wash. His name is Dwyer & Sen. Dan Evans wants him. I had wanted him named back in 85 or so but the Sen. adjourned without giving approval. He's a liberal. Ed is coming back tomorrow with recommendations for Dir. of FBI.

NSC—Heard about a meeting on Hill with Geo. S., Cap W., Sen. or Byrd & others on flagging Kuwaiti tankers. Sen. & others of course are opposed & want a delay. I'm unwilling. Then arms control & Soviets evidently wanting to re-introduce SDI as a chip. [. . .]

Doing homework was interrupted by news that John P. in hearings on the Hill had flatly denied ever telling me about the extra Iran funds & their diversion to Contras. This was the bombshell I've been waiting for 7 months. The day is brighter.

[Met with congressmen engaging in informing public on Nicaraguan situation; addressed Minority Groups Business Briefing; had lunch with Vice President Bush; met with Shultz and discussed ambassadorial appointment to Austria and strategy on negotiating nuclear arms agreement with Soviets; visit from former Cuban political prisoner, asked him to head U.N. Human Rights Team at upcoming meeting, commented, "We've never been able to get the Latin Am. nation to vote with us against Castro. We think Armando might make the difference."]

Thursday, July 16

Usual start—Staff then NSC—discussion of John P.'s testimony. My problem now is that both John & Ollie were dropped by me from NSC because they had not told me of the extra money or the diversion to the Contras. Now it seems that Ollie did tell John with the expectation John would tell me. I don't see though how I can do anything until they close down this investigation.

[New poll shows mixed results; presented civic awards; cabinet meeting on budget cuts; answered mail; photo session with Mrs. Reagan's staff; heard news that Armando Valladares agreed to head U.N. Human Rights Commission; photo session with commander of Jewish War Veterans; videotapings.]

Friday, July 17

[Staff meeting on impasse with Congress over debt limit.]

NSC—Discussion of possible actions we take if Iran does something drastic when we escort Kuwaiti oil tankers through Persian Gulf. Again Cong. has worked our people up so that any casualties on our side would raise a public storm.

[Briefing and meeting with British prime minister Margaret Thatcher, discussed Middle East, arms control, noted, "And were in agreement on everything"; went to Camp David; had swim.]

Saturday, July 18

Another beautiful day. Called Bill Bennett—just back from Europe. Told him I was sure some one had apprised him of our son Ron's article on AIDS in *People* mag. Ron gave both of us h--l. He can be stubborn on a couple of issues & won't listen to anyone's argument. Bill volunteered to have a talk with him. I hope it can be worked out.

[Radiocast; ride and swim.]

Sunday, July 19

Watched Sun. talk show—Margaret Thatcher on Face the Nation was absolutely magnificent & left Lady Stahl a little limp.

[Returned to W.H.]

Monday, July 20

[Staff meeting to discuss ways to block Democratic legislative initiatives; NSC meeting on Persian Gulf and starting date for escorts for Kuwaiti ships.]

A. B. Culvahouse came in—just checking some more on what testimony of Geo. S. & Cap will be to Iran-Contra hearing. Geo. had debriefed his people on our meetings & committee has his notes taken on this.

[An issues-briefing lunch; Economic Policy Council meeting to discuss a high-tech program and possibility of rewards for breakthroughs; ambassadorial formalities; meeting with Webster to be weekly; he reported on a defector who identified a circle of double-agents; telephoned Prime Minister Margaret Thatcher regarding a recent television appearance.]

Tuesday, July 21

1st meeting on a hot morning—a little preparation for a 9:30 meeting with GOP leadership. Debt Ceiling will be an important item with what the Dems are attempting to do to it. Then the same individuals are trying to make our catastrophic illness proposal into a budget busting expanded program which will actually impose costs on senior citizens. Two trade bills House & Senate—I'd have to veto either one. Now 9:30 & the meeting with Jim Baker, Clay Yeuter, Sec. Bowen all pitching in we had a good presentation as to why we had to stick together & get concessions from the Dem's or I'd veto.

Desk time—well after a brief NSC session—mainly on Kuwait. Our 1st flagged tankers with Navy support will lift anchor at 2 A.M. our time. I hope nothing happens to make the phone ring.

[Presented awards in Take Pride In America ceremony, Clint Eastwood and Louis Gossett Jr. also attended; appointment with eye doctor.]

Forgot—this morning had a meeting with Col. Oleg Antonvich Gordiyevsky—the Soviet KGB officer who defected to England. His wife & 2 little girls were left behind. We've been trying to get them out to join him.

Wednesday, July 22

Debt limit & Dem. games main subject of 1st meeting. Then Catastrophic Health program & here too they are distorting it to be a costly budget buster. A little discussion of "hearings." Tomorrow it's Geo. S. and the committee has some briefing notes of his that might be embarrassing.

[. . .] Our first 2 flagged Kuwait tankers are on the way through the Straits of Hormuz into the Persian Gulf.

A discussion of covert actions & the demands of the Cong. they be notified. How do we do it without risking exposure?

[Heard description of superconductors at meeting of W.H. Science Council; addressed state and local officials; meeting with Shultz, commented, "He was great in N.Y. at getting U.N. Security Council to call on Iran & Iraq to halt the war."]

Closed out the day with another meeting with Charles Wick. He has a great outline of a plan to get me on WorldNet among other things to reach world audiences & undo the damage done by our press with regard to the Iran-Contra Affair.

Thursday, July 23

Staff Time—Again a sum up of our legislative battles on the Hill. Partisanship is the order of the day.

NSC—A review of choices—Chf's of Staff are weighing with regard to action we may have to take in Persian Gulf.

It appears that Gorbachev's show off on zero—zero INF may be held up by their intransigence about W. Germany's Pershing I A's.

[Presented awards to businesses and organizations; had lunch with Vice President Bush; delivered speech at National Institutes of Health, visited children with AIDS and cancer there and met with AIDS commission; received report of Drug Policy Board, commented, "We've made great progress"; photo session with Governmental Affairs employees, and with departing military nurse; noted that Mrs. Reagan's brother, Dr. Richard Davis, was helpful in finding candidates for AIDS commission.]

Friday, July 24

A kind of hurried up & mixed up short day.

Awakened by phone at 2 A.M. Frank C. calling to tell me one of Kuwait tankers (Bridgeton) hit a mine in the Gulf. No casualties & ship able to make port.

Staff meeting 9 A.M.—discussed Judge Sessions who is our choice for Dir. of FBI. Then on to Debt Ceiling Bill—good news—the Chiles Amendment was defeated. Just a chance we can get a clean bill. A. B. Culvahouse called Shultz to tell him "well done" on testimony.

NSC—Bridgeton—details. It happened near Farsi Island where an Iranian naval vessel is berthed. Some Q. to be resolved after we can get into the ship—was it a mine or an inside explosion.

Had a cable from Bonn with latest on Pershing I A's. We don't want to pressure Kohl but he faces an iffy pol. decision. Factions in W.G. will have at him either way he goes. I suggested maybe he should volunteer to eliminate them in interest of disarmament.

New Subject—In Oct. another group of Soviet agents are going to be sent home by us from U.N.

Met briefly with Judge Sessions & asked him formally to be Dir. of FBI—He accepted. Then an NSPG meeting for report on several Covert Op's I ruled we continue them.

Then off to Ukrainian Church for lunch & ceremony recognizing Captive Nations Week. Extremely well received. The Soviets will be unhappy. Back to W.H. changed clothes & off to Camp D. greeting "Boys Nation" on way out of W.H.

Camp D. a few degrees cooler than Wash. but not much. A swim & quiet evening.

Saturday, July 25

Radio show, then over to medical center to receive my break proof safety glasses—new military issue. Mine are for using the chain saw at the ranch. Left one hearing device [. . .]. It seems that some hearing devices can become source of bugging. A wave can be directed at my hearing aid (if it's one of these) while I'm wearing it & my conversations would be heard by the eaves dropper.

Rcvd call Mac Baldridge had been seriously injured in a horse back fall—horse came down on him crushing him. I called Midge—her word was that he was going to be alright. A short time later we received word that he had died.

Sunday, July 26

Out of Camp D. early—thunderstorms expected. Back to W.H. where flags are flying at half mast. Called George S. in Mass. to tell him how great he was at the hearings

[**Monday, July 27**: *flew to Wisconsin; visited factory; addressed civic leaders in West Bend; returned to W.H.; Maureen visiting.*]

Tuesday, July 28

A little talk about a successor to Mac Baldrige. We'll get down to it after the memorial service & funeral. Some discussion about Sen. Dole's statement that I could appoint Bork to Sup. Ct. during Aug. recess. Trouble is the new Cong. in '89 could cancel that & there might just be enough animus to bring that about. Then it was Debt limit time—Friday is the last day we have before declaring default.

NSC—Discussion of Kohl possibly volunteering to give up Pershing missiles. We can't ask him to but Frank C. is going to see him & feel him out on the effect his doing this might have on his own standing. It looks like at last Shultz/Shevardnadze meeting will take place in September—this could mean a summit in November. The Bridgeton starts loading 260,000 tons of oil tomorrow.

Panama still in civil strife. Noriega arrested former Defense Minister. Military my be ready to turn on Noriega.

[Reviewed future budget targets; spoke to Federal Conference on Commercial Applications of Superconductivity; National Security Planning Group (NSPG) meeting on problems with embassies in Moscow; haircut; went to doctor for "sneeze shot" (allergy); videotapings; attended farewell dinner for Paul Volcker.]

Wednesday, July 29

Took up matter of George Allen. He's heading up our physical fitness program & the Foundation for the same. We have to ask him to resign from Foundation—he's not allowed to do both in addition to maintaining his interests in private sector.

It looks favorable for us to get a temporary debt ceiling right away. Sen. Byrd I'm happy to say has volunteered that he'll get Judge Bork's nomination out of the committee to the floor of Sen. We provided the investigating committee (Iran-Contra) with 200,000 pages of documents & did not invoke Exec. privilege.

NSC—While things are looking pretty good for an agreement with Soviets on INF, it seems likely they will try to link SDI to the matter of ICBMs—our "START" proposal.

Frank C. & Howard B. went up on Hill yesterday to meet with Dem. leaders. They support our Nicaragua peace proposals but won't go for Contra aid which is only way to get Sandinistas to talk peace.

At 9:45 A.M. picked up Nancy and we went to the Cathedral which was packed for memorial service to Mac Baldridge. I did the Eulogy. Some of our intelligence warned that such a gathering could be a prime terrorist target. The entire government chain of command was present. Except for one. I had to ask Sec. Bennet not to come so at least one person in the chain of succession if President is eliminated would be available to assume the office. But everything went well—security was overwhelming.

[Addressed National Law Enforcement Council; had lunch with Vice President Bush; spoke to leaders of Future Farmers of America; met with Shultz on issue of West German chancellor Helmut Kohl and the Pershing 1As, but mainly on his concern that NSC was infringing on State Department functions and responsibilities; greeted seven Special Olympics athletes escorted by Eunice Shriver.]

Then over to Dr's office where they took a shaving off my nose to see if

I'm having another Basal Cell Carcinoma. I'll know tomorrow if I have & then they'll take it off.

Thursday, July 30

Well its tomorrow & yes I have to have another carcinoma removed but now it will be tomorrow—Friday afternoon at Bethesda—late afternoon so they'll be no Camp D. this week.

Now to start of the day—good news—Cong. sent me a temporary 8 day extension of the Debt Ceiling so we wont be in default tomorrow. It's confirmed Shultz & Shevardnadze will meet here on September 15 through 17. New subject, Jo Ann Gasper fired by Bowen (he really had to) has been hired by Bennet for Department of Education. I'm glad—Bowen had no choice but she is a good person. I'm glad Bill Bennet had a place for her.*

A glimmer of hope on Nicaragua—it's possible the Dem. Leadership on the Hill will push as their own, legislation about identical to our proposal. Let em take the bow—we'll get what we want.

NSC—We're talking about funding for Contras on a 2 year basis at $180 mil. a year. Also touched on Persian Gulf—there is a plan to use Kuwaiti tugs towing a cable between the bring mines to the surface.

[Met with scientist on relative strength of U.S. and Soviet space programs, commented, "they have nothing yet to equal our Saturn rocket for ability to put heavy loads into space"; address to gathering of Right to Life group; visit by media advisor Gary Schuster; meeting with Latin American students; greeted Multiple Sclerosis Mother and Father of the Year; Mrs. Reagan returned from trip to Maine; Maureen visiting.]

Friday, July 31

Opened day talking about an Exec. Order I'm going to do on covert operations. We believe it might soothe some of the wild ones on the Hill. A little talk about my meeting with Presbyterian group on Nicaragua. I'm the only one who thinks I should do this.

[Noted progress of trade bill and debt-limit extension in Congress.]

NSC—Geo. S. attended. He discussed our Embassy bldg, or I should say rebuilding problems in Moscow—a memo of recommendations is on it's way. Then a few words on Mid-East peace process. He had word that Hussein & Shamir had a secret meeting at the King's (Hussein) house in Eng. King Hussein was most considerate, he asked S. to come stay the night on Fri. so he wouldn't have to break the Sabbath by traveling on Saturday. He also had a complete menu of Kosher food.

---

* *Mrs. Gasper, deputy assistant secretary of the Department of Health and Human Services, was fired for refusing orders to restore funding for Planned Parenthood.*

[Met Christopher Shays, Republican candidate for Congress in Connecticut; meeting with President Omar Bongo of Gabon; visited by American Legionnaires who had toured Central America; taped radiocast.]

At hospital went into surgery almost immediately. A little divot was removed from the tip of my nose. Then we waited for more than a half hour while they froze the divot & checked it out in the lab. The verdict was I had to have another slice removed. Again a wait & this time the verdict was, they had gotten all of the carcinoma. Then Nancy & I waited while the Dr's huddled on what procedure should be followed in my healing. The verdict was for cosmetic surgery, skin graft etc. for fastest healing. So onto the table again for the plastic surgeons. Back to W.H. 4 hrs. later for an 8 P.M. dinner. Once in bed I did the stupidest thing. The Dr. has me putting oil in my ear twice a day for a minor infection from swimming. This time I put the oil in my left nostril. I had a h--l of a time trying to get rid of it and rid my throat of the awful taste.

Saturday, August 1

Slept well, late breakfast—then something I haven't done since I was a kid—I upchucked. It must have been a follow on to the oil. A quiet day at home. Some homework & a lot of reading. After dinner ran a Fred Astair & Ginger Rogers movie.

[Sunday, August 2: reading, including as noted, "a book on the N.Y. Times & its bias & distortion of news to benefit Dem. regimes & denigrate Repub's"; dressing changed on nose.]

Monday, August 3

[Had discussion of debt ceiling–extension bill in Congress; NSC meeting pertained to U.S. company possibly providing technology to USSR and news of ship movement in Persian Gulf; received new dressing for nose; addressed Project Economic Justice group; issues-briefing lunch; meeting to discuss Shultz idea for peace plan in Nicaragua, commented, "I approve. We think maybe Cong. would too so we're going to see if Jim Wright will claim it as theirs & I'll approve it."]

A lot of desk time. Then Ed Meese came by regarding my answer the other day on no evidence of law breaking in Iran-Contra Affair. He had guessed right—I was thinking only of us—the Admin. He says there is evidence of possible legal problems with some individuals who did things I wasn't aware of.

[Meeting with Weinberger on Defense Department budget; received new dressing on nose.]

Tuesday, August 4

Spent our time in talking about who will replace Mac B. Then a Bipartisan approach to Central Am.—a diplomatic solution while continuing to fund the

Contras. A false story in Wash. Times about our policy on covert programs. I asked for someone to bring this to Arnold De Borchgrave's attention. Then it was our problem with Dem. Leadership & the debt limit also catastrophic health insurance. Our Repubs got 190 votes on a substitute which means we could count on a veto standing up if Dem. version is passed. Then it was Welfare Reform—again there are 2 bills—one bad & one possible and a costly Housing bill.

[Discussion of pending bills with Republican congressional leadership.]

Then desk time & lunch. An NSPG meeting regarding the Panama situation. Noriega head of Nat. Guard is a bad egg but right now has support of the people. At 3 P.M. a meeting with Cap W., George S., Colin & Howard. Cap is upset about Nicaragua peace plan we want with Dem. help. There was some merit in several of Cap's criticisms. I proposed we do an annex explaining my interpretation of the questionable terms in the compromise. At 4 P.M. Bill Webster briefed me on Contras progress inside Nicaragua—14,000 of them. Then my session with biographer Edmund Morris.

[Dressing and some stitches removed; noted report in earlier congressional leadership meeting that Robert Bork was "scoring very high with personal visits to individual Sen's." **Wednesday, August 5:** *staff meeting to discuss new secretary of Commerce; met with Soviet pilot who defected in 1976; met with congressional leaders on Nicaragua peace plan; spoke to press about same; met with six Nicaraguan resistance leaders; new polls indicate slowly rising approval rating; regular meeting with Shultz, who requested military transportation for Howard Baker and Frank Carlucci, which the president approved; received report on Soviet defector to Britain, discussed arms reduction, and commented, "It looks like a complete treaty is on the table in Geneva for the INF's. George thinks we're getting close on Nuc. Test Treaties & of course we must follow through on the ICBMs—a 50% reduction. Shevardnadze is due in mid-Sept. We hope to have some issues on reductions ready by then"; meeting with conservative senators to assure them that Nicaraguan peace plan is not a sell-out; meeting on secretary of Commerce vacancy, decided to look closely at Bill Verity for the job; visited doctor and had nose plaster removed.]*

Thursday, August 6

Before we were out of bed the phone rang. It was Geo. S. to tell us the Russian Pianist Feltsman was out of Moscow & in Vienna with his wife & child. We've been working on that for about 2 years. Into the office for a quick meeting. It seems Jack Kemp is blasting our peace proposal & even went so far as to call Pres. Duarte of El Salvador & try to persuade him to oppose our plan. Then we discussed Cong. & the possibility they'll send a bill I'd like to sign but with an amendment calling for a tax increase. That I can't sign. We've agreed on Bill Verity for Sec. of Commerce but at day's end still hadn't been able to reach him.

[NSC meeting regarding news that General John Vessey was on his way home from Vietnam; learned of unrest in Panama, commented, "It's hard to tell who the players are & who to root for."]

In the P. Gulf we're delaying flagging Kuwaiti tankers for a couple of days & we've tightened up on revealing plans & schedules. Our Turkish friends are nervous. The Cong. is again considering a bill demanding the Turks take blame for the Ottoman Empires persecution of Armenians when it was in power.

[Met with Democratic congressmen on Nicaraguan peace plan and then with some Republican ones on welfare reform and others on housing bill; had, as noted, "our regular Mexican lunch" with Vice President Bush; met with Meese on appointments; presented awards to Puerto Rican Americans, including a vice-admiral and performer Rita Moreno; photo sessions; last stitches removed from nose.]

Friday, August 7

[Noted positive employment figures.]

Howard had reached Bill Verity. He's coming in Monday—sounds as if he'll take the Sec. of Commerce job. Brad Holms has been confirmed as head of FCC. I'm getting a call from Dick Cheney tomorrow at 1:30.

NSC—Frank C. is back from Europe—had good meeting with Helmut & Margaret. We have some reason to believe the Soviets will back down on the German Pershing 1A missiles. In the P. Gulf we'll probably move 3 more tankers on the 9th. I've written letters to Sen.'s Boren & Cohen—Chairman & Vice Chrmn. of the Intelligence Committee on policy regarding covert actions. We still protect my right to defer notification of such actions if secrecy is necessary to protect human life etc. Frank C. is a little on edge about Geo. S. & Cap W. being a part of morning NSC meetings. I'll let them work this one out.

[Met with congressmen on their return from Central America; greeted chefs group; had NASA briefing comparing Soviet and U.S. space programs; had meeting with Shultz; presented letters to Senators David Boren (D-OK) and William Cohen (R-ME) in public ceremony; went to Camp David.]

Saturday, August 8

A quiet day—televised my radio address so Charlie Wick can play the TV tape for Europe. Subject was the Berlin Wall. A call from Dick Cheney about upcoming speech—he feels the public is fed up with the whole subject of Iran-Contra. I think he's right.

Calls from Frank C & Geo. S. on Nicaragua statement. The Presidents meeting in Guatemala have agreed to a cease fire plan—including Ortega.

[Sunday, August 9: returned to W.H.; meeting with Howard Baker, Senator John Tower (R-TX), and speechwriter Landon Parvin to discuss Wednesday speech.]

Monday, August 10

Again the schedule got moved around. It seems if I remember correctly this happens every year in those last days before our August vacation. Well anyway we started the day with our usual meeting—discussing Bill Verity & a possible foul up over some business deal years ago. Howard is going to check it out with Ed Meese who is in S.F. Former Chf. Justice Burger is going to be a witness for Bork in the Sen. hearings. A little talk about the Central American Peace proposal & some loopholes adverse to the Contras.

NSC—Report our 3 tankers & escort in the P. Gulf are laying back for a day off Bahrain. This was pre-planned to keep the Iranians off balance. The Central Am. Pc. plan is lousy. We've made a point by point comparison with one we & the Cong had agreed to. Some of our Central Am. friends were looking to their own wants. Salvador & Guatemala have guerilla problems & their plan helps them with that—never mind what it does to the Contras.

[Briefing from General John Vessey on return from Vietnam, reported that progress was made on MIA issue; issues lunch canceled in favor of more work on speech; cabinet meeting on Central America and Persian Gulf; signed savings and loan bill; met with new head of Securities and Exchange Commission; announced William Verity as new secretary of Commerce; met with Citizens Network, group supporting administration programs on foreign aid; saw doctors.]

Tuesday, August 11

NSC—Discussed verification problems if INF Treaty is signed. Our Mil. leaders believe implementation of the missile elimination should take longer than we would like. Their point is readjusting so we have deterrent power after missiles are gone.

After lunch I met with Joint Chiefs of Staff. Gen. Galvin gave quite a presentation of NATO planning—very informative. Over to East Room for swearing in of Alan Greenspan as Chairman (the 13th) of Fed. Reserve Board. Vice President swore him in—I made a welcoming speech. George S. came by for meeting on Central Am. & how we must plan to plug the plans in the Nicaragua peace agreement which are out & out loopholes for the Sandinistas. We must back the Contras. Then the longest taping session I think I've ever done—after Nancy put some make-up on my still unhealed nose. Mermie is here.

Wednesday, August 12

Some discussion of last night's TV news. They have jumped on me as selling out the Contras—acting as if the Central Am. Presidents plan is mine. Well it isn't. I'm determined we will continue to support them. Then I OK'd an appointment to Pub. Liaison of Rebecca Range.

NSC—Central America again. Got word Pres. Arias of Costa Rica is boasting he topped big Uncle Sam with his peace plan. Well his own plan has a loop-hole where the Contras are concerned. We'll plug it.

Then I met with some of the Presbyterians who went to Central Am. & came back with a report to the Annual Nat. Congress of the church that was adopted by the church. It has the church assailing the Contras & us as villains and Ortega & his thugs are the freedom loving heroes. We tried to give them some of the facts but I don't think we budged them. I was prepared for them to be left wing zealots but now I don't think so. They are sincere & honestly believe the things told them in Managua.

[Watched videotape of congressmen's trip to Nicaragua; meeting with Shultz, who had suggestion for a commission on fair trade to settle the question between Canada and U.S., and he also expressed concern about effect of budget cuts on embassies and foreign policy, the president commented, "he's right. It's short sighted of Cong. to say the least"; Shultz also reported possible Syrian efforts on freeing hostage in Lebanon; interviewed by Hugh Sidey; made speech on national television. **Thursday, August 13:** *flew to North Platte, Nebraska, visited a ranch there; delivered speeches on balance of term in office; flew to California and ranch.]*

Friday, August 14

A foggy morning. Barney, Dr. Hutton & I cleaned guns. I have 14 rifles & almost as many handguns & they were all in need of oiling & cleaning. Then the fog began lifting about lunch time & I decided to ride—which I did about 2 p.m. Nancy decided against it because of the uncertain weather. She was wrong—the clouds began to break & we soon had some sunshine on the ride. Back to the house & some homework as usual. Nancy got a heart break call—Joan Rivers' husband of 30 years or so committed suicide in a hotel in Phil.

[Conference call with Howard Baker and Frank Carlucci regarding adjustments in Central America policy. **Saturday, August 15:** *radiocast; rode; tree cutting.* **Sunday, August 16:** *rode; tree cutting.* **Monday, August 17:** *same schedule.* **Tuesday, August 18:** *received report that Lebanon hostage escaped; afternoon tree work.]*

Wednesday August 19

Fog burned off by 10 A.M. so a very pleasant ride. First a couple of phone calls—to Chris Shea who won the special election in Conn. for a Cong. seat. Then to Bill Brock—in hospital for detached retina. Got word our ships in P. Gulf escorted 4 Kuwait tankers thru Straits of Hormuz during the night. Phone Joe Coors re Citizens for America.

After lunch back to our exterior decorating—pruning trees etc.

[Thursday, August 20: rode; cut trees; attended press barbecue. **Friday, August 21:** *guests arrived; riding; preparations for next day's party.* **Saturday, August 22:** *radiocast; rode; afternoon party for seventy guests, including Ron and Doria.]*

Sunday, August 23

A morning ride on another beautiful day. Howard B. called to say we were issuing a disclaimer on a front page story—*Los Angeles Times* re visit by Gorbachev. I believe & so does Howard that it was a fishing trip by the Soviets.

[Afternoon spent pruning a trail.]

Monday, August 24

Another beautiful day. Nancy elected not to ride because of some phone call & homework. I ride with our usual troop of S.S. Agents & Mil. Aide. After lunch Dennis & the Dr. & I cleared out a trail that had become overgrown. A message came that by end of afternoon our convoy in the P. Gulf would be clear of the Hormuz Strait. Wound up our trail clearing early. A late afternoon call from Frank C. Our convoy made it through the Straits of Hormuz. One incident—2 small boats approached a Tanker—machine gun bullets, off their bows made them rear off.

Dinner & off to bed.

*[*Tuesday, August 25: *left for L.A.; looked at house to buy as home; homework; dinner with friends.* Wednesday, August 26: *addressed Town Hall of California; met Governor Guy Hunt (R-AL); had ear examination, no problems detected; dinner at home of Betsy Bloomingdale.]*

Thursday, August 27

Another busy day. Nancy off to see the land proposed for the Presidential Library. I stayed behind & met with George Scharfenberger & Mort Janklow re a book I should write after I leave the W.H. I agreed.

Then after lunch a meeting here with George Shultz, Howard B., F. Carlucci, Marlin F., Elliot Abrams, & Jose Sorzano. This was a preparatory meeting down stairs with the Nicaraguan Resistance leaders. They went down ahead of me & I went down a half hour later. The leaders were Col. Eurique Bernendez, Adolfo Calero, Pedro Jeaquin Charerro, Equcena Ferrey Escheuerry, Alfonso Robelo Callejas, & Arestideo Shachez Herdicia. A good meeting—preparing them to meet & be careful of Speaker Jim Wright. After meeting went up to the Suite with George S., Howard B., & Frank C. A good discussion of arms negotiations, Central America—agreed that Elliot A.'s man Busby should go there as an regional Ambassador. Also some planning about Middle East.

[Checkup with allergy doctor; received word of attempted military coup in Philippines, easily put down; dinner with Mr. and Mrs. Lew Wasserman. Friday, August 28: *met with law-enforcement leaders supportive of Robert Bork's Supreme Court nomination; photo session with twelve-year-old suffering terminal leukemia, with* Life *magazine.* Saturday, August 29: *radiocast; went to ranch; starting cutting up dead oak tree.* Sunday, August 30: *received word that two tankers made it through Strait of Hormuz in Persian*

Gulf; worked on oak tree. **Monday, August 31:** *lunch with Stu Spencer; finished bringing oak tree to wood pile; called Jim Brady with birthday greetings;* **Tuesday, September 1:** *rode; finished splitting wood from oak tree.]*

Wednesday, September 2

Still very warm. A nice ride—a trifle short but pretty. Word came—Iranians had attacked 5 ships in the P. Gulf—of different nationalities—S. Korean, Greek, Spanish etc. An intercepted message revealed they were under orders not to attack any ships with American Navy escort.

After lunch we shifted operations to the woods along Penn. Ave. towards El Refugio road beginning at our front entrance. We're back to pruning & it's going to be a great improvement.

[Thursday, September 3: unrest in Persian Gulf, but no attacks on U.S.-escorted ships; shuffle in leadership of Department of Commerce due to illness of acting secretary; rode; pruned trees; noted that Dr. Hutton had returned from San Francisco. **Friday, September 4:** *rode; pruned an oak tree.* **Saturday, September 5:** *rode; continued pruning oaks.]*

Sunday, September 6

Up at 6:30 A.M. and off & away at 7:40. Some haze in the air, not sparkle clear as it has been every day. Landed at Nat. Guard Field in Topeka, Kansas—greeted by Gov. & Mayor and then on to Alf Landon's home. He has some difficulty hearing but is sharp as a tack. He'll be 100 yrs. old on Wednesday. His wife had a 91st Birthday last Wed. A nice but short visit with Sen. Nancy Kassebaum (his daughter) & Sen. Dole also on hand. Then out to the porch to greet a crowd of friends & relatives of the Landons. The U. of Kansas marching band was on hand to play. It was a pleasant occasion topped off with Birthday cake, then back to A.F.1 & on our way to Andrews A.F. base in Wash. For some reason it seemed as if we'd been away from Wash. longer than on any of the other times we've been away. Ted Graber came with us. Rex barked a loving welcome to Nancy.

[Monday, September 7—Labor Day: slept late; caught up on desk work. **Tuesday, September 8:** *meetings to plan upcoming work; address two hundred appointees, commented, "a pep session for up coming battles"; National Security Planning Group (NSPG) meeting on arms-reduction talks, no decisions made; photo session with legislative candidates from various states; haircut; meeting with Senator Paul Laxalt (R-NV), Howard Baker, and Duberstein, Laxalt having bowed out of presidential race; Maureen arrived for visit.]*

Wednesday, September 9

Staff meeting—discussed Justice Thurgood Marshal's interview with Carl

Rowan to be aired on Sun. The Justice took me on as being a racist—the worst in that since Herbert Hoover. George B. is going to call him about a meeting with me. My record & the record of this admin. on such matters is out ahead of any other admin. in our history. A suggestion was made also to contact Justice Powell about endorsing Bork who I've nominated to take his place.

Some discussion about Independent Council Walsh & his wanting to Q. me on Iran-Contra affair. I've said OK.

[Briefing for visit by Prime Minister Ingvar Carlsson of Sweden; visit and meetings in morning; discussion with Vice President Bush, Shultz, and others of arms reduction, commented, "Our biggest difference is whether to try to block the Soviets from going mobile on their missiles or not"; state dinner with entertainment by Marilyn Horne.]

Thursday, September 10

A late start 10 A.M.—briefing for interview with several journalists for *USA Today*. I think the interview went well. They were all friendly & their Q's were fun.

[Spoke to student winners of essay contest on the Constitution; boarded Air Force One.]

We went to Miami, Fla. to greet Pope John Paul II on his arrival for a tour of the U.S. There were 5,000 people on hand at the airport. Then later in the afternoon we joined him again at an art museum where he & I had a one-on-one—joined later by Nancy for a 2 on 1. He's truly a great man. We did our farewell speeches before the press & went our separate ways. Ours was back to A.F.1 & Wash. where we arrived about 10 P.M. It was a most rewarding day.

*[***Friday, September 11:*** *meeting with Howard Baker on appointments for top posts at Council of Economic Advisors and Office of Management and Budget; NSC meeting; briefing for interviews, then had the interviews, first with James J. Kilpatrick, commented, "We're old friends & I enjoyed it," and then with* U.S. News & World Report; *signed proclamation for Hispanic Heritage Week; met with Shultz on upcoming meeting with Soviet Minister of foreign Affairs Eduard Shevardnadze; videotaping, including segment with Mrs. Reagan for ABC show* Good Evening America, *featuring Ron; went to Camp David.* **Saturday, September 12:** *radiocast; short walk with Rex.* **Sunday, September 13:** *returned to W.H.]*

Monday, September 14

Staff time began with Beryl Sprinkel resigning as Ec. Advisor. I couldn't ask him to stay—he had family reasons & so it's back to Illinois for him. Then Liz Dole came by & she is leaving the Cabinet in order to campaign for her husband.

NSC—Gen. Powell reported on a successful sting operation we brought off in the Mediterranean. A participant in 2 hijackings that took American lives was enticed to leave Beirut & go to Cyprus. There he was told of a Mr. Big in the drug business who had a money making plan for him was offshore on a luxury yacht. He fell for it. The crew turned out to be our FBI men & he's now on his way to America to be charged for his crimes.

We also have another Kuwait convoy on it's way through the Persian Gulf. Libya is sending planes to Iran by way of Turkey air space. We think they're carrying mines.

[Visit from South Korean ruling party's presidential candidate Roh Tae Woo; swearing-in ceremony for James Billington, new Librarian of Congress; addressed National Alliance of Business; met with leaders of state Junior Chambers of Commerce; photo session with congressmen and guests; sat for thirty minutes of serious portrait photography by Norman Parkinson.]

Tuesday, September 15

A very brief staff meeting as I met with Repub. Cong. leaders at 9:15 A.M. except for those Senators who are on the Judiciary Committee. The hearings on Bork's confirmation started today. It was a good meeting with much agreement about why I should veto some of the bills now before Cong. A suggestion was made that I should veto any "continuing resolutions" & insist that appropriations bill be sent to me individually. This would be like giving me line item veto in a way. A short NSC time & then briefing for visit with Shevardnadze. He came in at noon—we had a short time cluttered up by 4 waves of press & photographers. Then into Rose Garden for signing by him & Geo. S. of agreement to have a center in each country to minimize chance of accidental hostilities taking place. From there it was a Plenary meeting in Cabinet Room followed by a brief one-on-one & then a luncheon meeting in State Dinning Room. They were good meetings & free of the hostility we used to see even if we were disagreeing on some things.

[Desk work; greeted major donors to National Republican Congressional Committee; received word that Howard Baker was hospitalized with diverticulitis, would probably return to work next day.]

An early dinner (alone). Nancy is in Calif. on a drug thing with the Pope. At 7:20 I left for brief appearance at a reception—5th anniversary of USA Today. It was nice—a short address & home to bed.

Wednesday, September 16

Staff time—no great issues. Then NSC—a report on how the talks with Shevardnadze are going. Apparently going well—a far different tone than in earlier times before changes in Moscow. Then about 9:50 A.M. Pres. Diouf came by with a few people. We discussed problems of Tropical Africa in conquering hunger.

[Spoke to conference of U.S. Advisory Commission on Public Diplomacy; participated in Celebration of Citizenship rally; telephoned brother on his birthday; greeted poster child for Juvenile Arthritis, W.H. staffer back after cancer treatment, Airmen of Year, three of whom were female; quit work at 3 pm.]

Thursday, September 17

[Staff and NSC meetings; phone call from Shultz, commented, "He seemed upbeat about progress he was having with Shevardnadze"; flew to Philadelphia for celebration of bicentennial of Constitution; attended party fund-raiser.]

Back to plane—rain started again just as we boarded for take off. Approaching Wash. slowed down—Andrews AFB was having a thunderstorm. As we arrived & landed—sun came out. Back to W.H. about 4 P.M. Sen. Baker told me Geo. S. & Shevardnadze would like to come over & brief me on meetings at 5 P.M.—Well 5 P.M. became 5:45—the meeting grew to a dozen upstairs in our living room. Then Nancy was delayed. I changed dinner time 3 times. Now the briefing meeting is over—we've agreed to meet the press on it at 9 A.M.—they are gone & my room mate is at the door. Thank Heaven.

A call from Frank—In East Germany Soviets fired on a patrol of American soldiers wounding one but not seriously. A h--l of an incident to happen at this time. Also the bad guy in our sting operation in the Med. arrives tonight—he'll be charged & tried here.

Friday, September 18

In at 8:45 & at 9 A.M went into the Press Room to make a statement about the 3 days of meetings between Geo. S. & Soviet F. Minister Shevardnadze. Took a few Q's & then turned it over to Geo. Back in office for NSC—we mainly talked about the meetings. Then at 10 A.M. Lee Thomas came in to brief me on agreement we have negotiated with 23 nations to reduce fluorocarbons—in our effort to stop reducing the Ozone Layer. This is an historic agreement. At 11 A.M. an NSPG meeting—an update on P. Gulf situation. There has been a great change in world thinking about that and there is general approval of what we are doing.

After lunch met with & was photographed with 35 top Repub. leaders of La. who are supporting Cong. Livingston's run for Gov. Then it was usual meeting with Geo. S. He brought Ambas. Jack Matlock with him who reported on changes Gorbachev is trying to make in Soviet U. George has an idea that perhaps this change can be used to involve Soviet U. in our effort to bring peace to Middle East. We never could have accepted that idea under previous Soviet leaders.

[Photo session with Hispanic Scholarship Fund representatives; went to Camp David; received list of senators voting against administration to limit SDI.]

Saturday, September 19–Sunday, September 20

Two days of looking out the window at solid bank of fog & sitting by the fire. Did our usual short hike after my radio broadcast & stopped in to see the newly completed S.S. bldg with all its security measures. Very impressive. Saturday, P.M. called Judge Bork after he finished his testimony before Senate Judicial Committee. They never laid a hand on him.

Sunday—still deep fog—back to W.H. by about 2 P.M.

[Monday, September 21: flew to NYC for appearance at U.N.; meeting with Secretary General Javier Peréz de Cuéllar, and leaders of Pakistan, Japan, and Guatemala, finally with eighteen foreign secretaries on nuclear disarmament; returned to W.H.; received word of Iranian attack on British ship.]

Tuesday, September 22

An early morning call; our Navy has boarded & seized the Iranian ship. Three of the Iranians were dead, 2 missing, 4 wounded & 26 taken prisoner. This was subject of our brief staff meeting. Then we discussed Pres. Arias of Costa Rica who was going to meet with me prior to addressing the House of Repr's. That continued for a few minutes of NSC meeting then he arrived. He admitted to being concerned that Sandinistas would try to cheat & agreed the Contras should be supported. Then he went up to Cong & reversed himself.

Then a very disturbing meeting with Jim Miller, OMB, Will Ball, Cap Weinberger, Jim Baker & of course our staff. The subject was the bill to extend the debt ceiling which expires tomorrow. They have loaded that simple bill with a cut in Defense so great we'd have to kick 400,000 out of the service. I'd have to agree to bil's of $ in new taxes & on & on. Jim Baker says if I veto we could conceivably have a world crisis financially because we'd be defaulting on $24 Bil. worth of bonds. I have to come up with a decision on—do I sign or do I veto. Everything in me cries out to veto.

Then a meeting with group of Conservative leaders who through Paul Weyrich had appealed for a meeting with me at intervals to discuss issues. They've been off on belief that I'm violating my conservative principles. It went fairly well & I believe we should meet fairly regularly.

[Desk work; photo with Swedish editor; haircut; farewell visit with departing staffer.]

Wednesday, September 23

[Discussion of congressional opposition; NSC meeting with reference to president of Iran's speech at U.N. General Assembly; photo with Mr. & Mrs. Anatoly Shcharansky; greeted Boys Club Youth of the Year finalists.]

A meeting on whether I should sign or veto the debt extension bill. I'm really torn on this one. I'm sure my gang think I should sign but everything in me says veto. Watched a while—the Bork Hearing on TV—then lunch & some desk work. Three top officers of Farm Bureau came by—they are supporting us on Bork, taxes & even our effort to get rid of farm subsidies world wide by year 2000.

Eddie Serrano came by with his family & his son who was sworn in as a Navy Ensign. Meeting with Cap Weinberger—he's for a veto. Interview with Edmund Morris my biographer. He's been talking to old friends in Hollywood, Des Moines etc.

[Congressional picnic at W.H., with entertainment by Marine Band playing music of Meredith Willson.]

Thursday, September 24

What turned out to be a hectic day started with staff meeting. I have decided to sign the debt extension bill but with a hard statement to the effect that it's probably the worst legislation ever sent to me. But I must sign to keep us from defaulting on $24 Billion worth of Bonds. Then NSC—most of time taken up with a stop by of Egyptian Foreign Minister.

[Met with Republican congressional leadership, found they supported decision to sign debt extension; signed proclamation for Historically Black Colleges and Universities; visit from Congressman Christopher Shays (R-CT); had lunch with Vice President Bush; greeted hockey teams in town for exhibition, noted, "I shot a goal past a uniformed goal keeper on my 3rd try"; new poll numbers positive; cabinet briefing on Canada fair-trade problem; greeted Asthma poster child from Nebraska, officers of Associated Builders and Contractors group, leaders of the Sons of Italy; visit from the president of European Parliament.]

Friday, September 25

Ken D. & I were the staff meeting. Howard B. is in Tennessee & the V.P. is in Europe. I've discovered the Soviet Poetess who has been released from the Gulag is living in Chicago so we are making contact to bring about a meeting with Irina Ratushrinskaya. That will be in October. I'm told the Sen. may confirm Judge Session as new Dir. of FBI. We have some dissension on the AIDS Commission—some has to do with the Gay Dr.

[NSC discussion of possible coup in Fiji.]

The Sen. is battling over a bill that would undo the flagging of Kuwait Tankers in the P. Gulf. They are insane. Word has come we are going to sink the Iran mine layer ship possibly this afternoon.

Just learned that "60 Minutes" Sunday will have a story that Bill Casey before his death gave Woodward of the *Washington Post* 40 interviews for a book that will charge among other things that we planned assassinations in Lebanon with my approval.

At 10:40 went over to Va. to address the 4th Annual Convention of the

"Concerned Women of America." There were about 1,300 (Nat. membership is around 600,000). It was inspiring. They are all for Bork, also aid to the Contras. They were on their feet cheering for most every line of my speech.

[Dinner with King Juan Carlos and Queen Sophia of Spain, considered them "good friends & truly nice people."]

Saturday, September 26

Did my radio show on Dem. blackmail in connection with debt limit extension. A quiet day at home. Phoned Justice Powell & discussed the politics of the Sen. committee on confirming Judge Bork to take Powell's place. He's going to have a talk with Justice (Chf) Rehnquist. So far Powell has thought it improper for him to endorse Bork. I hope he can see his way clear to do it. Things liven up after dinner. Ron & Doria & Nancy's brother Dick & wife Patty arrive.

[Sunday, September 27: concert by Soviet pianist Vladimir Feltsman, noted, "Ron had met him in Russia a few years ago. That's how we became aware of his plight."]

Monday, September 28

Staff meeting started with Bob Woodward's claim in his book & his interview last night on "60 Minutes" that he'd interviewed Bill Casey just before Casey's death. He's a liar & he lied about what Casey is supposed to have thought of me. Then Howard B. told me Bill Brock wants to retire as Sec. of Labor. Some talk about possible replacements—I suggested possibly our 1st Sec. of Labor Ray Donovan. Some concern it might cause a rumpus over his trial even though he was not guilty.

NSC—More talk about Casey & his wife Sophia. Then the UNSC & Resolution 598. Iran is apparently building another Silkworm launching pad. Sen. Byrd has another bill that would amount to de-flagging the Kuwait Tankers. Savimbi has freed the Swedish prisoners he was holding. Then we talked of the arms Saudi Arabia wants to buy from us. As usual there is resistance in Congress.

[Desk work; appeared at gathering for Presidential Library Foundation lunch; photo with W.H. groundskeepers; interviewed by Arnaud de Borchgrave for the Washington Times.*]*

Upstairs—no Nancy, she is overnight at the Grunwalds place. As usual I'm lonesome & so is Rex.

Tuesday, September 29

Meeting began with Bill Brock coming in to tell me he wanted to resign as Sec. of Labor toward the end of the year. Then it was Bork & our concern that the Sen. might do us in. I'm starting to meet with some of those Sen's tomorrow. I told Howard & Ken if the Sen. didn't confirm him I'd favor letting the

Court stay at 8 judges til I'm gone. Some talk about Woodward's vicious attack on Bill Casey & his lies about having a death bed interview. Ed Meese who with Don Regan did visit Bill confirm that he was incapable of understanding or communicating.

NSC—Word that after 6 weeks of Gorbachev's silence & rumors that he might be ill—he has appeared in Moscow. We have a letter from 64 Sen's opposing our sale of weapons to Saudi Arabia. This is disgraceful & a blow to our diplomacy. Then some talk about Canada & the free trade negotiations. They are back in the meetings again.

[Addressed joint meeting of World Bank and International Monetary Fund (IMF); NSC meeting on nuclear plants; signed bill extending debt ceiling; report from Webster on trip to Mideast; videotapings. **Wednesday, September 30:** *staff meetings on various efforts, including Robert Bork's Supreme Court confirmation, Persian Gulf, and defense funding; met with Commission on Privatization; addressed several hundred supporters; photo session with W.H. chef and departing Secret Service agent; telephoned Senator John Stennis (D-MS) on Bork confirmation, noted, "He's not decided yet. Has material he wants to study over the weekend"; greeted 450 Republican Eagles at W.H.]*

Thursday, October 1

This was a day that shouldn't happen. Got an early phone call that the swearing in ceremony for Judge Sessions as FBI Director was off. He's in hospital with a bleeding ulcer.

Then staff meeting—a lot of talk about Bork & my talking to some Sen's. It seems Specter has announced he's against Bork—things look bad. Then Howard said Jim Miller had gone to the hill to propose some tax increase we might go for. Howard had him in & told him from now on Jim Baker would deal with Congress on budget affairs. A U.S. District Court upheld my proposal for drug testing employees.

NSC—Soviets tested a new monster missile near Hawaii & are said to have another which they'll shoot right over the islands. This is unprecedented & we've told them no.

Now some controversy with defense—they want to send the Battleship New Jersey into the P. Gulf. We're a little concerned about that.

Sen. Warner is suggesting a change in the War Powers Act. Cong. would have to vote to disapprove our action & I of course would have a veto.

Howard B. brought Ed Meese in while I was having lunch. More troubles on Bork. They had been up on the Hill button holing Sen's without much success.

[Photo session with Legal Task Force from Iran-Contra hearings.]

Personnel time with Bob Tuttle then Cap W. to tell me he was going to resign as Secretary of Defense—his problem is his wife's bad health. I can't argue about that.

[Photo session with departing staff; meeting with CEOs working on behalf of Robert Bork; met Senator Howell Heflin (D-AL), still undecided on Bork; more photo sessions; visited Judge Sessions at hospital. **Friday, October 2:** *staff meeting to plan Bork strategies; NSC meeting to discuss Persian Gulf situation; meetings with four senators regarding Bork; attended farewell party for Elizabeth Dole; met with Senator Jesse Helms (R-NC), noted, "he's going to get [out] of the way and allow Varity to be confirmed as Sec. of Commerce"; signed proclamation for German-American Day; interviewed by Fred Barnes of New Republic; went to Camp David; ran* Knute Rockne All-American *(1940), featuring Ronald Reagan as George Gipp.* **Saturday, October 3:** *radiocast; ongoing discussions about free-trade talks with Canada.* **Sunday, October 4:** *agreement set with Canada; watched NFL game with strikebreakers playing.]*

Monday, October 5

[Clarification of unchanged stand against tax increases; progress on seating new secretaries of Transportation and Commerce; NSC meeting on Iranian flotilla in Persian Gulf.]

Speaker Wright has deserted his & our agreement on how to make peace in Nicaragua & charges us with being guilty of breaking up the peace efforts. Lies, Lies, Lies. A visitor, Pres. Chissano of Mozambique. A half hour visit but worthwhile. I urged him to meet with their insurgents the RENAMO.

Then in the Rose Garden—271 High School Principles—winners of the nationwide recognition for excellence in Ed. On the way across the grass on my way back to Oval Office Sam Donaldson yelled a question as to was Bork out as Supreme Ct. nominee—I said "Over my dead body." The Principles cheered & then move in on the press giving them h--l.

[Issues-briefing lunch, commented, "as always worthwhile & fun"; desk work and calls to senators on Robert Bork vote; met with Suzanne Massie, who reported mixed feelings in USSR about glasnost; haircut; videotapings. **Tuesday, October 6:** *news that Bork confirmation was defeated in committee; met with Republican congressional leadership; photo session with Crown Prince Akihito of Japan; planned to keep Weinberger resignation secret for the time being; greeted Associated Press board of directors; went to doctor for allergy shot, blood pressure recorded as 128 over 84; dinner party for Prince Akihito and Princess Michiko, entertainment by cellist Yo-Yo Ma.]*

Wednesday, October 7

Went into office at 10 A.M. At 10:15 Bork came in. I told him I'd go all the way in trying to get the Sen. to confirm in a floor vote. He'd been talking to some in the Sen. who have him pretty convinced he can't make it. He's going to talk to some more but I have a feeling he's leaning toward taking himself out of the race. We won't know for a few days. Meanwhile I met with Ted Stevens he's going to vote yes & he seems to think there may be some who'll join him.

[NSC meeting included Soviet secretary of Agriculture Viktor Nikonov, who talked about collective farming; presented Minority Enterprise Development Awards; addressed Organization of American States (OAS) regarding Nicaragua; presented Medal of Freedom to Judge Irving R. Kaufman, presided over Rosenberg spy case; meeting with Shultz, on his way to Middle East hoping to induce Israel to approve international group to work on peace treaty.]

Later in day asked Admiral Watkins to take over as Chairman of our AIDS Commission. He said yes.

Upstairs to a lonely old house. Nancy is in Chi.

*[**Thursday, October 8:** staff and NSC with routine discussion; met with bipartisan congressional leadership to speak out for money for Freedom Fighters in Nicaragua; and for arms sales to Saudi Arabia, noted, "Before day was over we made a settlement with Cong. & the sale goes forward"; met with President Kenneth Kaunda of Zambia; announced nominee for secretary of Transportation, James Burnley; received bronze replica of Navy monument; visited GOP candidates in special elections; photo session with chairman of black mayors organization; Mrs. Reagan home, noted, "Nancy's home & faces—next week a biopsy to determine if she has a cancer in her breast."]*

Friday, October 9

Opened day with news that Sen. Cohen had announced <u>for</u> Bork—also Sen. Boren. The opposition still has enough votes to win unless we can persuade some to change. A little talk about Len Garment who has been a little evasive about the affair. Senator Specter has 2 candidates for Fed. Judgeships—after his performance I'll not reward him for his no vote on Bork.

[Discussion of War Powers Act, added, "which I claim is unconstitutional"; and of defense budget; report on incident in Persian Gulf; met with Deng Pugang, son of Chinese premier, in a wheelchair and working on behalf of handicapped; interviewed for C-SPAN; attended United States Information Agency (USIA) luncheon.]

Then upstairs to a meeting with Judge Bork who has decided to stay nominated for the Sup. Ct. At 3 P.M. he went on nat. TV & announced his decision. It's an uphill fight all the way but I'm with him. Then off to Camp David.

*[**Saturday, October 10:** radiocast; rode. **Sunday, October 11:** read by fireside. **Monday, October 12:** returned to W.H.; photo session with portrait painter Aaron Schikler; attended birthday party for Charles Wick.]*

Tuesday, October 13

Up early—Nancy had to give the Dr. & Nurse some blood for her upcoming surgery. Staff meeting without the V.P.—now that he's a declared candidate he's out on the trail. Talked mainly about the Supreme Court & possible replacements for Bork. Howard brought up the name of a lady. Brock had decided to announce his resignation as Sec. of Labor to handle Doles campaign.

NSC—discussion of INF Treaty as it appears to be & question as to whether we can get it ratified. The Dem's in the Sen. are feeling their oats.

[Flew to New Jersey for tour of Somerset Technologies plant; spoke at forum of CEOs; addressed Chamber of Commerce at luncheon; attended party fund-raiser; met with blind little girl, a cancer victim; returned to W.H.]

Wednesday, October 14

A wide ranging discussion of several subjects—Bork, Brock's resignation & possible successor, also successor to Cap W. when he leaves etc.

[Briefing for state visit of President José Napoleón Duarte of El Salvador, who kissed U.S. flag in arrival ceremony; discussion of Contra Aid; National Security Planning Group (NSPG) meeting on nuclear arms proposals; made speech on Robert Bork's nomination.]

Then—the State Dinner for the Duartes. It turned out to be a wonderful evening. My dinner partner was Madam Nien Cheng—the woman who was imprisoned & tortured for 7 years during the Chinese Cultural Revolution. She is an amazing woman—soon to be an American citizen.

Thursday, October 15

Day started at 10 A.M. a half hour staff go around. A little talk about Bork & my broadcast yesterday. The Dem. response, Sen. Terry Sanford was the height of demagoguery. We also talked about offices to be filled—Sec. of Labor etc. Suggestion was made that we should look for women & Hispanics etc.

[NSC briefing on budget problems for defense and Central American strategy; met with Senate Steering Committee, commented, "I assured them I would not soften my requirements for a Supreme Ct. Justice & please the lynch mob"; met with Shultz before his trip to Mideast and Moscow; desk work; met with Roy Brewer, who had collected copious documentation of the domestic communist scene; signed proclamation for use of seat belts; met new commander of the VFW, family of late fraternity brother from Eureka College; videotapings; addressed party fund-raiser.]

Friday, October 16

Again a general discussion of candidates for openings on Cabinet. I've reversed myself—new Sec. of Defense will be Frank C.—I'll make Gen. Colin Powell Dir. of Nat. Security. Had an NSPG meeting on latest P. Gulf incident—Iran Silkworm missile hit an American owned tanker at anchor. Our discussion was—action we should take. I think we agreed we're generally agreeable on a single target response—minimum risk to personnel—theirs & ours.

[Greeted Soviet high school students; sent legislation for three-step war on crime to Congress; meeting with Alan Greenspan, Jim Baker, Beryl Sprinkel, and others on the stock market plunge, commented, "I'm concerned about money supply—has Fed been too tight. Alan doesn't agree & believes this is only an overdue correction."]

Did my radio tape because tomorrow I'll be at hospital with Nancy. She faces a possible mastectomy. This morning she was in N.H. for a program linking—Foster Grand Parents & the Anti-Drug program. Early dinner & then to Bethesda to get her bedded down for tomorrow's surgery. Her brother Dick was with us.

Saturday, October 17

Up at 6 A.M.—then off to hospital by 6:45 A.M. Had to drive—fog made helicopter impossible. We got there in time to see her go down to the operating room. Then Dick & I buried ourselves with newspapers & some sessions with assembled Dr's keeping us posted on progress of surgery. The biopsy turned out to be traces of what they called a non-intensive carcinoma—very tiny. Decision was to perform moderate mastectomy. After lunch were able to see her in her room. She was barely conscious from anesthesia. As can be expected she's feeling bad about losing a breast. We did our best to let her know that was nothing compared to fact the cancer was gone. The Dr's are delighted with the operation—it went so well and was so effective. There won't be any chemotherapy or radiation treatment at all.

Dick & I stayed around dropping in now & then to see her but didn't want to worry her or tire her. Then I had to go back to W.H. for an NSPG meeting & Dick went back to Phil. In our meeting we settled on a plan (a target) for retaliating to the Iranian missile attack on an Am. owned Tanker in the P. Gulf. Tentative date set for response—Monday P.M.

Rcvd calls—one from Margaret Thatcher re Nancy. Gave a good nite call to Nancy myself.

Sunday, October 18

Began day with a phone call to Ida Nudel in Israel—recently released "refusnik". Then off to hospital. Nancy is doing fine—she got out of bed to have lunch with me at table—then I took her for a walk around the halls. I had to come back to the W.H. about 4 P.M. to preside over the 4th concert—last in the series of programs that go on public TV as "concerts from the W.H." It was great show & will be on next month. Then at 8 P.M. our gang & the Cong. leadership met up in the oval Yellow Room. We briefed them on our retaliatory strike against the Iranians for their attack on the Am. owned tanker of Kuwait. We intend to take out a radar & command control tower in the Gulf with Naval gun power. This will be done early tomorrow morning.

Monday, October 19

At 7 A.M. our time 4 U.S. destroyers warned men on 2 adjoining platforms in the gulf which served as radar & command & control posts to get off—most if not all did. We then loosed 1,000 rounds of 5 inch shells on the platforms. Mission successful. Most of our meeting time taken up by reports on this & our dealings with Cong. Cong. is for most part supportive of our action.

[Photo session with new secretary of Commerce William Verity and thirty-seven members of his family; issues-briefing lunch; presided at swearing-in ceremony for Verity; received report from Commission on Merchant Marine; short meeting with President Yoweri Museveni of Uganda.]

Then a haircut, a meeting with 10 children from all over the country who have been honored for performing heroic deeds. Then left (by helicopter) for Bethesda & dinner with Nancy. She has astounded the medical profession with her rapid recovery. She'll be home soon.

By days end the news was a 508 fall in the Dow Jones. Some people are talking of panic—the Dow is down 800 points in just over a week.

Tuesday, October 20

First meeting discussed mrkt. Howard said we'd meet with Alan Greenspan, Jim Baker & Beryl Sprinkel later in day. The Fed announced they would add a little liquidity to the money supply & 2 major banks reduced their interest rates. Chrmn. of Stock Exchange is acting very upset. Sale of Long Range Bonds is up. Mrkt in Hong Kong shut down for a week. (At 2:35 in the afternoon 490 million shares had been traded & the Dow was up 59.83 points.)

We also learned a list of possibles for Sup. Ct. will be along shortly. Incidentally 3 ladies are possibles for Sec. of Labor—Mrs. McLaughlin, Carla Hills, & Connie Horner.

Then a briefing for the upcoming meeting with P.M. Gandhi of India. And the meeting—I've gotten us on a 1st name basis. It was a good session including the working lunch. I believe we have a good personal relationship that carries over to our 2 nations. After his departure I met 6 incoming Ambassadors & their families. Then went up to the W.H. & had the meeting I mentioned.

It was a good one—we are going to meet with the Dem & Repub. leaders of Cong. to work out a process to balance the budget among other things. At 4 P.M. the market closed at 108.27 up & 640,000,000 shares traded.

Out to the hospital—Dick arrived also & we had dinner together—Nancy is fine.

Wednesday, October 21

Our 9 A.M. meeting had to do with our possible summit with Cong. leadership. Jim Wright has stormed that he won't meet if we bring Jim Miller (Dir. of OMB). My inclination is to tell him to go to h--l. Some talk also of replacements for Bork & Bill Brock.

[NSC discussion of military action in Persian Gulf; short meeting with King of Leso-

tho; brief meeting with President José Azcona of Honduras; addressed gathering of Network on Foreign Affairs Funding; two hours' rehearsal for next day's press conference; visited hospital; noted that market was up.]

Thursday, October 22

Left at 9:15 A.M. for hospital—Nancy up & waiting. Arrived back at W.H. at 10:35. A large group with signs, a Marine band, kids from "Just Say No" clubs, Foster Grandparents, & Exec. office staff to welcome her. It was very heartwarming. Rex is happy—so am I. Then over to the office for staff meeting—10:45 A.M. Had a phone call from Geo. S. & Frank C. in Moscow. Meetings with Shevardnadze, constructive & business like—covered Human Rts., Regional problems, & of course arms control. They said they believe the Pershing 1A's issue is settled.

Tomorrow they meet with Gorbachev. Then in our meeting we had at the stock mkt situation. This morning it started nose diving again. At one point it was down 200 or so points. By end of the day it closed only 55 pts. down.

We talked plans—I opted for an afternoon announcement that I will personally meet with Cong. leadership to negotiate on reducing the deficit. After that announcement the market improved.

[NSC heard report on Iranian attack on Kuwaiti oil facility; greeted conservative German politician; met with U.S. Savings Bonds Volunteer Committee.]

Lunch—then photo with departing S.S. agent & wife—off to San Jose. Then met with our "market" group—Alan Greenspan, Beryl Sprinkel, Jim Baker, et al. Discussed more things we can do. All believe market was over priced & this was largely a correction but believe some activity re the deficit can help. Into Family Theatre—more briefing for press conf. And so upstairs to my napping bride.

Mkt was down over 100 pts. When my announcement was released that I would meet Cong. regarding deficit. Mkt closed at 77 (down)—8 P.M. Press Conf. went well

Friday, October 23

Usual meetings—began with a phone call from Geo. S.—a few words about meetings. Looked over a list of 15 possibles to replace Judge Bork. Trimmed it down to 3 and in my mind it's down to 1—Judge Douglas H. Ginsburg. We need to talk to our Repub. Sen's. & ABA.

[Addressed Employment Expansion Forum and Department of Labor employees; interviewed by foreign press.]

Then some Bob Tuttle time—photo with departing S.S. Agent & a taping session. Then off to Camp David. Weather there was beautiful. Sen. had rejected Judge Bork. I phone him and expressed my very real regret & reiterated my great respect & admiration for him.

Saturday, October 24–Sunday, October 25

On Saturday did my radio talk on ec. & need to balance the budget. Also mentioned that in Moscow meeting Gorbachev brought SDI back into picture & linked it to a summit meeting. I think he feels Cong. has me on ropes & I need a summit. I told my radio audience I would not give anything away to bring about such a meeting. A couple of easy walks with Nancy. Phone calls to 2 top men of Securities Exchange. During movie Saturday night—Ron called Nancy. He's home from Moscow. Changed our clocks back to Standard Time. Sunday was another beautiful day. Howard B. called to tell me Geo. S. & Frank C.—both home now would come to the W.H. at 5 P.M. for a report on Moscow trip. We left Camp D. at 2:30. George reported on meeting with Gorbachev who sort of pounded the table at one point over a State Dept. paper he thought was critical of him. George pounded the table back & the air cleared. We still believe he wants a summit but is playing a game thinking I want a summit so bad I'll pay a price regarding SDI. He's wrong.

Monday, October 26

Reversed our morning schedule—NSC 1st, George told of his Meeting with Mrs. Nudel—just released from Soviet prison. She related a talk with Sakharov. He says Soviets are experimenting regarding their system, glasnost & all. She advises us to milk the experiment.

[Decided to nominate Ann McLaughlin for secretary of Labor; met with congressional leadership regarding budget; brief meeting with Pope of the Armenian Church; met with Shultz, heard reports of his trip; lunch with Republican Inner Circle.]

Back to the W.H. & an interview with 5 foreign TV persons. Right after that—schedule was cancelled. Kathy told me word had come that Nancy's mother had died this afternoon. I came home & told her the news. It was heartbreaking. We've cancelled my schedule for tomorrow. I'm taking her to Phoenix and then I'll fly back here. She'll stay there & arrange for a church service Fri. or Sat. I'll then fly out & return with her.

Tuesday, October 27

Left at 9 A.M. for Phoenix on our sad journey. Upon arrival we went direct to the mortuary. We saw Deede looking calm & peaceful in her red robe. This was too much for Nancy who broke down sobbing & telling her how much she loved her. I told her Deede knows that now & that she really wasn't in that room with her body but would be closer to her when we get to her apartment where her long time friends were waiting for us.

And they were John Chauncey, Mich Wood, Vick Zepp the Dr.

After a couple of hours I had to get back to the airport but by that time Nancy was in a better state of mind hearing all of us talk about Deede & our love for her. At 2:30 I left & got back to Wash. & the W.H. about 9 P.M. Tomorrow I go to West Point & Friday after a meeting with Shevardnadze I go back to Phoenix. Sat. will be the service & then we'll come back here Sun. A.M. I should have written that at 8:30 this morning I was in a conference call. The Soviets blinked. Shevardnadze speaking for Gorbachev is arriving Thurs. for meetings on INF and plans for the summit.

[Wednesday, October 28: *meeting with Howard Baker on his contact with congressional leadership on budget; flew to West Point; had lunch with cadets; returned to W.H.; heard from Mrs. Reagan, getting along well as possible.*]

Thursday, October 29

Opened day with NSC meeting. Date for summit may cancel our Thanksgiving vacation according to one Soviet source who said they might propose the last week in Nov. Then we discussed some of the things Cong. is talking about, are all kind of ridiculous legislation about arms treaties. They are actually sitting on the Soviet side of the negotiating table. We have to hold off on asking Cong. for the $270 mil. for the Contras. Fortunately they still have some arms in the pipeline. Later we can try for non-lethal aid. A Sandinista Major & his wife have defected. We're holding it secret while they try to get their young son out of Managua.

[*Report of encouraging meetings with Congress on deficit; greeted champion baseball team Minnesota Twins, all observed moment of silence for Mrs. Reagan's mother, Mrs. Davis; spoke to small business representatives.*]

Then Howard & Ed brought in Judge Ginsburg. We went over to the East Room—about 200 in attendance. There I revealed that the Judge was my nominee for Sup. Ct. It was well received by all present. Then some desk time & finally Admin. time. I was 1st President to ever be made an honorary member of the Telephone Pioneers. Next a photo of American winners of the Motocross motorcycle event—50 nations participate. Then S.S. brought in Painter of my pictures of the ranch & his son. Followed that with reception in the East Room for "Citizens for America."

Upstairs exercise—phone call from Nancy & days end.

Friday, October 30

Usual staff time & NSC meeting. Howard is taking Ginsburg up to the Hill to meet with some Senators. I taped my Sat. radio cast. An early lunch—then a briefing for meeting with Shevardnadze & then the meeting. George & Frank had been with them for some time & both sides had agreed to a communiqué to be released simultaneously. Shevardnadze had brought me a letter from Gorbachev. It was statesmanlike & indicated a real desire for us to work out any differences. The outcome of the letter & meetings was a summit to start here on

Dec. 7. Purpose to sign INF Agreement and set in motion START Treaty to reduce ICBM's by 50% & to finalize that at a summit in Moscow next Spring. On that note of optimism we lifted off the S. Lawn for Andrews AFB & Air Force One. With me on trip to Phoenix—all of Dicks family. Arrived around 5 P.M. at Arizona Biltmore & met a tired Nancy who has been making all arrangements & closing down Deedee's dwelling. Nine of us for dinner here in the suite—that included Ron & Doria. A pleasant reunion & so to bed.

Saturday, October 31

Slept in a bit—then stayed in bed & did some homework—the P.D.B. & preparation for my eulogy at the service.

About 1 P.M. the family joined us & at St. Joseph's Church Mike, Maureen & Dennis joined us. The service was very nice & the Lord was good to me, my Eulogy was very well received. Then everybody moved to the Boich lovely home for a reception. Friends from all over the country were on hand. It was most heart warming.

Back to the hotel & dinner in bed.

[Sunday, November 1: returned to W.H.]

Monday, November 2

Our 2 morning meetings were combined & total discussion was on our attempt to negotiate with the Democrat Congress on the deficit. It isn't going well—they are just stubborn that the answer must be a tax increase.

[Telephoned captain of Kuwaiti ship Sea Isle City *wounded in Iranian attack in Persian Gulf; cabinet meeting on Douglas Ginsburg nomination; swearing-in ceremony for Judge William Sessions as director of FBI; farewell party for Bill Brock; interviewed by David Frost.* **Tuesday, November 3:** *continuing discussion on Congress and budget; met with Republican congressional leadership regarding ways to reduce the deficit; met with Shultz about prospects of getting Senate ratification of INF Treaty, commented, "Believe it or not there are elements who are hinting it would be a bad treaty"; NSC meeting; desk work; videotaping; introduced Ann McLaughlin as secretary of Labor; had allergy shot; photo sessions; attended reception for Friends of Art and Preservation in Embassies.]*

Wednesday, November 4

[Meetings on debt reduction.]

Then a meeting with Soviet émigré—poetess Irina Ratushinskaya. She was released from Labor Camp about 2 days before our Iceland meeting with Gorbachev. She has a remarkable story & is the author of the miniature letter smuggled out of the camp to me a few years ago. It is written on a slip of paper about 5/8 of an inch by about 3½ inches. Ten of the inmates signed it. Too tiny to be read by the naked eye.

At 11:45 regular meeting with Bill Webster—CIA—report on defector—Major in Sandinista Army. We're not sure he isn't a plant.

[Had lunch with Vice President Bush; meeting with Nicholas Brady and his commission on stock exchange.]

Tom Gibson & fiancé came by—he's leaving to become a cartoonist. Then meeting with Geo. S.—2 subjects—one the market. He thinks we're taking it too lightly—that is really threatens a recession. Other was Gorbachev & the INF Treaty. A list of items yet to be worked out.

Over to EOB to address group supporting Canada Free Trade. Met with Edmund Morris—usual thing for biog. Upstairs for exercise & then redress for an early dinner & drop by at RNC reception Decatur Carriage House. This was installation of RNC board of directors. It was a very pleasant time. Brief speech, receiving line & mix & mingle.

Thursday, November 5

Staff meeting—a problem with our nominee Judge Ginsburg. Some one opposing him has revealed that in his younger day he smoked pot on several occasions. It was a brief thing back when a lot of people experimented a bit and that's all it amounted to. I think the fellow who leaked it was a friend at the time who smoked it with him. I don't see any reason why I should withdraw his name. Then there was more on the deficit. Frankly I think the Demos want a recession for the '88 Campaign.

NATO Sec. Gen. Lord Carrington in for a brief meeting. He's most supportive of our getting the INF Treaty signed.

Then to Rose Garden—a ceremony for Cap W.'s resignation & replacement Frank C. & Colin Powell to replace Frank at NSC.

[Received report on child-safety partnership; telephoned Bill Stout, who was gravely ill with heart ailment; photo sessions; met with leaders of Alzheimer's Disease Association, including Princess Yasmin Aga Kahn; greeted leaders of American Asian Society. **Friday, November 6:** *had vigorous debate with Republican congressional leadership over deficit; meeting with Valerio Zanone, defense minister of Italy; interviewed by five journalists; met with Shultz on upcoming summit; appearance to, as noted, "brief Ethnic, Minority, & Repub. leaders on Judge Ginsburg situation since his revelation that he smoked some pot when he was young"; went to Camp David.]*

Saturday, November 7

A morning call Bill Bennett wants me to ask Ginsburg to withdraw. I told him that was absolutely impossible. Later a call from Ginsburg that he wanted to withdraw. Well this I couldn't say no to. It looks increasingly that he can not be confirmed & he's taken enough abuse. Today at 2 P.M. in a very dignified manner he announced his withdrawal in the W.H. Press Room. That then became the top weekend story.

Now the good news—the day was beautiful & I got a horse back ride. Nancy must wait a bit longer before getting back in the saddle.

[Sunday, November 8: returned to W.H.]

Monday, November 9

Back to the old routine. More discussion of next nominee to Supreme Court. Then NSC & a charming lady dropped by—the Foreign Minister of Switzerland Mrs. Kopp who will be the next President.

[Toured United Way headquarters; lunch meeting with Organization of American States (OAS) leaders, commented, "Wednesday they'll have Ortega of Nicaragua—he was a good part of my speech"; met with Senator Strom Thurmond (R-NC), who had suggestion for Supreme Court; photo session with guests brought by congressmen.]

Then upstairs to a meeting with Howard Baker, Ed Meese, & Judge Kennedy—Sacramento, Calif. He is going to be my nominee for Supreme Court.

Tuesday, November 10

Howard B. is going to talk to Gary Bauer—word has it that he's been leaking to the press. Paul Laxalt has agreed to call Orin Hatch who has suddenly gone off the reservation & is throwing rocks at me & the exec. staff. Then we get down to the budget & Judge Kennedy. I've decided to nominate him for Sup. Ct. His wife is coming in from the coast & tomorrow at 11:30 A.M. I'll present him in the Press Room.

NSC became a briefing for visit of Pres. Herzog of Israel. Ceremony had to be in the East Room—rain. Arrival at 10 A.M. The Pres. of Israel has a touch of Irish brogue. He was born in Dublin—his father was a Rabbi. Our meeting in Cabinet Rm went very well. We did suggest Israel shouldn't be selling arms to Iran.

Shortly after meeting I returned to Cabinet Rm. where heads of all Vet's org's & several Sens. & Cong. men were present & with press covering—announced I was supporting legislation to make Veterans Admin. head a member of the Cabinet—loud applause.

[Photo session with family of Howard Baker, also with family advocate Dr. James Dobson; spoke to employees in anti-pornography and obscenity field.]

Some desk work then over to Dr. He had a specialist in to see my Dupreton Contraction. I think I'll have an operation to straighten my 3rd finger—my left hand. But not 'til next year sometime unless it progresses further—sooner.

[State dinner, entertainment by Roberta Peters.]

Wednesday, November 11

Veteran's Day & a heavy snow fell—about 3 inches on ground by 10 or 11 o'clock. Over to the Office—a short meeting with Dir. Sessions of FBI—a couple of his men, Howard B., A. B. Culvahouse etc. Report on FBI check on Judge Kennedy—clean as a whistle. Then he & his family—lovely wife, beautiful daughter—Soph. at Stanford, son who is a Sr. at Stanford, & son who is out in business. Then into Press Room & announcement of his nomination. So far we've heard encouraging words that he has bipartisan support from Dem's & Repub's.

Back upstairs for the day. Nancy cancelled trip to N.Y.—Dr's orders. I'm not at all unhappy about that. Maureen is with us.

Howard called—says making progress for a change on the deficit matter. I'm concerned that an element in the Dem. majority would like to see a recession with the '88 election coming up.

Thursday, November 12

Something like a foot of snow on the ground—but the sun is out.

A brief meeting with Bill Bennett. He was worried the he was out of favor about his phone call to Doug Ginsberg. The press was playing that I had encouraged him to call & urge Doug to withdraw his name. I reassured him that I didn't believe the stories & we were all OK. Then Jim Baker & Jim Miller along with Howard B. & Ken D. This was about negotiations re the deficit. They wanted instructions from me as to what they could do to get an agreement. I want them to hold out for more spending cuts in domestic programs.

NSC—presented possible schedule for 3 day meeting with Gorbachev. I OK'd showing it to Soviets for their approval. Learned John Poindexter was operated on for tumor in his sinuses. OK'd—flowers & wine.

Colin reported Iranians seemed to be acting a little better—possibly because of Arab League meeting. Almost all of the Arab nations have decided to restore relations with Egypt. He also reported that Jewish groups are planning big demonstration when summit is on. However it is not billed as protest against Soviet U. but as support for us.

Then a meeting with half dozen Afghan leaders who have come together to try unify various elements against Soviets. I presented them to an audience of Reps & Sen's, & public leaders in Roosevelt Room.

After lunch a meeting with Rep. Frank Wolf, Sen. Helms & others about removing "Most Favored Nation" status from Romania. I can do that on basis of violation of human rights. We have given them that status because of their apparent independence from Soviet U. Apparently that is a sham and their violations of human rights are worse than in Soviet U. I'll take this up with Geo. S. tomorrow.

[Meeting with Senators Joseph Biden (D-DE) and Strom Thurmond (R-SC) regarding hearings on Kennedy confirmation; greeted officers of American Legion; launched American Lung Association Christmas Seal campaign; farewell party for staffer.]

Friday, November 13

Staff time—I put in a bid to ask Justice Thurgood Marshall to come to the W.H. next week for a visit. He's made it plain that he thinks I'm a racist—but the Justice has only heard the case as presented by the prosecution. I intend to speak for the defense. Everyone agrees. More time spent on deficit. Demos. want about $9 bil. in new taxes but want us to agree without telling us what kind of taxes.

NSC—We have more proof that Marcos is cooking up a coup aimed at Pres. Aquino. More planning on our part going on for summit. I've proposed (on another subject) to drop Romania's most favored nation status until they clean up their human rights act.

[Presented medals to youthful heroes.]

Meeting with George Shultz. We've heard from Soviets—they want Gorbachev to address a joint session of Congress. It's OK with me. I think he has some supporters in Cong.

[Birthday party for Howard Baker; went to Camp David.]

Saturday, November 14–Sunday, November 15

[Radiocast; short walk.]

Sun. another bright day & even warmer. Howard B. called me in morning. He'd been called by Jim Wright from Texas who's panic'd a little by press treatment of his monkey shines with Ortega. Howard rode him a bit and so he'll be in to see me & George S. at 10 A.M. tomorrow.

Arrived back at W.H. around 3 P.M. Tonite Black Tie & over to the Edmund Morris home for dinner. It was a wonderful evening. Ten of us around the table—all of the others writers. We really enjoyed ourselves.

Monday, November 16

Staff—taken up a lot with the Jim Wright situation & the need to take him to the woodshed. NSC—Discussion of INF & the need to work out a verification with regard to Soviets using SS20's as 1st stage for the SS25's. I think it can be worked out.

At 10 A.M. Jim Wright came down at my invitation. Our whole team including Geo. S., we outnumbered him. We let him have it pretty good. I think we narrowed his field of action. Then I left for the Sheraton Wash. Hotel where I addressed about 1,000 members of Am. Council on Life Insurance. They were most receptive.

Back to W.H. for issues lunch. Then some desk time. At 2:30 a haircut. At 3:00 P.M. a taping session in the Library and at 3:30 the Dentist. For a couple of days I've been having a tooth ache. In the chair for 1 ½ hours. He thinks I have an inflamed nerve due to one tooth in the upper jaw being hit by one in the lower. He ground a lump off the sore one—then said if the inflammation didn't go away in a few days—it's a root-canal job for me. Home & Mother.

Tuesday, November 17

[Optimistic reports on budget talks; attended farewell ceremony for Weinberger.]

Then at 11:30 a meeting in the Cabinet Room with Repub. Senate Leadership. I called for support for my proposed veto of Housing Bill which is a pork laden budget buster. It was a good meeting & they were with me.

Lunch and a short meeting with A. B. Culvahouse. He reported on forthcoming report of Joint Committee investigating Iran-Contra. The majority report kicks me around pretty good. Minority report repudiates the full report & gives me a clean bill of health.

[Meeting with Congressman Bob Walker (R-PA) on involving private sector in the space program; met with Representative Robert Dornan (R-CA) and others just back from Nicaragua, commented, "Their report confirms our own findings that Sandinistas are playing games & have no intention of keeping their word on the peace plan—they just want to use the plan to get rid of the freedom fighters."]

From there to a meeting of Jewish leaders & 3 refusniks we succeeded in getting out of the Soviet U. I told them of how we intended to get more Jews released & hopefully better living conditions & freedom for all Soviet Jews. Then my sneeze shot & upstairs to a meeting with Justice Thurgood Marshall. I'd asked for the meeting because of his public statement to Carl Rowan that I was a racist. I literally told him my life story & how there was not prejudice in me. I gave examples of my relation with Minorities as a young man in school, as a sports announcer & as Gov. I think I made a friend.

Howard Baker called to tell me I wouldn't have to veto the Housing Bill—the Sen. killed it.

Wednesday, November 18

[Talk about deficit.]

NSC was brief—talk of summit. Soviets had suggested Gorbachev speak to Joint Session of Congress. We'd passed the word to the Hill. Looks like there is resistance to the idea on the Hill.

[Addressed a teachers gathering.]

Back to lunch & watched TV on presentation of finding of the Joint Committee on Iran-Contra. It's obvious the report is aimed at discrediting me. Repub. Members—have presented a minority report repudiating the majority report.

[Spoke to employees of Trade Representative's Office—observing twenty-fifth anniversary of that office; met with Shultz on summit; farewell party for Weinberger; further meeting on deficit with Representative Bob Michel (R-IL), Senator Bob Dole (R-KS), Secretary

of the Treasury James Baker, and Howard Baker, commented, "We don't have agreement yet, but revolution in the works. We may have a meeting here at W.H. That's as far as we could go."]

Thursday, November 19

Again we talked deficit & Dem's refused to commit themselves. It's getting frustrating. I sent over team back with a message that I would not ask for a delay in the sequester which is due tomorrow.

NSC—Heard from Max Kampleman in Geneva—he thinks INF agreement might require Geo. S. & Colin to go to Geneva this weekend.

[Spoke at U.S. Chamber of Commerce; photo session with departing staffer; addressed women's business group called Business Partners; cabinet meeting with general updates on current affairs; met with World Soccer Cup president on holding tournament in U.S. in 1994; photo with Epilepsy poster child, and with Notre Dame football stars and Gene Schoor, author of 100 Years of Notre Dame Football; *signed proclamation for Adoption Week; greeted boxer Tommy Hearns.]*

Finally upstairs. Billy Graham & Ruth our guests for dinner. A wonderful evening with Billy & Ruth as always. God Bless Them.

Friday, November 20

[Met with Senators Sam Nunn (D-GA) and John Warner (R-VA) regarding their defense bill; NSC briefing for meeting with Israeli prime minister Shamir; then meeting, commented, "I assured him of our continued friendship and told him of my plan to address Gorbachev on matter of Soviet Jewry."]

Then a meeting with Geo. S. who is leaving for Geneva to try & pin down the details on the INF Treaty.

The schedule go turned inside out—the 2 P.M. NSPG meeting was moved up to 11 A.M. It was a discussion on START Treaty which will be feature of Summit. There are still differences among us on some of the goals we should go for.

After lunch—A. B. Culvahouse deposited a pile of reading on my desk—Q's & suggested answers for Judge Walsh who is investigating Iran-Contra. Then at 2:45 Howard B. brought in Jim B. & Jim M.—We have an agreement on the deficit.

At 3 P.M. the Cong. leaders came in & at 3:15 we all joined the negotiating teams in the Cabinet Room. A discussion & decision that we were united on the agreement. Then at 3:30 we went into the Press Room & I made a statement. We took questions & everything went well. Some desk work & upstairs.

Saturday, November 21–Sunday, November 22

Both days cold but beautifully clear—the view from our windows was spectacular. Did my radio broadcast Sat. on the Deficit. For once the usual Dem. rebuttal—endorsed what I was saying.

Sun. continued the homework that occupied me on Sat. afternoon—The written questions the Independent Council sent on Iran-Contra plus written answers provided by A. B. Culvahouse & staff. I've gone thru them—added a few items by for most part they were accurate. Then some letter writing.

Sun. Nite—to Kennedy Center for our refusniks piano concert. Vladimir Feltsman—2 hours. It was magnificent, he took curtain calls several times & encores for an audience that just wouldn't stop clapping. Our guests were the Laxalts, Michels, Chris Buckley & wife, & Helen Hayes.

Monday, November 23

[Mrs. Reagan left for California.]

Staff time was brisk—talk about summit & Geo. S. & Colin's negotiations in Geneva. Still some unsettled problems.

Then NSC time & more of the same plus talk of how we must work on the Senate to see that a treaty (INF) if we get it will be ratified.

Then A. B. Culvahouse & 2 of his helpers came in to go over with me the written interrogatory. This went on for an hour & then it was Dick Wirthlin time. His figures were mixed some up, some down, reflecting the confusion of the people with the way the media is mis-stating some of all that's going on, what with Iran-Contra, the deficit, & the Summit.

[Swearing-in ceremony for Frank Carlucci as secretary of Defense; spoke to activists and supporters; accepted live turkey in annual rite of Thanksgiving; discussion with business and labor leaders on Congress and deficit effort.]

Finally a return of A. B. Culvahouse & his aides to complete our briefing on the written Q's. Nich Ruis stopped by & then it was upstairs.

Tuesday, November 24

Downstairs for the sneeze shot. Then the Map Room to sign notarized statement that I was telling the truth about the Iran-Contra Affair. And then off to Calif. by way of Denver where I was making a speech. Actually the visit was more than that.

[Visited Martin-Marietta Astronautics Facility in Colorado; went to ranch; received word en route from National Security Advisor Colin Powell that Shultz and Eduard Shevardnadze had worked out a full agreement on the INF Treaty.]

Wednesday, November 25

Bright, Sunny, & Windy as h--l today. No morning ride. After lunch Howard B. arrived with Colin—just back from Geneva. He reported on what took place. He said George S. was just great. We've achieved verification rules with

Soviets better than ever been achieved before. It looks like we'll sign on INF Treaty at the Summit.

Then out in the woods with Barney & Dennis, cutting up a giant Oak that tipped over. Afterward called George S. in Wash. to congratulate & thank him. He just got home this afternoon.

Thursday, November 26—Thanksgiving Day

Again a bright beautiful day & this time less wind. Rode again. Nancy can't ride yet because of her operation. Yesterday we rode Snake Lake Road. Today was around the Omni Station & back down Pa. Ave. About a 1½ hour ride. At 4 P.M guest started arriving just after Dennis & I finished pruning the Sycamore tree in the yard. Mike, Colleen & the children—then Maureen & Dennis. Dinner at 5 P.M. then a pleasant couple of hours to visit & early to bed.

[Friday, November 27: rode; had lunch with Ron and Doria; pruned trees; called members of Congress on budget bill. Saturday, November 28: radiocast; rode; ranch work. Sunday, November 29: returned to Washington.]

Monday, November 30

Back in the routine again. We opened talking about the Budget deal & speculating that maybe we'd have more trouble in the Sen. than in the House. Then came work the Demo's may pull a fast one & include the bill in a Continuing Resolution which I might have to veto. This of course would automatically put into effect the Sequestor. I brought up our need to conceive of an ongoing plan to follow to balance the budget once & for all. I changed the subject & asked to see if the thousands of high school students tomorrow in Fla. after my speech couldn't ask a few Q's. We'll see.

NSC—Mainly about the tragedy in Haiti. Then a visitor—King Olaf of Norway. A pleasant 15 min. visit.

[Appeared with two hundred business leaders to press for budget bill; addressed Heritage Foundation luncheon; videotapings; haircut; Maureen visiting. Tuesday, December 1: brief meeting regarding new poll that shows poor approval rating; flew to Jacksonville, Florida, for party fund-raiser; spoke to high school students; returned to W.H.; watched presidential candidates debate.]

Wednesday, December 2

Brief staff meeting—pretty much concerning my next event which was a meeting in the Cabinet Room with Repub. Cong. Leadership of both houses. Main subject there was the deficit plan & getting it approved. Then back to the Oval O. for the NSC brief. Ken Adelman came in & gave us some suggestions about the whole nuclear power issue & how we should approach it so as not to upset our European Allies etc. It was a worthwhile presentation. Then we discussed Soviet violations of existing treaties. I sent a report on these to Congress before the day was over. Colin reported that Frank C. in Europe had sent a message that our allies were in great support of the INF Treaty. Some talk also of when Israel will drop the other shoe. Word has it they are planning some reprisal against Syria for the Hang Glider attack that killed 6 Israeli soldiers.

[Desk work; meeting with fund-raisers for annual inaugural anniversary dinner; meeting with Shultz on upcoming summit; greeted University of Maryland champion women's college field hockey team.]

Thursday, December 3

[Staff meeting; Treasury Secretary James Baker reported on tax provisions related to budget bill; telephoned Washington Times columnist Bill Rusher for good piece.]

Then it was NSC time. A report on travel expenses of Soviet summiteers. They've demanded 50 we've said 20. But at Madison Hotel they empty the mini-bars in their rooms into their suitcases every morning in addition to running up big bar bills—all of which we have to pay. The St. Dept. has already had to ask Congress for a supplemental appropriation.

On another subject—Att. Gen. has asked Defense for 4 more helicopters in Atlanta (the prison). I said yes. Cong. voted no in Committee on Stinger sale to Bahrain—we have to get that changed. I called Sen's Inouye & Kasten for help—they are both on my side. We are waiting to see if there will be S. Korean retaliation for N.K.'s bomb in the S.K. airliner.

Then I went over to EOB to speak to Human Rights advocates. Outside door I met 5 individuals who have wives, husbands or fiancés in Soviet U. who've been held there for years. They asked me to appeal directly to Gorbachev & I will.

[Swearing-in ceremony for Secretary of Transportation James Burnley; interviewed by four anchormen (three major networks and CNN); short meeting with Victor "Brute" Krulak, Marine veteran; photo session with poster child for Better Hearing and Speaking.]

Upstairs to Nancy. We saw the interview on NBC. It was OK.

Friday, December 4

Staff time—Learned House has passed a Continuing Resolution I'll have to veto if it comes to my desk. Senator Byrd claims he has made a pledge to me—there will be no amendments on the Bud. Deficit Bill. I'm more than happy (& surprised) at this.

[NSC meeting received report on incident in Persian Gulf involving Iraqis and Saudis.]

At 9:30 A.M. a meeting with Bipartisan Cong. Leadership. We briefed them on upcoming summit. Bob Dole spoke once—very feisty re Senates responsibility to confirm. It was apparent he was taking off on me for remarks I made last night on the networks TV interview. I think really he was irritated because I said something nice about Vice President Bush. I understood Howard B. sort of took him to task.

[Signed defense-authorization bill; swearing-in ceremony for new archivist of the United States; spoke to Anti-Crime Coalition; meeting with Shultz to discuss Gorbachev and the summit; met in Situation Room for discussion of remaining details on START, commented, "Most had to do with SDI."]

Upstairs & to Dr. Hutton's office for a check up on a spot on my nose—apparently it is innocent—no trouble. Ran back to the office to deliver B.D. gift to Kathy. Then Home & Mother.

Saturday, December 5

Ted Graber here—also Maureen & Dennis. Did my radio bit in Oval O. Lunch upstairs in Atrium then while Ted & Nancy went into a session about Xmas decorations I went down to the study. I spend the afternoon with a 3 inch thick briefing book on the summit & the Army-Navy game which Army won 17 to 3. Later the Heisman trophy winner, Tim Brown of Notre Dame was announced. I phoned him to congratulate him. Later Howard B. called & requested I phone Bob Dole who has gotten over his feisty mood & who wrote me a nice letter in addition to a nice public statement. Mission accomplished—peace is restored. Dinner & Bed.

Sunday, December 6

Nancy lunched in the beauty parlor—Ted, Dennis, Maureen & I in the Atrium. Spent afternoon with briefing book on summit. Then a reception at 5:20 for the attendees & nominees of Kennedy Center Honor Gala. They were Sammy Davis, Jr., Perry Como, Bette Davis, Mrs. Nathan Milstein & Mr. Alwin Nikolais.

[Photo session with W.H. medical team; attended Kennedy Center gala.]

Monday, December 7

Pearl Harbor Day. I signed a proclamation that said it was. A busy staff session at 10 A.M. Jim Baker came in to tell me there was a proposal that we issue some of our U.S. Bonds in Yen & Marks. He's against it & so am I. Then he left & we talked about the summit. Then Howard brought up a bill changing the way of appointing an Independent Counsel. We agreed I should sign it but with a statement pointing out its shortcomings. The Reconciliation Bill is in the Sen. it may turn out OK but the Continuing Resolution out of the House is bad. We must get money for the Contras there.

NSC—Colin had a full agenda. Report on P. Gulf. Iran fired a Silkworm missile at Sea Isle loading station in Kuwait & missed. However their speed boats hit 2 ships—one Danish & a man killed, 2nd was a Singapore registered ship which they set on fire. Our Geneva negotiations started a walk out on their Soviet counterparts who then blinked. Colonel Lenhart is doing a great job for us. I signed an NSDD calling for agreement on summit issue we'd agreed on last week at NSPG meeting. It had to do with warhead count for START.

Colin warned that Gorbachev may suggest a zero-zero agreement on short range tactical nuclear weapons—that can only happen after we agree to parity on conventional weapons.

We should get together on banning Chemical weapons but that's a tough one to verify.

[Telephoned British prime minister Margaret Thatcher; met with Joint Chiefs of Staff, all support INF treaty.]

Then an Ec. briefing, Jim Baker, Beryl Sprinkel, & Alan Greenspan. They do not believe we face a recession in '88.

[Attended Christmas tree lighting.]

Tuesday, December 8

This is the big day. Staff meeting was small talk and actually so was NSC because at 10 A.M. a full ceremony on the S. Lawn for Gorbachev's arrival. He & Raisa arrived in a limo made in Russia that's bigger than anything we have. After the usual routine he & I went to the Oval O. All our talk must be through interpreters. A good rousing meeting—we got into a debate about human rights. He thinks we have fewer of those than they do. After a while we brought in additional member of our teams. It was a good meeting & it's plain he really wants more reduction of nuclear weapons. I think we'll make progress on the START Treaty.

After lunch Nancy & I met he & Raisa and went up to the East Room for the signing of the INF Treaty. It was TV'd live and really was an historic moment.

Back to the Cabinet Room for another session—with a full house—34 people in the room. Not nearly as good a meeting as this morning's.

Tonite the State Dinner—Van Cliburn entertaining.

Everything turned out fine—a very enjoyable evening. Then came word that the Market on Wall Street zoomed up 56 points right after we signed the treaty.

Wednesday, December 9

Had a half hour to review points for today's meeting. Then briefing for meeting. At 10:30 went out to the drive to meet Gorby—(I should say Mikhail). We held still in the Oval O. for 5 waves of press & photos. Then I took him into my study. We had a brief talk then joined our teams in the Oval O. for a 2 hr. meeting. I led off on the 50% cut in ICBM's—George S. & Frank C. added some remarks. Then the Gen. Sec. responded. We seem to be doing well on the 50% deal but then he brought up SDI and while he didn't link it to the treaty he still made an issue of it and suggested a 10 year abiding by ABM Treaty & then we should negotiate on whether we could deploy. Things got a little heated. We switched to regional problems—Afghanistan. I asked for a date certain for their leaving Afghanistan. He said he'd leave when we stopped helping the Mujahdeen. I pointed out we couldn't do that unless the puppet government laid down their weapons. Well we agreed to put our teams to work on 50% deal & time was up. I took him over to the Dip. Room to meet Raisa who was with Nancy. They took off for lunch at St. Dept. I went back to office for lunch.

After lunch a briefing for interview with 4 columnists. Interview went pretty well. Some desk time then meeting with several Repub. Sen's re the Budget plan & INF. I let them know if some of the games re the plan went through I'd veto. We want the plan we agreed to. Phil Gramm was bright spot—he came out for the 1st time in support of INF.

Then it was home to clean up & go to Soviet Embassy for dinner. A very pleasant evening but dinner was pretty much the entire evening. I lost count of how many courses but they just kept coming. Brief entertainment after dessert—a Soprano from the Bolshoi Opera—Moscow, then home.

Thursday, December 10

1st meeting—brought word I may not be able to do my annual trip to Palm Springs for New Year's. If the Sen. keeps playing games with the Bud. Deficit Bill I may veto & Cong. won't be able to have it's Xmas recess. (Please don't let that happen.)

NSC—urged me to hit Gorby with need to halt arms flow to Nicaragua & to urge him to get Vietnamese out of Cambodia. I got him to agree on Nicaragua when he & I took a walk across the Lawn.

Our Teams have been meeting literally around the clock & great progress has been made. One trouble spot had to do with ship & submarine launched cruise missiles & how you can tell nuclear ones from conventional.

Then it was meeting time 10:30 A.M. But a call came he was in a meeting at their embassy & would be 15 min.'s late. He finally arrived at 12 noon. We met in the Oval O. for 15 min's then I walked him around the lawn & into the W.H. for a working lunch. And we did work—while others on our team were back in the W. Wing putting finishing touches on Joint Communiqué. Farewell ceremony was slated for 2 P.M. on S. Lawn. Geo. S. left lunch & went back to West Wing. At a few minutes to 2 Gorby & I went into Red Room to join Raisa & Nancy.

Then word came the West Wing workers hadn't finished & needed to see us—by now 5,000 spectators were jammed on South Lawn waiting for us.

Well finally 40 min's late we went out on the lawn where it was raining. If we'd been on time we'd have missed the rain. Well at least it's over, they've departed, & I think the whole thing was the best summit we'd ever had with Soviet Union.

I finished up in the Oval O. signing a proclamation making this Human Rt's Day, in the presence of 15 refugees from all over the world. This is an annual affair.

Upstairs for rest of the afternoon & dinner then back to Oval O. for TV speech on summit—to the Nat. Speech went on at 9 P.M. Just as Gorby's plane was lifting off at Andrews A.F. Base.

Friday, December 11

Had an 8:30 A.M. breakfast in Cab. Room with entire Dem. & Repub. Cong. Leadership. When I walked into the room I got an ovation. The spirit of bipartisanship flavored the entire meeting.

Followed with NSC meeting in Oval O.—Colin led off & most of discussion was post mortem on summit.

[Press conference with regional journalists, commented, "They ask good questions aimed at getting news—not trying to trap me"; telephoned Prime Minister Takeshita of Japan to report on summit; went to Camp David; telephoned Prime Minister Margaret Thatcher of Britain and President François Mitterrand of France. **Saturday, December 12:** *radiocast; photo session with new military personnel at Camp; read.* **Sunday, December 13:** *new poll results show dramatic rise in approval rating; returned to W.H. to see Christmas decorations.]*

Monday, December 14

Staff time—talk about Ortega & his backing away from Contras—canceled talks & ignored Cardinal Brevo.

Then it was Kennedy & his chance for confirmation (good) for Sup. Ct. Then Budget talk & state of Continuing Resolution & Reconciliation Bill & whether Cong. can make it in time to go home for Christmas.

NSC—Busy time. Contra funding—Bud. Conf. will face Sen. proposal to appropriate $9 Bil. in non-lethal aid.

Our defector from Sandinistas—a General has said Sandinistas are planning a 600,000 man military with Soviet help. Sandinistas also intend to continue aid to guerillas in El Salvador. Prince Bandar visited Colin to tell him King Fahd is very concerned about Persian Gulf & situation worsening.

Colin briefed Ex-Pres's Ford & Carter on summit. Both were very supportive of what we're doing.

[Made speech at Center for Strategic and International Studies; cabinet meeting with discussion of summit and budget; interviewed by biographer Edmund Morris; haircut; videotaping.]

Tuesday, December 15

Staff time—Told to expect Deaver verdict today or tomorrow. I'm praying. Bill coming down on changes in special investigation—discussion on veto or sign. Right now the very things I object to are being litigated in court. Well it came to me & I signed—but with a message that I was awaiting the court action & thought it would be out of place to veto.

Then word Dole was going public tomorrow in favor of INF Treaty. I'll accompany him to Press Room.

[Met with Republican congressional leadership to press for ratification of INF treaty.]

NSC—Discussion of Gen. Ortega (brother of Pres. O.) revealed that with Soviet help their mil. was going to vastly build up. Now his brother is trying to play it down. Gorbachev had told me they would withhold arms shipments to help in peace plan.

[Meeting with Republican Task Force on INF; answered questions from Hugh Sidey on summit; had lunch with Vice President Bush.]

An NSPG meeting for a report on our covert activities. A good report. [. . .]

CIA Director Bill Webster came up to office after meeting & gave me a report on some new technology that was fabulous.

[Greeted Rabbis from Friends of Lubavitch; allergy shot; photo session with W.H. aides; attended Congressional Ball at W.H.]

Wednesday, December 16

An update on the Continuing Resolution & Reconciliation Bill. Things seem a little better on about half of those 13 objectionable items. Then some talk about Jim Wright since the defecting Major from Nicaragua who has revealed the Sandinistas planned military build up. I'm also told our PFIAB group is a little upset about our verification policy's. Then I learned Judge Kennedy has finished his testimony before the Sen. Committee on ratifying his appointment. They didn't lay a glove on him. He & his family stopped by the Oval Office.

NSC—The ASEAN Summit came off very well we're told. Japan has offered $2 billion in aid of various kinds to S.E. Asia. Walter Pincus of the W. Post has a story that we don't know how many INF weapons Soviets have. Well there could be a far fetched possibility that's true but actually our ways of estimating prove the Soviets are telling the truth. And if we find even 1 more than they've claimed, the treaty is violated.

We're still waiting to hear about Korean election. Sen. Cohen is demanding access by the Cong. to all information regarding covert operations.

[Briefing and then visit from Prime Minister Giovanni Goria of Italy.]

Then it was some desk time—a report on Mike Deaver Jury. They found him guilty on 3 of the 5 counts. Sentencing is set for Feb. 25. His lawyer has announced he'll appeal.

[Attended first of two holiday receptions for press.]

Thursday, December 17

Howard had quite an agenda for the opening meeting. First A. B. Culvahouse with all the legality's about Mike Deaver situation. For example he has 30 days to file an appeal & we should not make any contact at this time. There is still a case in court challenging the Constitutionality of the Independent Counsel. On other subject we're not out of the woods on the Continuing Resolution or the Reconciliation Bill. The Housing Bill blew up in the House of Rep's who refused to go along with the sensible Senate version I had approved.

[NSC meeting with positive report on attitude of allies from Shultz; report on South Korean election; noted, "With all the Soviets talk about leaving Afghanistan—they are planning a major road construction program there"; visited W.H. mail office to express holiday greetings, learned that in seven years, they handled 40 million letters; swearing-in ceremony for Ann McLaughlin as secretary of Labor, commented, "only 2nd woman to ever hold the job—1st was Ma Perkins during FDR's term"; participated in Senator Bob Dole's announcement of support for INF treaty; photo session with W.H. operations staff, George Allen's family, officers of FFA; presented Maritime History Award to the USS Constitution; attended second press holiday reception.]

Friday, December 18

No Camp David. Staff time—some conversation about Coach George Allen. I've asked to see if we can't give him a little of some kind because of his work abroad on physical fitness for children. Then we prepped ourselves for the meeting with Repub. Congressional meeting at 9:30. It was a good meeting & I made it clear that I would veto both the Reconciliation Bill & the Continuing Resolution if they were passed as they are now in the House. The Senate version seems OK & they are in conference.

NSC—We picked up a spy in the Soviet UN delegation. We're shipping him back to Moscow. It will be interesting to see if they retaliate.

[. . .] Maybe that will get the Soviets to join us in eliminating chemical warfare.

[Positive report on political situation in South Korea; economic briefing; interviewed by BBC for show on Prime Minister Margaret Thatcher; photo session with departing staffers.]

Some phone calls—judge appointments etc. And finally upstairs for the day. Whoops! George the Butler just let me know I was expected downstairs at the household help party. Nancy was scheduled to go but she isn't back from the Children's Hospital. So I went—35 minutes of hand shaking & picture taking. But all worthwhile. Nancy didn't get back until 5:25.

*[***Saturday, December 19:*** radiocast; still concerned with budget battle with Congress,*

noted, "I've served notice—what I'll veto if they send some of the things the Demos are push-
ing"; greeted W.H. military and their families in Christmas reception; did a lot of reading.]

Sunday, December 20

Nancy interviewing household help for when we go back to Calif. I'm due downstairs to greet Open House for more staff. These were all connected with management of the W.H. I got carried away and took Q's. It was more fun than I've had in a long time. Stayed about 20 or 25 min's.

Now it's after dinner—about 9 P.M. Howard B. just called from his car. He's on his way home from the Hill. Cong. just passed a 14 hour extension of the Continuing Resolution. Howard thinks they are very close & will probably wind it up satisfactorily tomorrow. I'm now waiting for the C.R. to get here for my signature. When I sign it will keep us from having to send workers home. And here is the usher now with the bill. Now it's signed & on it's way back & I'm going to bed.

Monday, December 21

A lot to talk about in 9 A.M. meeting. The business of signing the bill if the Congress ever gets around to passing one, is very complicated. We can't just sign without knowing what's in the fine print. The bill is 1,400 pages long. Congress is still talking about going home tomorrow but still won't remove the veto bait from the C.R.

Then at 9:30 I met with Bob Dole & several Sen's for more talk. They wanted to hear from me once more that I would absolutely veto the bill if we didn't remove the "Fairness Bill." They heard me say h--l yes I'll veto.

The balance of the morning was desk time—then an issues lunch in the Cab. Room. I always enjoy those. Then a long stretch of desk time. I even got some drawers cleaned out & as usual finding a few documents I'd forgotten I had. Photos with 4 staff members.

[Received report of Private Sector Initiative commission; ambassadorial formalities;
had allergy shot; attended another holiday reception.]

Tuesday, December 22

A wild day—the schedule was reversed about umpteen times. Ed Meese came in on the staff meeting. He's concerned about some of our Cong. & Sen's having a misperception about me—that I've changed etc. He suggested some meetings with those individuals to straighten them out. I think he's right. We didn't have word until mid-afternoon about the C.R. & the Reconciliation. The Housing Bill I learned had been passed & they had met our figures. A msg. from Geo. Shultz that I should veto the Authorization for St. Dept. because it called for using the Polygraph on St. Dept. employees. But if I did I'd also be vetoing the money to run the St. Dept.

NSC—Soviets allowed our verification team to see the new radar site—but hold out on some things. Team is convinced they are violating the ABM Treaty. Ambas. Matlock says Dobrynin was pumping him as to what we'll do in Afghanistan. They (the Soviets) are showing some willingness on 2nd UN Resolution in Iran. They also still want to talk about SDI. Then we talked about verification problems in connection with possible START Treaty.

Then I had some desk time until 11 A.M. when I had to go to the dentist. It seems I woke up with evidence of an abscess. Well I spent 2 hrs in the chair having that lanced but mainly having a root canal job.

I got back to the office just in time for a call from Tom Foley & Bob Michel to tell me the bills had been passed & they were reporting that the House had adjourned for the Holidays. Finally I had the boys put a bowl of soup on my desk. At 2 P.M. A. B. Culvahouse & Howard B.—re Ed Meese problem with special investigator.

Then a photo with John Rogers who is leaving government. Then a meeting with both Bakers, Jim Miller, & Ken D. This was a report that 300 our people had gone through the 1,000's of pages of the bill & they were OK for me to sign. So at 3:45 I signed both bills—following which Bob Byrd & Bob Dole called me with their msg. that the Sen. was adjourning.

[Brief meeting with Frank Carlucci on defense budget; photo session with W.H. schedule staff; attended holiday reception for senior staff party, entertainment by Rich Little.]

Wednesday, December 23

With Congress on its way home—the heavens smiled, blue sky & sunshine. Also the day was light. Our staff meeting was really general conversation. NSC much the same. Colin reported on Philippines ship collision—ferry boat & tanker—1,500 dead. One 5 yr. old boy found floating on a piece of driftwood. Then a day of mostly photos such as with Stewards (Eddie's gang) Staff had a small Christmas Tree put in study off the Oval O. Then a taping—for Rose Bowl Parade—NBC—then a speech to Soviet people & finally my radio casts for next 2 Saturdays. A lot of picture signing. George Bush & I lunched together. Tommy Thomas came by.

Some bill signing—including the one George S. had wanted veto. I think he changed his mind when he found it would cut off Defense [State] Dept. appropriation. Stopped by Dr's office to have my sore elbow looked at. No diagnosis as yet.

Home for the rest of the day & opening of Xmas gifts. That gets to be a chore—at same time it is heartwarming to bask in affection & friendship of so many generous people.

Thursday, December 24

Over to the office 1st to see A.B. re some Q's in writing from the special investigation. Then call to 6 overseas mil. personnel representative of 5 branches to wish them a Merry Christmas.

Back upstairs for photos in front of the tree. There was quite a lot of bill signing before that however. Lunch in the solarium. More gifts opened. Dick, Pat, Ann & Dan & Geo arrived & we all went over to the Wicks for our usual Xmas Eve dinner with their family & a few friends on hand. As always a wonderful time was had by all.

Friday, December 25—MERRY CHRISTMAS
Slept in 'til 9 or so and then a gathering around the tree for more gift opening. That's finally done & a good time was had by all. A late lunch—some time off & ready for dinner party at 5 P.M. It was a great success—family & friends—N.C.R. & son, George Will & the Laxalts. Early to bed.

Saturday, December 26
The day after Xmas & some of our time spent packing. We'll be off to Calif. tomorrow morning.

*[***Sunday, December 27***: had allergy shot, as noted, "so I'll be ready for Dr. Bookman tomorrow"; flew to L.A.; received word sailor died of injuries sustained in terrorist attack at USO club in Barcelona, Spain.]*

Monday, December 28
A haircut & then allergy tests by Dr. Bookman. Press account of Sailor's death (Ronald Strang) reveals he has a 19 year old pregnant wife—new widow. I'm trying to get her number so I can call. Call is made to Widow & his Parents—not a joyful task. Nancy is out seeing our future home. Tonite Doria & Ron for dinner. A nice visit & early to bed.

Tuesday, December 29
And this is off to Palm Springs day. But 1st some office work & a few things to sign. Then at 10:45 Nancy headed for the hair dresser. At 11 A.M. Mike Abrams came by to check on my exercises etc. He told me our gym equipment at the W.H. belongs to us not the W.H. I'll have to look into this—if so then our exercise room at Belair house we're looking at is all set.

As he left Dr. House & his crew plus our doctor & nurse came by to make ear molds for my new hearing aids. They are actually going to be smaller than the present ones and with a remote control device so I won't have to take one out for a phone call.

Lunch—some more desk work. The wind & rain storm has broken so we'll helicopter to the airport at 3 PM.

We took A.F.1 to Palm Springs & motored to Sunny Lands. We are 1st guests there—others coming tomorrow morning. An early dinner with Lee & Walter & to bed.

Wednesday, December 30

Weather clear—a little chilly but a sweater took care of that. I went over to Walter's driving range & hit a few—and strangely enough hit them perfectly & it's 1st time I've swung a club since last year. The Wilsons, Jorgensens, & Smiths arrived also the Deutchs. After lunch we golfed—2 threesomes. I wasn't as good as on the range. Some good shots—but didn't bother to keep score.

This was the night of the annual Wilson-Jorgensen dinner party at the Country Club. As usual it was great to see so many old friends—most of whom will also be at the Annenberg's tomorrow night.

Thursday, December 31

Some morning bill signing & a meeting about our future plans with George Sharfenburger who has been managing our blind trust. Then a buffet lunch with other additional guests who have arrived such as Ambas. Charles Price, Wicks etc.

After lunch a golf game with Walter, Geo. Shultz, & Charley Price. I was terrible & again did not keep score. George & Charley had a match—one dollar a hole going.

And then the annual New Years Eve Ball. It was great—better than ever. Dolores Hope, Dinah Shore & Frances Bergen sang some songs of our time. Bob Hope told some jokes & then it was midnight & Old Lang Syne.

CHAPTER 8

—

1988–1989

Friday, January 1—Happy New Year

Slept late—didn't get in bed til 2 A.M. Then a little desk work. Read Gorbachev's New Years greeting to the American People to our friends. Spent P.M. watching Rose Bowl game. Mich. St. beat U.S.C. 20 to 17. Charley Wick told me he wants to resign after Moscow Summit next summer.

Tonite a dinner party for about 40—very informal—but also a lot of fun. Singing around the piano.

Saturday, January 2

A few shots on the practice tee, then over to the tennis court to watch mixed doubles—Howard Baker & Jean Smith against Bill Webster & Betz Bloomingdale. Before lunch a phone call to Brian Mulroney (P.M. Canada) and then the signing of new trade treaty between the U.S. & Canada—an historic occasion.

A pleasant lunch in the game room & out for 9 holes of golf. I got one par & some good drives. In to pack & off to the Cultural Center to present an award to Bob Hope. It was a fun evening—good entertainment by Lucille Ball, John Forsyth and a number of others. Then home to a late supper & to bed.

Sunday, January 3

Get away day—up at 7:30 & airborne a little after 9 A.M. Day cloudy & chilly. Had a talk with Howard B. about plans for year ahead. Then a meeting with him, Geo. S., Colin P., Bill Webster on a letter I was handed in L.A. by a Filipino Dr. & his Dr. wife. It was a charge that President Acquino was playing footsie with the Reds & that we should be helping Marcos return to power. [. . .] We are now trying to find out how the Filipino's make contact with me.

Monday, January 4

The bomb fell early—the Dr. in the hall told me I should see the dentist in the afternoon. So at 3 P.M. there I was in the chair for 2 hrs & 15 min. having another go at the previous root canal job.

[Staff meeting: planned for budget meeting and upcoming legislative schedule, discussed, as noted, "a possible travel schedule for me about which I am not enthused."]

NSC—Colin gave us an update on Israel, Gaza Strip & W. Bank. Max Fisher has told Howard that he has notified the Israelis not to send trouble makers out of the country. Colin is going to check. Then Nicaragua & a proposed peace process for Jan. 15. We're talking about asking Congress for Contra aid after my St. of the Union Address. Colin is off to Central Am. end of this week.

Armitage is meeting Gen. Noriega in Panama—not our favorite person. We'll give the INF Treaty to the Sen. after January 25. We think we'll make Ambas. Rowell to Bolivia our interim appointee to Portugal.

[Telephoned Ambassador Mike Mansfield, in hospital for heart surgery; met Sports Illustrated's six honored athletes; met with Howard Baker and James Baker on stabilizing dollar; long-duration dentist appointment.]

Tuesday, January 5
Staff meeting—discussed outline on planning major speeches for coming year. We're thinking of sounding out Ken Khachigian as editor—if he can take a year off from his private life. Then some happy news. In early Feb. I'm slated to do a fund raiser in Calif. for Pete Wilson and a meeting (one day) in Mexico with Pres. De La Madrid—then about 4 or 5 days at the ranch.

NSC Time—Colin brought in our expert on Soviet U. He sees a split developing between Gorbachev & Ligachev. We'll soon see an Ec. plan to make Soviet enterprises self supporting. In June the once in every 4 yrs Soviet Cong. will meet. There should be some hint as to division in Soviet U. under the Glasnost plans.

[Desk work; farewell photo with staffer; presented awards to government executives; budget meeting; allergy shot; reception for W.H. Fellows. **Wednesday, January 6:** *Howard Baker will be doing interview with Hugh Sidey; positive news on markets; NSC meeting with discussion of son of Zimbabwe diplomat victimized by abuse; report of Taiwan producing nuclear weapon; talk of upcoming visit of Prime Minister Noboru Takeshita of Japan and of bilateral agreement with USSR on drugs; decided to attend NATO summit in Brussels in March; desk work; signed farm-credit bill; photo session with Dr. Hutton; met new military aide; meeting with Shultz on foreign policy plans for upcoming year; taped ABC special with children asking questions.* **Thursday, January 7:** *report on causes of October '87 stock market crash; discussion of Howard Baker's travel plans; NSC meeting on Takeshita visit; message from Frank Carlucci regarding Bahrain and from Undersecretary of State for Political Affairs Mike Armacost on Pakistan; word from National Security Advisor Colin Powell on Haiti, commented, "They now have 13 cand's seeking the presidency"; desk work; met with INF treaty negotiators; photo sessions.]*

Friday, January 8
The V.P. in Iowa—Howard in Florida so meeting was Ken & myself. We just visited. Then NSC—Negroponte reported that Pakistan P.M. had sent a warm "thank you" message to me for our continuing aid to Pakistan. We have

word 3,000 more Cuban troops on way to Angola. N.Y. Times did a story based on a leak that revealed [. . .] Lybia—they are making chemical weapons. Duarte (El Salvador) worrying about Contra (Freedom Fighter) radio on El Salvador territory. We're hoping he wont try to close it down. [. . .] Then it was desk time all the way to lunch. My appearance at Army-Navy Club dedication postponed 'til Tuesday.

After lunch met with George Shultz who delivered a pitch for all of us getting on the speaking circuit to sell our plans for the country's future now before the '88 campaign really gets underway. Made a call to Cong. man Pepper—just finished a cancer operation. Then dropped by a birthday celebration for Dr. Edward Teller's 80th B.D. Followed by a meeting with Nick Brady's task force which has been studying the stock mkt. problem & now has presented report.

[Left for Camp David. **Saturday, January 9:** *received news that former aide Ed Hickey died suddenly; called Richard Nixon on his birthday.* **Sunday, January 10:** *telephoned Hickey's widow; returned to W.H.]*

Monday, January 11

[Flew to Cleveland, Ohio; Mrs. Reagan in NYC; addressed City Club; attended fundraiser; returned to W.H.; telephoned Frank Kuehl in hospital for surgery.]

At 5 P.M. George Bush came by—he had called yesterday about a private meeting. We talked about campaign & the latest Wash. Post effort to portray Geo. as having some untold secrets about Iran-Contra & that I was ordering him to remain silent. It was a good talk & we're agreed on how to handle this canard. Then my exercise, dinner & an evening in front of the TV missing Nancy.

Tuesday, January 12

[Noted that Howard Baker returned from Florida where he was visiting his wife, ill with emphysema; discussed budget and economic commission specified in reconciliation bill, commented, "I don't think I'll go for that. We came here to cut down on commissions."]

NSC—Colin met with 4 Pres's in Central Am. They want us to keep supporting the Freedom Fighters but none of them want to come out & say so. Arias continues to ignore the Sandinistas intransigence. Colin told the 4—it was show time for the F.F.'s if Congress is to give us the money for them.

In the P. Gulf we retrieved one of the Iranian boats our helicopters sank some time ago. It was armed with a Stinger we had given to the Mujaheddin including the missile.

At 9:50 we had a short meeting in the Cab. Room about the Soviets setting dates for leaving Afghanistan.

[Attended opening of rebuilt Army Navy Club; visit from James J. Kilpatrick and family; signed proclamation for a Reverend Martin Luther King Jr. holiday; met with

board of governors of Hoover Institution; NSC and E.P.C. meetings on future plans; economic meeting with H. Baker, J. Baker, Sprinkel, and Greenspan; ambassadorial formalities; undertook cardiac stress test, commented, "I was proclaimed healthy as a horse."]

Wednesday, January 13

At about 3:30 A.M. I awoke & realized I had to upchuck—which I did in Nancy's bathroom. At 6 A.M. I repeated. Now the problem became—can I go to Ed Hickey's funeral & run the risk of having to run out of my 1st row pew & try to get to the growler while the Funeral Mass is going on. At 7 A.M. we called Dr. Hutton.

He was already up & dropped in at 7:30 A.M. Decision was made that Nancy would do the honors for the both of us. The Dr. tends to think that it's a bug that's been going around. I'm inclined to believe that I ate something last night that disagreed with me. Temperature, blood pressure even Electro-cardiogram were all on the nose.

Schedule was switched—I went at 10:30 A.M. for briefing on visit of P.M. of Japan. My breakfast was a glass of Ginger Ale. Everything went well—the 1 on 1 (brief) with the P.M. then the Plenary meeting in the Cabinet Room & over to the WH. for working lunch, the farewell in the East Room & the goodbye at the Dip. Room Entrance. Lunch was Consomme, Ginger Ale & Sorbet.

Then the problem was can I go to the dinner tonite where I'm supposed to present Roger Stevens with Medal of Freedom as he retires from the Kennedy Center directorship—well actually he's Chairman of the Board.

At 3 P.M. Dr. Hutton came up—my blood pressure is normal and actually I feel fine. So decision is I go to the dinner—same menu as for lunch.

If between now & evening I get a bout of barfing—we'll change signals.

Nope—I felt fine & we went to the dinner & a good time was had by all. Home about 11 P.M.

Thursday, January 14

A discussion of how to reach the people during the year about our goals & achievements. Then a talk about the continuing belief by so many that we had been less than candid about Iran-Contra.

NSC—a Korean spy story—the 24 yr. old woman arrested in Bahrain for the bomb plot against KAL plane has confessed her guilt & that she was a N. Korean agent ordered to sabotage the coming Olympics. Talk about the death of Pres. Chaing (Taiwan). We're going to ask Ford to be representative at funeral.

Ed Meese came in with letter from Bork who is going to resign as an Appeals Judge. We made an immediate statement. Met with Ann McLaughlin our Sec. of Labor. She's getting into – child care problem for women workers in a very intelligent way.

Lunch and then meeting with Sec. Carlucci on his Middle East visit. Kuwait is solidly supportive as is Saudi Arabia. They can't understand our Congress resistance to selling arms to them.

[Domestic Council meeting regarding acid rain; farewell meeting with staffer; met new commander of American Legion; videotaping; took tea with Prime Minister Noboru Takeshita and his wife.]

Friday, January 15

Dr. Hutton & nurse came by at 8 A.M. While I was still in bed. Drew some blood & started me on my every 15 minute schedule of drinking 10 oz. of Golightly. No breakfast. First meeting discussion of who should represent us at Chiang's funeral in Taiwan. I suggested Paul Laxalt & possibly Warren Burger—found out later they couldn't do it.

NSC—Have learned about a luggage search the French have been doing on Americans including govt. personnel on official visits. They do this secretly & collect a lot of information.

Our son Mike stopped by for a short visit. He's in town on business having to do with his upcoming book.

[Telephoned assembly at local school.]

At 10:30 a short meeting about upcoming St. of the Union Address. Some desk time then at 11:30 Geo. S. meeting. He left me with "Gorby's" book "Perestroika." A quick lunch after completing my imbibing 1 gallon of Golightly. Then at 1 P.M. Nancy & I departed on S. Lawn for Bethesda Hospital—her mammogram & my checkup (6 mo.) after surgery. It was the whole works—colonoscopy, electrocardiogram, CAT scan, stress test on treadmill, chest X-ray and prostate. Everything was judged as way above par—say like good for someone half my age. So on to Camp D. after 4 hours. Usual movie etc. I phone Mrs. Roger Stevens in Fla. to have her remind me of something she had mentioned to me I'd said I would look into it. It had to do with Japan still killing whales.

*[*Saturday, January 16: *radiocast; read* Perestroika. **Sunday, January 17**: *read* Perestroika; *exercised; watched football.* **Monday, January 18**: *walked; finished book; returned to W.H.; Maureen visiting.* **Tuesday, January 19**: *discussed political ramifications of appointing regional housing director in New York; NSC meeting, noted, "Report on Haiti election which is a scandal," also discussed Philippines, Central America; attended annual gathering of appointees; meeting on Department of Transportation budget appeal, commented, "Quite a hassle—I'll make a final decision tomorrow but wont be easy"; photo session with March of Dimes poster child, Japanese journalists; National Security Planning Group (NSPG) meeting on Persian Gulf, noted, "We're doing great there & not about to retreat"; presented Citizen's Medal to philanthropist Brooke Astor; participated in practice dinner for state dinner; Maureen still visiting.]*

Wednesday, January 20

A dreary, foggy & rainy day. First meeting—V.P. present, talked about budget appeal—dept. of transportation. I opted for reducing Rapid Transit money & using it for air safety etc. The Budget summit we agreed to included a 12 man commission—Nat. Ec. Commission, 14 members to take broad view of economy. I disapprove.

[Noted Senate committee vote; took personal action on case of family faced with $1 million medical bill.]

NSC—Japan fudging on Internat. Whale Protection Agreement. If continues we can deny them fishing rights in our waters. False story in the N.Y. Times that we're changing trade policy toward Soviet U.

Sam Nunn made a speech proposing a limited ground based ABM defense against possible accidental launch of a missile. Team that visited Soviet Test Site for nuclear tests seem very optimistic. Soviets are due here—same purpose shortly.

Scheduled an NSPG meeting Fri. for plans on supporting Freedom Fighters in Nicaragua.

[Made appearance at local high school; meeting with Vice President Bush, then Shultz, reporting on Middle East and Israel; addressed gathering of supporters of Central America and Freedom Fighters.]

Thursday, January 21

Began the day talking about St. of U. address. We're trying to boil it down a little but putting in a declaration that I get a budget. If I get another Continuing Resolution—I veto it & serve notice on this in the St. of U. We're going to set dates aside that Cabinet members & Sr. Staff can bring guests & run movies in the W.H. theatre. We're also asking Bob Gray to put together small stag dinners with business leaders etc. at W.H.

A. B. Culvahouse informed us we needed to have me photographed signing answers to interrogatory of special prosecutor—notarized—later in day. It was done in afternoon.

NSC—Sen. Helms is building up a straw man that I.N.F. destroys missiles not warheads. It is a straw man.

[Report from Lord Carringon that NATO partners are enthusiastic about March summit; briefing on and then meeting with Vice Chancellor Genscher of West Germany; meeting with Senator John Tower on ratification of INF; lunch with William F. Buckley, commented, "I tried to ease some concerns he has over I.N.F."; photo session with senior staff; attended reception of Republican National Committee; noted Maureen out to dinner.]

Friday, January 22

Started out a gray day. A little more discussion of St. of the U. speech. I'm going to do some work on this draft—it's too wordy & too long. Discussed an Evans & Novak column re N.J. Principal Joe Clark & our Gary Bauer. Gary offered Clark a job if school board fires him. He's a fine Principal & should not be fired. But Gary shouldn't have done what he did without checking with anyone.

NSC—Colin mentioned some press confusion about summit & the ABM treaty. press portrays us & Soviets having some kind of disagreement. Truth is after signing I.N.F. we & they agreed to a joint statement where we agreed to disagree about Space & Defense. We also agreed that we would have to negotiate on S. & D.

I explained to Colin where Bill Buckley was concerned about I.N.F. He feels Soviets can violate treaty by substituting SS 25's for SS 20's. We don't think that's something we can't handle. One other subject we mentioned was the Soviet proposal that we have a meeting—several states about Middle East.

At 11:00 am an N.S.P.G. meeting—discussed how we can convince Cong. we must appropriate money to help the Freedom Fighters in Nicaragua. Then I did a 10 minute phone call to the 10's of 1000's of Right to Lifers on the Mall—their annual demonstration.

[Photo with GOP candidate in special election; met with leaders of Right to Life movement; met with Shultz; addressed gathering on Central America; left for Camp David. **Saturday, January 23:** *noted snow; radiocast; exercised; read work-related material; watched movie.* **Sunday, January 24:** *watched panel shows; telephoned Senator Ted Stevens (R-AL) regarding his appearance on David Brinkley's show; returned to W.H.; dinner with Maureen and Dennis.]*

Monday, January 25

A gray, snowy day. Opened the day with talk of the Legislative schedule re I.N.F.—aid for the Contras & a new subject—the Cong. has proposed an idea for a Nat. Ec. Commission. They would weight it a little in favor of the Dems. Frankly I don't like the whole idea but it may be hard to block. On another subject I suggested bringing Jim Brady into some meetings—maybe once a week. He's not mentally crippled & he's always had good ideas.

NSC—A letter from Margaret Thatcher who is trying to sell some Tornado fighter planes to Jordan. She needs our approval because of some American parts in those planes. She's trying to head off sale of MIG 29's by the Soviets. I'm on her side. Results of the Haiti election are in & while it was totally manipulated, the winner is probably the best choice for a reasonable govt.

We're going to ask Cong. next week for aid to the Contras. Non-lethal to begin with—then lethal in an escrow with a date certain for sending it if Nicaragua has fouled up the peace process.

Some optimistic talk about getting ratification of I.N.F. treaty but it will take a lot of work. Then rest of morning desk time until after lunch a brief Cabinet meeting about need for Cab. members to join in helping us get Cong. moving.

[Haircut; signed photos; left office early, noted, "upstairs to think about St. of the U. address tonite."]

The 7th & best St. of the U. I've never had such a reception with even the Dem's. clapping. I was interrupted 37 times by applause. The speech ran 43 min's. because of it. I surprised Nancy by singling her out in the gallery & praising her for her anti-drug activity.

Tuesday, January 26

Only a brief meeting before a session in Cabinet room with Repub. Congressional Leadership. Subject—the legis. on aid to the Contras we will deliver to the House tomorrow. They felt it was nip & tuck as to whether we could get it passed but they seemed to feel it was worth trying. Then back to the Oval O. for NSC. Colin reported on Contra supply plane shot down by the Sandinistas. There were 12 Nicaraguans aboard. One was captured alive. There were 4 bodies in the wreckage & 6 parachutes were found in the jungle. That accounts for 11 with 1 missing.

[Met with meeting of Republicans and Democrats on vote regarding aid to Contras, commented, "They made some useful suggestions, one of which I'll carry out"; photo with Egyptian newspaper, already did written interview; meeting with CIA Director Webster with report on ability to detect Soviet missiles; meeting with president of Iceland, Vigdís Finnbogadóttir; National Security Planning Group (NSPG) meeting on upcoming visit by Egyptian president Hosni Mubarak; meeting with staff on Contra aid.]

Wednesday, January 27

This became a day that shouldn't happen to a dog. The schedule sort of went out the window. There was a very brief staff meeting & then at 9:30 NSC went out the window so I could meet with some Congressmen on the Contra aid bill. Then at 10:15 I met with one of the 5 Soviet Soldiers who went over to the Mujahedeen in Afghanistan who wrote me asking for shelter in America. We got them out by way of Canada. This one is named Vladislav Novamov. He's a young man in his 20's & believes in God. He told me many young people in Russia do. He was with Sen. Gordon Humphrey & a lady from Freedom House in N.Y. They brought a list of others who have deserted the Soviet Army & want to come to America. Then it was more Congressional meetings always on same subject—Contra aid. They ran right up to lunch—some in Cabinet room, some in Oval O. Then right after lunch it was Congressional photo time. This means individual Congressmen bring in individuals or groups they want to have photos with me.

At 1:30 it was Geo. Shultz's regular meeting. This time on Israel & Middle East—he's working on a plan for a multi-nation conference on the Palestinians in Israeli occupied territory, Gaza & the West Bank etc. It sounds good & tomorrow morning I'm to phone King Hussein.

[Addressed Reserve Officers Association; meetings with congressmen; desk work.]

Thursday, January 28

A report that Jim Wright has yielded to a couple of his left liberal reps. & is against us on Contra aid. Tony Kennedy has been approved for Sup. Ct. 14 to 0 in the Committee. Howard Baker is going to meet with Dir. Sessions of the F.B.I. & Ed M. on charges of racist conduct by some F.B.I. agents regarding a Black agent.

A Civil suit has been brought against a Forest Ranger exec. on charges of discrimination against women in hiring Foresters. NSC was briefing for visit of Pres. Moubarak of Egypt. He arrived at 10 AM & the ceremony was held in the East Room because of cold. The usual receiving line then over to the Oval O. where he & I had a one on one for about 10 min.—then into the plenary in Cabinet Room for a 45 min. meeting. Subject for entire time was middle east & getting Shamir to agree to a conference. It was a good meeting & we heard some good suggestions re Iran.

After lunch I met with some Repub. Cong. leaders—only one is doubtful on Contra aid.

A photo with Office of Communications staff & another dept. I had lots of desk time for a change. I phoned King Hussein re the idea of a conf. He agreed to meet with Phil Habib who we are sending over. Then at end of afternoon we ran Doug Morrows TV spots featuring numerous celebritys promoting the spin offs we get from our space program. Upstairs early to exercise & get ready for State Dinner. And the State Dinner was like—the best yet. Everyone had a good time. Patti Austin sang—more guests danced than at any previous State dinner.

Friday, January 29

A late start, 10 A.M. 1st meeting some unhappy news about Ed Meese. An independent counsel found some papers re an oil pipeline near Israel & pvt. sector involvement that might show Ed knew of something that would be a violation of the ethics bill. We'll have to wait but from what I've heard so far I think Ed is more victim than violator of anything.

NSC—On our Contra aid problem we may be about 15 Dems. short of what we need. Then some talk as to whether we can jiggle our proposal so as to win some of those votes. I like our proposal the way it is.

[Caught up on reading; visit from W.H. spokesman Mark Weinberg and family; addressed meeting of state legislators; meeting with Shultz concerning ways to elicit cooperation from Shamir; decided to send Habib to see King Hussein; greeted college football

champions from University of Miami, noted, "I was the smallest man on the platform &
I'm 6'1" & weigh 190"; update on Meese situation from Culvahouse.]

Saturday, January 30

A bright, warm (50 degrees) day. Did radiocast at Oval O. on Contra aid. A
quiet afternoon & then off to the annual Alfalfa dinner. Don R. was the Alfalfa
nominee for Pres. Everything of course was comedy. As usual I closed the show.
My routine was very well received—more & bigger laughs than I had anticipated. Home at 11 P.M. & bed.

Sunday, January 31

A late breakfast then our date with Meet the Press & This Week. They've
become standard TV viewing for us. After a quiet afternoon—down to the Family Theatre on the ground floor at 5:30 for the Super Bowl game on big screen
TV. About 38 friends on hand & a buffet dinner between halfs. It was a lot of
fun & a good time had by all. It didn't hurt that the Red Skins won 42 to 10.

Monday, February 1

Back to the routine. Naturally we talked about the game. Then sadly some
further reports on the latest attack on Ed Meese (The Israeli—pipe line deal).

[Received notice of upcoming congressional votes.]

NSC—Frank Carlucci & Gen. Odom have recommended I appoint Gerald R. Young as Deputy D. of Nat. Security—I said yes. Word has come from
Soviets—Gorby misspoke when he claimed Soviets have a way of identifying
number of nuclear missiles, ships & subs carrying. There ain't no such device.
Roy Brewer—called Kathy to tell her Sandinistas freed our citizen Denby in
return for his pledge that he would come up here & attack Contra Aid. He's told
our people whole deal including that they are holding his plane until he does.
Well some Americans have pd. him the cost of the plane & he'll tell the truth
publicly. It will be interesting to see if press will carry it.

Now we're hearing that Marcos—claiming he's dying wants Aquino to let
him come home to Philippines. We're trying to find out state of his health because we know he has continued to use phone trying to organize a coup. Habib
had his meeting with Hussein.

[Timed speech for following night; issues lunch; addressed National Religious Broad-
casters Convention; desk work; visit from friends of brother Neil; meeting with Frank
Carlucci on budget.]

Tuesday, February 2

Ground Hog Day & Puxatawny Phil didn't see his shadow—Spring will
be early.

At our staff meeting some talk of coming Korean Inaugural. V.P. can't go & I certainly can't but George S. & Frank C. are booked elsewhere. I've suggested H. Kissinger.

NSC—Plans for mid east peace talks. Habib met with Hussein who expressed great pleasure at my phone call. Discussion rages among all parties about transitional plans being accompanied by permanent goals—but the plans become the permanent settlement. Hussein wants an international conference.

[Series of meetings with congressmen on Contra Aid bill; met Virginia governor Chuck Robb.]

After lunch more Cong. meetings including 2 that were with single Dem. individuals. All the time of course I was also making phone calls to Dem. Congressmen & Sens. Finally at 3:30 went over for a sneeze shot & then upstairs. Still taking calls. Howard called me & we agreed to change our proposal to allow the Cong. to vote on final disposition of lethal aid when time comes. This is a concession by us but we believe we need to do that if we are to have a chance of winning the vote on the package. An early dinner then over to the Oval O. to broadcast speech on Contra Aid. Only CNN carrying it—the 3 majors refused which is an unprecedented affront & reveals where their sympathies lie—right with the Communist govt. of Nicaragua.

[Attended dinner for Republican Eagles.]

Wednesday, February 3

[Received word of unanimous confirmation vote for Kennedy.]

Told of a controversial bill—the Alaskan Claims Bill. There is sentiment in several depts. that I should veto it. It passed by an overwhelming voice vote—our Alaskan reps. want it signed & I think it should be too (Late afternoon I signed it). I also signed letters to Congress telling them of my switch to allow them to vote on Contra lethal aid.

NSC—Colin spoke to House Foreign Affairs Committee on the Contra situation & publicly took on Rep. Hamilton who had followed my speech last nite on TV with a totally dishonest diatribe. Colin let him have it.

Possible indictments may be served on Noriega in Panama. Could happen tomorrow. They are drawn up by U.S. Attorneys. Ambas. Dick Murphy is calling on Thatcher & others on Middle East situation. We have a problem that could create trouble with Margaret. Israel wants to sell F 4 planes to Argentina—it would require our permission. Argentina & Eng. are still in a state of war. Then at 11 AM started Congressional meetings & phone calls. These were all 1 on 1—no groups.

At lunch continued phone calls.

After lunch Jim Brady came to Oval O. with 2 lunch guests of his—both English film producers of fine movies.

[Meeting with Shultz; greeted Super Bowl champion Redskins; made pass to receiver, commented, "Thank Heaven I threw a perfect pass which he caught on the run."]

Another meeting with a Congressman & then upstairs. Then watched debate & vote in House on Contra aid. We lost 219 to 211. High spot was Claude Pepper closing the 10 hours of debate & on the side of "yes." He was just great & must have changed some votes in our favor.

Thursday, February 4

[Attended Prayer Breakfast; Mrs. Reagan went to Indianapolis, Indiana, for "affair on drugs."]

Our 1st meeting was a discussion on Contra vote & now will Sen. vote on it or declare it dead. I phoned Claude Pepper to thank him—also Bob Michel. Then NSC & actually we continued on same subject. I had desk time until lunch. Rt. after lunch a meeting with Repub. leaders & Sens. plus 2 Demos. who are with us on Contra, Sen. Boren & Rep. Stenholm. After a lot of talk—we decided on going for it in the Sen.—this was what I wanted all along.

[Telephoned senators; photo sessions; had dinner with Maureen; watched Senate vote in favor of Contra aid bill; call from Mrs. Reagan.]

Friday, February 5

[Discussion of plans after Contra aid vote.]

Then Colin came in NSC time. We are putting NSC to work keeping watch on Ortega & the Sandinistas to see if they are going to play games. We'll have rat traps set for them.

We have worries about King Fahd of Saudi Arabia. [. . .]

Our Rep. at Haiti inaugural will be our Ambas.—Sunday Noriega of Panama will be indicted today. And in Japan—P.M. Takeshita seems to be trying to do something about Beef Quotas & construction contracts for Am. co's. Then desk time & receipt of gifts.

At 11:25 AM over to EOB to speak to 1988 Class of U.S. Senate Prog. This is a program instituted by Hearst Papers. Randy was on hand. Then back to West Wing for signing ceremony of Housing bill. At noon Warren Burger came in & presented me with a solid silver bust of Ben Franklin—he made the original when he was 15 yr. old. Then Dick Wirthlin came by with good polling figures. My approval rating is 57%.

Then it was time to go over to the WH. for lunch. Howard Baker & Jim insisted on helping me with my gifts. When the elevator doors opened—there was Nancy waiting. I thought it was because she'd been away in Indianapolis. I didn't even notice we were on the 1st floor. Then she opened the door to the St. Dining Room & there were about 100 or so of Cab., Staff, Friends etc. singing Happy B.D. to me. I was completely surprised. What a job Nancy had done.

Marvin Hamlish on 1 weeks notice had written a stirring march—The "Ronald Reagan March."

When lunch was over we went up & changed clothes for Camp D. When we came down to leave there were more than 1000 people on S. Lawn, a podium & mike & the Army Band. Another surprise—Howard Baker made some remarks & I responded, then everyone sang Happy B.D. & the Band played the new March.

Finally off to Camp D. & that night the movie was "Santa Fe Trail."

Saturday, February 6

The radiocast—some exercises in the gym. Some phone calls & a lot of reading in the icy afternoon. Then movie time—but 1st another surprise—they ran a part of a retraining film I'd made during the war. Then gifts from our guests—gifts from the S.S., Mil. aides, Medical group, Camp D. people etc. Finally got to the movie—Spencer Tracy in "Bad Day at Black Rock."

[Sunday, February 7: returned to W.H.; had massage.]

Monday, February 8

Usual staff time—spent some time discussing possible leverage we might have on the Contra situation in the House in view of the Sen. Votes. Then the matter of George Bush & the NSC meeting of Jan. 7. This is the one I believed he wasn't present because Shultz was in his chair. Now it seems the record shows he was there.

[NSC meeting with report that French president François Mitterrand can't make up his mind about seeking reelection, word that Contras are slowing down in Nicaragua and rotating troops in from Honduras, awaiting plan from Representive Wright, discussion of Noriega appearance on 60 Minutes, report from India that Mikhail Gorbachev thinking of withdrawing troops from Afghanistan; desk work and responding to birthday greetings; flew to North Carolina for drug forum at Duke University; returned to W.H.]

Tuesday, February 9

Well the Iowa Caucus is over (thanks be.) On the Dem. side it's Gephardt, Simon & Dukakis & for us Dole, Robertson & Bush.

Downstairs to the Dr. at 8:39—Dr. House was here & my new hearing aids. They are supposed to reduce background noise & I carry a small slim control that regulates volume etc. Then to the office—some talk about Gephardt's campaign trade bill. Jim Baker agreed to represent us at the Korean inaugural. Then talk about the idea of a Nat. Ec. Commission. We don't think it should be appointed 'til after the election. I'm not sure it's a good idea at all.

NSC—Brf. on F. Minister of Saudi Arabia due in to office in 20 min's. Then talk of Panama. Noriega went on TV to say our military should be removed from the Canal Zone. Pres. of Panama has announced he's asking Noriega to take a leave of absence. We have found out that Noriega's shouting that Admiral Poindexter had asked him to invade Nicaragua has been proved a lie.

[Report that Israel was planning new settlements in West Bank, commented, "We are going to try & talk them out of that"; short meeting with Saudi foreign minister; desk work; greeted Boy Scout leadership; interviewed by Edmund Morris for biography; National Security Planning Group (NSPG) meeting on START treaty and, as noted, "a call to action by everyone"; videotapings; attended reception for ambassadors.]

Wednesday, February 10

Of course the 1st subject was Gen. Noriega who is spouting lies all over the news.

Several depts. of course have a hand in foreign affairs, Commerce, Justice Dept., Defense & of course State. We discussed setting up a channel so that before any one of them goes off on it's own it goes through me.

Speaker Jim Wright has publicly announced that I'm invited to join him in working out a new Contra aid bill. I've replied that he should send his plan over to me to look at & I'll give him my ideas.

Discussion of my part in Nat. Repub. Convention. We all agree maybe I should make a speech opening day.

N.S.C.—Pres. of Panama announced he was going to ask Noriega to take a leave of absence—then he postponed the Cabinet meeting where this was supposed to take place.

U.S. attorneys in Miami are drawing up papers demanding extradition. In Haiti things going well with new President but we're a little curious—there seems to be no sign of Namphy. A letter is coming from P.M. Mulroney (Canada) re our mutual problem of Acid rain. This is a touchy problem.

[News of Irani-Iraqi action in Persian Gulf; visit from vice president of Switzerland.]

Desk time—then lunch with V.P. Bush. He told me he had called & squared tif with Dan Rather—both are keeping it quiet. Meeting with C.I.A. Director Bill Webster—briefed on some Soviet naval advances. Then meeting with Geo. S.—He spoke of mid-East peace plan & transition in W. Bank & Gaza. Most countries are very pleased but P.M. Shamir is a hold out. Then he talked of his upcoming trips & my coming visit to NATO.

An Ec. briefing—nothing much new. And finally a visit by publisher & several cartoonists & cover artist of the New Yorker magazine. They gave me a cartoon book of those that had appeared in New Yorker.

[Worked on correspondence.]

Thursday, February 11

Some talk of Q's I might get in Cabinet meeting when press would be there for photos etc. Well when time came & they shouted their Q's. I refused to answer.

Talked also about Cab. members following my lead & being neutral on our Presidential candidates.

[Discussed nominees for National Economic Commission; briefing for upcoming meeting with President Miguel de la Madrid of Mexico; taped radiocast; worked on backlog of correspondence; cabinet meeting on budget; photo session with Culvahouse and legal staff, advised not to hold reception for Archie Roosevelt, author of new book, in case of appearance of promoting it, decided to meet in office instead; greeted leadership of Masons; attended Conservative Conference dinner.]

Friday, February 12

Up 7:30—no breakfast—fog—had to drive to Andrews A.F.B. then A.F.1 to Calif. By having breakfast on plane we got our schedule onto Calif. time. Arrived L.A. 11:50 Calif. time. Lunch in suite at Century Plaza. Then some down time until 4 P.M. when we left for David Murdock's home for a Calif. State Repub. fundraiser. About 50 people at $10,000 each. Total raised $630,000. Gov. & Mrs. Deukmejian & Pete Wilson & wife were there. We did a receiving line—mixed & mingled for a bit & then went back to the hotel for early dinner, the TV news (Al Haig quitting the campaign & signing on in support of Dole. He did it with a blast at George Bush).

Saturday, February 13

Up early in morning & breakfast on the plane. Flew down to Mazatlan for summit with Pres. de la Madrid. We're both in our last year in office. He & I met 1st with only note takers & interpreters. They were good meetings on our mutual drug problem, on trade & on the threatened increase of Soviets in Mexico. He's aware of that threat & refused to allow Soviets to set up consulates along our border.

[Plenary meeting; went to ranch, Mrs. Reagan waiting there. **Sunday, February 14—Valentine's Day:** *rode; tree pruning.* **Monday, February 15:** *participated in filming of documentary by Ray Stark on Mrs. Reagan, commented, "I'm a supporting player."]*

Tuesday, February 16

[Rode; did tree pruning.]

Rcvd. a call from Ken D. with news of George Bush's win in the New Hampshire primary—that made my day even if I do have to be neutral publicly.

Tomorrow we leave the house at 8:35 for the flight back.

Wednesday, February 17

[Returned to Washington; Maureen visiting.]

We did get bad news—Lt. Col. Higgens, a U.S. member of the U.N. force in Lebanon has been kidnapped. It occurred in the vicinity of a Palestinian refugee camp. The Amal militia has surrounded that camp—beyond that we have no info.

Thursday, February 18

Started day at 10 A.M.—Some talk about budget & I deliver it ('89) to the Congress today. Then some talk about Contras & Jim Wrights proposal on Contra aid. He wanted us to join him—we told him to present his version & we might provide one of our own.

Then it was time to receive the top 6 from the Hill—Byrd, Wright, Foley—Bob Michel, Al Simpson & John Stennis. I gave each of them signed bound volumes of budget. Then we all went into Cabinet room where full leadership was gathered. Jim Miller gave a run down on budget which conforms to the agreement we arrived at last November & no new taxes. Then it was desk time & lunch.

[Visit from Bill Stout, commented, "who I believe is terminal with Cancer"; farewell photo with Ken Adelman, arms negotiator; NSC meeting to discuss Higgins kidnapping; telephoned Higgins's wife; tea party with Lucky and Archie Roosevelt; swearing-in ceremony for Anthony Kennedy as Supreme Court Justice.]

Friday, February 19

[Telephoned Drew Lewis to invite him to join National Economic Commission.]

Several terrorist org's. are each claiming they have our Lt. Col. Higgens. Nabbi Berri's Amal Militia is in the hunt to try & find him. Our Gen. Walters faced Pres. Mitterrand & P.M. Chirac separately about French security going into hotel rooms of pvt. Am. business people as well as govt. representatives searching for & stealing business & technical information. Both denied knowledge of this & said they'd look into it.

Efforts are going forward to set up an interim govt. in Afghanistan. Geo. S. & Colin leave for Moscow this afternoon & will bring this up with Soviet ministers.

Word from Canada's P.M. Mulroney that as host of Ec. Summit this June he's going to try & loosen up the format & avoid a set hard & fast structure. I'm for that.

[Received copies of economic report; briefing for and then meeting with West German chancellor Helmut Kohl; brief meeting with Shultz and Powell; went to Camp David; watched Yankee Doodle Dandy *(1942).* **Saturday, February 20:** *radiocast; exercised; took short walk; watched* The Treasure of the Sierra Madre *(1948).]*

Sunday, February 21

A pretty day to look at through a window.—It was cold & bright. Watched a couple of the talk shows. Lunch & at 2:30 wheels up for Wash. Back to the W.H. & some desk work. Then at 7:15 downstairs—(Black tie) for the annual dinner for the Nat. Gov.'s Conf. It was a great evening—the Gov.'s dinner always is. Dave Brubeck Quartet entertained magnificently.

Monday, February 22

Announced I'll be doing a press conf. this Wed. 8 P.M. Then got a call from Geo. S. & Colin. They were somewhat upbeat about their meetings in Moscow—3½ hours alone with Gorbachev. We learned Jim Wright is suggesting the Dem. Contra Aid bill may come up for a vote Thurs. It will of course not provide for my lethal material. Bob Michel is going to try for a bill that will provide for such.

[Advised of date of state dinner in June; appointed acting director of arms control; NSC meeting with discussion of Higgins appearance on videotape; noted Human Rights Commission declined by one vote to condemn Cuba, commented, "This will be brought up in current meeting in Geneva. This time—new head of our commission is Veladores—22 yrs. a pol. prisoner in Cuba"; swearing-in ceremony for chairman of the Commodities Futures Trading Commission; issues briefing lunch; addressed employees of Office of Management and Budget.]

Back to office for Cabinet meeting. Beryl S. briefed Cabinet on economy. Photo with office of legislative affairs. Dir. Will Ball is going to become Sec. of Navy. Present Sec. Webb resigned over Navy budget cuts. I don't think Navy was sorry to see him go. A haircut then a long taping session & finally upstairs to exercise, dinner & bed.

Tuesday, February 23

Began day at 9 AM in map room downstairs on live TV—the World Network & Voice of America—speaking to all of Europe etc. on arms situation & assuring world we would not retreat from our position of solidarity with NATO. Then over to office for staff meeting. It seems Jim Wright is not sure he'll be submitting his version of a Contra Aid bill to Cong. for a vote Thurs. as he'd promised. We're told he may not have the votes. Howard is going to call him & tell him to send the pckg. or we'll send one of our own. Deadline is Feb. 29.

[Rehearsed for upcoming press conference; report that Prime Minister Brian Mulroney of Canada wanted private meeting, decided to try and comply.]

NSC—Iraq's planes continue to harass our ships—we don't know whether they just don't know what they're doing or not. Some of our top mil. are in Baghdad trying to find out. Afghan resistance has come up with a proposed interim govt. for when Soviets leave. They propose 28 ministers—7 from present govt. 7 from refugees & 14 from various factions of the freedom fighters.

Pres. of Panama is still talking that Noriega must retire. On Feb. 22 (yesterday) Noriega's force arrested & detained 28 Ams. mainly our soldiers on their way from home to post. They were charged with wearing their uniforms in Panama & detained for a time. Curiously on Feb. 21 Noriega's daughter & her husband were arrested by our mil. police for a traffic violation in one of our compounds.

[Desk work; greeted law-enforcement officers.]

At 1:30 Lady Olga Maitland of Eng. friend of Margaret T. & Charles Price came in—she did an interview. She's a journalist—very pleasant. She also heads up a women's org. that is in opposition to the peacenik bunch in Eng.—very supportive of I.N.F. etc.

[Allergy shot; rehearsal for press conference.]

Wednesday, February 24

A little talk about Jim Wright's Contra aid bill which is a stinker. Then Colin—just back from Soviet U. came in for NSC—There was some talk of Q's. I might get at press conf. Then George S. came in for a brief report on his Soviet Trip. We're not going public—but the Soviets spoke about dismantling Krasnoyarsk Radar (big violator of A.B.M. Treaty). There was progress on Human Rts. & nuclear testing. Also on our effort to get U.N. Security Council to come down hard on Iran regarding their refusal to end war. Iraq has already said yes to ending it.

Shevardnadze will be here March 23rd. George is off again tonite—this time for Middle East.

[Met with congressmen opposed to Wright plan on Contra aid; briefing about and then meeting with Prime Minister Aníbal Cavaco Silva of Portugal, commented, "he's very much along our way of thinking"; addressed members of Export Now team; held press conference, considered that it was successful.]

Thursday, February 25

Well the Jim Wright Contra aid bill will not be voted on in the House today. It is clear Jim Wright knows he doesn't have the votes. Some of his people are looking favorably upon to be counted on. I assured him he was absolutely right & I'd be saying the same thing when I go to Brussels & meet with NATO next week.

[Desk work; photos with departing staff; briefing about and then interviewed by Lou Cannon; greeted Easter Seal poster child and adult.]

Finally Armand Hammer came in & presented report from "President's Cancer Panel"—he has chaired this since 1981. That ended office time. I went up—did my exercise & had dinner with Mermie. On ABC News Sam Donaldson had somehow gotten a sheet of my talking points for one of the meetings & used it to show (his interpretation) that I could no longer be trusted to attend meeting without having everything I was to say written out for me.

Friday, February 26

Had a letter from our Ambas. to Ireland Peggy Heckler. She says Irish are all for us on I.N.F. Some talk with Howard & Ken about Noriega. That really is a mixed up mess in Panama. Then we talked about possibility of trying a line item veto just to get the problem before the Sup. Ct.

We are having a problem with Jesse Helms on our appointment of Gen. Burns to A.C.T.A. Right now we have several unfilled vacancies at Dept. of Commerce. Verity's man Bruce Smart is leaving.

N.S.C.—Colin reports that Pres. of Panama has been ousted. He denies he is out but the Assembly passed a measure declaring he was no longer Pres.

Ceau Ceauceson of Romania has asked us not to extend "most favored nation" to Romania.

Soviets & Armenians are having trouble. Armenians want independence.

[Taped message for congressional candidate; meeting on Noriega; lunch with GOP state chairmen; attended gathering of Crusade for a Drug Free America; photo with departing staff member.]

Then upstairs. Nancy is down in Dr's. office having a little tuck taken where her mastectomy was done.

Saturday, February 27

Cancelled Camp D.—I did radiocast on Drugs (1st half) & Central Am. (2nd). Spent most of day wood shedding on coming NATO Summit. Rcvd. word Pres. of Panama has gone into hiding, his wife is in Presidential residence & his children are in our embassy. Noriega had ordered the Presidents arrest.

Maureen & Dennis are here.

Sunday, February 28

Quiet at home until 5 P.M. then another concert at the W.H. Theme, "Salute to Broadway." Artists—Marvin Hamlisch, Pearl Bailey, Jerry Herman, Larry Kert, Judy Kuhn, Pamela Myers, & Jerry Orbach. It was the best show yet.

*[**Monday, February 29:** discussed staff appointments; NSC meeting with reports on Panama and South Africa; addressed W.H. Conference on a Drug Free America; received annual report from Office of Management and Budget, noted $125 billion in savings due to administration program; lunch with Vice President Bush; desk work; photo with members of Boone & Crockett Club; photo with Mr. and Mrs. Mike Mansfield; meeting with Frank Carlucci; packed for trip. **Tuesday, March 1:** made speech about NATO summit; flew to Brussels, stayed at Chateau Stuyvenberg.]*

Wednesday, March 2

Breakfast in bed at 9 AM Then up & a briefing with Colin & George S. Then we left for NATO H.Q. & joined everyone at noon for opening ceremony

& family photo. And back to meeting. Lord Carrington called on me for a statement. At 1 P.M. broke for lunch. I ad-libbed one that was hailed by all. Then back to meeting—a number of speeches by heads of state. Finally a break & back to Chateau—then reassembled at Val Duchese for dinner—really a social time & a lot of fun. Back to Chateau & a good nights sleep.

Thursday, March 3

Up at 7:30—brkfst. in bed then up for a 9:15 brf. by Senior Advisors. And on our way to NATO H.Q. Working session began at 10 AM We broke at 11 for press session. I made a statement then turned it over to George S. for Q's. Then I had a brief meeting 1 on 1 with King Baudoin. He's a most gracious & unassuming man.

We had also a private bilateral with Margaret Thatcher, & one with P.M. (Belgium) etc.

Finally a farewell to our employees at NATO—then a farewell to several hundred Americans at airport & on board A.F.1 for 8 hr. & 20 min. ride home.

Nancy met me on lawn & here I am. Good news—Jim Wrights Contra plan was defeated while we were landing.

Friday, March 4—Our 36th Wedding Anniversary

[Slept late; greeted group of Mexican congressmen; had allergy shot; Howard Baker and staff organized anniversary party; desk work.]

Colin sent secret memo—our Frigate USS John Moore fired on Iranian Speed Boats in Northern Persian Gulf (2). No word as to any hits.

[Saturday, March 5: radiocast; farewell visit with departing speechwriter; helped sister of W.H. photographer with college project on speeches; massage; watched home movies.]

Sunday, March 6

A truly beautiful day—temp. in the 50's. Watched morning talk shows then lunch upstairs. Got a secret memo—Iranians fired on 2 of our helicopters. Fire came from an oil platform & some small boats. No damage & we didn't return fire. Well the afternoon passed slowly—then dinner in the study & the day was over.

Monday, March 7

The best day yet—sunny & in the 60's. Howard B. is in Tenn. so Ken D. stood in. We discussed the So. Carolina primary & George Bushes victory—also Super Tues. Some talk about Dems. attempt to increase the minimum wage. Every time it's been increased unemployment has gone up. I can support a certain amount of increase but not what the Dems. want.

We've decided Drew Lewis will be our choice to serve as Co-Chairman of the Presidents Ec. Commission along with the Dems. choice.

NSC—[. . .] Saudi Arabia is building a training center for ballistic missiles. We believe the missile will be Chinese. Wu says China will not sell nuclear missiles to Saudis.

[Discussion of Panama Canal and Noriega; noted congressional action on sugar imports; addressed VFW convention; then media executives on drug initiative; received latest poll results (did not note numbers); reception for media executives.]

Tuesday, March 8

Another bright day. Howard B. will be back around noon. Ken D. & I talked about the Grove City bill. It presents a problem. It passed overwhelmingly in both houses but it is a lousy bill I'd like to veto—but the veto would be overridden & this could adversely affect our candidates.

Ken talked about having some W.H. stag dinners for small groups of public leaders. Then I had some desk time. Colin NSC came in at 10:30 & brought in Ambas. Bandar (Saudi Arabia). We talked about missile base being built [. . .]. Then Chinese F. Minister Wu came for his meeting. It was a good meeting.

[Lunch with Senator Paul Laxalt (R-NV); spoke at reception for Winter Olympic team; brief cabinet meeting on NATO; economic briefing, commented, "I approved an outside committee to look at answers to Brady Commission report"; photos with officers of U.S. League of Savings Institutions; videotaping.]

Wednesday, March 9

A little time in the office. Howard & Ken & I discussed again the Grove City bill. There is no doubt my veto will be overridden. I must present at the same time specifics on what needs correcting in the bill.

We have a problem concerning Sam Pierce & an appointment in New York re his dept. Our Senators have mobilized in support of one candidate & Sam has a different choice. I have no choice but to give in to the Senators. Both candidates are top rate. Howard wants me to have a talk with former Sen. John Sherman Cooper. I've said yes. His subject is Nicaragua.

[NSC meeting with report of helicopter accident; also mounting trouble in Haiti; flew to South Bend, Indiana, for appearance related to Knute Rockne; flew to Washington.]

Thursday, March 10

Ann McLaughlin came in & we discussed the minimum wage bill. She's doing great as Sec. of Labor. We're going to try for a beginner's minimum lower than the regular.

Howard got Sec. Bennett & Ed Meese together on their differences on drugs. Then we discussed a pardon for John Howell of Texas. Pat Jacobsen is coming in today & she's been pleading for this for several years. He is no longer in jail & has been removed from probation but our legal staffs resist a pardon.

[NSC meeting with talk on situation between UK and Argentina, Afghanistan, in which Pakistan wanted temporary government in place when Soviets withdrew; word that Taiwan is building a nuclear weapon, concern that Contras were in short supply and needed help; briefing for interview with European journalists.]

Then lunch & afterward our NSPG meeting on Panama & Gen. Noriega. We're agreed we must not let down or lose on this. We must get rid of him.

[Photos with fund-raiser for presidential library; greeted Air Force Academy football team, departing staff member; dinner to raise funds for presidential library.]

Friday, March 11

Started a beautiful Spring day talking about Noriega. There are reports of Bulgarian & East German agents on the scene possibly with Soviet money on the scene. Then we switched to Contras—there are indications Sandinistas may be mounting a major assault across Honduran border.

In Geneva the U.N. Human Rts. meeting adopted a resolution to send a Human Rts. investigating team into Cuba—first time we've ever gotten a consensus & doing this.

NSC—Colin came in & we did some summing up on our troop situation in Panama. We agreed on anti-Noriega actions we're going to take. Then the subject turned to Shamir who'll be here next week & who is being bullheaded about our peace proposal. Suzanne Massey just back from Soviet U. came by. Her reports on attitudes in Russia are very interesting. She says my appearances on their TV what with the summit have made me pretty popular among the Russian people. Then desk time—lunch & over to E.O.B. to address Coalition on Trade. I blasted the protectionist proposals of Congress. Back to Oval O. & meeting with George Shultz. We covered subjects already talked about—But he is following release of my Noriega statement to press with a press conference. That took me upstairs for a clothes change & off to Camp D. A beautiful afternoon—a good movie that evening.

[Saturday, March 12: radiocast; exercised; finished reading Michael Reagan's book.]

Sunday, March 13

Partly cloudy—rained in middle of night but only a shower. A "secret phone call" from Colin—a request from Defense & State & NSC for permission to quietly send about 100 Marines trained in guard work to Panama to guard our oil facilities & aircraft. I said yes—we'll make no announcement of it. I'm phoning Mike R. about his book which is fine. Left Camp D. at 2 P.M.—back to W.H. before 3 & did some desk work for tomorrow. A phone call from both of us to Mike R. to praise his book.

Monday, March 14

Gray & cold. Our team (Advance) is back from Moscow. They report Soviets are different. They really want the summit & want it to succeed. They want the Start treaty to be ready for signing. Then NSC Colin brought Prince Bandar who delivered a reply to my letter to King Fahd (Saudi Arabia). [. . .] After he left the subject was Noriega. It's true Spain & Paraguay have indicated they will offer him sanctuary. Word is possibly Dominican Repub. will also. No word yet though as to whether he'll leave Panama.

Word from or regarding Afghanistan is that Soviets want accords signed in Geneva on March 15—tomorrow.

[Met with Manfred Woerner, new secretary general of NATO, considered him amiable; addressed Foreign Policy Analysis Conference on SDI; briefing lunch; met with Frank Carlucci on his way to Switzerland to meet with Soviet defense minister, noted, "He's going to have at them over such things as harassing our mil. etc."; went to dentist for teeth cleaning.]

Tuesday, March 15

Colin came in briefly to tell us S. Africa was going to execute 6 people—5 men & 1 woman who had been arrested quite a while ago in Sharpsville, S.A. P.M. Thatcher has notified she's protesting & I joined her. It seems there were some people killed by demonstrators. The 6 did not do the killing but under S.A. law they were equally guilty because they were part of the demonstration.

Then I was told that an Opera is opening at Kennedy Center & I should go but unfortunately it's the night of Gridiron. The Opera is about Nixon opening China.

[NSC meeting, reviewed schedule for European trip in May; briefing on meeting with congressional leadership, commented, "Got a little hot & heavy on subject of Contra Aid. We must do something quickly for Contras are in a bad way"; noted that signing over Soviet withdrawal from Afghanistan probably postponed due to Pakistan position, talk of Panama; National Security Advisor Colin Powell to go to Spain; addressed P. Ministers Club—foremost Jewish group; meeting with Swedish-American Chamber of Commerce; greeted Miss Teenage Oklahoma and other visitors; met with Shultz on his meetings with Prime Minister Yitzhak Shamir of Israel; addressed Inner Circle party fund-raising dinner, commented, "Surprise—2 really old friends were there as members—Joel McCrea & Frances Dee his wife."]

Wednesday, March 16

Bright but cooler. George B. back from campaigning so part of our meeting again. We arrived at decision—I will veto "Grove City Bill" even though it will likely be overturned.

[Planned to send Congress supplemental appropriations bill.]

We're facing a Congressional play to increase drastically the minimum wage. We're trying to get it changed—plus giving us a separate minimum for beginners.

[NSC meeting covered possible coup attempt in Panama, Sandanista-Contra military clashes in Nicaragua; desk work; briefing for and then visit by Prime Minister Yitzhak Shamir, ten-minute private meeting, commented, "I let him know we were determined to remain a close friend & ally but did want to find a way to bring peace to Middle East"; further meetings with others present.]

At 2 P.M. a meeting in Oval O. with Jim Stewart—he's back here doing some lobbying against the computerized coloring of old classic black & white pictures. He makes a good point. I'm going to look into it to see if I can be of help. Then brief visit with Japanese Painter who is doing a painting of me for Repub. H.Q.'s that is going to be named the R. R. bldg. Then upstairs for fitting of coat the tailor in Brussels made for me—it needed some alteration. And at 5 P.M. down to the Blue Rm. for a social type meeting with Repub. Cong. Leadership. It turned out fine & our people were aroused over the Sandinistas assault & crossing the Honduran border.

After dinner a call from Colin & Howard B.—they asked to come up & of course did. We've gotten a msg. from Pres. Azcona of Honduras. He asked for a show of support by us. He's getting it. I said yes to 2 Btns [battalions] of the 82nd airborne & 2 btns of light infantry plus a number of helicopters for moving troops.

Thursday, March 17—St. Patrick's Day
First talk was about 23 indictments against Col. North, John Poindexter, Gen. Secord & Harkim. Nothing much to say really.

NSC—The 1st plane load of our soldiers took off at 7:03 AM this morning for Honduras. By end of day tomorrow all 3200 will be in place. Last loads of 82nd will parachute in.

Noriega has given hints he'd like to speak to reps. of our govt. Our Chiefs want to send more M.P.'s to Panama—I've OK'd another 100 tomorrow.

[Noted Italy had new prime minister, Saudi Arabia enlarging and adding missile facilities; Irish officials arrived to celebrate holiday; took Jack Kilpatrick to Irish restaurant in Alexandria; stopped by Representative Wright's St. Patrick's party; made calls eliciting support for Grove City bill veto; photo session with departing staff; stopped in W.H. kitchen staff's St. Patrick's party.]

Friday, March 18
We've had an informal request—but no follow-up regarding our helicoptering Honduran troops to the battle zone. No action taken as yet.

[Discussion of appointment of deputy counsel; noted Congress was using administration budget as basis for negotiating.]

NSC—Panama—2 St. Dept. reps. are on way to Panama to meet with Noriega at his request. Later in day we were told they'd had a lengthy meeting & made plain we want his departure—soon—from Panama. He's meeting now with his lawyers & has asked for a second meeting maybe—tomorrow.

Several of Noriega's staff have defected including his personal pilots of 727 & helicopter. They've asked us for asylum. Pilots claim they flew 48 tons of Soviet weapons & ammunition from Cuba to Noriega. Defectors are now in Fla.

Word from Honduras is that Sandinistas so far have not gotten Contras store of supplies.

Last word from Colin was that Saudi's missile plans are going forward.

[Received report from commission on privatization; interviewed by reporter from Paris Herald Tribune; lunch with Vice President Bush; photos with departing staff; usual Friday meeting with Shultz, indicated he'd like to return to Middle East, reported on diplomatic meeting with Noriega; greeted members of Young Republican National Federation; made calls to congressmen regarding Grove City bill, commented, "It's just possible we might get the veto sustained even though I made it with the belief we had no chance." Saturday, March 19: read most of the day. Sunday, March 20: watched panel shows, read.]

Monday, March 21

Just Howard—Geo. is back on the campaign trail & Ken is taking his little boy to the hospital for some kind of surgery.

Howard & I talked about chance to get Sen. Kassebaum's Contra Aid bill through Cong. (we have a fair chance) & same goes for upholding my veto of the Grove City bill. We face possible trouble from Cong. on I.N.F.—they might delay ratification 'til after Moscow Summit just out of sheer orneriness.

[NSC meeting included discussion of location of U.S. troops in Honduras and Contra aid bill; desk work; telephoned President José Azcona of Honduras to congratulate him on joint operation; photo with Advance Office staff; issues-briefing lunch; signed proclamation for Afghanistan Day.]

Some more desk time—phone calls to nominate 2 Ambassadors & 3 calls to congressmen soliciting support for my veto. Then a meeting with the Sens. & Reps. who went down to El Salvador as witnesses of yesterdays election. The guerillas did their best to louse things up—death threats, shut off power in the entire country etc. A haircut, a taping session & upstairs to a lonely White House. Nancy is in N.Y. Did my exercises & now Rex & I are in front of T.V.

Tuesday, March 22

[Met with Republican Congressional Conference; staff meetings.]

Then NSC—Colin was tied up, John N. discussed our nuclear agreement with Japan & the coming Sen. Vote. Then Panama & the strike—it is 90% effective. French Bank is threatening to make money available to Panama which would louse up our plans. The Nicaraguan cease fire talk has started. Sandinistas want the cease fire to be accompanied by Contras disarming. That just won't work.

[Spoke to local GOP officials about importance of Federalism program; watched Senate override of Grove City bill veto; photo with editors of Defense Daily; *visit from Basque president José Antonio Ardanza; ambassadorial formalities; attended party fund-raiser at Shoreham Hotel.]*

Wednesday, March 23

This is a day that shouldn't have happened—but it turned out O.K. It was a jammed up schedule but fortunately with an early quitting time. The only break for desk time was ½ hour at 10:30 AM

[Reviewed new summit schedule submitted by Soviets.]

Then we had a little post mortem on the veto.

On I.N.F. Sam Nunn, a highly over rated Sen. is kicking a tantrum which could cause trouble in getting it ratified.

[Short NSC meeting; visit from ex-president Chun Doo Hwan of South Korea, commented, "I complimented him on how much change he had brought about that led to the peaceful transfer to a new Pres. by way of election"; spoke to American Business Conference; briefing on Eduard Shevardnadze's visit; met with Shultz, decided to proceed with travel advisory on Middle East.]

Panama is the only country that has an embassy in Israel. We believe Israelis have some connections with Noriega. We're going to ask them to urge him to leave Panama. We discussed some Ambas. problems also some concerns we have over cease fire talks in Nicaragua. There may not be a cease fire. Finally Afghanistan & the accords negotiated with Soviets by Pakistan. Sen. Byrd is upset & doesn't seem to have a grasp of the situation.

I went out on the lawn & had a photo taken for use in the Leukemia Fund. effort. Then back in for some bill signing etc. & finally upstairs. Nancy is back from N.Y. & all is right with the world.

Thursday, March 24

Now Jim Wright is all for Contra non lethal aid since the Sandinistas have announced a cease fire for 60 or 90 days. Some of us are afraid some games are being played.

NSC—Our meetings (hours of them) with the Soviets are ended. We didn't make much progress. Sticking points for Soviets—Cruise missiles—air launched & sea launched, plus right of inspections. S.D.I. suddenly became an issue again. Soviets seem to want a treaty using lines from the Dec. Joint Communique having to do with Space & Space Defense. Cong. is beginning to put together a Contra Aid Program.

[Met with leader of opposition party in West Germany; short economic briefing with positive figures.]

A half hour of desk time then lunch with V.P.—pleasant as always. When Dole drops out of race I want to go public & declare that Geo. B. has been telling truth about "Iran-Contra." He played no part in whatever went on.

[Visited school in Virginia; returned to W.H.; photos with senior NSC staff; GOP candidates; greeted National Republican Heritage Council Board; visit from Anna Chennault; exercised.]

Friday, March 25

[Discussion of opposition senators, including Sen. Wilson (R-CA.) and his attack on INF treaty; NSC briefing for and then visit with President Joaquín Balaguer of Dominican Republic; turned down people who wanted to join trip to Moscow, specified, "Paul Laxalt, Armand Hammer, Suzanne Massey etc.—We have to say no to each one. No one we could say yes to without having to take the others"; discussion of sugar program and of Nicaragua, noted, "evidence that Contras gave in to Sandinistas in agreement because no longer believe they can trust us. All due to Congress shutting off aid"; meeting with Prime Minister Balaguer; greeted Archbishop of Greek Orthodox Church; addressed Center for Study of the Presidency.]

Then Geo. S.'s regular meeting. He started with Nicaragua & feels strongly that we must have an aid program fast & a fast track clause that we can if called upon provide lethal aid. On Noriega he proposed additional things we can do. One would be to get Pres. Delvalle to announce a post Noriega govt. He asked me to O.K. another mid-east trip over Easter so he can explain to the people there what's in our peace plan.

Sat.—tomorrow he is meeting with 2 distinguished scholars who are Americans—but also Palestinians. Shamir is very upset with him. Last item is—for some reason King of Saudi Arabia is on the outs with our Ambas. who is a darn good man. We'll bring him home though & nominate a replacement. He speaks flawless Arabic & for some reason this upsets the Arabs.

[Photo with departing staff member.]

That did it—upstairs for exercise & evening. Part of eve. I'll be alone. Nancy has to do a publicity thing for "drugs" at the Hockey game. I watched it on TV—she appeared on the ice with a group of young people & the mascot Snoopy—spoke of the contribution to the drug fight the Hockey League is making. Then she took a hockey stick & sailed 2 goals past the goalie. It was very nice & the crowd loved it.

*[**Saturday, March 26**: slept late; radiocast; spent afternoon reading book on Channel Islands in California; attended Gridiron Club dinner. **Sunday, March 27**: watched TV and read; dinner with Maureen.]*

Monday, March 28

[Met with Republican congressional leadership regarding Contra aid.]

The Sen. Foreign Relations Committee is starting mark-up on I.N.F. today & tomorrow. It should be on the Sen. floor right after Easter week. We're expecting the trade bill to come out of conf. this week. Then some top secret info. on Meese investigation. Howard thinks it might call for his resignation. D--n he's done nothing wrong.

NSC—S.D.I.—papers put some false statements together & came up with a story that we were backing away from it. Taint so! Panama—our actions of 2 weeks ago are working or aren't they? Do we need additional action? One idea is to get 8 Panamanians of distinction here in Am.—putting them on our mil. base in Panama to lead opposition to Noriega—there in Panama. Then a short desk time & a meeting with Bill Webster—reports by C.I.A. He let me in on a secret of an information source by way of a country friendly to us. Then an early lunch & at 12:45 off for Richmond Va. On arrival (by helicopter) to a hotel where I met with a group of Repub. leaders—handshakes & photos. Sen. John Warner was with me on round trip. After that to Reynolds Aluminum plant. First a tour of the factory—it was fascinating. Then to H.Q. & a forum on trade etc. I addressed about 600 employees then about 2000 were listening outdoors. I dropped by & said a few words to that crowd & home & mother. Mermie still here for dinner.

Tuesday, March 29

Rcv'd. 2 resignations from the number 2 & 3 man in the Justice dept. Their resignations just said they wanted to return to civilian life. We think they are leaving because they believe that special investigator is going to hand down an indictment or a censure on Ed Meese. I hope not. Ed may have used bad judgment a couple of times but he is honest as the day is long.

[Meeting (does not specify with whom) on prospects for INF in Senate; meeting with Republican congressional leadership; short meeting with NSC; briefing for group interview with journalists not based in Washington, participated in interview, then turned it over to Howard Baker.]

Lunch on the patio with George Will. He's been critical of our tax policy & some other things. I hope I straightened him out.

[Participated in ceremony honoring cancer survivors; National Security Planning Group (NSPG) meeting on Noriega and to consider military options there; interview with biographer Edmund Morris; meeting with senators on budget, commented, "I've agreed to go public to explain need for more money for Space & science"; allergy shot; measured for bust for presidential library.]

Wednesday, March 30

Ed Meese on ticket again—Howard B. thinks we're in for a big fire storm.

NSC—Contra funding—they are coming close on the hill & we may be winning. Plan calls for our AID program to deliver to Latin Am. then some other means must be found to get the supplies to the Contras—we're talking about Humanitarian supplies. If Lethal aid is required—the Dems. want me to write a letter to the "Speaker" & he'll have to respond in 10 days. It's possible we may have a deal by end of the week. I.N.F. treaty seems to be coming along in the Sen. OK. I received a phone call from Pres. Zia of Pakistan. I think we may be close to a deal on Afghanistan.

[Greeted science students; swearing-in ceremony for Secretary of Navy William Ball; met with Shultz; filmed scenes for documentary to be shown at presidential library.]

Finally upstairs—no Nancy—she's in Calif. now. But I'm entertaining a dinner—some prominent citizens—a mix just to hear their ideas. Ms. Cathy Black, Howard B., John Connally, Col. Frank Borman (USAF Ret.), Robert W. Galvin, Daniel J. Boorstin & Walter Cronkite. The dinner turned out great—good conversations & humor also. An early evening over by 9.

Thursday, March 31

Brief talk about the House passing Contra Aid by a sizeable margin. Also change in Moscow trip schedule—we leave for Helsinki May 25 instead of May 26.

NSC—Les Aspin of House Armed SVC'S. Committee took $300 mil. away from M-X rail launchers & added it to the Midget Man missiles which only the Congress wants—the mil. doesn't want them. F. Carlucci had already raised them by $200 mil. Now the Cong. also removed $800 mil. from our figure for S.D.I.—this is a terrible setback.

At 9:45 down to the Situation room for an NSPG meeting. Main purpose was on problem of Philippines making noises like we should remove our bases at Clark Field & Subic Bay. Of course if we wanted to raise the rent a fantastic amount they'd reconsider. Then we got on Noriega again & cleared up a few items that had been suggested but that would have been very counterproductive.

[Lunch with Vice President Bush; interviewed by student journalist; launched Baldridge National Quality Award.]

A Cabinet meeting during which word came the Senate had passed the Contra aid bill 88 to 7.

[Photo sessions with former staff member from California; with representatives of W.H. athletic center; bill signing; videotapings; Michael Reagan visiting.]

Friday, April 1

Getaway day. Mike taped an interview & was back at the W.H. in time for a 10:15 AM takeoff in Marine 1. At Andrews we left on A.F.1 for Calif. It was an uneventful trip. Whoa! I forgot before we left the W.H. I signed the $48 mil.

Contra aid bill. Landed at Point Mugu & helicoptered to Ranch. Mike of course headed home by car from Point M.

[Relaxed at ranch, commented, "I got into Jeans & then met Nancy at the helipad when she came in from L.A. an hour later." **Saturday, April 2:** *rode; cut trees.]*

Sunday, April 3—Easter

And there was a rabbit on the lawn when we sat down for breakfast. It's also the first day of daylight saving.

Another ride—then Dennis & I (after lunch) went back up on Mt. Rhino—more clean up. Ken Duberstein is coming for dinner. I just called Sydney his wife to say Happy 39th Birthday & to tell her from now on no more birthdays, just anniversaries of this one. Ken arrived & it was a very pleasant evening.

*[***Monday, April 4:*** *helicoptered to Newport Beach for library fund-raiser at home of William Lyon.]*

Tuesday, April 5

The best day yet—warm & bright. A good ride with a beautiful view whichever way we looked. Barney is back. A call from Howard early to let me know Ed M. was doing a press conference today & that we believe we have some solid names as nominees to replace the quitters in the Justice Dept. Then Colin called on secret phone to tell me the Soviets seem to be going ahead with pulling out of Afghanistan even though they seem unwilling to sign the Geneva accords.

Also he reported that Jesse Jackson is denying he tried to get into the matter of Noriega—even though we know what was in Noriega's letter to him.

After Lunch, Dennis, Barney & I as well as Dr. Hutton who always joins in when he's here went back up on Mt. Rhino to continue the pruning & clearing we've been doing there. It's now 5:25—I'm showered & in P.J.'s already but with daylight saving it looks like high noon.

Wednesday, April 6

Same schedule but a really warm day. Our morning ride, then after lunch Roy Miller & his lady assistant came with our income tax forms. We signed Fed. & State forms. Then I hurried up to Mt. Rhino to join Barney & Dennis. It was a short session but we got some more pruning done. A call from Ken D.—he's on his way back to Wash. & tomorrow Howard Baker will be here with us for dinner tomorrow nite. A lot of paper work was sandwiched between meals, riding & pruning. I almost forgot—we had a visitor before the ride—a Western Artist drove 1500 miles from Oklahoma to deliver as a gift 3 of his paintings. One was of "No Strings," then "Gaualianko" & finally "El Alamein." They are beautiful & I'm excited about having them.

[Thursday, April 7: rode; finished major pruning job; Howard Baker arrived for dinner, no progress on Noriega, commented, "I've O.K.'d further financial moves for whatever they are worth."]

Friday, April 8

[Rode.]

After lunch it was next to final day at Mt. Rhino—more pruning & cleaning up. TV news had our inc. tax on showing we paid less tax than last year on greater earnings. So we have tax reform & it works even for Presidents. Word came Judge sentenced Lyn Nofziger to 90 days in prison & $30,000 fine. I still say he did nothing wrong. Dinner & Early to bed.

Saturday, April 9

[Rode; read radiocast; photo session with military staff; received honor from California Highway Patrol Association.]

Rcvd. an explanation of the ABC special event alleging—h--l they outright accused us (based on evidence from unnamed sources) of using planes in Contra aid in cahoots with Israel & those same planes then ran drugs into U.S. It's all a phony story & just never happened.

[Continued pruning. **Sunday, April 10:** *back into, as noted, "store bought clothes"; flew to Las Vegas; addressed National Broadcasters Association; flew to Washington.]*

Monday, April 11

[Started work at 10 am, rather than 9 am, due to jet lag; Howard Baker reported on recent stock market increase.]

Howard reported Sam Nunn & Sen. Helms are raising h--l about I.N.F.— trying to block its ratification by the Senate.

N.S.C.—Colin & John Negroponte came in for their meeting. Colin told me George S. would report in a later meeting on the Afghanistan situation. Sen. Gordon Humphrey is objecting to the Geneva accords which George S. will be signing—possibly this Thurs. The Sen. thinks we're selling out the Afghans. He's wrong as H--l. We are not deserting them at all & if Soviets continue arming the Kabul govt. We'll continue arming the Mujahadeen.

A problem with the French has come up—they have a high tech firm that is selling high tech. products to the Soviets similar to the Toshiba deal.

Honduras—quiet now but leftist students still trying to get riots going again.

A witness at Sen. Kerry's hearings on drugs claimed P.M. Seaga of Jamaica was involved & that it's been covered up. Seaga is naturally upset & has a right to be. The story is a phony. Ethiopia is fighting some internal problems & has pulled all relief agencies out of famine area which can mean no food for ¾ of a million people.

[Issues lunch; meeting with Shultz, Carlucci, and Webster on Middle East and Afghanistan; greeted men's and women's college basketball champions; dinner for King and Queen of Sweden, entertainment by Barbara Cole after dinner.]

Tuesday, April 12

Another 10 AM day—they're spoiling me.

[Discussion of budget; trade bill nearing acceptable state; rejected bill calling for harsher treatment of Japan due to Toshiba matter.]

NSC—Kuwait & Jordan have both volunteered to storm the hijacked 747 on Cyprus. It won't work without causing the slaughter of the hostages.

Believe it or not we've been approached by Iran to see if we can't have a better relationship. In Honduras Pres. Ascona is having some trouble over his release of drug king-pin Mattja. Our Sec. of Air Force has been invited to the Soviet U. to visit their space center. Of course they then want a visit to ours. At 11:30 another meeting with leaders of hard core Conservative leaders Paul Weyrich, Gen. Graham etc. Half hour meeting became an hour. As usual they had us on the wrong side in Afghanistan settlement, Mozambique, Chile & Angola. It's amazing how certain they can be when they know so d--n little of what we're really doing. A late lunch for me then my postponed meeting with Frank Carlucci. He reports Pakistanis are most pleased with us & eager to cooperate in helping us in the Afghan affair. Then down to the situation room for an hour of NSPG meeting. Subject Panama. We're all agreed we must not fail in the Noriega matter. A group is going to explore additional steps we might take.

[Desk time; visit from personal physician Dr. Burton Smith and his wife; attended reception for PBS series "Concert at the White House"; dinner with Ted Graber.]

Wednesday, April 13

Back on regular time. Nine o'clock meeting—1st A. B. Culvahouse with a message I had to O.K. It seems that Eastern Airlines is bankrupt & cannot be approved as safe to fly. Even our military has ruled Eastern cannot be used by mil. persons. Then Ken Dub. came in, Howard B. is still in Tenn. We are somewhat encouraged by progress in cleaning up trade bill—but it still isn't something I can sign.

[Hopeful of progress against Senator Sam Nunn's resistance of INF Treaty; approved Ken Khachigian as writer for three speeches; briefing for visit by Prime Minister Lu Kuan Yew of Singapore.]

Some strange antics have been going on at our oil storage base in Panama. Last night Marines were reported to be firing on invaders. We don't know yet if there was a real fire fight or something cooked up by Panamanian radio. No one was hurt. Confusion ended about 1 AM.

Three more Soviet defectors (deserters) in Afghanistan have arrived in U.S. by way of Canada.

We have a problem with Sen. Jake Garn who is determined we must take some very punitive action against Japan's Toshiba Company. We think it is mindless interference. He's trying to get in the trade bill. We think we've taken action already. At 11 AM went over to E.O.B. to address "Young Presidents Org." They are being briefed by a number of our people. Then it was Lee Kwan Yew time. A short meeting in Oval O. & then over to the W.H. for lunch. It was really good to see him again. He's been working to help Pres. Aquino in the Philippines.

[Addressed American Society of Newspaper Editors.]

Thursday, April 14

Dem. Leadership in Cong. has asked for meeting with Howard B. & Jim Baker. I think they want to find out where I'll decide enough has been done on Trade bill to get me to sign. Howard B. leans toward reducing numbers & having me attend the meeting.

NSC—I'm going to call (Colin's request) P.M. (of Pakistan) Junejo. Later in morning I did call. We were just recognizing each other's part in Afghan agreements George S. signed in Geneva today.

One of the hostages released from the Hijacked plane now in Algeria identified one of the terrorists who got on plane in Iran as the murderer Izz Al Dim who killed our Navy man on the TWA 847 hijacked in 1985. We have notified Algerians we have a warrant for his arrest.

Colin and Howard are going to the hill to meet with Sens. on I.N.F. A 2nd INF issue has arisen—we had agreed to ban conventional cruise missiles (ground launched) as well as nuclear cruise missiles. Some Sen's. are objecting.

[Received new polls numbers, approval steady at 55 percent, commented, "Some figures reflected, I believe, result of Dem. Presidential candidates fairy tales. For example that Dems. are better than we are on cutting Fed. spending"; met with party donors; lunch with Vice President Bush, running unopposed in remaining GOP primaries; met with former Senator John Cooper (R-KY) regarding Contra aid; presented award to Teacher of the Year, and outstanding students; received word of U.S. frigate damaged by mine in Persian Gulf and of a car bomb near a USO building in Naples, Italy; dinner with Ted Graber and Maureen.]

Friday, April 15

[Negotiation with Congress over Trade bill.]

Next subject was I.N.F.—This time Colin & Howard met with Bob Dole & Byrd. We want ratification before I leave for Moscow. Sen. Nunn is still the roadblock with some of our right wingers on his side. The issue is over conventional warheads on cruise missiles.

Then a little time on 2 of our Sens. Pete Wilson & Al D'Amato who want us to decertify Mexico on grounds of not doing enough to fight drugs. I won't do it. Mexico is trying I know but is dealing with a lot of corruption.

NSC—Report on our Frigate hitting a mine in the Persian Gulf. List of injured up to 10—only 1 serious but Dr. is convinced he will live. He's burned on 90% of his body. Mines had only just been laid—48 hours earlier ships had gone through on that same course. Later in day we had an NSPG meeting. Our mil. is now studying what retaliatory measures we should take.

There was a bombing near an installation of ours in Spain. In Naples the Am. sailor killed was a 21 year old woman. Colin sounded a little more optimistic about Sen. action on I.N.F.

Later in morning had a meeting with Jim Baker on the Trade bill. I'm afraid I'm going to have to veto. Then the NSPG meeting—we all agree on retaliation, new Chiefs of Staff are working on targets & details.

[Had lunch with Vote America group; met with large group of lieutenant governors, then with Shultz with report of Afghanistan and his upcoming meeting with Eduard Shevardnadze; videotapings; went to Camp David. **Saturday, April 16:** *radiocast; rode in cold; watched, as noted, "the usual Sat. nite movie."]*

Sunday, April 17

Came home early & had lunch at the White House. Colin had called me late Sat. & told me the Chiefs were ready to brief me on our P. Gulf retaliation. We are going to blow up two & possibly three oil platforms—3 of our ships doing job. Also one Iranian Naval vessel. In each case we'll give time for their people to get off. We seek no killing—just the destruction of the targets. So our people will come to the W.H. at 8:30 P.M. Then at 9 P.M. 5 Congressional leaders will join us & be notified. Then at 1 AM tomorrow morning our time the action takes place.

In the meantime I have to get dressed up for a sitting with the painter who is doing my W.H. portrait at 2:30 P.M. this afternoon. That only took an hour.

At 8:30 P.M. our gang from George B., George S., Howard, Ken, Colin, Frank C., Admiral Crowe & a few more came by & we had the final briefing on our retaliation for Iran's hostile acts in the P. Gulf (Our frigate hit an Iranian Mine). Then at 9:00 P.M. Congressional leaders—Jim Wright, Robert Byrd, Tom Foley—all Dems. & Bob Dole & Bob Michel joined us. We told them what we planned—made it clear I had given no order yet & so we consulted. They appreciated that I would have limited our strike—limiting us to oil platforms & no ships. At end of meeting I gave order to take out the platforms & the ship or ships after giving a chance for crews to abandon ship. We didn't want to kill—but to destroy those properties. At 1 A.M. our time the operation began.

Monday, April 18

Awakened at around 5 AM for call on secret phone. It was Colin reporting we'd taken out 2 of the 3 platforms—3 of our Naval vessels there while 3 more awaited Iran ship or ships to come out of harbor. A little while later Colin called again to get my O.K. on diverting ships bound for 3rd oil platform to ships off Iranian harbor. I did it.

[Staff meeting regarding attack, drug issue, trade bill, INF treaty; Meese reported on trip to Latin America on drug issue; NSC meeting regarding trip of envoy to Panama; unconfirmed report that Amal had killed Colonel Higgins in Lebanon, commented, "Actually Amal is on our side & has tried to help get him back"; intent to sign restatement of relationship with Israel; greeted Youth Volunteers; had lunch with GOP state chairmen; addressed Associated General Contractors; dressed for Paul Laxalt's annual "Lamb Fry" dinner.]

Reports coming in on P. Gulf. We have sunk or disabled 3 vessels now & disabled 2 smaller craft.

The Lamb Fry was as always a lot of ribald fun & I was home before 10 P.M.

[Tuesday, April 19: word on scheduling of INF vote; NSC meeting regarding Persian Gulf action, possible fall of Danish government due to parliamentary action to ban nuclear ships in ports; Republican leadership meeting with report on Persian Gulf action, noted, "Everyone is approving what we did," also discussed trade bill; attended ceremony honoring law-enforcement officers who died in the fight against drugs; economic briefing with James Baker, Howard Baker, Kenneth Duberstein, and Alan Greenspan, with consensus on positive outlook; greeted sixteen former Heisman Trophy winners supporting Republican Senatorial candidate in New Jersey, Pete Dawkins; photo sessions, greeted delegation from the California Agricultural Leadership Program; attended reception for Dawkins; addressed Electronic Industries Association dinner.]

Wednesday, April 20

[Staff meeting to discuss trade bill.]

As to I.N.F. we're worried they'll delay ratification 'til after the summit.

NSC—Dept. of Defense will make public—photo of our high tech newest plane, the B-2.

Iranians fired a missile at Kuwait—possibly a Scud. Tunisians are talking about taking Israeli attack on Abu Jihad to the U.N. [. . .]

[Signed proclamation for Law Day; lunch with Vice President Bush; briefing for British TV interview; had interview, commented, "It was very pleasant & I enjoyed it I hope I don't have to go to Eng. to see it."]

Well then I met with Burns & Weld about their resignations from Justice Dept. They professed great affection for Ed Meese & belief that he was innocent of any wrongdoing but claimed the dept. was dead in the water.

Then we met with Ed. He gave chapter & verse of the dept's. activities which completely rebutted their story.

[Telephoned congressmen seeking support of upcoming veto of trade bill.]

Thursday, April 21

A little review of yesterday's meetings with Burns, Weld & Ed M. Then the trade bill. Bob Michel got a rule passed that he could move to recommit the bill—to the conference to see if under my threat of veto the conf. would drop the proviso for plant closings.

[Word on congressional action.]

NSC—[. . .] Apparently the Soviets have already moved out.

Kozak, our emissary to Panama reports that Noriega would step down if we lift Ec. sanctions immediately & if law limiting Mil. commander to 5 years in office is made retroactive so it would apply to him. And he would live in Panama as a retired—former Commander.

Pakistan—the explosion of the ammo dump destroyed about $100 mil. of armaments. But now we know it won't affect the Mujahadin. All evidence now indicates it was an accident.

[Report of attack on oil tank base in Panama.]

Captors of Col. Higgens in Lebanon say they are going to put him on trial. V.P. suggests we remind U.N. he was there as part of the U.N. detachment.

We are engaged with Soviets in dividing line off Alaska in Bering Sea. I sent memo to Geo. B. authorizing him to negotiate on this in Moscow.

[Signing ceremony for proclamation of support for Israel; photo session with Soviet media and government representatives, commented, "They've been meeting with Charles Wick on better communication between our peoples"; flew to Springfield, Massachusetts, for talk to World Affairs Council; returned to W.H.]

At 7:35 P.M. departed for Wash. Hilton Hotel & annual W.H. Correspondents Dinner. But before that a call from John Negroponte on secure phone. Frank C. had notified him our forces in P. Gulf had discovered an Iran supply ship which we believe has been used to lay mines. Frank wants my permission to board the vessel. If any sign of mines, remove crew & scuttle. If it refuses to halt & put's up opposition, sink it. I approved wholeheartedly. The dinner at the Wash. Hilton was, as always, very pleasant & Yakoff Smyronov who entertained was very funny as always.

Friday, April 22

[Discussed prospects for sustaining veto of trade bill and ratifying INF treaty.]

A little talk of Ed M. & the press stories.

Lamar Alexander is writing a book—sent Howard a few lines from his manuscript about me to see if his memory was correct—it was.

NSC—Briefing on Thailand foreign minister visit.

Proof now that Iranian supply ship Frank called me about laid the mines.

Unfortunately under cover of darkness it got into their harbor. Rcvd. a cable from Geo. S. reporting on Moscow meetings.

Then the Foreign Minister & his team arrived. It was a short visit but we talked about Cambodia their problem with being the 1st stop for all the Vietnamese refuges etc. Then an NSC meeting in Cabinet Room—entirely on Canada & next week's visit by P.M. Mulroney.

[Addressed American Legislative Exchange Council; went to Camp David, noted, "Ran a Golden Oldie—Astaire & Rita Hayworth in 'You'll Never Get Rich.'" **Saturday, April 23:** *radiocast; ride canceled due to weather.]*

Sunday, April 24

[Returned to W.H.; Maureen and Dennis visiting.]

Five o'clock & downstairs for "Performance at the W.H." Some long time greats like Mary Martin on hand for a kind of memories of broadway show. It was great & Mary sang—"My Heart belongs to Daddy." Then upstairs & dinner with Mermie & Dennis.

Monday, April 25

[Received report of explosion onboard submarine.]

Ken conducted 1st meeting—Howard on his way back from Tennessee. We discussed coming meeting with P.M. Mulroney. He has some pol. difficulty in Canada. Nancy & I are having them for lunch while they are here. I suggested a W.H. photographer get a shot of that which we'll release to the press.

[Noted number of amendments proposed for defense bill, and that Senator Dan Quayle (R-IN) switched to support INF treaty; NSC meeting, effect of Danish ban on nuclear vessels.]

Colin reported how Soviets raised h--l about my speech in Springfield Mass. last week. "Tass" was particularly critical of me. Both sides in Moscow feel we can't pull "Start" treaty together for signing at summit but all agreed we go right on working on it. Progress has been made on Chemical Weapons Agreement. Ethiopia is letting some of the relief workers back into the famine area.

[Approved sale of British nuclear submarines (using U.S. technology) to Canada; met with clergyman petitioning Gorbachev to allow religious freedom in USSR; issues-briefing lunch; desk work; National Security Planning Group (NSPG) meeting to plan further steps in relation to Panamanian dictator Manuel Noriega.]

I phoned Frank Robinson, manager of Orioles who have lost 18 straight games beginning with 1st game. I told him to ask them to win one for the Gipper.

Then a haircut & upstairs to exercise & get ready for opening of remodeled Blair House. We have house guests for dinner & overnight at the W.H.—Marion Jorgensen & Buffy [Bunny]—Bonita Granville.

Tuesday, April 26

1st meeting—Sen. Gordon Humphrey—upset, thinking we weren't taking care of Mujahaden. I convinced him he'd been misinformed. Some discussion of trade bill. Our effort is to get Sen. votes to sustain the veto I'm going to deliver.

I.N.F.—a clause bearing on futuristic weapons is being used to block ratification. Colin & Dave Abshire are working with Sen. leadership.

NSC—Saudis refused our request for cover over one of our barges when our navy forces were elsewhere in the gulf. Well nothing happened. But now Saudi Arabia is breaking relations with Iran & seem to be offering us a better relationship. We are formalizing somewhat our diplomatic relations.

We've learned the Soviets are having a problem with one of their nuclear powered space gadgets & are fearful it may crash as one did some years ago. Finished with some talk about P.M. Mulroney & our Acid rain problems.

[Meetings with senators on support for veto of trade bill; greeted outstanding educators; met with Shultz regarding his visit with Gorbachev; more meetings with senators; videotapings; dinner with Maureen and Dennis.]

Wednesday, April 27

Woke up with a sore ear or maybe a jaw. It feels like my ear but I have to eat on the left side of my mouth it hurts to bring my teeth together on the right.

[Met with Senator David Karnes (R-NE) on trade bill; National Security Advisor Colin Powell reported on situation with Canada regarding acid rain; noted that envoy had returned to Panama.]

We have a report that Iraq used Nerve Gas in driving the Iranians out of the Al Faun peninsula.

[Briefing for Canadian prime minister Brian Mulroney's visit and then official greeting, private talk on acid rain and then general meeting.]

After that Jack Hume & Jerry Carmen came in & I had Howard B., Ken D., A. B. Culvahouse etc. This was over my agreeing to be Honorary Chairman of "Citizens For America" after I leave office. Our legal beagles don't think as Pres. I can announce such things before I leave office. I've told them this was the only group I'd say yes to. I had to leave all of them to work this out Sen. Jim McClure & others from St. of Wash. brought in a couple who have been waiting 26 years for $700,000 owed them by the Gov. He had built a bldg. for the govt. at the Seattle World Fair. There was some bureaucratic mix-up that stopped payment. Jim McClure got into this & had a bill passed. I signed it at our meeting.

Then a late lunch with the V.P. Yesterday in Pa. he went over the top—winning enough delegates to make him Candidate at the Convention.

[Met blind Puerto Rican girl who had scholarships to U.S. colleges; state dinner.]

Thursday, April 28

The Trade bill is veto bound & we have the votes to sustain. Rostenkowski has told us he'll put a new bill out of the Ways & Means committee immediately minus the objectionable amendments. A report to me on Jack Hume & Jerry Carmen. Apparently yesterday's meetings turned out all right.

NSC—Acid Rain. I'm asking Geo. S. to get together with Canada's Joe Clark & work up an agreement.

Lebanon—it is possible there will be a commando operation by the Lebanese mil. against some S. Beirut bldgs. where it is believed the Hisballah is holding some hostages. Then a meeting with Mayor of W. Berlin Eberhard Diepgen. I assured him I intended to make Berlin a subject for discussion with Gorbachev.

[Had lunch with Mrs. Reagan and the Mulroneys; signed education bill; meeting with National Drug Policy Board, commented, "We've made great progress in intercepting drugs but the drug lords just increase their effort"; greeted Muscular Dystrophy poster child, escorted by Jerry Lewis; photos with Marriott hotel executives; with GOP congressman from Louisiana, with Defense Advisory Committee for Women in the Service.]

Then over to the East Room for ceremony presenting Congress. Gold Medal to Lady Bird Johnson. A very nice affair with her family & many friends in attendance.

Friday, April 29

Get away day for Nancy—she's off for Calif.—be back Mon. Howard B. is away so Ken D. & I had the first meeting. There was a discussion about what I do now with regard to Geo. B. & the endorsement of him now that he's the nominee—or will be at Convention time. His people have suggested an appearance by me at a campaign rally here in Wash. We're inclined to think it should be at a non political type of gathering where we go as Pres. & V.P.

[. . .]

N.S.C.—We've established the Iranians did not fire any silkworm in our recent shootout.

We're worried that Pakistan in spite of denials may be dickering for nuc. missiles. China has become arms mkt. to the world.

I.N.F.—ratification of treaty is being held up on issue of whether we & Soviets have an agreement on "futuristic" weapons. Kozak is back from Panama. There seems to be some give on Noriega's part. He wants indictment killed, does not want to leave Panama but would do some extensive traveling while regular govt. took over.

Later in morning we had an NSC meeting. Much of it was taken up on Noriega & how to continue negotiations. Other half of meeting was on P. Gulf & our future course. Cong. is trying to block us in our plan to send 2 Coast

Guard vessels to the Gulf. They are better equipped than some of our larger Naval vessels with regard to Mines.

Had a lot of dark time & then lunch in the study. Early afternoon a meeting with Bob Tuttle to O.K. a sizeable group of appointees. Then my meeting with Geo. S. A lot of our time was spent on Middle East & Shamir's objections to a peace plan. Also some early talk on upcoming Moscow summit.

A short visit then by Hal Roach (over 90 now). Then it was upstairs—where there's no Nancy. I picked out a dozen sweaters to give away. Then some desk time, my exercises & dinner with Mary Jane & Charley Wick.

It was a fun evening like old times but we missed Nancy.

[Saturday, April 30: radiocast; read.]

Sunday, May 1

More reading—finally finished the 2 vol's. of "The Presidents House." It was fascinating reading—well at least for someone who's living in it. I made a tour of the 2nd floor really looking at each room after I finished the 2nd vol. this morning.

Lunch on the balcony—a beautiful 70 degree day. Nancy called from Calif. Her problem with the leaking incision under her arm has worsened & she's in great pain. Then John Hutton called & said he & an L.A. doctor are going to treat her. A few hours later he called & said things went well & he thinks Nancy is going to be fine.

Well a rubdown, dinner & the day was over.

Monday, May 2

House is here in session but Sen. has gone home for the week. I.N.F.—the "party line" so to speak is correct language in Treaty plus changes which could cost $5 bil.

Last week's "Issues Lunch" leaked to the press. I'll make a pitch & a threat at this one today.

A little discussion also about when & how should I endorse the V.P. for Pres.

[NSC meeting with talk about Manuel Noriega of Panama, INF; meeting with Prime Minister Harri Holkeri of Finland on 350th anniversary of Finnish colony in present-day Delaware; spoke to National Chamber of Commerce annual meeting; desk work; issues lunch; meeting with Bill Webster.]

Bill said more about Frank C. coming by tomorrow with space proposal. [. . .] Belief is they have a nuclear capability & could put it on missiles in a few weeks. Same is true of Pakistan. [. . .]

At 3 P.M. dropped by Roosevelt Rm. with Bob Michel to meet with C. of C. members from Illinois. A good meeting & I took Q's. Then my trip to Hades—2 hrs. in the dentist's chair for more Root Canal business.

Upstairs & wait for Nancy—due in a little after 7 P.M.—Head Winds. She arrived—the place brightened up & Rex & I are happy. So dinner & off to bed.

Tuesday, May 3

The press have a new one thanks to Don Regan's book. We make decisions on the basis of going to Astrologers. The media are behaving like kids with a new toy—never mind that there is no truth in it.

Ed Meese has a candidate for Deputy Att. Gen. This time he'll keep it quiet while we run a background check.

[NSC meeting regarding Israeli movement of troops in Lebanon; envoy returning to Panama, some reason for hope of a deal; met with Premier John Swain of Bermuda; signed proclamation for Asian, Pacific-American Heritage Week; NSC briefing on Noriega situation and on proposed space program; met four Soviet defectors; appeared at W.H. seminar on influencing religious rights in the USSR; photo session with cardinal of Ukrainian Catholic Church; met with Jim Martin of North Carolina regarding project for islands around Cape Hatteras; exercised.]

Wednesday, May 4

A short meeting—some talk about this astrology mess Don Regan's book has kicked up. Some gal in L.A. claims she's a visitor to the W.H. & that she gives us frequent readings. She even claims she advised me on choosing Geo. B. for V.P. We've never seen her in our lives & don't know her at all.

[NSC meeting: approved letter to Representative Bob Michel (R-IL) protesting budget cuts in SDI, report on Israeli invasion of south Lebanon, noted that development of new technology on space information satellites facing budget cuts; flew to Chicago for speech to Foreign Affairs Council; appeared at GOP fund-raiser; returned to W.H.; received news of explosion at Nevada plant making solid rocket fuel; did exercises. **Thursday, May 5:** *telephoned radio personality Alden Aaroe in Atlanta with birthday greeting; also Senator Chic Hecht of Nevada offering help in wake of explosion; noted that Health and Human Services sent 100 million copies of flyer on AIDS, noted, "We did no tampering with it in compliance with the law"; NSC briefing for NSPG meeting later in day, report on invasion by French security officials of hotel rooms occupied by Americans in France, French government said it would stop; Israelis pulling out of Lebanon, noted, "Polish govt. is getting rough on the strikers"; heard details on cause of explosion in Nevada; greeted Kilgore Junior College women's basketball team; NSPG meeting with discussion of issues to be faced in Moscow; greeted NCAA hockey champs; received as honor season's first Atlantic salmon; signed bill for National Day of Prayer; then met head of GOP in Delaware, commented, "it's possible he is a terminal cancer case"; then Hispanic business leaders; Boeing executives; Admiral James Stockdale (ret.), chairman of Commission on W.H. Fellows; did exercises.]*

Friday, May 6

Word that unemployment is 5.4% & 62.6% of the total potential pool is employed.

There is some argument about my veto message of the trade bill. Geo. S. & a few others think I should veto on basis whole bill is bad instead of just the plant closing feature. I'm afraid I can't agree.

Don Regan is doing a press conf. Sun. to plug his book which is a vile attack on Nancy. I expressed that today in answer to a press Q.

N.S.C.—As of 7:15 AM Poland had not turned police etc. loose on ship-yard strikers in Gdansk.

Tass has commented on my Ch. speech & said I confessed to human rts. violations in the U.S.

[Noted dispute over elections in El Salvador; meeting with Alfred Dregger, majority leader of West German bundestag; met with Josip Vrhovec, Yugoslav leader, commented, "that once Communist country is really moving into a free enterprise system"; cabinet briefing by Secretary of Housing and Urban Development Samuel Pierce on housing; photo session with organizers of Public Service Week; telephoned GOP fund-raising volunteer; staff photo, commented, "with whole press office staff including Jim Brady"; usual meeting with Shultz; reviewed plans for summit; went to Camp David. **Saturday, May 7:** *radiocast; watched Russian movie recommended by NSC; received call from National Security Advisor Colin Powell and approved possible terms for deal with Noriega.]*

Sunday, May 8—Mothers Day

Another bright day but temp. only around 60 degrees. Watched TV talk shows. They are driving Nancy up the wall with their gossip that we both (but especially Nancy) live by our horoscopes & won't make any decisions without checking the stars. It's a phony but the press is making a big phony thing out of it. Left Camp D. at 2:30 & back to the W.H.

Monday, May 9

A little conversation about Don Regan's book & how we're going to handle this.

NSC—There seems to be a 50/50 chance that the plan Kozak & others are working on with Noriega might work. If it doesn't come together this week it's probably dead.

I.N.F.—Maybe Dan Quale [Quayle] (Sen.) is softening on his beef about "futuristic weapons." Colin thinks he's coming around after making some other gains.

Then Ed Meese joined us—this too was about Noriega. He thinks dropping the drug indictment as part of the plan would kick up a fuss but he's for it on grounds—our choice is Nat. Security.

At 11:30 AM a truly fine old friend came by—Yasu Nakasone—former P.M. of Japan. He's staying active as an elder statesman. It was good to see him.

Then Lunch with Howard B. & Carl Rowan. I'd asked him for this when we sat together at the Gridiron. It wasn't an off the record meeting. He brought his tape recorder & it was on the record. I think I did alright.

Then some desk time & finally a ceremony in the Rose Garden—awarding plaques for the 2 top rated small businessmen in the U.S. Most of the audience were state winners. At the end Sam Donaldson yelled a Q. about the Regan book saying I made decisions by Astrology & was I going to continue. I said I can't because I never did.

[National Security Planning Group (NSPG) meeting regarding arms treaties in preparation for Moscow summit; met with Frank Carlucci regarding cuts in defense budget; photo session with David Rockefeller and Council of the Americas; telephoned François Mitterrand during the day to congratulate him on winning the French presidential election.]

Tuesday, May 10

Most of meeting was spent on Don Regan. Another falsehood corrected— book says I sent Geo. B. to ask Don to resign. George came to me & said he wanted to talk to Don & tell him he should go but wouldn't do it if I was against it. I gave him permission to do it on his own.

Then we talked about games Sen. is playing with I.N.F. Tricks also by Dems. re the Trade Bill—they seem to be stalling about sending bill up for my veto. "Defense Authorization" is due up this week also.

[Meeting with Senators Bob Dole (R-KS), Richard Lugar (R-IN), Simmons (D-NC), and Ted Stevens (R-AK) regarding INF; NSC meeting on positive outlook on Noriega situation; visit from Vice-Premier Tian Jiyun of China, commented, "A pleasant half hour—no major issues to discuss"; lunch with Lord Carrington, secretary general of NATO; meeting with Shultz, on his way with Colin Powell to Geneva to tend to details on INF; interviewed by biographer Edmund Morris, as noted, "my regular session"; videotapings; allergy shot.]

Wednesday, May 11

Started day talking about Trade Bill. Jim Wright is staging a circus as he signs order sending the bill on to me. He's having some Union members around him talking about getting fired etc. All this has to do with my objection to the plant closing part of the bill. Then it was drugs. Cong. is talking legislation to bring military into drug war as a major factor & the appointment of a Drug Czar.

NSC—We have a problem getting help to Contras. Sandinistas have gone back on their word to allow neutral delivery of food & such to them. [. . .] Head of Soviet C.W. program was recently in Syria.

[Meeting with Republican congressional leadership on INF, trade bill, commented, "First meeting with Congress people that didn't go longer than scheduled—wound up 5 min. early"; stopped by W.H. mail office to thank volunteers; had lunch with eight experts on USSR.]

Later in office Tommy Thomas brought some relatives & friends in. Then photos with about 50 major donors to tonite's fund raising meeting. Upstairs for a time—then at 5:30 down to East Room for a reception with a larger group of major donors. Upstairs to get into Black tie & off to dinner which has raised $5½ mil.

The evening was just great—very successful & a most enthusiastic crowd—several thousand people & I concluded my remarks by announcing I'd be campaigning for Geo. B. for Pres.

Thursday, May 12

Started the day at 10 AM talking about Don Regan. Then discussed my Sat. Radio show & the Trade bill & what game is Cong. playing. Then Drugs & a possible summit with Cong. to come up with a plan. Ed Meese agrees. A little talk about my graduation speech to Coast Guard Academy. There is also talk of Blue Angels doing a fly by as we leave for Moscow. Then Nancy & I are both going to get a briefing on security.

NSC—Agreement has been reached in Geneva on futuristic weapons & 9 other disputed points. Negotiations in Panama still going on. We're still waiting to hear from Japan on some idea of helping they had. July 1st will be the 20th anniversary of opening a treaty on Nuc. testing.

[Desk time; meeting with leaders regarding plight of Soviet Jewry; photo session with staffer; met with leaders of Goodwill Industries; with officials of Non-Commissioned Officers Organization; attended dinner for donors to presidential library.]

Friday, May 13

Bob Byrd wants a report from me on I.N.F. before he takes it to the floor. There is a firestorm in the House about the news reports on Noriega & the deal we are trying to make. They don't seem to realize the only alternative would be a military invasion of Panama.

On drugs some on the hill are demanding the military be given the power to arrest. We are totally opposed. For 200 yrs. it has been against our every principle to give the mil. that power.

NSC—Colin is back from Geneva. Evidently George & Shevardnadze worked out an agreement on I.N.F. that should eliminate all the Senate reservations about ratification. In addition we'll have several bilateral agreements to sign on various things like fishery rights etc. Some progress was made on Angola & nuc. testing but none on S.L.C.M.'s. A little on A.L.C.M.'s

We're hoping for an agreement on Noriega next week.

[Received personal financial report ready for filing, commented, "Frankly I think it's an invasion of individual rights"; signed proclamation for National Safe Kids Week.]

A last min. addition to schedule—a meeting with Ed M., Ken D., Jim Baker, Colin P., Elliott Abrams, Negroponte, & a few others about Noriega & our using cancellation of drug indictment against him to get him out of Panama—but again there is no alternative & we can't use the indictment anyway because Panama law does not allow extradition.

[Went to Camp David.]

Finally Camp D. & my first swim of the season. Watched a John Wayne movie.

Saturday, May 14

Did radio broadcast on Trade bill. Lunch outside & I finally got in a ride—it's been a long time. Then I hit a few golf balls. A call from Colin—our negotiators are coming back from Panama. Evidently there is no Noriega deal. We'll meet on Monday George S. will be back then. Tonite we run an Astaire-Rogers golden oldie.

[Sunday, May 15: watched panel shows, noted misinformation on Noriega situation and that "Don Regan was on 'Face the Nation.' He's apparently trying to cool down a little but not much"; returned to W.H.; Maureen back.]

Monday, May 16

I sounded off about Noriega & the media's treatment based on leaks & unfounded gossip. Then I.N.F. hearings we're told will start this afternoon. Soviets may fudge a little on futuristic weapons. I meet Sen's. Dole & Boren tomorrow morning. Recent word about N. Korea moving ground to air missiles near the border. [. . .]

[NSC meeting regarding plan to get money to Contras for food.]

At 11:00 AM I met with Geo. Shultz who gave me some outlines & proposals about the upcoming summit—there'll be a lot of that from now on. Then at 11:30 a larger meeting with Att. Gen., Sec. of Treasury, Geo. S. plus one of our negotiators from Panama, Colin Powell & a few others. Subject was Noriega & what to do about him. Finally I interrupted the debate & told the negotiator to go back to Panama—lay down our terms & if Noriega doesn't go along, tell him we'll send in our military. That ended the meeting.

[Discussion of drug policy; desk work; addressed students in U.S.-Soviet exchange program, commented, "Why don't we let the young people run the world—there wouldn't be any wars"; met with leaders of National Federation of Republican Women; received report from Intermarket Working group on improvements for stock market.]

Tuesday, May 17

Panama on 1st—Dole has introduced a resolution as an amendment to Defense bill—that resolves we must not kill the indictment on Noriega. Before day was over it passed overwhelmingly—they just don't know what they are talking about.

I'm enclosing some drug talk in my Coast Guard Academy commencement address tomorrow.

Howard told Ed M. he has to support our Panama policy.

NSC—Kozak is waiting to see Noriega today. He'll tell Noriega we must have an answer on our proposal.

A very brief meeting & then a meeting with Sens. Cohen & Boren. This had to do with funding space science—they went away happy. A little desk time then Vice P.M. Peres of Israel came by. We had a good but brief meeting. He is all out for our Mid East peace proposal. I wish Shamir felt that way.

[Photos with GOP candidate; met with Americans working for human rights in USSR; National Security Planning Group (NSPG) meeting with, as noted, "nothing exciting"; met with Meese about staff departure; rehearsed presentation for W.H. photographers dinner; received positive news on poll numbers; videotaping; allergy shot. **Wednesday, May 18:** *flew to Coast Guard Academy in Connecticut; delivered address; returned to W.H.; did exercises; watched broadcast of concert at the White House on PBS with Mrs. Reagan and Maureen.]*

Thursday, May 19

As of last night Kozak in Panama had no reply to the ultimatum he delivered to Noriega—he plans to leave at 10 AM our time today & come home.

[Howard Baker gives positive chance of INF ratification; decided to hold up veto of trade bill as long as possible.]

I'm facing a bill on Vets remuneration for Cancer if they were in the forces at the testing in Nev. or based near Hiroshima & Nagasaki in WWII.

NSC—Colin commented on the anti INF measure proposed by Jesse Helms—it was defeated 92 to 6 in the Sen.

Word came at 10:25 AM our time that Kozak was on his way to the plane when a messenger gave him word that Noriega has accepted our terms.

This was a cut up day—first I'm working out of the study, the Oval office is torn up for TV interviews I'm doing. A brief meeting on what to do about Noriega. Geo. Bush is back & feels very strongly that we shouldn't quash N's. indictment for drug running even if it does result in his stepping down & out of Panama. My position is that since Panama law prohibits extradition the indictment is meaningless & if we don't follow through N. can stay in Panama as mil. dictator & we can't touch him.

[Interviewed for foreign television outlets.]

Lunch turned out to be a hastily called meeting in Cab. room to go on about Noriega. A division of ranks supporting my position & the VP's. Shultz was with me & I stuck by my decision.

Then after lunch an interview between me & 3 news magazines. A photo for a Finnish newspaper with which I've done a written interview. A photo for a Soviet News magazine. A meeting with Sen's. Murkowski & Al Simpson—they prefer a veto on a veterans bill—Cancer again. I heard them out but believe I'll sign.

[Photo session with Blue Angels team scheduled for fly-over on president's departure for Moscow summit; attended W.H. photographers annual dinner and showed slides in humorous speech; received news a horse at ranch broke a bone and would need a long rest.]

Friday, May 20

Very short staff & N.S.C. meetings. Discussed news that a bank in Colombia was providing bank in Panama with a $27 mil. deposit. Bank in Colombia is reported as one that laundered drug money. If this is part of our deal we are really in trouble at home. But as proving went on we learned bank wasn't in the dirty money business but also that wasn't connected in any way with our Noriega deal.

[Photo session with Johnny and June Carter Cash; interview with Soviet journalists; lunch with Vice President Bush; addressed group on Cuban Independence Day.]

Another 20 mins. at desk then a meeting with George Shultz followed by meeting with Joint Chiefs of Staff. They briefed us on their views of arms reduction treaties. Then a meeting hastily called by Geo. S., Colin P., Frank Carlucci, V.P., Howard B., Ken D. & me. It was on deal which looks like it's coming together in Panama. I made decision that we delay consummating the deal just while Kozak flies back here for weekend to bring his report & then goes back to give N. our decision. We will then announce our decision before we leave for Summit.

[Dinner party at home.]

Mermie is here for dinner. After dinner an argument over Noriega. Both Mermie & Nancy are against me on Quashing his indictment. "Looks to public like I'm giving in to a drug dealer."

Saturday, May 21

Armed Forces Day—that was theme of my noon radio cast. At 10:30 AM a meeting—10 of us in Yellow Oval room. Subject same as last night's—Noriega & his indictment. Kozak flew back from Panama with terms agreed on in negotiations with Noriega & his lawyers. He is willing to announce now that he's retiring as Commander of the nat. military on Aug. 12 & shortly thereafter he'll leave Panama 'til after the 1989 election. We in turn are to Quash the indictment Aug. 12. My position is it's a good deal. The indictment is meaningless because the Panama Const. prohibits our extraditing him. H. Baker, Jim Baker, the V.P. & Ed Meese want us to refuse the deal because of the politics—how it will look giving in to a drug dealer. I say what will it look like backing down & letting him continue as the absolute Dictator of Panama & still in the drug business. I had to duck out at noon & do my radio cast but was back in 15 min. & we went on 'til about 1:15. Geo. Shultz, Colin & I are practically alone—well no—include Kozak. I finally moved for thinking it over 'til Mon. & meeting again.

So Nancy (she disagrees with me) & I had lunch on the balcony.

An afternoon of reading then a Black Tie evening at the "Preservation Ball" of the "Am. Film Institute" at the Grand Hyatt Hotel. It was a wonderful evening—a tribute to Fred Astaire. Bunny Wrather was chairwoman & there were a lot of old friends—Marge Champion, Nanette Fabray, Betty Furness, Joan Leslie, Cyd Charise & a newer friend, Mikhail Baryshnikov. Film clips of Fred's movies were shown & finally a clip from "This is the Army" in which I starred & then I said a few words & we came home.

Sunday, May 22

A quiet day—caught up on some reading—getting ready for Moscow. Then a phone call from Colin & I end up the day with an 8 P.M. meeting in the Oval Yellow room. The same group—all 10 of us—on the Noriega matter. No one's minds had been changed so I made the decision. Kozak left for Panama & when we get the word from him, I announce that Noriega is stepping down, leaving Panama & we are dropping the indictment against him. Even Nancy is against me on this one but I'm absolutely convinced it's the only way to go.

Monday, May 23

A general discussion of INF—Noriega, the Congress & it's expensive staff. H. Baker brought me regards from Jimmy Carter. He saw him at a meeting of a commission meeting on which J. C. serves.

Howard has suggested that my appointees to a proposed bipartisan Drug commission might be J. Baker, Ed Meese, Colin P. & himself.

[NSC meeting on Soviet withdrawal in Afghanistan; videotapings; presented awards to businesses; NSC briefing on summit; NSPG on START.]

Bob Tuttle with some appointments to pass on & down to a haircut. Stopped in Dr's. office to have my Dupreton Contraction on my hand checked. And upstairs.

Tuesday, May 24

A talk about Veto which I'm signing later this morning & did. INF Helms & others still blocking. Howard B. is going to stay here 'til end of week to keep pushing. Discussed & approved some Marine Corps promotions.

NSC—Kozak in Panama is dealing with some intermediaries & has been unable to get a closure on the Noriega deal.

A shift in Moscow schedule. Mon. 3 to 4:15 meeting has been cancelled. Then some talk about the communiqué & or possible joint statement.

[Meeting with bipartisan representatives and senators on INF treaty, drug program; interviewed by foreign press; lunch with Vice President Bush; meeting with President's Foreign Intelligence Advisory Board (PFIAB) regarding START; cabinet meeting on same subjects.]

Jesse Helms, Gordon Humphrey, Jim McClure, Steve Symms & our gang in a straight plea to quit blocking INF & let me take it ratified to Moscow. We'll see what happens. Upstairs to pack & we're off in the morning.

*[***Wednesday, May 25:*** flew to Helsinki; noted, "George stayed behind because it looked like a Noriega deal was upcoming. Then we received word on the plane the deal had collapsed & Geo. will be joining us in Helsinki."]*

Thursday, May 26

Awakened at 12:10 after several hours of solid sleep. Our press is chewing the Noriega deal & using the word failure (applying it to us) every other min.

An afternoon walk at around 3 P.M. then back to the Guest House which is something of a hotel. Dinner—watched 90 min. of our 3 network news programs & so to bed.

Friday, May 27

Up at 9:30 this time—no hiking clothes this time. We dressed up & were driven to the Palace. The streets were lined with people—they are warm & friendly. We were greeted by Pres. & Mrs. Koivisto. Then a brief meeting with him & his team & ours. Then all of us went into a pleasant lunch—they served Calif. wines. After lunch it was back to the guest house. An early dinner & more news reels—then word came our Sen. passed or I should say ratified the I.N.F. treaty. I called Sen's. Byrd & Dole to thank them & invited them to bring the papers & join me for the ceremony on Wed.

After our early dinner Nancy & I took turns having a massage by a masseuse recommended by Geo. Shultz. And then to bed. Finally got to sleep.

Saturday, May 28

Breakfast at 9:30 AM after a fair night's sleep. Then at 11: AM a meeting in the "box"—that's the leak proof room—no bugs etc. Geo. S., Colin P., Ken D., Tom Griscom & Roz Ridgeway plus our Ambas. to the Soviet U. Jack Matlock. We discussed the 1st meeting tomorrow which will be a 1 on 1—me & Gorby. I'm going to tackle him on religious freedom—not as a deal with us but as a suggestion to him as an answer to some of his problems.

Then lunch & back to the box. Same cast & a wider range of discussion on what we hope to accomplish.

Then some homework with the manual they've prepared for me. Nancy had another massage.

Dinner in front of the TV again watching the tape of Am. TV news & then to bed.

Sunday, May 29

Get away day—this time to Moscow, our 1st time ever to see Russia. We drove to the Presidential Palace where Pres. & Mrs. Koivisto joined us for the drive to the airport. A brief goodbye visit in the airport V.I.P. then a goodbye message to our embassy people & on to A.F.1. We lunched on the plane & landed in Russia at 2 P.M.—now we're 8 hrs. different from Wash. time. We were met by Ambas. & Mrs. Matlock & Pres. & Mrs. Gromyko—also our Ambas. staff plus family. Then the ride to the Ambas. house "Spaso House"—a one time castle. On the 20 min. drive 1st to the Kremlin where we were greeted by Gen. Sec. Gorbachev & Raisa—a review of the troops—the Nat. songs of both nations then his welcome to us & my reply. Nancy & Raisa took a tour of the Palace while he & I had a one on one visit. I introduced my favorite pitch—why he should give his people religious freedom. It was a good session & a nice way to launch the summit. Then we went on to Spaso House.

Our people had an idea about us going out on the street to be seen by the people—our goal a kind of set up where children could be photo'd with Disney type animals. It was amazing how quickly the street was jammed curb to curb with people—warm, friendly people who couldn't have been more affectionate. In addition to our S.S. the KGB was on hand & I've never seen such brutal manhandling as they did on their own people who were in no way getting out of hand.

Back to Spaso House for the night.

Monday, May 30

2nd meeting with the Gen. Sec. Touched on other subjects—congratulated him for his courage in leaving Afghanistan. Then tried to show him how some of things were urging on him would actually help bring about his perestroika. There is no question in my mind but that a certain chemistry does exist between us.

Then joined Nancy & we visited the Danilov Monastery which is being refurbished for use as an Orthodox Church. We were shown the restoration of icons that is going on. I addressed the assembled monks about the need for freedom of worship.

Then back to Spaso house for meeting with group of dissidents & Refusniks we're trying to get released & allowed to emigrate. One man who has been trying to join his wife in Fla. for 8 yrs. got word last Sat. that he can go. It was a heart moving experience.

Then a quiet dinner by ourselves at Spaso House—NO—it was the big dinner at the Kremlin.

Tuesday, May 31

Nancy off early on her visit to Leningrad. My usual beginning in the ice house—that's what I call the bug proof box we can hold our meetings in. Then my 3rd meeting with Gorbachev & again we got along well—it was a 1 on 1. Then he took me for a brief walk around the Kremlin grounds before we went back for the plenary meeting. Following meeting we attended the Shultz, Shevardnadze signing ceremony of some bilateral agreements—we're up to about 47 of those now. Then it was a luncheon with a group of cultural & art community members at the House of Writers Assn. I spoke to them about the need for freedom of the arts. Then a couple of hours of free time (homework) and over to St. U. of Moscow where I addressed & took Q's. from more than 1,000 students. Later met with 35 American Faculty members there on exchange.

Before day was over Gorbachev walked me out into Red Square. It is quite a sight—the expanse is so great it is really something to see. We stopped & talked with several groups of people who were there.

Tonite our dinner for the Gorbachevs—very pleasant & entertainment by our famous American jazz artists quartet.

Wednesday, June 1

Usual staff time in the ice box then on to the 4th plenary meeting with Gorbachev. A little hassle—the joint statement was read. He wanted an addition we couldn't buy. I used argument his addition would give us trouble at home. He gave in. Then we went into signing ceremony of INF treaty ratifications. They had been brought here by Sen's. Robt. Byrd & Bob Dole who were present.

Then lunch with our team in the ice box getting ready for my press conf.

And at last the press conf. including foreign & Soviet as well as our entire Wash. press Corps. What can I say except "Thank The Lord." According to our people I did just fine. Now it's on to the Bolshoi Ballet & then a private dinner at Gorbachev & Raisa's country place with Shultz's & Shevardnadzes.

It was a nice ride out in the wooded countryside to a really lovely home & a good time was had by all. On the way back we drove through Red Square so Nancy could see it. Naturally the press was on hand & wanted photos. Believe it or not there were hundreds of people behind a rope there to see & wave at us. I don't know how they find out where we'll be. It was almost midnight.

Thursday, June 2

Up at 7:30 to pack & do a farewell to the embassy staff. Then over to the Kremlin for a farewell to Gorby & Raisa & off to the airport for a formal departure with the Gromykos & parading the troops etc. All was done & we boarded A.F.1 for the 3½ hour ride & the 3 hour time change.

We had a big breakfast & that turned out to be lunch for the day—well not really. Charlie & Carole had a quiet lunch for us at Winfield House. Then we had some down time—about 4 hours before we left for Buckingham Palace where we had a pleasant tea with Queen Elizabeth. From there it was over to 10 Downing Street for a tête-à-tête with Margaret Thatcher—I gave her a report on our summit in Moscow aided by Colin, George S. etc. Then a pre dinner reception & dinner at 10 Downing & a brief view of the mounted troop "Beating The Retreat." Over to Winfield House for a good nights rest.

Friday, June 3

Downstairs after 8:30 breakfast in bed & finishing packing. A staff meeting—or briefing I should say with Sec. Shultz, H. Baker, K. Duberstein, M. Fitzwater, Colin P., G. Sigur & P. Stevens—ended at 10:20 AM for arrival of Japan's Prime Minister Noburo Takeshita. The Prices joined in greeting him. The usual photos with him & several members of the Diet. Then a bilateral meeting mainly on our trade matters—particularly on beef & citrus fruits. Meeting over at 11 AM & we all motored to Guildhall—a very formal gathering led by The Lord Mayor of London & P.M. Thatcher—an audience of several hundred. I made about a 30 min. speech that was very well received. Then our farewells to the Thatchers & others & back to Winfield House for ceremonial farewell to embassy staff & familys & departure by Marine 1 with Prices & out to the airport where we boarded A.F.1 & headed home—a 7½ flight & a five hr. time change.

[Arrived at Andrews A.F.B.]

Geo. B. & Barbara were waiting to escort us in to the hangar where we greeted our Cabinet members, were led to a podium. The Nat. Anthem was played by the Army band. Geo. introduced me & I spoke of our trip & our thanks to the assembled people. Then we marched out the other side of the hangar to Marine 1 & on to the W.H. where there were more people to welcome us—also a small brown & white dog—Rex.

*[***Saturday, June 4***: overcame jetlag already; radiocast; Maureen and Dennis visiting; watched taped coverage of trip; read. ***Sunday, June 5***: relaxed with books, TV, massages.]*

Monday, June 6

Howard Baker called—he's in Tenn. with Joy who has had more health problems. He'll be here tonite. Ken D. & I held our morning meeting & it was mostly a re-hash of the trip. He thinks I should do some impromptu mixing with the people here as I did in Moscow.

Then Colin for NSC—he was reporting on Geo. S. in the Middle East. He just can't move Shamir. He has King Hussein & Egypt's Moubarak going for him but not the hold out. He's also met with Pres. of Syria but no word on progress with our hostages who are held by the Hezbollah. Wife of one—Joan Sutherland is in Lebanon—has been there working her heart out for a couple of years. She's made some top contacts. I hope we can help her get her husband out.

[Addressed World Gas Council; desk work; haircut; videotaping; telephoned President José Napoleón Duarte of El Salvador, commented, "he's at Walter Reed Hospital for Cancer surgery—prospects aren't good"; telephoned widow of Clarence Pendleton, chairman of the U.S. Commission on Civil Rights, who died the day before; Dennis and Maureen still visiting.]

Tuesday, June 7

Howard still not here. Ken & I carried on without him. Ken tells me Tommy Griscom is leaving—possibly by end of month. He's been offered a Vice Presidency in a business concern in Tenn. The Sen. will be voting on my Trade Bill veto tomorrow. I made phone calls to Mike Deaver & Bud McFarlane's wife Jonny to thank them for their letters on the Summit.

NSC—Colin brought in an NSC board member Ambas. Cohen to give me a rundown on Mozambique & the Renamo. He's really an expert & he convinced me we should not look kindly on Renamo. They are well organized & in some sectors are very kind to the people but in others they pillage & massacre. Afghanistan—have been rumors Soviets will halt their withdrawal in response to Mujahadeen activity. There is no evidence as yet to back the rumor.

[Received word that Contras and Sandinistas were resuming talks; Representative Mickey Edwards (R-OK) requested meeting; prepared for upcoming economic summit in Toronto; attended fund-raising dinner for Senator Trent Lott (R-MS), commented, "he & Tricia are deserving of everything we can do for them."]

Wednesday, June 8

George B. on hand for meeting—took the "Red Eye" in from Calif. Howard brought Tommy Griscom in to tell me he (Tommy) was resigning effective end of this month. He'll be a V.P. of an Insurance Company in Chattanooga. I learned this morning from Geo. B. that Dukakis as Gov. had vetoed a bill to have students in Mass. recite the Pledge of Allegiance. This P.M. Sen. votes on my veto of trade bill.

[NSC meeting with reports by National Security Advisor Colin Powell on resumption of talks between Contras and Sandinistas in Nicaragua, and by Deputy National Security Advisor John Negroponte on Panama.]

Mort Abramowitz went down to Panama last week—is back with some proposals for building pressure against Noriega. One idea is removing mil. dependents & bringing them home. After some desk time I was briefed at 11:30 for meeting & back with about 90 press, radio & TV media from all over the country.

[Received honor from Tau Kappa Epsilon (TKE) fraternity; greeted, as noted, "the lovely little girl who is Arthritis poster child"; meeting with Polish-American Congress.]

Went upstairs with Geo. B. & spent an hour talking about his campaign. Now I'm waiting for Nancy who went to Fla. today for young people's meeting on drugs.

Thursday, June 9

Howard is back in Tenn. with Joy. I have a feeling he'll have to make that a permanent move.

Yesterday the Sen. sustained my veto. Then Byrd pulled a political trick. He switched his vote to support the veto which now permits him when he wants to—to call for another vote.

We discussed the Catastrophic Health bill—it looks as if I'll sign it.

A judge has ruled North & Poindexter must be tried separately.

The House ethics Committee is deciding on whether to bring charges against Jim Wright.

Cong. is thinking of asking me to formally submit the Canada free trade bill to them for approval.

NSC—Soviet Ambass. & other officials visiting other countries are hailing the summit as a great success. Zia has officially cancelled his trip to the U.S.

[Noted that Eduard Shevardnadze raising issue at U.N. that U.S. was blocking progress on START over banning or not banning nuclear weapons on ships; met with President John Mobutu of Zaire, commented, "He's a good man & is helpful to us on Savimbi in Angola"; addressed National Conference on a Drug Free Workplace; briefing for and then interview with Life *magazine; photo sessions; telephoned members of Congress regarding starting work on a new trade bill.]*

Friday, June 10

A lot to talk about. Chf. Justice supports bill eliminating mandatory lie detector tests now before Cong. It's suggested that I send letters to Dole & Byrd re protecting Presidential prerogatives on treatys. A great speech line from Ken Crib—"Liberals don't care what you do as long as it's compulsory."

NSC—Nicaragua talks cancelled again. Sandinistas claim Contras made impossible demands. Both promise not to resume fighting. Contra leaders are coming to Wash. for consultations. There is talk of an arms sale—40 F 18's to Kuwait?? [. . .] Some Arab states are helping P.L.O. in Israel.

Norm Dicks (Cong.) is threatening to bring legislation to face us to eliminate a lot of our nuc. subs.

*[Briefing for and then visit to meeting of Self Help Leadership; planning session for
economic summit; presented Volunteer Action Awards; went to Camp David, watched
'Crocodile' Dundee II (1988).* **Saturday, June 11:** *radiocast; exercise; ride; heard by
telephone that writer Louis L'Amour died.* **Sunday, June 12:** *returned to W.H. in morning;
attended Mrs. Reagan's celebrity tennis tournament, with Fess Parker as emcee.*

Monday, June 13

A busy day—began with a call from Howard B. in Tenn. His 87 yr. old
mother has just been taken to the hospital so he's staying there. She may require
surgery. Ken & the V.P. started the day. We discussed 1st the sale of F 18's to
Kuwait—none of the Hebrew orgs. are opposing this sale. Learned Geo. S. is
meeting later in the day with the Contra leaders here.

Then it was Welfare Reform—some on the Hill are talking welfare reform
but they are proposing welfare increase. We need a bill that will provide work
for welfare recipients.

Cong. defeated a budget busting health care bill. We have a good chance of
getting a catastrophic bill I can sign.

NSC—Colin reported on Somalia—the horn of Africa & the continued
strife between Govt. forces in the S. & the Somalia people in the N. This is based
on ancient tribal rivalries & it looks like the govt. forces are the recalcitrants.

*[Discussion of Contras and then Greece, which refused U.S. request for extradition
of terrorism suspect; brief visit by Prince Sadruddin Aga Khan, head of U.N. commission to help Afghans; desk work; telephoned Mrs. Louis L'Amour with condolences; photo
session; meeting with Shultz on Mideast; addressed Atlantic Council; received report on
NATO from North Atlantic Assembly; received presidential gifts, including bronze buffalo
sculpture, bas-relief of president and Gorbachev; filmed campaign footage with GOP candidates; greeted gathering of Hispanic businessmen from New Mexico.]*

Tuesday, June 14

Howard B. came in to tell me he was announcing his resignation today.
With his mother & wife both hospitalized he feels his duty is in Tenn. & I agree
with him. We also announced Ken D. was replacing him.

[Discussion of speechwriter for convention speech.]

NSC—A short one. Jim Kelly NSC came in to report Philippine legis.
is giving us trouble on renewing our mil. bases there. And Guiliani (U.S. Attorney) is talking of drawing up an indictment against Marcos. I think he's
crazy.

We sold 3 planes—patrol types to China—they evidently delivered them to
N. Korea & one has already crashed.

Frank Carlucci is talking of having all D.O.D. suppliers swear they have drug free work places. Then a 9:30 AM meeting with Repub. Legislative leadership. We reported on Summit Nicaragua & discussed new trade bill.

[Visit from Prime Minister Ciriaco De Mita of Italy, commented, "We have a good solid relationship with Italy. They are taking in our 401st fighter squadron formerly based in Spain"; meeting on economic summit; desk work; regular meeting with Frank Carlucci; then with Bill Webster regarding Noriega.]

Wednesday, June 15

Learned Senate is working to get 2 appropriation bills down to me & intends to send 13—one at a time. This is a response to my St. of the U. address in which I stated I would not accept another 1000 page package of such measures as one bill.

[Noted that Frank Carlucci requested joint session of Congress; NSC meeting with news that Contra leaders visiting Washington consider aid essential to negotiations.]

Press is full of F.B.I. investigation at Pentagon which has been going on for 2 yrs. It is tracking down reports of kickbacks etc. in Defense Industry procurement. Then at 9:30 AM a congressional meeting (Dems. & Repubs.) in Cabinet room. These were all reps. who want to help Contras. I had a good meeting & lots of suggestions.

Desk time—then a meeting with Dick Wirthlin. My approval rating is up to 60%. It is 80% regarding dealing with the Soviet U. Other figures very good. People's greatest concern is drug problem.

[Had lunch with GOP state chairmen from fifteen states where Democrats were strong; interviewed by group of predominantly foreign journalists; met with people making boyhood home in Dixon, Illinois, a tourist attraction.]

Thursday, June 16

Howard on hand but leaving tonite for Tenn. 'til after the Summit in Toronto. Tom Griscom delivered his letter of resignation effective July 1. Report from hill 8 appropriation bills have been passed. An unhappy report on drought situation. I've asked our people to check on govt. owned surplus in storage—is there a way to offset crop loss for farmers & supply the mkt.

Ed Meese & Bill Sessions (F.B.I.) came in to report on D.O.D. investigation. It has been completely covert operation for 2 yrs. Some 85 contracts are involved. All started by one Navy man who went to Naval Investigation force with some information. Naval Force started, then asked F.B.I. for help.

[NSC meeting with talk of Soviet protester Yuri Zieman denied emigration papers.]

Zia has declared Islamic Law is law of Pakistan. That puts them into the Fundamentalist Revolution with the Ayatolah & Quadaffi.

[Noted that Mari Masing agreed to return as director of Communications; desk work;

greeted high school honor students; meeting on economic summit; spoke to CEOs on U.S.-Canada trade; photos with departing staff and with Lorraine Wagner, commented, that she "had been a member of my fan club 40 yrs. ago"; met with Ken Khachigian about speech for convention.]

Friday, June 17

[Summoned Secretary of Agriculture Richard "Dick" Lyng for meeting regarding drought; noted congressional progress on welfare reform, trade, and fair-housing bills; NSC meeting with report by Shultz on South America; received word that Soviets punishing dissidents supporting Reagan, then later that Yuri Zieman was released for emigration.]

Cap W. called Frank about changing S.D.I. plans—he's violently against it & so am I.

Our Nicaragua visitors will settle for non-lethal aid plus our right to deliver.

N.S.P.G. meeting in Situation room. [. . .]—help for Contras in Nicaragua. I've asked for specific plans on which I can base my decision.

[Briefing for and then interviews on summit with CNN and PBS; addressed United States Information Agency (USIA) International Council Conference; meeting with Shultz, commented, "He's fired up about some flaws in new trade bill"; haircut, videotapings. **Saturday, June 18:** *radiocast; hours of reading for summit; received confidential message regarding unsuccessful assassination attempt on Prime Minister Özal of Turkey.* **Sunday, June 19:** *flew to Toronto, staying at Royal York Hotel, changed into blue suit for welcoming ceremony; private meeting with Prime Minister Brian Mulroney; brief talk with French president François Mitterrand; then three-hour group meeting with heads of state, commented, "A good free-wheeling discussion of problems we must solve"; dinner with spouses present, commented, "More discussion & much agreement."]*

Monday, June 20

Brkfst. in bed at 7:45 AM Then up for a brief staff meeting—learned of coup in Haiti—Pres. Margat thrown out by Gen. Namphy. Then at 9:50 over to MTCC—part of L. hotel. Summit heads only for 15 min's., then a 2 hr. plenary meeting with Finance & Foreign Ministers. This became working lunch. An additional meeting & then back to Royal York for a meeting with P.M. Takeshita of Japan & several diet [Diet] members. He & I signed before an audience & press our Science & Technology Agreement. Upstairs—Nancy on way to another dinner. I showered & was off to Hart House at U. of Toronto for customary photo & an informal heads only meeting.

It worked just fine & our meeting before dinner & at dinner turned out to be the most fruitful of all. I came up with an idea about subsidys & the farm problem I'd never thought of before. How about subsidizing farmers to give up farming?? Got back to the Hotel before Nancy.

Tuesday, June 21

Get away day. Brief staff time. Ken & Colin—nothing very important. Talked a little about the coup in Haiti. Then to MTCC for plenary session—scheduled 2 hrs. This was for agreement on communiqué. Mitterrand gave us a battle over a couple of words & the 2 hours became almost 3 before he could be satisfied. Then back to the hotel to get Nancy & on our way to Ontario Art Gallery. Formal meeting & reception with Gov. Gen. who is a lovely friendly woman. We had our final summit luncheon as her guests. Then we were on our way to Roy Thernson Hall for reading of communiqué by Brian Mulroney. Back to Hotel for like 10 min's. then a briefing for my press conf. which took place in ball room of hotel. It went well but there were virtually no questions on the summit. The Mulroneys came to get us to take us to dinner meeting (no dinner for us). I spoke to Empire Club & Toronto Club—1st Am. Pres. to ever do so. Then it was off to the airport & a 1 hr. 5 min. flight to Andrews A.F.B. & Marine 1 to the W.H.

Wednesday, June 22

Howard back—in fact he met us on the S. Lawn last night. This morning he told me about Congress. Duncan's death by Cancer. I called Lois his widow & then son Jim who is running to replace his father. We then talked a little business in our meeting. What kind of Contra aid pkg. to put before the Cong. It must be one that will pass. The fair housing bill looks like it will get to my desk & be one I can sign. Everybody & his brother seem to be talking about new drug programs & more spending on them. Well let's see if that is the answer. I have one of my own I'm going to try. I'm calling Lane Kirkland, AFL-CIO to see if Labor could help by voluntarily drug testing in Unions like Railroad, trucking etc. Tomorrow Dick Lyng & the V.P. will be back from their tour of the drought country. On the Hill we face a fight over funding for the Space Station.

NSC—Frank Carlucci is pushing a plan to put 3rd phase of S.D.I. in place before going any further on 1. & 2. I'm against it. Phase 3 is merely ground based A.B.M.'s to hit any strays that get through the main defense line. I'm afraid Congress would halt spending on 1 & 2 maybe permanently. I granted Colin authority to withhold classified information from court ordered surrender of it. Colin tells me Saudi's are now (and a 1st for anyone) partners in Texaco Oil Co.

Last month 1000 Visas for Jews leaving Soviet U. were changed for U.S. when they reached Vienna—1000 out of 1100 total.

[Noted opposition in Congress to sale of F-18s to Kuwait; desk work; review of styles in welcoming ceremonies for visiting heads of state; photo with Turkish newspaper; meeting with Shultz, who was against trade bill; discussed plans for upcoming trip to Central America, commented, "Then we talked about Sandinistas. We're trying to come up with

further action against them"; met new Air Force aide; exercises; addressed Annual Meeting of Business Round Table; dinner with Mrs. Reagan and Maureen.]

Thursday, June 23

[Discussion with Howard Baker on trade bill; NSC meeting with National Security Advisor Colin Powell reporting on his meeting with pro-Contra senators and on possible path in Angola ultimately deemed unfeasible; prepared for meeting later with Prime Minister Robert Hawke of Australia; met with Puerto Rican Americans honored with achievement awards.]

Then it was Hawke time. He's been here 4 times. We had a good meeting & lunch at the WH. We have a great relationship between our 2 countries. I think I reassured him that our summit negotiations weren't going to affect their Wheat exports. Lunch turned into a story telling session.

[Cabinet meeting on summit, drought, and privatization.]

Some photos then the Gershwin family to whom I presented medals struck by order of Congress. One to George G's sister & one to Ira's widow.

Upstairs—exercise then into Black tie, on to dinner for Sen. John Stennis. It was a moving affair to a deserving, remarkable man. I announced we were naming a nuc. powered aircraft carrier the U.S.S. John Stennis. Home & Mother.

Friday, June 24

Didn't look at my schedule carefully. Arrived at office at 9 AM—wasn't due til 10 AM. Well I had an hours desk time. Then at 10 started meeting with Ken, V.P., & A. B. Culvahouse. He reported on Judge Gesell—on the North & Poindexter case. Had to do with classified info. & whether it can be used as evidence. I'm just not sure what it's all about.

Then another case involving a Sup. Ct. decision. Move on to appeal to Sup. Ct. for reconsideration of something called the Runyan decision. We are not participating. Some discussion of gun control & my statement that I approved what Calif. has done about a waiting period for anyone buying a gun—until their background has been checked.

NSC—Colin briefed me on coming visit by Patriarch of Lebanese Catholic Church. He is addressed as "Your Beatitude." Talk also about possible visit to Pres. Duarte who has begun Chemotherapy treatment.

Mujahadeen fired rockets at Kabul Airport & destroyed 7 Soviet fighter planes.

Rioting in Yugoslavia against Communist party.

Then it was time for his Beatitude. A pleasant meeting—he was accompanied by Papal D'Nuncio.

[Met with California Young Presidents group; photos with departing staff members;

short meeting with Shultz, as noted, "he was late—little time for business"; attended ceremony honoring former POWs.]

A reception preceding tonite's Ford Theatre Gala—a receiving line—upstairs for dinner & off to the theatre. The show was the best of the 8 we've seen—a lot of old friends: Jimmy & Gloria Stewart, Rosemary Clooney, Mike Feinstein, Shirley Jones, Sid Caesar, Fabian & others.

Saturday, June 25

A nice day—did my morning radiocast on drugs & our discussion of the subject at the Toronto summit. Lunch on the Truman balcony.

At 2 P.M. Doug Wick brought a friend to see us. It was a young lady with a tragic story. She has a husband & 2 children—a girl & boy. At the girl's birth 7 years ago (it was a Caesarean birth) she hemorrhaged & was given a transfusion—7 pints of blood. At age 3 her daughter became ill. To sum up, she has AIDS from breast feeding. Her mother had been made a carrier by the blood transfusion. Discovery of this meant her 2nd child, the boy is also infected. Her story is one of great courage & tragedy & she wanted me to know in connection with the report I'm getting this week from our AIDS Commission. Actually I feel better prepared after hearing her to deal with the report & decisions I must make.

Spent afternoon catching up on reading plus my exercises. Then a quiet dinner & bed.

[Sunday, June 26: watched panel shows, read Revolution by Marty Anderson, commented, "it's great. No kiss & tell in this one"; quiet dinner with Librarian of Congress Jim Billington, his son, and the Shultzes.]

Monday, June 27

First discussion was of the Polygraph Bill. Ed Meese is against it but everyone else is for it & if it comes to me as is, I believe I'll sign it. Still haven't gotten the Catastrophic bill. Tomorrow I get a briefing on D.O.D. investigation.

NSC—Ambas's of several countries calling about the Space Station. They are countries we asked to join us in a joint venture. Now Congress is balking & says only way they'll provide for the program is if we take the money out of Defense budget. We can't do that. Soviet Support for Sandinistas is off by 60%. The Middle East continues to dabble in nuclear weapons. [. . .]

[Briefing about and then visit by President Kenan Evren of Turkey.]

Desk time, lunch outdoors—a couple of photos—a meeting with Howard & Jim Baker. Jim was giving his side of differences between him & Geo. S. on Trade bill. I think I have to go with Jim & sign the Bill.

A Cab. briefing—it really was presentation to me of our AIDS Commission Report. I'm going to work with Dr. Ian MacDonald on implementing it.

[State dinner, entertainment by guitarist Christopher Parkening.]

Tuesday, June 28

[Staff meeting on pending legislation; approved appointments.]

NSC—I brought up matter of Holocaust Museum. It seems someone has approved a room dedicated to 1915 massacre of some Armenians by the Turks. I'm against it but don't know what we can do. Latest intelligence has 10 more divisions on the border in N. Korea than we had estimated. They're playing games that may threaten the Olympics in S. Korea. We have the Soviets pledge they won't let anything like that happen. [. . .] a bomb plot in Panama City—an apartment bldg. housing about 40 of our mil. officers. They've been moved out & on to the base.

Met with our Sen. & Cong. leadership—Dole & Simpson, Bob Michel & Trent Lott. A discussion of pending bills & the pol. games the Dems. are playing trying to get campaign issues. Then a meeting with Meese & Carlucci regarding the investigation of the Pentagon. Nothing much yet to report.

[Interviewed by biographer Edmund Morris; meeting with Israeli defense minister Yitzhak Rabin; went to dentist for teeth cleaning; received report on Conference for a Drug Free America; photos with GOP candidates; received allergy shot and, as noted, "a little doctoring of my hearing aid."]

Wednesday, June 29

[Flew to Miami for party fund-raiser and meeting with leaders of Cuban American Society, commented, "I spoke to them & reassured them we were not getting soft on Castro.]

Then back to A.F.1 & to the W.H. by about 5:25. After dinner received a call from Howard B. that 3 of our Uniformed Secret Svc. & 2 now ex-employees of NSC will be front page in the Wash. Post tomorrow for dealing in Cocaine. The 3 have been banned from the W.H. grounds. I'll find out more tomorrow. Also had a call from Ron—the first in a long, long time.

[Thursday, June 30: discussed allegations against agents; noted that House had passed all thirteen appropriations bills; talked about states and counties that refused to provide plans for removing citizens in case of nuclear power plant accident, commented, "In doing this the plant can't be licensed to operate"; NSC meeting regarding upcoming visit by Jonas Savimbi, then extradition cases related to Cuba and Mexico, some talk about Gorbachev and the opposition he faced in the Communist Party Conference; noted that European allies were organizing commercial airline service to Berlin; visit from Savimbi; photo session; meeting with South Carolina members of Congress promoting their state for tritium-producing plant; signed proclamation for National Safety Belt Use Week; meeting with congressional leaders on drug policy; lunch with basketball coach Pat Riley; desk work; cabinet briefing on nuclear power stations, commented, "some were shut down because countries & or states refused to submit plans for emergency evacuation. Bil. $ plants sitting there unable to operate"; farewell visit with staffer; greeted Eureka College students; photo

with Multiple Sclerosis Mother and Father of the Year; appeared at gathering of Republican conference leaders; exercised.]

Friday, July 1

Word rcv'd. that Stu Spencer is coming back for a meeting re the Bush campaign Thurs. I was told I should say a few words of greeting from the W.H. balcony to all our people who will be on the S. Lawn for fireworks on July 4th.

Cong. has gone home & will be back next Wed.

I approved Billy Graham for benediction at Repub. Convention on night I speak.

N.S.C.—Some talk about reps. at Soviet Conference calling for ouster of older people in office such as Gromyko etc. all of whom were present to hear.

[Heard that Italian parliament approved presence of U.S. F-16s; message from Shultz reflecting positive reception in Central America; signed bill providing catastrophic-illness coverage for Medicare patients; received interim report on drought, commented, "In some areas it has broken all records. It is a serious problem"; videotapings; went to Camp David.]

Saturday, July 2

My radio cast, then groundbreaking for new Chapel at Camp David. Meet & thank service personnel who rounded up & captured new recruit who armed himself & set out threatening Camp personnel. Then a meeting with new Camp Commander. Photo with Jim Broadus & family, our last trip with him as Commander. Sat. P.M. a horse back ride. But main event was a 5 AM phone call that one of our cruisers, the USS Vincennes in the Persian Gulf accompanied by 2 Frigates had been engaged by a flotilla of Iranian missile carrying speed boats. The Vincennes sank 2 & disabled a 3rd. While this was going on radar picked up a blip on a plane taking [off] in Iran & on a course directly toward the action. Radio warnings were sent on both mil. aircraft frequency & commercial. There was no reply. Then radar revealed the plane was descending right on line with the Vincennes. The Captain ordered firing of 2 missiles. One brought the plane down. Later it was learned the plane was an airbus carrying according to Iran, 290 people. It's a terrible tragedy & I so stated but while I've ordered a Naval investigation, I don't believe the Captain had any other choice but to fire on the plane.—This happened on Sun.

Sunday, July 3

A quiet day except for the phone calls—one a conference call re today's tragedy. We are all in agreement & approved statements to that effect.

Later in the afternoon—my first swim of the season.

[Monday, July 4: returned to Washington; lunch on balcony; homework all afternoon; informal dinner with friends; W.H. staff celebrating on the lawn.]

Tuesday, July 5

A 10 A.M. day. We of course discussed the Gulf plane tragedy & press coverage. Tomorrow expecting a House vote on plant closing. Judge McKay has submitted his findings on Ed Meese to Ed. & to the 3 judge panel. Nothing in it charges crime or calls for indictment. Ed is busy reviewing their study on drug policy.

NSC—Colin brought in an amazing personality. He's Wm. Harrington, 93½ years old—a former Sgt. Major in the 9th & 10th Cavalry. These were the 2 black Cavalry regiments. He really is an amazing man with a great personality. Then we got down to business. We're sending request to Cong. to allow us to sell Kuwait 40 F-18's. Saudis are buying arms, $30 Bil. from the U.K.

Max Kampelman & Elliott Abrams came in & reported on George S's. trip to Central Am. We can have some optimism about the results.

Later in morning Ken Khachigian came in & we talked about my convention speech. Then a 90 yr. old lady was brought in. She had gotten an invitation to attend last weeks Repub. Leaders conference. They hadn't included the card that called on her to contribute $2500 so she couldn't get in. She doesn't even have enough money to get home until her Soc. Security check comes. Thank Heaven our people found out about her & brought her in for a visit, & they're arranging to get her back to Calif.

[Desk work; addressed conservative political leaders; ambassadorial formalities; taped interview on inaugural process.]

Then Geo. Bush & I met at Walter Reed hospital to visit Pres. Duarte of El Salvador. Just before we left I got a call from Ed Meese telling me about the report & now that he's cleared he wants to resign effective in early August. While we were at the hospital he held a press conf. & announced he was stepping down.

Wednesday, July 6—"Happy Birthday Nancy!"

First meeting—Ken D & B. Oglesby. Talked about U.S.A. Today poll showing great support for action taken in shooting down plane. Dems. in House talking about an increase in minimum wage but only $4.35 not the original $4.55.

Ken is trying to set up a lunch meeting tomorrow with Geo. B. & the others—Paul Laxalt, George's campaign people, Stu Spencer etc. to talk campaign strategy.

A. B. Culvahouse came in briefly to report on Ed Meese situation & the McKay report. Then we did a little talking about possible replacements for him.

NSC—An argument between State & Treasury over Soviet exports to us of Tea & Chocolate. Our law prohibits American import of products produced by convicts or slave labor. It seems Tea & Chocolate are made in Prison Labor Camps in the Soviet U. I've asked to see the evidence.

[. . .] Then some talk about the Iran plane shoot down. There is a Question as to whether plane wasn't actually inside the Commercial Aircraft Corridor.

Some desk time—then Ed M. came in briefly to make some suggestions on his replacement & an offer to help. An interesting successor he suggested was Sec. of Interior Don Hodel.

Then a double N.S.P.G. meeting—50 min's. on changing our charge that Krasnoyarsk Radar in Russia was a violation of ABM treaty to a term in treaty making it a material act outside the treaty which would call for some action on our part. Good arguments for both positions & now it's up to me to decide. For 1st time in Defense Dept. that I can recall—Chiefs of staff were opposite side from Sec. Carlucci. Second meeting was a 10 min. CIA presentation of [. . .] action to rid Panama of Noriega [. . .]. Then lunch—signed a bunch of pictures & upstairs. Tonite dinner with Paul & Carol Laxalt & Merme & Dennis then over to Kennedy Center to see Les Miserables. It was like something I've never seen before, a fascinating new idea in theatre.

Thursday, July 7

A 10 AM day & the day Nancy leaves for N.Y. for about 10 days. Started the day talking about Wash. Post poll showing great support for the Navy, et al in the gulf tragedy. Also speculated on possible replacements for Ed Meese. Some whose names have been mentioned took themselves out of the running—Bill Smith, Paul Laxalt, Libby Dole. An energy & water bill (appropriation) coming down to me. It's supposed to be alright & eligible for a signing ceremony—but we don't have final word yet. Sen. passed the plant closing bill 72 to 23. It will go to the House next week.

NSC—Chf. of Staff—Soviet mil. forces on visit to U.S. visiting a number of mil. bases.

Soviets are talking of withdrawing 68,000 Soviet mil. from Hungary.

[Noted President Eric Delvalle of Panama came out of hiding for his mother's funeral.]

Desk time 'til a photo with Connie Heckman. Then our lunch meeting with V.P. & his campaign people. On our team Ken D., Stu Spencer, Paul Laxalt & a few others. A good meeting—we're going to work together on campaign.

[Desk work; signed photos; received award from Salvation Army.]

Then a drop-by at farewell reception for Tom Driscoll [Griscom]. And over to East Room to speak to "Gopac" group. Up to exercise & dinner with Merme & Dennis.

*[*Friday, July 8: *unemployment figures show improvement; discussion of farm program to relieve drought-stricken areas; received word that Office of Management and Budget*

deficit projections were as expected; NSC meeting; National Security Advisor Colin Powell reported that he and Soviet ambassador discussed release of refusniks, heard that Soviets released text of most of Reagan toasts offered during Moscow summit; plan for Soviet chief of military to visit Oval Office, noted that "Question still up in air about whether to label Krasnoyarsk a 'Material Breach' or not"; presumed president of Mexico was Carlos Salinas; desk work; National Security Planning Group (NSPG) meeting discussion of whether to compensate victims of airplane incident in Persian Gulf; greeted Rick Mears, Indy 500 winner; desk work; photos with GOP candidates; hosted stag dinner with Dennis and seven members of Congress from both parties, noted, "A great evening, one story after another. We should have been doing this more often"; planned attendance at Baltimore Orioles game the next night canceled due to security reasons.]

Saturday, July 9

Woke up early. I don't like being alone. Some homework—phoned Nancy & told her I wouldn't be going to the ball game tonite. She was delighted.

Did my radiocast mainly on drugs plus a paragraph or two on Ed Meese's announcement that he was stepping down. Merme got back from Atlanta. She & Dennis & I are going to have dinner on trays in the Study.

And so we had dinner after I had another call from Nancy. Then I ran a tape of a TV show I did with Gizelle MacKenzie back in the '50's—I actually sang in it—with her. But I can't remember one thing about when or where I made it. And so to bed.

[Sunday, July 10: read; had lunch with Maureen and Dennis; dinner alone in front of TV.]

Monday, July 11

NSC time began at 9:05 AM & was a meeting with Marshal Akromeyer— Chf. of Soviet armed forces. He has toured our country with Admiral Crowe. He was much impressed by what he saw—including our mil. bases, ships etc.

Then it was staff meeting time. Cong. is in session 'til near end of week. Trade Bill in House has none of provisions that made me veto it. Bill Verity suggests that if plant closing bill comes down I let it pass without my signature. We'll see. Today was 1st meeting of our combined campaign team for Geo. Bush. A. B. Culvahouse came in with list of names being pushed for Attorney Gen. job. I asked for record of Dick Thornburg, former Gov. of Pa. I'm leaning toward him.

[Farewell visit from Salvadoran president José Napoleón Duarte, arrangements made to fly him and his guests home to El Salvador; met with Senators Strom Thurmond (R-SC), Orrin Hatch (R-UT), and Bob Dole (R-KS), who wanted to suggest names for attorney general, commented, "I didn't reveal my choice."]

Then N.S.P.G. meeting—a review of [. . .] covert operations [. . .].

A meeting with N. H. Congressional meeting re the SeaBrook nuclear plant which is unlicensed because of the rule they must have a plan for moving people away from plant in an emergency. We have to do something about that ridiculous rule. A haircut. Then photos with about 70 candidates for State legislative seats.

[Met bipartisan congressmen about drought measures; allergy shot; attended fund-raiser for GOP candidate; dinner with Maureen.]

Tuesday, July 12

Ken D. informed me former Pa. Gov. Dick Thornburgh would accept appointment as Attorney Gen. if I asked him.

[Received conference report on defense authorization.]

Dick Thornburg came in—I asked him to take the job & he enthusiastically said yes.

N.S.C.—Ambas. Dobrynin (Soviet U.) told Colin the Ziemans were told July 5 they could come to America. It's possible the hold up is money for an airline ticket.

Gorby in Poland talked of taking mil. planes out of Poland if we'd bring our NATO planes—F-16's home—they're slated to move from Spain to Italy. Not a fair deal. His planes only move the short distance to Russia—ours come 1000's of miles across the Ocean. Nicaragua kicked our Ambassador out of Managua along with 7 embassy officials. We're going to kick his Ambas. plus 7 comrades out of Wash.

[U.N. debate scheduled on Persian Gulf airplane tragedy, Vice President Bush to represent U.S.; received news that no Americans hurt in terrorist attack on Greek ferry boat; briefing for and visit by Prince Saad of Kuwait.]

Then a Cabinet meeting—briefs by Dick Lyng on drought, other subjects reported (drugs, Contra aid, education. etc.) by CABINET MEMBERS. Real aim of meeting was to turn everyone on for V.P.'s campaign.

[Attended dinner for presidential library.]

Wednesday, July 13

[Started day at 10 am; NSC meeting with discussion of defense appropriations bill, orders Nicaraguan ambassador and seven others leave U.S.; positive reception to Dick Thornburgh, except anti-abortion rights Senator Gordon, commented, "thinks Dick is pro-abortion. He's not"; discussion of bill making Veterans Administration director cabinet-level post; met with Prime Minister Edward Adami of Malta, noted, "He gave me credit for Malta turning against socialism"; lunch with Vice President Bush; visit from actress Nell Carter.]

A departure photo & then signing ceremony in Rose Garden for Captive Nations Day Proclamation. Went well. We've done this every year. Some desk time then addressed 48 state chairmen of Farm Bureau in Roosevelt Rm. And upstairs where I did a phone interview with Adell Specht—writer for Moline Ill. paper—I'll be going there tomorrow. A phone call from Ron who'll miss tribute to Nancy at convention but will be there for mine. I wish it could be the other way around.

Exercise—dinner & bed.

[Thursday, July 14: went to Marion, Illinois, to see drought damage; went to Davenport, Iowa, to visit radio station WOC, first employer in 1930s; returned to W.H.]

Friday, July 15

Started day talking about yesterday's trip. Everyone who went along thinks it was a great success. N.Y. Times had a front page photo. Wash. Post had story on P. 14.

The D.O.D. authorization passed in both houses. It may present a problem—to sign or veto.

Ambas. Tunnerman who we are sending back to Nicaragua got no support from Org. of Am. States. He's due out at 5:30 P.M. Still says he's not going. We say he is. If he takes shelter in the embassy we'll have a problem because we can't set foot in an embassy. Senate moving fast on Thornburgh approval—have said it could happen by Aug. 5.

Both houses passed resolution denouncing Sandinista govt. That is a turnaround. Now if they'll finally stop being stubborn & let us give aid to the Contras. It's possible I'll be scheduled for an address at Gov.'s Conference in Cincinnati Ohio.

N.S.C.—Pres. Botha of S. Africa has granted a permanent stay of execution for 6 Blacks under death sentence. We had asked him to do that.

We're encouraged by progress in talks about Angola & getting the Cuban military out of there.

The problem of whether to declare the Krasnoyarsk radar is a violation of the A.B.M. as it is now or take the next step by calling it a "Material Breach" still faces us. Sec. of Defense says yes but Joint Chfs. of Staff say no.

[Presented National Medals in science; met with U.N. Secretary General Javier Peréz de Cuéllar, discussed many issues, including U.S. payments, which were late; met with new surgeon general of the Army; photo session; phoned Max Fisher on his birthday.]

Then a taping session & a drop-by in Dr. Hutton's office. He had 2 specialists there to look at my lip. He has felt there was a spot which might be a beginning Carcinoma. Now they've decided maybe not so they'll look again on Aug. 5.

Tonite in family theatre about 35 assorted guests to see "The Winning Team"—my picture where I play Grover Cleveland Alexander. It was a great evening. We all had dessert & coffee in the hall outside the theatre—then inside for the movie. It was very well received & the audience included a few ex sports announcers & big league ball players of a previous time. After the pic. was over I told them a few anecdotes about "Alex" & the making of the movie. Then beddy bye.

Saturday, July 16

Brkfst. at 8 A.M. A morning of reading then—my radio cast which dealt with the Dems. coming convention & their distorted attacks on the economy & what we've accomplished. A brief meeting in the study off my office with Ken D. & A. B. Culvahouse. It had to do with top secret classified matters that Judges Walsh & Gesall wanted for the "North" trial. We determined we could not make these available because they are essential to Nat. Security.

[Mrs. Reagan returned; read; ran tape of Gisele MacKenzie program. **Sunday, July 17:** *received word of gunfire attack on U.S. soldiers in Honduras, noted, "No word as to their—the terrorists identity. But Iran has been calling for an attack on American military"; flew to ranch; hung a mirror in guest room.* **Monday, July 18:** *rode; received call from National Security Advisor Colin Powell regarding movement by Iran at U.N. indicating new hope for peace in Iran-Iraq War; pruned trees all afternoon.* **Tuesday, July 19:** *tree pruning, noted, "Nancy got a call from George Will in Atlanta telling us not to watch the Dem. Convention—it was terrible."* **Wednesday, July 20:** *rode; received cable from Shultz visiting Marshall Islands, reporting good relations there; cut down a tree; telephoned Giselle MacKenzie.]*

Thursday, July 21

[Rode; did chores around house; pruned oak trees all afternoon; telephoned Governor Bill Clements (R-TX) to offer good wishes after very mild stroke.]

After dinner watched Dukakis make acceptance speech at Dem. Convention in Atlanta. He pushed all the buttons including premises that sounded as if our country was in the Ec. doldrums instead of a continuing Ec. expansion. And, of course he charged me with the deficit spending & the big debt. Presidents can't spend a dime, only Congress can spend the money & the Cong. is Democratic.

Friday, July 22

A little muggy again but not really bad. Before the ride called Marty Anderson about his book "Revolution" which is about me largely but as Gov. & now President. It's no kiss & tell. It's a factual account of what we've done. Then a call from Ken D. about what we do in the remaining 6 months. He & about 14 or 15 of our gang had a lengthy meeting. I'm waiting for their report but from what he said I believe I'll agree with what they talked about. Then the ride—started early because we're going exploring. We rode down to the old

Eckert ranch & through the property down to the spring fed lake we've known about but never seen. Round trip was about an hour & 40 minutes. I had to get back for lunch with Ken Khachigian who is working on my speech for the convention. It was a working lunch & progress was made.

Then I caught up with Barney & Dennis & we went at the Oak trees again. Late in day after shower & clean clothes Ron & Doria arrived for dinner. A pleasant evening.

Saturday, July 23

[Radiocast.]

Then our final ride for this trip. I forgot my spurs which I've never done before.

Phoned Sam Donaldson re the death of his mother. She was 93 & lived alone on a farm doing the necessary chores. He told me she voted solidly Republican all her life.

[Cut firewood; Mrs. Reagan received word of suicide of Gloria Vanderbilt's son Carter. **Sunday, July 24:** *returned to W.H.; Ted Graber visiting for dinner.]*

Monday, July 25

A 10 A.M. day & a busy one. First meeting a kind of general discussion of the legislative situation & problems we can look forward to with necessary authorization & appropriation bills.

NSC—Our young (18-19-20) crewmen of the C-5F's that delivered our inspectors for treaty verification. Report is the Soviets were fascinated by them & we were showing them everything there was to see.

Morocco wants to buy F 16's from us.

Jim Baker suggests banning import of tea & chocolate from Soviets on basis of makers being convicts in production of those items. We don't agree. First of all our importation is minuscule & 2nd we can't confirm whether products are made by convicts.

[Desk work; audiotaping for convention documentary; met with Boy's Nation participants; received report on drought; signed letters on U.S.-Canadian free-trade bill; met with Senator James McClure (R-ID) regarding location of nuclear fuel production plant; videotapings; allergy shot; dinner with Graber and Maureen; ran documentary on inaugural of Jimmy Carter Presidential Library.]

Tuesday, July 26

A little fun hearing about the Belly Dancer at Bob Doles Birthday party last night. Then to business. It looks like Thornburgh will come up for clearance Aug. 5. It appears deficit will be $140 Bil. instead of the estimated $146 Bil.

I have to make a correction in yesterdays item about our young air-crews in

U.S.S.R. They were on 6 missions carrying Dept. of Energy drilling equipment to Semipalatinsk near Chinese border. This is in support of U.S—U.S.S.R. joint nuclear testing agreement.

[Addressed gathering of Take Pride in America; meeting with Shultz on his return from extensive travels, noted, "His report was very interesting & also he said the U.S. was at the highest level of popularity world-wide that it has ever been."]

[Photos with departing employees and with party donors.]

Then a photo session with donors to Repub. Senatorial Committee. An N.S.C. briefing—Dr. Edward Teller & a distinguished group of scholars displayed a new & most secret means of destroying nuclear missiles. It is technology that is almost unbelievable. After some desk time a meeting with Korean War Veterans group who are working to create a Korean war memorial. Tomorrow is anniversary of end of that war. Gen. Von Fleataway with them. He is 96 yrs. old—his grandfather fought in the revolutionary war.

Cong.man Bill Archer came in to talk about legislation. Bill Webster report on covert operations etc. Congressional Photo time—photos with friends & family of several Reps. And now—I forgot to report that day began with a meeting in Cab. room with Repub. Cong. leadership. It was a good session. I listened to each one of them make proposals about what should we do in these 6 mo's. to insure our achievements would be carried on after I'm gone.

[Mrs. Reagan on one-day trip to NYC for funeral of Gloria Vanderbilt's son.]

Wednesday, July 27

[Positive report on economy from Beryl Sprinkel.]

Cong. working on some supplementals & calling them "Dire" supplementals. Have to do with things like Black Lung etc. Drought bill is getting the Christmas tree treatment—an airy subsidy etc. Our leaders think they can eliminate some of the tree ornaments.

We discussed a definite date of departure for Ed. Meese. A group of Repub. legislators are meeting with me tomorrow to discuss D.O.D. authorization bill on whether to sign or veto. Jesse Jackson has asked for Secret Sve. protection. We'll do it for 30 days—then he's on his own. Some talk for my campaign appearances.

NSC—Talk of Covert operations in general. [. . .]

Then briefing for visit with P.M. Grosz of Hungary. He is also Gen. Sec. of the Communist Party of Hungary Then the meeting—brief 1 on 1 in Oval O. then plenary in Cabinet R. & lunch at the WH. He's an Hungarian Gorbachev. He's putting in reforms & encouraging pvt. investment etc. We got along fine. A half hour meeting with Shultz. He reported on his trip.

Some photos—daughter of my 2nd cousin Lea Wilson—with her family—husband, daughter & son. Some S.S. departure photos. A meeting with A.

B. Culvahouse who is meeting with Ollie North's lawyer tomorrow. He wants me to be a witness—A.B. says no—so do I. It would set a precedent for future Presidents. A Pres. cannot testify or make public classified material to which he is privy.

[Early dinner.]

Thursday, July 28

The morning was a little hurried. Ken reported that Speaker Wright & Rostenkowski say they will pass the U.S.-Canada Free Trade Bill before the Repub. Convention. No word yet from Byrd on the Sen. side.

Got word that Lee Greenwood whose great song was played at the Dem. Convention much to his annoyance will sing at the Repub. Convention on the date of tribute to Nancy & me. Beryl Sprinkle & Jim Miller were in with more good Ec. news. Earnings are up & so is Disposable Income.

[NSC meeting, with report that Soviets wanted review of ABM treaty in September, Iraq wanted direct talks with Iran, noted, "Both sides are now using chemical warfare"; met with congressional leadership representatives regarding defense authorization bill; spoke to Student Congress on Evangelism.]

Back to W.H. & lunch with George B. We talked possible V.P. cand's. Both of us are without a firm choice.

[Photo with departing staff member; addressed Future Farmers of America.]

Dr. Ian Macdonald reported on Aides [AIDS] findings of commission I'd appointed. He has the situation well in hand.

Then the Bush campaign spot ads & finally upstairs to await visit by Richard Nixon. As always, he had some campaign suggestions that make great good sense.

Friday, July 29

Another hot day. We started by talking leaks to the press—suddenly there is a new wave. One about covert operations is so destructive to what we're trying to do with regard to Noriega could actually get a few people killed. I think it's disgraceful the press would use it.

[Meese requested meeting for following Monday; commented that congressional bills on drought relief have costly additions attached.]

Dukakis has accepted our offer to provide intelligence briefings during the campaign. Plans are going forward with regard to both cand's in regard to where their personal residence would be in the event of becoming Pres. etc.

NSC—Leadership of P.O.W.-M.I.A. organization is concerned about whether the new administration will continue our work in seeking information & remains of the MIA's. We will urge that it be carried on. Laos & Vietnam are proposing a joint operation. We anticipate some problems when P.M. Prem of

Thailand steps down. We're not happy about some of his possible replacements. Things are stirring in Cambodia. We're hopeful Prince Sihanouk will play a major role.

[Visit from Franz Josef Strauss, president of Bavaria; spoke to National League of POW/MIA Families; photo session with Pakistani foreign minister Yaqub Khan; meeting with Shultz who described upcoming Central American trip; photos with party donors; went to Camp David. **Saturday, July 30:** *radiocast; exercised; telephoned Elizabeth Dole with birthday greetings.]*

Sunday, July 31

Watched the TV shows "Meet the Press" & "This Week." Lunch outside— a little cooler. Back to the W.H. by about 3 P.M. Then Jim Baker came by—he'd asked for a meeting. Jim, Ken D., Nancy & I met about half an hour. Jim is going to resign as Sec. of the Treasury to be George's campaign chairman. I think it's great & George needs him. We'll ask Nick Brady to take Jim's job & hope we can make announcement next Fri. P.M. Then some exercises for me because I can't tomorrow night. Dinner & so to bed.

Monday, August 1

Another muggy, steam bath day.

Ed Meese came in for a part of the 1st meeting. He wants his successor, Gov. Thornburgh to be a part of N.S.C. as he is. It's fine with me. Then he made a pitch for Gov. Sununu's recommendation on Nuc. Power Plants emergency regulation & an executive order. On S.D.I. being gutted in Defense authorization bill I should veto bill & make the S.D.I. an important part of my veto message. Again I agree. Then he left & meeting went on. Tomorrow I'll meet with Cong. leaders on Plant closing bill. I think I should let it become law without my signature—no way it wouldn't be over ridden if I vetoed. We're releasing Aides [AIDS] report today (my task force).

NSC—Sec. Herrington has made a wise decision on locating new nuc. material plant. There'll be 2—1 in S. Carolina using "heavy water" system. The other in Idaho using gas.

It looks good for getting Congressional approval of sale of F.18's to Kuwait.

[Meeting with President George Vassiliou of Cyprus, commented, "He's quite a man & determined to wipe out the Greek-Turkey split & reunite the island"; lunch with Republican leaders in state legislatures; met with members of student choir.]

Back to Cabinet Room for a Domestic Policy Council meeting. Subject was about international program calling for reduction of pollutants—Nitrates & Oxides etc. We are ahead of most nations in eliminating or reducing emissions now.

[Briefing on Yellowstone park fires; Nick Brady agreed to replace James Baker; went to doctor for prostate exam, noted, "verdict excellent"; attended party fund-raiser.]

Tuesday, August 2

Decided I'd let the Plant Closing bill become law without my signature. Marlin made a press announcement this morning. Then a summary of outstanding bills & whether we should push some to be acted on before the Aug. recess & which should be left for later. Actually the situation is up in the air.

NSC—There is consideration as to whether we & the USSR should destroy our chemical weapons. We're considering approaching them on a proposal that we do it & each side watch the other one do it.

We're making progress on F 18 sale to Kuwait. It's up to the Sen. now.

A short meeting in Roosevelt Rm. with 4 Repub. Cong. leaders. Bob Michel absent but Dick Cheney took his place. We were all pretty much agreed on Defense Authorization Bill (I should veto) & the plant closer.

[Desk work; presented awards to government employees.]

Back to a lunch with Suzanne Massie (just back from Russia). Nancy, Colin & Ken D. on hand. It was a strange one. We expected her usual insider type of information but instead it was almost like a travelogue. She did say the Russians have a real feeling of friendship for Nancy & me.

Then I went into Oval O. & did TV footage for 16 of our cand's.—14 for Sen. & 2 for Gov. That over, Nancy joined me in Roosevelt Rm. for reception for Jim & Sarah Brady & the Brady Foundation donors.

[Photo session; ambassadorial formalities; allergy shot.]

Wednesday, August 3

Made it public I was vetoing the Defense Authorization bill. On this Saturdays radiocast I'll talk about the veto & why. More legislation discussed—the compensation for the Japanese interred in W.W.II.—Aides [AIDS] discrimination—O.K. for Fed. employees but having Justice Dept. look into legalities of applying it to Pvt. sector. Then faced with leaks of report on Iran plane shoot down. Press asking questions—but neither military high command or my office have received the report as yet.

NSC—Took up that same matter re the shoot down. Then desk time until I signed the veto message on the Defense authorization—it THE BILL was 100's of pages long. I went into the press room & reported on my veto & why.

[Briefed press on Central America; had lunch with Citizens for America; sat during refinements on bust for presidential library.]

Thursday, August 4

Defense has always pouted because the budget for building a space station came out of Defense. Well now it's coming out of H.U.D. We need to officially name Nick Brady for Treasury Sec. so F.B.I. can do their field check on him. I have O.K.'d for new Ed. Sec.—the Pres. of Texas Tech. He is an Hispanic, Larry Cavazos. Twice faculty voted a no confidence in him. Once when he agreed with Bd. of Regents that spending should be reduced. Second when he voted against tenure for entire faculty. He's our kind of guy.

Sen. has passed a trade bill we support 85 to 11—it includes repeal of windfall profit tax which I've been trying to get for 5 years.

N.S.C.—Aug. 24 is date for review of ABM treaty. Question is do we declare Krasnoyarsk Radar a "Material Breach" of treaty or leave it as a "serious violation." If now we were to make the change—wouldn't it raise an inquiry as to why now? Chiefs are opposed to changing it & so am I. This is not a time to kick up a storm. Word from Sov. U. is that Ziemans family are leaving next Wed. for the U.S. I've got my fingers crossed. Contra Aid. Dem's. put an impossible package on table yesterday. Only weapons they'll permit & that will be questionable are the $18 mil. worth stored in warehouses awaiting delivery.

Ed M. wants us to appeal judges ruling that we can't close the P.L.O. information agency in N.Y.—part of U.N.

There is a news story out of Managua that a boatload of people including religious group of Americans were fired on from the shore—a number wounded. We know there are no Contras in that area. My belief is the Sandinistas did it & so as to deliberately frame the Contras.

I sneaked up to the 2nd floor this morning to say Happy B.D. to Bob Tuttle.

[Meeting with Senator Tribble (R-VA) with idea for summit of all Western Hemisphere nations on drug policy; meeting with defense experts, commented, "They are all behind my veto of the Defense Authorization bill"; photo sessions; exercised.]

Friday, August 5

Drought aid bill cleaned up in Conference.—the $5 & $6 Bil. Xmas tree was brought down to $3.9 Bil. That I can sign.

[Congressional Democrats fighting Contra aid; Senate holding up approval of appointees to Federal Regulatory Commission.]

NSC—We've learned that over the years the Soviets have used the Russian Orthodox Church outside of Russia for spying. This is very hush, hush—[. . .].

[Possible progress on negotiated withdrawal of Cuban troops from Angola; woman kidnapped in Panama released; Secretary of Agriculture Richard "Dick" Lyng reported on drought aid; photos with departing staffers; received new poll results, commented, "It appears that Dukakis has a big lead on George but much of it was direct result of the Dem. Convention"; received gift of Boehm bird sculptures; greeted Meese's daughter, but did not attend farewell party for Meese.]

Then it was over to Dr. Hutton who had a couple of specialists in to look at my lip—suspicion of possible Carcinoma. They decided it had gotten better since last look so things are going right. Upstairs to work on Convention Speech. Then back to Oval O. to meet Jim Baker & Nick Brady. Today is day to announce Jim's resignation & Nick as replacement. We went into the press Room at 4:30. I made the announcement, then Jim said a few words followed by Nick & the show was over.

[Rehearsed appearance for PBS show In Performance at the White House. **Saturday, August 6:** *slept late; radiocast; read all afternoon; appeared in PBS show.* **Sunday, August 7:** *quiet day; caught up on homework.]*

Monday, August 8—The 8888 Day

[Flew to Cincinnati, Ohio; made speech at U.S. Precision Lens factory; had lunch with bipartisan group of governors.]

After lunch into the Gov's meeting (46 of them), plus an audience of 800. My speech on Federalism was very well received. Then back to the chopper & to Ky & A.F.1. Got word Geo. Shultz's motorcade in Bolivia was target of a bombing. But while there was some damage to cars—particularly the one Obie was riding in—there were no injuries. Back to the W.H. at about 3:30—upstairs & days end. Exercises dinner & bed time. Oh, I did stop on the way at the Dr.'s office & got my sneeze shot.

Tuesday, August 9

A sum up of bills still to be brought up & which ones we might expect before Cong. takes off in a week or so. A little talk about my proposal to do an explanation of the Iran Contra affair in my speech (Convention) Mon. Some are against it.

[Received news of legislative progress on drought, trade and fair-housing bills.]

NSC—The Contra Aid bill looks like we'll have to compromise with Byrd to get anything. I favor doing that rather than having the Contras face a complete rejection.

[GOP leadership arrived to make legislative plans.]

At noon Nancy & I went into the East Room for the Medal of Arts lunch. We handed out 12 medals—1 to Helen Hayes I'm happy to say.

[Photos with departing staffers; saw Secretary Wall, just back from visit to Persian Gulf; announced new director of Education; photos with presidential exchange executives; haircut; did exercises.]

Wednesday, August 10

Everybody is overruling me on the Iran-Contra insert I wanted to put in the Convention speech. They may be right that it would take away from the rest of the speech & be the only thing the media would latch on to.

Claude Pepper & Hatfield have been promoting our F.D.R. memorial for a long time. They have come up with an idea that seems to have caught on—cost $5 mil. Ed Meese has announced tomorrow will be his last day. Sen. Judiciary Com. has passed Thornburgh & he'll probably be voted in this week. House passed U.S.-Canada free trade bill but there is a question as to whether Sen. will get to it before this recess or in Sept.

NSC—Dole & Byrd each have Contra aid bills. Byrd has offered his if Dole will let his be defeated. It's lousy politics but even Dole says Byrd is better than nothing.

[Powell suggested tentative schedule for U.N. visit.]

Sen. Helms is off again—this time a letter charging that Gen. La Joie stood by while Major Neilson bled to death after being shot by Soviets. Colin took Jesse on but good. Gen. La Joie is a fine officer & did no such thing.

[Briefing for and visit by Prime Minister Thorsteinn Palsson of Iceland; signed Bill 442 (providing restitution for Japanese Americans interred during WWII); convention speech completed; photo with departing National Park Service official; did exercises; watched broadcast of "In Performance at the White House."]

Thursday, August 11

[Received word that Soviet dissidents, the Ziemans, due to arrive; commented, that congressional Democrats holding up appointments; received positive news on auto sales; Congress would start clearing ambassadorial appointments, commented, "Jesse Helms was blocking them—now he's relented."]

Heard a story of Jim Wright throwing a tantrum in front of annual "Gym Dinner" when Bush tried to say hello to him.

Looks like we may get U.S.-Canada Free Trade bill 1st week in September.

NSC—George S. got in at 6 AM this morning from his Central Am. trip U.S. will contribute to U.N. Peace keeping force in Middle East.

Frank Carlucci has decided to retire our SR 71's. These are fast (1800 miles an hour) high flying spy planes. Operation costs are $350 mil. a year. Now our space satellites are doing the job. Frank is working with Hill to improve D.O.D. appropriation bill.

[Signed drought bill; briefing for and then interview with Hugh Sidey, commented, "In Sat. talk show he is always the most fair when some of the media hatchets are slashing at me"; photo sessions; met with Meese on his last day; met with Shultz on travels and ongoing relations with USSR; meeting with Joint Chiefs of Staff on START, which they support.]

A call from Ken D. just before dinner—Sen. confirmed Thornburgh—85 to 0. I'll swear him in tomorrow. Dinner, ran a tape of Larry King interviewing Kirk Douglas who has written a book. Then to bed.

Friday, August 12

Lots of things to talk about—speeches—Sun. Mon. noon & night. A phone

call to be made to the Ziemans tonite. Cong. has gone home 'til Sept. Sen. has
cleared almost all the Ambas. appointments they've been stalling on.

The so-called Dire supplemental is progressing. This is funding for things
like Black Lung etc. Welfare reform is put off 'til Sept. Strom Thurmond for
some reason is blocking clearance of the other 3 justice dept. appointments.

[Noted schedule for future vacations.]

NSC—Things are quiet on Foreign Affairs Front. [. . .] They're making a
little fuss about it. Soviets look like they'll make good on getting ½ their forces
out of Afghanistan by Aug. 15.

Contras re-crossing border out of Nicaragua into Honduras reveal a problem—many are under nourished, sick & some wounded.

[Addressed administration officers; photo sessions.]

Lunch & an NSPG sleep inducing meeting. A discussion of Mobile nuc.
missiles & some of points to work out with Soviets on verification.

Then a photo with 26 members of Council on the Humanitys. A lengthy
taping session for 6 different org's. & affairs. Upstairs a date with a tailor—to
make a little improvement of coat of suit that was a gift when I was in Brussels.

Saturday, August 13

[Radiocast.]

Got a phone call about 2 P.M. today—Admiral Zumwalt's son died of cancer. He was one of the Vietnam group that was exposed to Agent Orange. It's
believed this was the cause of his Cancer. Tomorrow we're off for New Orleans
& then Calif.

*[Sunday, August 14: flew to New Orleans, Louisiana; made campaign appearances
for Vice President Bush; Mike and Colleen Reagan, Maureen and Dennis arrived.]*

Monday, August 15

This is my surprise day. A great luncheon tribute is on for Nancy which
I'm not supposed to attend. I will, however. After she left I got dressed & then
over at the N. O. Convention Center I waited behind a curtain & when she
finally appeared to respond to the tribute that had been addressed to her & she
responded I walked out & on stage spoke my piece. It was a complete surprise
to everyone. For once there was no W.H. leak.

Then it was back to the Hotel. Ron & Doria are due in any min. Nancy
can't dine with us. She has to go over to the Convention for some interviews.
Ron arrived without Doria, she has a touch of the flu.

We all had dinner—Mike, Colleen, Cameron, Ashley Marie, Ron, Dick
& the family, Maureen & Dennis. Then it was on to the Convention where
we joined up with Nancy. There were speeches by Paul Laxalt, John McCain,

a touching movie as a tribute to me, Jack Kemp spoke, Liz Dole was Chairwoman. After the movie I spoke—well Nancy had earlier said a few words. My speech was well received & the session ended with the song "I Love the U.S.A." sung by Lee Greenwood.

Back to the Hotel & a reception for a gang of our friends & cabinet members. We didn't get to bed until about 1 A.M.

Tuesday, August 16

I had to get up early & speak to the Eagles Breakfast. That too went well—then a receiving line & photos with the 150 of them. Back to the hotel for Nancy then out to the airport where we met Geo. B. & Barbara. We each said a few words to the crowd. The press was clamoring for word about who George would name for the V.P. spot. Just before we parted Geo. whispered his choice was Dan Quayle—Ind. Sen. On plane with us was Mike & family, the Wicks, Dennis LeBlanc & Betsy Bloomingdale. Before we landed, got word Geo. B. has made his V.P. choice known.

Beautiful at the ranch—spent afternoon moving in & doing some picture hanging as some now at ranch are going into our new home in Bel Air. Then dinner—with Barney & end of day.

Wednesday, August 17

At 7:30 AM phone call from Colin Powell. One of our C-130's in Pakistan—the Punjab where a mil. demonstration of one of our new tanks was taking place crashed on takeoff killing all on board. This included Pres. Zia of Pakistan, our Ambas. & 2 of our Generals. We're waiting further word as to others & possible cause of crash. We took our 1st ride this morning—Nancy too. Then some more picture hanging—lunch & back to pictures. Afternoon—back to our Oak tree pruning. Then a call from Colin & Bea O.—more details on C-130 plane crash. Apparently an explosion. Trying to see if it was sabotage. Thirty seven dead. We lost our Ambas. & 1 Gen. Seven of their 15 top mil. were killed. Geo. Shultz leaves tomorrow with a delegation for funeral ceremony. We are appointing immediately a new Ambas. who will accompany George.

[Thursday, August 18: telephoned widows of U.S. officials killed in crash in Pakistan, commented, "not the happiest task but certainly necessary"; rode; spent time with Edmund Morris for biography; watched Vice President Bush deliver convention acceptance speech, commented, "he was just great. I phoned him to tell him so." **Friday, August 19:** *rode; watched preparations for luncheon party Saturday; received account of successful rally in Indiana with Vice President Bush and Senator Dan Quayle.* **Saturday, August 20:** *radiocast; photo session; lunch party; Ron and Doria present, commented, "I tried to give Ron a pair of boots but they didn't fit him."]*

Sunday, August 21

Again we awoke to fog. I switched our ride to 2 P.M. & joined Barney & Dennis in our pruning work. By lunchtime the fog was blowing away so we had our ride on what turned out to be a beautiful day We phoned Maureen to tell her how great she'd been with all her work on the convention & especially the luncheon for Nancy.

Then it was dinner—TV—"60 minutes" & bed.

[Monday, August 22: rode for over an hour; received new poll, steady approval rate at 58 percent, Vice President Bush strong versus Governor Michael Dukakis. Tuesday, August 23: Mrs. Reagan went to new house in Bel Air; flew to Long Beach to sign trade bill; attended party fund-raiser; went to Century Plaza Hotel; met with doctors on site of future exams, settled on Mayo Clinic. Wednesday, August 24: read Presidential Daily Briefing (PDB); desk work; talked to loyal supporters, then at rally, Bob Hope appeared too; haircut and manicure; desk work; attended dinner party at Betsy Bloomingdale's home.]

Thursday, August 25

Bright, hot & humid. An 11:30 AM meeting with Ken D., Kathy & Fred Ryan. This was about my post President set up. We have taken offices—the top floor of a new building across the street from the Century Plaza Hotel in Westwood. I was amazed to discover I'll have a staff of 12 & Nancy will have 4. This is in addition to Secret Svc.

There was also discussion of the Presidential library which will have a groundbreaking in November.

Then lunch—alone. Nancy is at a luncheon where our friends are giving her a kitchen shower.

[Photos with General Services Administration (GSA) staff setting up post-presidency offices.]

Then later Dr. Bookman & his team to check on my allergies. Found one new one—cats.

Then a meeting with George Scharffenberger. This was a discussion of our financial set up after January 20, also the book I'm to write etc. etc.

Then at 6:40 P.M. off to the Tom Jones's for a dinner party with our friends—some of them the same as last night & for that matter tomorrow night. The Wicks, Annenbergs, Jorgensons, Wilsons, Jimmy & Gloria Stewart, Marty Manulis, Betsy Bloomingdale. A good time was had by all.

[Friday, August 26: met with personal trainer; had hearing aids checked, commented, "My hearing remains the same—there has been no decline"; desk work; dinner party hosted by Annenbergs at bistro. Saturday, August 27: went to two party fund-raisers, first one with entertainment by the Judds; went to ranch. Sunday, August 28: rode; pruned trees; watched Emmy Awards.]

Monday, August 29

A warmer day—morning ride—this time with Nancy. Lunch—Barney, Dennis & I decided we'd try to get in a little work before we had to clean up for the big party & dinner at Barney Klinger's home. Sarah & Jim Brady came out for it & Sam Donaldson left his assignment with Dukakis to attend. It is the annual party for all the press & our staff.

But 1st we got in an hour & ½ hauling brush from yesterdays pruning to the dump. Then at 3 P.M. we headed for the showers to clean up for our 4:55 departure.

A phone call from Ken D.—Dick Wirthlin has taken a new poll—started last Sat. to check changes now that the Convention surge is over. My job rating is 59%. Handling of Ec. 54% approval. Foreign Policy handling 62%. Handling of U.S.S.R. relations 78%. Handling Persian Gulf 57%. Dukakis & Bush are tied at 47 to 47%. On the downside only 33% favor lethal aid to the Contras. On handling the deficit is 64% negative to 34% approve.

Left for Heli pad at 4:55. Arrived at Klinger's ranch. Did a receiving line of almost 500 people. Then dinner. A few words by me & recognition of Jim Brady's 48th birthday. Then we were given a gift of a 7 ft. Redwood sapling to plant. Back to Marine 1 & quite a pckg. of homework & then to bed.

Tuesday, August 30

[Discussion with Scharffenberger about easement on part of ranch being sold.]

A nice ride—Nancy too. Lunch & then with Barney & Dennis back to pruning. Quit about 4 P.M. & I faced some paperwork. Not too much & no really great decisions. Well anyway—clean up, dinner & a little TV. Heard Dukakis on TV claiming that Bush was part of a program of selling weapons to Iran. I've got to find the right occasion & tell the truth about Iran-Contra. Then to bed.

*[***Wednesday, August 31***: rode; pruned trees; watched television news, commented, "press trying to portray Geo. B. as breaking away from me on environmentalist. This is an image the press has tried to paint me as against environmentalism."* **Thursday, September 1**: *received letter from President Hosni Mubarak on his dealings with International Monetary Fund (IMF); rode; spoke for two hours with coauthor of Mrs. Reagan's memoir; called Lew Wasserman about environmentalists protesting site selected for presidential library.* **Friday, September 2**: *rode; planted tree; pruned trees; homework.* **Saturday, September 3**: *radiocast; photo session; tree pruning.]*

Sunday, September 4

Bright, hot & Nancy decided against riding. Maybe the rest of us should follow her lead. We didn't & when we got there in the hills there was a gale blowing that countered the heat of the sun. It was a nice ride as it turned out.

Lunch—pleasant as always with Barney & Dennis. Then about 2 hrs. of pruning & a cool drink (soft) in the "Bide A Wee," our name for the Dr. & Mil. Aides quarters. Signed some photos for Dennis & Barney & home for a shower. Then dinner & some T.V. watching—Jerry Lewis Telethon for Muscular Dystrophy. I phoned him—or tried to but couldn't get him. Show was coming from Las Vegas. So I called L.A. number & made a pledge of $1000. Not too long thereafter, Jerry called me. He'd gotten word that I'd tried to reach him. Then on air he told of our conversation & had the nicest things to say about me.

[Monday, September 5—Labor Day: rode; pruned trees; called friends whose son committed suicide, and Michael Reagan, who had been in Ireland; watched Broken Arrow *(1950).* **Tuesday, September 6:** *left ranch for east; Mrs. Reagan to return two days later; flew to Nebraska for speech at Hastings College and at party fund-raiser; flew to Louisville, Kentucky, and spoke to American Legion national convention; returned to W.H., commented, "in the AM I'm going to raise a holler. My box with all my cuff links & tie tacks etc. is missing."]*

Wednesday, September 7

Breaking me in gently they set starting time at 10 AM. A lovely day—temp. around 70 degrees. Started with the usual staff meeting. Talk about confrontations in Cong. re various legislative matters. Truth is the Dem's. are playing outright politics with their approach to bills. The idea is to keep us from getting a bill that will help in the election or to pass one I'll have to veto & then alienate some voters. Typical of this are amendments they're talking about in the drug omnibus bill.

Canada Free Trade should clear the Sen. in the next couple of weeks. Tomorrow is date set for our destruction of our 1st nuc. weapon under I.N.F. Treaty.

NSC—Colin took us in to see the whole new lighting plan in the Roosevelt room. It actually makes the room look bigger. New lighting in Oval office is a big improvement.

[Received news of unrest in Burma.]

China has gone into business with long missiles launching satellites into space commercially. Some interest in our own country about using Chinese launchers. Contra funding on agenda again. Bob Byrd has a bill calling for $27 mil. non lethal aid but no provision for weapons. It isn't good enough.

Next a meeting in Cab. Room with Repub. Cong. leadership. A good session on much of the same problems we'd first been discussing. Al Symms brought up Yellowstone fire & said we are not doing all we could to control it. I've told Ken to look into that—we must make every effort to put that fire out.

Lunch on the patio then a short but pleasant meeting with a Polish couple. He is top aide (press wise) of Lech Walesa & Solidarity. They think very much like us & are chafing under the Soviet control.

Then meeting with Geo. Shultz—mostly a report on the Middle East situation. Both sides are at fault.

A haircut, a sneeze shot & found out I'm up about 4 pounds. That has to be muscle, these last 3 weeks have been hard work. There is no sign of increased flab.

Some phone calls—to Bill Buckley & Al Schwabacher. Maureen here for dinner. A sleepless night. I miss Mommie.

Thursday, September 8

[Staff meeting at 9 am, with talk about Yellowstone fire; discussion of appointments.]

NSC—Still Q's. as to whether Gorbachev is coming to the U.N. opening. We're going to quietly begin bringing back the 2000 military police we sent to Panama. We'll also start reducing our fleet in the Persian Gulf but only after we're sure the Iran-Iraq peace is for real.

Frank C. is still in China & talking to them about reducing their sale of weapons world wide.

Later in A.M. Don Hodel, Dick Lyng & Will Taft (Defense) came in for a meeting about the fires. They agreed to go to Yellowstone this weekend & come back with a plan for advance fire fighting training, particularly for our military.

Then an N.S.P.G. meeting about our delinquent payments to U.N. I'm to make a final decision on making good on payments soon.

Lunch on the Patio.

[Addressed Executive Women in Government; photo session with members of Hispanic-American Chamber of Commerce; visit from Japanese philanthropist Sasakawa; saw positive poll results; photo session with George O'Neill at work on a book on leading conservatives.]

A taping session in the Library & upstairs—exercise, shower etc. And at 9:15 P.M. Nancy arrived from Calif. Rex got to her 1st but I was right behind.

[Friday, September 9: staff meeting with discussion of confirmation of nominee for secretary of Education, Lauro Cavazos, and of remaining speeches, including farewell address; report that diplomat dependents leaving Burma; Angola negotiations include South Africa.]

George Shultz really laid it on Iraq Ambas. about Iraq gassing civilians—the Kurds. He tried to deny it but George really laid it on him. We have the evidence.

[Ambassadorial formality; signed bill on commercial fishing safety; signed proclamation for Youth 2000 Week; heard about system for notifying specialists in any city to which the president travels; telephoned Drug Enforcement officer wounded in the line of duty; addressed lawyers in conservative Federalist Society; desk work. Saturday, September 10: reported good night's sleep; radiocast; reading; did exercise; watched TV. Sunday, Septem-

ber 11: *slept late; watched panel shows; had lunch with Mrs. Reagan and Maureen; spent afternoon on homework; attended dinner party with friends.]*

Monday, September 12

Up early & unnecessarily so for me. Hadn't looked at my schedule which had me due in office at 10 AM instead of 9. Well I had a whole hour to clear up some paperwork. Then came staff meeting. Some talk of the campaign. Talk of legislation. Ken says the so-called labor bill is an argument about language not dollars. The Dems. are so mixed up on the Drug bill we may not get one at all. There may be a floor fight on the South African sanctions bill. The textile bill is a veto candidate—pure protectionism. There is a so called Meese report that says nothing. My schedule for campaign is taking shape. Wed. I go to Missouri, Texas on the 22nd—Baylor U., then over to Fla. & back on the 23rd to meet with Shevardnadze. On Sept. 30 I'll do some events in ILLINOIS Wrigley field for one.

[NSC meeting with news from Lebanon of kidnapped American who escaped captors; Kampelman and others reported on ABM treaty review with Soviets; work on U.S. speech ongoing; desk time; read the Presidential Daily Briefing (PDB); briefing for and then interview with Warren Brooks, commented, "my favorite Ec. columnist."]

A Mr. Blackburn, wife & granddaughter came by for a photo. He was on the bill with me as one of the Blackburn twins at the "New Frontier" in Las Vegas—1954. My only night club performance.

[Issues-briefing lunch followed by communications planning meeting.]

Geo. Shultz came by—gave a rundown on Middle East. Then a long desk session, got a lot of work done. A meeting with Bill Webster & some agents status information. These were agents we've recruited in Soviet U. A brief personnel time with Bob Tuttle & over to Library to film an interview for a documentary on John Wayne. Upstairs to exercise—dinner & bed.

[Tuesday, September 13: Stu Spencer requested a meeting with Senator Dan Quayle (R-IN); planned radiocast on Olympics; Senate to discuss abortion language in the Labor–Health and Human Services appropriation bill; reviewed questions about nominee for administrator of General Services Administration (GSA); considered that "The Dem's. are playing games on the minimum wage bill. We want a training wage for beginners & will accept a modest increase on the minimum regular wage"; notified Congress of decision to pay dues withheld from U.N., noted, "They have corrected some of the practices that caused us to withhold"; NSC meeting report of Cuban diplomat expelled from Britain; visit from Philippines foreign minister for negotiations over payment for bases; meeting with Senator Bob Dole (R-KS) and Representative Bob Michel (R-IL) to plan legislative strategies; brief meeting with President Desmond Hoyt of Guyana; desk work; signed fair-housing bill; signed Hispanic Heritage Proclamation; photo with uniformed Secret Service agents; report from cabinet secretaries on fire situation in Yellowstone; autographed pictures; photo session

with GOP candidates; received report from United Way; allergy shot; did exercise; had dinner with Mrs. Reagan and Maureen.]

Wednesday, September 14

A busy day. 1st meeting discussed Sen. vote on Labor-H.H.S. bill. They voted our language & money.

Next was defense authorization bill. Sen. Nunn has become a real pain. He's acting as if he's the supreme authority on defense matters. Nick Brady received a unanimous committee vote for Sec. of Treasury. Textile bill vote delayed awhile. Heads off the game Byrd is playing. Sen. has agreed it will take up Canada-Free Trade bill Mon.

NSC—Iran Contra is a subject right now of Jack Anderson's column. He's more correct than the rest of the media. Discussion of Hurricane moving on Jamaica.

[Ambassadorial formalities; flew to Missouri for political rally and reception for major donors.]

Back to the airport & off for St. Louis. Met by Gov. John Ashcroft & wife Janet, also Geo. B.'s brother. I holed up for a couple of hours in the Omni hotel. Then at 5:40 Central time, went downstairs & did a reception—photos & all with major donors. Then into ball room where fundraising dinner was being held. Gov. introduced me (before dinner). I spoke & again the reception was overwhelming. Then it was back to A.F.1 & on our way back to Wash. dinner on the plane.

Arrived at W.H. at 10:05. Maureen was arriving same time from some meeting or other. Upstairs to Nancy—end of a good day.

Thursday, September 15

Into office at 10 AM—Ken told me he'd had a good day Wed.—meeting with Sen. Steering Committee.

Sen. is voting on Textile bill today. Rostenkowski wants it to go to Conf. I think he's stalling to prevent me from having a pocket veto.

Sen. Dems. are against our proposal for a minimum wage for training jobs.

House is working on a drug bill & probably will run out of time.

Then some discussion of a ridiculous book by Jane Mayer. She quotes from some staff members (Don Regan's men) to the effect that at time of Don's leaving I was so depressed & confused they were advocating I resign under the 25th amendment as no longer capable of doing job. It is all a complete lie but the press is being its usual pain in the rear.

NSC—We have a No. Korean defector [. . .].

Soviets have completed their nuc. test explosion with our observers there.

This is carrying out terms of test ban treaty. Hurricane has hit Jamaica—most powerful ever recorded. 300,000 people homeless. I have suggested we name someone—say a David Rockefeller to organize a pvt. sector fund raising to provide help.

[Briefing and then meeting with Hispanic Media; usual Thursday Mexican food lunch; addressed elementary school educators; report from Frank Carlucci on China trip; desk work; quit work early in preparation for barbecue for Congress, with entertainment by Capitol Steps.]

Friday, September 16

1st meeting did a little discussion of B.B.Q. & Cong.man Foley's remarks were very nice. We agreed I should write him a note which I did.

Then some talk about the disgraceful book by Ms. Mayer & McManus of the L.A. Times. It's based on a flurry of outright lies.

House defeated Sarah Brady's plan to have gun buyers checked for criminal records etc.

Yesterday Sen. passed Textile bill but apparently we have the votes to sustain the veto I'll deliver.

Sen. Committee voted approval of Cavazos as Sec. of Ed. Justice dept. is demanding indictment of Marcos on fraud charges.

NSC—Papandreau of Greece divorcing his wife to marry his Mistress after he has quadruple heart by-pass surgery. Gorbachev at Krasnoyarsk made a speech proposing we get our bases out of Philippines and he would close Soviet base at Camrahn Bay. He also volunteered to make Krasnoyarsk Radar a space tracking system open to international use. Catch is it could be transferred back at any time.

[Report on health problem for ambassador to Cyprus; swearing-in ceremony for Nicholas Brady as secretary of the Treasury; meeting with new secretary general of NATO, Dr. Manfred Wörner; also met with Shultz.]

Then a visit by Dorothy Ferguson who is mother of a young man who went into the Army a few years ago. She wrote me a beautiful letter about sitting in his bedroom after he left. We've kept in touch since. He's served his enlistment term & is now entering college in Iowa.

[Went to Camp David, watched The Girl from Jones Beach *(1949).* **Saturday, September 17:** *canceled ride because of rain; radiocast; homework; after dinner, as noted, "watched a long & not very entertaining movie about a man fighting drugs";* **Sunday, September 18:** *returned to W.H.; sat for last details on official portrait with artist Aaron Schikler, commented, "I'm happy with how it turned out."]*

An early dinner with Dick, Patty & Jeff then down to the family theater & about 40 guests for dinner, coffee & a movie. It was Jim Cagney in "Yankee Doodle Dandy." It was great & everybody enjoyed it. A.C. & Marf were here for it.

Monday, September 19

Started day at 10 AM—Saw Maureen & Dennis off for their week in Africa. Then a meeting about legis. & where it stands. We expect the drug bill this week. Sen. is expected to approve Cavazos today. If so, we'll have him take the pledge tomorrow. Sen. is expected to vote on Canada Free Trade bill today. Thurs. on my trip to Texas I'll meet the astronauts. Cong. is passing more appropriation bills. I'll meet with the leaders tomorrow. Decided to make next Saturdays Radio script a preview of my U.N. speech.

N.S.C.—Colin says the violence in Burma is because Gen. now in charge is for retention of socialism.

In the coup in Haiti—Gen. Avril top man seems to be supporting democracy but concern is if he's going to put Col. Paul in charge of the military. Paul is under indictment for drug running.

The press reported in detail our planks for reducing forces in Persian Gulf. They are totally irresponsible & could throw a monkey wrench in all we're trying to do.

We're considering bringing all the nations that signed the 1925 pact against Poison gas together again. Before week is out I'll be talking to Shevardnadze about Krasnoyarsk Radar.

Talk of a Basic Agreement with Soviets but no one can define what is definition of Basic Science.

[Photo session for Soviet publication, as noted, "printing an article by me on differences etc"; met with Commission on Workforce Quality.]

Then lunch & some desk time. At 1:45 met with Sec. Bennett. He's leaving us but he's done such a job I now believe there is a legitimate place for a Fed. Ed. program.

[Greeted new W.H. Fellows; ambassadorial formalities; attended farewell reception for William Bennett.]

Tuesday, September 20

First topic—Canada Free Trade Bill passed the Sen. last night. Sad news—Deaver is to be sentenced Fri. Just rcvd. news Emperor Hirohito (Japan) may not last out the week. If he dies I will attend the funeral—but according to their custom that comes 35 days after death. Feelers have come about a 1 on 1 meeting in Europe between a top level man from Iran & one of those from Wash. My feeling is that we should say such a meeting could only take place if our hostages are released.

Dan Crippen joined us at this point & I met with a half dozen Cong. leaders—both partys. We discussed appropriation bills. I made it plain I wanted them all on my desk before Oct. 1. If it happens it will be first time since 1948.

[Signed Labor/Health and Human Services/Education appropriation bill.]

NSC—Lenhart came in to tell me about a Soviet proposal. They want a free standing agreement—an exchange of letters while we continue negotiating on "Start". We'd agree on how many warheads we could put on existing missiles. It presents some problems.

Japanese Crown Prince has been named acting Emperor. There is a deadlock in Lebanon on the Presidential selection. Syria is holding out for their candidate.

New Cabinet in Haiti—all but 1 are civilians. Armed fighting has broken out between Azerbaijan & the Armenians. Ortega is coming to N.Y. for U.N. opening. He says he'll visit Wash. New leader at U.N. has come down to a choice between Argentina's foreign minister & Dame Nita Barrow of Barbados. I favor the lady.

Back to lunch in the study—a photo with Silvio Conte & a Rabbi. Then Geo. S. came in for 10 minutes. He doesn't agree with me about asking for our hostages before we have meeting with Iranians. An N.S.P.G. meeting to discuss Middle East.

[Swearing-in ceremony for Lauro Cavazos, secretary of Education; photo session; allergy shot; videotapings; did exercise; telephoned Prime Minister Brian Mulroney on Senate approval of Canada free-trade bill.]

Wednesday, September 21

Ken D. not on hand for 1st meeting—due in at noon—the Jewish Holiday. Bea conducted meeting. I learned Nick Brady will be off to Europe for meetings having to do with World Bank & International Monetary Fund. Jim Wright sounded off on a supposed C.I.A. operation in Nicaragua—this is a grave violation of policy regarding intelligence operations. Our Repub. Reps. are after him. Some talk about my speeches to come in Texas—tomorrow.

NSC—Well it didn't do me any good to be for Dame Nita Barrow for top spot at U.N. The Argentine Foreign Minister won. I had to rule on an agreement between us & U.S.S.R. having to do with Basic science. I gave it a go ahead but with a couple of changes to prevent it involving Applied science.

We'll talk tomorrow & Fri. on Warheads on missiles but feel Krasnoyarsk Radar should come down 1st.

[Report on diplomacy surrounding Lebanese elections.]

The proposed (by the Soviets) Human Rts. Conf. in Moscow. I feel we should insist Soviets clear up their lack of Human Rts. first—then we say yes to a conference & after we've taken it up with our allies. Geo. S. would like to tell Shevardnadze that we agree on a conf. if U.S.S.R. meets some requirements.

[Met with President's Council on Integrity and Efficiency, then with Joint Chiefs of Staff, commented, "They gave me a report on how much the Dem. budget cuts have hurt our

mil."; *had lunch with Vice President Bush; met with gathering of bankruptcy judges, then with Shultz about upcoming meetings with Eduard Shevardnadze; Howard Baker stopped in, Mrs. Baker improving; greeted football teammate from Eureka, noted, "He was Left Guard at Eureka when I was R. Guard"; went to dentist for final root canal work; back at W.H.; signed off on basic-science proposal.* **Thursday, September 22:** *flew to Waco, Texas, for speech at Baylor University; campaign appearances; flew to Houston to meet astronauts; addressed NASA gathering; hour-and-a-half rest at Four Seasons Hotel; spoke at campaign dinner; flew to Boca Raton, Florida.* **Friday, September 23:** *made campaign appearances; returned to Washington; briefing for and then meeting with Shevardnadze, commented, "No great decisions reached but both sides agreeing that much progress has been made & we'll continue to meet"; went to Camp David; watched* Eight Men Out *(1988).]*

Saturday, September 24

Up for radio cast. It was about the speech I'm doing on Mon. at the U.N. Back to "Aspen" for lunch & then on a dark day we went for a quiet horse back ride. Rest of afternoon—homework for Mon. & Tues. meetings with heads of state in N.Y. Dinner & the movie "Gorillas in the Mist"—the story of Dian Fossey played by Sigourney Weaver. Very good.

Sunday, September 25

A wet, foggy day—left for W.H. at 11:30 (have to pack for N.Y. tomorrow). Arrived about noon—a 12:30 lunch & then to desk work & packing. Late in the afternoon a phone call from Billy Graham. He was present last night at a tribute dinner for Paul Harvey. I had taped a speech for the occasion. Billy was most gracious in his praise for my effort & said it was in the top ten of all the speeches I've ever made. He made my day.

Monday, September 26

Up & on our way—Marine 1 to Andrews, A.F.1 to Newark airport. Another Marine 1 to helipad in NY. harbor & motorcade to U.N. Bldg. Met by Chief of protocol & Sec. Gen. Perez de Cuellar. A brief stop in his office with Shultz & Vernon Walters. Then to courtesy call on U.N. Pres. (brand new) Dante Caputo, Pres. of Gen. Assembly. Then into Gen. Assembly & my speech. Walters said most applauded speech he's ever heard in the Assembly. Apparently I did strike a chord or two. Then Nancy left for Waldorf Astoria. I met with Sec. Gen. Then on to Am. Mission & lunch with Walters, Shultz, Ken D., Bea. Oglesby & Colin Powell. Up to 12th floor for a brief talk to our U.N. delegation. Over to Waldorf Astoria & meetings—1st with Foreign Minister Maguid of Egypt & also F.M. Shimon Peres of Israel. Next was Pres. Mugabe of Zimbabwe—he took most of meeting attacking Southern Africa. Then it was Sheikh Jabu Al-Ahmed—Amir of Kuwait.

From there we went to a brief meeting with the Gulf Cooperative Council members chaired By His Royal Highness Prince Saud, Foreign Minister of Saudi Arabia. Then upstairs in time to see Nancy off—to the Opera with the Annenbergs. For me it was shower & homework time. Then at 6 P.M. downstairs for reception for Sec. Gen. de Cuellar & heads of state. It went well. I shook a few more than 200 hands—had a few one on ones afterward around the room— Pres. of Venezuela, F.M. of Thailand etc. Then back to the suite for dinner alone & so to bed. Got word a few cliffhanger bills had gone our way—F18 sale to Kuwait for example.

Tuesday, September 27
Up fairly early—Ken D. & Bea & Colin came in for a little routine work & a bill to sign. All this at the Waldorf Astoria of course. Then at 9:30 down stairs for a gathering of NATO & Allied Foreign Ministers. A round table discussion of our future & it turned out unanimous agreement on the importance of preserving our unity.

Back upstairs for a brief time then into another room for a meeting first with Foreign Minister Khan of Pakistan. A worthwhile meeting to make contact with the new admin. there since Pres. Pias [Zia's] tragic death.

Then a meeting with F. Min. Rao of India—this, too was worthwhile. We've had a good relationship with P.M. Ghandi & a meeting like this helps.

Back to the suite & saw Nancy off to her luncheon date with friends, then lunch myself. And a couple of hours of reading. Met Nancy downstairs—a little late—N.Y. traffic held her up. Motorcade to Wharf—into Marine 1 & back to Newark airport. Then A.F.1 to Wash. & back at the W.H. a little after 4 P.M. Exercises, shower & dinner with Merme & Dennis who are just back from a week in Africa. Then we ran the tape of my U.N. speech.

Wednesday, September 28
Not into office until 10 AM. Actually it was the study. The Oval O. was being set up for some filming with candidates We had some talk about schedule. I'll do some Calif. campaigning toward the end of Oct. which will mean a couple of days at the ranch. Then A. B. Culvahouse came in to explain why I must turn down prosecutor in Ollie North's case who wants to call me as a witness. It would set a bad precedent. I have agreed to provide Defense with some written answers to Q's.

[NSC meeting regarding chemical warfare and desire for meeting of nations, also military bases in Spain, and Noriega aggression; Eduard Shevardnadze's schedule; vote in Chile on General Augusto Pinochet; signed U.S.–Canada Free Trade Agreement, commented, "Now it only remains for the Canadian Sen. to sign on"; photo session with Boys

Clubs Youth of the Year, and with representatives of Historically Black Colleges; received Alexander Graham Bell medal for contributions to the hard of hearing; attended appropriations bill ceremony; filmed campaign commercials.]

Thursday, September 29

[Meeting to discuss congressional blocking tactics in legal services; legislative plans on drug bill; NSC meeting regarding asking David Rockefeller to head drive for disaster help for Jamaica; Shultz secured France's approval of plan for conference on chemical warfare, commented, "It's beginning to look as if we can close out the C.S.C.E. conf. in Vienna & move on."]

Carlucci has suggested we should break the secrecy on the Stealth fighter & announce it next week. We have more than 50 now & we should put them into our regular air force program. Took a farewell photo with Marian Dicks of NSC. Then desk time & some bill signing 'til 11:45. Then it was a Rose Garden ceremony—handing out 30 C. Flag Crystals to org's. & businesses who have been outstanding in private sector programs.

[Received Irish medals from ambassador to Ireland.]

A briefing for Mitterrand meeting. He arrived. We had the usual 1 on 1 then a walk around the Rose Garden & into the Cabinet Rm. for Plenary meeting. Things went very well.

[Dinner for the Mitterrands, entertainment by Peggy Lee. **Friday, September 30:** *flew to Chicago, Illinois, met family of policewoman slain in the line of duty; made campaign appearance at Wozniak's Casino beer garden; threw out first ball at Wrigley Field, joined in play-by-play with announcer Harry Carey; attended party fund-raiser at Mercantile Exchange; returned to Washington; received call just after midnight with news that Congress passed appropriations bills with two minutes to spare.]*

Saturday, October 1—Begins the '89 Fiscal Year

A meeting at 11 AM with Sen. Quayle our VP cand., Stu Spencer, Ken & Bea. I was supposed to give him some debating tips & did. He's a fine person & has been given a bum rap by the media. Then down to the Oval O. for signing of the appropriation bills. Did my radiocast & took photos with Nancy Robert's family all 21 of them. Received a plaque from 2 police officers—St. Louis county police. I'm now an honorary policeman.

Then upstairs—lunch on the balcony & changed clothes for appearance at Georgetown U. where I'm to get an honorary degree—Dr. of Humane Letters, & make a speech.

Well I had to don an iron vest for the outdoor event where somewhere close to 10,000 attended—most of them students. I was warmly received.

[Security scare at W.H. was investigated without incident; watched news clips from Chicago appearance.]

Sunday, October 2

At noon Nancy off to the Redskins F.B. game where something goes on involving her with regard to the anti-drug campaign. So I watched the game on TV—Redskins lost 24 to 23.

[Desk work; exercise; dinner with friends.]

Monday, October 3

[New Time *magazine poll shows Bush leading; presidential approval rising; NSC meeting regarding hostages in Beirut, noted that President Straus of Bavaria died; greeted exchange students from Yale and University of Moscow, commented "Nice looking kids & you couldn't tell the Russians from the Americans. All could speak the others language"; desk work.]*

Then a ceremony in East Room for signing of Columbus Day proclamation. It's 500 years since he discovered Am. A Cabinet luncheon—just a session to talk about things & greet our 3 new Cabinet members—Brady, Thornburgh & Cavazos. In Roosevelt rm. after lunch, spoke to meeting of Nat. Drug Policy Board now chaired by Dick Thornburgh.

[Authorized commemorative silver dollar for Eisenhower; meeting with Prime Minister Edward Seaga of Jamaica, looked at hurricane pictures; haircut; desk work; attended two GOP fund-raisers.]

Colin called—we have further information but no proof on hostage. According to information it would be Singh the Indian Professor at Am. U. in Beirut. He may be on his way to American Embassy in Damascus.

Tuesday, October 4

Nancy off to a luncheon in N.Y.—Anti-drugs again. My 1st meeting—at 9 A.M. Learned Sen. had put the death penalty we want in the drug bill. Polls in Canada look good for P.M. Mulroney in coming election He was way down but apparently has recovered.

I'm faced with a cliff hanger—the votes to sustain or override my veto of Textile bill is going on in the House.

NSC—Discussion about Nicaragua & the Contras—doesn't look like we have the votes to get military aid for them.

In Iran Fadlallah says more hostages not coming out. Then the 9:30 meeting with Repub. Leadership. It was a good meeting & ended up with almost every one of them paying a great tribute to me for what they called my leadership to the world. I was touched.

[Met with President Joao Vieira of Guinea-Bissau; desk time; presented awards in ceremony for Minority Enterprise Development Week; had lunch with 114 party donors of $100,000 or more; greeted pageant winners; met with Senator Bill Armstrong (R-CO), who suggested executive order banning pornographic material on federal property (includ-

ing bases), commented, "I'm in favor but must check out some constitutional problems"; allergy shot; telephoned congressmen seeking support for veto of textile bill; went to party event, learned veto had been sustained.]

Wednesday, October 5

A beautiful Fall day—clear, sunny, cool & no humidity. First meeting 9 AM. Talk about the veto vote of course. Dan Rostenkowski (Dem.) Chmn. House Ways & Means Comm. was of tremendous help to us. He's been that way on a number of important issues. He is truly bipartisan.

Some discussion of my use (in speeches) of the figure 62.7% of Ams. male & female from age 16 & up employed is not completely clear to the audience. I think I can add a little explanation that will clear things up. It is the highest percentage in our history.

Colin brought A. B. Culvahouse in to explain a situation that was distorted in the press. Charge has been made that I have decided on exec. order & will now approve assassinations as part of anti-terrorist programs. Truth is I have not. My order against such acts is still in force.

[Problem with ice breaker caught in Canadian territory; visit from President Joseph Momoh of Sierra Leone; desk work; laid cornerstone for Holocaust Memorial Museum; photo with Secretary of Transportation James Burnley and Governor Jim Martin (R-NC) to promote program against drunk driving; photo with candidate from Wisconsin; videotapings; photo session for Vanity Fair; *greeted Ziemans, formerly of USSR; watched vice presidential debate with Maureen and Mrs. Reagan, commented, "I think Quayle did very well & Bentsen got away with some blatant falsehoods."]*

Thursday, October 6

First meeting—we spent some time re-hashing the debate. Then got report Sen. Judiciary Committee had voted to send 11 of our judge appointees to the floor. That leaves 14 still waiting. Sen. & House conferees made it easier to close obsolete mil. bases. I was told Helen Thomas wrote a good column about me.

NSC—More talk about closing bases. Then a rumor that our Lt. C. Higgens—kidnapped in Lebanon, is dead. But we can't get any confirmation.

One of Gorbachev's new appointees, Medveded is making speeches that sound like he's pushing for free enterprise.

[Visit from President Moussa Traore of Mali; desk work; met with Frank Carlucci, who was pleased with new military authorization bill; attended ceremony for German-American Recognition Day.]

At 2:00 P.M. the Grand Duke of Luxembourg for a brief visit. He was here in 1985 for a State reception & dinner. Bob Dole & Bob Michel came in to get my urging that we get the drug bill before Cong. goes home. And that did it until the State Dinner for Pres. Traore.

Again, a wonderful evening. Some old friends on hand but all in all a good mix. After dinner entertainment was Pete Fountain, the great clarinet player & his group of 6. Then some dancing & by midnight upstairs to bed.

[**Friday, October 7:** *flew to Detroit, Michigan; spoke at high school; at Chamber of Commerce lunch, commented,* "*actually made my speech a partisan campaign speech for Bush & Quaale [Quayle]. It was received as if they wanted a Repub. rally*"; *spoke at party fund-raiser; returned to Washington; went to Camp David.* **Saturday, October 8:** *slept late; radiocast; walked; rode; homework; movie in the evening.* **Sunday, October 9:** *watched panel shows; rode; homework; movie.*]

Monday, October 10—Columbus Day

A bright day. Finished my book by Larry Mc Murty—"Anything For Billy." Some homework—lunch & off to the W.H. Arrived there about 2:30. Watching playoffs game between Dodgers & Mets for awhile after arrival. Dodgers won—now one to go in L.A. tomorrow. Dinner & bed time. Ted Graber was on hand for dinner.

A call from Ron. I don't know who he's been talking to but we had a debate. He's convinced I'm not protecting the environment.

Tuesday, October 11

[*Discussion of Canadian election; of prospects for drug bill.*]

We need to provide the Contras with some hope. A proposal has been made that I threaten to call the Congress back in a special session to get them to act now.

Some talk about Marcos. Justice dept. want to indict him—I believe we should suggest a deal—he gives Philippine govt. a bundle of money & we lay off.

[*Talk about remaining speeches.*]

NSC—Colin brought up Contra matter. He supports idea of me threatening to call Cong. back.

Duarte (Pres. of Salvador) back in hospital here. I called him. He says it was to check & that his liver cancer is reduced by half.

In keeping with I.N.F. treaty we've notified U.S.S.R. we'll be taking space photos of SS25 missiles in its garage. They are to open the door so we can see it's not an SS20. Treaty lets us do this 6 times a year.

[*Report on progress of icebreaker in Canadian waters; short meeting with Prince Sihanouk of Cambodia; photo with Reserve Officers Association; visit by cousins, Virginia and Harry Smith; NSC meeting briefed on security measure [. . .]; interviewed by biographer Edmund Morris; taped ads for candidates; heard that friend Bunny Wrather (Bonita Granville) died of cancer; attended party fund-raiser.* **Wednesday, October 12:** *flew to Philadelphia, Pennsylvania; spoke at high school; spoke at party fund-raiser; paid visit*] *to Archbishop Bevilacqua; met with family of police officer slain in the line of duty; flew to Newark, New Jersey; spoke at Italian American society; meeting with major donors; returned to W.H.]*

Thursday, October 13
Started day at 10 AM. Talked about legislation permitting some closing of obsolete mil. bases. We've tried for a long time to get this.

Also talked of N.Y. Times poll shows my approval rating at 60%. It was 45% last Nov. Then the Drug bill—about 100 amendments are kicking around & holding up progress.

NSC—Word on Duarte isn't as good as he led me to believe. Apparently he's trying to stay through next election in March.

France has agreed to go to a Human Rights meeting in Moscow. Thatcher is saying no.

[Photo sessions with staff members; visit from delegation of Asian-Indian Americans; then delegation of Ukrainian religious leaders; signing ceremony for welfare-reform bill; farewell reception for departing director of Office of Management and Budget Jim Miller; taped commercials for candidates; exercised; watched debate.]

Friday, October 14
[Proposal for Mrs. Reagan to speak at U.N. on drugs.]

Then talk of doing a retirement event (me) for Jack Kemp. Some progress is being made on drug bill in the Sen. A statement approved by me about aid to Contras if Sandinistas attack.

A Prof. (History) has refuted Sen. Bentsen's claim to have been a close friend of J.F.K.

NSC—Report that our soldiers in Korea fired on an N. K. soldier who was seen in the demilitarized zone. He was seen to fall down but nothing further is known.

Pakistan about to realize & report their version of plane crash that killed Pres. Zia. We have no more information than that.

Our negotiations regarding war in Angola have come down to, how long Cuba can take to remove their troops. Angola govt. is holding out for 30 months. Strangely enough the U.S.S.R. is asking for a shorter time—which is what we also want.

[Noted that Mexico's minister of finance was meeting with Secretary James Brady regarding currency exchange rates; pollster Dick Wirthlin optimistic regarding Bush campaign; attended ceremony for astronauts; photo session; signed alternative-motor-fuels bill; greeted officials of Congress of Independent Unions; greeted rabbi brought out of USSR with help of two senators; regular meeting with Shultz, most of discussion on negotiations on Angola.]

Some more photos in Oval O. & Dip. Rm. then upstairs to a lonely W.H. Nancy is in N.Y.—due back for dinner. Did my exercise. Surprise, surprise, Nancy arrived about 5 P.M. I'm trying to get her to toss me a baseball a few times. Tomorrow night she'll be in Los Angeles tossing out the ball for the 1st World Series Game at Dodger Stadium. Well she gave in & we did it in the Center Hall. She did very well. Dinner, a quiet evening & so to bed.

Saturday, October 15

And Nancy is off again, this time to L.A. to throw out the 1st ball tonite at the World Series starts. A leisurely morning here & then my radiocast—this time on who is responsible for deficits in Fed. Spending—who? The Congress.

Well—spent the day reading & checking over things for the Library. Then exercise—shower & dinner. Watched 1st game of World Series. Saw Nancy throw out the 1st ball—she done good. Also told crowd & TV audience about the need to step up the anti-drug campaign. Dodgers won 6 to 5 with an exciting come from behind finish. And so to bed.

[**Sunday, October 16**: *read newspaper; watched panel shows; watched football game; Mrs. Reagan expected back, noted, "Nancy due back at 6:25—but she arrived at 6:10. Hooray!"*]

Monday, October 17

In the 1st meeting a little summing up as this is last week Congress is in session. The drug bill has to go to conference—differences in House & Sen. bills. That may mean a delay 'til after recess. Then some talk about the lousiest pol. trick ever played. A discharged employee of Designer Jimmy Galanos has hit the press by way of "Time mag." Claims he's writing a book on the Reagans. Then he claims Nancy has bought $1.4 mil. worth of gowns at special favor prices. That she's borrowed & not given back scores of others etc. If what he says was true she could be in for tax evasion etc. None of this crud is true & her assailant is a Greek—way out in front for Dukakis. Does anyone say this isn't a political trick.

[*Talk of schedules and campaign.*]

NSC—We're waiting for word on the plant [base] closing negotiations with the Philippines. I've another letter on the way to Pres. Aquino. Another meeting here with the Mexicans. I think we've arranged a 3½ bil. $ loan to help them out.

Pakistanis have made official their claim Zia's plane was sabotaged. Some talk about Korean & Pres. Roh's efforts to bridge the gap with N. Korea.

Assad (Pres. of Syria) made a secret visit to Moscow. Our information is that Moscow is now demanding cash payment for military supplies. A brief desk time then over to the high school. Cardinal Hickey took me up to the chapel 1st

to see the altar used by Arch Bishop Carroll more than 200 yrs. ago—the first
Catholic altar in the U.S.

[Addressed students; returned to W.H.; luncheon for recipients of Medal of Freedom;
desk work; mostly photo signing; exercised.]

Tuesday, October 18

Some talk about Nancy then newest poll on Bush in Wall St. Journal—he's
jumped way ahead of Dukakis. Something like that has happened in Canada—
Mulroney is in first place with 43%. Then we decided I'll attend the dinner
where Nancy is being honored Thursday. And some talk about my campaigning.

NSC—An attempted coup in Haiti—3 Lt's. taken captive. More about the
Mexican loan. Papandreau, P.M. of Greece—open heart surgery—now a relapse
& infection set in. After NSC meeting—Ken D. stayed behind to recommend
some promotions for a number of my people including Kathy & Jim K. After a
desk session—bill signing etc. A photo & short visit with Patriarch Sabbah—of
Philistine Catholics in Israel. Then lunch & at 12:45 photos with 8 West wing
guards. At 12:50 Dr. Hutton—at his office. Sneeze shot & molds were taken of
my ears for new hearing aids.

At 1:30 Kathy & Fred Ryan for some talk about my office in L.A. We'll
have a staff of 13 & Nancy will have 4 of her own. Then a meeting with George
S. Talked mostly about Central Am. But 1st a brief meeting with Sen. Stennis
about a promotion for Admiral presently assigned to NASA. I'm in favor. A session of tapings—7 all told. Then upstairs for exercise & shower. Nancy is very
upset about the fuss raised over her wardrobe. Dinner & bed.

[Wednesday, October 19: flew to Columbus, Ohio; spoke at Ohio State University;
went to Bowling Green; spoke at rally; went to Cincinnati, attended campaign dinner;
returned to W.H., commented, "It was an overwhelming day. The warmth of my reception
at every stop was very moving."]

Thursday, October 20

Started day at 10 AM Some talk of budget for 1990. Under the law I submit such a budget & the succeeding Pres. can change it. I decided we should do
our best to come up with the budget I'd want if I were going to administer it. At
same time confer with Bush so he would have some change he'd like to make.

Then some talk about the pending immigration legislation. It's on hold
because my friend Al Simpson (Sen.) objects to a couple of points that are important to me. Ken is working on him.

[NSC meeting with details of suicide bombing along Lebanon-Israeli border; former
Philippine president Ferdinand Marcos, defying subpoena, was charged with contempt
of court; visit from Bettino Craxi, former prime minister of Italy; visit by President Roh

Tae Woo of South Korea, commented, "Discussed our trade problems, his efforts & ours to establish a relationship with N. Korea"; greeting George Nash with new book on Herbert Hoover, along with Hoover descendants; photo session; reviewed appointments.]

Then over to the Dr's. office to be fitted with a form of magnifying glass mounted on a pair of spectacles.

Upstairs for exercise & shower & dress up for testimonial dinner for Nancy—a really big nationwide affair. Kiwanis, Lyons, Sentorna, Scents & a host of other service & fraternal org's. joined in a salute to Nancy for the fight against drugs. Bob Hope was a surprise guest, speaker after speaker gave her praise & named her as the leader in the war on drugs. So to bed.

[Friday, October 21: flew to Raleigh, North Carolina, and Bowling Green, Kentucky, for campaign appearances, including one at Western Kentucky University basketball arena, commented, "I went down some steps from the dais to a toe marker & shot—with 12,000 watching a basket—missed on 1st try & sank the next one"; went to Camp David with the Dubersteins. Saturday, October 22: cold day, but rode anyway; watched That's Entertainment, Part II (1976).]

Sunday, October 23

Wouldn't you know—the sun was shining & it was a beautiful day. We boarded Marine 1 at 11:30 AM & were at the W.H. a few mins. after 12. So lunch—an afternoon of TV (The Red Skins won), a lot of reading. Nancy & I called Sandra Day O'Connor in the hospital. She's just had a mastectomy. Nancy was most helpful to her. Then I called Lew Wasserman to thank him for all he's done to help us on the Library & my office. Dinner & to bed—well not quite that simple. Nancy had to appear at the Nat. Horseshow—but she was back around 10 P.M.

Monday, October 24

Dave Gergen had a column in U.S. News & Report. For a change he was quite nice to me. John Henderson replied to Ohio's Gov. Celeste for me. He had written a letter about the leaking nuc. plant in Ohio. John explained it would not reopen 'til it was absolutely safe. Some discussion of the bills we got & didn't get before Cong. went home. One we got was V.A. becoming a Cabinet post & the plant closing. We did not get U.S.I.A.'s World Net. expansion, nor did we get U.N. Peace Keeping appropriation. Did get Bermuda tax treaty & the Immigration bill. The Nat. Adoption bill was killed by Jim Wright.

NSC—More on Marcos & some of the wild rumors in the media—that he contributed to my campaigns & that we're indicting him to get a settlement on the mil. bases in the Philippines.

Shultz is asking permission to tear down our Embassy (half built) in Moscow & start over—question is where does the money come from. Soviets want

a meeting on Krasnoyarsk. We want them to tell us what they are thinking of
1st.

[Haitian president requested help with major drug bust; desk work; greeted 550 American Olympians; photo to launch United Federal Fund Campaign; signed base-closing bill; greeted twelve-year-old essay winner from Vote America Foundation; Mrs. Reagan in NYC for address at U.N., commented, "Slept lousy—at 1 AM I tried reading for an hour."]

Tuesday, October 25

A lot of press about my meeting with the Olympic team. Some talk about a radio speech on competence & ideology in campaign. I say we talk about both. Tomorrow I'm to call Lou Cannon for 3 Q's. he's doing a campaign article. Congress is working on an ethics bill.

N.S.C.—A letter from Gorbachev—he is now President as well as Gen. Sec. Soviet Space Shuttle is going up without a crew. We never did that. E. Germany had paid no reparations to Jews who went through the Holocaust. Mujahadeen are laying siege to Kabul. We have 17 Americans in our embassy there. Japan is upset by our wanting to sell rice to them. Then brief meeting with Pres. Conte of Guinea. He's making progress on the economy. He spoke eloquently about how great our Peace Corps volunteers are. Then some desk time & at 10:30 AM I left for Ft. McNair where I addressed a large crowd of mil., veterans & officials on the 5th anniversary of our landing on Grenada. Then I signed legislation making dept. of Veterans Affairs a Cabinet position.

[Bill signings; photo sessions; received gift from refugees from USSR.]

A 1 hr. N.S.P.G. meeting on E. Europe. Those countries are restless under the Soviet domination.

[Desk time; farewell party for departing staffer; videotaping; allergy shot; Mrs. Reagan home.]

Wednesday, October 26

Connie Mack wants me for Fla. again. I may have to try & offer a tape or something. The schedule looks like Fla. is impossible. I approved a replacement for Gary Bauer—Fran Marie Keal. We discussed a new Ambas. for Morocco—Mike Ussery. discussed a transition team to get under way Nov. 9th. Word on a Whistle Blower bill. When it comes I'll pocket veto & ask for a better one in Jan.

N.S.C.—Sakkarov is coming to the U.S. for a visit. I'll see him. He's representing something called—"A Fund For the Survival of Humanity." Colin P. reports on a N.Y. Times reporter who has done an article revealing terms of an N.S.S.D. I signed off on (& which is top secret) last week. John Negroponte reported on a valley in Peru that produces 40% of all the Cocaine in the world. D.E.A. wants us to double our mil. involvement & presence in Peru. Also increase our financial help from 10 to $20 mil. They suggest a survey team to go & work with Peru's Nat. Guard.

[Flew to Baltimore, Maryland, for party fund-raiser luncheon; returned to W.H.; greeted L.A. Dodgers, world champion baseball club; Mrs. Reagan left for California; economic meeting on positive outlook for economy; short meeting with Shultz [. . .]; met with Howard Baker on his campaign efforts; short interview with Lou Cannon; packed for trip. **Thursday, October 27**: *flew to Little Rock, Arkansas, for airport rally; flew on to Springfield, Missouri, and San Diego, California, for campaign appearances; met Mrs. Reagan at Beverly Wiltshire Hotel in L.A.; received telephone call from Vice President Bush.* **Friday, October 28**: *addressed World Affairs Council; went to ranch.]*

Saturday, October 29

[Rode; changed into work clothes, commented, "First chore cutting up a dying Oak tree that had gone down"; further cutting; quiet evening.]

A call from Walter Annenberg. He had seen the Video Tape of the "Reagan Days" which Lew Wasserman is putting on the market—all proceeds going to the "Library." Walter said he was in tears for much of it.

Sunday, October 30

[Rode; received gift of photo of ranch as it was in 1920s.]

Lunch, some picture signing for Barney, packing, a shower & all dressed up for trip to L.A. & drop by dinner for Simon Wiesenthal—80th Birthday. Mainly it was to announce the start of the Wiesenthal museum on the Holocaust. It was a pleasant evening with a number of our old friends present. Arnold Schwarzenegger made a speech that was most generous about me. Then Nancy received the 1st of their "Museum of Tolerance Awards." My turn & I was presented the 1988 Humanitarian Award." And back to the Beverly Biltmore to bed.

Monday, October 31

Up fairly early. Nancy was off to our new home at 9:15 AM for a day of opening boxes & arranging. My 1st was a meeting—Ken D., Colin etc. here in the suite—Marlin. First topic was campaign schedule for me. We have decisions to make. Jim Baker wants Ill. & Texas—one or both etc.

[Canadian campaign at an even split, with Prime Minister Brian Mulroney having lost lead in polls.]

Learned I've got another 20 bills to sign today—possibly 1 veto.

NSC—Colin brought up video tape of Terry Anderson supposedly criticizing us for blocking rescue attempts. Obviously he was taking orders from his captors.

We're having Soviet Ambas. called about Soviets moving heavy bombers close to the Afghanistan border & within range of Pakistan. Then downstairs for signing of the Berne Pact legislation. This is legislation to protect Am. copy-wrights, books, movies & Computer software. A.C. Lyles had stage full of ce-lebritys, long time friends of ours. Lew Wasserman gave me 1st printing of E.T. videotape. Then a lot of photos & upstairs after a barrage of press Q's. most on Terry Anderson.

[Meeting with Kathy and Fred Ryan to plan L.A. office and presidential library; then with Ventura County Supervisors; signed thirty bills, with one veto; showered and changed into pajamas; telephoned Secret Service to advise that Mike Reagan and his wife wanted to end agent protection in mid-November.]

Nancy returned—worn out by the work she'd been doing. We had dinner in bed with the TV news.

*[*Tuesday, November 1: *left for campaign speeches in Orange County, California, Reno, Nevada, and Milwaukee, Wisconsin.* Wednesday, November 2: *flew to Ohio for campaign appearances in Cleveland; returned to W.H.]*

Thursday, November 3

Went to office at 10:00 AM. First subject was about a N.Y. Times story about me—surprisingly in the Times. Then K. told me that I'll be signing about 300 bills the Cong. left for me. Signed about 40 of them before the day was over. Polls in Calif. went up 3 points after my trip so—I'm going to Calif. Monday. In upcoming transition I'll have to work out a resignation plan for many of my appointees. Also, I have to submit a budget for 1990.

[NSC meeting regarding Israeli election; attempted coup in the Maldives; Soviet warn-ing to Mujahadeen in Afghanistan; Geneva talks with no breakthrough on Krasnoyarsk Radar, commented, "which we see as breach of ABM treaty; Soviet military producing new SS18 missile more powerful than present model"; desk work; spoke at party fund-raiser luncheon but left before the meal; had lunch alone in study at W.H.]

At 1:30—Dick Wirthlin's new figures—our polls are up all along the line. Then Geo. Shultz meeting—no startling news—some about Afghanistan etc. Haircut at 3 PM, then a visit & photos with Courtney Trissler (our Ranch Fore-man) & his two daughters. Bob Tuttle & some appointments—my sneeze shot & a taping in the Library. Upstairs—some desk time & Nancy arrived. Merme is also here. Dinner & bed.

*[*Friday, November 4: *announced that unemployment rate was down to 5.2 percent; flew to Chicago, Illinois; signed Genocide Implementation Act; made campaign speech at community college; flew to Philadelphia, Pennsylvania, for further campaigning in New Jersey; returned to W.H.* Saturday, November 5: *made campaign appearances in Michi-gan and Texas; returned to W.H.]*

Sunday, November 6

A lazy day—beautiful, sunny & no humidity. I got some bothersome correspondence out of the way, also some reading. Exercised, showered—a few hours after a lunch in the Solarium & early to bed.

[Monday, November 7: campaign appearances in Long Beach and San Diego, California; returned to W.H. at 11:10 pm, commented, "Joined Nancy in bed & the campaign is over."]

Tuesday, November 8—Election Day

No office schedule. Edmund Morris at 2:30 (after Nancy & I had lunch in the solarium). Then Ken & Dick W.—report on exit polls, they think the news is good. Then a rub down & up to shower & dress for our dinner party. Phone call from Ambas. Price & asking about election. A photo of us watching TV—returns by photo pool. Dick & Patti have arrived. (Phoned George & Dan with congratulations.)

Then 28 friends here for buffet & to watch the returns on TV. A great evening as almost from the beginning it was evident that Bush-Quale were winners. About 11 P.M. to bed—happy.

Wednesday, November 9

Into the office at 10:00 AM The Rose Garden was jam packed with all our staff cheering me. A very touching moment for me. Into the morning session. Some talk about the election of course. Planned for George's afternoon arrival here from Houston. I have appointed Ken as Chmn. of our transition team. Dan Crippen joined us & we talked about '89 budget I must submit by Jan. 9.

N.S.C.—Dep. Foreign Minister of Kabul's Afghan Puppet govt., ABDUL GHAFFAR LAKANIHL was in N.Y. to address the U.N. He came down to Wash. & turned himself in to State dept. asking for Asylum. We're keeping it secret 'til his family can get out of Afghanistan. Col. Paul (involved in drugs in Haiti) dead of heart attack. Rumor has it he was poisoned. Arafat has asked for a visa to come to the U.N.

A man—no hostage—was in jail in Iran for something during Pres. Carter's term. Then he was freed. He committed something or other & is back in jail in Iran. Ex Pres. Carter wants him freed again. Libya is building a chemical weapons plant that threatens the entire Middle East. [. . .] We, too are looking at that. Kohl (W. Germany) has been of help to Libya. I'll have to take this up with him when he comes.

[Bill signing; gave National Security Medal to General William Odom, director of the National Security Agency (NAS); photo session; met with member of the Afghanistan Resistance Alliance; regular meeting with Shultz, commented, "nothing of great importance"; ambassadorial formalities.]

Then at 3:50 Nancy came down—we greeted the Bushes & Quayles. Made an appearance—the 6 of us—before the press in the Rose Garden. George & I had a little one on one & I asked him to continue my custom of returning military salutes. He's for it. Then upstairs, exercised (finally)—dinner & bed time.

Thursday, November 10

Regular meeting only with the V.P. present after a long absence. Talk of transition & early staff & Cabinet meetings. All appointees are asked to resign but to agree to stay until replacements are in place.

Told my radio cast will be about Veterans day which is tomorrow. A. B. Culvahouse came in. There is a disagreement between the Attorney General & Sec. of Transportation. Burnley wants testing for drugs in travel industry. Dick T. objects & claims it violates the Const. On the Oliver North case—"Walsh" says Cong. could deny the President the right to order any covert operations.

NSC—Today Defense announces existence of Stealth Fighter planes. We've had 50 in the desert operating only at night. Now they are going to be operational.

Pres. Carter sent a note that anything to do with his former pilot now in an Iranian jail is called off.

Canada is presenting a problem. P.M.'s opponent Turner looks like he's going to be the winner.

Our Sec. of the A.F. is resigning to take a position with an aircraft firm. Sec. Shultz wants us to deny a visa for Arafat. [. . .] Our one open channel to Iran's govt. brought a message of congratulations to Geo. B.

[Desk work, read Presidential Daily Briefing (PDB); received award, as did Gorbachev, from the Beyond War Foundation; attended luncheon for all living recipients of Medal of Freedom; cabinet meeting to hear Joe Wright on the budget for 1990.]

After that we swore Joe in as new Dir. of O.M.B. Usual meeting with Bill Webster, dir. of C.I.A. He reported on new satellites we have [. . .].

[Gave POW medal to World War II veteran.]

Upstairs—exercise—shower. Dinner—Merme here. After dinner Dick Davis, Patti & Geoff arrived. At 8:30 I phoned Yasu Nakasone—returning his call. Just congratulations. Early to bed.

Friday, November 11—Veteran's Day

At 10:40 AM off to Arlington—all our guests along. I laid a wreath at the Tomb of the Unknown soldier. Then into the amphitheater where I gave a brief speech. From there over to the Vietnam Vets. Memorial—some 20,000 people there. I spoke—message was well received. Then I read a note Nancy & I had signed & we put it at the base of the memorial. It's a tradition that has grown of people leaving letters, gifts etc. at the memorial.

[Went to Camp David with family and friends; watched Winchester '73 *(1950).*
Saturday, November 12: *radiocast; lunch with guests, noted, "Eddie Serrano (steward)*

showed the gals how those special brown W.H. crackers are made"; rode; after dinner movie was Cocoon: The Return *(1988).* **Sunday, November 13:** *watched ABC's* This Week; *returned to W.H.; read and watched TV; dinner party with friends.]*

Monday, November 14

Into the office at 10 AM—discussed random drug testing by dept. of transportation. Then Gorbachev coming to U.N. to address Gen. Assembly. He'll be available for a get-together with the V.P. & me in N.Y.

NSC—More on Gorbachev & N.Y. Then Colin said Soviet Shuttle is supposed to go up at 10 P.M. our time tonite.

Then it was time for meeting with Andrei Sakkarov in Oval O. A half hour meeting—talked mainly of Perestroika & his view that part of that should involve human rights.

[Read Presidential Daily Briefing (PDB); presented Malcolm Baldrige Quality Awards to three firms; addressed junior high students.]

A haircut then over to Dr. Hutton to check my allergies—he believes I may have picked up a virus. I'm having post nasal drainage problems.

Upstairs to exercise & shower. I should have pointed out that our morning discussion of drug testing was because Dept. of Transportation has issued order to entire transportation industry for drug testing.

Quiet evening & early to bed.

Tuesday, November 15

My Biographer, Edmund Morris wants to stick with me an entire day. I've said yes. We've decided on a press conf. Dec. 8. A luncheon will be held in Wash., 2nd week in January to raise money for F.D.R.'s Presidential Library. They want me to be Honorary Chmn. How do you say no to that? Gorbachev & Raisa are going to visit Eng. on way home from U.N. speech.

Noted misinterpreted letters sent by Federal Agency.

NSC—Lots of Q's. about P.L.O. statement on Israel. We still don't know all the details.

Colin is going to see Sec. of St. about denial of visa to Arafat who wants to come to U.N.

[Discussed Angola, details not yet known on Cuban withdrawal; desk work; briefing for and then visit by West German chancellor Helmut Kohl, commented, "He's a good friend"; bill signing; received first copy of Constitution *magazine; greeted heart transplant recipient.]*

A meeting with Dennis (Merme's husband). Merme is feeling lost & left out now that we are leaving. She wants to stay involved in govt. & politics. I had to tell him I can't ask George to find a place for her.

[Videotaping; allergy shot; exercised; Maureen and Dennis at dinner.]

Wednesday, November 16

Trade deficit figures. We have highest level of exports in history & the deficit is lowest it's been in years.

Resignations are coming in. George has announced he's turning down Nick Brady's & keeping him on as Sec. of the Treasury.

Same discussion of helping resolve the situation of so many "Savings & Loans" that face financial ruin. We're going to make some appts. "Recess"—people Sen. has not acted on.

N.S.C.—Tomorrow I have to make certification of no nuclear weapons in Pakistan.

Colin, Bill Webster & Geo. S. met with Kohl on Chemical warfare. Some German businesses are participating in construction of a nuc. weapon plant in Libya. He says he'll put a stop to this. After the P.L.O. statements that looked like they were going to recognize Israel as a nation—now spokesmen are saying they have not made such agreements.

Then it was briefing time for Thatcher visit. At 10 AM we meet her & Dennis on S. Lawn. The regular ceremony—19 gun salute—Mil. on hand & we inspected the troops. Her speech was a eulogy to me & our admin. She praised me as having changed the whole world. Then the receiving line & the meeting in the Oval O. She really is a great stateswoman. A bit of desk time—then lunch in Study. Out to Oval O. to meet Mayor Teddy Kollek—Jerusalem. He presented me with a 15th century map of the middle East. Then I received the P.S.I. in Ed. Report. The pvt. sector has gone from 40,000 business & school partnerships to 142,000 & they are still growing.

Then over to the East Wing—1st Ladys Garden for a photo session—the P.C. of Nancy & me we'll use for picture requests once we leave here.

Tonite—State Dinner for Margaret Thatcher. It was a wonderful evening & a great mix of people—from Oral Hershiser the great Dodger pitcher, to Billy Graham, Loretta Young & on & on. There was an air of nostalgia—our final state dinner etc. We got upstairs about midnight.

Thursday, November 17

Day started at 10 AM in the office. On Mon. I leave for Calif. And ex-Pres's. Ford & Carter came to see George B. I await a bill that Commerce wants me to veto but everyone else wants me to sign (which I think I'll do). It has to do with NASA. Geo. S. is going to the Inaugural in Mexico. It looks like my Sat. Radio talk will focus largely on Thanksgiving.

NSC—A kidnap in Lebanon of a Swiss member of the Internat. Red Cross. Three armed men—possibly Hizbollah. In Canadian election Mon. looks like Mulroney has gained some ground. Looks like Ms. Butto is winner (not official yet) in Pakistan. Soviets may have fired a SCUD missile into Pakistan. Soviets have suddenly & seemingly opened the door for a new long term grain deal with us.

[Addressed gathering of business people and diplomats on free trade; stopped in at gathering of Advisory Committee on Presidential Libraries; lunch with Vice President Bush; noted that biographer Edmund Morris sat in on meetings all day; received report from Committee on the Arts and Humanities; photo session; signed one hundred photos and more than a dozen bills; reviewed recess appointments; met with 1968 campaign manager Cliff White; received membership card number 1 in Republican National Committee; doctor came to W.H. residence to perform nose and throat exam; attended black tie dinner in honor of Charley Wick.]

Friday, November 18

Into the office at 10 AM. Discussed bills I'll be signing later in the day. Ken D. suggested that while at the ranch I might try putting down some thoughts for the farewell address. Ken also reminded me that 11 AM I would have a full briefing on the budget I must submit. In the meantime Dan Crippen gave me a little advance on it. A. B. Culvahouse came in to discuss who I'd turn to as legal counsel when I left office. I told him I already retained a firm of which Bill Smith was a member.

N.S.C.—Sofear of St. Dept. is going to the Hague about the Iranian funds we still have tied up & which we need to turn over to them. A B-1 Bomber crashed last night. Crew all safe. Apparently it was pilot error, not a failure of the plane. Last word on Persian Gulf is that we can make another reduction of our forces there. Then some desk time & bills to sign but only briefly. The Pres. of Bangladesh (Ershad) came by for a brief visit. We've been helping them with their flood situation. Then the budget briefing. More desk time & lunch. A photo with a staff members family. Then over to E.O.B. for ceremony—10 awards for Preservation of Historic sites. Then meeting with George Shultz after photos with his entire family—18 in all. Occasion is a daughter getting married. Then George's regular meeting. This time he had Chet Crocker reporting on the Angola situation & the schedule for Cuban troops going home. Out to the Rose Garden for annual presentation of Turkey. The bird isn't the one they give us to eat—it's a 50 lb. monster. Finally up to get Nancy then down to East Room for signing of Drug Bill—The Anti-drug Abuse Act of 1988. Quite a turnout. Congressmen, Cabinet, the father & mother of a young policeman Dave Byrne—murdered by a drug pusher, & the mother of a girl brutally raped & murdered by 3 Pornographers. Upstairs more bill signing & paperwork, then exercise, shower, dinner & early to bed.

Saturday, November 19

Slept in, then up & at the desk for P.D.B. & radio script timing. And then a packet of bills arrived for signature or veto. No vetoes, all signed & delivered when I went to Oval O. for radiocast. The afternoon was a mix of packing things for Nancy to take to Calif. Mon. More bill signing, more mail & reading. Then dinner & early to bed.

[Sunday, November 20: watched panel shows; packed papers for California office, commented, "Finally tired of that & took a shot at TV"; Patti, Dick, and Geof arrived; attended black tie dinner thrown by Katharine Graham. **Monday, November 21:** *flew to California; motored to site of presidential library for ground-breaking ceremony; Mrs. Reagan went to home in Bel Air; went to ranch, watched football; commented, "Then bed time & for me the 1st time to go to bed without Nancy ever at this ranch."]*

Tuesday, November 22

Mulroney re-elected P.M. of Canada—phoned him this morning. Then a horse back ride on a very beautiful day. Forgot to say yesterday I sorted wines & put together 8 cases to go down to L.A. & our new house. Barney, Dennis & I had lunch then went back to our pruning. Then waited for news about Nancy's arrival. Call came she was 15 min's. out. Up to landing pad to meet her helicopter. She's home at last. I put 2 more cases of wine together for the new house—cleaned up, dinner & finally bedtime.

Wednesday, November 23

Up & into riding clothes but no go. First fog rolled in & before breakfast was over the rain came. A load of homework arrived—mainly bill signing plus 3 vetoes. One was the much discussed "Ethics Bill." It's a lousy piece of legislation. Every Cabinet member said veto & a few Dem. members of Congress have gone on record favoring veto—Dingell & Rostenkowski the 1st 2. Looks like an indoor day. And it stayed that way. Dennis here for lunch. Rain set in for most of afternoon. So it was homework, bill signing, vetoed the controversial Ethics legislation. Kathy sent me a bunch of speeches I'd done back in the 60's & 70's when I didn't have speech writers & a Xerox of my handwritten script of 1980 inaugural address.

A couple of calls from Ken D. & one from son, Mike. Then dinner in front of TV. Watched "War & Remembrance," 8:30 to 11 P.M.—then to bed.

Thursday, November 24—"Thanksgiving"

A cold, windy day—so cold I said no to a ride. A gray cloud cover—no rain—it's too cold to do that. Late morning there were some signs of the clouds breaking. Did some chores, turned on the heat in the guest house in case our guests might want a look this afternoon. At 3 P.M. Dennis & Merme arrived. She & I had a most productive talk about her future. Then at 4 P.M. Jane E., Elaine Crispen & her daughter arrived. Cocktails & a 5 o'clock turkey dinner. It was a most enjoyable evening. They left around 7:30 to drive home in the rain. Nancy & I watched "Mary Poppins" on TV & then to bed. The Johnny in the bathroom had a mechanical problem—flusher wouldn't work. I had to flush by lifting the plunger in the water box.

[Friday, November 25: *wet weather again precluded riding; homework; Ron and Doria arrived for visit; hung pictures.*]

Saturday, November 26

Up early to do broadcast—subject "protectionism." Blue sky, bright sunshine & a cold wind that kept the temperature in the 40's. All of that plus the sloppy ground kept us from riding. Finished up with the picture hanging which involved the house, guest house & tack room. We took down things that will go to the Library, Eureka or Dixon House & replaced them with things we had in storage. Lunch with Dennis & Barney. Ron & Doria left for home about 10:30 AM.

After lunch we split the wood we had cut on Tues. Then we all gathered in the bldg. of the Dr. & the Mil. Aides & watched the finish of the N.D. game—U.S.C. N.D. 27, U.S.C. 10. Then a ceremonial picture of all of us in front of the door with the Bide A Wee sign evident. The last time we'll ever have those sessions. The Bldg. will be disassembled & returned to the mil. They presented me with a home made plaque with all our names on it. I've hung it in the Tack Room. Dinner—Barney present. Then Nancy & I watched TV for awhile—"Philadelphia Story." And so to bed. Left a 7:15 AM call.

[Sunday, November 27: *returned to Washington; read* Final Flight *by Stephen Coonts en route.*]

Monday, November 28

Staff time at 10 AM—after months of negotiation—a Grain agreement with the Soviet U. Extends present agreement (now expiring) until 1990—thru 1990 harvest. Then we talked about reaction to my veto of the ethics bill. Some flurry in Press but already running down. Then subject of Ollie North & possible pardon. I can't do that before the trial. It would leave him under a cloud of guilt for the rest of his life.

NRA [NSC]—George Shultz's decision to deny Arafat a visa to come to U.N. Israel is very happy. Arab States are upset about it. Our law is very clear on the subject. George is right & the visa should be denied.

Then a little planning talk re Gorbachev's visit to U.N. George & I will spend about 2 hrs. at lunch with him at residence on Governors Island in N.Y.

[*Session with portrait painter; quick hello from James Baker; greeted Dick and Tommy Smothers escorting officials of American Lung Association; haircut; visit with W.H. operator facing surgery for brain tumor; photo session with W.H. staff of ninety-three people;*

Betsy Bloomingdale staying as houseguest, noted "She & Nancy are buried in closets—I think eliminating clothes that will not go back to Calif. with us. A brief interlude for dinner—then back to the closets."]

Tuesday, November 29

Up early & had breakfast in Study. Then woke Nancy before I left for the office. First meeting—Ollie North case—Judge Gesell has made it plain that if certain secret information can not be used in trial because of Nat. Security—some charges may have to be dropped. We have notified Judge & prosecuting attorney Walsh that we can't allow some facts to be made public. Rev. Falwell has sent a petition with 2 mil. signatures demanding a pardon for Ollie.

Ken D. is having a meeting with Sununu to show him the ropes. Sen. Burdick of No. Dakota was taken to the hospital last night. Their law calls for election if Senator is disabled. Nick Brady is working on a plan to rescue the Savings & Loan banks. Last night Cong. John Dingell told Ken I was absolutely right in vetoing the Ethics bill.

N.S.C.—I received a letter from Egypt P.M. Hosni Moubarak—friendly but pleading with me to allow Arafat to have a visa & come to the U.S. Sorry, I can't do that. A meeting is taking place between Soviets & Mujahadeen regarding P.O.W.'s. Another meeting is scheduled.

Rcvd. a message from Pres. Botha (S. Africa) explaining things that need to be done if they are to sign off on an Angolan settlement.

Then desk work 'til Dick Wirthlin came in with latest polls My approval rating is 63%. I'm 80% on handling Soviets. Almost everything is in that range or thereabouts. Lunch, after lunch calls to Bob Dole & Sen. Mitchell—both elected to Sen. leadership positions. Then a photo with Phil Dusenberry—he's retiring. More desk time. Then A. B. Culvahouse came in. I have a plea from Art Linketter about a pardon for his son-in-law who faces trial or sentencing December 7 for some mix-up with a Texas Savings & Loan for whom he was brokering a loan to a Mexican group. He has admitted he did not reveal to authorities some violation of Texas laws on interest rates by the bank. A.B. has convinced me I can't do what Art asks. It's plain however that an injustice is being done. We'll keep watching it.

Then a group including Mayor Barry came in & presented me with a sledge hammer. Has to do with the successful transfer of ownership in a Wash. housing development from govt. to the residents. We've started this & now about 13 more in different cities are moving to do the same thing.

Then a sneeze shot & up to shower & change for a busy evening. First we go downstairs to a reception for about 250 people involved in the "Art in Embassies" program. Then over to Library for Cong.—The Repub. Senators dinner.

At the 1st event we were presented with a silver box holding tapes of favorite songs. Then a receiving line for about 270 people. Some mix & mingling then over to the Library of Cong. & another receiving line—a few hundred people. Dinner & some ceremony about departing Senators. Then Bob Dole did a beautiful tribute to me—I responded. Bob toasted me & gave me a gift—a weed chopper for the ranch. Homeward at 10 P.M. A wonderful evening.

Wednesday, November 30

Another 10 AM day. The Wash. Post mis-stated the case of Ollie North. A. B. Culvahouse gave me the legal details & corrected the story that had suggested I would be a witness. Budget Review Board meets today to receive budget proposals for various dept's.

Ken reported on his meeting with his successor Gov. Sununu. The Gov. wants a detailed report on our staff duties & performance. Fred Ryan came in & reported on Presidential Library progress. I was pleased to learn we've been shipping stuff to Eureka College for their museum.

[NSC meeting regarding Canadian prime minister Brian Mulroney's request for meeting over trade; Swiss government confirmed for Philippine government that Marcos's accounts contain $350 million, also doctors say Marcos is healthy enough to travel; received news of gun accident on USS Nimitz.]

We are losing further our Persian Gulf protection. Ships will patrol zones instead of accompanying ships. Also a barge is being taken away. It carried supplies & served as a landing pad for helicopters.

Word is Bhutto will be named Pakistan P.M. tomorrow. First woman to serve as such in any Islam state. Met with Ricardo Bofill—leader of Cuban Rights Committee. He's campaigning worldwide on subject of human rights. Castro had him imprisoned for 11 of his 45 yrs.

[Session with portrait artist; lunch with administration officials; photos with Savings Bond Volunteer Committee; Shultz submitted letter of resignation; attended black tie dinner for supporters of presidential library.]

Thursday, December 1

And again a 10 AM start. A.B. came in with more on the North case. Judge has asked independent council Walsh to tell him what the info. blocked from use in trial is. This may force a drop of 2 of the charges.

U.N. voted 151–2 against us on Arafat—England abstained. Bud. Review Board rejected NASA's request for further increase. We've already proposed a 21% increase plus a 22% increase next year. Mil. is asking for 4% pay raise—we've offered 2%.

NSC—Soviets have stopped jamming radio broadcasts such as Radio Liberty & all the others. They have also announced they are releasing 120 pol. prisoners.

Then on P.F.I.A.B. (Anne Armstrong) & several others came in to report. They have been working with CIA on a matter that goes back 25 years. [. . .]

[Visit from Foreign Minister Sousuke Uno of Japan, commented, "We really have established a good relationship with Japan"; then annual visit of rabbis with menorah for Hannukah; lunch with Vice President Bush, commented, "He's really torn about the choice for the Sec. of Defense. I couldn't help him"; photos with appointees; met with publisher Robert Anderson about program dedicated to the American horse on a 23,000-acre preserve in New Mexico; trade briefing; filled appointments to commissions; photos with Secret Service agents; went to doctor for exam of hand, commented, "Dupreyton Contraction. It should be operated on—I suggest early Jan."; made before-dinner speech at dinner for Jack Kemp; returned to W.H. for dinner.]

Friday, December 2

Sec. Cavasos is new in the job—he asked for a budget increase of $70 mil. He didn't know Congress had already added a bil.$ to what we had asked for. Ann McLaughlin—Sec. of labor is asking for a $400 mil. increase. This is in the budget overview we were having. Energy has had a $21 mil. increase, they want more. Leaders of our American Indians want a meeting with me. 61 Sen's. & 8 Senator-elects have signed an endorsement of Geo. S's. ruling on Arafat.

[Noted upcoming schedule of speeches and TV appearances; NSC meeting noted meeting of Caribbean and Latin American countries in Miami; Soviet hostage crisis; NATO meeting; birthday lunch with secretary; met with David Rockefeller about his private initiative programs; received report from Advisory Commission on the Arts; telephoned Caribbean meeting and made remarks.]

Then the whole gang of stewardesses, flight crews etc. from Pan Am who carry our press on trips but who always gather as a greeting group for me when A.F.1 lands came into the Oval O. It was great to see them & really meet them.

Then Lyle Parker—Teke from Eureka when I was there dropped by with his family. Then upstairs to change for Camp David. On way to Helicopter—supposed to choose some wood carvings as gifts from the carver. We chose a large pair of Swans & a Santa Claus. Photos in Dip. room & outside with about 40 Interns. Then on to helicopter & Camp D. Temperature at the camp, 34 degrees. The movie after dinner—Jim Stewart in "The Way The West Was Won."

[Saturday, December 3: radiocast; homework; talk with Dr. Hutton regarding upcoming hand operation; received news of sudden death of longtime housekeeper; watched football.]

Sunday, December 4

Dawned bright & cold. Up at 9, watched "This Week with David Brinkley." George Shultz was solid & good. Then at 11:30 AM left for W.H.—there at 12 noon (had a tailwind). Both of us brought back most of clothing we had at Camp David. Only one or two more weekends left up there. I spent the afternoon at my desk. Then the Kennedy Center Honors reception, 5:20 P.M.—W.H. Honorees—Alvin Ailey, George Burns, Myrna Loy, Alexander Schneider & Roger Stevens.

1st the reception & a receiving line of about 350—many of them old friends—Bob Hope, Claudette Colbert—well the list is too long. Then upstairs, a fast dinner with Merme & Dennis & Ted Graber. Then over to the Kennedy Center—a great show honoring the 5 & a surprise finale when they did a tribute to Nancy & me with Auld Lang Zyne.

Monday, December 5

[Reviewed week's scheduled speeches; noted new congressional leadership.]

Budget meeting—main problem—military pay. We're $1.1 Bil. over the Gramm, Rudman $100 Bil. ceiling. Dan Crippen reported on this one.

N.S.C.—Some talks about the Wed. meeting with Gorbachev. We have to find out what title to use. He is now Gen. Sec., Chmn. & President.

Venezuela election—John Perez who was Pres. in 70's was elected again. We can work with him. The radical Left did poorly. The brief mil. revolt in Argentina has been controlled. The Heritage Foundation revealed its true nature. They are against an Angola settlement getting Cubans out charging we are selling out Savimbi. No such thing—we strongly support Savimbi. Desk time 'til lunch. At 12:45 up from lunch & a photo with Lindy St. Cyr. Then photos with a group of Jim's friends. Then a photo with a 6 yr. old girl & her mother. Meeting me was the child's request. She only has weeks to live. She has a brain tumor.

At 1:30 a signing ceremony for Proclamation designating 1989 as "Year of the Young Reader."

[Meeting with George Shultz, as noted, "mostly on Middle East, the P.L.O. & upcoming meetings"; did exercises; dressed for Capitol reception for congressmen, old and new; returned to W.H. for dinner with Mrs. Reagan and Graber, who had plans for the new office in California.]

Tuesday, December 6

Again the day began at 10 AM A. B. Culvahouse began the meeting with talk about pardons for North, Poindexter & McFarlane. Jon (Mrs. McFarlane) wrote me a letter on her own asking for a pardon for Bud. He awaits sentence, he pleads guilty to making a couple of statements that weren't true on the basis of things he said later. I don't think he deliberately lied at all. I lean very much toward a pardon for him.

Woodward of the Wash. Post sprang a story that Israel's Nir had told him there was a secret intelligence agreement between Israel & the U.S. There is no such thing. Woodward wrote the story after Nir turned up dead in Mexico—a plane crash. It looks like a repeat of his story about Bill Casey—after Casey died.

[NSC meeting with word that commander of U.S. F-16 base in Spain was killed in crash of one of the jets; South Africa said it would return to Angola talks; discussed Canadian trade problem.]

There are rumors that Gorbachev is planning a shake-up of the military. Tomorrow is the day Geo. B. & I lunch with him in N.Y. Nancy has left this morning for N.Y. but we won't see each other while we're there. Another rumor has Gorby creating something like our N.S.C. Another rumor has it that someone had planned a coup while Gorby was in the far East.

[Briefing for Wednesday meeting; photo session.]

Dr. Hutton brought in Dr. Hummel of St. Johns Hospital in L.A.—he has set up a team for any medical needs we may have after January 20. Then the Board of the Alzheimer's Disease Assn. came in &, among them Yasmin Khan, daughter of Rita Hayworth who died a victim of Alzheimers. I had declared November as Alzheimers month. Today I received a plaque. Then it was a photo with the top brass of the Air Force Assn. I'm a charter member.

Then an unscheduled meeting. The P.L.O. were meeting in Sweden re the middle east situation. Arafat made a statement that if it remains as told to us for his public pronouncement tomorrow, 6 A.M. our time we can grant him the visa for the U.N. & we are the winner.

Then over to the family theater for rehearsal of Thurs. nite's Press Conf. Upstairs, exercise, shower & then dinner with Ted Graber.

Wednesday, December 7

10 AM COLIN in our meeting. I signed a photo of Gorby & me in our first meeting in Geneva when he & I agreed to more meetings. I'm taking it to him in N.Y. We have information that Gorbachev will tell the U.N. he is reducing his military unilaterally. Learned Soviet U. had an earthquake in the Caucasus—great damage & many dead. Arafat's news conf. in Sweden—he recognized Israel's right to exist as a nation & rejected terrorism.

[Senator Bob Dole (R-KS) requested that subject of opportunity for physically handicapped be raised with Gorbachev; flew to NYC and residence on Governors Island.]

Geo. B. joined us there, Shultz & others arrived & briefed us on Gorby's U.N. address. In it he declared he wanted mil. reduced by 500,000 men, 10,000 tanks, 8000 artillery pieces & 800 front line planes. Then Geo. B. & I met Gorby on his arrival. The 3 of us met for a short meeting then it was lunch with about 20 of us including translators & note takers. Meeting & lunch ended at 3 PM. I think the meeting was a tremendous success. A better attitude than at any of our previous meetings. He sounded as if he saw us as partners making a better world. Then he &

Geo. B. & I motored to the waterfront where we faced Lady Liberty, then a view of the N.Y. skyline. George left us there & we went on to the Ferry dock where Gorby & I parted. Marine 1—then A.F.1 to Andrews & Marine 1 to the W.H. Nancy is home I'm glad to say. But I hurried through a shower & clothes change & at 6:25 P.M. left for "American Enterprise Inst." for annual dinner. I spoke before dinner & returned to W.H. for 8 P.M. dinner with Nancy & Ted G.

Thursday, December 8

First meeting upstairs here with Colin, John, Ken & Bea. Main thing was that the Gorbachevs are returning to Russia today because of the tragic earthquake yesterday. He has said the deaths might well go into the 1000's. We phoned him in N.Y.—I did my goodbyes etc. & offered to be of help if we were needed. Then Nancy got on for a farewell to Raisa.

Then a pre-news conf. briefing in the family theater. Desk time in the office & lunch. My lunch was clear broth & jello & tea. Breakfast was a ginger ale & tea—Dr's. orders. Last night I upchucked my dessert—the last time at 3 AM I feel very much OK now but I obeyed the Dr's. orders.

[Met with NAM executive board; signed proclamation of Human Rights year; briefed for news conference.]

Over to Dr. Hutton's office—other Dr's. there. We got a bit of tomorrow's annual check-up done. Then Nancy & I did photos for News Week & upstairs—desk time 'til dinner then the press conf. in the East Rm. It turned out very well. They seemed more friendly & all our gang thought it was great.

Friday, December 9

10 AM start—no breakfast—10 ounces of Golightly & a repeat every 10 minutes for a total of 13 glasses. Dressed for Camp David. First meeting a little post mortem on last night. Then told Stu Spencer had suggested after Jan. 20 that I might think of contributing a couple of speaking fees to the United Negro College Fund. Ken told me there would be no appeals for more money in the budget by any of our Cabinet.

N.S.C.—Colin reported Israel had moved into Lebanon again militarily on land, sea & in the air. We all think it's stupid. We've discovered the French are back to stealing from hotel rooms of American businessmen in Paris. Our A-10 that crashed in a German village has stirred up more German demands that may stop our flying there. Apparently Arafat who signed the paper we thought would meet our demands has fuzzed things up again & says it must be approved by some other group there in the middle east.

[Greeted head of Paramount Pictures; photos with Mrs. Reagan, including one with dog breeder who raised Rex; received sculptures for W.H.; went to Bethesda for checkups; went to Camp David; watched Destry Rides Again *(1939).]*

Saturday, December 10

Bright & cold. Did radiocast—had to do a little writing—to get an insert in on meeting with Gorbachev. After radiocast at Laurel took our places before fireplace for greeting & photos with each of 135 newcomers to the Camp D. crew since last Xmas. Then lunch at the mess with the enlisted men & women. I did a little Q&A with them. It was fun.

Spent afternoon mainly on homework I'd brought along. Dinner & at 7:45—Camp D. detail Christmas Caroled us. Then the movie, "North by Northwest"—Cary Grant. A call from Ken D.—we are going to stop by the Russian Embassy tomorrow afternoon & sign the condolence book re the earthquake.

Sunday, December 11

[Watched panel shows briefly; returned to W.H.; received word of deadly crash of Soviet plane headed to earthquake zone with relief supplies.]

Lunch in the Atrium—then a short afternoon of cleaning out drawers. Then downstairs for photos with beautiful Xmas decorations—over to the Soviet Embassy—signed the condolence book for earthquake victims. Then on to the annual "Xmas in Wash." Jimmy Stewart, Shirley Jones, Vicki Farr [Carr] & others—a wonderful Xmas program. Then back to W.H. for dinner in front of the T.V.

Monday, December 12

10 AM—Q's. are being raised again about whether something should be done about people with incomes (this time above $90,000) collecting Soc. Security. Talk of Wed. schedule—speech at the exec. forum & Fri. to Va. on Foreign Policy. I've suggested Q&A because this one is at a University. Thurs. I get the commission's recommendation on pay increases. Kranowitz is being hired away from us by Bob Michel & the Repub. Cong. leadership. Cong. resumes on Jan. 3. Tomorrow I get the budget.

NSC—Colin in Detroit making a speech. John took over. He gave me or I should say an A.I.D. man gave me an update on our aid to Soviet's on Earthquake disaster. Another plane crash there. This was a Yugo Slav plane—mistook a lighted highway for a runway—7 dead.

Swedes are telling us Arafat is going to make his statement to the U.N. in Geneva (5 A.M. our time tomorrow). It will, we're told, be the statement we've asked for. Moubarak wrote me a letter that Arafat is coming around. Argentina—Noriega helped finance the attempted mil. rebellion. Pres. Alfonsin has broken off relations with Panama. Polisario now admits they shot down our plane believing it was a Moroccan mil. plane. Some desk time—P.D.B. & so forth. Then my Commission on Physical fitness brought me the annual report.

There were 18 of them including our friend Pam Shriver. Then into Lunch. Two staff members for the 2nd time smuggled a Xmas Tree into the study where I lunch. At 2 P.M. I met in Cabinet Room with 18 heads of Indian tribes. I think meeting went well. They want less bureaucratic control of tribes & more freedom. I'm for that.

[Desk work; haircut; photo with employee who processes photos; dropped in on Mrs. Reagan's tea with Sammy Davis Jr. and his wife; more photos; attended Congressional Ball at W.H., entertainment by Peter Duchin.]

Tuesday, December 13

10 AM & Arafat still speaking in Geneva to the U.N. The 5 AM was postponed to 9. He's been going for an hour & a half. Colin was listening to him on CNN. I don't think Arafat has met our demand. Our 4th plane is on it's way to Armenia. Our W.H. staff exceeded it's quota on the Fed. govt. charity drive. Carlucci is proposing a new idea in Defense long term ordering of weapons that would eliminate the every year squabble over budget. A.B. came in & clarified some of the issues over the North trial and classified info. that can't be used.

NSC—Angola, Cuba & S.A. have signed an agreement calling for Cuba to get it's troops out of Angola. We're trying to contact Savimbi to make sure he is involved. Out near Indian Ocean one of our ships testing a harpoon missile. An Indian ship came into target area in spite of our warnings & was hit by missile—1 man killed. We discussed the Higgens matter. We're still not sure he isn't already dead. In Armenia, 18,000 bodies have been retrieved from the rubble.

[Addressed three thousand administration employees on economic policy of the two terms; greeted friend from Sacramento days; economic briefing, noted, "Our growth continues & the predictions of a possible recession don't ring true"; spoke to Business-Government Relations Council; met with Webster about new satellite, the Lacrosse; had allergy shot; went to dentist for tooth cleaning; exercised; Maureen and Dennis at dinner.]

Wednesday, December 14

Back to 9 AM—News clippings carried a statement Bob Hope made about me in Miami, Fla. It was most complimentary. Some talk of remaining schedule then Danny told me budget report I'm getting this afternoon will be showing a surplus by 1993.

NSC—Nations are divided about Arafat's speech to U.N. Our own view is that he did not come through with clear cut statement we need. U.N. is to get together December 22 on Angola agreement. Need in Armenia is for a concerted plan as to how best we can help.

Turmoil in Soviet U. as a great shifting & changing of leaders takes place.

Then 10 AM Ambas. from China—Hau Xu in for a visit I asked for. It is

10th anniversary of China-U.S. establishing of relations. Desk time but a brief hello to Henry Kissinger who was here to see Ken D. Then a pleasant time in Roosevelt room with 30 or 40 of advisors to me on "Pvt. Sector Initiatives." They gave me a report. They've done a great job. Lunch with V.P. Then—entire crew of A.F.1 came by for photos—about 35 all told. Then photos with the 3 groups Nancy Risque heads up. George Shultz's regular meeting. He showed me a miniature railroad built in to a carrying case—a present from our NATO leaders to him. He tells me details of verification of Angola withdrawal has yet to be worked out. Vienna meeting is supposed to report on a variety of things having to do with conventional weapons on Jan. 5. Egypt says P.L.O. should be talked to now. Israel adamantly says no. Then photos with 58 Admin. officials, 1 at a time.

[Presented medal to longtime employee; budget briefing, commented, "Joe Wright has done a great job. We'll present a budget to Cong. Jan. 9 that is within the Gramm Rudman target by billions of dollars"; received list of recommended raises from Commission on Executive, Legislative and Judicial Salaries; exercised; had dinner; attended holiday reception for press; noted that "Tonite's news (TV) featured George S. announcing that at last Arafat had made the statement we've been waiting for & our Ambas. in Tunis will begin meeting with him. I issued a written statement to the press. We mustn't say it out loud but we outlasted him & we won."]

Thursday, December 15

10 AM again. I'm told our budget includes a small start on the Savings & Loan problem.

A follow up on my meeting with the Indian leaders. The Bureau of Indian affairs has a $45 mil. fund for loans to Indians (guaranty) for Ec. development.

NSC—Reactions are showing up re our announcement of a dialogue with Arafat. Israel, of course is dead set against it. A number of Am. Jewish groups are rather noncommittal & we are being praised in a number of circles here & abroad. As we see it—a new game begins today.

We're still watching the Conf. on Security & Cooperation in Europe (C.S.C.E.). Geo. S. may be a little too optimistic in his belief that its job is finished & we move on to other areas of negotiation & dialogue.

A meeting with Turkish P.M. Turgut Ozal. A short but pleasant meeting. We are on good terms. He indicated a need for more help on military support. I hope we can but we've trimmed down on help to a number of other countries.

A little desk time then our 2nd visitor, Italian P.M. Ciriaco De Mita. He was late so our meeting was short but again it was pleasant. Italy has agreed to take our F-16's once based in Spain. Spain refused to extend the lease. Finally—15 min's. later—lunch. Then 6 photo ops. for S.S. agents being transferred. Then photo op as I received a plaque from the new officers of the Future Farmers of America. Bob Tuttle with appointments to a number of boards &

commissions—followed the P.M. A meeting with Att. Gen. Thornburgh for a rundown on the indictment of the Marcos's. After him, Peggy Noonan— another session on our creation of my Farewell Address. Then a photo op with Ambas. Montgomery & his daughter. He's been our man in Oman for 3 yrs. & a great help in the Persian Gulf business.

[Videotaping, including a broadcast to the Soviet people; attended Christmas tree lighting; dinner with Maureen and Dennis. **Friday, December 16:** flew to Charlotte, Virginia, for foreign-policy speech at University of Virginia, commented, "The old school abounds with tradition & the spirit of Thomas Jefferson who founded the U."; photo sessions at school; returned to W.H.; more photo sessions; Christmas reception for W.H. staff; noted that "Nancy is so tired we're having dinner in bed."]

Saturday, December 17

Nancy slept in—I got up & went out to the Park Police Stables. Did photos with them in the riding arena & got a thank you plaque from them. They trained our S.S. agents in riding & always brought the horses up to Camp David. I should have been giving them a plaque. Back to the W.H. & to the Oval O. for the radiocast—then up to the solarium for lunch—just the 2 of us. Then at 2:15 down the Grand Staircase to greet the W.H. & E.O.B. staffs—the whole downstairs was filled with them. Then upstairs to exercise & shower before a rubdown Nancy had scheduled. Dinner in study & then to bed.

[**Sunday, December 18:** watched panel shows regarding deficit and Arafat situation; packed.]

Monday, December 19

A 9 A.M. start. A little fuss with dept. of Energy. They wanted a budget increase to restore nuc. plants. We pointed out they are holding excess funds & those must be used for the nuc. safety measures. They hold $900 mil.

Just rcv'd. a fire management report on handling of natural fires (lightening etc.). Rule is to let them burn themselves out. We agree with rule but need provisions in case of certain weather conditions etc. A dispute also with task force looking into Fetal tissue research. They say using aborted fetuses is not immoral. Well I say it is because abortion is immoral.

[Discussed pay raises for judges and other executives; NSC meeting with news of bombing of Peace Corps building in Honduras, of kidnapping of two people identified as Americans in El Salvador; resolution of problems regarding Chinese launch of civilian satellites; report that Soviets were waging an all-out air war on Mujahadeen in Afghanistan; report from aid workers on Armenian earthquake relief; photo sessions; addressed Grace Commission; many photo sessions; telephoned Governor George Deukmejian (R-CA) urging him to run for a third term; taped a greeting to CBS reporter Terry Drinkwater who was in the hospital with cancer; attended Christmas party for W.H. employees.]

Tuesday, December 20

Well, George will have to do the honors if our friend Emperor Hirohito leaves this mortal soil. If he does anytime in the next 30 days—the services wouldn't take place 'til after Jan. 20.

[Received positive word on inflation and that kidnap victims in El Salvador were not Americans but Salvadorans; NSC meeting with report on Israeli political situation; agreed to requested meeting by President Eric Delvalle of Panama; lengthy briefing for interview with David Brinkley following day.]

Lunch—then photos shaking hands with 55 dept. people & 74 military. Another session with Peggy Noonan on my farewell address. Over to E.O.B., another photo with Uniformed Security. Then to East Room to greet Volunteers at their Xmas party. Back down to Dr. Hutton's office for check on my eyes by Dr. Glamp & his assistant. And upstairs—I'll exercise, shower & get dressed again for tonite's Sr. Staff Party.

It was a lot of fun, first the handshakes—about 300 & photos, then entertainment in the East Room by Yakov Smyrnoff—very funny. And so to bed.

Wednesday, December 21

A 10 AM start—talked about schedule. I present the budget to Congress Jan. 9. Don Hodel has endorsed the Quad Commission's proposed pay raise. Our people have put in a request that the signal phone system we have be continued for 6 months.

NSC—Geo. S. will be present tomorrow in N.Y. when the Angola, Cuba, S. Africa, Navimbi signing takes place.

Top Secret the Chfs. of Staff have drawn up a plan for taking out the Libya poison gas plant if we should decide to go that route. We have sent a cable in my name to P.M. Shamir reaffirming our continued support for Israel.

Then a briefing for my upcoming TV interview. Everything this morning has been in the study. The Oval O. is being set up for A.B.C. to rig it for my interview with David Brinkley. Then it was zero hour. Everyone says it went very well. I'll wait 'til I see it tomorrow night. It's a segment of an hour show on my presidency. Then lunch & after lunch a photo with Dale Petrosky who left us a short time ago.

A Cabinet meeting—2 subjects—Montreal meeting on free trade. We got 11 out of 15 things we waned. Agriculture still a sticking point with our European trade partners. 2nd subject was a report on budget by Joe Wright.

[Photo sessions with more than 130 people.]

Then a meeting with Geo. Shultz. He gave his views on the Angola settlement even though we are not a signatory. Then made a pitch for me to pardon North & Poindexter now on those charters that involve secrets having to do with Nat. Security. Let them stand trial on the indictments having to do with possible breaking of the law. It was a hell of a presentation & I've ordered that we pursue this further.

[Photos and Christmas reception for Secret Service.]

Upstairs—dinner & beddy bye. A phone call from Colin re the Pan Am plane crash in Scotland—258 aboard & all dead. We believe a great many were Am. mil. on their way home for Xmas.

Thursday, December 22

Ken D. heard from Dick Wirthlin who said a poll had shown 59% favoring our talking with Arafat & 19% against it. We decided I'd pre-tape my radio script in Palm Springs. We've decided I'll give one more Medal of Freedom before I leave office. It will go to George Shultz. We're invoking a 301 against Taiwan. They've banned our shipment of Turkey parts to Taiwan. A.B. came in. I'd sent him letters from Ollie North's lawyers asking me to pardon Ollie. Now Judge has stated Nat. Security secrets must be released. We've said no.

NSC—Air crash of Pan Am plane over Scotland began with some kind of explosion at 31,000 ft. New figure on dead—273 on the plane. Ambas. Charles Price & Carol went to scene of crash last night. Margaret Thatcher went early in the morning. I talked to both of them later in the day. Israeli's 2 parties voted on yes on a unity govt.

Basic Science agreement with Soviets came to me for approval. Our ranks are divided—some say sign, others want to go back & get some word changes. I haven't made up my mind yet. A brief meeting with Pres. Delvalle of Panama. He went away pleased when he found George B. is going to continue our anti-Noriega campaign when he takes office.

Photos with 5 of our Ambas's. & familys. At 10:30 Thornburg came in with a well reasoned argument against pardons for North & Poindexter. Then new Sen. designate Dan Coats & family came by for a picture. He is the Congressman Gov. Orr of Ind. appointed Senator to replace Quayle. Then a brief address to about 50 Vol. rescue workers just back from Armenia & lunch with V.P. He told me of some other appointments he's going to make & then asked if he & Barbara could use Camp D. over New Years. I said yes.

[Taped radiocast for Saturday; many more photos; dinner party for friends in W.H. residence; concluded, "Then to bed & watched my interview with David Brinkley. O.K. I guess." **Friday, December 23:** *made statement to the press regarding the Pan Am tragedy; flew to L.A.; went to new home in Bel Air, commented, "Nancy worried all the way that I wouldn't like it. It's beautiful"; got to work hanging pictures and unpacking boxes.]*

Saturday, December 24

Up a little early considering we're in Calif. & it's Sat. A dismal, rainy day but O.K.—we're indoors still surrounded by cardboard boxes. But as day went on we are finally just about finished. Ted, Peter & Nancy concentrated on picture hanging. At 10 AM I made calls to overseas forces & an individual in each of the mil. forces. They were stationed in Panama, Puerto Rico, Egypt & a little island in the Bering Sea. I made one to our Navy Man in Japan at 3 P.M. We've learned that Ron & Doria won't be here tonite. They are going up to Lake Arrowhead. Well they actually showed up at the Wicks after dinner. It was the usual Wicks Xmas eve party with Charlie at the piano. As usual it was a time of old friends & most enjoyable. Santas arrival is always a big moment. This time it was a well padded young lady—CZ's fiancé. Home & to bed nearing midnight.

Sunday, December 25—Merry Xmas

Rain all gone & a beautiful sunny day—a little cool. We're still unpacking & opening Xmas gifts. The basement is the scene of action for the 1st. We've discovered we have a set of those shipboard high powered binoculars we can install out on the patio. An easier day than we've had. Ken D. called—[. . .] has found some fragments of the Pan Am plane that do indicate—it was a bomb explosion. An early dinner with Ron & Doria, Merme & Dennis, & Mermies friend Jeanne—a lovely person—totally blind. A pleasant evening & then to bed.

Monday, December 26

Another beautiful day—not a cloud in the sky. We're waiting for Ted G. He was due at 9 AM. Now we learn (at 10 A.M.) he was on time & the S.S. wouldn't let him through the gate. Nancy got on the phone & got things straightened out. It had to do with his car. The boys just wanted the bomb dogs to sniff it. He wasn't suspect—the security is concerned that someone coming to see us might not know that someone slipped a bomb under the car attached to a magnet.

Well Ted & Peter are here & once again it's picture hanging etc. It was an unpack plus desk work day. Come shower time & we couldn't. The hot water was shut off because of a faucet's failure that wouldn't shut off. Dinner & bed.

Tuesday, December 27

Still no hot water. Sunny day. About 11 AM I went out on the patio in slacks & blazer for photos by Pete & MaryAnn. After lunch Bea O. came by with some desk work & to accompany me to the Fox Plaza to see our suite of offices. I'm overwhelmed—they are of course not completed. It's the top floor of a new bldg. but they are going to be magnificent. Nancy has an office as well as mine & then space for a staff of 13. Back to our home & desk work. Called Lew Wasserman who was responsible for us getting that office space to thank him. We now have hot water.

Out to Chasens for dinner with Wicks & their 41st Wedding anniversary. All their family including CZ's fiancé & Ron & Doria. We had a table for all of us in a new addition to which I had never been. The old Chasens is bigger. A good time was had by all.

Wednesday, December 28

Bright day—57 degrees. A call from Colin before breakfast. The British have completed their examination of the Pan Am 103 crash & are announcing today that it was a bomb explosion.

Barney & Aggie came by—also Ron & Doria—twice. Ron took slippers he'd given me for Christmas & exchanged them for a bigger size. There was more moving in & unpacking. Also some packing. Tomorrow we go to Palm Springs. Back to desk work. Then dinner & early to bed.

Thursday, December 29

Up for breakfast at 8:30. Some reading—P.D.B. etc. Then up. Not much "moving in, work." It's go to Palm Springs day & it's a beautiful one—a little warmer than we've had but still below Calif. normal. Got some telephoning done. Called Peewee & Nancy Wms. also Lee Clearwater's son. His son who I last saw 8 years ago is now a Marine. A late lunch then off to the airport. Before getting on A.F.1 I told the press about our position on Pan Am 103. It was a bomb & we are going to do all we can to track down the bomber. At 4:15 P.M. we landed at Palm Springs. A 20 min. drive & we were at Sunny Lands—met by Lee & Walter. It's beautiful as always. Ardie Deutsch had gifted me with a massage—a new kind. It was great & I feel great. Then it was dinner with Lee & Walter & an early bed time.

*[*Friday, December 30: *lunch with friends; golfed and noted, "I've really forgotten all the things that should be automatic in a golf swing. This once a year golf game has left me a real duffer"; dinner party at El Dorado Country Club.]*

Saturday, December 31

A sunny day to start with—warmer than we've been having. Afternoon clouds came in probably because of rain in L.A. Did my radiocast, then photo of me signing one agreement with Canada Free Trade act. Phoned P.M. Mulroney. He & his family are in Fla. They were taking their children to Disney World. Geo. Bush called me from Camp D. to thank us & say Happy New Year. Two great golfers of the pro tournaments were here—Lee Trevino & Watson, George S. & I made a 4 some with them for 9 holes before lunch. It was quite an experience. Then after a luncheon we did another 9. We had a gallery—all the house guests. We all had a lot of fun. Then it was time for a little desk work & clean up for the New Years ball. It was grand as usual. Dinah Shore, Mrs. Bob Hope

& Mrs. John Gavin sang after dinner. Bob Hope entertained with some funnies. Geo. S. spoke & said some nice things about Nancy & me—so did our host Walter A. I responded to the toast—then it was 12 & Happy New Year. Bed about 1:30 A.M.

Sunday, January 1, 1989

Up but not too early. Sun & scattered clouds—somewhat warmer than it has been. Hit a bucket of balls on the driving range—did pretty well. Then at noon a meeting with George Scharffenberger. He handles our blind trust. That of course will end on January 20 then lunch—about 40 of us & as always a lot of fun. A quiet afternoon, some homework—then TV & the last half of S.F. defeating Minn. in the playoffs. Nancy had a massage. Cleaned up & dressed (no necktie), dinner for all the house guests & a few others in the dining room & again it was a very pleasant time. Nancy was so tired we left right after dinner & into bed.

[**Monday, January 2, 1989:** *went back to L.A.; desk work; watched football.*]

Tuesday, January 3, 1989

Scattered clouds but warmer than it's been. Got right down to desk work. Interrupted once to pick places for planting a peach tree & an apple tree—both gifts. Problem is to keep them from blocking out our magnificent view when they grow up. Right now they are about 3 feet tall. Back to my desk—then lunch & a drop by of Ken D. & Colin. They brought me up to date on my decision & George Shultz's to tell Gorby we'll go for the human rights conference which will wind up in Moscow in 1991. We've worried about P.M. Thatcher—if she'd go along. Well I have a message from her that she will. Then 2 P.M. Dr. Bookman & staff to test me for some different allergys. Apparently I still just have the old ones. I got my usual sneeze shot. Betts B. came by to see Nancy. I went back to my desk to work on my farewell address—Peggy Noonan has sent a 1st draft.

Now it's shower time. Dinner & bed—while TV experts try to discover why all of our sets are suddenly haywire. Now we learn—other people are having the same problem. It must be at the sending end.

Wednesday, January 4, 1989

At 2:45 AM Colin phoned to tell me 2 of our F14's from the USS Kennedy shot down 2 Mig 23's from Libya. There have been signs that Quadafy is antsy about a possible attack on his chemical factory. This time the ship & plans were on their way to Israel & 600 miles East of his factory when they fastened on to our F.14's & turned on their afterburners, our pilots decided they were hostile & knocked them down. So back to sleep. Dr. Hutton brought 4 local Doctors in he has lined up for us. Then I had Pearl & Sam for a manicure & haircut. Then a photo outdoors with men who installed our new house security provisions & their wives.

More homework including some changes in Peggy N's. draft of the Farewell address. Sent it over to Ken D. for any suggestions he might have. Nancy over to rehearse for tonight's program. It's a big evening in support of the "N.R. Drug Center." 1000 people at $1000 each. A receiving line for a special group of fundraisers then a film clip on scenes from the last 8 yrs. Dinner & a program of Merv Griffin singing—Mary Martin, Don Rickles—we both said a few words. All in all a very pleasant evening with a lot of old friends—Maureen & Mike. Ron is in Yosemite. Home to bed.

[Thursday, January 5, 1989: *returned to Washington; decided to submit to Congress recommended pay raises for judges and others; dinner with Graber.*]

Friday, January 6, 1989

Into Study near Oval O. at 10 A.M.—Oval O. being set up for my interview with Mike Wallace on "60 minutes." Meetings with Ken & Colin very short—a little talk about my approval of wage increase & a little mention of our shoot down of 2 of Quadafi's Migs. Then a brief for my interview. It & the interview started at 11:30—Mike ran over so lunch was at 12:20 back in the study. Then at 1:15 I taped tomorrow's radio cast because tomorrow I'll be at Walter Reed Hosp. getting my finger operated on. Did some desk work 'til about 3:20 then upstairs to Join Nancy in more 60 min's. Finished & off for the Hosp. Had dinner at Walter Reed with Nancy then to bed early.

Nancy went back to the W.H.

Saturday, January 7, 1989

Up at 6 A.M.—shower & scrub left arm & hand with special disinfecting soap. No breakfast. In fact, nothing, not even a drink of water after midnight. A lot of briefing by doctors etc. then into the operating room for 3 hrs. 2½ of them actual surgery. My arm was totally deadened plus some mild intravenous anesthetic which made me a little woozy. Operation over & back to room for lunch with Nancy & Dick who came down from Phil. He left in the afternoon but Nancy stayed for dinner—I ate in bed. I was in bed for the afternoon with my left arm propped up & hanging from a pole. Dinner in bed—I'm learning how handicapped one is with the use of only 1 hand. After dinner Nancy went back to W.H. I watched some TV—read awhile & then lights out.

[Sunday, January 8, 1989: *received smaller bandage; returned to W.H. after lunch; watched football; Graber a guest at dinner; watched "60 Minutes" and "Murder She Wrote."*]

Monday, January 9, 1989

Into the office—my bandage improved on way by Dr. Hutton. Ken D. told me Peggy was fired up & against some of suggested changes in my Farewell address. Met with her later in morning & smoothed things over.

NSC—Still trying to get Vietnamese out of Cambodia. A new development—it looks like China is willing to meet with Soviet U. on jointly getting V's. to leave.

Japan—funeral has been set for 47 days from now. Geo. B. will attend. Geo. S. is back from Chemical Warfare symposium in Europe. Quadaffi invited world press to come to see his bldg. which we say is making poison gas & he says it isn't. Well the press arrived & had a picnic in desert several miles from plant for them. Kept them there 'til dark then drove them past the plant whose lights were all out & took them back to town. Soviets have resumed troop withdrawal from Afghanistan. After Peggy—Lunch. Then Personnel with Bob Tuttle.

[Met with business group regarding budget; photos with staffers.]

A session with Edmund Morris—probably last as Pres. Stopped by Dr's. office. He notified me he's been put on list to get his Generals star. Upstairs to shower & dinner with Merme & Ted G.—Nancy's in N.Y. getting an award from Stylists. Back tomorrow P.M. Forgot to note—this afternoon I went over to the Japanese Embassy & signed the Condolence book re the death of Emperor Hirohito. And so to bed.

Tuesday, January 10, 1989

Into the office after a stop at the Drs., at 9 AM. I really had a bad night—woke up at one point—4 AM & read for awhile.

Dennis (our son-in-law) called Ken to tell him his company had signed on to represent Uganda. The Dr. Koop's report is not really a report. It is a letter to me expressing a need for research on how to evaluate the abortion set up. The matter of a pardon for North is now being looked at as—the judge should drop the charges. I signed the annual Ec. report.

NSC—Colin told me 450 Cubans are leaving Angola. Also, Soviets are stepping up their withdrawal from Afghanistan. A former Navy Petty Officer who was kicked out of Navy has been picked up for trying to sell info to Soviets.

[Addressed W.H. Workshop on Choice in Education.]

Then desk time & a budget meeting. I presented bound volumes of budget to a few who helped on budget. A short lunch ½ hour then on our way for a drop-by at Fundraiser lunch at Nat. Archives for F.D.R. Library. F.D.R. started the idea of Presidential Libraries.

[Dropped by meeting of old friends from Sacramento days; met with President José Azcona of Honduras; allergy shot; Mrs. Reagan suffering from sinus infection; dinner party for cabinet and spouses.]

Wednesday, January 11, 1989

To the Dr. at 9:30 AM—bandage changed. A much smaller one & the splint—made of piano wire on back of hand & finger. Then to the office by 10:10. Ken told me the Justice Dept. will file a motion today to quash the subpoenas for Geo. & me in the North Case.

Nancy Risque has ordered "Council on Environmental Control" to kill a proposal re the greenhouse effort. It would have made practically everything illegal. I'm to call Trent Lott this afternoon.

NSC—Libyans have agreed to return the body of American flyer they've been holding since 1986.

Greeks & Turks are working out a compromise re a Turkish Port the Greeks thought should be under some kind of international control.

A very hush, hush problem [. . .].

[Presented medals at ceremony to honor Mr. and Mrs. Anatoly Shcharansky; meeting with Economic Commission on the budget; more photo sessions.]

Regular meeting Geo. S. reporting on Chemical Warfare meeting in Europe. Things are looking pretty good. Also spoke of easing some sanctions on Libya because they are hurting some Americans worse than they are hurting Quadafi. Then into Cabinet Room for official photo with Cabinet. They presented me with my cabinet chair. They bought it for me—$1200.

[Received report of the Export Now Commission.]

Then in Oval O. received printed report—1st copy of W.H. publication of the Reagan Admin. record—what we promised & what we delivered.

Some desk time—phoned Trent Lott who has a candidate for position having to do with sport & commercial fishing in Mississippi. Over to Dr. to check on a little bleeding in my hand. Nothing serious. Shower, dress, dinner & at 9 P.M. farewell TV address in Oval O. to the people. It was well received. Our exec. office staff stayed on & listened to the speech in the Roosevelt Room. They had a buffet. I dropped in for a few mins. after the speech. They kept the party going for quite awhile. Then upstairs & bed.

Thursday, January 12, 1989

Stopped at Dr. for a look at my bandage etc. Then over to office. Ken told me about Bill Bennett who retired to get out in the world. He has repented & Geo. B. is making him drug Czar. A lot of talk about last night's speech & the transition to new admin.

NSC—Geo. S. is going back to Vienna this weekend. The resolution at U.N. against us on the Libyan shoot down failed. We're having problems on our refugee allotment program. They all want to come to the U.S.

Then a visitor—P.M. Charles of Dominican Republic. Good to see a long time friend. She is a real leader in the Caribbean states.

[Attended armed forces salute in president's honor at Andrews A.F.B., commented, "I puddled up when the Marine Band played 'Auld Lang Syne'"; lunch with Vice President Bush; received keys for new limousine built by Ford; briefing for interview in CNN; photos with congressmen; went to doctor for new bandage; attended presidential library dinner.]

Friday, January 13, 1989

Another 10 AM start. 1st meeting some talk about Dr. Koop & the mix-up of his letter on abortion. It seems he never wanted it made public. Then A.B. came in with regard to the Ollie North trial. The Judge asked that the Attorney Gen. make a formal request to drop charges 1 & 2. The Rep. of Justice Dept. will have in his possession a signed statement by Attorney Gen. in his pocket. Dan Crispen [Crippen] came in with some replys to Dem. charges that I have increased taxes 13 times.

NSC—Very brief—a little talk about our [. . .] artillery based in Germany.

[Greeted leaders of Layman's National Bible Association; Joint Chiefs of Staff arrived to say good-bye and presented lighted world globe.]

Lunch & over to E.O.B. for meeting with Russian & Am. students, representatives of schools (30 each) in the 2 countries. It is my proposal to Gorbachev for a youth exchange fulfilled.

Then a photo with W.H. Science Council. And scattered through day were numerous photos of departing govt. employees & their familys & friends of friends etc. Day became quite busy.

[Flew to NYC for Knights of Malta dinner; talked first with Peter Grace and Cardinal O'Connor; returned to W.H.]

Saturday, January 14, 1989

I slipped out of bed early & left Nancy sleeping—she's worn out between the flu & her packing. Brkfst. in the study—timed my radio broadcast & did some desk work. Then over to Oval O.—did broadcast—back to the W.H. & a couple of photos in dip. room with family & friends of staff. Upstairs Nancy & Ted G. were packing odds & ends—I joined in some odds & ends of my own. P.D.B. arrived late but after lunch I got it read. Back to packing until our 3 P.M. departure for Camp D. More photos on departure. A different kind of chopper—not very comfortable. It's something new they wanted us to see. It was 21 degrees at Camp D. & then it started to rain. Usual routine—showered, washed my hair & then dinner & movie—"Cattle Queen of Montana." And to bed.

[Sunday, January 15, 1989: watched "Meet the Press"; bade farewell to staff and rangers; returned to W.H.; packed; Graber on hand to help; went to doctor for new bandage; dinner included Graber and Maureen; watched "60 Minutes" with segment filmed previous week at W.H.]

Monday, January 16, 1989—Martin Luther King Day

Brkfst. around 9 AM, some more packing—this time my White Tie & Tails. To my desk for some backed up mail. Nancy out to lunch with the girls. Ted & I lunched in the West hall. Back to the desk & then some more packing. Nancy got home from lunch with the press ladies about 3:30. In the meantime Jim K. was up with a bunch of pictures for me to sign. Then Ken D. came in for a 20 min. session on things hanging fire—like pardons for several people. Amanda—Mike Deaver's daughter called him about a pardon for Mike—yet Mike has passed the word he wouldn't accept a pardon. In the evening into Black tie & over to the Laxalt's home for a dinner party—about 30 people. It was a most enjoyable evening with old friends. Charley Wick played piano & we had a little song fest after dinner. Back to the W.H. about 11 P.M.

Tuesday, January 17, 1989

A 10 AM start—in the study. NBC setting up in Oval O. for a TV interview with Tom Brokaw.

We talked in the first meeting about the bureaucrats in Justice dept. who gave a report on Ed Meese that had him some kind of a Monster. They gave their report to the Wash. Press. Ed has already been cleared. Even the Special Prosecutor refused to indict him.

We're after H.H.S. to quit stalling on a directive regarding use of Fetal tissue on experiments etc.

3 of Bushes appointees will be examined by Congressional Committee.

Colin has reported that our continued sanctions re oil industry in Libya is now hurting American businesses to the benefit of Quadafy.

Project Hope is engaged in a big program of aid to the Armenian children who are victims of earthquake.

There is some objection by North's lawyer against Quashing subpoenas for Geo. B. & me.

NSC—Colin late—John Negroponte opened meeting. Colin was at a farewell ceremony by Joint Chiefs.

March 6 is date for resumption of Geneva talks & March 9 it is agreed will be meeting on conventional weapons.

Shevardnadze is in Kabul. There are great shortages there—oil only a 1 days supply. State dept. is urging us to bring our embassy people home. Some European countries have already done that.

Talks between Chinese & Vietnamese are going on in Beijing re Cambodia.

Then time for interview briefing. Tom says he wants to stick to personal things about my life beginning back in Dixon Ill.

The interview proceeded & everyone thinks it went very well. Lunch—then back in the Oval O. Rcvd. Timmy Awards for Nancy & myself from officials of The Wash. Touchdown Club.

Long stretch of desk time—a jillion P.C.'s. & letters to sign.

Then another parade of people from a variety of depts. for individual photos. Got word Pres. Seaga setting date for (JAMAICA) election Feb. 9. Dick Wirthlin came in with final figures. Approval rating of 63%—highest for any Pres. in this spot. Rating for handling the Soviets 81%.

Then over to Dr's. office. Stitches were removed from my hand. Upstairs to clean up. Dinner tonite—here 42 of us for unveiling of official portrait. It was another great evening. Old friends, Nancy's brother & his family & after dinner the unveiling of our portraits. There seems to be a feeling mine could be better. Even the artist, Shiller feels that way & is going to make some changes.

Wednesday, January 18, 1989

Another 10 AM day.

Some talk about the proposal for an exec. order re Fetus Tissue use from aborted babies.

Last night Bill Plante reported in CBS poll my approval rating was 68% to 75% on P.L.O. decision.

NSC—I'm to check on Contingency guidance for Defense. I'm to check on it. It was a proposal in Packard Commission Report. I've said yes. [. . .] Soviets are continuing to allow divided familys to rejoin—20 now.

Again decision called for on Krasnoyarsk Radar. We view it as a serious breach of ABM treaty. We do not have time to change that to a "Material Breach."

[Presented medals to Senators Dole (R-KS) and Strom Thurmond (R-SC) and Representative Bob Michel (R-IL); briefing for interview by visiting journalists; photos; met with Shultz, commented, "kind of a review of his 84 foreign trips to meetings with 74 heads of state"; greeted champion Notre Dame football team, which presented as a gift the letter sweater of George Gipp.]

After that single photos with about 86 members of V.P.'s staff. A haircut—my last in the W.H. Then Ron's agent & his fiancé came in for a photo. Then over to East Room where some 700 members of W.H. staff were gathered for a goodbye. Presents to Nancy & me & a dog house made like the W.H. for Rex. The Marine Band played Auld Lang Syne. Lots of tears. Upstairs—packing & desk work. Dinner with Merme & Dennis.

Thursday, January 19, 1989

9 AM Started day with photos—2 Marine W. Wing W.H. Guards. Then our usual meeting—some late schedule changes. Then into the Oval O. came a horse—the kind you see in vaudeville—2 human beings, one the front & one the rear end. Dan Crippen was in front & Fran Marie Kennedy Keel brought

up the rear. The horse was wearing the bridle etc. that was given to me yesterday. Some discussion of tomorrow's schedule then it was NSC time.

Colin reported on S. Africa's Pres. Botha having a stroke. No details on prognosis on his future.

Geo. S. is considering reducing embassy personnel down in Kabul down to 11. If trouble threatens the plan is to move our people into the British embassy. Kuwait is going to take the American flags off 6 of it's 11 tankers & leave them on 5 but the 5 will also go as American vessels with American crews.

Soviets propose reducing it's nuclear tactical battlefield weapons. This may be a ploy to win approval of German people who are aware their country would be the battlefield for an exchange of such weapons.

Vernon Walters came into the office to receive the medal he was to get yesterday but couldn't be there. Then a pre-brief for the 11:30 interview with Wire Service Reporters—It went O.K. I guess. Then over to W.H. to get Nancy for the East Room Medals of Freedom lunch. It was awarded to Ambas. Mike Mansfield & George Shultz. A nice affair.

[. . .]

[Photos with another hundred staffers.]

Dick Thornburg came in to see me about pardons. He doesn't believe I should pardon Patty Hearst, North, Poindexter or McFarlane. I'm afraid he's right. Nick Brady came by to say goodbye & thank me. Same with Alan Greenspan. Then it was the V.P.'s turn. All of my W.H. staff was gathered in the Roosevelt room for him to thank & say a goodbye. I went in with him. He said very nice things about me.

Then over to the Dr's. office—upstairs to shower. John H. came up & put a new bandage on my hand. Then it was final packing time. Dinner with Ted. Merme & Dennis went out to attend one of the Inaugural Gala's. And after signing a whole stack of pictures & letters brought by Jim Kuhn to bed.

Tomorrow I stop being President.

Friday, January 20, 1989

Up fairly early. A little before 10 AM I went over to the Oval O. now looking pretty bare. Took a picture with the photo pool. Then I took a look at the desk—for their photos & walked out the door. Over at the W.H. Nancy & I went to N. entrance for mag. photos of us awaiting the Bushes. Then into the State Dining room for farewell to W.H. staff. It was everyone from Ushers to gardeners, plumbers etc. Then back to the No. entrance to meet the Bushes & Quayle's. Into the Blue room for coffee with the Cong. & Sen. members of the Inaugural Committees & spouses. At 11 AM we departed for Capitol. First limo Geo. & I & Jim Wright & Sen. Ford. Behind us, Barbara B. & Nancy & Bob Michel & Sen. Mitchell.

Some waiting in Capitol with Cong. Committee then out to the platform for Swearing in. George is now Pres. & I'm ex. Through the Capitol Dome to E. side where Nancy & I boarded helicopter for Andrews A.F. Base. There I reviewed the troops. There was a crowd of a few thousand. On board A.F.1 & off to Calif. At L.A. airport, several hundred friends & supporters. Speeches by Mayor Bradley, Bill Smith, Rich Little & Bob Stack. I responded & the motorcade took us to our home. Should mention also—several bands including U.S.C. band at airport. Barbara Logan sang God Bless Am.

Then home & the start of our new life.

ACKNOWLEDGMENTS

Shortly after I wrote *The Boys of Pointe du Hoc: Ronald Reagan, D-Day and the U.S. Army 2nd Ranger Battalion*, Fred Ryan telephoned me. He liked the book. As chairman of the Ronald Reagan Foundation, he had a staggeringly interesting project for me to consider: editing our fortieth president's White House diaries. I had heard about the diaries for over a decade from my friend Pete Wilson, former governor of California. They had, in fact, taken on an almost mythological status for anybody involved in U.S. presidential studies. Obviously, I was extremely interested in studying them. Upon special invitation, I flew from New Orleans to Simi Valley and read all five volumes in their entirety. Clearly they were a treasure trove of the utmost historical significance. I wanted the job of editing them. The final decision to turn them over to me was made by former First Lady Nancy Reagan herself, who is smart, candid, and extremely loyal to her husband's legacy. Getting to know Mrs. Reagan was the most memorable part of this engaging project. A truly deep appreciation is in order to all three of these individuals for trusting my instincts as a historian.

To facilitate the editorial process, I moved my family to Thousand Oaks, California, a wonderful community of 117,000 only eleven miles from the Reagan Presidential Library. Our leased home was adjacent to Paradise Falls, which includes a maze of trailheads that leads to the Pacific Ocean. Every day, after working long hours at the Reagan Library, I'd jog down to the falls and hike. My half-year spent around the Santa Monica National Recreation Area will never be replicated. It's one of the most beautiful geographical wilderness areas in the world. Equally important, while living in Thousand Oaks, my wife, Anne, gave birth to Cassady Anne (our third child) at the local hospital. We call her our California baby.

Words won't ever fully explain how much help I received from the Reagan Library in transcribing and editing the diaries. Duke Blackwood (executive director of the Reagan Library and Foundation) was my all-purpose advisor. Deeply committed to Reagan's legacy, he kept a close eye on the editorial process I undertook to cut the diaries in half for publication. Imbued with an offbeat sense of humor, and a heroic work ethic, he essentially rode shotgun over me for six months. We became good friends. His personal assistant, Shawna Adolph, was a delight to work with. She put in extra hours helping me meet publishing deadlines and fact-check the manuscript. She was always on call and is a real gem of a human being.

Whenever I had a question pertaining to NSC concerns, I worked with the indomitable Joanne Drake (chief of staff of the Reagan Foundation) to iron out the difficulties. The Reagans—both Ronald and Nancy—were lucky to have somebody as loyal and dedicated as Joanne in their daily lives. Blessed with keen editor's eyes, she caught many missed commas and misspelled words while vetting the manuscript. Her intense razorlike professionalism was truly something to behold. Nobody knows the diaries better than Joanne. Without her tireless efforts this book wouldn't have come to fruition.

Others helped me transcribe and decipher much of the diaries, especially Mike Duggan, supervisory archivist. With the exception of Joanne Drake, nobody knows the diaries better than he does. An employee at the Reagan Library since 1993, Mike was responsible for dealing with the national security classification redactions. We became good friends on the project, and I look forward to working with him again down the line. Steve Branch (AV archivist) and Ray Wilson (archivist) also put in a lot of hard work. With very little guidance from me, they tracked down all the photographs found in this book. As a devotee of presidential libraries, I can say that these diligent archivists truly rank among the best in the business.

A number of Reagan Foundation board members and former White House staffers took an active interest in the diaries. Catherine Busch, Mark Weinberg, and Kathy Osborne were all sources of great wisdom and fine-tuning as the words went from handwritten copy to the typed version and then moved into the marketing phase. A special thanks is due to current and former Reagan staffers Kirby Hanson, Marguerite Jagard, and Peggy Grande, who lent a hand on everything from photos to forewords and all else in between.

At HarperCollins I worked closely with editor Tim Duggan. "Old school" in the best way, he attacks manuscripts with a blue-pencil vengeance. He took a deep interest in the Reagan Diaries, convinced it was both a historical classic and a *New York Times* bestseller. Because HarperCollins is eventually going to be publishing the entire diaries in a multivolume set, he urged me to make sure my manuscript wasn't two thousand pages long, which it could easily have been. Meanwhile, Kate Pruss was responsible for public relations, and John Jusino for copyediting and formatting. Everybody at Harper worked hard to make sure this book maintained impeccable integrity.

Two of the best agents in America—Bob Barnett and Lisa Bankoff—did a terrific job of bringing the Reagan Library, HarperCollins, and me together for this project.

At Tulane University I was assisted by Andrew Travers in innumerable ways. In California both Jennifer Smith and Justin Ho helped out with typing. My historian friends Robert Dallek and Julie Fenster were full of good advice. The History Department and the James A. Baker III Institute at Rice University allowed me to present diary material in their public forums in October 2006. From the minute the Baker Institute heard that the Reagan diaries might be published, they opened every door imaginable to help me lessen my work load. Rice is committed to have *all* the Reagan diaries eventually published in a full, unedited HarperCollins set. Particularly helpful at Rice was Professor Alan Matusow, one of the greatest authorities on U.S. presidential history, and President David Leebron, a true intellectual visionary.

Luckily for me I was able to use four Reagan scholars for regular editorial advice: Tony Dolan, a former speechwriter for Ronald Reagan who knows everything about how his old boss wrote; Lou Cannon, the leading Reagan biographer who is living in Santa Barbara; and the husband-and-wife team of Martin and Annelise Anderson, who are planning a multivolume biography of our fortieth president. They were always available to talk and swap ideas. I am greatly indebted to each of them.

Finally, my parents often took over babysitting duties while Anne and I were off-to-the-races in Southern California. I thank them profusely for their love and support. And, last but not least, my in-laws, Jerry and Lynn Goldman, as always, read an early version of this manuscript and offered insightful criticism. It's wonderful to have my core family always at my side.

GLOSSARY

Abshire, David (1926–)—NATO Ambassador from 1983 to 1987, appointed as Special Counselor to the President in 1986, serving as the White House Coordinator for the Iran-Contra inquiry.

Alexander, Lamar (1940–)—Republican Governor of Tennessee from 1979 to 1987.

Allen, Richard (1936–)—National Security Advisor from 1981 to 1982.

Anderson, Martin (1936–)—Senior Policy Advisor in Reagan's 1976 and 1980 presidential campaigns; member of White House advisory committees, including the Economic Policy Advisory Board from 1982 to 1989.

Arafat, Yasser (1929–2004)—Chairman of the Palestinian Liberation Organization from 1969 until his death.

Baker, Howard, Jr. (1925–)—Republican Senator from Tennessee from 1967 to 1985; Minority Leader from 1977 to 1981; Majority Leader from 1981 to 1985; White House Chief of Staff from 1987 to 1988.

Baker, James, III (1930–)—White House Chief of Staff from 1981 to 1985; Secretary of the Treasury from 1985 to 1988; served as chairman of Reagan's Economic Policy Council and was a member of the National Security Council.

Baldrige, Malcolm (1922–1987)—Secretary of Commerce from 1981 to 1987.

Begin, Menachem (1913–1992)—Prime Minister of Israel from 1977 to 1983.

Bennett, William (1943–)—Chairman of the National Endowment for the Humanities from 1981 to 1985; Secretary of Education from 1985 to 1987.

Bentsen, Lloyd (1921–2006)—Democratic Senator from Texas from 1971 to 1993.

Boren, David Lyle (1941–)—Democratic Governor of Oklahoma from 1975 to 1979; Senator from Oklahoma from 1979 to 1994.

Bork, Robert (1927–)—Circuit Judge of the U.S. Circuit Court of Appeals for the District of Columbia from 1982 to 1988; nominated by President Reagan as Associate Justice of the U.S. Supreme Court in 1987 but was denied Senate confirmation.

Brady, James S. (1940–)—Assistant to the President and White House Press Secretary from 1981 to 1989; paralyzed in the assassination attempt on President Reagan in 1981.

Brady, Nicholas F. (1930–)—Republican Senator from New Jersey from April to December 1982; Secretary of the Treasury from 1988 to 1993.

Brezhnev, Leonid (1906–1982)—General Secretary of the Soviet Communist Party from 1964 to 1982.

Brinkley, David (1923–2003)—Host of ABC's *This Week with David Brinkley* from 1981 to 1996.

Brock, William (1930–)—Chairman of the Republican National Committee from 1977 to 1981; U.S. Trade Representative from 1981 to 1985; Secretary of Labor from 1985 to 1987.

Burger, Warren (1907–1995)—Chief Justice of the U.S. Supreme Court from 1969 to 1986.

Burns, Arthur (1904–1987)—Former economics professor; Ambassador to the Federal Republic of Germany (West Germany) from 1981 to 1985.

Bush, George H. W. (1924–)—Republican U.S. Congressman from Texas from 1967 to 1971; Ambassador to the United Nations from 1971 to 1973; Director of the Central Intelligence Agency from 1976 to 1977; Vice President to Ronald Reagan from 1981 to 1989; U.S. President from 1989 to 1993.

Byrd, Robert (1917–)—Democratic Senator from West Virginia from 1958 to the present; Senate Majority Leader from 1977 to 1980 and from 1987 to 1988; Minority Leader from 1981 to 1986.

Carlucci, Frank (1930–)—Deputy Director of the Central Intelligence Agency from 1978 to 1981; National Security Advisor from 1986 to 1987; Secretary of Defense from 1987 to 1989.

Casey, William (1913–1987)—Chairman of Reagan's presidential campaign committee in 1980; Director of the Central Intelligence Agency from 1981 to 1987.

Castro, Fidel (1926–)—President of Cuba from 1959 to the present.

Chirac, Jacques (1932–)—Prime Minister of France from 1974 to 1976 and from 1986 to 1988; President of the French Republic from 1995 to the present.

Chun Doo Hwan (1931–)—Military officer who staged a coup in South Korea; President from 1980 to 1988.

Clark, William "Bill" P. (1931–)—Associate Justice of the California State Supreme Court from 1973 to 1981; Deputy Secretary of State from 1981 to 1982; National Security Advisor from 1982 to 1983; Secretary of the Interior from 1983 to 1985.

Cronkite, Walter (1916–)—Anchorman of the *CBS Evening News* from 1962 to 1981.

Culvahouse, Arthur B., Jr. (1948–)—Counsel to the President from 1987 to 1989.

D'Amato, Alfonse (1937–)—Republican Senator from New York from 1981 to 1999.

Darman, Richard (1943–)—White House advisor on economic and political strategy from 1981 to 1985; Deputy Secretary of the Treasury from 1985 to 1987.

Davis, Patti (1952–)—Daughter of Ronald and Nancy Reagan; used her mother's maiden name while establishing herself as an actress.

Deaver, Michael (1938–)—White House Deputy Chief of Staff from 1981 to 1985.

Delehanty, Thomas (1934–)—District of Columbia police officer wounded in the assassination attempt on Reagan in 1981.

Dole, Elizabeth (1936–)—Secretary of Transportation from 1983 to 1987; Secretary of Labor from 1989 to 1990; Republican Senior Senator from North Carolina since 2003; married to Senator Robert Dole in 1975.

Dole, Robert (1923–)—Republican Senator from Kansas from 1969 to 1996; Majority leader from 1985 to 1987; Minority Leader from 1987 to 1995; married to Elizabeth Dole.

Duarte Fuentes, José Napoleón (1925–1990)—President of El Salvador from 1984 to 1989.

Duberstein, Kenneth (1944–)—Assistant to the President for Legislative Affairs from 1981 to 1983; Deputy Chief of Staff in 1987; White House Chief of Staff from 1988 to 1989.

Fahd, bin Abdul Aziz Al Saud (1923–2005)—King of Saudi Arabia from 1982 to 2005.

Fraser, Douglas (1916–)—President of the United Auto Workers union from 1977 to 1983.

Gergen, David (1942–)—White House Director of Communications from 1980 to 1984.

Gorbachev, Mikhail (1931–)—General Secretary of the Communist Party of the Soviet Union from 1985 to 1990; President of the Soviet Union from 1990 to 1991; negotiated with President Reagan on historic nuclear arms reduction.

Greenspan, Alan (1926–)—Chairman of the National Commission on Social Security Reform from 1981 to 1983; Chairman of the Federal Reserve Board from 1987 to 2006.

Gromyko, Andrei Andreyevich (1909–1989)—Foreign Minister of the Union of Soviet Socialist Republics (USSR) from 1957 to 1985.

Habib, Philip (1920–1992)—Under-Secretary for Political Affairs from 1976 to 1978; Special Envoy to the Middle East from 1981 to 1983, to the Philippines in 1986, and to Central America in 1986.

Haig, Alexander (1924–)—Four-star General; Adviser to Secretary of State Henry Kissinger from 1969 to 1973; Secretary of State from 1981 to 1982.

Hinckley, John, Jr. (1955–)—Attempted to assassinate Reagan outside the Washington Hilton Hotel in 1981.

Hooks, Benjamin (1925–)—Executive Director of the National Association for the Advancement of Colored People (NAACP) from 1977 to 1993.

Hussein bin Talal (1935–1999)—Hashemite King of Jordan from 1952 to 1999, worked toward peace in the Middle East.

Hussein, Saddam (1937–2006)—President of Iraq from 1979 to 2003, perpetrated war with Iran from 1980 to 1988.

Jackson, Jesse (1941–)—Baptist Minister; Democratic presidential candidate in 1984 and 1988.

John Paul II (1920–2005)—Born Karol Wojtyla, led the Roman Catholic Church as Pope John Paul II from 1978 to 2005; combated Communism in his native Poland and elsewhere.

Kemp, Jack (1935–)—Republican Representative from New York State from 1971 to 1989.

Kirkland, Lane (1922–1999)—President of the AFL-CIO from 1979 to 1995; strong supporter of Solidarity Movement in Poland.

Kirkpatrick, Jeane (1926–2006)—Ambassador to the United Nations from 1981 to 1985, the first woman to hold that post.

Koch, Edward (1924–)—Democratic Mayor of New York City from 1977 to 1989.

Kohl, Helmut (1930–)—Chancellor of Germany from 1982 to 1998 (West Germany between 1982 and 1990) during Reagan's "Tear Down This Wall Speech" and his visit to Kolmeshöhe Cemetery near Bitburg.

Laffer, Arthur (1941–)—Member of Reagan's Economic Policy Advisory Board from 1981 to 1989.

Laxalt, Paul (1922–)—Republican Senator from Nevada from 1974 to 1987; Chairman of Reagan's 1976, 1980, and 1984 presidential campaigns.

Lewis, Andrew (1931–)—Secretary of Transportation from 1981 to 1983, consulted with Reagan on the air traffic controllers strike.

López Portillo, José (1920–2004)—President of Mexico from 1976 to 1982.

McFarlane, Robert (1937–)—President Reagan's National Security Advisor from 1983 to 1985, he entered a guilty plea in 1998 on charges of withholding information from Congress in the Iran-Contra crisis.

Mansfield, Michael (1903–2001)—Democratic Senator from Montana from 1953 to 1977; Ambassador Extraordinary and Plenipotentiary to Japan from 1977 to 1988.

Marcos, Ferdinand (1917–1989)—President of the Philippines from 1965 to 1986, was forced to flee the country after attempting to claim victory in a reelection bid he did not win.

Meese, Edwin, III (1931–)—Counselor to the President from 1981 to 1985; U.S. Attorney General from 1985 to 1988.

Michel, Bob (1923–)—Republican Representative from Illinois from 1957 to 1995; Minority Whip from 1975 to 1981; Minority Leader from 1981 to 1995.

Mitterrand, François (1916–1996)—Socialist President of France from 1981 to 1995, was a strong proponent of the European Union.

Mondale, Walter (1928–)—Vice President of the United States from 1977 to 1981; Democratic presidential nominee in 1984, lost to Reagan.

Morris, Edmund (1940–)—Commissioned to write President Reagan's authorized biography, *Dutch: A Memoir of Ronald Reagan* (1999).

Mubarak, Hosni (1928–)—Vice President of Egypt from 1975 to 1981; after the assassination of President Anwar Sadat in 1981, assumed presidency and has remained in office.

Mulroney, Brian (1939–)—Prime Minister of Canada from 1984 to 1993, he became one of Ronald Reagan's closest allies and friends.

Nakasone, Yasuhiro (1918–)—Prime Minister of Japan from 1982 to 1987, representing the Liberal Democratic Party, he was regarded as a conservative leader.

Negroponte, John (1939–)—A career diplomat who held several posts in the Reagan administration, including Deputy Assistant to the President for National Security Affairs from 1987 to 1989.

Nitze, Paul (1907–2004)—Served as head of arms negotiation talks with the Soviet Union from 1981 to 1984; chief arms negotiator for the Intermediate-range Nuclear Forces Treaty, signed in 1987.

Noonan, Peggy (1950–)—Special Assistant to President Reagan from 1984 to 1986, known for her speechwriting skills.

Noriega, Manuel (1936?–)—Military dictator of Panama from 1983 to 1989. President Reagan cut off U.S. military and economic aid to the country. In 1989 Noriega was deposed by a U.S. military invasion and has been serving a prison sentence on drug charge in Florida since 1992.

North, Oliver (1943–)—U.S. Marine Corps Colonel; worked for the National Security Council starting in 1981; operated the process of weapons transfer to Iran and the channeling of funds to Contras in Nicaragua; convicted on charges related to the Iran-Contra affair in 1989; the conviction was later overturned.

Nunn, Sam (1939–)—Democratic Senator from Georgia from 1972 to 1997; Chairman of Senate Armed Services Committee from 1987 to 1995.

O'Connor, Sandra Day (1930–)—Judge on the Arizona Court of Appeals from 1979 to 1981; Associate Justice of the Supreme Court from 1981 to 2006; the first woman to serve on the Supreme Court, she was unanimously confirmed by the Senate.

O'Neill, Thomas "Tip," Jr. (1912–1994)—Democratic Representative from Massachusetts from 1953 to 1987; Speaker of the House from 1977 to 1987.

Orr, Robert (1917–2004)—Republican Governor of Indiana from 1981 to 1989; Ambassador to Singapore from 1989 to 1992.

Petroskey, Dale A. (1955–)—White House Assistant Press Secretary from 1985 to 1987.

Poindexter, John (1936–)—Vice Admiral in the U.S. Navy; White House military advisor from 1981 to 1983; Deputy National Security Advisor from 1983 to 1985; National Security Advisor from 1985 to 1986; convicted of conspiracy, perjury, obstruction of justice, and other felonies in relation to his involvement in the Iran-Contra affair in 1990, verdicts which were later overturned.

Powell, Colin (1937–)—As a senior military aide, General Powell helped orchestrate attacks on Grenada and Libya; Deputy National Security Advisor from 1986 to 1987; National Security Advisor, 1987-1989.

Qaddafi, Mohammar Abu Minyar al- (1942–)—President of Libya from 1969 to the present.

Reagan, Maureen (1941–2001)—Daughter of Ronald Reagan and Jane Wyman.

Reagan, Michael (1945–)—Adopted son of Ronald Reagan and Jane Wyman.

Reagan, Nancy (1921–)—Former actress, married Ronald Reagan in 1952. First lady of California from 1967 to 1974; supported Foster Grandparents Program and advocated for POWs and MIAs. First lady of the United States from 1981 to 1988, known for her "Just Say No" anti-drug campaign. Recipient of the Presidential Medal of Freedom and the Congressional Gold Medal of Honor.

Reagan, Ronald Prescott (1958–)—Son of Ronald and Nancy Reagan.

Regan, Donald (1918–2003)—Secretary of the Treasury from 1981 to 1985; White House Chief of Staff from 1985 to 1987; architect of Reagan's 1981 tax cuts and the Tax Reform Act of 1986; criticized for his role in the Iran-Contra affair.

Rostenkowski, Daniel (1928–)—Democratic Representative from Illinois from 1959 to 1995; Chairman of the House Ways and Means Committee from 1981 to 1994.

Rostow, Eugene (1913–2002)—A Democrat, Rostow was appointed by President Reagan as Director of the Arms Control and Disarmament Agency, a position he held from 1981 to 1983.

Rumsfeld, Donald (1932–)—Served the Reagan White House from 1981 to 1989 in a variety of posts, including as an envoy to the Middle East from 1983 to1984, and a member of the General Advisory Committee on Arms Control from 1982 to 1986.

Schmidt, Helmut (1918–)—Chancellor of Federal Republic of Germany (West Germany) from 1974 to 1982.

Shamir, Yitzhak (1915–)—Prime Minister of Israel from 1983 to 1984 and from 1986 to 1992; cooperated with President Reagan on Middle East peace efforts and signed trade agreements with the United States.

Shevardnadze, Eduard Amvrosiyevich (1928–)—Foreign Minister of the Union of Soviet Socialist Republics (USSR) from 1985 to 1991; implemented the more open and humanitarian initiatives of the Gorbachev era.

Shultz, George (1920–)—President Reagan's Secretary of State from 1982 to 1989; key negotiator with the Soviet Union; one of Reagan's closest foreign policy advisers.

Smith, William French (1917–1990)—A Los Angeles lawyer, Smith was one of Reagan's earliest political supporters; U.S. Attorney General from 1981 to 1985.

Stockman, David (1946–)—Republican Representative from Michigan from 1977 to 1981; Director of the Executive Office of Management and Budget from 1981 to 1985.

Suzuki, Zenko (1911–2004)—Prime Minister of Japan from 1980 to 1982; strengthened relations with the United States.

Teller, Edward (1908–2003)—Contributed to the development of atomic and hydrogen bombs from 1941 to 1952; closely advised President Reagan in support of the Strategic Defense Initiative.

Teresa, Mother (1910–1997)—Born Agnes Gonxha Bojaxhiu in Mecedonia; Roman Catholic nun, known for her work with the sick and poor; founded the Order of the Missionaries of Charity in India in 1948.

Thatcher, Margaret (1925–)—Prime Minister of England from 1979 to 1990, the first woman to hold that post; was a trusted Cold War ally of President Reagan.

Thornburgh, Richard (1932–)—Republican Governor of Pennsylvania from 1979 to 1987; U.S. Attorney General from 1988 to 1991; helped craft the Americans with Disabilities Act.

Vessey, John, Jr. (1922–)—Four-star Army General; Chairman of the Joint Chiefs of Staff from 1982 to 1985.

Volcker, Paul (1927–)—Chairman of the Federal Reserve from 1979 to 1987.

Walters, Vernon "Dick" (1917–2002)—Lieutenant-General in the U.S. Army; Deputy Director of the Central Intelligence Agency from 1972 to 1976; appointed by President Reagan as U.S. Ambassador to the United Nations, a position he held from 1985 to 1989.

Watt, James (1938–)—President Reagan's Secretary of the Interior from 1981 to 1983, resigned in the wake of public outcry against bigoted remarks made about a Senate advisory panel.

Webster, William (1924–)—A former judge in Missouri; Director of the Federal Bureau of Investigation from 1978 to 1987; Director of the Central Intelligence Agency from 1987 to 1991.

Weinberger, Caspar (1917–2006)—Budget advisor during Reagan's California governorship; Secretary of Defense from 1981 to 1987; oversaw the largest peacetime military buildup in U.S. history.

Wilson, William (1914–)—Presidential envoy to the Vatican from 1981 to 1984, when Congress raised the post to Ambassadorial status; Wilson served until 1986.

Wirthlin, Dick (1931–)—Political strategist and pollster for President Reagan.

Wright, James (1922–)—Democratic Representative from Texas from 1955 to 1989; Speaker of the House from 1987 to 1989.

Zia-ul-Haq, Mohammad (1924–1988)—President of Pakistan from 1978 to 1988, until his death in an airplane crash.

INDEX

Mon. March 30.

My day to address the Bldg. & Const. Trades Nat. Conf. A.F.L.-CIO. at the Hilton Ball room – 2 P.M. Was all dressed to go & for some reason at the last min. took off my really good wrist watch & wore an older one.

Speech not riotously received – still it was successful.

Left the hotel at the usual side entrance and headed for the car – suddenly there was a burst of gun fire from the left. S.S. agent pushed me into the floor of the car & jumped on top. I felt a blow in my upper back that was unbelievably painful. I was sure he'd broken my rib. The car took off. I sat up on the edge of the seat almost paralyzed by pain. Then I began coughing up blood which made both of us think – yes I had a broken rib & it had punctured a lung. He switched orders from W.H. to Geo. Wash. U. Hosp.

By the time we arrived I was having great trouble getting enough air. We did not know that Tim McCarthy (S.S.) had been shot in the chest, Jim Brady in the head & a policeman Tom Delahanty in the neck.

I walked into the emergency room and was hoisted onto a cart where I was stripped of my clothes. It was then we learned I'd been shot & had a bullet in my lung.